Wildfowl
Special Issue No. 4

"Ecology and Conservation of Waterfowl in the Northern Hemisphere"

Proceedings of the 6th North American Duck Symposium and Workshop
Memphis, Tennessee, 27–31 January 2013

Editors
Eileen Rees, Rick Kaminski & Lisa Webb

Editorial Board
Brian Davis, Mike Eichholz, Gary Hepp, Rick Kaminski,
Dave Koons, Tom Nudds, Jim Sedinger & Lisa Webb

Published by
Wildfowl & Wetlands Trust
Slimbridge, Gloucestershire
GL2 7BT, UK
Registered Charity No. 1030884

Wildfowl

Editor
Dr Eileen C. Rees
Wildfowl & Wetlands Trust
Slimbridge, Gloucestershire GL2 7BT, UK.

Associate Editor
Prof Anthony D. Fox
Department of Bioscience, Aarhus University
Kalø, Grenåvej 14, DK-8410 Rønde, Denmark.

Editorial Board

Prof Jeff Black
Department of Wildlife
Humboldt State University
Arcata, California 95521, USA.

Assoc Prof Bruce D. Dugger
Department of Fisheries & Wildlife
Oregon State University
104 Nash Hall, Corvallis, OR 97331, USA.

Prof Andy J. Green
Wetland Ecology Department
Doñana Biological Station-CSIC
C/ Américo Vespucio s/n
41092 Sevilla, Spain.

Dr Matthieu Guillemain
Office National de la Chasse et de la Faune
   Sauvage
La Tour du Valat
Le Sambuc, 13200 Arles, France.

Cover photographs: King Eider, by Danny Ellinger/Minden Pictures/FLPA; Bewick's Swan and White-fronted Goose, by Paul Marshall/WWT. The cover images were selected to represent a duck, goose and swan species of the northern hemisphere, consistent with the title of this special issue of *Wildfowl*.

Cover design by Paul Marshall

Published by the Wildfowl & Wetlands Trust, Slimbridge, Gloucestershire GL2 7BT, UK

Wildfowl is available by subscription from the above address. For further information call +44 (0)1453 891900 (extension 257), or e-mail wildfowl@wwt.org.uk

ISBN 0 900806 65 6
Print ISSN 0954-6324
On-line ISSN 2052-6458

Printed on FSC® compliant paper by Berforts Information Press Ltd.

Recommended citation: Rees, E.C., Kaminski, R.M. & Webb, E.B. (eds.), 2014. *Ecology and Conservation of Waterfowl in the Northern Hemisphere. Wildfowl* (Special Issue No. 4).

Contents

Dedication 1
Dr. Guy A. Baldassarre
Rick Kaminski

Foreword 2
Bruce Batt

Preface 5
R.M. Kaminski & J.B. Davis

PLENARY PAPERS

Habitat use and selection

An introduction to habitat use and selection by waterfowl in the northern hemisphere 9
R.M. Kaminski & J. Elmberg

Habitat and resource use by waterfowl in the northern hemisphere in autumn and winter 17
J.B. Davis, M. Guillemain, R.M. Kaminski, C. Arzel, J.M. Eadie & E.C. Rees

Spring migration of waterfowl in the northern hemisphere: a conservation perspective 70
J.D. Stafford, A.K. Janke, M.J. Anteau, A.T. Pearse, A.D. Fox, J. Elmberg, J.N. Straub, M.W. Eichholz & C. Arzel

Nest site selection by Holarctic waterfowl: a multi-level review 86
M. Eichholz & J. Elmberg

Waterfowl habitat use and selection during the remigial moult period in the northern hemisphere 131
A.D. Fox, P.L. Flint, W.L. Hohman & J-P.L. Savard

Demography, cross-seasonality and integrated population management

Drivers of waterfowl population dynamics: from teal to swans 169
D.N. Koons, G. Gunnarsson, J.A. Schmutz & J.J. Rotella

Conspecific brood parasitism in waterfowl and cues parasites use *H. Pöysä, J.M. Eadie & B.E. Lyon*	192
The effects of harvest on waterfowl populations *E.G. Cooch, M. Guillemain, G.S. Boomer, J-D. Lebreton & J.D. Nichols*	220
Cross-seasonal effects and the dynamics of waterfowl populations *J.S. Sedinger & R.T. Alisauskas*	277
Managing harvest and habitat as integrated components *E.E. Osnas, M.C. Runge, B.J. Mattsson, J. Austin, G.S. Boomer, R.G. Clark, P. Devers, J.M. Eadie, E.V. Lonsdorf & B.G. Tavernia*	305

SPECIAL SESSION PAPERS AND EVALUATION OF THE SYMPOSIUM
Ecology and Conservation of North American Waterfowl

Implementing the 2012 North American Waterfowl Management Plan: people conserving waterfowl and wetlands *D.D. Humburg & M.G. Anderson*	329
Wetland issues affecting waterfowl conservation in North America *H.M. Hagy, S.C. Yaich, J.W. Simpson, E. Carrera, D.A. Haukos, W.C. Johnson, C.R. Loesch, F.A. Reid, S.E. Stephens, R.W. Tiner, B.A. Werner & G.S. Yarris*	343
Opportunities and challenges to waterfowl habitat conservation on private land *W.L. Hohman, E. Lindstrom, B.S. Rashford & J.H. Devries*	368
Estimating habitat carrying capacity for migrating and wintering waterfowl: considerations, pitfalls and improvements *C.K. Williams, B.D. Dugger, M.G. Brasher, J.M. Coluccy, D.M. Cramer, J.M. Eadie, M.J. Gray, H.M. Hagy, M. Livolsi, S.R. McWilliams, M. Petrie, G.J. Soulliere, J.M. Tirpak & E.B. Webb*	407
Annual variation in food densities and factors affecting wetland use by waterfowl in the Mississippi Alluvial Valley *H.M. Hagy, J.N. Straub, M.L. Schummer & R.M. Kaminski*	436

Atmospheric teleconnections and Eurasian snow cover as predictors of a weather 451
severity index in relation to Mallard *Anas platyrhynchos* autumn–winter migration
M.L. Schummer, J. Cohen, R.M. Kaminski, M.E. Brown & C.L. Wax

Waterfowl populations of conservation concern: learning from diverse challenges, 470
models and conservation strategies
J. Austin, S. Slattery & R.G. Clark

Waterfowl in Cuba: current status and distribution 498
P.B. Rodríquez, F.J. Vilella & B.S. Oria

Assessment of the 6th North American Duck Symposium (2013): "Ecology and 512
Conservation of North American Waterfowl"
L.P. LaBorde, Jr., R.M. Kaminski & J.B. Davis

Sponsors of NADS 6/ECNAW and this Publication

Designed by Justyn Foth

Dedication

Unfortunately and sadly, there comes a time in life when it ends despite individual desires to continue onward and to contribute professionally and personally. Dr. Guy A. Baldassarre surrendered life on 20 August 2012 at the age of 59 years after a courageous battle against leukaemia (Moorman 2013, *The Auk* 130: 194–195). The 6th North American Duck Symposium and Workshop (NADS 6) and this special issue of *Wildfowl* are dedicated to Guy for his decades of contributions to waterfowl science and conservation. Guy inspired countless people internationally through his teaching, research, writing, mentoring, service, humour and friendships. He had an extraordinary ability to amass and synthesise technical information on waterfowl and wetlands into comprehensive texts of international acclaim; most notably the two editions of *Waterfowl Ecology and Management* (1994, 2006; with Eric G. Bolen) and *Ducks, Geese, and Swans of North America* (2014). These treatises will serve as an information resource on waterfowl, and thus our profession, indefinitely. Indeed, Guy so wanted to attend NADS 6 to learn, visit with colleagues and students, and autograph copies of his books, but that trip and symposium were not part of his destiny. We thank Guy for his dedication to family, whom he considered first and foremost in his life, the wildlife profession, humanity, waterfowl, and all the birds he loved so dearly. Guy's contributions are timeless; his legacy lives on among all who knew this gentleman!

Rick Kaminski

Photograph: Guy A. Baldassarre 1953–2012, by Eileen Baldassarre.

Foreword

"What a great idea." That was my immediate reaction when Rick Kaminski told me about collaboration between the Wildfowl and Wetlands Trust (WWT) in the United Kingdom and the team that conducted the 6th North American Duck Symposium and Workshop (NADS 6), *Ecology and Conservation of North American Waterfowl* (ECNAW). The event occurred in Memphis, Tennessee in January 2013. A key part of that great idea was to broaden the content and title of this publication to include waterfowl from across the Northern Hemisphere. Indeed, since 1997, the NADS have stimulated intercontinental relationships and efforts among colleagues to understand, conserve and sustain waterfowl and their habitats throughout the northern part of the globe.

The NADS 6/ECNAW symposium was built on a rich history of progress in waterfowl and wetlands research and conservation, as is nicely explained in the Preface to follow. Briefly, NADS 6/ECNAW combined efforts of three groups of specialists that had previously established separate conferences in North America: the NADS, the North American Arctic Goose Conference, and the International Sea Duck Conference. Each separate gathering had assembled many of the world's specialists to review advancements in waterfowl and wetlands science; this progress has helped steer ongoing management, conservation and future scientific efforts for their focal waterfowl taxa. These conferences each had an excellent reputation as a "must attend" event for academic, government, and non-government scientists and students. Thus, NADS 6/ECNAW was assured to be an unprecedented assemblage of the world's waterfowl specialists and, from the level of knowledge exchanged, this proved to be the case for all those attending the meeting.

This special issue of *Wildfowl* is the first published proceedings from any NADS. The specialists in the field will appreciate the effort required to conduct a symposium of this magnitude and then follow with a peer-reviewed proceedings. This outcome was accomplished through an extraordinary level of engagement by all who were involved and they are acknowledged in the following Preface. All those folks also recognise that Rick Kaminski's leadership was the driving force of the symposium and publication of this volume – a vision he has pursued for over half a decade.

Dedication of the symposium to the memory and contributions of Guy Baldassarre was an inescapable decision in view of all participants' recognition of Guy's huge lifetime commitments to waterfowl science, conservation and education. His untimely death in 2012 was an overwhelmingly sad event. Guy had made personal contact and was friends with a very large portion of those in attendance at NADS 6/ECNAW, as a result of his superb career. He wrote, with Eric Bolen, the 1994 original and the 2006 second edition of *Waterfowl Ecology and Management* – the waterfowl profession's first comprehensive text book. Many conference participants have a tag-eared edition of both treatises on their reference shelf. I cherish the personalised copy of the second edition.

Not one to seek praise or credit, Guy would nevertheless surely have been proud of the conference and this proceedings – for a moment at least, before he sought conversation with whoever was willing to talk about waterfowl and the latest findings from new studies, especially if the research resulted from students' efforts. Indeed, Guy had planned to attend the symposium and sign copies of his monumental 2014 revision of *"Ducks, Geese, and Swans of North America,"* as he did for his and Eric's book at previous NADS. Francis Kortright and Frank Bellrose, the first and second authors of *"Ducks, Geese, and Swans of North America,"* respectively, would be thankful to Guy for bringing new life to this classic waterfowl treatise.

The WWT brings its own record of excellence in waterfowl research and conservation that extends historically back to the 1940s. They originally focused on the status of waterfowl in Great Britain, on introducing the general public to the beauty of waterfowl, and on emphasising the need for conservation to assure the future of these remarkable species. Since then they have established a singular position of leadership in a multitude of waterfowl and wetlands conservation matters, extending from their local education centers and across the world through engagement in many venues (including the United Nations) and the development of international research and conservation projects, a record that continues to mount to this day. For most North American waterfowl biologists, arrival of the annual *Wildfowl Trust Report* (which evolved into *Wildfowl*) was a highlight that provided a window for many subjects that extend from propagation of endangered species to detailed analyses of courtship behaviour and to population inventories of waterfowl throughout the world. The WWT's record of accomplishment is unequalled. I urge the world's waterfowl enthusiasts to make the WWT Centres in England, Scotland, Wales and Northern Ireland high priority destinations if they ever find themselves in the United Kingdom.

The WWT actually engaged in North American waterfowl conservation from their earliest years of population monitoring, because a portion of Light-bellied Brent Geese, known as Atlantic Brant in North America, fly across the Atlantic Ocean from breeding areas in eastern Arctic Canada to winter in coastal Ireland. My predisposition is to label these geese as "North American" because that is where they breed. I suspect WWT thinks of them as European geese that migrate to Canada to breed before they return "home" to Ireland. Both perceptions are correct and augur well for the future of these birds. On a broader basis, it is critical that people throughout the flyways of waterfowl assume the responsibility of caring for the birds and the places in which they live whether they are breeding, migrating or wintering.

Also, Sir Peter Scott, the founder of WWT, conducted the first ever field investigation of the then endangered Ross' Goose in the central arctic of Canada in 1949. The conservation of the entire world's waterfowl clearly has been their mission from the onset. Collaborations have since been much more active and ongoing, so this publication should be seen as an important example of the richness of today's pattern of international cooperation and accomplishment in waterfowl science and conservation.

The life histories of most northern hemispheric waterfowl encompass annual passages that extend between northern and southern latitudes and breeding and wintering habitats. The ecology of waterfowl throughout their annual cycle and the occurrence of cross-seasonal, carry-over effects on their survival and recruitment are now a core paradigm guiding waterfowl science, conservation and management. However, few researchers are able to work closely with the birds year round amongst latitudes. Bringing students, managers and scientists together to share and review their findings at gatherings like NADS is one of the most significant benefits to the birds and to the scientific basis for their future. This partnership between WWT and NADS in the production of these proceedings is a natural and welcome step in the continued quest for intelligent and effective application of science to the future understanding of waterfowl ecology and management in the northern hemisphere.

Bruce Batt
Chief Biologist (retired), Ducks Unlimited, Inc.

Photograph: Lesser Snow Goose (blue morph) on Wrangel Island, Russia, by Sergey Gorshkov.

Wildfowl Special Issue No. 4: Preface

This special issue of *Wildfowl* is the proceedings from the 6th North American Duck Symposium and Workshop (NADS). Archives of NADS 1–5 are filed at http://www.northamericanducksymposium.org/index.cfm?page=home. Because this special issue is the first NADS proceedings to be published in a scientific journal, a brief history of NADS seemed warranted, given that relatively little has been documented on the development of NADS to date. Additionally, this preface reports the goals and themes of NADS 6, the contents of the symposium, and acknowledges people who aided or sponsored NADS 6 and this special issue.

Being inspired by the North American Arctic Goose Conference in the 1990s, the conceptual founders of NADS – Alan Afton of the Louisiana Cooperative Fish and Wildlife Research Unit and Robert Helm of the Louisiana Department of Wildlife and Fisheries – envisioned a periodic forum (*c.* every three years) for waterfowl biologists, managers, researchers, conservationists and especially students to present and discuss current research and management related to ducks worldwide but with an emphasis on North America. They and particularly Dave Ankney, Michael Johnson and Bob McLandress pioneered NADS. Collectively, they believed that research questions and management issues related to sustaining duck populations and maintaining wildfowling traditions and ecological study of the birds in North America were of paramount importance. Moreover, they believed that waterfowl ecologists were among leaders in avian research and sought to ensure that this legacy was perpetuated, in part by creating a venue for discussion of topics significant to waterfowl and their habitats. Therefore, NADS was established in 1997 as an independent meeting free of agency and organisational politics and charged the scientific and local organisational committee of each subsequent symposium with authority to ensure that no group could use the forum for promoting personal agendas without consideration of either alternative viewpoints or scientific-based resolutions to management or other issues. In 2009, NADS, Inc. was created to formalise the organisation, obtain non-profit status, and ensure that NADS continued in perpetuity. The primary mission of NADS, Inc. is to advance the science, management and outreach that guide waterfowl conservation in North America and beyond by convening symposia at regular intervals to present and discuss information related to ducks and other waterfowl. The stated vision of NADS, Inc. is that well-trained and informed educators, researchers and managers will promulgate effective strategies to sustain duck habitats and populations for scientific study, wildfowling, observation, ecosystem services and conservation, and other ecological and societal benefits.

Six NADS have convened since inception: NADS 1 at Baton Rouge, Louisiana (1997); 2 at Saskatoon, Saskatchewan (2000); 3 at Sacramento, California (2003); 4 at Bismarck, North Dakota, (2006); 5 at Toronto, Ontario, (2009); and 6 at Memphis, Tennessee (2013). NADS 7 is planned for February 2016 in Annapolis, Maryland, the first NADS to assemble in the

Atlantic Flyway. When possible, symposium locations were alternated between the United States and Canada and among Flyways in North America.

The NADS 6 organisational committee and the board of directors of NADS, Inc. agreed in 2009 at NADS 5 that NADS 6 would include all taxa of Anatidae (*i.e.* ducks, geese and swans). Thus, the scientific committee dubbed NADS 6 as "Ecology and Conservation of North American Waterfowl (ECNAW)". Consistent with this taxonomic expansion to all waterfowl, the committee invited the North American Arctic Goose Conference and the International Sea Duck Conference as joint partners of NADS 6/ECNAW (http://www.northamericanducksymposium.org/).

In 2010, a 20-member scientific planning committee was formed with representation from universities, agencies and organisations across North America and Europe with expertise in ecology and conservation of ducks, geese and swans (http://www.northamericanducksymposium.org/index.cfm?page=committees). Richard M. Kaminski and J. Brian Davis (both of Mississippi State University) served as co-chairs for local planning, fund raising and logistics of NADS 6/ECNAW, hosting the event at the Peabody Hotel in Memphis, Tennessee, from 27–31 January 2013. A total of 450 conferees attended NADS 6/ECNAW, and a majority of those attending responded to a survey to evaluate the symposium (see Laborde *et al.* 2014 in this volume). The grand theme for NADS 6/ECNAW was "Science and Conservation: Sustaining Waterfowl Forever." Although this theme may seem grandiose, it provided an ageless, challenging and inspiring goal for waterfowl scientists and stewards presently and into the future.

The scientific committee's first challenge was to identify broad, prominent topics for plenary sessions consistent with the mission of NADS. By vote, the committee selected the following topics in sequential order occurring Monday–Thursday (28–31 January 2013) of the symposium: 1) habitat use and selection; 2) annual-cycle and biological carry-over effects; 3) life-history strategies and fitness; and 4) population and community ecology and dynamics. These topics were chosen because the committee believed use of habitats and intrinsic resources by waterfowl, relative to myriad exogenous influences, shape annual-cycle and life-history adaptations of individuals and ultimately influence biological outcomes for individuals, populations and communities. Leaders of the plenary sessions were also selected by vote; they and invited presenters represented colleagues with worldwide reputations in the topical areas of the plenaries.

In addition to the morning plenary sessions, afternoons of the symposium were filled with concurrent contributed and special sessions wherein professionals and students made oral presentations. There were numerous presentations that spanned theoretical and applied ecology and conservation of waterfowl and wetlands in North America and Europe. Abstracts from presentations are archived at http://www.northamericanducksymposium.org/index.cfm?page=agenda. The final afternoon of the symposium included a novel "Syntheses and Futures" session wherein senior and junior colleagues presented thoughts on major issues revealed during the symposium and visions for future research and conservation to sustain and advance science and conservation of northern hemispheric waterfowl. Many of these progressive notions are presented in the articles in this special issue. Moreover, there were two evening poster sessions and a special session for student mentees to meet and interact with professionals. A special session of the North American Waterfowl Management Plan 2012 (NAWMP) also convened before the official opening of the symposium to discuss implementation and advancement of the revised plan (http://nawmprevision.org/). A synthesis paper of NAWMP 2012 is included in this volume (see Humburg & Anderson 2014).

From the wide array of presentations, the scientific committee deliberated and concluded that this volume would be composed of manuscripts from plenary and special sessions of the symposium. As mentioned, the committee selected broad topics for plenary sessions because of their fundamental importance to individual survival, reproductive performance and fitness, as well as how individual biological outcomes impact collectively on vital rates, population dynamics and community ecology of waterfowl. Indeed, this sequential acquisition of knowledge from individual to population and community levels is essential for holistic understanding of waterfowl ecology and guiding effective management and conservation for abundant and rare species, populations, and communities of waterfowl worldwide. Additionally, the committee did not envision the proceedings to be a mere compilation of selected "souvenir" papers from NADS 6/ECNAW but a contemporary compendium of knowledge related to waterfowl and their habitats in the northern hemisphere. Not since 1987 (in Winnipeg, Manitoba, Canada), have scientists, students and managers of waterfowl assembled for an international symposium focused on all waterfowl. The Winnipeg symposium was followed by a seminal publication entitled *"Ecology and Management of Breeding Waterfowl"* (Batt *et al.* 1992, University of Minnesota Press), which synthesised knowledge of breeding waterfowl ecology and management from the 20th century. Since this conference and publication, ecologists have greatly advanced understanding of waterfowl ecology and conservation throughout the birds' annual cycle and range in the northern hemisphere (*e.g.* Baldassarre & Bolen 2006 in *"Waterfowl Ecology and Management"*, Krieger Publishing Company; Rees *et al.* 2009, *Wildfowl* Special Issue 3; and Baldassarre 2014 in *"Ducks, Geese and Swans of North America"*, The John Hopkins University Press). Thus, NADS 6/ECNAW served as a cornerstone for this special issue with its own primary goals: 1) synthesising classical and contemporary information related to waterfowl ecology and conservation throughout the northern hemisphere and the birds' annual cycle and range; 2) comparing this knowledge across taxa of Anatidae with diverse habitat use, annual ecologies and cross-seasonal carry-over effects, life-history traits and fitness strategies, and population and community ecologies, for the purpose of archiving current knowledge from species to communities and enabling cross-taxa generalisations; and 3) using science-based information from the first two initiatives to adapt and advance local, regional, and intra- and intercontinental management and conservation of waterfowl (*e.g.* NAWMP 2012). Fulfilment of these goals, coupled with presentation here of papers from special sessions at NADS 6/ECNAW, are intended to make significant contributions toward understanding and sustaining northern hemispheric waterfowl and other migratory birds during and after the 21st century. Moreover, because ecologists studying waterfowl have contributed greatly to understanding the ecology of migratory avifauna since the 20th century, this special issue should be useful to ornithologists in general and provide a 21st century standard for guiding and advancing science and conservation for waterfowl and other migratory birds and their habitats in the northern hemisphere. Indeed, we hope our mission has been accomplished.

Clearly, our mission would not have been successful without the dedicated efforts of all leaders of the plenary and special sessions and especially the authors contributing manuscripts to this special issue of *Wildfowl*, the 20-member scientific planning committee for NADS 6/ECNAW, the co-editors of this volume (Eileen Rees, Rick Kaminski and Lisa Webb), the editorial committee (Brian Davis, Mike Eichholz, Gary Hepp, Rick Kaminski, Dave Koons, Tom Nudds, Jim Sedinger and Lisa Webb), the many external peer reviewers of manuscripts, Bruce Batt for penning the Foreword, Jeanne Jones for drawing the logo for NADS 6/ECNAW, the local logistics committee for the symposium (Amy Alford, Bruce Batt, Karen Brasher, J. Brian Davis, Jim Feaga, Justyn Foth, Charlsie Halford, Steve Jones of the Mississippi State University Extension Service, Rick and Loretta Kaminski, Molly Kneece, Jennifer Kross, Joe Lancaster, Joe Marty, Kira Newcomb, Tom Peterson, Jessica Myers, Jessie Schmidt, Clay Shipes, Jake Straub, Lisa Webb and Matt Weegman), Laurie Grace and Justyn Foth for designing and producing wood plaques for student presentation award recipients, session moderators, student poster and oral presentation award judges, student organisers of the mentor/mentee session (Elizabeth St. James, Justyn Foth, Jessi Tapp, David Messmer, Matt and Mitch Weegman), the student mentors, vendors, and the Peabody Hotel staff – especially Shannon Williams and Betsy Wilson. We also sincerely thank Eileen Rees (Editor-in-Chief of *Wildfowl*) and the Wildfowl & Wetlands Trust for accepting our request to publish the proceedings in *Wildfowl*, and her editorial staff for reviewing and editing the manuscripts herein. Moreover, we thank the NADS, Inc. Board of Directors who assisted with and supported NADS 6/ECNAW (Brian Davis, John Eadie, Michael Johnson, Rick Kaminski, Tom Nudds, Scott Petrie, Ron Reynolds, Mike Szymanski and Dan Yparraguirre). Rick Kaminski extends sincere thanks to George Hopper, Dean and Director of the College of Forest Resources and the Forest & Wildlife Research Center, Mississippi State University, for providing fiscal and moral support before, during, and after the symposium and for allowing him time to focus work on completion of this special issue.

Lastly but not least, we are deeply indebted to the sponsors and conferees who defrayed costs of NADS 6/ECNAW and publication of this special issue (see sponsors' logo page). This international, premiere symposium and publication are due largely to your support. If we have omitted anyone deserving acknowledgment, we accept full responsibility for the non-intentional oversight and extend our sincere thanks to you now.

Rick Kaminski & Brian Davis
(Local co-organisers of NADS 6/ECNAW)

An introduction to habitat use and selection by waterfowl in the northern hemisphere

RICHARD M. KAMINSKI[1]* & JOHAN ELMBERG[2]

[1]Department of Wildlife, Fisheries and Aquaculture, Mississippi State University, Mississippi State, Mississippi 39759, USA.
[2]Division of Natural Sciences, Kristianstad University, SE-291 88, Kristianstad, Sweden.
*Correspondence author. E-mail: rkaminski@cfr.msstate.edu

Abstract

This introductory article aims to provide a theoretical framework to the topics of habitat use and selection by waterfowl (*i.e.* family Anatidae) in the northern hemisphere during the four stages of their annual cycle: autumn migration and winter, spring migration and pre-breeding, nesting and brood rearing, and post-breeding and moulting. Papers addressing each of these seasonal sectors of the annual cycle, which follow this introduction, were presented at the 6th North American Duck Symposium, "Ecology and Conservation of North American Waterfowl" in Memphis, Tennessee in January 2013. Here, we consider the theory and selected empirical evidence relevant to waterfowl habitat and resource use and selection that may affect individual survival and fitness of waterfowl in Nearctic and Palearctic ecozones. Additionally, where possible, a comparative taxonomic approach is attempted in the following papers to identify and generalise patterns in habitat and resource use and selection across waterfowl taxa that may influence biological outcomes for individuals, populations and species through space and time. Each of the subsequent papers use accumulated science-based information to recommend future opportunities and strategies for research and for habitat and population conservation. Collectively, our goals in synthesising information on waterfowl are to help sustain harvestable populations of waterfowl and to protect rare species amid worldwide changes in climate, landscape, economics, socio-politics and growth of human populations.

Key words: Anatidae, annual cycle, biological outcome, conservation, fitness, habitat, habitat use, habitat and resource selection, management, survival, waterfowl.

Habitat is a principal unifying entity and concept in wildlife ecology and conservation, because wildlife would not exist in the absence of habitat and associated resources (Block & Brennan 1993). The English word "habitat" is derived from the Latin word "*habito*," meaning "to live or inhabit." Herein, we define habitat as environmental space occupied by living individuals of a species for any amount of time during their life cycle and which contains resources for individual survival and reproduction (Block & Brennan 1993; Jones 2001). All waterfowl species (*i.e.* ducks, geese and swans: Anatidae) of the northern hemisphere use three-dimensional habitat space (*i.e.* aerial, terrestrial and aquatic habitats) and the resources therein for physiological maintenance during their annual cycle and ultimately to gain fitness (*i.e.* individual survival and reproductive success resulting in genetic representation in subsequent generations; Mayr 1970). Generally, use of only two-dimensional space (terrestrial and aquatic) by waterfowl has been quantified using modern analyses and technologies (Belant *et al.* 2012; *cf.* O'Neal *et al.* 2010). Nonetheless, much literature exists which provides evidence that habitats and associated resources impose critical influences on individuals and evolution of adaptive strategies (Hildén 1965; Lack 1968; Stearns 1976, 1992; Southwood 1977, 1988; Kaminski & Weller 1992; Block & Brennan 1993; Jones 2001).

Block and Brennan (1993) and Jones (2001) provided excellent reviews of habitat use and selection and related concepts, primarily based on the ornithological literature. To resolve and reduce future ambiguity in terminology, they clarified and defined terms such as habitat, niche, habitat selection (or preference) and habitat suitability, and we have adopted their terminology here for consistency. The simplest and most common measure of habitat use by individuals of a species is their occurrence within habitats. Habitat associations (or correlations) are demonstrated quantitatively when animals' presence or abundance varies with measured habitat features (Wiens 1976, 1985). However, habitat associations should not be construed as habitat or resource selection, because selection is influenced by availability including accessibility and procurement (hereafter, availability) of habitats and associated resources (Block & Brennan 1993; Garshelis 2000; Jones 2001). Although unlikely to happen in nature, Fretwell and Lucas (1970), Fretwell (1972) and Wiens (1976, 1977, 1985) hypothesised that "true" habitat selection occurs when animals use habitats disproportionately to their availability and resource quality without constraints from extrinsic factors such as predation, competition, territoriality, density dependence, anthropogenic effects or a combination of these and other factors. Likely, habitat selection has evolved through increased immediate or short-term benefits to individuals using specific habitats and resources, leading to long-term rewards in survival, reproduction, and fitness accrued through natural selection and possibly other agents of evolution (*e.g.* gene flow; Lack 1968; Wilson & Bossert 1971; Jones 2001). Demonstration of causal linkages between cross-seasonal and habitat-resource use and fitness is difficult, especially for migratory waterfowl because of disconnects in daily to seasonal use of resources to meet physiological and behavioural needs during the annual cycle. Therefore, researchers generally infer habitat and resource selection when these are used disproportionately in relation to their estimated availability (Johnson 1980; Mulhern *et al.* 1985; Alldredge & Ratti 1986; Jones 2001). For example, Mulhern *et al.* (1985) proposed and tested three theoretical models of habitat use by Mallard *Anas platyrhynhos* and Blue-winged Teal *Anas discors* in relation to habitat availability, based on structural characteristics of used and available wetlands: 1) no selection when use of habitats "mapped" availability, 2) "plastic" selection when use was statistically different from available wetlands but use was temporally dynamic, and 3) "stenotopic" (*i.e.* specialistic or static) selection when use was statistically different from available wetlands but was consistent through time despite varying availability. Therefore, here and in subsequent papers in this section on seasonal habitat use, the term "habitat selection" is used when referring to results of studies that have demonstrated disproportionate habitat or resource use, or relationships between one or both of these and measures of biological outcomes (*i.e.* metrics of individual condition or performance (fitness correlates); Chalfoun & Martin 2007; Ayers *et al.* 2013).

Habitat use and selection by migratory birds, such as most waterfowl, may be envisioned as a multi-stage, hierarchical process from macro- to micro-spatial scales throughout the birds' annual cycle and geographic range (Johnson 1980). However, spatial distributions of waterfowl and other gregarious animals may not be completely influenced by individuals or habitat and resources. For example, habitat use by male waterfowl is often influenced by female philopatry to natal or other areas in their annual range, and juvenile birds, without prior migratory or dispersal experience, are likely influenced by co-occurring conspecifics or closely related taxa with similar niches (Brennan & Block 1993; Elmberg *et al.* 1997; Thomson *et al.* 2003). Despite possible social effects on habitat and resource use, migratory waterfowl seemingly make an initial "first-order" selection of geographic regions, such as those important to the birds during breeding and non-breeding seasons (Johnson 1980; Baldassarre & Bolen 2006). Within first-order occupied regions, waterfowl make "second-order" selections of wetland systems (Cowardin *et al.* 1979) and possibly associated landscapes (*e.g.* nesting habitats or arable fields for foraging). Next, they make "third-order" selections of local, site-specific wetlands or other locations and finally "fourth-order" selections of microhabitats where individuals specifically may roost, nest, forage or engage in other activities to acquire food or other resources including mates (Wiens 1973; Johnson 1980; Kaminski & Weller 1992, Baldassarre & Bolen 2006). Considering non-migratory species that do not move seasonally among geographic regions (*e.g.* subtropical species; Mottled Duck *Anas fulvigula*), their local-regional, annual home range and inclusive habitats constitute their "first-order" selected habitats, followed hierarchically by second- to fourth-orders of selected habitats within their home range as described above for migratory species (*sensu* Jones 2001). Additionally, the reversal of this process from micro- to macro-habitats also can be envisioned when birds depart microhabitats to disperse or migrate to different areas and habitats therein.

Ideally, habitat use and selection should be investigated at all relevant spatio-temporal scales to identify the scale(s) at which possible limiting factors may have greatest impacts on survival, reproduction and fitness. The outcome of habitat use and selection analyses can be influenced by the spatio-temporal scale of the investigation and associated environmental variation (Wiens 1985). Moreover, habitat use and selection should be studied among individuals of different ages, sexes and social status to yield accurate inferences and effectively guide conservation at population and species levels, always recognising that some habitat use by individuals may be related to presence and abundance of related or unrelated species (*e.g.* Götmark 1989; Jones 2001). Additionally, Buskirk and Millspaugh (2006) stated that an informative metric of habitat use may be risk to fitness (*e.g.* the probability of individual mortality; Lima & Dill 1990). Nonetheless, Ayers *et al.* (2013) and Lancaster (2013) recommended that researchers should measure individually based biological outcomes resulting from habitat and resource use, because the population-level approach common in wildlife-resource studies limits ability to make inferences about the importance of habitats and resources by failing to link these with fitness metrics or masking effects through sampling error or model averaging. Indeed, population-level habitat use and selection processes merely reflect the sum of individual responses; thus, individual-based approaches should be emphasised because natural selection operates at this level and individuals vary (*e.g.* Block & Brennan 1993; Goss-Custard *et al.* 1995; Stillman 2008).

Given that habitat quality or suitability (*sensu* Fretwell & Lucas 1970) influences habitat and resource use, fitness and species' life-history strategies, ecologists have hypothesised and measured relationships between indices of habitat suitability (*i.e.* biological outcomes) and covariates predicted to influence these outcomes. Fretwell and Lucas (1969) and Fretwell (1972) equated suitability among habitats to the capacity of habitats to promote fitness of individuals relative to variation in density of coexisting individuals. Theoretically, in a state of "ideal-free distribution" of animals, fitness prospects should decrease with increasing intra- and interspecific density of and competition among co-existing organisms and ancillary negative interactions (Fretwell & Lucas 1969; Fretwell 1972). Alternative models have been proposed and tested (*e.g.* ideal despotic model, Fretwell 1972; ideal pre-emptive model, Pulliam & Danielson 1991), but support for these and others has been inconsistent (Kaminski & Prince 1981; Rosenzweig 1985; Kaminski & Gluesing 1987; Pöysä 2001). Thus, researchers have cautioned that positive associations between indices of habitat suitability and population density should not be inferred without supporting data on biological outcomes (Van Horne 1983; Kaminski & Gluesing 1987; Kaminski & Weller 1992; Elmberg *et al.* 2006; Ayers *et al.* 2013).

Habitat use and selection vary seasonally, especially for migratory species. Anderson and Batt (1983) stated: "Obviously, a comprehensive understanding of ecology and evolutionary relationships of any species requires an appreciation of selective forces that act upon individuals of it during all seasons and throughout its range." During most of the 20th century, research conducted to understand ecology and habitat use of waterfowl has focused on the breeding grounds and season (Batt *et al.* 1992; Kaminski & Weller 1992; Baldassarre & Bolen 2006). However, since the 1970s, ecologists have recognised and investigated the importance of biological events experienced and habitats and resources used by waterfowl throughout the birds' annual cycle and range (*e.g.* Weller 1975; Fredrickson & Drobney 1979; Heitmeyer & Fredrickson 1981; Kaminski & Gluesing 1987; Tamisier & Dehorter 1999; Baldassarre & Bolen 2006). Nevertheless, ecology and habitat use of vernal and autumnal staging waterfowl remain understudied compared to other seasonal sectors of the annual cycle (Arzel *et al.* 2006). Nonetheless, increased knowledge of waterfowl ecology throughout their annual cycle and range has been paramount in shaping waterfowl habitat conservation plans in North America and Europe (Canadian Wildlife Service & U.S. Fish and Wildlife Service 1986; U.S. Department of Interior, Environment Canada, & Environment and Natural Resources, Mexico 2012; Kadlec & Smith 1992; Baldassarre & Bolen 2006; Elmberg *et al.* 2006).

As researchers continue studies of waterfowl habitat use and selection, they would be wise to heed concerns and advice of Jones (2001). To paraphrase Jones (2001), "… ornithologists tend not to consistently evaluate the behavioural and fitness contexts of their findings. That can be ameliorated by recognizing that (1) habitat selection refers to a process (by individuals) and perhaps less so a pattern, (2) that there are many extrinsic factors that influence habitat selection, and (3) that a complete test of habitat selection involves an assessment of whether or not the documented habitat preferences are adaptive. A second concern was that ornithologists do not consistently use habitat-related terminology. That lack of consistency can be remedied by providing operational definitions to limit misunderstanding. A third concern was that methodologies commonly employed to document habitat selection do not account for the hierarchical nature of habitat selection and do not generate accurate representations of habitat availability. Comparisons of used habitat with available habitat are more appropriate than comparisons of used and unused habitats. Definitions of habitat availability ought to be informed by the natural and life-history characteristics of the species."

Given this background, the papers following in this section of the journal, which were presented during the 6th North American Duck Symposium, "Ecology and Conservation of North American Waterfowl," in Memphis, Tennessee in January 2013, will synthesise knowledge on habitat and resource use and selection across ducks, geese and swans, when possible, throughout the birds' annual cycle and range, focusing primarily on published studies conducted in Nearctic and Palearctic ecozones. The synthesis is organised cross-seasonally to understand carry-over biological effects related to individual fitness and population dynamics in the following sequence: 1) autumn migration and wintering habitats, 2) spring migration and pre-breeding habitats, 3) nesting habitats, and 4) post-breeding and moulting habitats. Additionally, authors will identify new questions deserving future research and suggest how advances in technology and analytical approaches may enhance understanding of waterfowl habitat use and selection. Finally, the synthesis of empirical information will be used to recommend future scientific investigations and strategies for habitat and population conservation, to help sustain harvestable populations of waterfowl and to protect rare species in the northern hemisphere amid global changes in climate, landscape, economics, socio-politics and human population growth. As we have increased our knowledge of waterfowl of the northern hemisphere throughout their annual cycle and range during the 20th and early 21st centuries (*e.g.* Batt *et al.* 1992; Baldassarre 2014), the future is rich in opportunities for collaboration of scientists and managers to advance our understanding and conservation of Anatidae and their habitats worldwide.

Acknowledgements

This introduction benefitted greatly from reviews by Chris Ayers, Jerry Belant, J. Brian Davis, Tony Fox, Andy Green, Matt Guillemain, Joe Lancaster, Aaron Pearse and Josh Stafford. Kaminski was supported by the James C. Kennedy Endowed Chair in Waterfowl and Wetlands Conservation and the Forest and Wildlife Research Center (FWRC) and Department of Wildlife, Fisheries and Aquaculture (WF), Mississippi State University. Elmberg was supported by a grant from the Swedish Environmental Protection Agency. This manuscript was approved for publication by the Director of the FWRC as a peer-reviewed journal article WF-388.

References

Alldredge, J.R. & Ratti, J.T. 1986. Comparisons of some statistical techniques of analysis of resources selection. *Journal of Wildlife Management* 50: 157–165.

Arzel, C., Elmberg, J. & Guillemain, M. 2006. Ecology of spring-migrating Anatidae: a review. *Journal of Ornithology* 147: 167–184.

Ayers, C.R.,. Belant, J.L, Millspaugh, J.J. & Bodinof, C.M. 2013. Directness of resource use metrics affects predictions of bear body fat gain. *Polar Biology* 36: 169–176.

Anderson, M.G. & Batt, B.D.J. 1983. Workshop on the ecology of wintering waterfowl. *Wildlife Society Bulletin* 11: 22–24.

Baldassarre, G.A. 2014. *Ducks, Geese and Swans of North America.* John Hopkins University Press, Baltimore, Maryland, USA.

Baldassarre, G.A. & Bolen, E.G. 2006. *Waterfowl Ecology and Management, 2nd edition.* Krieger Publishing Company, Malabar, Florida, USA.

Belant, J.L., Millspaugh, J.J., Martin, J.A. & Gitzen, R.A. 2012. Multi-dimensional space use: the final frontier. *Frontiers in Ecology and the Environment* 10: 11–12.

Block, W.M. & Brennan, L.A. 1993. The habitat concept in ornithology: theory and applications. *Current Ornithology* 11: 35–91.

Buskirk, S.W. & Millspaugh, J.J. 2006. Metrics for studies of resource selection. *Journal of Wildlife Management* 30: 358–366.

Canadian Wildlife Service & U.S. Fish and Wildlife Service. 1986. North American waterfowl management plan. Canadian Wildlife Service and U.S. Fish and Wildlife Service Report, Ottawa, Ontario, Canada and Washington D.C., USA.

Chalfoun, A.D. & Martin, T.E. 2007. Assessments of habitat preferences and quality depend on spatial scale and metrics of fitness. *Journal of Applied Ecology* 44: 983–992.

Cody, M.L. (ed.) 1985. *Habitat Selection in Birds.* Academic Press, Orlando, Florida, USA.

Cowardin, L.M., Carter, V., Golet, G.C. & Laroe, E.T. 1979. Classification of wetlands and deepwater habitats of the United States. Biological Service Program Report No. FWS/OBS-79/31. United States Fish and Wildlife Service, Washington D.C., USA.

Elmberg, J., Pöysä, H., Sjöberg, K. & Nummi, P. 1997. Interspecific interactions and coexistence in dabbling ducks: observations and an experiment. *Oecologia* 111: 129–136.

Elmberg, J., Nummi, P., Pöysä, H., Sjöberg, K., Gunnarsson, G., Clausen, P., Guillemain, M., D. Rodrigues, D. & Väänänen, V.-M. 2006. The scientific basis for new and sustainable management of migratory European ducks. *Wildlife Biology* 12: 121–127.

Fretwell, S.D. 1972. Populations in a seasonal environment. Princeton University Press, Princeton, New Jersey, USA.

Fretwell, S.D. & Lucas, H.L., Jr. 1970. On territorial behavior and other factors influencing habitat distribution in birds. I. Theoretical development. *Acta Biotheoretica* 19: 16–36.

Fredrickson, L.H. & Drobney, R.D. 1979. Habitat utilization by postbreeding waterfowl. *In* T. A. Bookhout (ed.), *Waterfowl and Wetlands – an Integrated Review*, pp. 119–131. LaCrosse Printing Co., LaCrosse, Wisconsin, USA.

Garshelis, D.L. 2000. Delusions in habitat evaluation: measuring use, selection and importance. *In* L. Boitani & T.K. Fuller (eds.), *Research Techniques in Animal Ecology: Controversies and Consequences*, pp. 111–153. Columbia University Press, New York, USA.

Goss-Custard, J.D., Caldow, R.W.G., Clarke, R.T., le V. dit Durell, S.E.A. & Sutherland, W.J. 1995. Deriving population parameters from individual variations in foraging behaviour. I. Empirical game theory distribution model of oystercatchers *Haematopus ostralegus* feeding on mussels *Mytilus edulis*. *Journal of Animal Ecology* 64: 265–276.

Götmark, F. 1989. Costs and benefits to eiders nesting in gull colonies: a field experiment. *Ornis Scandinavica* 20: 283–288.

Heitmeyer, M.E., & Fredrickson, L.H. 1981. Do wetland conditions in the Mississippi Delta hardwoods influence mallard recruitment? *Transactions of the North American Wildlife and Natural Resources Conference* 46: 44–57.

Hildén, O. 1965. Habitat selection in birds. *Annales Zoological Fennica* 2: 53–75.

Johnson, D.H. 1980. The comparison of usage and availability measurements for evaluating resource preference. *Ecology* 61: 65–71.

Jones, J. 2001. Habitat selection studies in avian ecology: A critical review. *The Auk* 118: 557–562.

Kadlec, J.A. & Smith, L.M. 1992. Habitat management for breeding areas. *In* B.D.J. Batt, A.D. Afton, M.G. Anderson, C.D. Ankney, D.H. Johnson, J.A. Kadlec & G.L. Krapu (eds.), *Ecology and Management of Breeding Waterfowl*, pp. 590–610. University of Minnesota Press, Minneapolis, USA.

Kaminski, R.M. & Gluesing, E.A. 1987. Density- and habitat-related recruitment in mallards. *Journal of Wildlife Management* 51: 141–148.

Kaminski, R.M. & Prince, H.H. 1981. Dabbling duck and aquatic macroinvertebrate responses to manipulated wetland habitat. *Journal of Wildlife Management* 45: 1–15.

Kaminski, R.M. & Weller, M.W. 1992. Breeding habitats of Nearctic waterfowl. *In* B.D.J. Batt, A.D. Afton, M.G. Anderson, C.D. Ankney, D.H. Johnson, J.A. Kadlec & G.L. Krapu (eds.), *Ecology and Management of Breeding Waterfowl*, pp. 568–589. University of Minnesota Press, Minneapolis, USA.

Lack, D. 1968. *Ecological Adaptations for Breeding in Birds*. Chapman & Hall, London, UK.

Lancaster, J.D. 2013. Survival, habitat use, and spatiotemporal use of wildlife management areas by female mallards in Mississippi's Alluvial Valley. M.Sc. thesis, Mississippi State University, Mississippi, USA.

Lima, S.L. & Dill, L.M. 1990. Behavioral decisions made under the risk of predation: a review and prospectus. *Canadian Journal of Zoology* 68: 619–640.

Mayr, E. 1970. *Populations, Species and Evolution*. Harvard University Press, Cambridge, Massachusetts, USA.

Mulhern, J.H., Nudds, T.D., & Neal, B.R. 1985. Wetland selection by mallards and blue-winged teal. *Wilson Bulletin* 97:473–485.

O'Neal, B.J., Stafford, J.D. & Larkin, R.P. 2010. Waterfowl on weather radar: applying ground-truth to classify and quantify bird movements. *Journal of Field Ornithology* 81: 7–82.

Pöysä, H. 2001. Dynamics of habitat distribution in breeding mallards: assessing the applicability of current habitat selection models. *Oikos* 94: 365–373.

Pulliam, H.R. & Danielson, B.J. 1991. Sources, sinks, and habitat selection: a landscape perspective on population dynamics. *American Naturalist* 137 (Supplement): 50–66.

Rettie, W.J. & Messier, F. 2000. Hierarchical habitat selection by woodland caribou: its relationship to limiting factors. *Ecography* 23: 466–478.

Rosenzweig, M.L. 1985. Some theoretical aspects of habitat selection. *In* M.L. Cody (ed.), *Habitat Selection in Birds*, pp. 517–540. Academic Press, Orlando, Florida, USA.

Southwood, T.R.E. 1977. Habitat, the templet for ecological strategies? *Journal of Animal Ecology* 46: 337–365.

Stearns, S.C. 1976. Life history tactics: a review of the ideas. *Quarterly Review of Biology* 5: 3–47.

Stearns, S.C. 1992. *The Evolution of Life Histories*. Oxford University Press, Oxford, UK.

Stillman, R.A. 2008. MORPH – An individual-based model to predict the effect of environmental change on foraging animal populations. *Ecological Modelling* 216: 265–276.

Tamisier, A. & Dehorter, O. 1999. Camargue, canards et foulques: fonctionnement et devenir d'un prestigieux quartier d'hiver. Centre Ornithologique du Gard, Nîmes, France.

Thomson, R.L., Forsman, J.T. & Mönkkönen, M. 2003. Positive interactions between migrant and resident birds: Testing the heterospecific attraction hypothesis. *Oecologia* 134: 431– 438.

U.S. Department of Interior (DOI), Environment Canada & Environment Natural Resources, Mexico. 2012. North American Waterfowl Management Plan: People conserving waterfowl and wetlands. U.S. DOI, Washington D.C., USA.

Van Horne, B. 1983. Density as misleading indicator of habitat quality. *Journal of Wildlife Management* 47: 893–901.

Weller, M.W. 1975. Migratory waterfowl: a hemispheric perspective. *Publicaciones Biologicas Instituto de Investigaciones Cientificas* 1: 89–130.

Wiens, J.A. 1973. Patterns and process in grassland bird communities. *Ecological Monographs* 43: 237–270.

Wiens, J.A. 1976. Population responses to patchy environments. *Annual Review of Ecology and Systematics* 7: 81–120.

Wiens, J.A. 1977. On competition and variable environments. *American Scientist* 65: 590–597.

Wiens, J.A. 1985. Habitat selection in variable environments: shrub-steppe birds. *In* M.L. Cody (ed.), *Habitat Selection in Birds*, pp. 227–251. Academic Press, Orlando, Florida, USA.

Wilson, E.O. & Bossert, W.H. 1971. *A Primer of Population Biology*. Sinauer Associates, Inc. Publishers, Sunderland, Massachusetts, USA.

Habitat and resource use by waterfowl in the northern hemisphere in autumn and winter

J. BRIAN DAVIS[1]*, MATTHIEU GUILLEMAIN[2], RICHARD M. KAMINSKI[1], CELINE ARZEL[3], JOHN M. EADIE[4] & EILEEN C. REES[5]

[1]Department of Wildlife, Fisheries and Aquaculture, Mississippi State University, Mississippi State, Mississippi 39762, USA.
[2]Office National de la Chasse et de la Faune Sauvage, CNERA Avifaune Migratrice Le Sambuca, 13200 Arles, France.
[3]Section of Ecology, University of Turku, 20014 Turku, Finland.
[4]Department of Wildlife, Fish and Conservation Biology, University of California, One Shields Avenue, Davis, California 95616, USA.
[5]Wildfowl & Wetlands Trust, Martin Mere, Burscough, near Ormskirk, Lancashire L40 0TA, UK.
*Correspondence author. E-mail: bdavis@cfr.msstate.edu

Abstract

A particular aim of avian ecologists, especially those studying waterfowl *Anatidae*, in the 20th and early 21st centuries has been to elucidate how organisms use habitats and intrinsic resources to survive, reproduce and ultimately affect fitness. For much of the 20th century, research was mainly on studying species during the breeding season; however, by the 1970s, the focus had changed to understanding migratory waterfowl throughout their annual cycle and range in Europe and North America. Autumn and winter are considered the non-breeding seasons, but habitat and resource use through these seasons is crucial for completing spring migration and subsequent breeding. Here we review the literature on autumnal and winter habitat use by Nearctic and Palearctic waterfowl to determine characteristics of important landscapes and habitats for the birds during autumn migration and in winter. Selection of habitats and resources is discussed (when literature permits) in relation to Johnson's (1980) model of hierarchical habitat selection. Habitat use by selected species or groups of waterfowl is also reviewed, and important areas for future research into habitat ecology are identified. We suggest that the greatest lack of understanding of waterfowl habitat selection is an ongoing inability to determine what habitats and intrinsic resources, at multiple scales, are truly available to birds, an essential metric in quantifying "selection" accurately. Other significant challenges that impede gaining knowledge of waterfowl ecology in the northern hemisphere are also described. Nonetheless, continued technological improvements and engagement of diverse interdisciplinary professional expertise will further refine understanding of waterfowl ecology and conservation at continental scales.

Key words: autumn, habitat use, migration, selection, waterfowl, winter.

Understanding how wildlife and especially birds use habitats and resources to survive and reproduce (*i.e.* promote fitness; *sensu* Kaminski & Elmberg 2014) has long been the subject of ecological research (Darwin 1859; Lack 1944; Morrison *et al.* 1992). Studies of waterfowl habitat use and selection are well represented within the substantial avian literature (Block & Brennan 1993; Kaminski & Elmberg 2014). David Lack's (1966) early reference to habitat selection remains valid today, and visionaries such as Lack and also Fretwell (1972) further hypothesised that non-breeding habitats and resources may be important limiting factors for birds of the northern hemisphere, especially migratory species such as waterfowl. Conditions at non-breeding habitats (*e.g.* winter wetlands) correlate with waterfowl recruitment (Heitmeyer & Fredrickson 1981; Nichols *et al.* 1983; Kaminski & Gluesing 1987; Raveling & Heitmeyer 1989; Guillemain *et al.* 2008). However, understanding habitat use and selection by seasonally mobile waterfowl remains challenging, because technology, logistics, economics and other constraints impede monitoring and assessment of resource availability, exploitation and biological outcomes for individuals and populations, from local to flyway scales and cross-seasonally (Elmberg *et al.* 2014; Kaminski & Elmberg 2014; Sedinger & Alisauskas 2014).

The number of waterfowl species and different populations, and their abundance and geographic distribution in the Holarctic, makes waterfowl dominant fauna of aquatic and terrestrial systems in the northern hemisphere (Raveling 2004). Many waterfowl species are largely tied to freshwater systems but several use agricultural, estuarine and marine environments (Bellrose 1980; Baldassarre 2014). Some waterfowl habitats are relatively stable and seasonally predictable relative to hydrology (*e.g.* estuarine and lacustrine wetlands; Cowardin *et al.* 1979), whereas other habitats provide food and other resources temporarily but are characteristically dynamic, such as harvested agricultural lands, riverine and palustrine wetlands (Tourenq *et al.* 2001; Fredrickson 2005; Baldassarre & Bolen 2006; Mitsch & Gosselink 2007; O'Neal *et al.* 2010).

Here, classic and contemporary literature that revealed habitat and associated resource use by Holarctic waterfowl during autumn and winter is reviewed, with emphasis on the latter season of the annual cycle. The review does not provide an exhaustive summary of habitat and resource use by each species or group of waterfowl, but gives an overview focusing on habitat use by non-breeding waterfowl from macro- to finer spatial scales, when available information permitted such coverage (*sensu* Johnson 1980; Kaminski & Elmberg 2014). Space limitations required us to review a selected group of waterfowl species and tribes, but planning is underway to address non-breeding seasonal ecology of lesser known taxa (*e.g. Cairini* sp., *Dendrocygnini* sp. and *Anas fulvigula*) and better known or more widely distributed Nearctic species in a future publication (*e.g. A. americana, crecca, clypeata, strepera, rubripes* and *Branta canadensis*). We begin with a conceptual overview of autumn migration applicable to Nearctic and Palearctic waterfowl, followed by a review of selected eco-regions important to non-breeding waterfowl in the Holarctic and the aforementioned review of selected species or groups of ducks, geese and swans. Finally, currently perceived challenges in studying habitat selection by non-breeding waterfowl are conveyed to stimulate further research and conservation of these birds and their habitats in the northern hemisphere and worldwide.

Hierarchical habitat use and selection

Kaminski & Elmberg's (2014) conceptual review of hierarchical habitat selection (*sensu* Johnson 1980), indicated that habitat use and selection by migratory birds, such as most waterfowl, can be envisioned as a multi-stage, spatio-temporal process from macro- to micro-scales throughout the birds' annual cycle and range. Migratory waterfowl seemingly make 1st order selection of geographic regions, such as those important to and used by the birds during breeding and non-breeding seasons (Johnson 1980; Baldassarre & Bolen 2006). Within 1st order occupied regions, waterfowl make 2nd order selections of wetland systems (Cowardin *et al.* 1979) and possibly associated landscapes for some species adapted to terrestrial habitats (*e.g.* arable lands). Next, waterfowl make 3rd order selections of local, site-specific wetlands or other locations in their seasonal home range, and finally 4th order selections of microhabitats where individuals may roost, forage or engage in other activities to acquire food or other resources, including mates (Wiens 1973; Johnson 1980; Kaminski & Weller 1992; Baldassarre & Bolen 2006). A reversal of this process from micro- to macro-habitats also can be envisioned, as birds depart micro-habitats to disperse or migrate to different regions.

Autumn migration

Avian migration involves complex physiological, behavioural, genetic and ecological influences at individual and flock levels, which can influence population dynamics and demography (Dingle & Drake 2007). Numerous publications focus on avian migration (*e.g.* Dingle 1996; Dingle & Drake 2007; Newton 2007; Stafford *et al.* 2014), but a disproportionate number address passerines, while relatively few consider waterbirds. This reality is surprising given the well-known migratory nature of most Holarctic waterfowl (Arzel *et al.* 2006).

Migration involves large-scale movements from breeding to non-breeding grounds and vernal returns to breeding grounds (Salewski & Bruderer 2007; Zink 2011). Autumnal migration may be considered endogenously and exogenously influenced seasonal movements of birds between breeding and non-breeding areas (Alerstam & Lindström 1990; Dingle 1996; Salewski & Bruderer 2007). A perplexing aspect of autumn migration in waterfowl is that timing of departure in birds is especially complicated (O'Neal *et al.* 2010; Krementz *et al.* 2012). Long-migrant passerines typically exhibit a time-minimisation strategy (Dänhardt & Lindström 2001; O'Neal *et al.* 2010), and although geese and swans refuel at staging sites for shorter periods in autumn than in spring (Madsen 1980; Luigujõe *et al.* 1996; Beekman *et al.* 2002), some ducks, such as larger-bodied species like Mallard *Anas platyrhynchos*, may remain at mid-migration stopovers for weeks or longer despite harsh weather conditions that seemingly would stimulate migration (Bellrose & Crompton 1970; O'Neal *et al.* 2010; Schummer *et al.* 2010; Krementz *et al.* 2012; Dalby 2013). Moreover, autumn migration and winter habitat use are further complicated by habitat availability and quality and human-related disturbance (*e.g.* Väänänen 2001; Roshier *et al.* 2006; Legagneux *et al.* 2009; O'Neal *et al.* 2010; St. James *et al.* 2013).

Life histories of waterfowl vary considerably among species and confound simple explanations of migration patterns. For instance, although body size influences migration and habitat use (Raveling 2004), American Black Duck *Anas rubripes* (1,100 g; Zammuto 1986; Baldassarre 2014) overlaps in time and space with American Green-winged Teal *A. crecca carolinensis* (318 g; Zammuto 1986), the smallest dabbling duck species, during migration and winter (Bellrose 1980, Baldassarre & Bolen 2006). Conversely, Blue-winged Teal *A. discors* (363 g; Zammuto 1986), although ~12% heavier than Green-winged Teal, winter at more southerly latitudes (≤ 30°N; Thompson & Baldassarre 1990). Clearly, waterfowl migration patterns do not strictly follow ecological generalisations such as Bergmann's Rule (Bergmann 1847).

Many Palearctic waterfowl converge from Fenno-Scandian and Russian breeding grounds toward the Baltic Sea, where they use various habitats as staging sites before gradually moving south during winter. Some birds such as Eurasian Teal *A. crecca crecca* move by successive small flights in early autumn, whilst Mallard lag behind and move later in less numerous but longer flights (Dalby 2013). Others, such as Northern Pintail *A. acuta*, may be nomadic and seek newly flooded but ephemeral habitats in autumn (Bellrose 1980), whereas Mallard may have protracted migrations (Bellrose 1980; Krementz *et al.* 2012).

Movements, site fidelity and turnover rates of waterfowl during autumn-winter are likely to reveal patterns of habitat suitability and trade-offs made by waterfowl during these periods of the annual cycle (Rodway 2007). Winter site fidelity is known to be strong in geese and swans (Owen 1980) but of lesser importance in ducks, which exhibit greater spatio-temporal plasticity in habitat use (Mulhern *et al.* 1985; Robertson & Cooke 1999). Moreover, interspecific comparisons of winter philopatry are confounded by vast differences in the size of regions investigated (Robertson & Cooke 1999). In Europe, studies of individually-marked Eurasian Teal highlighted significant wintering site fidelity among and within winters (Guillemain *et al.* 2009; Guillemain *et al.* 2010a), suggesting that birds were able to evaluate site quality and adapt their use of traditional wintering areas, perhaps resulting in increased individual fitness. Of course, such traditions may be jeopardised if abrupt habitat changes occur. Indeed, the ecology of waterfowl migration in the northern hemisphere remains a frontier for future scientific investigation (Arzel *et al.* 2006).

Selected important Holarctic regions for non-breeding waterfowl

Eastern United States

The eastern U.S. historically has been an important region for migrating and wintering waterfowl, particularly lacustrine and estuarine coastal wetlands and deep-water habitats (Cowardin *et al.* 1979; Bellrose 1980). The region of the Atlantic Coast Joint Venture (ACJV) encompasses 17 states in the Atlantic Flyway and is the most densely human-populated area in the conterminous U.S., wherein about 35% of the population resides (ACJV 2009).

Landscape diversity in this region includes ~22% agricultural land and 25% wetlands, which together support ~37 native species of waterfowl (ACJV 2009). Considering 2nd order habitat selection within this region, estuarine systems of coastal Maine are important to wintering American Black Duck, Common Eider *Somateria mollissima* and scoters *Melanitta* sp. that use sheltered ice-free areas for foraging and loafing (ACJV 2005), while fringes of saltmarshes and mudflats are important to Mallard and other dabbling ducks (Jorde *et al.* 1984). Barrier beaches, back-barrier coastal lagoons and salt marshes of Long Island and New Jersey provide additional important winter habitats for American Black Duck and Brent Geese *Branta bernicla* (ACJV 2005; Plattner *et al.* 2010). Farther south exists the Chesapeake Bay, the largest estuary in the conterminous U.S. with a watershed that drains 165,760 km², along with North Carolina Sounds, natural and artificial lakes and reservoirs, flooded bottomland hardwoods, Carolina bays and estuarine and salt marshes that provide habitat for a diversity of ducks, geese and swans (Hindman & Stotts 1989).

Additionally, South Carolina and Georgia provide habitat for wintering dabbling, diving and sea ducks (Gordon *et al.* 1989; ACJV 2005). South Carolina alone winters ~30% of all dabbling ducks in the Atlantic Flyway including Green-winged Teal, Northern Shoveler *Anas clypeata*, Mallard, American Wigeon *A. americana* and Northern Pintail (Gordon *et al.* 1989). In Florida, the St. John's and Indian Rivers basins provide important waterfowl habitat, supporting nearly 400,000 ducks during winter (ACJV 2005). Freshwater lakes, such as Lake Okeechobee, also provide important wintering habitats for many waterfowl, including Lesser Scaup *Aythya affinis*, Ring-necked Duck *A. collaris*, American Wigeon, and Blue-winged Teal (Johnson & Montalbano 1989).

Mississippi Alluvial Valley

Largely forested prior to settlement by Europeans in the 19th century, flood control for agriculture and human inhabitation influenced a nearly 80% loss of lowland forests in the Mississippi Alluvial Valley (MAV) by the late 20th century, with only highly fragmented tracts remaining today (MacDonald *et al.* 1979; Klimas *et al.* 2009). The MAV contains flooded croplands, wetlands, deep water habitats and aquaculture ponds that are important to migrating and wintering ducks and geese (Cowardin *et al.* 1979; Christopher *et al.* 1988; Reinecke *et al.* 1989; Stafford *et al.* 2006; Kross *et al.* 2008; Feaga 2013). Swans (*e.g.* Trumpeter Swans *Cygnus buccinator*) are rarely sighted in winter in the MAV (R.M. Kaminski, pers. obs.; MAV Christmas Bird Counts unpubl. data).

Within the flooded agricultural landscape (including the aquaculture ponds), migrating and wintering waterfowl use 2nd order lacustrine (*e.g.* oxbow and watershed lakes, reservoirs), palustrine (*e.g.* forested and moist-soil wetlands) and riverine systems in the MAV (*e.g.* Mississippi River and tributaries; Cowardin *et al.* 1979; Mitsch & Gosselink 2007). Considering 3rd order habitat use of agricultural lands and wetlands within 2nd order systems, Reinecke *et al.* (1992) reported that over half of the Mallard observed during aerial surveys across most of the MAV used flooded rice and soybean fields during winters 1987–1990. Subsequently, during the early 2000s, Pearse *et al.* (2012) reported that greatest densities of Mallard in the Mississippi portion of the MAV during winter were observed in habitat complexes composed of 50% flooded cropland, 20% hardwood or scrub-shrub wetlands, 20% moist-soil and other emergent wetlands and 10% permanent water bodies (*e.g.* rivers, lakes, ponds). Greatest densities of other dabbling duck species were also associated with a similar habitat composition (Pearse *et al.* 2012).

Waterfowl associations with flooded cropland might be expected given that the MAV is now largely an agricultural landscape. Despite losses of natural wetlands in the MAV and continentally (Mitsch & Gosselink 2007), migrating and wintering waterfowl have adapted to flooded agricultural lands and make significant use of them in the MAV to meet nutritional and other physiological needs (Delnicki & Reinecke 1986; Reinecke *et al.* 1989; O'Neal *et al.* 2010). Indeed, ricelands in the MAV are critical for meeting seasonal requirements of waterfowl using this region (Stafford *et al.* 2006). In the late 1970s and early 1980s, Delnicki & Reinecke (1986), studying food use and body weight,

estimated that rice represented > 41% of total food intake by Mallard. However, because rice, soybean, and other seed crops are planted and harvested earlier nowadays in the MAV than during the 20th century, deterioration of waste seed occurs because of germination, decomposition and consumption by non-waterfowl species after harvest but before major wintering flocks arrive in the MAV (Stafford *et al.* 2006; Foster *et al.* 2010; Petrie *et al.* 2014). Reduction in waste rice from harvest through late autumn–early winter in the MAV is estimated at 71–99% (Manley *et al.*; Stafford *et al.* 2006). Despite reduced availability of waste rice in harvested fields in the region, flooded rice fields however have structural characteristics similar to natural wetlands (Elphick 2000; Huner *et al.* 2002; Marty 2013). The mid-winter population goal for the Lower Mississippi Valley Joint Venture of the North American Waterfowl Management Plant (LMVJV) is > 7.8 million dabbling ducks, and winter-flooded rice fields provide ~11% of all food energy available to dabbling ducks in flooded habitats in the LMVJV (Petrie *et al.* 2014). Approximately 20% of the 748,668 ha of ricelands is winter-flooded in the LMVJV (Petrie *et al.* 2014). If the LMVJV rice fields were able to produce a second harvested crop intra-seasonally as in Louisiana and Texas (*i.e.* ratoon crop, Marty 2013), the amount of food available to dabbling ducks from the flooded fields in the LMVJV would increase 12-fold (Petrie *et al.* 2014). Development of rice varieties and other crops with ability to ratoon at latitudes within the MAV would increase substantially the abundance of waste grain following harvest and benefit migrating and wintering waterfowl (Wiseman *et al.* 2010; Petrie *et al.* 2014; Marty 2013).

Despite dominant coverage of agricultural land in the MAV, Mallard and other waterfowl use 3rd and 4th order wetland sites in the MAV (Reinecke *et al.* 1989). Reinecke *et al.* (1992) reported that Mallard used forested wetlands (3–11%) and moist-soil wetlands (3–29%) within and among winters. Additionally, Davis & Afton (2010), working in the Louisiana portion of the MAV, reported that radio-marked female Mallard selected forested wetlands and suggested that continued restoration and establishment of these habitats should benefit females. However, they did not report any relationships between Mallard winter survival or other correlates of fitness that might implicate benefits resulting from female use of forested wetlands. Subsequently, Lancaster (2013), working in the Mississippi portion of the MAV, investigated habitat-related survival of radio-marked female Mallard. Greatest rates of winter survival (≥ 75%) were exhibited by females that used habitat complexes composed mostly of forested and emergent wetlands (86% combined) and 12% cropland, which was notable considering that most of the MAV landscape now is cropland (Lancaster 2013; Kaminski & Davis 2014). Thus, although Mallard may be considered habitat generalists, they also use certain habitats disproportionately, affording increased fitness prospects consistent with the concept of habitat suitability (*sensu* Fretwell 1972; Kaminski & Elmberg 2014).

Considering 4th order microhabitats, Mallard and Wood Duck *Aix sponsa* differentially used flooded hardwood bottomlands in the Interior Flatwoods and MAV in Mississippi during winter. Mallard used microhabitats that contained less woody understory cover, whereas Wood Duck were associated with microhabitats of increased understory vegetation (Kaminski *et al.* 1993). Within moist-soil wetlands in the MAV, dabbling ducks of several species foraged in experimental plots with water depths ranging from 3–16 cm (Hagy & Kaminski 2012). Such a range of depths may facilitate forage acquisition by a diversity of species using a common habitat, at least until food depletion occurs (Greer *et al.* 2009; Hagy *et al.* 2014).

In addition to flooded croplands and natural wetlands in the MAV, aquaculture ponds for production of Channel Catfish *Ictalurus punctatus* and bait fish have become important staging and wintering habitats used by dabbling and diving ducks since their construction in the 1970s (Christopher *et al.* 1988; Reinecke *et al.* 1989; Wooten & Werner 2004). Species of waterfowl commonly using catfish ponds include Lesser Scaup *Aythya affinis*, Ruddy Duck *Oxyura jamaicensis* and Northern Shoveler, along with lesser abundances of Mallard, Gadwall *A. strepera*, and introduced resident Giant Canada Geese *Branta canadensis maxima* (Christopher *et al.* 1988; Dubovsky & Kaminski 1992; Vest *et al.* 2006, Feaga 2013). Dubovsky & Kaminski (1992) estimated that 150,000 ducks used catfish ponds in Mississippi, with an average of 100,000 individuals using ponds weekly in the mid-1980s. Wooten & Werner (2004) collected Lesser Scaup from Arkansas baitfish ponds and reported scaup primarily ingested *Chironomidae* larvae, but ~25% of collected birds contained fish biomass or bones.

Because of competition from foreign markets, infrastructural and other costs, catfish aquaculture has declined in the MAV (U.S. Department Agriculture 2010). There were 64,000 ha of ponds in Mississippi, Louisiana and Arkansas in 2001, but only 25,000 ha remained in operation in those states by 2012 (Lehnen & Krementz 2013). Feaga (2013) reported that migrating and wintering waterfowl and other waterbirds occurred in densities on catfish production impoundments (~130 birds/ha) similar to idled impoundments (~120 birds/ha). However, different bird communities existed in production *versus* idled production ponds, the latter now managed to provide emergent vegetation, mudflats and shallow wetland areas < 30 cm during summer–winter wetland birds (Feaga 2013; Kaminski & Davis 2014). Diving and dabbling ducks and American Coot *Fulica americana* were primary users of production aquaculture impoundments (Dubovsky & Kaminski 1992; Feaga 2013), whereas idled impoundments were used by over 40 species of ducks, shorebirds, waders and other waterbirds (Feaga 2013; Kaminski & Davis 2014).

Louisiana-Texas Gulf Coast

The coastal tallgrass prairies of Louisiana and Texas once covered over 1 million ha (Chabreck *et al.* 1989; Hobaugh *et al.* 1989). They have slight topography, relatively impervious soils and thus seasonal wetlands (Smeins *et al.* 1991; Petrie *et al.* 2014). Winter rains and tropical storms in summer–autumn periodically inundate basins and provide habitat for numerous migrating and wintering waterfowl (Petrie *et al.* 2014).

Fresh and intermediate brackish marshes have been among the greatest wetland losses in the coastal prairies; ~100,000 ha of non-farmed freshwater wetlands have been lost in the coastal plains of Texas since the mid-1940s (Moulton *et al.* 1997). Conversion of rice agriculture to cotton and soybean production has further reduced important habitats for waterfowl (Anderson & Ballard 2006). Gulf coastal wetlands are critical to several guilds of wintering waterfowl (Weller 1964; Chabreck *et al.* 1989; Hobaugh *et al.* 1989; Marty 2013), and an estimated 19% of all waterfowl wintering in the U.S. use marshes in the Louisiana Gulf Coast (Michot 1996; Bolduc & Afton 2004). The Texas Mid-Coast once wintered 78% of the Northern Pintail in the Central Flyway (Ballard *et al.* 2004). Contemporary estimates of midwinter population goals for the Gulf Coast JV region include > 5.6 million dabbling ducks (Petrie *et al.* 2014).

Considering 2nd and 3rd order habitat selection, freshwater and intermediate marshes along the Gulf of Mexico are perhaps the most important wetland habitats for waterfowl in the region (Chabreck *et al.* 1989; Batzer & Baldwin 2012). Brackish marshes are the most extensive habitat and considered historical habitats for wintering Snow Geese *Anser caerulescens* (Chabreck *et al.* 1989; Batzer & Baldwin 2012), but salt marsh habitats are generally regarded as less favourable to waterfowl in Gulf coastal systems (Williams III & Chabreck 1986; Batzer & Baldwin 2012). In addition to these, lakes (*e.g.* Grand, White), bays (*e.g.* Atchafalaya, Terrebonne) and off-shore habitats have been important historically for scaup and other diving and sea ducks in the Gulf region (Harmon 1962; Afton & Anderson 2001).

Scaup wintering off-shore in Louisiana have comprised 50–86% of the total wintering population and were much more abundant off-shore than in in-shore habitats in January (Kinney 2004). Kinney (2004) flew transect surveys and determined that only about 15% of scaup were detected in some years by traditional Midwinter Waterfowl Surveys. One hypothesis for scaup wintering farther off-shore is that Surf Clams *Mulinia lateralis* were historically a preferred food for the species (Harmon 1962; Kinney 2004) and recent increases in hypoxic areas in the near-shore waters of the Gulf may be causing scaup to venture farther off-shore for food.

Along the Texas Gulf Coast, the Laguna Madre is a large shallow lagoon that contains ~80% of the seagrass communities along the Texas coast (Ballard *et al.* 2010). The dominant species is Shoal Grass *Halodule wrightii* and ~80% of the continental Redhead *Aythya americana* population winters in the region, primarily because of seagrasses (Division: Angiospermae) and associated habitats (Weller 1964; Mitchell *et al.* 1994; Michot *et al.* 2006; Ballard *et al.* 2010). Several studies have documented the importance of proximate inland freshwater ponds to Redhead and other ducks including Lesser Scaup (Adair *et al.* 1996; Michot *et al.* 2006; Ballard *et al.* 2010). The proximity of coastal ponds to seagrass foraging areas on the Gulf Coast is important, as Redhead were never observed using ponds > 5.7 km from the shoreline or > 8.1 km from the nearest foraging area (Ballard *et al.* 2010). Thus, proximity of freshwater ponds to seagrass beds in the Laguna Madre is an example of a critical synergistic habitat association, particularly in drier winters (Ballard *et al.* 2010).

United States Great Plains

The Playa Lakes Region (PLR) contains 60,000–100,000 playa lakes or shallow wetlands that generally occur at the bottom of large watersheds and are formed by wind and water dissolution processes (Smith 2003; Venne *et al.* 2008). Playa wetlands range in size from < 1 ha to > 300 ha, extend from Wyoming and Nebraska to Texas and New Mexico, and are habitat to a wide diversity of life forms including waterfowl (Playa Lakes Joint Venture 2014). Historic native grassland has largely been replaced with arable crops, and subsequent erosion of topsoil has contributed to sedimentation of ~90% of all playas in the Southern High Plains (SHP; Venne *et al.* 2008). Moreover, ~80,000 playas throughout the Great Plains states are currently incapable of recharging the Ogallala aquifer (Playa Lakes Joint Venture 2014). Historically, one-third of the Central Flyway Northern Pintail population (~300,000 birds) used playa lakes in the SHP, but this population has declined 47% since 1977 (Bellrose 1980; Luo *et al.* 1997; Haukos 2004; Moon *et al.* 2007). Concomitantly, body condition of pintail in the PLR has declined considerably since the mid-1980s (Moon *et al.* 2007).

The SHP is a southern extension of the PLR and is a critical region to waterfowl, once containing 25,000–30,000 wetlands (Smith 2003; Baldassarre & Bolen 2006; Venne *et al.* 2008). Obenberger (1982) studied several species of dabbling ducks from autumn–late winter 1980–1982 and reported that ducks generally had a bimodal migration. Migration phenology of Northern Pintail and Green-winged Teal peaked in November, and autumn abundances were at least double their greatest numbers during vernal peaks. Nearly 30 years later, Baar *et al.* (2008) conducted similar research in the SHP and observed that duck use of playas was much more intermittent, protracted or less intensive compared to previous decades. Baar *et al.* (2008) offered two possible explanations for these patterns. First, abundance of playa wetlands, irrigation ponds and tailwater reservoirs were greatly reduced, and playas have become more rainfall dependent (Smith 2003; Baar *et al.* 2008). Second, playas have been subjected to significant sedimentation, with negative impacts to hydrologic patterns and function (Smith 2003). Moon & Haukos (2006) attributed declining body condition of Northern Pintail to harassment and stress, resulting from increased movements by hunters pursuing waterfowl and Ring-necked Pheasant *Phasianus colchicus* (Baar *et al.* 2008).

Generally, evidence suggests that important waterfowl foods, such as waste agricultural or natural seeds, are becoming depleted in early winter in the SHP (Baldassarre & Bolen 1984; Bolen *et al.* 1989; Smith & Sheeley 1993; Moon & Haukos 2006). As a consequence, exploitation of these environments by dabbling and other ducks may be more limited during late winter and spring (Baar *et al.* 2008) compared with prior decades (Obenberger 1982). Dedicated conservation programmes have been championed and are needed in the SHP (Haukos & Smith 2003; Smith 2003).

Central Valley of California

California always has been one of the most important regions for wintering waterfowl in North America (Gilmer *et al.* 1982; Miller 1986; Heitmeyer *et al.* 1989; Fleskes *et al.* 2005; Miller *et al.* 2010). The state has lost ~95% of its historic wetlands (Central Valley Joint Venture 2006) but continues to support millions of non-breeding waterfowl. Within California, the Central Valley provides critical wetland and agricultural habitat for migrating and wintering waterfowl and was the focus of one of the original Joint Ventures of the North American Waterfowl Management Plan (NAWMP 1986). The Central Valley encompasses ~4.1 million ha, stretching 724 km north to south and 64 km east to west. The valley is dominated by two riverine systems – the Sacramento River and the San Joaquin River, which meet at the Delta then flow into the Pacific Ocean past the Suisun Marsh, one of the largest contiguous brackish marshes in the western United States.

The hydrology of the valley determines the main habitat types and influences seasonal and inter-annual patterns of waterfowl use (Fleskes 2012). However, hydrology has been altered drastically from agriculture and urban growth and caused considerable changes in distribution of waterfowl habitats. Before the 1849 Gold Rush, the valley contained > 1.6 million ha of wetland habitat (Central Valley Joint Venture 2006). Most of these wetlands were seasonal, inundated by riverine flooding in the valley, bordered by expansive riparian and grassland habitats, which may have supported 20–40 million waterfowl during migrations and winter.

Seasonal and permanent wetlands in the Central Valley are distributed in four sub-regions: the southern San Joaquin Valley (including Tulare Basin, which held the now dry Tulare Lake, once the largest freshwater lake west of Mississippi; Fleskes 2012), the northern Sacramento Valley, the Delta and the Suisun Marsh. Historically, many waterfowl wintering in California would migrate first to Tulare Lake, a vast shallow complex of seasonal and permanent marshes. As winter progressed birds moved north, through the San Joaquin Valley, Delta and Suisun Marsh into the Sacramento Valley. Prior to land conversion, ~40% of waterfowl habitat occurred in the San Joaquin Valley (including Tulare Basin), while the remaining 60% occurred in the Sacramento Valley, Delta and Suisun Marsh (Fleskes *et al.* 2005). By approximately 1900, the Tulare lakebeds were effectively drained by diversion of water for agriculture, and the lakebeds now remain dry in all but extremely wet years. Wetlands in the San Joaquin and Sacramento Valleys were also converted to agricultural land, leading to cotton, orchard, vegetable and rice production in the Sacramento Valley. In the Delta, islands were leveed to grow corn, barley and other grain crops, some of which have value to ducks and geese.

Brackish marsh wetlands in the Suisun Marsh historically were significant to wintering waterfowl, but populations of dabbling ducks and geese there have declined. The Suisun Marsh currently provides wintering habitat for > 60,000 waterfowl, of which dabbling ducks are the most numerous (55,000), followed by diving ducks, geese, sea ducks, and swans (Ackerman *et al.* 2014). Following decades of considerable landscape changes, the Central Valley is left with merely 162,000 ha of wetlands nested within a largely agricultural matrix.

Most existing wetland habitat in the valley is managed and comprises seasonal, semi-permanent and permanent wetlands. Seasonal wetlands are flooded in autumn for waterfowl and other waterbirds and drawn down in late winter. Many wetlands are managed as waterfowl hunting clubs or state and federal wildlife areas or refuges. Seasonal wetlands provide critical foraging habitat for non-breeding waterfowl. These wetlands are managed annually using several methods (*e.g.* disking, irrigation and water management) to promote moist-soil plants such as Watergrass *Echinochloa crusgalli*, smartweed *Polygonum* sp. and Swamp Timothy *Crypsis schoenoides* (Heitmeyer *et al.* 1989). Semi-permanent wetlands are flooded from autumn to early July, while permanent wetlands are flooded throughout the year (Central Valley Joint Venture 2006). Semi-permanent and permanent wetlands produce less food, but provide important roosting and brood habitat for locally breeding ducks, mostly Mallard and Gadwall.

The most significant change to waterfowl habitats in the Central Valley over recent decades has been the development of rice agriculture, particularly in the Sacramento Valley. Planted rice acreage has increased from nearly 41,000 ha (1930s) to almost 243,000 ha, and now averages > 202,000 ha (Petrie *et al.* 2014). Waste grain remaining in fields after harvest provides a valuable food source for wintering waterfowl (Eadie *et al.* 2008). Along with the increase of planted rice, there has been a significant change in management of residual rice straw after harvest. Before the 1990s, fire was the primary method for rice straw disposal.

However, with air quality concerns, the Rice Straw Burning Reduction Act of 1991 mandated that burning of straw be reduced and currently less than 10% of all harvested rice fields are currently burned. As an alternative, rice growers turned to post-harvest flooding, accompanied by disking, rolling or chopping of straw. The result was that flooded rice fields provided valuable foraging habitat to a diversity of dabbling ducks and geese. At the peak, > 141,000 ha of harvested rice fields were flooded in autumn, nearly 70% of the planted rice acreage (Central Valley Joint Venture 2006; Petrie *et al.* 2014).

Waterfowl wintering in the Central Valley have responded strongly to these changes at both 2nd and 3rd orders of habitat selection. Timing and distribution of 2nd order selection by waterfowl have been altered considerably with the draining of Tulare Lake and increase of rice agriculture in the northern reaches of the valley. Fleskes *et al.* (2005) reported that the total area of croplands intentionally flooded in winter increased by 157% in the Sacramento Valley and 58% in the Delta, but declined by 23% in the San Joaquin Valley between 1973 and 2000, leaving only 3% of the total winter-flooded agricultural land in the latter region. In response, birds have shifted winter distributions northward. Fleskes *et al.* (2005) conducted extensive surveys and radio-telemetry in 1998–2000 and compared results to data from 1973–1982 (Heitmeyer *et al.* 1989; Miller *et al.* 1993; Miller *et al.* 1995). The recent research indicated that the percentage of dabbling ducks using the Tulare basin and the San Joaquin Valley declined, especially in late winter, while use increased in the Sacramento Valley. Cinnamon Teal *Anas cyanoptera* were an exception and did not shift northward. In contrast to dabbling ducks, the percentage of diving ducks using the San Joaquin and Tulare Basins increased concurrently with a decrease in diving ducks using the Suisun Marsh and Delta. Use of the Suisun Delta and San Joaquin Valley declined for geese, with concomitantly large increases in the Sacramento Valley. Thus, the Central Valley has experienced substantial shifts in the distributions of all waterfowl, reflecting significant changes at the 2nd order level of habitat selection.

Most of these distributional shifts of waterfowl in the Central Valley have been driven by the large-scale changes in habitat availability and 3rd (and possibly 4th) order levels of habitat selection. Currently, dabbling ducks in the Central Valley rely on three major habitat types: 1) flooded harvested rice fields, 2) managed seasonal wetlands, and 3) flooded and unflooded harvested corn fields (Central Valley Joint Venture 2006). Geese in the valley also use unflooded rice fields and uplands. Petrie *et al.* (2014) estimated that winter-flooded rice fields provided 44% of all food energy available to dabbling ducks in flooded habitats in the Central Valley, while flooded and unflooded rice fields provided 49% of all food energy available to dark geese but 73% of all food energy for white geese. These results were corroborated by Fleskes *et al.* (2005); they reported the importance of agricultural habitat (relative to managed wetlands) for Northern Pintail, Mallard and Greater White-fronted Geese *Anser albifrons* was greater than 20–30 years ago, presumably as birds increased their use of flooded rice fields.

In addition to the above patterns, the importance of managed wetlands has increased in the Suisun Marsh. Most waterfowl that winter in Suisun Marsh are dabbling ducks, which primarily use managed wetland habitats provided by duck hunting clubs and state wildlife areas (Ackerman *et al.* 2014). Coates *et al.* (2012) radio-marked and relocated 330 female Northern Pintail in the Suisun Marsh to estimate resource selection during non-breeding months and found strong evidence for selection of managed wetlands. Ackerman *et al.* (2014) reanalysed Northern Pintail telemetry data to examine habitat selection. They compared spatial patterns of habitat use by ducks to availability of habitats at two spatial scales and found that Northern Pintail strongly selected managed wetland habitats at both small and large scales. Further, Northern Pintail avoided tidal marshes, bays, sloughs and some other habitats (Ackerman *et al.* 2014). These results have important implications for Northern Pintail given current efforts to restore large portions of the Suisun Marsh to tidal wetlands. The consequences for dabbling ducks using the marsh have not yet been thoroughly assessed, and loss of managed wetlands in the Suisun Marsh remain a concern for waterfowl managers (Ackerman *et al.* 2014).

Patterns of habitat selection by waterfowl in the Central Valley represent large-scale shifts in the area and type of habitats available; as a consequence, significant changes in 2nd and 3rd order habitat selection have occurred by many species of ducks and geese. Most remaining wetlands are intensively managed to produce seed-producing moist-soil plants. The decline of Northern Pintail has resulted in management of seasonal wetlands toward more densely vegetated marshes favoured by Mallard. This technique has reduced amount of sparse and short vegetation which is likely more representative of seasonal flooded wetlands sought historically by Northern Pintail. The greatest recent change in the Central Valley has been the considerable increase in rice acreage, especially in the Sacramento Valley. This change has led to a northern shift from the San Joaquin Valley by most species (2nd order habitat selection) and a substantial increase in use of flooded and unflooded rice fields as foraging habitat (3rd order). Indeed, rice landscapes have become so important to wintering waterfowl that decline or loss of this agriculture would seem catastrophic to Northern Pintail and likely other wetland-dependent birds (Petrie *et al.* 2014). Nearly half of all duck-use-days in the U.S. portion of the Pacific Flyway occur in the Central Valley, and loss of rice would have continental impacts on Northern Pintail and other waterfowl using ricelands (Petrie *et al.* 2014). However, the future of flooded rice as winter habitat for waterfowl is in question with recent record droughts, water requirements for in-stream flows to meet needs of several species of federally endangered fish, and ever-growing urban demands. Petrie *et al.* (2014) estimated that > 75,000 ha of additional managed moist-soil wetlands would be required to replace the waterfowl food value provided by existing ricelands in the Central Valley. While rice agriculture is unlikely to disappear from the valley, the total acreage and the way it is managed post-harvest are uncertain.

Understanding the shifting mosaic of available winter habitats and bird responses will be an ongoing research need to guide conservation initiatives.

Pacific Coast

San Francisco Bay is the largest estuary along the west coast of the continental U.S. and historically important migration and wintering grounds for sea and other diving ducks (Conomos *et al.* 1985; Hothem *et al.* 1998). More than 85% of the tidal wetlands of the Bay have been lost to agriculture and development in the 20th Century (Nichols *et al.* 1986; Hothem *et al.* 1998). Anthropogenic changes and impacts have affected numerous waterfowl and other birds, including Canvasback *Aythya valisineria* whose overwintering numbers dropped by 50% during the 1970s–1990s (Hothem *et al.* 1998). Despite habitat modifications, San Francisco Bay may harbour nearly 50% of the total population of several diving duck species during winter (Accurso 1992; Brand *et al.* 2014). Given the history of mining in California, the position of the San Francisco Bay makes it susceptible to accumulating contaminants such as mercury, cadmium and selenium (Heinz *et al.* 1989; Hothem *et al.* 1998).

Farther up the northern California coast, the coastal lowlands are important migration and wintering areas for > 20 species of waterfowl, with populations ranging from 25,000–100,000 birds per day from autumn through spring (Pacific Coast Joint Venture 2004). Humboldt Bay is particularly important for brant because of its extensive Common Eelgrass *Zostera marina* beds. An estimated > 40% of the Pacific Flyway population of brant use Humboldt Bay as a migratory stopover from late February through to mid-April.

Inter-mountain West and Great Salt Lake

The Inter-mountain West region comprises two regions of special importance to non-breeding waterfowl: Southern Oregon Northeastern California (SONEC), including the Klamath Basin, and the Great Salt Lake. The SONEC region covers approximately 10% of the Great Basin, although waterfowl habitat comprises a much lower percentage (Petrie *et al.* 2013). Historically, peak waterfowl abundance occurred during autumn and spring migration. Migrating waterfowl in autumn likely would have experienced dry conditions and were probably restricted to a few large complexes of permanent or semi-permanent wetlands (Petrie *et al.* 2013). Few birds remained over winter because of the below-freezing winter temperatures. Today, nearly all autumn and winter waterfowl habitat in SONEC occurs on public land. Two refuges are of particular significance: Lower Klamath National Wildlife Refuge (Lower Klamath) and the Tule Lake National Wildlife Refuge (Tule Lake). Although these refuges account for only a fraction of the region, they support a significant portion of the waterfowl that use SONEC in autumn and winter (Kadlec & Smith 1989; Fleskes & Yee 2007). In fact, the Klamath Basin is recognised as a region of continental significance to North American waterfowl populations (NAWMP Plan Committee 2004).

Management of waterfowl habitats on Lower Klamath and Tule Lake refuges depends on water supplies. Increasing demands for water within the Klamath Basin by farmers, native communities and endangered fish have hindered refuges from obtaining sufficient water for waterfowl. A recent analysis using bioenergetics models (TRUEMET) indicated that food resources at Tule Lake were adequate to meet energy needs of diving ducks and swans, but were insufficient for dabbling ducks and geese. Food for dabblers was exhausted in early autumn, well before traditional peak migration in November (Petrie *et al.* 2013). Thus, dabbling duck numbers at Tule Lake have declined significantly since the 1970s. The SONEC region is also critical during spring migration, especially for Northern Pintail. Over 70% of habitat use by radio-marked Northern Pintail in SONEC (outside of the Lower Klamath) occurred on privately-owned habitats, primarily flood-irrigated agriculture (Fleskes *et al.* 2013).

The Great Salt Lake (GSL) is one of the largest wetland complexes in western U.S. and is recognised internationally for its importance to migratory waterfowl (NAWMP Plan Committee 2004). As many as 3–5 million waterfowl migrate through the GSL annually (Petrie *et al.* 2013). The GSL is surrounded by >190,000 ha of wetlands maintained by fresh water from rivers that flow into the basin. The surrounding marshes are extensive and provide rich diversity of invertebrate and plant food resources (Petrie *et al.* 2013). Waterfowl use of the GSL is greatest during late summer – early autumn and also in spring. Peaks occur in September, with birds arriving from northwestern and mid-continent Canada and Alaska, and some from the Prairie Pothole Region. Banding data indicate that many ducks that migrate through the GSL spend the winter in the Central Valley of California and west coast of Mexico (Petrie *et al.* 2013). Use of GSL by waterfowl is lowest in mid-winter but increases during spring. Dynamic ebbs and flows of water and fluctuating lake salinities are significant in maintaining this productive wetland system (Petrie *et al.* 2013).

The Inter-mountain West Joint Venture estimated 17.4 million waterfowl-use-days of the GSL during winter of which dabbling ducks accounted for 74% (Northern Pintail = 39% of dabbling duck use-days; Green-winged Teal = 23%; Mallard = 21% and Northern Shoveler = 11%), while diving ducks comprised 19% of total waterfowl-use-days during winter, with Common Goldeneye *Bucephala clangula* representing 91% of all diving duck use (Petrie *et al.* 2013). Bioenergetics analyses of food supplies in the GSL needed to support migratory waterfowl suggested that seed resources required by dabbling ducks were depleted during autumn migration by late October (Petrie *et al.* 2013). Yet, there may have been > 1 million dabbling ducks alone in the GSL in October and November. These results suggest that dabbling ducks are obtaining unknown but critical energy supplies from perhaps aquatic invertebrates, submerged aquatic vegetation, tubers, or a combination of these (Petrie *et al.* 2013). Petrie *et al.* (2013) concluded that improved understanding and estimation of the spatiotemporal variability of wetland resources and waterfowl resource selection in the GSL system were needed to refine assumptions about the foraging guilds.

Europe

As in North America, substantial changes in land use and management have occurred in Europe since the early 20th century, where landscapes at staging and wintering areas for waterfowl are now a matrix of agricultural land and other habitats greatly transformed by humans (*e.g.* industrial and residential zones) which envelop small protected areas of remaining wetlands (Thomas 1976; Owen *et al.* 1986; Tamisier & Grillas 1994; Guglielmo *et al.* 2002). Autumn-migrating Western Palearctic waterfowl largely concentrate in a flyway corridor along the Baltic and North Sea coasts (*e.g.* Scott & Rose 1996; Söderquist *et al.* 2013; Calenge *et al.* 2010). Here, the global concerns of sea level rise and other loss of habitat associated with climate change are serious concerns for waterbirds in coastal wetland habitats (*e.g.* Clausen & Clausen 2014), which are further threatened by eutrophication (*e.g.* declines in seagrass beds, Clausen *et al.* 2012) and the encroachment of vegetation that is less nutritious for waterfowl (*e.g.* Common Cord-grass *Spartina anglica*; Percival *et al.* 1998). In contrast, climate warming and increased fertilisation of grasslands in northwest Europe may have enhanced terrestrial habitats for geese, where several populations are flourishing, and some are short-stopping or becoming partly non-migratory (*e.g.* Greylag Geese *Anser anser*, Voslamber *et al.* 2010; Barnacle Goose *Branta leucopsis*, Ganter *et al.* 1999). Hunting restrictions also have likely enhanced the abundance and influenced the distribution and timing of migration of swans and some goose populations. Further south along the flyway, wintering waterfowl, especially ducks (*e.g.* Eurasian Wigeon *Anas penelope*), have largely switched from using marine habitats to freshwater wetlands during daylight hours as the latter have increasingly been managed as nature reserves since the 1950s (*e.g.* Owen & Williams 1976; Guillemain *et al.* 2002). Reserves nowadays not only provide safety from hunting and other human disturbance, but habitats are managed specifically for waterfowl. Yet despite active habitat management, there is an increasing awareness that alien species (*e.g.* Red Swamp Crayfish, *Procambarus clarkia* and Water Primrose *Ludwigia* sp. and Swamp Stonecrop *Crassula helmsii*) are a threat to protected U.S. habitats and European wetlands (*e.g.* Dandelot *et al.* 2005; Meineri *et al.* 2014).

Along the Mediterranean coasts, primary wintering habitats of waterfowl are brackish lakes, lagoons and temporary wetlands. Wetlands of the Mediterranean region have been reduced by 80–90% by urban population growth and conversion to agriculture (Toral & Figuerola 2010). Fortunately, some of these are now rice fields which, as in North America, provide valuable resources to wintering waterfowl (*e.g.* Tamisier & Grillas 1994) and help compensate lost wetland habitats (*e.g.* Tourenq *et al.* 2001; Rendón *et al.* 2008). In the Camargue, southern France, portions of remaining natural wetlands are protected and most are on private estates, wherein temporary and seasonal wetlands are flooded beyond natural hydroperiods to attract waterfowl for hunting and observing.

This practice is detrimental to wetland biodiversity in general, but it has greatly promoted hydrophyte beds on which waterfowl forage (Tamisier & Grillas 1994). Such management is mostly beneficial to herbivorous species (*e.g.* Gadwall) but the other dabblers also benefit from seeds spread as bait in these properties (Brochet *et al.* 2012). Hunting management practices could likely be responsible for considerable improvement of wintering body condition of Common Teal (up to 12%) and other dabbling ducks in past decades (Guillemain *et al.* 2010b).

Habitat resources of selected northern hemispheric waterfowl

Dabbling ducks

Mallard

Mallard challenge clear distinctions of autumn migration and subsequent winter habitat distributions because of great seasonal and annual variation in settling by individuals or sub-populations within flyways. The breadth of habitats occupied by Mallard in North America is particularly fascinating. In the Sacramento Valley of California, Mallard use agriculturally dominated and largely treeless environments, where patches of seasonally flooded and emergent wetlands and flooded rice fields mostly occur, notwithstanding the Butte Sink wherein riparian wetlands consisting of willow *Salix* sp., California Sycamore *Platanus recemosa*, Buttonbush *Cephalanthus occidentalis* and other woody and herbaceous species exist (Gilmer *et al.* 1982; Heitmeyer *et al.* 1989; Eadie *et al.* 2008; Elphick *et al.* 2010). In Central U.S., Mallard use Gulf coastal and interior wetlands, cattle ponds, irrigation and flood-control reservoirs, playa lakes, seasonal wetlands, riparian and flooded forest wetlands, rivers and irrigation canals, plus flooded and dry agricultural lands including grain and legume crops within their geographic ranges from the Gulf Coast to southern Canada (Jorde *et al.* 1984; Chabreck *et al.* 1989; Miller *et al.* 2000; Link *et al.* 2011). In the Atlantic Flyway, Mallard use coastal and inland freshwater emergent marshes and managed wetlands developed from 18th century rice fields (Gordon *et al.* 1989, 1998). Perhaps most intriguing is the winter residency of some Mallard along the sandbar flats of the Missouri River in North Dakota, where these birds tolerate frequent inhospitable winter conditions while largely subsisting on Rainbow Smelt *Osmerus mordax* (Olsen & Cox, Jr. 2003; Olsen *et al.* 2011).

The MAV is considered the ancestral wintering grounds of North American Mallard (Nichols *et al.* 1983; Reinecke *et al.* 1989; Heitmeyer 2006). Nichols *et al.* (1983) examined winter distributions of Mallard and found support for the flexible homing hypothesis, given that Mallard wintered farther south in United States during wetter and colder winters (also see Green & Krementz 2008). Mallard typically migrate in autumn from latitudes of central Missouri after cumulative days of temperatures of ≤ 0°C, snow cover and ice conditions (*i.e.* weather severity index (WSI) of ≥ 8; Schummer *et al.* 2010). A quadratic and cumulative WSI model explained ≥ 40% of the variation in changes in relative abundance of Mallard and other dabbling ducks in Missouri during autumns–winters 1995–2005 (Schummer *et al.* 2010, 2014).

Recent capture-recapture results suggest similar patterns in Europe (Dalby 2013). Interestingly, satellite-marked Mallard in the Mississippi Flyway (Krementz et al. 2012) revealed patterns of incremental migrations similar to those described by Bellrose (1980).

Mulhern et al. (1985) investigated use and selection of wetlands by Mallard broods in Saskatchewan and found that broods used structurally different wetlands, but use was in proportion to availability of wetland types and thus not selective. How this apparent plastic habitat use by brooding ducks may ramify into habitat use subsequently during autumn and winter unearths interesting questions: 1) What drives individuals to seek and use diverse habitats? 2) What are survival and fitness outcomes related to these decisions? 3) What non-breeding habitat complexes are associated with greatest survival rates of individuals? 4) Where do these birds breed, and what are their reproductive outcomes? For example, do more competitive or fit Mallards occupy the MAV, the supposed region of greatest habitat quality for the species (Nichols et al. 1983), whereas other Mallard distribute to other regions? Alternatively, perhaps the regions occupied have little influence on fitness prospects, so long as adequate food, freshwater and potential mates are available. As previously mentioned, evidence exists that habitat complexes used by the greatest densities of Mallard and those individuals with greatest winter survival rates in the MAV differ in habitat composition (Pearse et al. 2012; Lancaster 2013; Kaminski & Davis 2014). Drivers of differential habitat use are not always clear but are likely related to foraging, weather, disturbance or a combination of these and other factors related to survival during winter. For example and relative to 3rd and 4th order selection, Mallard used irrigation canals in Nebraska agricultural landscapes over nearby natural riverine wetlands during harsh winters because canals were climatically more suitable than other habitats (Jorde et al. 1984). Additionally, Mallard may exercise trade-offs by selecting habitats of perhaps lesser foraging quality but prone to fewer disturbances which contribute to greater survival. Krementz et al. (2012) postulated that Mallard may forego wintering in the Grand Prairie region of Arkansas to avoid this area because of intense hunting pressure.

Northern Pintail

Similar to their reliance on rice in California's Sacramento Valley, ~52% of all locations (n = 7,022) of radio-marked Northern Pintail females were in rice habitats, which included active (18% use) and fallow rice fields (34% use) along the coast of Texas (Anderson & Ballard 2006). Many radio-marked female pintail that were located > 64 km from the Texas rice prairies flew to rice field habitats at some point during winter, which demonstrated the importance of flooded ricelands to pintail in this region (Anderson & Ballard 2006). In Louisiana, Cox, Jr. and Afton (1997) found extensive use of sanctuaries by radio-tagged Northern Pintail during hunting seasons, but less so before and after legal waterfowl seasons. Female pintail used flooded rice and fallow fields nocturnally where combined these habitats accounted for 68–93% of nocturnal use by the birds (Cox, Jr. & Afton 1997).

In California, Fleskes *et al.* (2007) attributed greater survival of Northern Pintail to increased area of flooded rice habitats. Other landscape factors important to pintail survival, such as the size and management of sanctuaries, types of feeding habitats (*e.g.* rice, wetlands) and the juxtaposition of these, may also have been important (Fleskes *et al.* 2007). Nonetheless, contemporary (1998–2000) survival estimates (87–93%) of adult female Northern Pintail in the Suisun Marsh and Sacramento and San Joaquin Valleys were greater than in any other region of North America (Fleskes *et al.* 2007). Clearly, sanctuaries adjacent to rice and other agricultural habitats are critical to survival and habitat use by Northern Pintail throughout their wintering range (Cox, Jr. & Afton 1997; Fleskes *et al.* 2007).

Wood Duck

The North American Wood Duck is the only *Aix* species in the Nearctic (Birds of North and Middle America Check list; http://checklist.aou.org/). Wood Duck are also unique among North American waterfowl, because they are the only species with migratory and non-migratory populations (Baldassarre 2014). Wood Duck have been widely studied in North America since their near extirpation in the early 20th century (Bellrose & Holm 1994). Migration routes of Wood Duck are not well defined, given the substantial overlap in breeding and winter ranges (Baldassarre 2014). Given their broad occupancy of geographic areas, Wood Duck use diverse freshwater wetlands, although they avoid brackish and marine systems (Bellrose & Holm 1994; Baldassarre 2014). Despite being a forested wetland specialist, wherein Wood Duck forage on red oak *Quercus* sp. acorns and aquatic invertebrates (Heitmeyer *et al.* 2005; Foth *et al.* in press), Wood Duck also use flooded croplands where they forage on waste agricultural seeds (Delnicki & Reinecke 1986; Bellrose & Holm 1994; Barras *et al.* 1996; Kaminski *et al.* 2003). Much of the non-breeding information about Wood Duck is derived from eastern populations and birds using the MAV and southern Atlantic Flyway (Arner & Hepp 1989; Reinecke *et al.* 1989; Peterson 2014), but much remains to be learned about non-breeding Wood Duck use and selection of unique habitats in regions such as the Central Valley of California and even xeric environments in Nevada that lack traditional expansive bottomland hardwood forests (Baldassarre 2014).

Diving Ducks

Ducks that are among the more ecologically pelagic have historically used estuarine or freshwater systems, usually along coastlines, shorelines of lakes and major rivers (Bellrose 1980). The significance to diving ducks *Aythya* sp. of myriad bays of North America, including Chesapeake and San Francisco Bays, has been recognised for centuries (Audubon 1840; Haramis 1991a,b; Perry *et al.* 2007). Unfortunately, these systems are often plagued by anthropogenic effects of shoreline development, boat traffic, increased sediments and nutrients and other factors (Perry *et al.* 2007; Lovvorn *et al.* 2013). Knowledge of niche overlap and "carrying capacity" of habitats by these ducks is necessary to understand relations between birds and potential invertebrate or other prey (Lovvorn *et al.* 2013).

Diving ducks wintering in Chesapeake Bay from 1950–1995 comprised 23% of Atlantic Flyway and 9% of North American populations of these ducks (Perry & Deller 1995; Perry *et al.* 2007). Some species wintering in Chesapeake Bay have been more adversely affected than others. For example, Redhead and Canvasback that feed on submerged aquatic vegetation (SAV), seeds and tubers have been impacted more than species that forage in slightly deeper water on invertebrates, particularly Lesser Scaup (Perry *et al.* 2007). Increased nutrients and sedimentation have lessened SAV in shallower reaches of Chesapeake Bay (Perry *et al.* 2007). Moreover, recently expanding hypoxic zones may be negatively impacting sessile prey of diving ducks (Perry *et al.* 2007) and have been linked to decreased body mass and survival in Canvasback (Haramis *et al.* 1986).

Pollutants and invasive species are thought to be especially problematic for diving ducks such as scaup and Canvasback (Lovvorn *et al.* 2013). In San Francisco Bay, Hothem *et al.* (1998) found that mercury and selenium levels in late winter had accumulated in scaup and Canvasback to levels that impair reproduction in game-farm Mallard (Heinz *et al.* 1989). Invasive species, such as Asian Clam *Potamocorbula amurensis*, which has displaced the former bivalve prey community (*e.g. Macoma balthica*), are considered a second primary concern for diving ducks in the Bay (Richman & Lovvorn 2004; Lovvorn *et al.* 2013). Asian Clams may harbour greater levels of selenium than other bivalve species (Richman & Lovvorn 2004), which could be especially problematic to Lesser and Greater Scaup as they comprised as much as 43–47% of all waterfowl in the Bay. Richman and Lovvorn (2004) collected Lesser Scaup in winters 1998–2000 and found that 98% of clams consumed by scaup were Asian Clams. Asian Clams apparently provide scaup with a profitable food source, because they mostly are distributed in the top 5 cm of sediments where scaup intake rates are greatest (Richman & Lovvorn 2004). Additionally, Lesser and Greater Scaup and Surf Scoter *Melanitta perspicillata* wintering in San Francisco Bay had decreased body mass and fat and increased foraging effort, causing them to disperse from upon food limitation. There also was substantial niche overlap and opportunistic use of dominant prey species by these ducks (Lovvorn *et al.* 2013). Lovvorn *et al.* (2013) concluded that scaup and scoter did not exploit a substantial fraction of food above local profitability thresholds before abandoning the habitat, and encouraged future research to better understand thresholds of energetic profitability for diving ducks.

Despite vast size and dynamics of San Francisco Bay, adjacent habitats in the region provide vital resources for some species using the Bay. Specifically, estuarine intertidal and subtidal mudflats and salt ponds provide additional food and water for diving ducks (Dias 2009; Brand *et al.* 2014). Brand *et al.* (2014) found that diked salt ponds, salt pans and managed seasonal wetlands in South San Francisco Bay collectively provided enough food energy to sustain 79% of the energy and nutrients required by diving ducks when birds were at maximum numbers, and basically 100% of the nutrients when average bird abundances prevailed. Managed ponds serve as important roosting and foraging habitats in this region. Ponds that intake, circulate or discharge water directly to or from the Bay or adjacent sloughs supported > 95% of the diving duck abundance (Brand *et al.* 2014). However, greater bird and invertebrate abundances and prey energy density occurred in meso-haline (*i.e.* 5–30 ppt) rather than low-hypersaline (*i.e.* 31–80 ppt) circulation ponds (Brand *et al.* 2014). Ruddy Duck *Oxyura jamaicensis* exercise dietary flexibility in these same wetland complexes, feeding on amphipods *Amphipoda* sp. or polychaetes *Polydora* sp. depending on prey occurrence or abundance among different wetland types (Takekawa *et al.* 2009; Brand *et al.* 2014). Thus, similar to identifying important habitat complexes for Mallard or other dabbling ducks (Pearse *et al.* 2012; Lancaster 2013), maintaining diverse foraging wetlands in ecosystems like San Francisco Bay is imperative for supporting waterfowl and other wetland dependent birds using this system (Brand *et al.* 2014).

A primary difference between historical and contemporary habitat use for some diving ducks, such as Ring-necked Duck in the U.S., has been a shift away from traditional winter habitats to open-water lakes because of a proliferation of invasive plants such as Hydrilla *Hydrilla verticillata* and other species that form dense floating mats (Johnson & Montalbano 1984; Roy *et al.* 2013). Some of the greatest wintering concentrations of Ring-necked Duck may occur in managed impoundments of coastal and inland Louisiana (Roy *et al.* 2013). Ring-necked Duck use small marshes adjacent to open water, whereas Canvasback, Redhead and scaup typically use open-water areas only (Korschgen 1989; Roy *et al.* 2013). Elsewhere herein, Stafford *et al.* (2014) provided a detailed account of scaup habitat use during late winter and spring migration. Diverse coastal and interior wetlands of south-central Louisiana are critical to diving ducks such as Redhead and Canvasback (Hohman & Rave 1990; Hohman *et al.* 1990). Canvasback in the Mississippi River Delta and at Catahoula Lake in Louisiana, both important wintering areas to these species (Hohman *et al.* 1990), consumed about 97% plant matter at each site, with below-ground plant biomass composing 94% aggregate dry mass (Hohman & Rave 1990). Mudflats with tubers or water that permitted Canvasback to tip-up and feed were important components of used habitats (Hohman & Rave 1990). Similar to plant-eating Canvasback, the importance of Shoal Grass *Halodule wrightii* to Redhead and several avian guilds has long been mentioned (Cornelius 1977; Michot *et al.* 2008). Redhead wintering in the Chandeleur Sound of Louisiana and Laguna Madre, Texas consumed as much as 74% dry mass of shoalgrass (Michot *et al.* 2008). Conserving *Halodule* beds arguably is the most critical conservation priority within the winter range of Redheads, particularly given that most of the North American population of the species winters along coastal habitats of Texas and Louisiana (Michot *et al.* 2008).

Sea ducks

North America

There are 15 species of North American sea ducks (Tribe: Mergini) and arguably they are the least understood taxa of waterfowl (Bellrose 1980; Goudie *et al.* 1994; Silverman *et al.* 2013). Evidence suggests that 10 of these species are in decline, including eight of 12 species that winter off the Atlantic coast of North America, a primary wintering area for this tribe (Sea Duck Joint Venture 2004; Zipkin *et al.* 2010). Eleven species of sea ducks commonly winter in Pacific coastal regions, nine of which commonly occur in the Puget Sound of Washington state (Faulkner 2013). Sea duck declines are occurring concomitantly with uncertainty about their habitat preferences (Zipkin *et al.* 2010). Shoreline development and associated pollution and climate change are potential negative influences on sea ducks in North America (Zipkin *et al.* 2010). Recent proposals for wind turbines along the Atlantic coast and threats from offshore energy development will also challenge sea ducks, so further understanding of habitat selection by these ducks is imperative (Zipkin *et al.* 2010).

Spatial distribution of sea ducks is generally determined by winter weather conditions and habitat diversity (Zipkin *et al.* 2010). At greater spatial winter ranges, food availability, local environmental conditions, habitat suitability, ocean depths and water temperatures influence sea ducks' use of habitats (Lewis *et al.* 2008; Zipkin *et al.* 2010; Dickson 2012). Northern seas are hostile during winter, with below freezing temperatures, wind, ice and limited daylight because the sun is below the horizon for two months (Systad *et al.* 2000). Sea ducks, however, remain in these rigorous environments during winter and forage on molluscs, echinoderms, crustaceans and other invertebrates. These foods are depauperate in energy density, so sea ducks must forage voraciously to maintain positive energy balances (Systad *et al.* 2000).

Surf Scoter *Melanitta perspicillata* and White-winged Scoter *M. deglandi* in the Pacific Flyway use soft-bottom habitats and forage on bivalves (Bourne 1984; Richman & Lovvorn 2003; Lewis *et al.* 2008). Scoters encounter considerable variation in clam densities and potentially face an exhaustible food supply (Lewis *et al.* 2008). However, Lewis *et al.* (2008) found that scoters in Baynes Sound (British Columbia) did not switch winter prey or move extensively to foraging sites, suggesting clam density was relatively high there (Kirk *et al.* 2007).

Sea ducks in the eastern U.S. have been monitored by the Atlantic Flyway Sea Duck Survey (AFSDS) in at least nine bays and sounds off of the Atlantic coast to quantify winter distributions and population indices (Migratory Bird Data Center 2009; Zipkin *et al.* 2010). Zipkin *et al.* (2010) modelled effects of bottom depths, monthly averages of sea surface temperature, and ocean floor topography for five species of wintering sea ducks. The North Atlantic Oscillation (NAO; *i.e.* fluctuation in sea surface pressure across the northern Atlantic Ocean between areas of high (Azores High) and low (Icelandic Low) pressure: Ottersen *et al.* 2001; Stenseth *et al.* 2002; Hurrell *et al.* 2003; Zipkin *et al.* 2010) was the only environmental covariate that had a significant influence on all five species; its effect was negative for the three scoter species and positive for Common Eider and Long-tailed Duck *Clangula hyemalis* (Zipkin *et al.* 2010). These results suggest that climatic conditions along the Atlantic coast during migration and winter may have direct or indirect influences on sea duck distributions, perhaps as prey are re-distributed (Zipkin *et al.* 2010). Scoters predominated inshore during cold, snowy winters and Common Eider and Long-tailed Duck were more abundant inshore during wet, mild winters (Zipkin *et al.* 2010). Sea surface temperature (SST) negatively affected Long-tailed Duck and White-winged Scoter abundance but positively affected Common Eider, although there was some interaction of effects between NAO and SST on birds' habitat distribution. Overall, sea ducks may respond to a combination of local habitat conditions and broader-scale weather patterns (Zipkin *et al.* 2010). Collectively, scoters used flatter bottom sites, which seemed consistent with knowledge that Black Scoter *Melanitta americana*, Surf Scoter and White-winged Scoter preferred sandier basins along the Atlantic shoreline (Stott & Olson 1973; Zipkin *et al.* 2010). In contrast, Common Eider used rugged substrates, but Long-tailed Duck have not yet been linked to bottom substrates (Perry *et al.* 2007; Zipkin *et al.* 2010).

Other important habitats for non-breeding sea ducks in central and eastern North America include the Great Lakes and Chesapeake Bay (Schummer *et al.* 2008). Mixed species of Bufflehead *Bucephala albeola*, Common Goldeneye and Long-tailed Duck use inshore areas of Lake Ontario and forage on energy-dense *Amphipoda* and larvae of *Chironomidae*, both abundant in the shallow-water zone near shore (Schummer *et al.* 2008). Despite concentrated mixed flocks of ducks, Schummer *et al.* (2008) did not detect declining abundances of macroinvertebrates during winter. They concluded that exploitative competition was likely not occurring and interference competition appeared below thresholds that would cause birds to spatially segregate. Overall, winter forage did not appear to limit habitat use of these species in Lake Ontario during winter (Schummer *et al.* 2008).

Chesapeake Bay is considered one of the most important areas for several species of scoters and Long-tailed Duck (Sea Duck Joint Venture 2004; Ross *et al.* 2009), but little is known about the birds' use of the system. Surf Scoter *M. perspicillata* is thought to forage preferentially in subtidal, sandy soft sediment habitats > 6 m deep (Ross *et al.* 2009), but will also use hard-substrates (Lewis *et al.* 2007; Perry *et al.* 2007). Long-tailed Duck in the upper Chesapeake Bay primarily consume bivalves (Perry *et al.* 2007), likely procuring food from soft-sediment areas (Żydelis & Ruskete 2005; Ross *et al.* 2009). Ross *et al.* (2009) suggested that limited availability of hard substrate bottom in Chesapeake Bay might dictate habitat use patterns among these sea ducks in the upper Chesapeake compared to other regions. Further concerns are linked to declining water quality since the 1960s in the lower region of Chesapeake Bay (Ross *et al.* 2009). Excessive sedimentation and nutrient loading have caused eutrophication and oxygen depletion, negatively affecting portions of the Bay's substrate, and are linked to dramatic declines in seagrass beds (Chesapeake Bay Program 2007; Ross *et al.* 2009). These consequences are problematic because seagrasses supply important substrates for bivalves compared to bare ground under the Bay (Peterson 1982; Peterson *et al.* 1984; Ross *et al.* 2009).

Europe

Recent count data indicate that most European sea duck populations, with the exception of Common Goldeneye, are now in decline (Hearn & Skov 2011; Skov *et al.* 2011). Common Goldeneye winter extensively in freshwater habitats along coastlines, whereas other sea ducks tend to have an offshore distribution. The Baltic Sea is the key wintering area for most European sea ducks, and it is a region of major concern. Recent surveys indicate that Long-tailed Ducks, Velvet Scoter and Steller's Eider have declined by 65%, 55% and 66%, respectively, with declines in Common Eider (51%), Common Scoter (47%), Red-breasted Merganser *Mergus serrator* (42%) and Greater Scaup (26%) also recorded (Skov *et al.* 2011). Declines have similarly been reported in other European countries, notably in Britain and the Netherlands, which are also important wintering grounds for European sea duck populations. Generally, wintering sea ducks aggregate in shallow coastal waters or over offshore banks where they can dive for food on the sea floor. In winter, > 90% of sea ducks use areas amounting to < 5 % of the Baltic Sea (Bellebaum *et al.* 2012), where they forage primarily on Blue Mussel *Mytilus edulis*.

Ecosystem changes that have a negative effect on habitat and food resources during the non-breeding season (*e.g.* extraction of sand and gravel, dredging of shipping channels or coastal development), are potentially the most important explanation for the decline in arctic-breeding sea duck populations (Skov *et al.* 2011). Moreover, shipping and offshore wind farms may permanently displace sea ducks from favoured feeding grounds (Petersen *et al.* 2006; Skov *et al.* 2011). Among sea duck species, the Long-tailed Duck is particularly sensitive to wind farms (Petersen *et al.* 2006), and plans for offshore wind farm construction exist in all Baltic countries. Traffic along the major shipping routes (which cross or pass close to Long-tailed Ducks wintering sites) is also predicted to increase (Skov *et al.* 2011). Oil illegally discharged from ships continues to kill tens of thousands of birds each year, despite enforcement of international regulations (Larsson & Tydén 2005; Skov *et al.* 2011; Brusendorff *et al.* 2013), and other hazardous chemicals are suspected of having a negative impact on Baltic wildlife (including sea ducks) when birds ingest bivalves or organisms that filter polluted sea water (*e.g.* Pilarczyk *et al.* 2012; *cf.* Skov *et al.* 2011). Additionally, sea duck food resources in the Baltic Sea have changed substantially in recent decades concomitantly with nutrient loading. Increase of nutrient loads after 1950 might explain rising bivalve biomass in shallow waters, which in turn may have stimulated sea duck population growth. But decreases in nutrient loads (nitrogen and phosphorous) have occurred in some coastal regions since the 1990s, whereas nutrient levels remain high in other parts of the Baltic Sea. Declines in nutrient loads along the coastline and subsequent effects on sea duck food quality need further investigation. Nevertheless, Skov *et al.* (2011) stressed the importance of eutrophication in spatio-temporal variability in food supply for and abundance of waterbirds in the Baltic Sea, with control of eutrophication being a plausible reason for the decrease of several benthic species in Danish waters.

Phytoplankton composition also has changed in the Baltic, perhaps through the increase in water temperatures in recent decades or overfishing leading to a decrease in food quality for filter-feeding bivalve mussels. In addition, in warmer waters mussels metabolise their own reserves during winter instead of hibernating, which could decrease the quality of mussels for bivalve feeders (Waldeck & Larson 2013). Lastly, overexploitation by commercial mussel fisheries (*e.g.* in the Wadden Sea) may cause food shortages for bivalve feeding species such as Common Eiders (Skov *et al.* 2011).

Concomitant with warming temperatures of the Baltic Sea, ice coverage has decreased and permitted access to new wintering areas for waterfowl. Common Goldeneye and some *Aythya* species are shifting northward in their wintering distribution in the Baltic Sea (Skov *et al.* 2011; Lehikoinen *et al.* 2013). The limited degree of northward shift in the distribution of seaduck feeding offshore suggests reduced food availability in the northern Baltic area, which is now partly ice-free in winter. Nevertheless, populations of some species including Common Eider have relocated to the southwest Baltic Sea from previous wintering quarters in northwest Denmark. Lastly, European sea duck populations also may be directly or indirectly affected by commercial fishing and the use of gillnets for fishing (Žydelis *et al.* 2009).

Geese

Geese and agriculture

As for ducks, habitat modifications influence distribution, movement and resource exploitation in geese. Geese are generally more adept at exploiting farm crops than most duck species (Owen 1980), so their autumn and winter habitat use is largely driven by and has changed markedly in response to variations in farming practices, both in North America and Europe, during the 20th and 21st centuries. For example, Pacific Flyway Greater White-fronted Geese commonly stage in the SONEC, and then migrate and winter in the Sacramento and San Joaquin Valleys (Ely 1992; Ackerman *et al.* 2006; Ely & Raveling 2011). Approximately 80% of foraging flocks of White-fronted Geese used harvested barley, wheat or oat fields from early September to mid-October in SONEC, 1979–1982, then switched to potato fields by mid-October–late November of those years (Frederick *et al.* 1992; Ely & Raveling 2011). When White-fronted Geese migrated to the Sacramento Valley in autumn and winter, they primarily used complexes of rice field habitats (Ely & Raveling 2011). After White-fronted Geese departed the Sacramento Valley for the San Joaquin Valley, green forage, waste corn and other grain and vegetable crops were available to the geese, but birds disproportionately used corn relative to its availability (Ely & Raveling 2011). The future of Greater White-fronted Geese in the San Joaquin Valley is uncertain because corn acreage declined there by 20%, largely because of urbanisation (Ackerman *et al.* 2006; Ely & Raveling 2011). Changes in agricultural practices and crops produced are commodity-market driven and largely beyond the control of wildlife biologists, thus challenging to conservation planning (Ely & Raveling 2011; Skalos 2012; Petrie *et al.* 2014).

Another striking example of dynamic habitat use by geese within agricultural landscapes comes from the North American Snow Geese and Ross' Geese *Chen rossii* (Ankney 1996; Abraham *et al.* 2005). White goose use of waste grain is well documented in the literature (Alisauskas *et al.* 1988; Ankney 1996; Alisauskas 1998; Abraham *et al.* 2005). Recent research has sought to identify winter origins of white geese migrating through Nebraska's Rainwater Basin, a region of continental significance to autumn and spring migrating waterfowl and Sandhill Crane *Grus Canadensis* (Krapu *et al.* 1984; Alisauskas & Ankney 1992; Alisauskas 2002; Stafford *et al.* 2014). Henaux *et al.* (2012) used stable isotopc analysis and found flexibility in diets and regional landscape use by Snow Geese. They determined origins of wintering Snow Geese harvested in the Rainwater Basin as follows: Louisiana (53% and 9% in 2007 and 2008, respectively), Texas Gulf Coast (38% and 89%, respectively), Arkansas (9% and 2%, respectively). However, no birds from the Playa Lakes region were detected. Beyond annual variability in their winter origins, differences in diet also helped to characterise their winter habitat use. Snow Geese relied on rice and wheat fields (C_3 plants isotopic signature) as well as corn and grain sorghum (C_4 plants). Geese collected from Texas and Louisiana were generally characterised by using estuarine and marsh habitats *versus* uplands typical of Arkansas and playa eco-regions (Alisauskas & Hobson 1993).

General plasticity of North American white geese in exploiting agricultural and marsh habitats (Bateman *et al.* 1988; Alisauskas 1998; Jefferies *et al.* 2004) creates complex challenges in arresting the growth of overabundant populations in the 21st century (Batt 1997; Jefferies *et al.* 2004; Abraham *et al.* 2005). However, dwindling rice acreage in Texas may influence white goose population levels. For example, rice acreage was ~203,152 ha and white geese numbered > 1.2 million in 1979; whereas ~378,000 geese were counted and only > 54,000 ha of rice existed in Texas in 2013 (K. Hartke, Texas Parks and Wildlife, unpubl. data). The contemporary estimate of rice acreage is the lowest ever for Texas since records originated *ca.* 1948 (K. Hartke, Texas Parks and Wildlife, unpubl. data).

Similarly, contemporary estimates of geese wintering in the Western Palearctic are 4.8 million, up from 3.3 million in 1993 (Fox *et al.* 2010). Most species exhibit signs of exponential increase, whereas others (*e.g.* the Greenland White-fronted Goose *Anser albifrons flavirostris*, Red-breasted Goose *Branta ruficollis* and Dark-bellied Brent Goose *Branta bernicla*) have declined in recent years (Fox *et al.* 2010). Although reduced hunting pressure on geese in some regions probably played an important role, increases in most species of European geese have likely resulted from exploitation of grains and root and grass crops, similar to patterns in North America (Abraham *et al.* 2005; Fox *et al.* 2010). Since the 1950s, wild geese wintering in the western Palearctic have partially or completely switched from feeding on natural vegetation to managed pastures and agricultural croplands (Madsen 1998; Jensen *et al.* 2008; Hake *et al.* 2010). Agricultural producers in Europe have been concerned with losses of wheat and oilseed as goose and swan populations have increased (Dirksen & Beekman 1991; Rees *et al.* 1997). Several measures have attempted to deter geese from crops, including providing supplemental feed in accommodation fields to influence movements of and use by geese, scaring of birds, fencing habitats and adjusting farming strategies, such as growing barley varieties that mature and are harvested before varieties used previously (Hake *et al.* 2010).

Black Brant and estuarine-marine systems

Besides agricultural lands, estuarine and marine wetland systems are critical to many waterfowl, including Black Brant in North America (named Brent Goose in Europe). Important autumn staging areas for brant include shallow marine waters along shorelines, within lagoons or behind barrier beaches (Shaughnessy *et al.* 2012; Lewis *et al.* 2013). Some of the important habitats for the nine-month non-breeding period of brant include the Northeast Pacific United States, the lagoons along the west coast of Baja California, areas of Mexico and Atlantic coastal habitats (Smith *et al.* 1985; Lewis *et al.* 2013, Martínez Cedillo *et al.* 2013). Pacific Black Brant solely use natural habitats during winter and avoid agricultural lands (Ward *et al.* 2005; Lewis *et al.* 2013). As mentioned, the unifying food resource for Holarctic brant is eelgrass (Moore *et al.* 2004; Moore & Black 2006; Shaughnessy *et al.* 2012; Lewis *et al.* 2013). Macrogreen Algae *Ulva* sp. beds also serve as important food in coastal areas in the Atlantic Flyway (Lewis *et al.* 2013). Brant exhibit different foraging strategies in Atlantic coastal states of New York, New Jersey and Virginia, where brant select eelgrass, cordgrass *Spartina* sp. or exploit grasses and clover in upland habitats (Smith *et al.* 1985). Smith *et al.* (1985) attributed diet switching by brant from eelgrass to other foods because of eelgrass declines. However, brant foraged on cultivated grass and clovers in New York, despite an increasing trend in availability of SAV in the state. They attributed differential feeding strategies among regions to the birds' winter philopatry and social organisation.

Brant have been negatively affected by loss of eelgrass habitats in the North American Atlantic Flyway and Europe (Vickery *et al.* 1995; Ganter *et al.* 1997; Ward *et al.* 2005; Shaughnessy *et al.* 2012). Brant in those regions use eelgrass where available, but birds also exploit salt marsh habitat. Moreover, European birds have moved inland to use golf courses and pastures with cattle (Vickery *et al.* 1995; Ganter *et al.* 1997; Ward *et al.* 2005; Shaughnessy *et al.* 2012). Lovvorn & Baldwin (1996) recognised the value of habitat complexes for wintering brant in Western Europe that include intertidal flats, bays and other permanent wetlands that provide sea grasses, as well as nearby farmlands containing waste grains and natural seeds. This complex of suitable habitats allow brant to move and forage among them and thereby enhance their survival (Lovvorn & Baldwin 1996). However, synergistic effects of climate change, possible negative effects on sea level rise and declining eelgrass communities are emerging concerns for waterfowl ecologists conserving brant (Shaughnessy *et al.* 2012).

Swans

Migratory swans

Of the five swan species and subspecies in the northern hemisphere, the Tundra Swan (a.k.a. Whistling Swan) *C. c. columbianus* and Trumpeter Swan of North America and the Bewick's Swan *C. c. bewickii* (conspecific with the Tundra Swan) and Whooper Swan *C. cygnus* in Eurasia are all migratory, whereas the Mute Swan *C. olor* is relatively sedentary in its native Europe and in North America where it has colonised (*e.g.* Petrie & Francis 2003). Trumpeter Swans were widespread in North America prior to 1900 (Rogers & Hammer 1998; Engelhardt *et al.* 2000), but hunting caused their numbers to drop nearly to extinction by the early 20th century, and use of established migration routes waned (Gale *et al.* 1987; Mitchell & Eichholz 2010). Legal protection from persecution (since the 1918 Migratory Bird Treaty) and more recent conservation measures (*e.g.* habitat protection and reintroduction programmes) saw Trumpeter Swan numbers recover to ~16,000 birds by 1990, and > 34,000 free-ranging swans were estimated in 2005 (Moser 2006; Mitchell & Eichholz 2010). Whooper Swan numbers have also increased in Europe in recent decades (Wetlands International 2014), and the Tundra Swan – the most numerous and widely distributed of North American swans – is likewise increasing. Indeed, agricultural foraging opportunities are thought to have contributed to a near doubling of Tundra Swan numbers (to > 200,000 birds) between 1955–1989, leading to regulated hunting of the species in some states (Serie & Bartonek 1991). In contrast, although the Northwest European Bewick's Swan population similarly rose from ~16,000 birds in the mid-1980s to a peak of ~29,000 individuals in the mid-1990s, its numbers are now in decline (Rees & Beekman 2010), with several poor breeding seasons in recent years probably a major contributing factor.

The Eastern Population of Tundra Swans, which breeds across northern Canada and north of the Brooks Range in Alaska, migrates to the U.S. eastern seaboard (allocating about half their time between boreal forest and northern prairie-Great Lakes habitats during autumn migration; Weaver 2013), whereas the Western Population, which breeds in coastal regions of Alaska south of the Brooks Range, migrates to western North America to winter mainly on the Pacific coast from Vancouver Island to central California, and the inland valleys of California (Bellrose 1980; Ely *et al.* 2014). The Northwest European Bewick's Swan population also migrates along a well-defined corridor, from breeding grounds in the Russian arctic along the arctic coast and across Karelia to autumn staging sites on the Baltic (particular Estonian wetlands) and wintering grounds in northwest Europe. Whooper Swans are thought to migrate on a broader front (Garðarsson 1991; Matthiasson 1991), but like the arctic-nesting swans they show strong fidelity to staging and wintering sites (Bellrose 1980; Black & Rees 1984; Rees 1987).

Historically, migratory swans fed on SAV during autumn and winter, often reflecting regional and seasonal variation in availability and dietary requirements. For Tundra Swans, this included Arrowhead *Sagittaria* sp., Sago Pondweed *Potamogeton pectinatus* and Wild Celery *Vallisneria americana* (Bellrose 1980), with Bewick's Swans also favouring pondweeds (*Potamogeton pectinatus* and *P. perfoliatus*) along with hornworts *Ceratophyllum* sp., watermilfoil *Myriophyllum* sp., stoneworts *Chara* sp. and other emergent vegetation (Rees 2006). However, wetland drainage and intensification of farming (including increased use of fertiliser on grasslands and more extensive planting of arable crops) has resulted in a large-scale movement of swans from wetland habitats to agricultural land. In Europe, Whooper Swans were recorded feeding on cereals and potatoes as early as the 19th century, but changes in agriculture saw an increase in their use of arable habitats during the second half of the 20th century (Kear 1963; Laubek *et al.* 1999). More recently, Tundra Swans were first observed in grain fields in the mid 1960s (Nagel 1965; Tate & Tate 1966; Munro 1981), and Bewick's Swans have been utilising arable habitats since the early 1970s (review in Rees 2006). Trumpeter Swans typically use freshwater marshes, ponds, lakes, rivers and brackish estuaries with abundant pondweed (Gale *et al.* 1987; LaMontagne *et al.* 2003; Mitchell & Eichholz 2010), but also forage on arable land in winter and early spring (Babineau 2004; Mitchell & Eichholz 2010), where they avoid soybean and prefer winter wheat and corn (Varner 2008). In the mid-west U.S., swans use reclaimed surface mine wetlands close to agricultural fields, which rarely freeze and are relatively undisturbed compared to reservoirs (Varner 2008; Mitchell & Eichholz 2010). The drivers of swan exploitation of arable lands remain unclear; however, historic and novel food availability, nutrition and foraging efficiency in croplands may be influences (Rees 2006). Several studies have described seasonal variation in the swans use of farmland, with birds generally moving from harvest waste (*e.g.* cereal stubbles, potatoes and sugar beet) to growing cereals (*e.g.* winter wheat) and then to pasture as the winter progresses, which has been attributed to a combination of food availability and changes in dietary requirements (*e.g.* Laubek 1995; Rees *et al.* 1997). Weaver (2013), studying habitat use by 63 satellite-tagged Tundra Swans, found seasonal differences in habitat selection. Tundra Swans selected open water over wetlands in autumn, but agriculture was used substantially less during autumn migration (despite representing 45% and 80% of Tundra Swan habitats in the Great Lakes and Northern Prairies, respectively, at this time) than in winter, when swans selected agriculture lands, and wetlands were used less than their availability. Weaver (2013) concluded that if adequate aquatic habitats were available, swans may not have made forays to agricultural fields, although agricultural seeds provided alternative foods of similar energy value (Kaminski *et al.* 2003), and recommended that wetland conservationists interested in managing non-breeding Tundra Swans should conserve and restore wetlands within agricultural landscapes < 8 km of known roosts and aim to protect open water habitats, especially those containing SAVs. Detailed studies of Bewick's Swan feeding ecology have also illustrated the importance of aquatic habitats for swans arriving in autumn, with swans feeding on below-ground Fennel Pondweed tubers at the Lauwersmeer, Netherlands, preferably in shallow waters and sandy sediments rather than in areas of deeper water, likely reflecting increased effort and energy costs (*e.g.* up-ending as opposed to head-dipping for food) required to feed in deeper waters or where the tubers are in clay (Nolet *et al.* 2001). Further analysis of the timing of the swans' switch from feeding on pondweed tubers to feeding on sugar beet in fields around the Lauwersmeer found that most swans switched habitats when the net energy gain from staying on tubers fell below that from feeding on beet alone. However, the swans would attain a substantially increased energy and total nutrient gain by feeding on both beet and tubers, and there was evidence from van Eerden (1997) that mixed exploitation of tubers and beet does occur in the Lauwersmeer area. Overall, swans seemingly switch to the beet fields long after they would first benefit from doing so due to energy gain alone (Nolet *et al.* 2002).

Mute Swans

Mute Swan movements tend to be relatively localised (< 50 km radius; Birkhead & Perrins 1986), although some long-distance flights have been recorded (*e.g.* those at more northerly latitudes heading south in cold winters). They frequent a wide range of lowland wetland habitats throughout the year, including freshwater lakes, estuarine wetlands, commercial fishponds, sea lochs and shallow coastal waters, where they feed primarily on SAV, and are also commonly found on rivers and canals in urban areas where they rely on bread and other provisions from humans (Birkhead & Perrins 1986; Sears 1989; Gayet *et al.* 2011). They also use farmland, for instance moving to agricultural fields and improved grasslands during winter (Birkhead & Perrins 1986), but tend to be more widely dispersed than the migratory species (Rees *et al.* 1997). In parts of the United Kingdom, where three swan species (Bewick's, Whoopers and Mutes) coincide in winter, segregation across habitats has been recorded, with Whooper and Mute Swans predominately using permanent inland waters and improved pasture, whereas Bewick's Swans were mostly on arable land (Rees *et al.* 1997), indicating a range of habitats are important for foraging by these swan species.

On studying effects of patch size and isolation on Mute Swan habitat use in France, Gayet *et al.* (2011) found that the swans' winter distribution and occurrence on fishponds was influenced by pond structure more than surrounding landscape and other features. Specifically, fishponds drained and cultivated for grain the previous year provided crop residues utilised by the swans the following winter. Understanding habitat selection of Mute Swans is important because they are perceived as having negative influences on other waterfowl, through territorial behaviour or intensive grazing on aquatic macrophytes, sometimes within their European range but particularly where they have been introduced to North America (Conover & Kania 1994; Petrie & Francis 2003; Gayet *et al.* 2011), with a Mute Swan control programme instigated in Maryland in 2005 (Hindman & Tjaden 2014; Hindman *et al.* 2014).

Future challenges and needs

Planning and implementing conservation strategies for waterfowl and their habitats are challenging because some species are declining or remain below long-term averages (*e.g.* Scaup, Northern Pintail American Wigeon), whereas others have become superabundant (*e.g.* Snow Goose) despite some using similar resources (*e.g.* agricultural fields used by Northern Pintail and Snow Geese) in autumn and during migration. Wiens (1989) discussed habitat quality in terms of "fitness potential", whereby habitat quality may be assessed through demographic, physiological and behavioural approaches. Nonetheless, Norris & Marra (2007) alluded to the difficulty in understanding habitat selection in migrating species, particularly in identifying spatio-temporal connectivity of individuals or populations among stages of the annual cycle. Indeed, there is strong research and conservation interest in determining the extent of migratory connectivity among birds occupying specific wintering and breeding areas (Norris & Marra 2007; Guillemain *et al.* 2014; Kaminski & Elmberg 2014). Here we consider some challenges hindering understanding of habitat use and selection by waterfowl during the non-breeding season and suggest future needs for research. We recognise there are other ecological, economic, bio-political and human dimensional considerations, but believe that addressing the following five issues will advance science and stewardship of waterfowl and their habitats in the Holarctic and worldwide.

(1) *Habitat and resource availability*. Resources available for migrating and non-breeding waterfowl are typically dynamic and unpredictable. Indeed, many migratory birds (*e.g.* Svalbard Barnacle Geese) seemingly cannot assess local resource conditions from afar and must "sample" habitats upon settling in them, though others (*e.g.* Svalbard Pink-footed Geese *Anser brachyrhyncus*) appear to use conditions at one site as an indicator of conditions that they might encounter at the next (Tombre *et al.* 2008). Habitat and other environmental dynamics may result in patchily distributed food and other resources within and across seasons, inter-annual site-specific changes in potential foraging areas (*e.g.* ploughed *versus* flooded field; 4th order selection), natural inter-annual droughts or flooding, weather that may dictate where birds winter and exploit resources, disturbance from hunting and other human-related factors, physiological and behavioural dynamics and other scenarios (Fig. 1). During winter, some species like Northern Pintail, Mallard, teal and diving ducks move inter-regionally, likely in search of suitable habitats (*sensu* Fretwell 1972; Cox, Jr. & Afton 1996; Heitmeyer 2006; Caizergues *et al.* 2011; Gourlay Larour *et al.* 2013). Interpreting true migration from movements to and fro (*i.e.* foraging flights) can be challenging (Dingle & Drake 2007) and documenting habitat selection across broad landscapes in brief intervals may be even more equivocal.

Arguably, one of the greatest current challenges waterfowl habitat researchers face relative to identifying true selection involves an inability to determine true habitat and resource availability at scales influencing biological outcomes for the birds (Kaminski

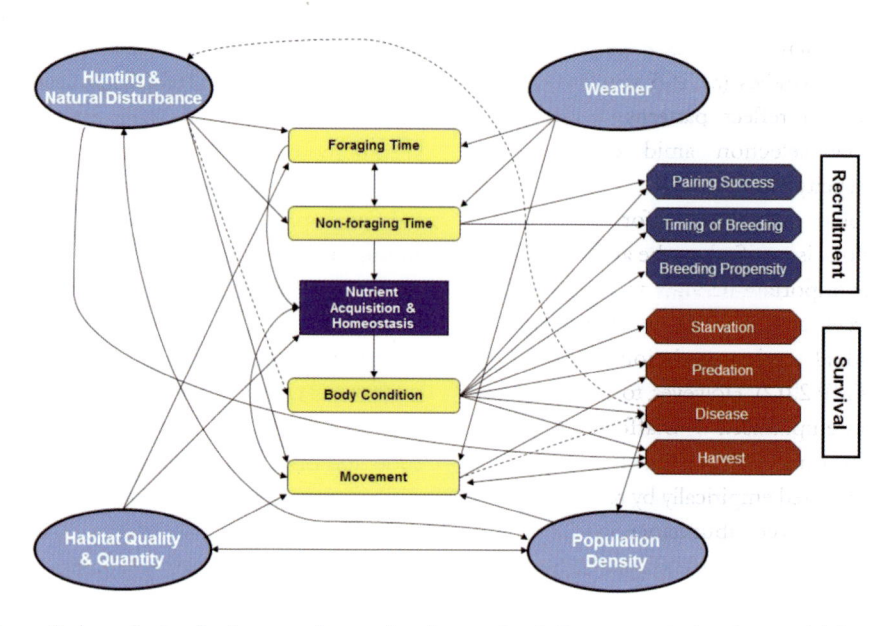

Figure 1. A synthesis of primary and secondary factors that influence survival and potential fitness of Holarctic waterfowl.

& Elmberg 2014). For example, non-breeding waterfowl that exploit agricultural environments (thousands of hectares of agricultural land in one region alone) may suddenly move from dry to shallowly flooded fields during autumn–winter (Reinecke *et al.* 1989). Mallard commonly feed in dry fields in southern Canada and the northern U.S. prairies, but not in the MAV where they utilise puddled fields. This typical scenario is further complicated during winters of below average temperatures; then, Mallard use dry agricultural fields in winter as wetlands freeze and foods become inaccessible. These and other scenarios create great resource variability across regions, temporal variability within regions, and basically constrain researchers' efforts to categorise and estimate available resources. We concur that recent analysis of habitat use by mid-continent Mallard (Beatty *et al.* 2013) is statistically robust, but may be ecologically tenuous because they could not estimate full availability of agricultural lands possibly accessible by Mallard. Despite broad spatial and temporal scaled information obtainable from satellite-tracked birds (Krementz *et al.* 2012; Beatty *et al.* 2013), sample sizes of marked birds are often small because of funding limitations (Lindberg & Walker 2007). This limitation constrains determining selection of habitats, because a small cohort of individuals is assumed to represent the greater population. Moreover, when making inferences of resource selection beyond one or two variables, sample sizes must be increased significantly (Lindberg & Walker 2007). Given the challenges in capturing environmental variability across vast landscapes, we suggest long-term studies (*i.e.* ≥ 5 years) should be invoked to reflect patterns of waterfowl resource selection amid environmental stochasticity.

Habitat conservation for non-breeding waterfowl is justified on the assumption that certain important habitats and intrinsic food resources are limited and thereby ramify individual and population implications (NAWMP 2012). However, to our knowledge (and as emphasised by Stafford *et al.* 2014), true resource limitation has not been demonstrated empirically by relating food or other resource abundance to biological outcomes for waterfowl. Indeed, further understanding these scenarios is required for assessing whether true resource limitation exists and is affecting individuals and populations (Neu *et al.* 1974; Johnson 2007; Stafford *et al.* 2014).

(2) *Populations important to study*. Individuals of some species (*e.g.* Mallard), are widespread in North America during autumn and winter (Bellrose 1980). Some Mallard winter along sandbars and adjacent agricultural lands along the Missouri River in North Dakota (Olsen & Cox, Jr. 2003), while others predominately occupy the southern U.S. (Nichols *et al.* 1983; Reinecke *et al.* 1989). We typically regard the former region as "breeding grounds", yet some Mallard remain there during winter. Although some resources (*e.g.* agriculture) in all these geographic regions get exploited by Mallard, basing habitat selection on a cohort of a species in one region may not reflect important resource components elsewhere in the species' range. Thus, what cohorts of birds should be studied? Comparative studies of conspecifics across geographic regions would be interesting and valuable; thus, studying non-breeding resource use and in regions with the greatest abundance of individuals of a species is a suggested approach. The genetic variability among individuals in these regions should reveal patterns of resource exploitation important to subsequent breeding success. The greater challenge and future research endeavour is to discover if population cohorts of a species that occupy ecologically disparate landscapes during non-breeding seasons contribute differently to population recruitment for the species. Conversely, analysis of bands recovered over a large geographical area have demonstrated that some population boundaries in western Europe were largely artificial (Guillemain *et al.* 2005), so that habitat selection studies should be conducted at much greater geographic scales.

(3) *Functional use of habitats*. Understanding the range of benefits that birds derive from different habitats is also a critical need. Time-budget studies have been conducted at sites across the Holarctic for decades, but new technology such as unmanned aircraft (drones) or GPS accelerometers would help to quantify the birds' activities at local and micro-habitat scales, which in turn would improve our knowledge of the functional values of habitats frequented by waterfowl.

(4) *Remoteness and difficulty in accessing habitats*. Inhospitable conditions and remoteness of habitats pose challenges to studying birds such as sea ducks (Silverman *et al.* 2013) and other arctic-nesting waterfowl. Establishing true habitat selection

among sea ducks in remote environments, especially when trying to link movements or habitat use in relation to food, is particularly problematic. Researchers hypothesise that serious challenges face wintering sea ducks, including marine (boat) traffic, wind-power development and aquaculture practices (Skov *et al.* 2011; Silverman *et al.* 2013). Despite inherent difficulties in investigating birds and habitats in marine environments, recent research has greatly advanced understanding of non-breeding ecology of sea ducks, albeit continued efforts are essential to sustain these birds (Faulkner 2013; Silverman *et al.* 2013).

(5) *Cumulative resource use*. Lastly, there exists a lack of understanding of how cumulative use of resources during the non-breeding period may influence reproduction and recruitment (*i.e.* Heitmeyer & Fredrickson 1981; Kaminski & Gluesing 1987). Indeed, body condition is an important factor in waterfowl survival and fitness. For example, Devries *et al.* (2008) found that female Mallard which arrive in better condition on breeding grounds in the Canadian prairie-parklands hatched eggs 15 days earlier than those in relatively poor condition. Guillemain *et al.* (2008) also observed more juveniles during autumn in southern France when body condition of females was greater at the end of the previous winter. Gunnarsson *et al.* (2005) used stable-carbon isotopes to demonstrate that Black-tailed Godwits *Limosa limosa* wintering in high quality sites in Europe were more likely to use higher-quality breeding habitats and have greater reproductive success than birds using poorer-quality habitats (see also Norris & Marra 2007).

These and related metrics are useful for understanding cross-seasonal carry-over effects (Harrison *et al.* 2011; Sedinger & Alisauskas 2014), but difficulty lies in the fact that autumn staging and migration immediately follow the breeding season, and are temporally furthest from the next breeding season. Hence, "back-dating" and identifying resources used by birds following their arrival on the breeding grounds, in relation to previous habitat use, are paramount needs. For example, if body condition of a cohort of Mallard in Nebraska in late March was known and these birds were subsequently sampled on the breeding grounds, linking March condition and breeding success seems reasonable (*i.e.* Devries *et al.* 2008). However, how should we consider body condition in relation to future fitness prospects in a cohort of birds examined months earlier, during autumn–winter?

No doubt, fitness is partly a result of some cumulative use of resources during an animal's annual cycle. The greatest uncertainty seems to be in understanding at what point in the non-breeding phase of the cycle a potential shortfall (or indeed windfall) of resources might influence future fitness prospects. There are likely bottlenecks or thresholds related to resource use during the year which could impose disproportionate impacts on subsequent fitness; these may vary considerably between years and across species, and deserve further investigation.

As an alternative to indexing body condition or some other fitness metric, perhaps coordinated inter-regional aerial transect surveys of waterfowl during autumn–spring migration could be conducted (*sensu* Pearse *et al.* 2008) to determine "hot spots" of waterfowl use, thereby identifying and characterising complexes of wetlands and uplands used by the majority of waterfowl (Pearse *et al.* 2012). Aerial survey data could be incorporated with GIS layers to illustrate habitat features and describe high and low priority habitats for North American waterfowl during winter and migration (*e.g.* Pearse 2007), analogous to the "thunderstorm maps" used by waterfowl breeding ground JV programmes (Loesch *et al.* 2012). Clearly, we must be creative in engaging diverse human expertise and reliable technologies to understand the ecology of waterfowl throughout their annual cycle and range, then use this knowledge to conserve important habitats for birds across the Holarctic region and worldwide.

Acknowledgements

We are indebted to Joe Lancaster and Justyn Foth, both Ph.D. students, and J. Czarnecki, Mississippi State University (MSU), for dedicating days to help us complete review and references for this manuscript. We thank the Forest and Wildlife Research Center (FWRC) and Department of Wildlife, Fisheries and Aquaculture, MSU for supporting J.B. Davis and R.M. Kaminski during writing of the manuscript. This manuscript has been approved for publication by the FWRC as FWRC/WFA publication 395.

References

Abraham, K.F., Jefferies, R.L. & Alisauskas, R.T. 2005. The dynamics of landscape change and snow geese in mid-continent North America. *Global Change Biology* 11: 841–855.

Accurso, L.M. 1992. Distribution and abundance of wintering waterfowl on San Francisco Bay, 1988–1990. M.Sc. thesis, Humboldt State University, Arcata, California, USA.

Ackerman, J.T., Takekawa, J.Y., Orthmeyer, D.L., Fleskes, J.P., Yee, J.L. & Kruse, K.L. 2006. Spatial use by wintering greater white-fronted geese relative to a decade of habitat change in California's Central Valley. *Journal of Wildlife Management* 70: 965–976.

Ackerman, J.T., Herzog, M.P., Yarris, G.S., Casazza, M.L., Burns, E. & Eadie, J.M. 2014. Chapter 5: Waterfowl ecology and management. *In* P.B. Moyle, A. Manfree & P.L. Fiedler (eds.), *Suisun Marsh: Ecological History and Possible Futures,* pp. 103–132 and maps 10 & 11. University of California Press, Berkeley, California, USA.

Adair, S.E., Moore, J.L. & Kiel, W.H., Jr. 1996. Winter diving duck use of coastal ponds: an analyses of alternative hypotheses. *Journal of Wildlife Management* 60: 83–93.

Afton, A.D. & Anderson, M.G. 2001. Declining scaup populations: a retrospective analysis of long-term population and harvest survey data. *Journal of Wildlife Management* 65: 781–796.

Alerstam, T. & Lindström, Å. 1990. Optimal bird migration: the relative importance of time, energy and safety. *In* E. Gwinner (ed.), *Bird Migration: Physiology and Ecophysiology,* pp. 331–351. Springer, Berlin, Germany.

Alisauskas, R.T. 1998. Winter range expansion and relationships between landscape and morphometrics of midcontinent Lesser Snow Geese. *Auk* 115: 851–862.

Alisauskas, R.T. 2002. Arctic climate, spring nutrition, and recruitment in midcontinent Lesser Snow Geese. *Journal of Wildlife Management* 66: 181–193.

Alisauskas, R.T. & Ankney, C.D. 1992. Spring habitat use and diets of midcontinent adult Lesser Snow Geese. *Journal of Wildlife Management* 57: 43–54.

Alisauskas, R.T. & Hobson, K.A. 1993. Determination of Lesser Snow Goose diets and winter distribution using stable isotope analysis. *Journal of Wildlife Management* 57: 49–54.

Alisauskas, R.T., Ankney, C.D. & Klaas, E.E. 1988. Winter diets and nutrition of midcontinental lesser snow geese. *Journal of Wildlife Management* 52: 403–414.

Anderson, J.T. & Ballard, B.M. 2006. *Survival and Habitat Use of Female Northern Pintails Wintering Along the Central Coast of Texas*. Final report to Ducks Unlimited, Inc., Ridgeland, Mississippi, USA.

Ankney, C.D. 1996. An embarrassment of riches: too many geese. *Journal of Wildlife Management* 60: 217–223.

Arner, D.H. & Hepp, G.R. 1989. *Beaver Pond Wetlands: a Southern Perspective*. Texas Tech University Press, Lubbock, Texas, USA.

Arzel, C., Elmberg, J. & Guillemain, M. 2006. Ecology of spring-migrating *Anatidae*: a review. *Journal of Ornithology* 147: 167–184.

Atlantic Coast Joint Venture (ACJV). 2005. North American Waterfowl Management Plan. Atlantic Coast Joint Venture Waterfowl Implementation Plan Revision: June 2005. Hadley, Massachusetts, USA.

Atlantic Coast Joint Venture (ACJV). 2009. Atlantic Coast Joint Venture Strategic Plan: Updated July 2009. Hadley, Massachusetts, USA.

Audubon, J.J. 1840. *The Birds of America*. J.J Audubon & J.B. Chevalier, New York, USA.

Baar, L., Matlack, R.S., Johnson, W.P. & Barron, R.B. 2008. Migration chronology of waterfowl in the Southern High Plains of Texas. *Waterbirds* 31: 394–401.

Babineau, F.M. 2004. Winter ecology of Trumpeter Swans in southern Illinois. M.Sc. thesis, Southern Illinois University Carbondale, Carbondale, Illinois, USA.

Baldassarre, G.A. 2014. *Ducks, Geese, and Swans of North America*. Johns Hopkins University Press, Baltimore, Maryland, USA.

Baldassarre, G.A. & Bolen, E.G. 1984. Field-feeding ecology of waterfowl wintering on the Southern High Plains of Texas. *Journal of Wildlife Management* 48: 63–71.

Baldassarre, G.A. & Bolen, E.G. 2006. *Waterfowl Ecology and Management*. Krieger Publishing Company, Malabar, Florida, USA.

Ballard, B.M., Thompson, J.E., Petrie, M.J., Chekett, M. & Hewitt, D.G. 2004. Diet and nutrition of northern pintails wintering along the southern coast of Texas. *Journal of Wildlife Management* 68: 371–382.

Ballard, B.M., James, J.D., Bingham, R.L., Petrie, M.J. & Wilson, B.C. 2010. Coastal pond use by redheads wintering in the Laguna Madre, Texas. *Wetlands* 30: 669–674.

Barras, S.C., Kaminski, R.M. & Brennan, L.A. 1996. Acorn selection by female Wood Ducks. *Journal of Wildlife Management* 60: 592–602.

Bateman, H.A., Joanen, T. & Stutzenbaker, C.D. 1988. History and status of midcontinental snow geese on their Gulf Coast winter range. In M.W. Weller (ed.), *Waterfowl in Winter*, pp. 495–515. University of Minnesota Press, Minneapolis, Minnesota, USA.

Batt, B.D. 1997. *Arctic Ecosystems in Peril: Report of the Arctic Goose Habitat Working Group*. Arctic Goose Joint Venture Special Publication. U.S. Fish & Wildlife Service, Washington D.C., USA and Canadian Wildlife Service, Ottawa, Ontario, Canada.

Batzer, D.P. & Baldwin, A.H. 2012. *Wetland Habitats of North America: Ecology and Conservation Concerns*. University of California Press, Berkley, California, USA.

Beatty, W.S., Kesler, D.C., Webb, E.B., Raedeke, A.H., Naylor, L.W. & Humburg, D.D. 2013. Quantitative and qualitative approaches to identifying migration chronology in a continental migrant. *PloS ONE* 8: e75673.

Beekman, J.H., Nolet, B.A. & Klaassen, M. 2002. Skipping swans: differential use of migratory stop-over sites in spring and autumn in relation to fuelling rates. *Ardea* 90: 437–460.

Bellebaum, J., Larsson, K. & Kube, J. 2012. Research on sea ducks in the Baltic Sea. Uppsala University – Campus Gotland, Visby, Sweden. http://seaducks.hgo.se/?q=system/files/dokument/Reserach%20on%20Sea%20Ducks.pdf.

Bellrose, F. 1980. *Ducks, Geese and Swans of North America*. Stackpole Books, Harrisburg, Pennsylvania, USA.

Bellrose, F.C. & Crompton, R.D. 1970. Migrational behavior of mallards and black ducks as determined from banding. *Illinois Natural History Survey Bulletin* 30: 167–234.

Bellrose, F.C. & Holm, D.J. 1994. *Ecology and Management of the Wood Duck*. Stackpole Books, Mechanicsburg, Pennsylvania, USA.

Bergmann, C. 1847. Über die verlhältnisse der wärmeökonomie der thiere zu ihrer grösse. *Göttinger Studien* 3: 595–708.

Birkhead, M.E. & Perrins, C.M. 1986. *The Mute Swan*. Croom Helm, London, UK.

Black, J.M. & Rees, E.C. 1984. The structure and behaviour of the Whooper Swan population wintering at Caerlaverock, Dumfries and Galloway, Scotland: an introductory study. *Wildfowl* 35: 21–36.

Block, W.M. & Brennan, L.A. 1993. The habitat concept in ornithology. *Current Ornithology* 11: 35–91.

Bolduc, F. & Afton, A.D. 2004. Relationships between wintering waterbirds and invertebrates, sediments and hydrology of coastal marsh ponds. *Waterbirds* 27: 333–341.

Bolen, E.G., Baldassarre, G.A. & Guthery, F.S. 1989. Playa lakes. *In* L.M. Smith, R.L. Pederson & R.M. Kaminski (eds.), *Habitat Management for Migrating and Wintering Waterfowl in North America*, pp. 341–365. Texas Tech University Press, Lubbock, Texas, USA.

Bourne, N. 1984. *Clam Predation by Scoter Ducks in the Strait of Georgia, British Columbia, Canada*. Canadian Technical Report of Fisheries and Aquatic Science No. 1331. Department of Fisheries & Oceans Fisheries (Research Branch), Pacific Biological Station, Nanaimo, British Columbia, Canada.

Brand, L.A., Takekawa, J.Y., Shinn, J., Graham, T., Buffington, K., Gustafson, K.B., Smith, L.M., Spring, S.E. & Miles, A.K. 2014. Effects of wetland management on carrying capacity of diving ducks and shorebirds in a coastal estuary. *Waterbirds* 37: 52–67.

Brochet, A.L., Mouronval, J.B., Aubry, P., Gauthier-Clerc, M., Green, A.J., Fritz, H. & Guillemain, M. 2012. Diet and feeding habitats of Camargue dabbling ducks: What has changed since the 1960s? *Waterbirds* 35: 555–576.

Brusendorff, A.C., Korpinen, S., Meski, L. & Stankiewicz, M. 2013. HELCOM actions to eliminate illegal and accidental oil pollution from ships in the Baltic Sea. *In* A.G. Kostianoy & O.Y. Lavrova (eds.), *Oil Pollution in the Baltic Sea,* pp. 15–40. Springer, Heidelberg, Germany.

Caizergues, A., Guillemain, M., Arzel, C., Devineau, O., Leray, G., Pilvin, D., Lepley, M., Massez, G. & Schricke, V. 2011. Emigration rates and population turnover of teal *Anas crecca* in two major wetlands of western Europe. *Wildlife Biology* 17: 373–382.

Calenge, C., Guillemain, M., Gauthier-Clerc, M. & Simon, G. 2010. A new exploratory approach to the study of the spatio-temporal distribution of ring recoveries: the example of Teal (*Anas crecca*) ringed in Camargue, Southern France. *Journal of Ornithology* 151: 945–950.

Central Valley Joint Venture. 2006. *Central Valley Joint Venture Implementation Plan-Conserving Bird*

Habitat. U.S. Fish & Wildlife Service, Sacramento, California, USA.

Chabreck, R.H., Joanen, T. & Paulus, S.L. 1989. Southern coastal marshes and lakes. In L.M. Smith, R.L. Pederson & R.M. Kaminski (ed.), *Habitat Management for Migrating and Wintering Waterfowl in North America,* pp. 249–277. Texas Tech University Press, Lubbock, Texas, USA.

Chesapeake Bay Program. 2007. Bay trends and indicators. http://www.chesapeakebay.net/indicators.htm.

Christopher, M.W., Hill, E.P. & Steffen, D.E. 1988. Use of catfish ponds by waterfowl wintering in Mississippi. In M.W. Weller (ed.), *Waterfowl in Winter,* pp. 413–418. Minnesota Press, Minneapolis, Minnesota, USA.

Clausen, K.K., Clausen, P., Fælled, C.C. & Mouritsen, K.N. 2012. Energetic consequences of a major change in habitat use: endangered Brent Geese *Branta bernicla hrota* losing their main food resource. *Ibis* 154: 803–814.

Clausen, K.K. & Clausen, P. 2014. Forecasting future drowning of coastal waterbird habitats reveals a major conservation concern. *Biological Conservation* 171: 177–185.

Coates, P.S., Casazza, M.L., Halstead, B.J. & Fleskes, J.P. 2012. Relative value of managed wetlands and tidal marshlands for wintering Northern Pintails. *Journal of Fish and Wildlife Management* 3: 98–109.

Conomos, T., Smith, R. & Gartner, J. 1985. Environmental setting of San Francisco Bay. *Hydrobiologia* 129: 1–12.

Conover, M.R. & Kania, G.S. 1994. Impact of interspecific aggression and herbivory by Mute Swans on native waterfowl and aquatic vegetation in New England. *Auk* 111: 744–748.

Cornelius, S.E. 1977. Food and resource utilization by wintering redheads on lower Laguna Madre. *Journal of Wildlife Management* 41: 374–385.

Cowardin, L.M., Carter, V., Golet, F.C. & LaRoe, E.T. 1979. *Classification of Wetlands and Deepwater Habitats of the United States.* U.S. Department of the Interior, Fish & Wildlife Service, Washington D.C., USA.

Cox, R.R., Jr. & Afton, A.D. 1996. Evening flights of female Northern Pintails from a major roost site. *The Condor* 98: 810–819.

Cox, R.R., Jr. & Afton, A.D. 1997. Use of habitats by female Northern Pintails wintering in southwest Louisiana. *Journal of Wildlife Management* 61: 435–443.

Dalby, L. 2013. Waterfowl duck distributions and a changing climate. Ph.D. thesis, Aarhus University, Aarhus, Denmark.

Dandelot, S., Verlaque, R., Dutartre, A. & Cazaubon, A. 2005. Ecological, dynamic and taxonomic problems due to *Ludwigia (Onagraceae)* in France. *Hydrobiologia* 551: 131–136.

Dänhardt, J. & Lindström, Å. 2001. Optimal departure decisions of songbirds from an experimental stopover site and the significance of weather. *Animal Behaviour* 62: 235–243.

Darwin, C. 1859. *The Origin of Species.* Penguin Books, New York, USA.

Davis, B.E. & Afton, A.D. 2010. Movement distances and habitat switching by female Mallards wintering in the Lower Mississippi Alluvial Valley. *Waterbirds* 33: 349–356.

Delnicki, D. & Reinecke, K.J. 1986. Mid-winter food use and body weights of Mallards and Wood Ducks in Mississippi. *Journal of Wildlife Management* 50: 43–51.

Devries, J.H., Brook, R.W., Howerter, D.W. & Anderson, M.G. 2008. Effects of spring body condition and age on reproduction in Mallards (*Anas platyrhynchos*). *Auk* 125: 618–628.

Dias, M.P. 2009. Use of salt ponds by wintering shorebirds throughout the tidal cycle. *Waterbirds* 32: 531–537.

Dickson, D.L. 2012. *Movement of King Eiders from breeding grounds on Banks Island, Northwest Territories, to moulting and wintering areas.* Technical Report Series No. 516, Canadian Wildlife Service, Edmonton, Alberta, Canada.

Dingle, H. 1996. *Migration: The Biology of Life on the Move.* Oxford University Press, New York, USA.

Dingle, H. & Drake, V.A. 2007. What is migration? *Bioscience* 57: 113–121.

Dirksen, S. & Beekman, J.H. 1991. Population size, breeding success and distribution of Bewick's Swans, *Cygnus columbianus bewickii*, wintering in Europe 1986–87. *Wildfowl* (Supplement No. 1): 120–124..

Dubovsky, J. & Kaminski, R.M. 1992. Waterfowl and American coot habitat associations with Mississippi catfish ponds. *Proceedings of the Annual Conference of Southeastern Association of Fish and Wildlife Agencies* 46: 10–17.

Eadie, J.M., Elphick, C.S., Reinecke, K.J. & Miller, M.R. 2008. Wildlife values of North American ricelands. *In* S. Manley (ed.), *Conservation of Ricelands of North America*, pp. 8–90. Ducks Unlimited, Inc., Memphis, Tennessee, USA.

Elmberg, J., Hessel, R., Fox, A.D. & Dalby, L. 2014. Interpreting seasonal range shifts in migratory birds: a critical assessment of 'short-stopping' and a suggested terminology. *Journal of Ornithology* 155: 1–9.

Elphick, C.S. 2000. Functional equivalency between rice fields and seminatural wetland habitats. *Conservation Biology* 14: 181–191.

Elphick, C.S., Parsons, K.C., Fasola, M. & Mugica, L. 2010. Ecology and conservation of birds in rice fields: a global review. *Waterbirds* 33 (Special Publication 1): 246 pp.

Ely, C.R. 1992. Time allocation by Greater White-fronted Geese: influence of diet, energy reserves and predation. *The Condor* 94: 857–870.

Ely, C.R. & Raveling, D.G. 2011. Seasonal variation in nutritional characteristics of the diet of Greater White fronted Geese. *Journal of Wildlife Management* 75: 78–91.

Ely, C.R., Sladen, W.J.L., Wilson, H.M., Savage, S.E., Sowl, K.M., Henry, B., Schwitters, M. & Snowdon, J. 2014. Delineation of Tundra Swan *Cygnus c. columbianus* populations in North America: geographic boundaries and interchange. *Wildfowl* 64: 132–147.

Engelhardt, K.A., Kadlec, J.A., Roy, V.L. & Powell, J.A. 2000. Evaluation of translocation criteria: case study with Trumpeter Swans *Cygnus buccinator*. *Biological Conservation* 94: 173–181.

Faulkner, H. 2013. Influence of aquaculture on winter sea duck distribution and abundance in south Puget Sound. M.Sc. thesis, The Evergreen State College, Olympia, Washington, USA.

Feaga, J.S. 2013. Winter waterbird use and food resources of aquaculture lands in Mississippi. M.Sc. thesis, Mississippi State University, Mississippi State, Mississippi, USA.

Fleskes, J. 2012. Climate change impacts on ecology and habitats of Central Valley waterfowl and other waterbirds. http://www.werc.usgs.gov/Project.aspx?ProjectID=204.

Fleskes, J.P., Yee, J.L., Casazza, M.L., Miller, M.R., Takekawa, J.Y. & Orthmeyer, D.L. 2005. *Waterfowl Distribution, Movements and Habitat Use Relative to Recent Habitat Changes in the Central Valley of California: A Cooperative Project to Investigate Impacts of the Central Valley Joint Venture and Changing Agricultural Practices on the Ecology of Wintering Waterfowl.* U.S. Geological Survey, Western Ecological Research Center, Dixon, California, USA.

Fleskes, J.P. & Yee, J.L. 2007. Waterfowl distribution and abundance during spring migration in southern Oregon and northeastern California. *Western North American Naturalist* 67: 409–428.

Fleskes, J.P., Yee, J.L., Yarris, G.S., Miller, M.R. & Casazza, M.L. 2007. Pintail and Mallard survival in California relative to habitat, abundance, and hunting. *Journal of Wildlife Management* 71: 2238–2248.

Fleskes, J.P., Skalos, D.A. & Farinha, M.A. 2013. Changes in types and area of postharvest flooded fields available to waterbirds in Tulare Basin, California. *Journal of Fish and Wildlife Management* 4: 351–361.

Foth, J. R., Straub, J., Kaminski, R.M., Davis, J.B. & Leininger, T. In press. Aquatic invertebrate abundance and biomass in Arkansas, Mississippi and Missouri bottomland hardwood forests during winter. *Journal of Fish and Wildlife Management* 00: 000–000.

Foster, M.A., Gray, M.J. & Kaminski, R.M. 2010. Agricultural seed biomass for migrating and wintering waterfowl in the southeastern United States. *Journal of Wildlife Management* 74: 489–495.

Fox, A.D., Ebbinge, B.S., Mitchell, C., Heinicke, T., Aarvak, T., Colhoun, K., Clausen, P., Dereliev, S., Faragó, S. & Koffijberg, K. 2010. Current estimates of goose population sizes in western Europe, a gap analysis and an assessment of trends. *Ornis Svecica* 20: 115–127.

Frederick, R.B., Clark, W.R. & Takekawa, J.Y. 1992. Application of a computer simulation model to migrating White-fronted Geese in the Klamath Basin. In D.R. McCullough & R.H. Barrett (eds.), *Wildlife 2001: Populations*, pp. 696–706. Elsevier Applied Science, New York, USA.

Fredrickson, L.H. 2005. Contemporary bottomland hardwood systems: structure, function and hydrologic condition resulting from two centuries of anthropogenic activities. In L.M. Smith, R.L. Pederson & R.M. Kaminski (eds.), *Ecology and Management of Bottomland Hardwoods Systems: The State of our Understanding*, pp. 19–35. University of Missouri-Columbia, Puxico, Missouri, USA.

Fretwell, S.D. 1972. *Populations in a Seasonal Environment*. Princeton University Press, Princeton, New Jersey, USA.

Gale, R.S., Garton, E.O. & Ball, I.J. 1987. *The History, Ecology and Management of the Rocky Mountain Population of Trumpeter Swans*. U.S. Fish & Wildlife Service, Montana Cooperative Wildlife Research Unit, Missoula, Montana, USA.

Ganter, B., Prokosch, P. & Ebbinge, B.S. 1997. Effect of saltmarsh loss on the dispersal and fitness parameters of Dark bellied Brent Geese. *Aquatic Conservation: Marine and Freshwater Ecosystems* 7: 141–151.

Ganter, B., Larsson, K., Syroechkovsky, E.V., Litvin, K.E., Leito, A. & Madsen, J. 1999. Barnacle Goose *Branta leucopsis*: Russia/Baltic. In J. Madsen, G. Cracknell & A.D. Fox (eds.), *Goose Populations of the Western Palearctic – A Review of Status and Distribution*, pp. 270–283. Wetlands International Publication No. 48, National Environmental Research Institute, Ronde, Denmark.

Garðarsson, A. 1991. Movements of Whooper Swans *Cygnus cygnus* neckbanded in Iceland. *Wildfowl* (Special Supplement No. 1): 189–194.

Gayet, G., Guillemain, M., Fritz, H., Mesleard, F., Begnis, C., Costiou, A., Body, G., Curtet, L. & Broyer, J. 2011. Do Mute Swan (*Cygnus olor*) grazing, swan residence and fishpond nutrient availability interactively control macrophyte communities? *Aquatic Botany* 95: 110–116.

Gilmer, D.S., Miller, M.R., Bauer, R.D. & LeDonne, J.R. 1982. California's Central Valley wintering waterfowl: concerns and challenges. *Transactions of the North American Wildlife and Natural Resources Conference* 47: 441–452.

Gordon, D.H., Gray, B.T., Perry, R.D., Prevost, M.B., Strange, T.H. & Williams, R.K. 1989. South Atlantic coastal wetlands. In L.M. Smith, R.L. Pederson & R.M. Kaminski (eds.), *Habitat Management for Migrating and Wintering Waterfowl in North America*, pp. 57–92. Texas Tech University Press, Lubbock, Texas, USA.

Gordon, D.H., Gray, B.T. & Kaminski, R.M. 1998. Dabbling duck-habitat associations during winter in coastal South Carolina. *Journal of Wildlife Management* 62: 569–580.

Goudie, R., Brault, S., Conant, B., Kondratyev, A., Petersen, M. & Vermeer, K. 1994. The status of sea ducks in the North Pacific rim: toward their conservation and management. *Transactions of the North American Wildlife and Natural Resources Conference* 59: 27–49.

Gourlay-Larour, M.-L., Pradel, R., Guillemain, M., Santin-Janin, H., L'hostis, M. & Caizergues, A. 2013. Individual turnover in common pochards wintering in western France. *Journal of Wildlife Management* 77: 477–485.

Green, A.W. & Krementz, D.G. 2008. Mallard harvest distributions in the Mississippi and Central Flyways. *Journal of Wildlife Management* 72: 1328–1334.

Greer, D.M., Dugger, B.D., Reinecke, K.J. & Petrie, M.J. 2009. Depletion of rice as food of waterfowl wintering in the Mississippi Alluvial Valley. *Journal of Wildlife Management* 73: 1125–1133.

Guglielmo, C.G., O'Hara, P.D. & Williams, T.D. 2002. Extrinsic and intrinsic sources of variation in plasma lipid metabolites of free-living Western Sandpipers (*Calidris mauri*). *Auk* 119: 437–445.

Guillemain, M., Fritz, H. & Duncan, P. 2002. The importance of protected areas as nocturnal feeding grounds for dabbling ducks wintering in western France. *Biological Conservation* 103: 183–198.

Guillemain, M., Sadoul, N. & Simon, G. 2005. European flyway permeability and abmigration in Teal *Anas crecca*, an analysis based on ringing recoveries. *Ibis* 147: 688–696.

Guillemain, M., Elmberg, J., Arzel, C., Johnson, A. & Simon, G. 2008. The income–capital breeding dichotomy revisited: late winter body condition is related to breeding success in an income breeder. *Ibis* 150: 172–176.

Guillemain, M., Fuster, J., Lepley, M., Mouronval, J.B. & Massez, G. 2009. Winter site fidelity is higher than expected for Eurasian Teal *Anas crecca* in the Camargue, France. *Bird Study* 56: 272–275.

Guillemain, M., Devineau, O., Brochet, A.-L., Fuster, J., Fritz, H., Green, A.J. & Gauthier-Clerc, M. 2010a. What is the spatial unit for a wintering Teal *Anas crecca*? Weekly day roost fidelity inferred from nasal saddles in the Camargue, southern France. *Wildlife Biology* 16: 215–220.

Guillemain, M., Elmberg, J., Gauthier-Clerc, M., Massez, G., Hearn, R., Champagnon, J. & Simon, G. 2010b. Wintering French Mallard and Teal are heavier and in better body condition than 30 years ago: effects of a changing environment? *Ambio* 39: 170–180.

Guillemain, M., Van Wilgenburg, S.L., Legagneux, P. & Hobson, K.A. 2014. Assessing geographic origins of Teal (*Anas crecca*) through stable-hydrogen (δ^2H) isotope analyses of feathers and ring-recoveries. *Journal of Ornithology* 155: 165–172.

Gunnarsson, T.G., Gill, J.A., Newton, J., Potts, P.M. & Sutherland, W.J. 2005. Seasonal matching of habitat quality and fitness in a migratory bird. *Proceedings of the Royal Society of London Series B* 272: 2319–2323.

Hagy, H.M. & Kaminski, R.M. 2012. Winter waterbird and food dynamics in autumn-managed moist-soil wetlands of the Mississippi Alluvial Valley. *Wildlife Society Bulletin* 36: 512–523.

Hagy, H.M., Straub, J.N., Schummer, M.L. & Kaminski, R.M. 2014. Annual variation in food densities and factors affecting wetland use by waterfowl in the Mississippi Alluvial Valley. *Wildfowl* (Special Issue 4): 436–450.

Hake, M., Månsson, J. & Wiberg, A. 2010. A working model for preventing crop damage caused by increasing goose populations in Sweden. *Ornis Svecica* 20: 225–233.

Haramis, G., Nichols, J., Pollock, K. & Hines, J. 1986. The relationship between body mass and survival of wintering Canvasbacks. *Auk* 103: 506–514.

Haramis, G.M. 1991a. Canvasback *Aythya valisineria*. *In* S. Funderburk, S. Jordan, J. Mihursky & D. Riley (eds.), *Habitat Requirements for Chesapeake Bay Living Resources, Revised 2nd Edition*, pp. 17.1–17.10. Chesapeake Research Consortium, Solomons, Maryland, USA.

Haramis, G.M. 1991b. Redhead *Aythya americana*. *In* S. Funderburk, S. Jordan, J. Mihursky and D. Riley (eds.), *Habitat Requirements for Chesapeake Bay Living Resources, 2nd ed. revised*, pp. 18.1–18.10. Chesapeake Research Consortium, Solomons, Maryland, USA.

Harmon, B.G. 1962. Mollusk as food of lesser scaup along the Louisiana coast. *Transactions North American Wildlife and Natural Resource Conference* 27: 132–138.

Harrison, X.A., Blount, J.D., Inger, R., Norris, D.R. & Bearhop, S. 2011. Carry-over effects as drivers of fitness differences in animals. *Journal of Animal Ecology* 80: 4–18.

Haukos, D.A. & Smith, L.M. 2003. Past and future impacts of wetland regulations on playa ecology in the Southern Great Plains. *Wetlands* 23: 577–589.

Haukos, D.A. 2004. *Analyses of selected mid-winter waterfowl survey data (1955–2004)*. U.S. Fish & Wildlife Service, Albuquerque, New Mexico, USA.

Hearn, R. & Skov, H. 2011. A brief overview of the status of European seaducks and actions required for their conservation. www.wwt.org.uk/blog/wp-content/uploads/2011/12/Seaducks-briefing-note-Dec-11.pdf.

Heinz, G.H., Hoffman, D.J. & Gold, L.G. 1989. Impaired reproduction of mallards fed an organic form of selenium. *Journal of Wildlife Management* 53: 418–428.

Heitmeyer, M. & Fredrickson, L.H. 1981. Do wetland conditions in the Mississippi Delta hardwoods influence Mallard recruitment? *Transactions of the North American Wildlife and Natural Resources Conferences* 46: 44–57.

Heitmeyer, M.E., Connelly, D.P. & Pederson, R.L. 1989. The Central, Imperial and Coachella Valleys of California. *In* L.M. Smith, R.L. Pederson & R.M. Kaminski (eds.), *Habitat Management for Migrating and Wintering Waterfowl in North America*, pp. 475–505. Texas Tech University Press, Lubbock, Texas, USA.

Heitmeyer, M.E., Cooper, R.J., Dickson, J.G. & Leopold, B.D. 2005. Ecological relationships of warm-blooded vertebrates in bottomland hardwoof ecosystems. *In* L.H. Fredrickson, S.L. King & R.M. Kaminski (eds.), *Ecology and Management of Bottomland Hardwood Systems: the State of Our Understanding*, pp. 281–307. University of Missouri-Columbia, Gaylord Memorial Laboratory Special Publication No. 10, Puxico, Missouri, USA.

Heitmeyer, M.E. 2006. The importance of winter floods to mallards in the Mississippi Alluvial Valley. *Journal of Wildlife Management* 70: 101–110.

Hénaux, V., Powell, L.A., Vrtiska, M.P. & Hobson, K.A. 2012. Establishing winter origins of migrating Lesser Snow Geese using stable isotopes. *Avian Conservation and Ecology* 7: 5.

Hindman, J.L. & Stotts, V.D. 1989. Chesapeake Bay and North Carolina Sounds. *In* L.M. Smith, R.L. Pederson & R.M. Kaminski (eds.), *Habitat Management for Migrating and Wintering Waterfowl in North America*, pp. 27–55. Texas Tech University Press, Lubbock, Texas, USA.

Hindman, L.J. & Tjaden, R.L. 2014. Awareness and opinions of Maryland citizens toward Chesapeake Bay Mute Swans *Cygnus olor* and management alternatives. *Wildfowl* 64: 167–185.

Hindman, L.J., Harvey, W.F & Conley, L.E. 2014. Spraying corn oil on Mute Swan *Cygnus olor* eggs to prevent hatching. *Wildfowl* 64: 186–196.

Hobaugh, W.C., Stutzenbaker, C.D. & Flickinger, E.L. 1989. The rice prairies. *In* L.M. Smith, R.L. Pederson & R.M. Kaminski (eds.), *Habitat Management for Migrating and Wintering Waterfowl in North America,* pp. 367–383. Texas Tech University Press, Lubbock, Texas, USA.

Hohman, W.L. & Rave, D.P. 1990. Diurnal time-activity budgets of wintering Canvasbacks in Louisiana. *Wilson Bulletin* 102: 645–654.

Hohman, W.L., Woolington, D.W. & Devries, J.H. 1990. Food habits of wintering Canvasbacks in Louisiana. *Canadian Journal of Zoology* 68: 2605–2609.

Hothem, R.L., Lonzarich, D.G., Takewaka, J.E. & Ohlendorf, H.M. 1998. Contaminants in wintering Canvasbacks and Scaups from San Francisco Bay, California. *Environmental Monitoring and Assessment* 50: 67–84.

Huner, J.V., Jeske, C.W. & Norling, W. 2002. Managing agricultural wetlands for waterbirds in the coastal regions of Louisiana, USA. *Waterbirds* (Special Publication 2): 66–78.

Hurrell, J.W., Kushnir, Y., Ottersen, G. & Visbeck, M. 2003. An Overview of the North Atlantic Oscillation. *In* J.W. Hurrell, Y. Kushnir, G. Ottersen & M. Visbeck (eds.), *The North Atlantic Oscillation: Climatic significance and nvironmental Impact,* pp. 1–35. American Geophysical Union, Washington D.C., USA.

Jefferies, R., Rockwell, R. & Abraham, K. 2004. The embarrassment of riches: agricultural food subsidies, high goose numbers and loss of Arctic wetlands a continuing saga. *Environmental Reviews* 11: 193–232.

Jensen, R.A., Wisz, M.S. & Madsen, J. 2008. Prioritizing refuge sites for migratory geese to alleviate conflicts with agriculture. *Biological Conservation* 141: 1806–1818.

Johnson, D.H. 1980. The comparison of usage and availability measurements for evaluating resource preference. *Ecology* 61: 65–71.

Johnson, F.A. & Montalbano, F. 1984. Selection of plant communities by wintering waterfowl on Lake Okeechobee, Florida. *Journal of Wildlife Management* 48: 174–178.

Johnson, F.A. & Montalbano, F. 1989. Southern reservoirs and lakes. *In* L.M. Smith, R.L. Pederson & R.M. Kaminski (eds.), *Habitat Management for Migrating and Wintering Waterfowl in North America,* pp. 93–116. Texas Tech University Press, Lubbock, Texas, USA.

Johnson, M.D. 2007. Measuring habitat quality: a review. *The Condor* 109: 489–504.

Jorde, D.G., Krapu, G.L., Crawford, R.D. & Hay, M.A. 1984. Effects of weather on habitat selection and behavior of Mallards wintering in Nebraska. *The Condor* 86: 258–265.

Kadlec, J.A. & Smith, L.M. 1989. The great basin marshes. *In* L.M. Smith, R.L. Pederson & R.M. Kaminski (eds.), *Habitat Management for Migrating and Wintering Waterfowl in North America,* pp. 451–474. Texas Tech University Press, Lubbock, Texas, USA.

Kaminski, R.M. & Davis, J.B. 2014. *Evaluation of the Migratory Bird Habitat Initiative: Report of findings.* Forest and Wildlife Research Center, Research Bulletin WF391, Mississippi State University, Mississippi, USA.

Kaminski, R.M. & Elmberg, J. 2014. An introduction to habitat use and selection by waterfowl in the northern hemisphere. *Wildfowl* (Special Issue 4): 9–16.

Kaminski, R.M. & Gluesing, E.A. 1987. Density- and habitat-related recruitment in mallards. *Journal of Wildlife Management* 51: 141–148.

Kaminski, R.M. & Weller, M.W. 1992. Breeding habitats of nearctic waterfowl. *In* B.D.J. Batt, A.D. Afton, M.G. Anderson, C.D. Ankney, D.H. Johnson, J.A. Kadlec & G.L. Krapu (eds.), *Ecology and Management of Breeding Waterfowl* pp. 568–589. University of

Minnesota Press, Minneapolis, Minnesota, USA.

Kaminski, R.M., Alexander, R.W. & Leopold, B.D. 1993. Wood duck and mallard winter microhabitats in Mississippi hardwood bottomlands. *Journal of Wildlife Management* 57: 562–570.

Kaminski, R.M., Davis, J.B., Essig, H.W., Gerard, P.D. & Reinecke, K.J. 2003. True metabolizable energy for wood ducks from acorns compared to other waterfowl foods. *Journal of Wildlife Management* 67: 542–550.

Kear, J. 1963. The history of potato-eating by wildfowl in Britain. *Wildfowl Trust Annual Report* 14: 54–65.

Kinney, S.D. 2004. Estimating the Population of Greater and Lesser Scaup during Winter in off-Shore Louisiana. M.Sc. thesis, Louisiana State University, Baton Rouge, USA.

Kirk, M.K., Esler, D. & Boyd, W.S. 2007. Foraging effort of Surf Scoters (*Melanitta perspicillata*) wintering in a spatially and temporally variable prey landscape. *Canadian Journal of Zoology* 85: 1207–1215.

Klimas, C., Murray, E., Foti, T., Pagan, J., Williamson, M. & Langston, H. 2009. An ecosystem restoration model for the Mississippi Alluvial Valley based on geomorphology, soils and hydrology. *Wetlands* 29: 430–450.

Korschgen, C.E. 1989. Riverine and deepwater habitats for diving ducks. *In* L.M. Smith, R.L. Pederson & R.M. Kaminski (eds.), *Habitat Management for Migrating and Wintering Waterfowl in North America*, pp. 157–180. Texas Tech University Press, Lubbock, Texas, USA.

Krapu, G.L., Facey, D.E., Fritzell, E.K. & Johnson, D.H. 1984. Habitat use by migrant Sandhill Cranes in Nebraska. *Journal of Wildlife Management* 48: 407–417.

Krementz, D.G., Asante, K. & Naylor, L.W. 2012. Autumn migration of Mississippi Flyway mallards as determined by satellite telemetry. *Journal of Fish and Wildlife Management* 3: 238–251.

Kross, J.P., Kaminski, R.M., Reinecke, K.J. & Pearse, A.P. 2008. Conserving waste rice for wintering waterfowl in the Mississippi Alluvial Valley. *Journal of Wildlife Management* 72: 1383–1387.

Lack, D. 1944. The problem of partial migration. *British Birds* 37: 122–130.

Lack, D. 1966. *Population Studies of Birds*. Clarendon Press, Oxford, UK.

LaMontagne, J.M., Jackson, L.J. & Barclay, R.M. 2003. Compensatory growth responses of *Potamogeton pectinatus* to foraging by migrating Trumpeter Swans in spring stop over areas. *Aquatic Botany* 76: 235–244.

Lancaster, J.D. 2013. Survival, habitat use and spatiotemporal use of wildlife management areas by female mallards in Mississippi's alluvial valley. M.Sc. thesis, Mississippi State University, Mississippi State, Mississippi, USA.

Larsson, K. & Tydén, L. 2005. Effects of oil spills on wintering Long-tailed Ducks *Clangula hyemalis* at Hoburgs bank in central Baltic Sea between 1996/97 and 2003/04. *Ornis Svecica* 15: 161–171.

Laubek, B. 1995. Habitat use by Whooper Swans *Cygnus cygnus* and Bewick's Swans *Cygnus columbianus bewickii* wintering in Denmark: increasing agricultural conflicts. *Wildfowl*, 46: 8–15.

Laubek, B., Nilsson, L., Wieloch, M., Koffijberg, K., Sudfelt, C. & Follestad, A. 1999. Distribution, numbers and habitat choice of the NW European Whooper Swan *Cygnus cygnus* population: results of an international census in January 1995. *Vogelwelt* 120: 141–154.

Legagneux, P., Inchausti, P., Bourguemestre, F., Latraube, F. & Bretagnolle, V. 2009. Effect of predation risk, body size, and habitat characteristics on emigration decisions in Mallards. *Behavioral Ecology* 20: 186–194.

Lehikoinen, A., Jaatinen, K., Vähätalo, A.V., Clausen, P., Crowe, O., Deceuninck, B., Hearn, R., Holt, C.A., Hornman, M. & Keller, V. 2013. Rapid climate driven shifts in wintering distributions of three common waterbird species. *Global Change Biology* 19: 2071–2081.

Lehnen, S.E. & Krementz, D.G. 2013. Use of aquaculture ponds and other habitats by autumn migrating shorebirds along the lower Mississippi River. *Environmental Management* 52: 417–426.

Lewis, T.L., Esler, D. & Boyd, W.S. 2007. Foraging behaviors of Surf Scoters and White-winged Scoters during spawning of Pacific herring. *The Condor* 109: 216–222.

Lewis, T.L., Esler, D. & Boyd, W.S. 2008. Foraging behavior of Surf Scoters (*Melanitta perspicillata*) and White-winged Scoters (*M. fusca*) in relation to clam density: inferring food availability and habitat quality. *Auk* 125: 149–157.

Lewis, T.L., Ward, D.H., Sedinger, J.S., Reed, A. & Derksen, D.V. 2013. Brant (*Branta bernicla*). *In* A.Poole (ed.), *The Birds of North America Online*. Cornell Laboratory of Ornithology, Ithaca, New York, USA.

Lindberg, M.S. & Walker, J. 2007. Satellite telemetry in avian research and management: sample size considerations. *Journal of Wildlife Management* 71: 1002–1009.

Link, P.T., Afton, A.D., Cox, R.R., Jr. & Davis, B.E. 2011. Daily movements of female Mallards wintering in southwestern Louisiana. *Waterbirds* 34: 422–428.

Loesch, C.R., Reynolds, R.E. & Hansen, L.T. 2012. An assessment of re-directing breeding waterfowl conservation relative to predictions of climate change. *Journal of Fish and Wildlife Management* 3: 1–22.

Lovvorn, J.R. & Baldwin, J.R. 1996. Intertidal and farmland habitats of ducks in the Puget Sound region: a landscape perspective. *Biological Conservation* 77: 97–114.

Lovvorn, J.R., De La Cruz, S.E., Takekawa, J.Y., Shaskey, L.E. & Richman, S.E. 2013. Niche overlap, threshold food densities and limits to prey depletion for a diving duck assemblage in an estuarine bay. *Marine Ecology Progress Series* 476: 251–268.

Luigujõe, L., Kuresoo, A., Keskpaik, J., Ader, A. & Leito, A. 1996. Migration and staging of the Bewick's swan (*Cygnus columbianus*) in Estonia. *Gibier Faune Sauvage* 13: 451–461.

Luo, H.-R., Smith, L.M., Allen, B. & Haukos, D.A. 1997. Effects of sedimentation on playa wetland volume. *Ecological Applications* 7: 247–252.

MacDonald, P.O., Frayer, W.E., Clauser, J.K. & Forsythe, S.W. 1979. *Documentation, Chronology and Future Projections of Bottomland Hardwood Habitat Loss in the Lower Mississippi Alluvial Plain*. Ecological Services, U.S. Department of the Interior, Fish & Wildlife Service, Vicksburg, Mississippi, USA.

Madsen, J. 1980. Occurrence, habitat selection and roosting of the pink-footed goose at Tipperne, Western Jutland, Dermark, 1972–1978. *Dansk Ornitologisk Forenings Tidsskrift* 74: 45–58.

Madsen, J. 1998. Experimental refuges for migratory waterfowl in Danish wetlands. II. Tests of hunting disturbance effects. *Journal of Applied Ecology* 35: 398–417.

Manley, S.W., Kaminski, R.M., Reinecke, K.J. & Gerard, P.D. 2004. Waterbird foods in winter-managed ricefields in Mississippi. *Journal of Wildlife Management* 68: 74–83.

Martínez Cedillo, I., Carmona, R., Ward, D.H. & Danemann, G.D. 2013. Habitat use patterns of the Black Brant *Branta bernicla nigricans* (Anseriformes: Anatidae) in natural and artificial areas of Guerrero Negro, Baja

California Sur, Mexico. *Revista de Biologia Tropical* 61: 927–935.

Marty, J.R. 2013. Seed and waterbird abundances in ricelands in the Gulf Coast Prairies of Louisiana and Texas. M.Sc. thesis, Mississippi State University, Mississippi State, Mississippi, USA.

Mathiasson, S. 1991. Eurasian Whooper Swan *Cygnus cygnus* migration, with particular reference to birds wintering in southern Sweden. *Wildfowl* (Special Supplement No. 1): 201–208.

McKelvey, R.W. & MacNeill, A.C. 1981. Mortality factors of wild swans in Canada. *In* G.V.T. Matthews & M. Smart (eds.), *Proceedings of the Second International Swan Symposium,* pp. 312–318. International Waterfowl and Wetland Research Bureau, Slimbridge, UK.

Meineri, E., Rodriguez Perez, H., Hilaire, S. & Mesleard, F. 2014. Distribution and reproduction of *Procambarus clarkii* in relation to water management, salinity and habitat type in the Camargue. *Aquatic Conservation: Marine and Freshwater Ecosystems* 24: 312–323.

Michot, T., Woodin, M. & Nault, A. 2008. Food habits of Redheads (*Aythya Americana*) wintering in seagrass beds of coastal Louisiana and Texas, USA. *Acta Zoologica Academiae Scientiarum Hungaricae* 54: 239–250.

Michot, T.C. 1996. Marsh loss in coastal Louisiana: implications for management of North American *Anatidae. Gibier Faune Sauvage* 13: 941–957.

Michot, T.C., Woodin, M.C., Adair, S.E. & Moser, E.B. 2006. Diurnal time activity budgets of Redheads (*Aythya americana*) wintering in seagrass beds and coastal ponds in Louisiana and Texas. *Hydrobiologia* 576: 113–128.

Migratory Bird Data Center. 2009. http://mbdcapps.fws.gov/mbdc/databases/afsos/aboutafsos.html.

Miller, M.R. 1986. Northern pintail body condition during wet and dry winters in the Sacramento Valley, California. *Journal of Wildlife Management* 50: 189–198.

Miller, M.R., Orthmeyer, D.L., Casazza, M.L., McLandress, M.R. & Connelly, D.P. 1993. *Survival, Habitat Use and Movements of Female Northern Pintails Radio-marked in the Suisun Marsh, California.* National Fish & Wildlife Foundation and California Waterfowl Association, Sacramento, California, USA.

Miller, M.R., Fleskes, J.P., Orthmeyer, D.L., Newton, W.E. & Gilmer, D.S. 1995. Survival of adult female northern pintails in Sacramento Valley, California. *Journal of Wildlife Management* 59: 478–486.

Miller, M.R., Garr, J.D. & Coates, P.S. 2010. Changes in the status of harvested rice fields in the Sacramento Valley, California: implications for wintering waterfowl. *Wetlands* 30: 939–947.

Miller, O.D., Wilson, J.A., Ditchkoff, S.S. & Lochmiller, R.L. 2000. Consumption of agricultural and natural foods by waterfowl migrating through central Oklahoma. *Proceedings of the Oklahoma Academy of Science* 80: 25–31.

Mitchell, C.A., Custer, T.W. & Zwank, P.J. 1994. Herbivory on shoalgrass by wintering redheads in Texas. *Journal of Wildlife Management* 58: 131–141.

Mitchell, C.D. & Eichholz, M.W. 2010. *Trumpeter swan (Cygnus buccinator). In* A.Poole (ed.), *The Birds of North America Online.* Cornell Laboratory of Ornithology, Ithaca, New York, USA.

Mitsch, W.J. & Gosselink, J.G. 2007. *Wetlands.* John Wiley and Sons, Inc., Hoboken, New Jersey, USA.

Moon, J.A. & Haukos, D.A. 2006. Survival of female northern pintails wintering in the Playa Lakes Region of northwestern Texas. *Journal of Wildlife Management* 70: 777–783.

Moon, J.A., Haukos, D.A. & Smith, L.M. 2007. Declining body condition of northern pintails wintering in the Playa Lakes Region. *Journal of Wildlife Management* 71: 218–221.

Moore, J.E., Colwell, M.A., Mathis, R.L. & Black, J.M. 2004. Staging of Pacific flyway Brant in relation to eelgrass abundance and site isolation, with special consideration of Humboldt Bay, California. *Biological Conservation* 115: 475–486.

Moore, J.E. & Black, J.M. 2006. Historical changes in Black Brant *Branta bernicla nigricans* use on Humboldt Bay, California. *Wildlife Biology* 12: 151–162.

Morrison, M.L., Marcot, B.G. & Mannan, R.W. 1992. *Wildlife-Habitat Relationships: Concepts and Applications*. Island Press, Washington D.C., USA.

Moser, T.J. 2006. *The 2005 North American Trumpeter Swan Survey*. Division of Migratory Bird Management, U.S. Fish & Wildlife Service, Denver, Colorado, USA.

Moulton, D.W., Dahl, T.E. & Dall, D.M. 1997. *Texas Coastal Wetlands; Status and Trends, Mid-1950s to Early 1990s*. U.S. Fish & Wildlife Service, Albuquerque, New Mexico, USA.

Mulhern, J.H., Nudds, T.D. & Neal, B.R. 1985. Wetland selection by Mallards and Blue-winged Teal. *Wilson Bulletin* 97: 473–485.

Munro, R.E. 1981. Field feeding by *Cygnus columbianus columbianus* in Maryland. *In* G. V. T. Mathews and M. Smart (eds.), *Second International Swan Symposium,* pp. 261–272. IWRB, Sapporo, Japan.

Nagel, J. 1965. Field feeding of Whistling Swans in northern Utah. *Condor* 67: 446–447.

Neu, C.W., Byers, C.R. & Peek, J.M. 1974. A technique for analysis of utilization-availability data. *Journal of Wildlife Management* 38: 541–545.

Newton, I. 2007. *The Migration Ecology of Birds*. Elsevier, London, UK.

Nichols, F.H., Cloern, J.E., Luoma, S.N. & Peterson, D.H. 1986. The modification of an estuary. *Science* 231: 567–573.

Nichols, J.D., Reinecke, K.J. & Hines, J.E. 1983. Factors affecting the distribution of Mallards wintering in the Mississippi Alluvial Valley. *Auk* 100: 932–946.

Nolet, B.A., Langevoord, O., Bevan, R.M., Engelaar, K.R. Klaassen, M., Mulder, R.J.W. & Van Dijk, S. 2001. Spatial variation in tuber depletion by swans explained by differences in net intake rates. *Ecology* 82: 1655–1667.

Nolet, B.A., Bevan, R.M., Klaassen, M., Langevoord, O. & Van Der Heijden, Y.G.J.T. 2002. Habitat switching by Bewick's Swans: maximization of average long-term energy gain? *Journal of Animal Ecology* 71: 979–993.

Norris, D.R. & Marra, P.P. 2007. Seasonal interactions, habitat quality, and population dynamics in migratory birds. *The Condor* 109: 535–547.

North American Waterfowl Management Plan. 1986. *North American Waterfowl Management Plan. A Strategy for Cooperation*. U.S. Department of the Interior, Washington D.C., USA and Environment Canada, Ottawa, Ontario, Canada.

North American Waterfowl Management Plan (NAWMP). 2012. People conserving waterfowl and wetlands. http://nawmprevision.org/.

North American Waterfowl Management Plan Committee. 2004. *North American Waterfowl Management Plan 2004. Implementation Framework: Strengthening the Biological Foundation*. Canadian Wildlife Service, Ottawa, Ontario, Canada, U.S. Fish & Wildlife Service, Washington D.C, USA and Secretaria de Medio Ambiente y Recursos Naturales, Mexico City, Mexico.

O'Neal, B.J., Stafford, J.D. & Larkin, R.P. 2010. Waterfowl on weather radar: applying ground truth to classify and quantify bird movements. *Journal of Field Ornithology* 81: 71–82.

Obenberger, S.M. 1982. Numerical response of wintering waterfowl to macrohabitat in the Southern High Plains of Texas. M.Sc. thesis, Texas Tech University, Lubbock, Texas, USA.

Olsen, R.E. & Cox, R.R., Jr. 2003. Body size and condition of male Mallards during mid-winter in North Dakota, USA. *Waterbirds* 26: 449–456.

Olsen, R.E., Cox, R.R., Jr., Afton, A.D. & Ankney, C.D. 2011. Diet and gut morphology of male Mallards during winter in North Dakota. *Waterbirds* 34: 59–69.

Ottersen, G., Planque, B., Belgrano, A., Post, E., Reid, P.C. & Stenseth, N.C. 2001. Ecological effects of the North Atlantic oscillation. *Oecologia* 128: 1–14.

Owen, M. & Williams, G. 1976. Winter distribution and habitat requirements of Wigeon in Britain. *Wildfowl* 27: 83–90.

Owen, M. 1980. *Wild Geese of the World: their Life History and Ecology.* Batsford, London, UK.

Owen, M., Atkinson-Willes, G.L. & Salmon, D.G. 1986. *Wildfowl in Great Britain – Second edition.* Cambridge University Press, Cambridge, UK.

Pacific Coast Joint Venture. 2004. Strategic Plan. https://projects.atlas.ca.gov/frs/download.php/15218/PCJV_StrategicPlan_2004.pdf.

Pearse, A.T. 2007. Design, evaluation, and applications of an aerial survey to estimate abundance of wintering waterfowl in Mississippi. Ph.D. dissertation, Mississippi State University, Mississippi State, Mississippi, USA.

Pearse, A.T., Dinsmore, S.J., Kaminski, R.M. & Reinecke, K.J. 2008. Evaluation of an aerial survey to estimate abundance of wintering ducks in Mississippi. *Journal of Wildlife Management* 72: 1413–1419.

Pearse, A.T., Kaminski, R.M., Reinecke, K.J. & Dinsmore, S.J. 2012. Local and landscape associations between wintering dabbling ducks and wetland complexes in Mississippi. *Wetlands* 32: 859–869.

Percival, S., Sutherland, W. & Evans, P. 1998. Intertidal habitat loss and wildfowl numbers: applications of a spatial depletion model. *Journal of Applied Ecology* 35: 57–63.

Perry, M. & Deller, A. 1995. Waterfowl population trends in the Chesapeake Bay area. *In* P. Hill & S. Nelson (eds.), *Proceedings of the 1994 Chesapeake Research Conference. Toward a Sustainable Coastal Watershed: The Chesapeake Experiment,* pp. 490–500. Chesapeake Research Consortium, Edgewater, Maryland, USA.

Perry, M.C., Wells-Berlin, A.M., Kidwell, D.M. & Osenton, P.C. 2007. Temporal changes of populations and trophic relationships of wintering diving ducks in Chesapeake Bay. *Waterbirds* 30: 4–16.

Petersen, I.K., Christensen, T.K., Kahlert, J., Desholm, M. & Fox, A.D. 2006. *Final Results of Bird Studies at the Offshore Wind Farms at Nysted and Horns Rev, Denmark.* National Environmental Research Institute, Ministry of Environment, Copenhagen, Denmark.

Peterson, C.H. 1982. Clam predation by whelks (*Busycon* spp.): experimental tests of the importance of prey size, prey density and seagrass cover. *Marine Biology* 66: 159–170.

Peterson, C.H., Summerson, H. & Duncan, P. 1984. The influence of seagrass cover on population structure and individual growth rate of a suspension-feeding bivalve, *Mercenaria mercenaria*. *Journal of Marine Research* 42: 123–138.

Peterson, T.G. 2014. Wintering waterfowl use of Delta National Forest, Mississippi. M.Sc. thesis, Mississippi State University, Mississippi State, Mississippi, USA.

Petrie, M., Vest, J. & Smith, D. 2013. Intermountain West Joint Venture Implementation Plan-Waterfowl. http://iwjv.org/resource/2013-implementation-plan-chapter-4-waterfowl.

Petrie, M., Brasher, M. & James, D. 2014. *Estimating the Biological and Economic Contributions that Rice Habitats Make in Support of North American Waterfowl*. The Rice Foundation, Stuttgart, Arkansas, USA.

Petrie, S.A. & Francis, C.M. 2003. Rapid increase in the lower Great Lakes population of feral mute swans: a review and a recommendation. *Wildlife Society Bulletin* 31: 407–416.

Pilarczyk, B., Tomza-Marciniak, A., Pilarczyk, R., Kavetska, K., Rząd, I., Hendzel, D. & Marciniak, A. 2012. Selenium status in sea ducks (*Melanitta fusca*, *Melanitta nigra* and *Clangula hyemalis*) wintering on the southern Baltic coast, Poland. *Marine Biology Research* 8: 1019–1025.

Plattner, D.M., Eichholz, M.W. & Yerkes, T. 2010. Food resources for wintering and spring staging black ducks. *Journal of Wildlife Management* 74: 1554–1558.

Playa Lakes Joint Venture. 2014. Priority habitats. http://pljv.org/about/habitats.

Raveling, D.G. 2004. *Waterfowl of the World: A Comparative Perspective*. University of Missouri-Columbia, Gaylord Memorial Laboratory, Puxico, Missouri, USA.

Raveling, D.G. & Heitmeyer, M.E. 1989. Relationships of population size and recruitment of pintails to habitat conditions and harvest. *Journal of Wildlife Management* 53: 1088–1103.

Rees, E.C. 1987. Conflict of choice within pairs of Bewick's Swans regarding their migratory movement to and from the wintering grounds. *Animal Behaviour* 35: 1685–1693.

Rees, E.C. 2006. *Bewick's Swan*. T. & A.D. Poyser, London, UK.

Rees, E.C. & Beekman, J.H. 2010. Northwest European Bewick's Swans: a population in decline. *British Birds* 103: 640–650.

Rees, E.C., Kirby, J.S. & Gilburn, A. 1997. Site selection by swans wintering in Britain and Ireland; the importance of habitat and geographic location. *Ibis* 139: 337–352.

Reinecke, K.J., Kaminski, R.M., Moorehead, D.J., Hodges, J.D. & Nassar, J.R. 1989. Mississippi Alluvial Valley. *In* L.M. Smith, R.L. Pederson & R.M. Kaminski (eds.), *Habitat Management for Migrating and Wintering Waterfowl in North America.*, pp. 203–247. Texas Tech University Press, Lubbock, Texas, USA.

Reinecke, K.J., Brown, M.W. & Nassar, J.R. 1992. Evaluation of aerial transects for counting wintering mallards. *Journal of Wildlife Management* 56: 515–525.

Rendón, M.A., Green, A.J., Aguilera, E. & Almaraz, P. 2008. Status, distribution and long-term changes in the waterbird community wintering in Doñana, south-west Spain. *Biological Conservation* 141: 1371–1388.

Richman, S.E. & Lovvorn, J.R. 2003. Effects of clam species dominance on nutrient and energy acquisition by spectacled eiders in the Bering Sea. *Marine Ecology Progress Series* 261: 283–297.

Richman, S.E. & Lovvorn, J.R. 2004. Relative foraging value to Lesser Scaup ducks of native and exotic clams from San Francisco Bay. *Ecological Applications* 14: 1217–1231.

Robertson, G.J. & Cooke, F. 1999. Winter philopatry in migratory waterfowl. *Auk* 116: 20–34.

Rodway, M.S. 2007. Timing of pairing in waterfowl II: Testing the hypotheses with Harlequin Ducks. *Waterbirds* 30: 506–520.

Rogers, P.M. & Hammer, D.A. 1998. Ancestral breeding and wintering ranges of the Trumpeter Swan (*Cygnus buccinator*) in the Eastern United States. *Bulletin of The Trumpeter Swan Society* 27: 13–29.

Roshier, D.A., Klomp, N.I. & Asmus, M. 2006. Movements of a nomadic waterfowl, Grey Teal *Anas gracilis*, across inland Australia – results from satellite telemetry spanning fifteen months. *Ardea* 94: 461–475.

Ross, P.G., Luckenbach, M.W., Wachapreague, V. & Bowman, T. 2009. *Distribution, Habitat Characteristics, Prey Abundance and Diet of Surf Scoters* (Melanitta perspicillata) *and Long-tailed Ducks* (Clangula hyemalis) *in Polyhaline Wintering Habitats in the Mid-Atlantic Region: a Comparison of Shallow Coastal Lagoons and Chesapeake Bay Environs*. U.S. Fish & Wildlife Service, Sea Duck Joint Venture, Anchorage, Alaska, USA.

Roy, C.L., Herwig, C.M. & Doherty, P.F. 2013. Mortality and refuge use by young ring necked ducks before and during hunting season in north central Minnesota. *Journal of Wildlife Management* 77: 947–956.

Salewski, V. & Bruderer, B. 2007. The evolution of bird migration – a synthesis. *Naturwissenschaften* 94: 268–279.

Schummer, M.L., Petrie, S.A. & Bailey, R.C. 2008. Interaction between macroinvertebrate abundance and habitat use by diving ducks during winter on northeastern Lake Ontario. *Journal of Great Lakes Research* 34: 54–71.

Schummer, M.L., Kaminski, R.M., Raedeke, A.H. & Graber, D.A. 2010. Weather related indices of autumn–winter dabbling duck abundance in middle North America. *Journal of Wildlife Management* 74: 94–101.

Schummer, M.L., Cohen, J., Kaminski, R.M., Brown, M.E. & Wax, C.L. 2014. Atmospheric teleconnections and Eurasian snow cover as predictors of a weather severity index in relation to Mallard *Anas platyrhynchos* autumn–winter migration. *Wildfowl* (Special Issue 4): 451–469.

Scott, D.A. & Rose, P. 1996. *Atlas of Anatidae Populations in Africa and Western Eurasia*. Wetland International Publication No. 41. Wetlands International, Waginingen, Netherlands.

Sea Duck Joint Venture. 2004. Sea duck information series. www.seaduckjv.org/ infoseries/toc.html.

Sears, J. 1989. Feeding activity and body condition of Mute Swans *Cygnus olor* in rural and urban areas of a lowland river system. *Wildfowl* 40: 88–98.

Sedinger, J.S. & Alisauskas, R.T. 2014. Cross-seasonal effects and the dynamics of waterfowl populations. *Wildfowl* (Special Issue 4): 000–000.

Serie, J.R. & Bartonek, J.C. 1991. Population status and productivity of Tundra Swans, *Cygnus columbianus* in North America. *Wildfowl* (Special Supplement No. 1): 172–177.

Shaughnessy, F.J., Gilkerson, W., Black, J.M., Ward, D.H. & Petrie, M. 2012. Predicted eelgrass response to sea level rise and its availability to foraging Black Brant in Pacific coast estuaries. *Ecological Applications* 22: 1743–1761.

Silverman, E.D., Saalfeld, D.T., Leirness, J.B. & Koneff, M.D. 2013. Wintering sea duck distribution along the Atlantic Coast of the United States. *Journal of Fish and Wildlife Management* 4: 178–198.

Skalos, D.A. 2012. Evaluating body condition and predicting lipid mass of wintering Pacific greater white-fronted geese (*Anser albifrons frontalis*). M.Sc. thesis, University of California-Davis, Davis, California, USA.

Skov, H., Heinänen, S., Žydelis, R., Bellebaum, J., Bzoma, S., Dagys, M., Durinck, J., Garthe, S., Grishanov, G., Hario, M., Kieckbusch, J.J., Kube, J., Kuresoo, A., Larsson, K., Luigujoe, L., Meissner, W., Nehls, H.W., Nilsson, L., Petersen, I.K., Mikkola-Roos, M., Pihl, S., Sonntag, N. & Stipniece, A. 2011. *Waterbird Populations and Pressures in the Baltic Sea*. Nordic Council of Ministers, Copenhagen, Denmark.

Smeins, F.E., Diamond, D.D. & Hanselka, C.W. 1991. Coastal prairie. *In* R.T. Coupland (ed.), *Ecosystems of the World 8A. Natural Grasslands: Introduction and Western Hemisphere,* pp. 269–290. Elsevier Science, New York, USA.

Smith, L.M., Vangilder, L.D. & Kennamer, R.A. 1985. Foods of wintering brant in eastern

North America. *Journal of Field Ornithology* 56: 286–289.

Smith, L.M. & Sheeley, D.G. 1993. Factors affecting condition of northern pintails wintering in the Southern High Plains. *Journal of Wildlife Management* 57: 62–71.

Smith, L.M. 2003. *Playas of the Great Plains*. University of Texas Press, Austin, Texas, USA.

Söderquist, P., Gunnarsson, G. & Elmberg, J. 2013. Longevity and migration distance differ between wild and hand-reared Mallards *Anas platyrhynchos* in Northern Europe. *European Journal of Wildlife Research* 59: 159–166.

St. James, E.A., Schummer, M.L., Kaminski, R.M., Penny, E.J. & Burger, L.W. 2013. Effect of weekly hunting frequency on duck abundances in Mississippi wildlife management areas. *Journal of Fish and Wildlife Management* 4: 144–150.

Stafford, J.D., Kaminski, R.M., Reinecke, K.J. & Manley, S.W. 2006. Waste rice for waterfowl in the Mississippi Alluvial Valley. *Journal of Wildlife Management* 70: 61–69.

Stafford, J.D., Janke, A.K., Anteau, M.J., Pearse, A.T., Fox, A.D., Elmberg, J., Straub, J.N., Eichholz, M.W. & Arzel, C. 2014. Spring migration of waterfowl in the northern hemisphere: a conservation perspective. *Wilfowl* (Special Issue 4): 70–85.

Stenseth, N.C., Mysterud, A., Ottersen, G., Hurrell, J.W., Chan, K.-S. & Lima, M. 2002. Ecological effects of climate fluctuations. *Science* 297: 1292–1296.

Stott, R.S. & Olson, D.P. 1973. Food-habitat relationship of sea ducks on the New Hampshire coastline. *Ecology* 54: 996–1007.

Systad, G.H., Bustnes, J.O. & Erikstad, K.E. 2000. Behavioral responses to decreasing day length in wintering sea ducks. *Auk* 117: 33–40.

Takekawa, J., Miles, A., Tsao-Melcer, D., Schoellhamer, D., Fregien, S. & Athearn, N. 2009. Dietary flexibility in three representative waterbirds across salinity and depth gradients in salt ponds of San Francisco Bay. *Hydrobiologia* 626: 155–168.

Tamisier, A. & Grillas, P. 1994. A review of habitat changes in the Camargue: an assessment of the effects of the loss of biological diversity on the wintering waterfowl community. *Biological Conservation* 70: 39–47.

Tate J., Jr. & Tate, D.J. 1966. Additional records of Whistling Swans feeding in dry fields. *The Condor* 68: 398–399.

Thomas, G. 1976. Habitat usage of wintering ducks at the Ouse Washes, England. *Wildfowl* 27: 148–151.

Thompson, J.D. & Baldassarre, G.A. 1990. Carcass composition of nonbreeding Blue-winged Teal and Northern Pintails in Yucatan, Mexico. *The Condor* 92: 1057–1065.

Tombre, I.M., Høgda, K.A., Madsen, J., Griffin, L.R., Kuijken, E., Shimmings, P., Rees, E. & Verscheure, C. 2008. The onset of spring and timing of migration in two arctic nesting goose populations: the pink-footed goose *Anser brachyrhynchus* and the barnacle goose *Branta leucopsis*. *Journal of Avian Biology* 39: 691–703.

Toral, G.M. & Figuerola, J. 2010. Unraveling the importance of rice fields for waterbird populations in Europe. *Biodiversity and Conservation* 19: 3459–3469.

Tourenq, C., Bennetts, R.E., Kowalski, H., Vialet, E., Lucchesi, J.L., Kayser, Y. & Isenmann, P. 2001. Are ricefields a good alternative to natural marshes for waterbird communities in the Camargue, southern France? *Biological Conservation* 100: 335–343.

U.S. Department of Agriculture. 2010. *Catfish 2010. Part I: Reference of Catfish Health and Production Practices in the United States, 2009*. U.S. Department of Agriculture – Animal and Plant Health Inspection Service, Fort Collins, Colorado, USA.

Väänänen, V.-M. 2001. Hunting disturbance and the timing of autumn migration in *Anas* species. *Wildlife Biology* 7: 3–9.

Varner, D.M. 2008. Survival and foraging ecology of Interior Population trumpeter swans. M.Sc. thesis, Southern Illinois Universy, Carbondale, Illinois, USA.

van Eerden, M.R. (ed.). 1997. Patchwork: patch use, habitat exploitation and carrying capacity for water birds in Dutch freshwater wetlands. Ph.D. thesis, Rijksuniversiteit Groningen and Rijkswaterstaat Directie IJsselmeergebied, Lelystad, Netherlands.

Venne, L.S., Anderson, T.A., Zhang, B., Smith, L.M. & McMurry, S.T. 2008. Organochlorine pesticide concentrations in sediment and amphibian tissue in playa wetlands in the Southern High Plains, USA. *Bulletin of Environmental Contamination and Toxicology* 80: 497–501.

Vest, J.L., Kaminski, R.M., Afton, A.D. & Vilella, F.J. 2006. Body mass of lesser scaup during fall and winter in the Mississippi flyway. *Journal of Wildlife Management* 70: 1789–1795.

Vickery, J., Sutherland, W., Watkinson, A., Rowcliffe, J. & Lane, S. 1995. Habitat switching by dark-bellied Brent Geese *Branta b. bernicla* (L.) in relation to food depletion. *Oecologia* 103: 499–508.

Voslamber, B., Knecht, E. & Kleijn, D. 2010. Dutch Greylag Geese *Anser anser*: migrants or residents. *Ornis Svecica* 20: 207–214.

Waldeck, P. & Larsson, K. 2013. Effects of winter water temperature on mass loss in Baltic Blue Mussels: implications for foraging sea ducks. *Journal of Experimental Marine Biology and Ecology* 444: 24–30.

Ward, D.H., Reed, A., Sedinger, J.S., Black, J.M., Derksen, D.V. & Castelli, P.M. 2005. North American Brant: effects of changes in habitat and climate on population dynamics. *Global Change Biology* 11: 869–880.

Weaver, K.H. 2013. Tundra Swan (*Cygnus columbianus columbianus*) habitat selection during the nonbreeding period. M.Sc. thesis, The University of Western Ontario, London, Ontario, Canada.

Weller, M.W. 1964. Distribution and migration of the redhead. *Journal of Wildlife Management* 28: 64–103.

Wetlands International 2014. *Waterbird Population Estimates*. http://wpe.wetlands.org.

Wiens, J.A. 1973. Pattern and process in grassland bird communities. *Ecological Monographs* 43: 237–270.

Wiens, J.A. 1989. *The Ecology of Bird Communities. Volume 1. Foundations and Patterns*. Cambridge University Press, Cambridge, UK.

Williams III, S.O. & Chabreck, R.H. 1986. *Quantity and quality of waterfowl habitat in Louisiana*. Research Report No. 8. Louisiana State University, Baton Rouge, Louisiana, USA.

Wiseman, A.J., Kaminski, R.M., Riffell, S., Reinecke, K.J. & Larson, E.J. 2010. Ratoon grain sorghum and other seeds for waterfowl in sorghum croplands. *Proceedings of the Southeast Association of Fish Wildlife Agencies* 64: 106–111.

Wooten, D.E. & Werner, S.J. 2004. Food habits of Lesser Scaup *Aythya affinis* occupying baitfish aquaculture facilities in Arkansas. *Journal of the World Aquaculture Society* 35: 70–77.

Zammuto, R.M. 1986. Life histories of birds: clutch size, longevity, and body mass among North American game birds. *Canadian Journal of Zoology* 64: 2739–2749.

Zink, R.M. 2011. The evolution of avian migration. *Biological Journal of the Linnean Society* 104: 237–250.

Zipkin, E.F., Gardner, B., Gilbert, A.T., O'Connell, A.F., Jr., Royle, J.A. & Silverman,

E.D. 2010. Distribution patterns of wintering sea ducks in relation to the North Atlantic Oscillation and local environmental characteristics. *Oecologia* 163: 893–902.

Žydelis, R.M. & Ruskete, D.R. 2005. Winter foraging of Long-tailed Ducks (*Clangula hyemalis*) exploiting different benthic communities in the Baltic Sea. *Wilson Bulletin* 117: 133–141.

Žydelis, R.M., Bellebaum, J., Österblom, H., Vetemaa, M., Schirmeister, B., Stipniece, A., Dagys, M., van Eerden, M. & Garthe, S. 2009. Bycatch in gillnet fisheries – an overlooked threat to waterbird populations. *Biological Conservation* 142: 1269–1281.

Photograph: A spring of Green-winged and Blue-winged Teal wintering in Louisiana, by Charlie Hohorst.

Spring migration of waterfowl in the northern hemisphere: a conservation perspective

JOSHUA D. STAFFORD[1]*, ADAM K. JANKE[2], MICHAEL J. ANTEAU[3], AARON T. PEARSE[3], ANTHONY D. FOX[4], JOHAN ELMBERG[5], JACOB N. STRAUB[6], MICHAEL W. EICHHOLZ[7] & CÉLINE ARZEL[8]

[1]South Dakota Cooperative Fish and Wildlife Research Unit, U.S. Geological Survey, Department of Natural Resource Management, South Dakota State University, Brookings, South Dakota, USA.
[2]Department of Natural Resource Management, South Dakota State University, Brookings, South Dakota, USA.
[3]U.S. Geological Survey, Northern Prairie Wildlife Research Center, Jamestown, North Dakota, USA.
[4]Department of Bioscience, Aarhus University, DK-8410 Rønde, Denmark.
[5]Aquatic Biology and Chemistry, Kristianstad University, SE-291 88 Kristianstad, Sweden.
[6]Center for Earth and Environment Science, State University of New York-Plattsburgh, Plattsburgh, New York, USA.
[7]Cooperative Wildlife Research Laboratory, Center for Ecology, Southern Illinois University Carbondale, Carbondale, Illinois, USA.
[8]Section of Ecology, Department of Biology, University of Turku, 20014 Turku, Finland.
*Correspondence author. E-mail: jstafford@usgs.gov

Abstract

Spring migration is a key part of the annual cycle for waterfowl populations in the northern hemisphere, due to its temporal proximity to the breeding season and because resources may be limited at one or more staging sites. Research based on field observations during spring lags behind other periods of the year, despite the potential for fitness consequences through diminished survival or cross-seasonal effects of conditions experienced during migration. Consequently, conservation strategies for waterfowl on spring migration are often only refined versions of practices used during autumn and winter. Here we discuss the current state of knowledge of habitat requirements for waterfowl at their spring migratory sites and the intrinsic and extrinsic factors that lead to variability in those requirements. The provision of plant foods has become the main conservation strategy during spring because of the birds' energy requirements at this time, not only to fuel migration but to facilitate early clutch formation on arrival at the breeding grounds. Although energy sources are important to migrants, there is little evidence on the extent to which the availability of carbohydrate-based food is limiting for many migratory waterfowl populations.

Such limitation is relatively unlikely among populations that exploit agricultural grain during migration (*e.g.* arctic-nesting geese), suggesting that conservation strategies for these populations may be misplaced. In general, however, we found few cases in which an ecological understanding of spring-migrating waterfowl was sufficient to indicate true resource limitation during migration, and still fewer cases where conservation efforts ameliorated these limitations. We propose a framework that aims to address knowledge gaps and apply empirical research results to conservation strategies based on documented limitations and associated fitness impacts on migrating waterfowl. Such a strategy would improve allocation of scarce conservation resources during spring migration and greatly improve ecological understanding of migratory waterfowl and their habitats in the northern hemisphere.

Key words: conservation, limitations, lipids, nutrients, spring migration, waterfowl.

Spring is a critical phase of the annual cycles of waterfowl *Anatidae* sp. in the northern hemisphere because of the physiological and environmental conditions encountered during migration, and the co-occurrence of pre-breeding life-history events. Maintenance or acquisition of nutrient reserves at staging areas is generally necessary in order to complete migration, and is often also a prerequisite for successful breeding (Ankney *et al.* 1991; Jenni & Jenni-Eirmann 1998). Individuals often experience diminished food availability as they await the thaw of wetland habitats or because of food depletion by autumn-migrating birds (Stafford *et al.* 2006; Greer *et al.* 2009; Straub *et al.* 2012). Moreover, in addition to migration, many species are undertaking energetically expensive activities such as courtship, pair-bond maintenance and moulting into breeding plumage at this time (Heitmeyer 1988; Lovvorn & Barzen 1988; Richardson & Kaminski 1992; Hohman *et al.* 1997; Barras *et al.* 2001; Anteau *et al.* 2011a). Adverse and unpredictable weather can kill birds directly or lead to starvation by making food resources temporarily unavailable (Trautman *et al.* 1939; Newton 2006, 2007). Further, migratory movements themselves can be dangerous and energetically costly, requiring individuals and flocks to exploit habitats and foods that promote survival (*sensu* Fretwell 1972; Kaminski & Elmberg 2014). The choice of migratory strategy therefore represents important trade-offs with lasting consequences for individual fitness and population dynamics, which may be sensitive to management strategies used by conservation organisations along migratory corridors in the northern hemisphere.

Habitat conditions encountered during spring migration also have potential to influence waterfowl populations through cross-seasonal (or carry-over) impacts on individual reproduction (Davis *et al.* 2014, Sedinger & Alisauskas 2014). The seminal works of Weller (1975), Fredrickson and Drobney (1977), Ankney and MacInnes (1978), and others (*e.g.* Heitmeyer & Fredrickson 1981; Kaminski & Gluesing 1987) prompted research on the nature and mechanisms for cross-seasonal effects on waterfowl in particular and migratory birds in general (Drent & Daan 1980; Harrison *et al.* 2011; Sedinger & Alisauskas 2014). This work has shown evidence for various cross-seasonal relationships in waterfowl, for populations both in Europe and in North America, and ranging across species from those using capital breeding strategies (*e.g.* Lesser Snow Goose *Chen c. caerulescens*; Alisauskas 2002) to those which mainly acquire the food resources needed for egg-laying on or near the breeding territories ("income breeders"; *e.g.* Eurasian Teal *Anas crecca crecca*; Guillemain *et al.* 2008). Although spring migration is widely recognised as being an important time both for individual survival and for subsequent breeding success, it remains largely understudied in comparison with other stages in the annual cycle (Arzel *et al.* 2006), and management strategies during this period are often refinements of practices intended for breeding or wintering populations (Soulierre *et al.* 2007).

In this paper we synthesise published information on the habitat requirements of waterfowl during spring migration and discuss potential applications of the knowledge for conservation initiatives at migratory stopover areas. Arzel *et al.* (2006) has made a comprehensive review of the current state of literature on spring-migrating waterfowl, so we do not intend to repeat their work here. Rather, we endeavour to assess available information and consider gaps in knowledge that have the potential to diminish the efficacy of conservation strategies aimed at enhancing habitat conditions for waterfowl on spring migration. Identifying knowledge gaps can inform the management and conservation of waterfowl at spring staging areas and help to set research priorities for improving our understanding of migratory species. Specifically, our objectives are to: 1) review the general requirements of waterfowl on spring migration and discuss intrinsic and extrinsic factors that influence these requirements, 2) discuss inter-specific differences in the requirements of migrating waterfowl and the limitations imposed by habitat conditions encountered during spring, and 3) propose a framework for evaluating limitations on waterfowl during spring migration, and for implementing habitat management and conservation that may alleviate or mitigate these limitations.

General requirements of spring-migrating waterfowl

Body reserves which can be converted into metabolic energy are the most recognised currency for avian migration (Jenni & Jenni-Eirmann 1998) and are also necessary for subsequent reproduction in many waterfowl species (Ankney *et al.* 1991). Lipids provide the most efficient means of storing energy for migration, and lipid metabolism therefore is considered a key factor influencing onwards migration and the selection of stopover sites. Individuals are expected to choose habitats where energy sources are readily available during migration and avoid energetically expensive staging areas (Bauer *et al.* 2008; Mini & Black 2009; Brasher 2010). Waterfowl gain energy and build lipid reserves from seeds, other plant material and invertebrates, with the relative contribution of each to the diet varying considerably among species. Similarly, the relative distribution of plant and invertebrate foods varies across different foraging habitats, which has species-specific implications for food availability at each site (Straub *et al.* 2012). Many North American species use abundant waste agricultural seeds in croplands as a carbohydrate source during spring migration (Krapu *et al.* 1995; Anteau *et al.* 2011b, Pearse *et al.* 2011). Others rely on invertebrates to build lipid reserves, which are likely to be more variable in abundance and distribution, and also are apparently declining in some regions (Anderson 1959; Wilson *et al.* 1995; Anteau & Afton 2008a,b; Anteau *et al.* 2011c; Straub *et al.* 2012).

Nutrients other than lipids are also required by waterfowl during spring, most notably protein, essential amino acids and minerals (*e.g.* calcium). Evidence from Snow Geese suggests that some waterfowl mobilise protein reserves gained during migration for subsequent reproduction (Ankney & MacInnes 1978; Gauthier *et al.* 2003). Moreover, moulting birds require protein to synthesise feather tissue (Heitmeyer 1988; Barras *et al.* 2001); some species consume protein-rich foods during contour feather moult (Fox *et al.* 1998; Anderson *et al.* 2000; Anteau *et al.* 2011a), whereas protein reserves may be related to contour feather moult intensity in other species (Lovvorn & Barzen 1988). Protein is also required for repairing muscles injured or catabolised during flight, similar to the way in which fat reserves are consumed and replenished during migration (Guglielmo *et al.* 2001; Piersma 2002). Earlier work has established the importance of a diverse diet for maintaining body condition during winter (Loesch & Kaminski 1989), suggesting that foraging decisions may be influenced by the specific amino acids to be found in food items (Heitmeyer 1988). However, our understanding of the role of specific nutrients (particularly at the essential amino acids and fatty acids level) for maintaining body condition at different times of year is still in its infancy.

The most basic requirement for all waterfowl (and indeed for most living organisms) is water. Water is gained primarily by drinking, but it can also be acquired in the diet or derived through metabolic pathways. Wetlands provide not only a water source but are important for a range of functions most notably foraging, roosting, pair formation (Anderson & Titman 1992), safety from predators, isolation from disturbance and protection from inclement weather conditions (LaGrange & Dinsmore 1989; Havera *et al.* 1992; Zimmer *et al.* 2010). Research in Europe revealed that many geese migrate toward breeding areas through agricultural regions, but that bird distributions may be constrained within a radius of a safe body of water or ice that can function as a predator-free overnight roost (*e.g.* the spring migration of Pink-footed Geese *Anser brachyrhynchus* within Britain is thus confined to particular areas; Bell 1988; Fox *et al.* 1994). Similar patterns have been shown with migrating Mallard *Anas platyrhynchos* (LaGrange & Dinsmore 1989) and geese (Anteau *et al.* 2011b) in agricultural landscapes in central North America.

Factors influencing waterfowl requirements during spring migration

Although these are generally universal for waterfowl during spring migration, the relative importance of each varies within and among species in response to conditions encountered *en route* (*e.g.* weather, disturbance) and in accordance with their migration and/or breeding strategies. Weather can influence individual requirements during migration, particularly among early migrants that may encounter physiologically demanding conditions on reaching high latitudes before the ice and snow has melted in the northern part of their range. For example, LaMontagne *et al.* (2001) reported differences in foraging activity among spring-migrating Trumpeter Swans *Cygnus buccinator* in response to cold temperatures encountered during migration. It is likely that early migrants exposed to wide variations in temperature and precipitation during spring would exhibit similar weather-dependent foraging and roosting behaviours, such as hyperphagia or seeking thermal cover. Weather affects habitat conditions along the migration route, and generally influences the availability of food and other resources throughout the year, as discussed further below.

Disturbance is another important factor influencing the relative importance of habitat requirements for migrating waterfowl (Madsen 1995), as it may affect the timing of migration strategies or individual body condition during stopover (Drent *et al.* 2003; Feret *et al.* 2003; Pearse *et al.* 2012). Variation in predation pressure during spring migration also may influence foraging ecology or the ability to exploit resources necessary for migration (Guillemain *et al.* 2007).

Variation in breeding and migration strategies leads to considerable variation in the conditions required by waterfowl throughout migration. Birds expected to adhere primarily to a capital breeding strategy (*e.g.* arctic nesting geese) need more resources from stopover locations than those using an income-breeding or local-capital strategy (*sensu* Klaassen *et al.* 2006), in which they acquire most breeding resources and nutrients from breeding habitats. Variation in migration strategies among species invoking an income-breeding strategy further differentiates requirements throughout migration. Income migrant waterfowl (*e.g.* Eurasian Teal; Arzel *et al.* 2007) rely especially on lipids acquired at staging sites to fuel subsequent flights, whereas other species may carry reserves to facilitate onward migratory movement (Krapu *et al.* 1995; Pearse *et al.* 2011). Across this gradient, from capital breeding species to income migrants, considerable variation in nutrient accumulation and storage rates have been documented throughout spring migration. For example, Garganey *Anas querquedula* in southern France effectively forgo nutrient reserve accumulation during stopover (Guillemain *et al.* 2004), whereas Greater White-fronted Geese *Anser albifrons* in the Rainwater Basin of Nebraska accumulate 11–22 g (dry) mass/day in the staging areas (Krapu *et al.* 1995).

European geese provide an example of variable requirements during migration that manifest as a result of variable migration strategies. These populations rely upon the new growth of grasses and sedges at higher latitudes following the emergence of the "green wave" of above-ground production following spring thaw (Drent *et al.* 1978; van der Graaf *et al.* 2006; van Wijk *et al.* 2012). Tracking this green wave is more easily undertaken in a series of relatively short flights, as is the case of Greater White-fronted Geese *Anser albifrons* in continental Europe. In contrast, Greenland White-fronted Geese *A. a. flavirostris* make long overseas flights from Britain and Ireland to staging areas in Iceland, and from there to breeding areas in west Greenland (Fox *et al.* 2003). Such stepping-stone migrants have to take calculated risks when moving onwards to staging areas, perhaps without adequate cues to predict meteorological conditions and the advancement of spring phenology further ahead (Fox *et al.* 2006; Tombre *et al.* 2008). Variation in the availability of spring staging areas has considerable effects on nutrient acquisition strategies adopted by the species with the same or similar body structure but in different parts of its range. The Greenland White-fronted Goose may deplete 800–900 g of fat when flying from winter quarters to spring staging areas in Iceland, and there it must acquire similar fat stores for the onward journey to breeding areas in west Greenland (Fox *et al.* 2003). Remarkably, the Greenland population now leaves the wintering areas on average three weeks earlier than 25 years ago (Fox & Walsh 2012), but because of a lack of warmer springs in Greenland it remains longer in Iceland (Fox *et al.* 2012) and fattens at a slower rate to arrive on the breeding areas at the same time as recorded in the 1860s (Fox *et al.* 2014). Such behaviour suggests considerable phenotypic plasticity in migration behaviour and ability to acquire fat stores in a fluctuating environment.

Distinguishing between requirements and limitations during spring migration

There is considerable variation between waterfowl of the northern hemisphere in the conditions that best match their social, ecological and physiological requirements at different stages of migration, creating many challenges for research and conservation along the flyway. However, in many cases the populations' requirements are met through large-scale habitat use and selection processes (*i.e.* adherence to flyways with necessary resources). For example, no studies of mid-continent goose populations in North America have documented nutrient deficiencies during migration or upon arrival on the breeding grounds, despite the importance of nutrient reserves to these populations having been established. In this case, adherence to the central flyway, where the availability of waste agricultural seed exceeds population needs, ensures nutrient accumulation and maintenance during migration and increases the likelihood of successful reproduction. Thus, although energy is the most important nutritional requirement among these migrating geese during spring, its availability (at least in the form of carbohydrates derived from agricultural seeds) is not a limiting factor during migration (Krapu *et al.* 1995; Jefferies *et al.* 2004; Stafford *et al.* 2006; Foster *et al.* 2010; Anteau *et al.* 2011b). This example raises the need for a distinction between requirements (resources that sustain migration and subsequent breeding) and limitations (resources that are not provided in sufficient supply to meet fully the needs of individuals at the time and/or thereafter) for spring migrating populations, which in turn should guide current and future conservation strategies developed for these populations. Such a distinction is fundamental for the effective implementation of conservation throughout the annual cycle, but is not yet explicitly recognised in conservation strategies for spring-migrating waterfowl. This is likely due to the uncertainty surrounding mechanisms regulating populations during the period and the aforementioned tendency to adapt wintering conservation strategies (*e.g.* provision of energy) for spring-migrating waterfowl. Misguided conservation strategies based only on requirements, rather than a limiting resource, may lead to ineffective conservation.

In populations where the availability of a necessary resource is limiting, observations of habitat use and distribution patterns for the birds during migration are likely to describe these limitation(s), which may be driving cross-seasonal effects on population productivity. The opposite is also likely; when resources are not in limited supply, populations may be freed from constraints on production originating during spring migration (Jefferies *et al.* 2004). Limitations of waterfowl during migration vary spatially and temporally, and the scale at which limitations are assessed is important and should be determined by management objectives. For example, food depletions at local scales may or may not be consequential for waterfowl populations, but regionally depressed food resources from drought or other impacts could influence annual recruitment (*e.g.* Davies & Cooke 1983).

In the former case, management may be ineffective at improving population productivity through spring migration, whereas in the latter, large-scale efforts to abate food limitation would likely have population-level implications. Therefore, research during spring migration should seek to identify limitations at appropriate spatial scales, through intensive study of migrant habitat use and behaviour, so as to identify important limitations on populations. This knowledge can then be applied to the development and delivery of conservation strategies for spring-migrating waterfowl. Additionally, identification of habitat limitations during spring migration might be investigated and used cautiously and insightfully to reduce populations of burgeoning species, such as Lesser Snow Geese, whilst ensuring no impacts to other waterfowl species.

Evidence for habitat limitations in other periods of the annual cycle exists. Availability of suitable breeding habitat in the North American prairies is a well-documented driver of annual population dynamics for a range of waterfowl (Johnson & Grier 1988; Bethke & Nudds 1995; Reynolds *et al.* 2001), and population-level implications of food shortages on the wintering grounds has similarly been documented for various species (Heitmeyer & Fredrickson 1981; Kaminski & Gluesing 1987; Raveling & Heitmeyer 1989). However, few studies have been conducted at an appropriate spatial and temporal scale to document population-level limitations during spring. Here we highlight a case where research has been conducted at appropriate scales to document limitations with the potential to inform conservation strategies for Lesser Scaup *Aythya affinis* on spring migration. Considerable research has been conducted on the ecology of the Lesser Scaup in spring following range-wide population declines throughout North America since the 1980s. We have therefore taken this body of work as an example of the benchmark necessary for achieving a reasonable understanding of populations-level limitations during migration, which may be applied to the conservation of other species during spring migration.

Research conducted along the Lesser Scaup's mid-continent spring migration corridors indicated that the females' lipid reserve levels had declined throughout the upper Midwest since the 1980s (Anteau & Afton 2004, 2009a), but not on wintering areas in the southern Mississippi Flyway (Anteau & Afton 2004; Vest *et al.* 2006). This work led to the spring-condition hypothesis, which predicted that females were unable to acquire energy or nutrients required during spring migration from stopover habitats through the Midwest. The limitation was predicted to result in decreased survival or diminished productivity from poor condition upon arrival in the breeding areas, reduced breeding propensity, delayed breeding or a combination of these (Anteau & Afton 2004, 2009a). Further research demonstrated that females were catabolising lipid reserves at spring migration stopover areas throughout the Midwest where they were expected to be storing lipids, lending support to the proposed link between habitat conditions during migration and diminished condition prior to the breeding season (Anteau & Afton 2011).

Research on wetlands used as stopovers during migration in the Midwest indicated that the availability of amphipods, an important, lipid-rich food item for migrating and pre-breeding Lesser Scaup (Arts *et al.* 1995; Lindeman & Clark 1999; Anteau & Afton 2009b), had declined in the region in conjunction with the documented declines in body condition (Anteau & Afton 2006, 2008a, b; Anteau *et al.* 2011c). Lesser Scaup select habitats with abundant amphipods (Lindeman & Clark 1999; Anteau & Afton 2009b); however, they likely use proximate cues (*e.g.* turbidity) to identify wetlands previously rich in amphipods, but which are now less numerous due to land use changes and invasions of fish into traditionally fish-free habitats (Anteau & Afton 2008a, 2009b). Spring is an energetically and nutritionally costly period for Lesser Scaup; thus, they clearly require both nutrient- and energy-rich foods. However, their adaptation to consuming an animal-based diet and the evidence reviewed above suggests that Lesser Scaup are likely limited by lipid availability during migration in the Midwest. Further research in the region suggested that amphipod abundance in stopover habitats may be subject to management (Anteau & Afton 2008a; Anteau *et al.* 2011c) and could focus on regions with high annual Lesser Scaup use in the Midwest (Anteau & Afton 2009b). Accordingly, implications of the research were to focus on identifying these key Lesser Scaup migration habitats in the region and to undertake work to improve the availability, quality and productivity of amphipods and other invertebrate food sources through wetland conservation practices, such as the implementation of upland vegetation buffers and manipulating fish densities (Anteau *et al.* 2011c).

This example with Lesser Scaup illustrates the role of large-scale research on ecology and habitat use of migrating waterfowl, for identifying limitations and focusing management of those limitations at relevant scales, with a view to improving conditions encountered during migration. Large-scale relationships between population productivity and limitations encountered during spring migration are one case in which an explicit focus on limitations in conservation is appropriate. A similar focus has applications at finer spatio-temporal scales for improving local management efforts and the value of habitat reserves intended for use by spring migrants. For example, detailed studies of Lesser Snow Goose stopover ecology in the Rainwater Basin of central Nebraska has shown that fine-scale habitat features rather than energy requirements during migration drive the birds' use of space at staging sites (Pearse *et al.* 2010; Anteau *et al.* 2011c; Sherfy *et al.* 2011). Cornfields dominate the landscape in the Rainwater Basin, resulting in an estimated 10-fold net surplus of energy for Lesser Snow Geese and other migratory waterfowl (Bishop *et al.* 2008), but their location in relation to wetland roosts appears to be more important than variability between fields in the availability of waste grain. Changes in the distribution and area of wetlands therefore would likely have the greatest influence on space use by Lesser Snow Geese and other waterfowl in the region (Vrtiska & Sullivan 2009; Webb *et al.* 2010). Although the main nutritional requirements for spring-migrating Lesser Snow Geese are energy and protein, this example illustrates that the habitat factor most appropriate for management is the distribution of the primary limiting resource, wetlands.

Conclusions

Conservation planning during spring has traditionally focused on ensuring adequate food energy at stopover locations because of the well-established importance of energy during migration and the importance of lipid reserves for subsequent breeding (*cf.*, Ankney *et al.* 1991). For example, the North American Waterfowl Management Plan, through its Joint Ventures (*i.e.* public and private partnerships that plan and implement conservation activities), has typically adopted an approach of estimating the energetic carrying capacity (ECC) for a region based on estimated waterfowl population levels and goals during the non-breeding periods of the annual cycle. In this scenario, management activities during spring migration target provision of food resources to meet the energetic needs of waterfowl, typically through wetland creation, enhancement or management to produce carbohydrate-rich plant foods. The ECC approach relies on the critical assumption that energy derived from wetland habitats is the main requirement limiting waterfowl populations in spring. However, this assumption is largely untested for the spring migration period, and the importance of energy may not equate with limitation should food availability exceed the requirements of the population. We contend that the assumption that energy is the primary limiting factor for waterfowl populations in spring may be untenable for many species that supplement their diet with residual agricultural food sources. Rather, any of the requirements of spring-migrating waterfowl discussed herein, or perhaps those yet undocumented, could limit spring migrants and have annual implications for survival or population productivity at various temporal and spatial scales.

In some cases, energy availability motivates habitat use and appears to limit population growth (*e.g.* the Lesser Scaup example described above). However, in other cases, energy during spring may not be limiting and can be in surplus (*e.g.* for mid-continent geese in North America). We suggest that the framework of ECC models could be reconsidered and perhaps restructured to evaluate whether energy is limiting for a species or guild within a conservation region. This would require a more comprehensive ECC that assumes that all sources of energy are equal if they are available to the species/group (*e.g.* agricultural *versus* wetland) and would require detailed information on the birds' diet and foraging behaviour given that all food sources – agricultural foods, wetland seeds and plants and invertebrates – must be considered as sources of energy. If careful evaluation indicates that available energy exceeds requirements for a given population or region, a focus on identifying or managing other possible limiting factors, if they exist, would be prudent. Such an approach may change the focus of conservation and management for some organisations (*e.g.* Joint Ventures, resource agencies); however, such an endeavour re-focused on the ecology of relevant species would lead to more efficient allocation of resources and be more likely to affect measurable impacts to populations.

Habitat use and selection, along with diet and behavioural studies, can provide the foundation for determining what might be limiting a certain species at a certain staging area, if at all (Callicutt *et al.* 2011; Hagy & Kaminski 2012). Some studies of this nature to date have identified cross-seasonal effects related to spring limitations, but many questions remain and adoption of novel research will be necessary to resolve them. Telemetry and other spatially explicit individual-based studies and local-scale surveys of waterfowl concentrations would help identify factors associated with improved individual fitness in response to conditions experienced during spring and identify key habitats for foraging and roosting, respectively. Similarly, such studies may yield insights into the relative role of specific stopovers during migration and assist in prioritising further research and conservation across the expansive landscapes transited between wintering and breeding areas.

Many factors drive hierarchical resource selection and knowledge of these factors can inform conservation strategies (Johnson 1980). For example, intensive research on Sandhill Cranes *Grus canadensis* during spring migration at a major stopover area in central North America revealed that access to protein constituents of their diet was a strong driver of fine-scale habitat selection during staging, despite accounting for only *c.* 3% of their diet in the region (Krapu *et al.* 1984; Reinecke & Krapu 1986). This research suggests that habitat selection and time investments among migrating waterbirds can be considerable in the acquisition of an apparently rare but important resource (*i.e.* a limiting resource). Similarly detailed studies for ducks and geese could assist in identifying other potential limiting factors.

Accomplishing a revised focus on limitations during spring conservation will require knowing the precise demands of the birds along the route of staging sites at different times during migration, and delivering the appropriate energy, protein, water and other resources such that birds may access them under a range of conditions (*e.g.* land use or climate change). Recognising the opportunities of expanded, individual-based cross-seasonal studies opens up a portfolio of research objectives that asks what birds need during spring, and how can we provide them most effectively, in a way that enhances condition, survival and preparation for the breeding season, regardless of species. The challenge for conservation will be to provide resources in adequate quantities to confer benefits to individuals of targeted populations and species. Such a task would be difficult given the dearth of information on factors truly limiting some waterfowl populations during spring. Nonetheless, until limiting factors are identified (or ruled out) it would be difficult to design and implement truly effective conservation programmes for populations of conservation concern.

Acknowledgements

We thank R.M. Kaminski, D.C. Kesler and an anonymous reviewer for comments that improved this manuscript considerably. Any use of trade, product or firm names is for descriptive purposes only and does not imply endorsement by the U.S. Government.

References

Alisauskas R.T. 2002. Arctic climate, spring nutrition, and recruitment in mid-continent lesser snow geese. *Journal of Wildlife Management* 66: 181–193.

Anderson, H.G. 1959. Food habits of migratory ducks in Illinois. *Illinois Natural History Survey Bulletin* 27: 289–344.

Anderson, M.G. & Titman, R.D. 1992. Spacing patterns. *In* B.D.J. Batt, A.D. Afton, M.G. Anderson, C.D. Ankney, D.H. Johnson, J.A. Kadlec & G.L. Krapu (eds.), *Ecology and Management of Breeding Waterfowl*, pp. 251–289. University of Minnesota Press, Minneapolis, USA.

Anderson, J.T., Smith, L.M. & Haukos, D.A. 2000. Food selection and feather molt by nonbreeding American green-winged teal in Texas playas. *Journal of Wildlife Management* 64: 222–230.

Ankney, C.D. & MacInnes, C.D. 1978. Nutrient reserves and reproductive performance of female Lesser Snow Geese. *Auk* 95: 459–471.

Ankney, C.D., Afton, A.D. & Alisauskas, R.T. 1991. The role of nutrient reserves in limiting waterfowl reproduction. *Condor* 93: 1029–1032.

Anteau, M.J. & Afton, A.D. 2004. Nutrient reserves of lesser scaup during spring migration in the Mississippi Flyway: a test of the Spring Condition Hypothesis. *Auk* 121: 917–929.

Anteau, M.J. & Afton, A.D. 2006. Diet shifts of lesser scaup are consistent with the Spring Condition Hypothesis. *Canadian Journal of Zoology* 84: 779–786.

Anteau, M.J. & Afton, A.D. 2008a. Amphipod densities and indices of wetland quality across the upper-Midwest, USA. *Wetlands* 28: 184–196.

Anteau, M.J. & Afton, A.D. 2008b. Diets of spring migrating lesser scaup throughout the upper-Midwest are consistent with the Spring Condition Hypothesis. *Waterbirds* 31: 97–106.

Anteau, M.J. & Afton, A.D. 2009a. Lipid reserves of lesser scaup (*Aythya affinis*) migrating across a large landscape are consistent with the Spring Condition Hypothesis. *Auk* 126: 873–883.

Anteau, M.J. & Afton, A.D. 2009b. Wetland use and feeding by lesser scaup during spring migration across the upper Midwest, USA. *Wetlands* 29: 704–712.

Anteau, M.J. & Afton, A.D. 2011. Lipid catabolism of invertebrate predator indicates widespread wetland ecosystem degradation. *PLoS One* 6: e16029.

Anteau, M.J., Anteau, A.C.E. & Afton, A.D. 2011a. Testing competing hypotheses for chronology and intensity of lesser scaup molt during winter and spring migration. *Condor* 113: 298–305.

Anteau, M.J., Sherfy, M.H. & Bishop, A. 2011b. Location and agricultural practices influence spring use of harvested corn fields by cranes and geese in Nebraska. *Journal of Wildlife Management* 75: 1004–1011.

Anteau, M.J., Afton, A.D., Anteau, A.C.E. & Moser, E.B. 2011c. Fish and land use influence *Gammarus lacustris* and *Hyalella azteca* (Amphipoda) densities in large wetlands across the upper Midwest. *Hydrobiologia* 664: 69–80.

Arzel, C., Elmberg, J. & Guillemain, M. 2006. Ecology of spring-migrating Anatidae: a review. *Journal of Ornithology* 147: 167–184.

Arzel, C., Elmberg, J. & Guillemain, M. 2007. A flyway perspective of foraging activity in Eurasian Green-winged Teal, *Anas crecca crecca*. *Canadian Journal of Zoology* 85: 81–91.

Arts, M., Ferguson, M.E., Glozier, N.E., Robarts, R.D. & Donald, D.B. 1995. Spatial and temporal variability in lipid dynamics of common amphipods: assessing the potential for uptake of lipophilic contaminants. *Ecotoxicology* 4: 91–113.

Barras, S.C., Kaminski, R.M. & Brennan, L.A. 2001. Effect of winter-diet restriction on prebasic molt in female wood ducks. *Proceedings of the Annual Conference of the Southeastern Association of Fish and Wildlife Agencies* 55: 506–516.

Bauer, S., Van Dinther, M., Høgda, K., Klaassen, M. & Madsen, J. 2008. The consequences of climate-driven stop-over sites changes on migration schedules and fitness of Arctic geese. *Journal of Animal Ecology* 77: 654–660.

Bell, M.V. 1988. Feeding behaviour of wintering Pink-footed and Greylag Geese in north-east Scotland. *Wildfowl* 39: 43–53.

Bethke, R.W. & Nudds, T.D. 1995. Effects of climate-change and land-use on duck abundance in the Canadian prairie-parklands. *Ecological Applications* 5: 588–600.

Bishop, A.A. & Vrtiska, M. 2008. Effects of the Wetlands Reserve Program on waterfowl carrying capacity in the Rainwater Basin Region of south-central Nebraska. Natural Resource Conservation Service, U.S. Department of Agriculture, USA. Accessible online at http://www.nrcs.usda.gov/technical/NRI/ceap/library.html (last accessed 20 August 2014).

Brasher, M.G. 2010. Duck use and energetic carrying capacity of actively and passively managed wetlands in Ohio during autumn and spring migration. Ph.D. thesis, Ohio State University, Columbus, USA.

Callicutt, J.T., Hagy, H.M. & Schummer, M.L. 2011. The food preference paradigm: a review of autumn-winter food use by North American dabbling ducks (1900–2009). *Journal of Fish and Wildlife Management* 2: 29–40.

Davies, J.C. & Cooke, F. 1983. Annual nesting productivity in snow geese: prairie droughts and arctic springs. *Journal of Wildlife Management* 47: 291–296.

Davis, J.B., Guillemain, M., Kaminski, R.M., Arzel, C., Eadie, J.M. & Rees, E.C. 2014. Habitat and resource use by waterfowl in the northern hemisphere in autumn and winter. *Wildfowl* (Special Issue No. 4): 17–69.

Drent R.H., Ebbinge B.S. & Weijand B. 1978. Balancing the energy budgets of arctic-breeding geese throughout the annual cycle: a progress report. *Verhandlungen der Ornithologischen Gesellschaft Bayern* 23: 239–264.

Drent, R.H. & Daan, S. 1980. The prudent parent: energetic adjustments in avian breeding? *Ardea* 68: 225–252.

Drent, R., Both, C., Green, M., Madsen, J. & Piersma, T. 2003. Pay-offs and penalties of competing migratory schedules. *Oikos* 103: 274–292.

Féret, M., Gauthier, G., Béchet, A., Giroux, J.F. & Hobson, K.A. 2003. Effect of a spring hunt on nutrient storage by greater snow geese in southern Quebec. *Journal of Wildlife Management* 67: 796–807.

Foster, M.A., Gray, M.J. & Kaminski, R.M. 2010. Agricultural seed biomass for migrating and wintering waterfowl in the Southeastern United States. *Journal of Wildlife Management* 74: 489–495.

Fox, A.D. & Walsh, A.J. 2012. Warming winter effects, fat store accumulation and timing of spring departure of Greenland White-fronted Geese *Anser albifrons flavirostris* from their winter quarters. *Hydrobiologia* 697: 97–102.

Fox, A.D., Mitchell, C., Stewart, A., Fletcher, J.D., Turner, J.V.N., Boyd, H., Shimmings, P, Salmon, D.G., Haines, W.G. & Tomlinson, C. 1994. Winter movements and site-fidelity of Pink-footed Geese *Anser brachyrhynchus* ringed in Britain, with particular emphasis on those marked in Lancashire. *Bird Study* 41: 221–234.

Fox, A.D., Kahlert, J. & Ettrup, H. 1998. Diet and habitat use of moulting Greylag Geese *Anser anser* on the Danish island of Saltholm. *Ibis* 140: 676–683.

Fox, A.D., Glahder, C.M. & Walsh, A.J. 2003. Spring migration routes and timing of Greenland white-fronted geese – results from satellite telemetry. *Oikos* 103: 415–425.

Fox, A.D., Francis, I.S. & Bergersen, E. 2006. Diet and habitat use of Svalbard Pink-footed Geese *Anser brachyrhynchus* during arrival and pre-breeding periods in Adventdalen. *Ardea* 94: 691–699.

Fox, A.D., Boyd, H., Walsh, A.J., Stroud, D.A., Nyeland, J. & Cromie, R. 2012. Earlier spring staging in Iceland amongst Greenland White-fronted Geese *Anser albifrons flavirostris* achieved without cost to refuelling rates. *Hydrobiologia* 697: 103–110.

Fox, A.D., Weegman, M.D., Bearhop, S. Hilton, G., Griffin, L., Stroud, D.A. & Walsh, A.J. 2014. Climate change and contrasting plasticity in timing of passage in a two-step migration episode of an arctic-nesting avian herbivore. *Current Zoology* 00: 000–000.

Fredrickson, L.H. & Drobney, R.D. 1977. Habitat utilization by postbreeding waterfowl. *In* T.A. Bookhout (ed.), Proceedings 1977 Symposium, Madison, Wisconsin, pp. 119–131. The Wildlife Society, Madison, Wisconsin, USA.

Fretwell, S.D. 1972. *Populations in a Seasonal Environment*. Princeton University Press, New Jersey, USA.

Gauthier, G., Bêty, J. & Hobson, K.A. 2003. Are Greater Snow Geese capital breeders? New evidence from a stable isotope model. *Ecology* 84: 3250–3264.

Greer, D.M., Dugger, B.D., Reinecke, K.J. & Petrie, M.J. 2009. Depletion of rice as food of waterfowl wintering in the Mississippi

Alluvial Valley. *Journal of Wildlife Management* 73: 1125–1133.

Guglielmo, C.G., Piersma, T. & Williams, T.D. 2001. A sport-physiological perspective on bird migration: evidence for flight induced muscle damage. *Journal of Experimental Biology* 204: 2683–2690.

Guillemain, M., Fritz, H., Klaassen, M., Johnson, A.R. & Hafner, H. 2004. Fuelling rates of garganey (*Anas querquedula*) staging in the Camargue, southern France, during spring migration. *Journal of Ornithology* 145: 152–158.

Guillemain, M., Arzel, C., Legagneux, P., Elmberg, J., Fritz, H., Lepley, M., Pin, C., Arnaud, A. & Massez, G. 2007. Predation risk constrains the plasticity of foraging behaviour in teals, *Anas crecca*: a flyway-level circumannual approach. *Animal Behaviour* 73: 845–854.

Guillemain, M., Elmberg, J., Arzel, C., Johnson, A.R. & Simon, G. 2008. The income-capital breeding dichotomy revisited: late winter body condition is related to breeding success in an income breeder. *Ibis* 150: 172–176.

Hagy, H.M. & Kaminski, R.M. 2012. Apparent seed use by ducks in moist-soil wetlands of the Mississippi Alluvial Valley. *Journal of Wildlife Management* 76: 1053–1061.

Harrison, X.A., Blount, J.D., Inger, R., Norris, D.R. & Bearhop, S. 2011. Carry-over effects as drivers of fitness differences in animals. *Journal of Animal Ecology* 80: 4–18.

Havera, S.P., Boens, L.R., Georgi, M.M. & Shealy, R.T. 1992. Human disturbance of waterfowl on Keokuk Pool, Mississippi River. *Wildlife Society Bulletin* 20: 290–298.

Heitmeyer, M.E. 1988. Protein costs of the prebasic molt of female mallards. *Condor* 90: 263–266.

Heitmeyer, M.E. & L.H. Fredrickson. 1981. Do wetland conditions in the Mississippi Delta hardwoods influence mallard recruitment? *Trans. North Am. Wildl. Nat. Resour. Conf.* 46: 44–57.

Herrmann, K.K. & Sorensen, R.E. 2009. Seasonal dynamics of two mortality-related trematodes using an introduced snail. *Journal of Parasitology* 95: 823–828.

Hohman, W.L., Manley, S.W. & Richard, D. 1997. Relative costs of prebasic and prealternate molts for male Blue-winged Teal. *Condor* 99: 543–548.

Jefferies, R.L., Rockwell, R.F. & Abraham, K.F. 2004. Agricultural food subsidies, migratory connectivity and large-scale disturbance in Arctic coastal systems: a case study. *Integrative and Comparative Biology* 44: 130–139.

Jenni, L. & Jenni-Eiermann, S. 1998. Fuel supply and metabolic constraints in migrating birds. *Journal of Avian Biology* 29: 521–528.

Johnson, D.H. 1980. The comparison of usage and availability measurements for evaluating resource preference. *Ecology* 61: 65–71.

Johnson, D.H. & Grier, J.W. 1988. Determinants of breeding distributions of ducks. *Wildlife Monographs* 100: 1–37.

Kaminski, R.M. & Elmberg, J. 2014. An introduction to habitat use and selection by waterfowl in the northern hemisphere. *Wildfowl* (Special Issue No. 4): 9–16.

Kaminski, R.M. & Gluesing, E.A. 1987. Density- and habitat-related recruitment in Mallards. *Journal of Wildlife Management* 51: 141–148.

Klaassen, M., Abraham, K.F., Jefferies, R.L. & Vrtiska, M. 2006. Factors affecting the site of investment, and the reliance on savings for arctic breeders: the capital-income dichotomy revisited. *Ardea* 94: 371–384.

Krapu, G.L., Facey, D.E., Fritzell, E.K. & Johnson, D.H. 1984. Habitat use by migrant sandhill cranes in Nebraska. *Journal of Wildlife Management* 48: 407–417.

Krapu, G.L., Reinecke, K.J., Jorde, D.G. & Simpson, S.G. 1995. Spring-staging ecology of midcontinent greater white-fronted geese. *Journal of Wildlife Management* 59: 736–746.

Lagrange, T.G. & Dinsmore, J.J. 1989. Habitat use by mallards during spring migration through central Iowa. *Journal of Wildlife Management* 53: 1076–1081.

LaMontagne, J.M., Barclay, R.M.R. & Jackson, L.J. 2001. Trumpeter swan behaviour at spring-migration stopover areas in southern Alberta. *Canadian Journal of Zoology* 79: 2036–2042.

Lindeman, D.H. & Clark, R.G. 1999. Amphipods, land-use impacts, and lesser scaup (*Aythya affinis*) distribution in Saskatchewan wetlands. *Wetlands* 19: 627–638.

Loesch, C.R. & Kaminski, R.M. 1989. Winter-body weight patterns of female Mallards fed agricultural seeds. *Journal of Wildlife Management* 53: 1081–1088.

Lovvorn, J.R. & Barzen, J.A. 1988. Molt in the annual cycle of Canvasbacks. *Auk* 105: 543–552.

Madsen, J. 1995. Impacts of disturbance on migratory waterfowl. *Ibis* 137: S67–S74.

Mini, A.E. & Black, J.M. 2009. Expensive traditions: energy expenditure of Aleutian geese in traditional and recently colonized habitats. *Journal of Wildlife Management* 73: 385–391.

Newton, I. 2006. Can conditions experienced during migration limit the population levels in birds? *Journal of Ornithology* 147: 146–166.

Newton, I. 2007. Weather related mass-mortality events in migrants. *Ibis* 149: 453–467.

Pearse, A.T., Krapu, G.L., Brandt, D.A. & Kinzel, P.J. 2010. Changes in agriculture and abundance of snow geese affect carrying capacity of sandhill cranes in Nebraska. *Journal of Wildlife Management* 74: 479–488.

Pearse, A.T., Alisauskas, R.T., Krapu, G.L. & Cox, R.R., Jr. 2011. Changes in nutrient dynamics of midcontinent greater white-fronted geese during spring migration. *Journal of Wildlife Management* 75: 1716–1723.

Pearse, A.T., Krapu, G.L. & Cox, R.R., Jr. 2012. Spring snow goose hunting influences body composition of waterfowl staging in Nebraska. *Journal of Wildlife Management* 76: 1393–1400.

Piersma, T. 2002. Energetic bottlenecks and other design constraints in avian annual cycles. *Integrative and Comparative Biology* 42: 51–67.

Raveling, D.G. & Heitmeyer, M.E. 1989. Relationships of population size and recruitment of pintails to habitat conditions and harvest. *Journal of Wildlife Management* 53: 1088–1103.

Reinecke, K.J. & Krapu, G.L. 1986. Feeding ecology of sandhill cranes during spring migration in Nebraska. *Journal of Wildlife Management* 50: 71–79.

Reynolds, R.E., Shaffer, T.L., Renner, R.W., Newton, W.E. & Batt, B.D.J. 2001. Impact of the conservation reserve program on duck recruitment in the US Prairie Pothole Region. *Journal of Wildlife Management* 65: 765–780.

Richardson, D.M. & Kaminski, R.M. 1992. Diet restriction, diet quality, and prebasic molt in female mallards. *Journal of Wildlife Management* 56: 531–539.

Sedinger, J.S. & Alisauskas, R.T. 2014. Cross-seasonal effects and the dynamics of waterfowl populations. *Wildfowl* (Special Issue No. 4): 277–304.

Sherfy, M.H., Anteau, M.J. & Bishop, A.A. 2011. Agricultural practices and residual corn during spring crane and waterfowl migration in Nebraska. *Journal of Wildlife Management* 75: 995–1003.

Soulliere, G.J., Potter, B.A., Coluccy, J.M., Gatti, R.C., Roy, C.L., Luukkonen, D.R., Brown, P.W. & Eichholz, M.W. 2007. Upper Mississippi River and Great Lakes Region Joint Venture Waterfowl Habitat Conservation Strategy. U.S. Fish and Wildlife Service, Fort Snelling, Minnesota, USA.

Stafford, J.D., Kaminski, R.M., Reinecke, K.J. & Manley, S.W. 2006. Waste rice for waterfowl in the Mississippi Alluvial Valley. *Journal of Wildlife Management* 70: 61–69.

Straub, J.N., Gates, R.J., Schultheis, R.D., Yerkes, T., Coluccy, J.M. & Stafford, J.D. 2012. Wetland food resources for spring-migrating ducks in the Upper Mississippi River and Great Lakes Region. *Journal of Wildlife Management* 76: 768–777.

Tombre, I.M., Høgda, K.A., Madsen, J., Griffin, L.R., Kuijken, E., Shimmings, P., Rees, E. & Verscheure, C. 2008. The onset of spring and timing of migration in two arctic nesting goose populations: the pink-footed goose *Anser brachyrhynchus* and the barnacle goose *Branta leucopsis*. *Journal of Avian Biology* 39: 691–703.

Trautman, M.B., Bills, W.E. & Wickliff, E.L. 1939. Winter losses from starvation and exposure of waterfowl and upland game birds in Ohio and other Northern States. *The Wilson Bulletin* 51: 86–104.

van der Graaf A.J., Stahl, J., Klimkowska, A., Bakker, J.P. & Drent, R.H. 2006. Surfing on a green wave – how plant growth drives spring migration in the Barnacle Goose *Branta leucopsis*. *Ardea* 94: 567–577.

van Wijk, R.E., Kölzsch, A., Kruckenberg, H., Ebbinge, B.S., Müskens, G.J.D.M. & Nolet, B.A. 2012. Individually tracked geese follow peaks of temperature acceleration during spring migration. *Oikos* 121: 655–664.

Vest, J.L., Kaminski, R.M., Afton, A.D. & Vilella, F.J. 2006. Body mass of lesser scaup during fall and winter in the Mississippi Flyway. *Journal of Wildlife Management* 70: 1789–1795.

Vrtiska, M.P. & Sullivan, S. 2009. Abundance and distribution of lesser snow and Ross's geese in the Rainwater Basin and central Platte River Valley of Nebraska. *Great Plains Research* 19: 147–155.

Webb, E.B., Smith, L.M., Vrtiska, M.P. & LaGrange, T.G. 2010. Effects of local and landscape variables on wetland bird habitat use during migration through the Rainwater Basin. *Journal of Wildlife Management* 74: 109–119.

Weller, M.W. 1975. Migratory waterfowl: a hemispheric perspective. *Publicaciones Biologicas Instituto de Investigaciones Científicas, U.A.N.L.* 1: 89–130.

Wilson, D.M. Naimo, T.J., Wiener, J.G., Anderson, R.V., Sandheirich, M.B. & Sparks, R.E. 1995. Declining populations of fingernail clam *Musculium transversum* in the upper Mississippi River. *Hydrobiologia* 304: 209–220.

Zimmer, C., Boos, M., Petit, O. & Robin, J.P. 2010. Body mass variations in disturbed mallards *Anas platyrhynchos* fit to the mass-dependent starvation-predation risk trade-off. *Journal of Avian Biology* 41: 637–644.

Nest site selection by Holarctic waterfowl: a multi-level review

MICHAEL W. EICHHOLZ[1]* & JOHAN ELMBERG[2]

[1]Mailcode 6504, Cooperative Wildlife Research Laboratory, Center for Ecology, Department of Zoology, Southern Illinois University Carbondale, Carbondale, Illinois, USA.
[2]Division of Natural Sciences, Kristianstad University, SE-291 88, Kristianstad, Sweden.
*Correspondence author. E-mail: eichholz@siu.edu

Abstract

Because of birds' mobility, behaviour and many species' migratory nature, they select repeatedly and spatially among habitats and have been central figures in studies of avian breeding habitat selection during the 20th and 21st centuries. The scientific literature on habitat use by breeding waterfowl has origins dating back to the writings of Charles Darwin in *The Voyage of the Beagle*, wherein he described the distribution and habitat differences of two species of geese on the Falkland Islands. Since that time, waterfowl ecologists have gone from descriptive studies of nest site characteristics used for planning waterfowl conservation and management to comparing nest site use in relation to potential habitat availability and determining selection for a wide array of ecological correlates. Waterfowl ecologists most recently have been investigating the adaptive significance of nest site selection by associating the latter with individual fitness and demographic measurements to assess the birds' adaptability under environmental conditions at multiple scales of selection. While little direct assessment of 1st and 2nd order nest site selection has occurred (*sensu* Johnson 1980), available information is most consistent with the hypothesis that selection at these scales is driven by food availability. At the 3rd and 4th order of selection, data are consistent with hypotheses that both food availability and predator avoidance drive nest site selection, depending on the species and type of nesting aggregation. We also identify understudied areas of nest site selection important for the conservation and management of waterfowl and suggest that the large-scale influence of current anthropogenic and natural effects on the environment indicates that greater emphasis should be directed toward understanding waterfowl nest site selection at the 1st and 2nd orders of selection and how nesting habitat selection interfaces with community ecology of sympatric breeding waterfowl. Moreover, because habitat selection of pre-fledging waterfowl is inherently linked to breeding habitat selection, we suggest an updated review of brood habitat selection should ensue from our synthesis here.

Key words: hierarchal habitat selection, nest, nesting habitat, nest site selection, waterfowl.

Elucidation of use and selection of habitat by animals for breeding and other life-history segments during their annual cycle is essential for answering basic ecological and applied questions at the individual, population and community levels (Fretwell & Lucas 1969; Lack 1971; Cody 1981). Because of birds' mobility and many species' migratory nature, they select repeatedly and spatially among habitats and have been central to studies of breeding habitat selection during the 20th and 21st centuries (Grinnell 1914; Lack 1933; Hildén 1965; Cody 1985; Block & Brennan 1993).

The scientific literature on habitat use by breeding waterfowl has origins dating back to the writings of Charles Darwin in The Voyage of the Beagle wherein he described distribution and habitat differences of the "Upland Goose" *Anas magellanica* and the "Rock Goose" *Anas antarctica*, although it appears he was describing what are now considered subspecies of upland geese *Chloephaga picta leucoptera* and *C. p. picta* on the Falkland Islands. Oberholser & McAtee (1920) reported on population declines in waterfowl in North America, exhorted the benefits of eliminating spring harvest in North America for its resultant increase in waterfowl populations. Displaying considerable insight for their time, they recognised the importance of habitat by suggesting that agricultural land development was negatively affecting duck populations through decreased availability of crucial wetlands and other habitats. Although early conservationists were beginning to recognise the importance of preserving breeding habitat for the conservation of wildlife populations (Leopold 1933), only sporadic reports of waterfowl nest sites can be found in the scientific literature until a synthesised description of the nesting habitats for waterfowl in North America was published by Bent (1923). Although Bent reviewed the available information, data were lacking on species-specific use of nesting habitats. This lack of understanding was also apparent in Pirnie's (1935) book on waterfowl ecology and management in Michigan. He recommended planting junipers *Juniperus* sp. and pine *Pinus* sp., or other evergreens and other shrubby vegetation, as nesting cover for upland game birds including Mallard *Anas platyrhynchos*, and other ducks nesting in uplands in Michigan. He later noted, however, an abundance of successfully nesting ducks in flooded meadows that provided dense herbaceous cover. Likely in response to the decreasing abundance of waterfowl in North America during the drought stricken 1930s and early 1940s, numerous descriptions of waterfowl nesting habitat were published in the mid 20th century (*e.g.* Bennet 1938; Low 1941; Gross 1945; Leitch 1951). Indeed, recognition became prevalent in the 1950s that factors including habitat type, social avoidance and attraction, and predator avoidance may influence species-specific nest site selection (*e.g.* Earl 1950; Kossack 1950; Glover 1956). As biologists recognised that nesting habitat use varied by species and even within species, and that habitat choices influenced reproductive success, considerable effort was spent during the 1960s–1980s describing species-specific variation in nesting habitat and how such variation was influenced by predation pressure (*e.g.* Keith 1961; Duebbert & Lokemoen 1976; Weller 1979;

Livezey 1981; McLandress 1983; Kaminski & Weller 1992). These studies emphasised the applied aspects of the data in identifying habitat use patterns for conservation and management purposes, but rarely used the data specifically to advance understanding of causative drivers of waterfowl habitat selection (*cf.* Kaminski & Prince 1984; Cody 1985) despite the fact that the early habitat selection models were developed and widely tested at that time (Fretwell & Lucas 1969; Fretwell 1972). Furthermore, because dabbling and diving ducks tend to breed in temperate, lower latitude regions than arctic nesting swans, geese and sea ducks, their habitat has undergone increased anthropogenic modifications. However, even the more northern boreal regions that were once relatively immune to human developments are now influenced by both direct (forest harvest, mining, extraction of fossil fuels and wind power development) and indirect (atmospheric warming, airborne pollution and eutrophication) anthropogenic forces. Indeed, the rate of anthropogenic modification has recently increased dramatically in some boreal regions in the northern hemisphere (Murphy & Romanuk 2014).

In general, factors that influence waterfowl reproduction are thought to have the greatest influence on populations of generally *r*-selected species (dabbling and diving ducks; Flint *et al.* 1998; Hoekman *et al.* 2002; Coluccy *et al.* 2008), while factors that influence post-fledging survival tend to have a greater influence on population dynamics of more *K*-selected species (sea ducks, geese, and swans; Nichols *et al.* 1976; Cooch *et al.* 2001; Schamber *et al.* 2009). Managers have recognised these distinctions and accordingly often tend to focus their effort toward managing habitat that influences reproductive success for dabbling and diving ducks while managing non-breeding habitat and harvest for sea ducks, geese and swans.

Thus, because reproductive success has strong influence on population dynamics, understanding breeding habitat use and selection is fundamental for wise decisions regarding habitat conservation and management. This recognition has led to numerous detailed studies of how management actions influence habitat selection and nest success of boreal and temperate nesting dabbling and diving ducks. Additionally, possibly because of influential writings by Romseburg (1981) and Walters (1985), waterfowl biologists have begun to view breeding habitat selection in basic and applied ecological contexts (*e.g.* Nudds 1983; Clark & Shutler 1999; Fast *et al.* 2007). This change coincided with recognition that spatial scale is an important factor in the habitat selection process (Johnson 1980; Hutto 1985; Wiens 1989). Thus, issues of spatial and temporal scale have been influential in framing and articulating theories on which research about breeding habitat selection in waterfowl is based (Johnson 1980; Hutto 1985; Forbes & Kaiser 1994). Selection of habitat by waterfowl during breeding and other annual cycle events may be viewed as a hierarchal process. For example, selection of nest sites is a fine-grained hierarchical process with individuals utilising large-grain characteristics for initial (first order) selections of geographic regions, then using characteristics that can be distinguished at more reduced grain size as the selection process continues (Johnson 1980; Wiens 1989; Kristan 2006). Grain size for nesting waterfowl generally will be smaller than other habitat components of the annual cycle, except perhaps individual foraging and rest sites, because nests occupy microhabitats and their location is such an important aspect of reproductive success (O'Neila *et al*. 1986; Wiens 1989).

In this contribution, we review the characteristics waterfowl use to make decisions at various spatial levels leading downward to nest site selection, the selective pressures that affect those characteristics, and the resulting fitness of those decisions. We review how research on waterfowl nest site selection has influenced our understanding of basic ecological theory and how this information is used when making management decisions. To facilitate comparison of cross taxa variation, this paper is partitioned into four scales of selection: 1st order – general region or latitude, 2nd order – landscape type (biome) within a region, 3rd order – location, wetland, or upland within a landscape and 4th order – specific nest site within the location (Kaminski & Elmberg 2014). By organising the review in this manner, we hope to articulate benefits of life-history comparisons. We conclude by identifying areas where information is inadequate to answer basic ecological questions and make reliable conservation decisions.

First-order selection

Habitat selection at this scale is best addressed by considering studies asking questions associated with diversity, distribution, and abundance of organisms at the continental scale. Brown & Maurer (1989) coined the term "macrocecolgical approach" to describe ecological approaches of this scale. They stated, "Our goal is to understand the assembly of continental biotas in terms of how the physical space and nutritional resources of large areas are divided among diverse species." Two areas of study that often use a macroecological approach to address associations among space and resource availability and species distribution, abundance and diversity are migration and community ecology; thus, these should provide insight into 1st order nest site selection of waterfowl are migration and community ecology (*e.g.* Davis *et al.* 2014).

Most waterfowl species occurring in the Holarctic are long- to medium-distance migrants. A general pattern among these waterfowl is they nest in temperate, sub-arctic and arctic regions too inhospitable to support them during the non-breeding period, and then spend the non-breeding period in locations with more moderate climates (Bellrose 1980; Kaminski & Weller 1992). Thus, if we assume pursuit of nesting habitat begins when birds begin to transition from non-breeding to breeding periods, the initial consideration for 1st order selection is really a question of migration (*i.e.* Does an individual breed at the same location it spent the non-breeding period or move to a different location?). Considerable debate exists whether migration evolved from northern and temperate breeding populations migrating south during the non-breeding season as winter ensued or from sub- and tropical breeding locations as species expanded their range into more seasonal, higher latitudes after Pleistocene glaciation (Cox 1968; Chesser & Levey 1998; Alerstam *et al.* 2003). Regardless of origin and mechanistic stimuli of migration, the Cordillerin and Laurentide ice sheets in North America and the Scandinavian ice sheet in Europe during the most recent glaciation (from *c.* 120,000–8,500 years ago) generally limited the distribution of waterfowl to regions that serve primarily as wintering areas under current climatic conditions (*e.g.* Hawkins & Porter 2003; Hortal *et al.* 2011). Thus, most species of Holarctic waterfowl since glaciation have evolved strategies to migrate north to breed. A number of hypotheses have been proposed to explain northern latitude breeding, including migration as a method to exploit rich seasonal food resources, avoid predation, reduce exposure to disease and parasites, exploit latitudes with long day lengths during the growing season, or reduce intra- or interspecific competition (Cox 1985; Fretwell 1980; Alerstam & Högstedt 1982; Piersma 1997; Chesser & Levey 1998; Rappole & Jones 2002).

The tendency for Holarctic waterfowl to migrate from southern wintering regions to more northern breeding regions has created a somewhat unique pattern of species richness. While species richness of most organisms, including most bird taxa, decreases with increasing latitude, species richness of Holarctic waterfowl tends to peak between 40°–65° latitude, declining rapidly north and south of that range (Dalby *et al.* 2014). Within this range of latitudes, waterfowl inhabit all the major biomes including grassland, temperate rain forest, tundra, taiga, eastern deciduous forest and desert when adequate water is present (Bellrose 1980; NAWMP 2012). The northern range limit for species is most likely driven by the length of the ice-free period being too limited to provide adequate time for reproduction (Schmidt *et al.* 2011), while the factors that influence the southern range limits are generally undefined for Holarctic waterfowl.

As the unique association between species richness and latitude in Holarctic waterfowl demonstrates, 1st order selection for Holarctic waterfowl seems a question of migratory behaviour, because it can influence species distribution which can drive spatial variation in species richness. Thus, an additional ecological concept to gain insight into the selective forces of 1st order selection and the potential driving force behind the southern range limit in Holarctic waterfowl is the concept of community ecology. Nudds (1992) thoroughly reviewed theories on species richness in waterfowl communities. In general, the discussion has changed little since that review. Species richness is dependent on three processes: speciation, extinction and dispersal. The speciation and extinction process influence variation in richness among clades over evolutionary time and potentially large scale (continental) spatial variation, while the dispersal process likely has the greatest influence in ecological time and on more local (regions and less) variation for mobile organisms such as waterfowl (Hulbert & Stegen 2014). Because of the close association between species distribution, migration and species richness, some of the same selective forces posited to influence migratory behaviour (*i.e.* disease, predation pressure, resource availability and intra- and interspecific competition) as well as other factors such as the species-area relationship, habitat heterogeneity and time for or rate of diversification have been proposed as primary mechanisms for spatial variation in species richness (Arrhenius 1921; Willson 1976; Wright 1983; Evans *et al.* 2005). While none of these mechanisms have been excluded, the roles of time, habitat heterogeneity and resource availability have been accepted as the most likely mechanisms.

A commonly proposed mechanism to explain the decline in species richness with increasing latitude observed in most groups of organisms other than waterfowl is that the former have had a longer period of time to diversify, diversify at a faster rate and have greater niche conservatism in the tropics (*i.e.* long-term stability of the environment conserving the niche); thus, the tropics support greater species richness (Brown 2014). Using similar logic, more recent large scale climatic factors such as glaciation have also been proposed to explain patterns of species richness at the continental scale within the Holarctic (Hawkins & Porter 2003; Hortal *et al.* 2011). If time were the primary mechanism driving the observed relationship between Holarctic waterfowl species richness and latitude, we would predict greater species richness in the more southern latitudes, given that vast areas of the more northern latitudes were glaciated as recently as 8,000 years ago. This prediction is opposite of the present pattern, indicating time is an unlikely explanation.

Total habitat area and heterogeneity also have been found to be a strong predictor and causative agent of species richness for some taxa (Roth 1976; Elmberg *et al.* 1993). If total wetland area or wetland heterogeneity were driving latitudinal variation in waterfowl species distribution, we would predict greater wetland area and heterogeneity in more northern latitudes. We were unable to locate latitudinal data reflecting wetland heterogeneity, but the wetland trends data provide an estimate of total wetland acreage for each of the United States (Dahl 2011). If wetland area is driving the latitudinal variation in waterfowl species richness, we predict an increase in wetland area with latitude. When we regressed total wetland acreage for each state in the United States against its geographic midpoint, we found no relationship between latitude and total wetland acreage for all states ($r^2 = 0.18$, $P = 0.24$) or those states between 80–105° longitude where most nesting waterfowl occur ($r^2 = 0.15$, $P = 0.29$). Furthermore, if we assume the relationship between wetland heterogeneity and wetland area at regional scales is similar to the relationship at more local scales (Elmberg *et al.* 1993, 1994), there should be a strong correlation between total wetland abundance and wetland diversity. Thus, if wetland heterogeneity is driving the relationship between species richness and latitude, we should again detect a positive relationship between latitude and overall wetland acreage, a relationship that contradicts our observation.

The final explanation of latitudinal patterns in waterfowl species richness is resource availability. Resource availability has been found to be strongly correlated to species richness at multiple scales for a variety of taxa. Latitudinal variation in per capita resource availability could be due to reduced intra- or inter-specific competition in more northern latitudes (Ashmole 1963), more wetlands (or other requisite habitat) in more northern latitudes, or higher productivity in more northern latitudes.

Ashmole (1963) argued that because the inhospitable winter climate of more seasonal environments limited the number of birds breeding in temperate regions, there was greater per capita resource availability. This argument does not appear to hold true for waterfowl, however, in that both abundance and richness peak at higher latitudes.

In recent large scale studies, resource availability seems to be the best predictor of species richness for the majority of taxa (Hawkins *et al.* 2003; Hulbert & Haskell 2003; Evans *et al.* 2005). As the analysis described above indicates, relative surface area of wetlands, assuming a correlation between surface area and heterogeneity, wetland heterogeneity does not appear to increase with latitude in North America but our analyses currently are restricted to this continent. Thus, if resource availability is the driving mechanism for latitudinal variation in waterfowl species richness, a latitudinal trend in per area resource availability may be the driving mechanism. Resource availability per area may vary due to greater exploitation by waterfowl in more southern wetlands or greater per area productivity in higher latitude wetlands.

During winter, waterfowl congregate in regions that provide predictable food availability (*i.e.* remain unfrozen) exploiting leafy vegetation, seeds, tubers and agricultural seeds produced during the previous growing season, as well as aquatic invertebrates. The high concentration of waterfowl exploiting these resources likely reduces their availability to waterfowl as winter progresses and transitions to spring (Davis *et al.* 2014). Thus, it's possible the exploitation of resources by wintering waterfowl reduces the resources to migrating and breeding waterfowl, reducing the species richness and abundance in more southern latitudes.

Alternatively, a similar amount of wetlands in more northern latitudes could provide more resources if vegetation in more northern regions provides greater benefit to secondary consumers (Coley *et al.* 1985; Moles *et al.* 2011; Morrison & Hay 2012). Vegetation with greater nutritional value would directly influence resource availability for geese which are herbivorous throughout their annual cycle and produce young that require nutrient rich plants during early post-hatch growth (Coley *et al.* 1985; Sedinger 1992). Nutritional quality of vegetation could also influence ducks that are primarily carnivorous during the breeding season by providing substrates and food for aquatic invertebrates that ducks consume (Krapu & Rcinecke 1992). Because most vegetation in higher latitude wetlands continues to grow until freezing temperatures cause senescence, annual rate of plant decomposition is much lower in more northern wetlands (Webster & Benfield 1986; Magee 1993; Holt 2008). Thus, as opposed to more southern wetlands where plant decomposition and nutrient turnover continues throughout the winter with little remaining by spring, there is considerable organic material and a large food base for invertebrates when spring arrives at more northern latitudes. Moreover, seeds and invertebrates are not exposed to predation by birds during the winter, providing returning breeders with a less depleted food base than is the case in more southern latitudes where wetlands are used year round by resident and seasonally occurring species of waterbirds. Additionally, a number of studies have demonstrated that invertebrates prefer to forage on plant material in more northern latitudes (*e.g.* Pennings *et al.* 2001, 2007). Some have suggested this strategy is due to fewer chemical defences from plants from more northern latitudes while others have suggested it is due to higher concentrations of nitrogen (Coley *et al.* 1985; Moles *et al.* 2011; Morrison & Hay 2012). Regardless of the mechanism, vegetation from more northern latitudes appears to be of greater nutritional value to invertebrates and herbivorous waterfowl, potentially producing greater nutritional resources.

Finally, as Willson (1976) suggested the shorter growing season in more northern latitudes may itself lead to an increase in standing biomass of aquatic invertebrates. The briefer period for reproduction may cause more species of invertebrates to reproduce simultaneously, leading to a higher spike in overall invertebrate biomass in more seasonal northern environments. High-latitude wetlands may thus offer a high standing biomass and higher per capita food resource availability when waterfowl nest as a result of higher productivity, concurrent food peaks or a long period without food depletion (Danell & Sjöberg 1977).

Although other mechanisms for explaining the more northern latitudinal peak in species richness of waterfowl cannot be excluded, per capita nutrient availability appears to have the greatest level of support (Dalby *et al.* 2014). While at a more regional scale, waterfowl richness appears to increase with both nutrient availability and habitat heterogeneity (2nd order level selection; Elmberg *et al.* 1993). Although the nutrient availability hypothesis appears to be most consistent with data currently available, direct tests with empirical data have not been conducted. For example, no one has tested for general latitudinal variation in aquatic invertebrate biomass (*cf.* Arzel *et al.* 2009 for a one-species all-flyway example), and although evidence exists indicating vegetation from more northern areas may be preferred by invertebrates over the same species of vegetation from more southern latitudes, studies have not isolated confounding effects of chemical or physical defence and concentration of nutrients for the more northern vegetation.

Second-order selection

Often waterfowl have multiple options for choice of biome (*e.g.* grassland, forest, tundra) to select for nest sites even after selecting a geographic latitude in which to breed. Unlike latitudinal variation in species richness, there appears no clear pattern in species richness across various northern hemispheric biomes. In fact, although some species appear to be habitat specialists in that most individuals are found almost exclusively in one biome (*e.g.* Blue-winged Teal *Anas discors* in grassland and Green-winged Teal/Eurasian Teal *A. crecca* in boreal environs), most species appear to be habitat generalists, selecting specific nest site characteristics regardless of the biome and thus can be found in multiple biomes (*e.g.* Nicolai *et al.* 2005; Safine & Lindberg 2007). Furthermore, with the exception of cavity nesting species, which largely require forested habitat, or sea ducks that specialise in coastal waters for brood rearing, almost all species can be found nesting in multiple biomes. Even species like the Common Eider *Somateria mollissima*, which depend on marine environments for foraging and are considered a tundra nesting species in North America, will nest in island forests when such forests are in close proximity to brood habitat (Öst *et al.* 2008a).

Species that demonstrate a clear preference also have likely adapted to specialise on certain characteristics of habitat (Mulhern *et al.* 1985). The two selection factors that appear to be acting most at the scales of 3rd and 4th order selection are food availability and predator avoidance, making them good candidates for proximate cues that may drive 2nd order selection. While nest predation pressure potentially varies with latitude (Hanski *et al.* 1991; Elmberg *et al.* 2009), it doesn't appear to vary among biomes so selection of biome may be dictated by food availability (Grand & Flint 1997; Fournier & Hines 2001; Walker *et al.* 2005; Schamber *et al.* 2009; *cf.* Elmberg *et al.* 2009). Currently, increasing exploitation of natural resources of the tundra and boreal regions emphasises the need to better understand the requirements of habitat specialists for future management and conservation. Studies addressing the question of whether the limited breeding ranges overlap between Blue- and Green-winged Teal or Gadwall *A. strepera* and American Wigeon *A. americana*, species that breed in different biomes but appear closely related genetically and ecologically, is due to variation in habitat requirements or competition may help elucidate questions associated with 2nd order selection.

Third-order selection

Third-order selection is the level at which they select a specific local habitat(s) within a biome. A number of nest site selection characteristics discussed here also could be considered 4th order characteristics. We consider them 3rd order characteristics for the sake of this discussion due to the grain size at which the characteristic may have been measured. For example, if a characteristic was measured at a scale that was relevant to more than one female (*i.e.* a field or patch of trees), we considered it 3rd order selection; whereas, if a characteristic was measured at a scale relevant to one female (*e.g.* density of vegetation surrounding a specific prospective nest or size in the opening of a tree cavity), we considered it 4th order selection. In the following sections, we review potential ecological, environmental or social influences of 3rd order selection of nest habitats.

Predator and coexisting prey densities

Nest predation is the primary cause of nest failure in waterfowl (Sargeant & Raveling 1992; Stephens *et al.* 2005). Egg predators are distributed heterogeneously across landscapes concentrating in habitats that provide efficient foraging (Kuehl & Clark 2002; Phillips *et al.* 2003; Elmberg & Gunnarsson 2007; Klug *et al.* 2009) and protection from higher trophic level predators (Crabtree *et al.* 1989; Dion *et al.* 2000; Chalfoun *et al.* 2002). High nest predation and heterogeneous distribution of predators invoke selective pressure for waterfowl to adapt strategies of selecting nest habitats and sites with fewer predators or more coexisting prey than other potential sites, leading to a reduction in nest predation (Holt 1977; Ackerman 2002; Eichholz *et al.* 2012).

Predator avoidance appears to be the primary selective force for colonial nesting species, apparently having less impact on dispersed nesting species (Schmutz *et al.* 1983; Bousfield & Syroechkovskiy 1985; Fox *et al.* 2009). This inconsistency may be related to the level of feeding that occurs by females during incubation and the distance young can travel after hatch. Most colonial nesting species of waterfowl, with the possible exception of Black Brant *Branta bernicla nigricans* and Ross's Goose *Chen rossii*, feed little, if any, during incubation and often travel long distances from nests to brood rearing locations; thus, there appears to be little pressure to nest near high concentrations of food for females and goslings. Although Black Brant often travel a substantial distance from nest sites to brood rearing areas, females spend as much as of 20% of the incubation period off the nest in maintenance activities such as feeding (Eichholz & Sedinger 1999; Sedinger *et al.* 2004). Brant colonies typically are located near the coast where nutritious foods are available and Arctic Fox *Vulpes lagopus* numbers are reduced due to fall flooding from storm surges (Mickelson 1975; Raveling 1989).

Researchers have long assumed nest site selection was influenced by predator avoidance, but only recently has there been evidence that birds could assess local predator density and modify their behaviour accordingly. Fontaine & Martin (2006) found numerous species of passerines modify their reproductive investment by increasing their feeding behaviour when predator abundance was reduced, but provided no explanation as to the mechanism parents used to assess predator density. Similarly, Dassow *et al.* (2012) found evidence that ground nesting ducks modify reproductive investment based on density of predators (*cf.* Duebbert & Kantrud 1974). A number of studies have now provided evidence that birds are able to use various mechanisms to assess predator abundance and modify reproductive strategies (*e.g.* Lima 2009; Eichholz *et al.* 2012; Forsman *et al.* 2012). Additionally, researchers conducting predator exclusion and reduction studies have observed increases in nesting densities in areas where predators were reduced. Although greater adult philopatry and an increase in the abundance of breeding yearlings associated with increased nest success and natal philopatry are typically proposed as the mechanisms for this increase (Duebbert & Kantrud 1974; Duebbert & Lokemoen 1980; Garrettson & Rohwer 2001), these results also are consistent with the idea that waterfowl select sites with reduced predator abundance, thus immigrating into experimental areas. Although results from studies consistent with the idea that waterfowl have developed a mechanism for assessing predator abundance and avoiding areas of high predator density, further empirical evidence is needed to test this hypothesis.

In addition to predator density, coexisting prey density may play a role in nest site selection. A number of studies have found a correlation between coexisting prey abundance and nest success in tundra, taiga and temperate grasslands (Pehrsson 1986; Summers *et al.* 1994; Ackerman 2002; Brook *et al.* 2008; Iles *et al.* 2013). For some tundra nesting species, the relationship between nest success and coexisting prey appears adequate to cause certain populations of waterfowl to modify nesting distribution or forgo nesting in years when prey abundance is low (Underhill *et al.* 1993; Sittler *et al.* 2000; Quakenbush *et al.* 2004). Researchers cannot explain whether decreased reproductive investment is due to a lack of coexisting prey (Bêty *et al.* 2001, 2002; Gauthier 2004; Iles *et al.* 2013) or "protective umbrella" species – species of predatory birds that inadvertently defend other birds' nest from mammalian predators while defending their own nest (Dyrcz *et al.* 1981). The mechanism(s) by which waterfowl assess abundance of coexisting prey is also unclear. A potential mechanism may be use of mammalian urine similar to that demonstrated for raptors. Evidence consistent with the hypothesis that predatory birds use UV light reflecting off phosphorous in mammal urine to locate areas of high prey density is well documented (*e.g.* Viitala *et al.* 1995; Koivula & Viitala 1999; Probst *et al.* 2002). Ducks can also see into the UV light spectrum (Jane & Bowmaker 1988) and may use a technique similar to that described by Eichholz *et al.* (2012) to assess indirectly coexisting prey abundance. This hypothesis, however, has not been tested empirically. Furthermore, while a pattern of increased nesting effort and success in years of greater small mammal abundance is well established in arctic and subarctic regions, the relationship is less clear in temperate regions, potentially due to greater abundance of generalist predators (Hanski *et al.* 1991). Perhaps an increase in abundance of coexisting prey, such as other waterfowl or bird eggs, arthropods or small mammals, would produce a functional response by satiating or decreasing movement of predators, thus decreasing susceptibility of nests to predation (Crabtree & Wolfe 1988; Crabtree *et al.* 1989; Larivière & Messier 2001; Ackerman 2002). In contrast, increased abundance of coexisting prey may produce a numerical response by concentrating predators into areas of high abundance of coexisting prey, decreasing waterfowl nest success (Holt & Lawton 1994). In the only known experimental study of nesting ducks belonging to different guilds (*i.e.* tree cavity *versus* ground nesters), Elmberg & Pöysä (2011) found that adding ground nests near cavity nests did not increase predation risk for the latter in an area where the main nest predator (Pine Marten *Martes martes*) was a genuine generalist. Clearly, the interrelationships between waterfowl nest success, predator abundance and coexisting prey abundance are complex and unresolved. A lack of consistent results between nest success and coexisting prey may be due to the variability in the balance between the strength of functional and numerical responses associated with varying species of predators and abundance of coexisting prey (Ackerman 2002; Brook *et al.* 2008), making predictions uncertain about how coexisting prey distribution should impact nest site selection by waterfowl.

Food availability

Non-breeding or abandonment of reproductive attempts have been observed in Northern Pintail *Anas acuta* (Derksen & Eldridge 1980), Mallard (Krapu *et al*. 1983) and Lesser Snow Geese *Chen caerulescens* (Ankney & MacInnes 1978), indicating that securing adequate resources is an important component of reproductive success. Thus, in contrast to colonial nesting species, a number of studies have found that dispersed nesting waterfowl nest in areas where food for adults during incubation and post-hatch young is available (Swanson *et al*. 1974; Derksen & Eldridge 1980; Haszard & Clark 2007; Fox *et al*. 2009). This phenomenon may be because dispersed nesting waterfowl tend to be smaller bodied than colonial nesting species, thus are required to feed more during incubation, or tend to nest amid more structurally complex vegetation that limits overland movement of young. For example, a number of studies have found reduced survival associated with increased overland movement of dispersed nesting females and broods (Rotella & Ratti 1992; Pearse & Ratti 2004; Simpson *et al*. 2005; Krapu *et al*. 2006; Davis *et al*. 2007); however, other studies have found no relationship (Talent *et al*. 1983; Dzus & Clark 1997; Pöysä & Paasivaara 2006). In contrast, colonial nesting species, such as Snow and Barnacle Geese have adapted to travel long distances from nest sites to brood sites to maximise fitness, indicating little cost to overland travel (Larsson & Forslund 1991; Sedinger 1992; Aubin *et al*. 1993; Cooch *et al*. 1993).

Interspecific Associations

Multiple studies have found evidence that subarctic and arctic nesting species nest in association with large avian predators even though the same predators prey on young waterfowl (Young & Titman 1986; Underhill *et al*. 1993; Summers *et al*. 1994; Quakenbush *et al*. 2004; van Kleef *et al*. 2007). In the process of deterring mammalian predators from their own nests, these avian predators inadvertently deter mammalian predators from nearby waterfowl nests. Thus, waterfowl selecting nests within a protective umbrella of predatory birds may gain benefits of egg protection that outweigh potential predation of hatchlings (Vermeer 1968; Young & Titman 1986; Bird & Donehower 2008). Some have suggested the extent of this protection is so important for some species that certain individuals will forgo breeding in years when predatory birds are not present to provide protection (Underhill *et al*. 1993; Summers *et al*. 1994; Quakenbush *et al*. 2000). In addition to nesting in association with predatory species, smaller waterfowl may enjoy fitness benefits by nesting near large waterfowl that deter small and medium-sized predators (McLandress 1983; Baldwin *et al*. 2011). For example, Canada Geese are known to reduce predation on and increase species richness of co-nesting ducks (Fabricius & Norgren 1987; Allard & Gilchrist 2002). To date, such relationships between predatory birds, colonial birds and dispersed nesting waterfowl have been reported mainly from arctic biota, but Fabricius & Norgren (1997) observed diving and dabbling ducks nesting close to geese on islets in archipelagos in the temperate biome. We see no reason why this should not be a widespread phenomenon, suggesting it should be investigated more thoroughly.

Nesting congregations

Lack (1965) suggested that birds have evolved two primary forms of nest distribution, colonial and dispersed nesting. In waterfowl, however, there appears to be a gradient from dense to loose colonies for some species to species generally considered dispersed nesters, but congregate into nest "clumps" or nest in high densities on islands. Here, we partition the discussion of nesting congregations into three sections: (1) coloniality – which pertains to species that generally congregate when nesting, (2) clump nesting – pertaining congregations of typically dispersed nesters in contiguous upland habitat, and (3) island congregations – typically dispersed nesters are found nesting in congregations on islands.

Coloniality

Coloniality appears to be the evolved trait from the ancestral condition of dispersed nesting (Coulson & Dixon 1979; Wittenberger & Hunt 1985; Rolland et al. 1998) and may have evolved multiple times due to a variety of selective pressures (Siegel-Causey & Kharitonov 1990; Rolland et al. 1998). One reason this topic has garnered substantial attention is the few measured benefits (advantages linked to predation and enhanced food finding) relative to costs of colonial nesting (competition for food, nest sites and mates, increased conspicuousness, transfer of disease and parasites, cannibalism; Siegel-Causey & Kharitonov 1990; Rolland et al. 1998). The most recent discussions of the evolution of colonial nesting suggest it evolves through: (1) a "limitation of breeding site" framework where a lack of nesting sites force individuals to nest in aggregation with no net benefit (Wittenberger & Hunt 1985), (2) an "economic" framework where a cost-benefit tradeoff of specific habitat conditions favour coloniality (Alexander 1974; Wittenberger & Hunt 1985; Sachs et al. 2007), or (3) a "by-product" framework where individual habitat selections or sexual selection leads to aggregation and colonial breeding results from these individual selection decisions not as a direct result of being aggregated (Wagner et al. 2000; Wagner & Danchin 2003; Sachs et al. 2007).

To our knowledge, no studies have been conducted directly to address theories on the evolution of coloniality in waterfowl; however, a number of studies appear to be consistent with factors described under the "economic" framework. For Holarctic waterfowl, evolution of coloniality has been limited to species that generally breed in open tundra, although admittedly it is a matter of definition whether intra- and interspecific aggregation of nests on islands in prairies and archipelagos in temperate and boreal regions should be construed as colonial nesting. Because open tundra makes nest concealment difficult, this observation appears consistent with the hypothesis that open habitat favours coloniality over dispersed nests as a means of predator avoidance. Indeed, positive correlations between colony size and density and nest success appear consistent with this hypothesis (Bousfield & Syroechkovskiy

1985; Raveling 1989). Furthermore, the species of waterfowl that commonly nest colonially generally feed little during incubation; thus, shared information of feeding sites appears unlikely to exert considerable selective pressure for coloniality (Milne 1976; Korshgen 1977; Parker & Holms 1990; Erikstad & Tveraa 1995).

Food availability during brood rearing, however, may be influential for evolution of coloniality in waterfowl. A number of studies have now documented reduced growth rate of goslings with increased colony size, observations consistent with a cost associated with colonial nesting (Cooch et al. 1991; Larsson & Forslund 1991; Sedinger et al. 1998). In the case of Black Brant, however, grazing pressure from high densities of colonial geese appeared to maintain quality grazing lawns; thus, colonial behaviour appears to impact nutrient availability positively for young brant (Person et al. 2003; Nicolai et al. 2008). Hence, with currently available data, factors associated with the "economic" framework seem most likely to explain the evolution of coloniality in waterfowl; however, the specific mechanism(s) is still unclear and may vary among species.

Clump nesting

With exception of a few species of sea ducks and geese, waterfowl are generally dispersed nesters (Anderson & Titman 1992). An unusual phenomenon often described by researchers, but yet to be explained, is the clumping of nests in relatively uniform habitat (Duebbert & Lokemoen 1976; Hines & Mitchell 1983; Fowler et al. 2004; Fowler 2005). In theory, this behaviour contradicts the strategy of dispersed nesting, because nests become concentrated, allowing for possible functional or numerical responses by predators (Tinbergen et al. 1967; Holt 1977). Studies have found evidence for negative density dependence, no density dependence, and positive density dependence of nests in both artificial and natural nest studies (Duebbert & Lokemoen 1976; Andrén 1991; Major & Kendal 1996; Larivière & Messier 1998; Sovada et al. 2000; Ackerman et al. 2004; Gunnarsson & Elmberg 2008). Inconsistency in results among studies likely is due to variation in the numerical (Holt 1977) and functional response behaviour of predatory species (Holling 1965; Tinbergen et al. 1967), variations in the response behaviour of prey, in the scale of the studies and in the habitat condition in which predators and prey exist (Grand & Flint 1997; Flint et al. 2006; Ringelman et al. 2012). Thus, the adaptive costs and benefits of nesting within close proximity of heterospecifics and conspecifics are not well understood.

Clump nesting may be adaptive and due to multiple individuals selecting nest sites in locations with fewer predators (Eichholz et al. 2012; Forsman et al. 2012), selection by multiple individuals of a yet unidentified nest site characteristic that provides appropriate nest microclimate or safe habitat such as sites that more adequately disperse scent of nests and hens (Conover 2007), rate of homing by successful hens (Greenwood 1982; Hepp & Kennamer 1992; Blums et al. 2003; Öst et al. 2011), natal philopatry of young (Hines & Mitchell 1983; Lindberg & Sedinger 1997; Coulton et al. 2011), or social attraction and transfer of information (Hines &

Mitchell 1983; Pöysä 2006; Valone 2007). Alternatively, in the case of negative density dependence and survival, clumped nesting may even be maladaptive behaviour, being due to environmental change outpacing the ability of birds to adapt, creating a false signal for appropriate nest site selection (Dessborn *et al.* 2011). Because clumping behaviour may appear obtuse evolutionarily and with regard to conservation ramifications, a better understanding of the mechanistic characteristics of this behaviour and its adaptive significance is needed.

Natural islands and island congregations

A number of species prefer islands as nest sites (Ryder 1972; Gosser & Conover 1999; Traylor *et al.* 2004; Öst *et al.* 2011). Island nesting is thought to be beneficial because important nest predators such as skunks, badgers and foxes generally avoid water (Ryder 1972; Mickelson 1975; Thompson & Raveling 1987; Petersen 1990; Zoellick *et al.* 2004). An interesting aspect of island nesting is that a number of species of ducks and geese tend to nest at densities as much as two orders of magnitude greater than densities observed on the mainland (Hammond & Mann 1956; Dwernychuk & Boag 1972; Duebbert *et al.* 1983; Willms & Crawford 1989). This occurrence is especially surprising for species such as Canada geese that typically are extremely territorial, maintaining territories as large as ≥ 100 m around the nest on mainland. The mechanism(s) allowing extremely high nesting density in territorial species may be due to decreased predation pressure leading to fewer individuals attempting to maintain widely dispersed nests or the inability of early nesters to defend territories and maintain dispersed nests because of an overwhelming drive of individuals to nest on islands (Mack *et al.* 2003). The latter explanation appears most likely based on the extreme number of pursuit flights emanating from islands during early nesting (Duebbert 1966). An additional likely prerequisite for dense nesting congregations on island is adequate food resources to support high densities of adults and young.

Wetland proximity

Primary and secondary productivity during summer dry seasons in the Holarctic is often concentrated around wetlands (Greenwood *et al.* 1995; Larivière & Messier 2000). Greater primary productivity within and immediately adjacent to local complexes of wetlands is thought to increase secondary productivity, concentrating higher trophic organisms, including predators, near wetlands (Greenwood *et al.* 1995; Larivière & Messier 2000). In theory, birds should nest away from wetlands, where predators are less abundant (Robb & Bookhout 1995; Pasitchniak-Arts *et al.* 1998a; Phillips *et al.* 2003; Traylor *et al.* 2004; *cf.* Keith 1961). For wetland dependent precocial species, such as waterfowl, shorter travel distance from the nest to brooding habitats may induce selective pressure to nest close to wetlands (Duncan 1987). Furthermore, most ducks make daily or multiple feeding bouts per day, leading to an additional energetic constraint for nesting far from wetlands (Shutler *et al.* 1998). Thus, most waterfowl species face a trade-off between nesting farther from wetlands, where hatching success may be increased, with nesting closer to wetlands where duckling survival is maximised (Dzubin & Gollop 1972; Ball *et al.* 1975; Duncan 1987; Pöysä *et al.* 1999). The nest site distance from wetlands that maximises fitness likely varies among species with different life history traits (Duncan 1987). This prediction is supported by various studies finding Northern Shoveler *Anas clypeata* and Blue-winged Teal nesting closer to water than other species or random sites while Northern Pintails nest farther from water than other duck species (Dzubin & Gollop 1972; Ball *et al.* 1975; Livezey 1981; Shutler *et al.* 1998). Studies comparing nesting distance from water between mainland and islands also have found results consistent with this tradeoff. Individuals nesting on islands secluded from mammalian predators selected nest sites nearer to water, suggesting the threat of predation associated with different landscapes and nest substrates may affect the distance that females build nests to water (Kellet & Alisauskas 1997; Bentzen *et al.* 2009).

Habitat fragmentation

Historically, pristine nesting landscapes for waterfowl, whether in temperate forests, grasslands, or sub-arctic boreal and tundra, were vast mosaics of upland and wetlands. However, agriculture, forestry, damming for hydroelectric power and other human development have fragmented these landscapes, especially temperate uplands, making habitat patch size a potentially recent evolutionary nest site selection characteristic (Clark & Nudds 1991; Reynolds *et al.* 2001).

To become a selected trait, patch size would need to influence fitness predictably for an adequate period of time (Clark & Shutler 1999). Because of benefits associated with widely distributed nests, a generally accepted paradigm is large patches of habitat are better for production and fitness of birds than small patches (Ball *et al.* 1995; Greenwood *et al.* 1995; Reynolds *et al.* 2001). Smaller patches are thought to be less productive because they increase foraging efficiency of predators by providing proportionally more edge habitat, increase density of foraging predators (*i.e.* concentration of enemies hypothesis), force birds to nest in greater density, impact species composition of predator communities, provide more homogeneous vegetation facilitating movement of predators, or increase dispersal inhibiting maintenance of higher concentrations and more intact communities (Higgins 1977; Clark & Nudds 1991; Stephens *et al.* 2004; Bayard & Elphic 2010).

Selection of larger patches should lead to a positive relationship between nest density and patch size, termed area sensitivity (Robbins *et al.* 1989; Bender *et al.* 1998; Conner *et al.* 2000). When considering breeding density and patch size relationships for a wide diversity of fauna, a neutral relationship due to equilibrium theory of biogeography tends to be most supported (Bender *et al.* 1998; Conner *et al.* 2000). For avifauna, area sensitivity due to resource concentration (more resources in larger patches) or concentration of enemies (higher concentration of predators in smaller patches) often garners greatest support (Raupp & Denno 1979; Conner *et al.* 2000). For example, Ribic *et al.* (2009) reviewed statistically rigorous studies of 32 species of obligate grassland passerines and noted evidence for area sensitivity in half of those species. The review by Ribic et al. (2009), and work of other researchers, suggested that area sensitivity is strongest for forest dwelling species relative to grassland species (Conner et al. 2000; Bayard & Elphick 2010), and proposed that area sensitivity appears to ignore the potential negative impacts of negative density dependence on nest success discussed earlier.

For upland nesting ducks breeding in the Prairie Pothole Region in North America, where habitat fragmentation has been most dramatic, available evidence is inconsistent with the area sensitivity hypothesis (Clark & Nudds 1991; Arnold et al. 2007; Haffele 2012). There are a number of reasons why one may not expect to observe a nest density-patch size relationship for upland nesting ducks when considering mechanisms typically invoked to explain area sensitivity in passerines. In forested landscapes and tall grass prairies adjacent to forests, edges typically result in an intermediate scrub-shrub habitat that decreases nest success by concentrating predators; thus, birds should select nest sites farther from habitat edges (Johnston & Odum 1956; Root 1973; Gates & Gysel 1978; Vickery et al. 1992). Because larger more uniform patches would allow birds to select nest sites farther from edge, larger patches are thought to be beneficial (Gates & Gysel 1978; Pasitchniak-Arts & Messier 1996; Clark & Shutler 1999). Although edge effects on nest success appear intuitive and have been well documented in some landscapes (Root 1973; Gates & Gysel 1978; Whitcomb et al. 1981; Sliwinski & Koper 2012), empirical evidence supporting the relationship among birds nesting in mixed and short grass prairies, where most temperate upland nesting waterfowl nest, remains inconsistent (Sargeant et al. 1984; Krasowski & Nudds 1986; Clark & Nudds 1991; Horn et al. 2005). In short and mixed grass prairie grasslands, patch edges tend to be very distinct, with no transitional zones, and perhaps these abrupt edges don't attract predators (Pasitchniak-Arts & Messier 1998; Phillips et al. 2003; Horn et al. 2005). Additionally, grasslands tend to support both core and edge predators, making predation more distributed across the landscape (Bergin et al. 2000; Chalfoun et al. 2002; Winter et al. 2006).

An alternative mechanism for area sensitivity in grassland birds is the "concentration of enemies" hypothesis (Conner et al. 2000). This mechanism hypothesises that predators respond numerically or behaviourally to fragmentation in the landscape, leading to a positive relationship between patch size and fitness of birds. Smaller patches may allow a predator to modify its behaviour by concentrating foraging effort in remaining grassland, increasing the likelihood a predator encounters a nest regardless of nest density. Additionally, smaller patches may be more attractive to predators, thus increasing their abundance and concentration within the patch (Root 1973; Sovada et al. 2000; Kuehl & Clark 2002). Although this theory appears intuitive, empirical support for the hypothesis is conflicting, with results of studies being both consistent (Fritzell 1975; Oetting & Dixon 1975; Cowardin et al. 1985; Johnson & Shaffer 1987) and inconsistent with this hypothesis (Duebbert & Lokemoen

1976; Livezey 1981; Vickery *et al.* 1992; Esler & Grand 1993; Phillips *et al.* 2004).

The remaining alternative explanations of area sensitivity in grassland birds are associated with greater resource availability in larger patches. Resource availability is often found to influence habitat selection and breeding distribution of wildlife (Stephens & Krebs 1986). Studies often find evidence for area sensitivity with no corollary relationship in productivity, suggesting that resource availability is the probable mechanism (Van Horn 1983; Bock & Jones 2004; Winter *et al.* 2006). While this explanation is intuitive for passerines, which acquire most resources from upland landscapes, most resources for ducks are gleaned from wetlands during breeding seasons, whereas grassland and other uplands merely provide cover for nests. In fact, for grassland nesting species of waterfowl, which may not be as susceptible to edge effects and acquire resources from wetlands, inverse area sensitivity often is predicted between nesting density and patch size (MacArthur *et al.* 1972; Pasitchniak-Arts *et al.* 1998; Sovada *et al.* 2000; Donovan & Lamberson 2001). Similar to the mechanism proposed for predators in the concentration of enemies hypothesis, habitat fragmentation hypothetically forces birds to nest densely in remnant cover, potentially leading to "ecological traps" (Clark & Nudds 1991; Ball *et al.* 1995; Reynolds *et al.* 2001). This hypothesis assumes bird population abundance is at some level independent of grassland abundance, so when grasslands are reduced, birds modify distributions but do not exhibit an isometric decline in abundance, leading to inverse area sensitivity (Braun *et al.* 1978; Clark & Nudds 1991;

Greenwood *et al.* 1995). Such an outcome may be especially relevant to upland nesting waterfowl in grasslands and many authors have suggested this mechanism has led to a decline in reproductive success of these birds (Greenwood *et al.* 1995; Sovada *et al.* 2000; Reynolds *et al.* 2001). Assuming wetland resources drive waterfowl distribution, even if waterfowl have adapted to select large patches, fragmentation and loss of grasslands have potentially forced them to nest in remnant patches near wetlands that may not be the preferred patch size, obscuring any relationship between patch size and nest density.

Finally, habitat fragmentation may not have been enacting selective pressures on ducks for an adequate period of time to evolve an adaptation. The majority of the fragmentation of prairie grassland has occurred since the 1950s and fragmentation of boreal forest is even more recent. Although selective pressures that directly influence species demographics tend to evolve quickly, 60 generations may not be adequate time for a behavioural adaptation such as the selection of large patches to occur especially considering that breeding individuals may not be reproductively successful annually during their longevity.

In summary, unlike forest birds, evidence for area sensitivity or a positive relationship between reproductive success and patch size for grassland birds in short or mixed grass prairies is equivocal. Furthermore, the few studies that have tested for a breeding density:patch size relationship were not consistent with the prediction of area sensitivity or inverse area sensitivity typically proposed in studies associated with habitat fragmentation. The limited number of studies that address this question appears surprising considering patch size plays such an important role in habitat conservation and restoration.

Fourth-order selection

This level of selection has been the focus of most waterfowl nesting studies (Kaminski & Weller 1992). While predator avoidance and food availability for adults and young likely play a strong role in 1st–3rd orders of nest site selection, predator avoidance seems the predominant selective force for 4th order selection. Nest site selection likely influences predation rate and most female annual mortality occurs during the breeding season (Ricklefs 1969; Southwood 1977; Hoekman *et al*. 2002).

Nest-site characteristics are well documented for many species of birds including waterfowl and are often found to be significantly different from characteristics at randomly selected locations (Bellrose 1980; Clark & Shutler 1999). This non-random distribution is usually assumed to be caused by habitat preference and thus adaptive (Martin 1998; Clark & Shutler 1999). Preferred nest site characteristics have been described in numerous studies, but evidence of their adaptive value is inconsistent and limited (Hines & Mitchell 1983; Crabtree *et al*. 1989; Clark & Shutler 1999; Durham & Afton 2003). This inconsistency is at least partially due to species-specific variation; geographic, temporal and design variation of studies; focus on restricted components of the ecological community (*i.e.* only waterfowl and predators, waterfowl and vegetation, or predators and vegetation); and short study durations (2–3 years) preventing inference regarding short- versus long-term variation (Clark & Nudds 1991). Additionally, dramatic anthropogenic changes to natural environments since the 20th century may have caused some nest sites selection characteristics to become maladaptive traits as environmental change outpaced adaptive ability of birds (Dwernychuk & Boag 1972; van Riper 1984; Schlaepfer *et al*. 2002). Herein, we review characteristics proposed as 4th order nest site characteristics of waterfowl and their level of empirical support.

Vegetation structure

Evidence supports the notion that physical structure of the vegetation is an important criterion of nest site selection for almost all species of ground and over-water nesting waterfowl (McLandress 1983; Miller *et al*. 2007; Safine & Lindberg 2008; Haffele *et al*. 2013). Two characteristics that have been identified as important, especially for many ground-nesting ducks, are height and density of vegetation. In general, they appear to select taller and thicker vegetation (Schrank 1972; Martin 1993; Clark & Shutler 1999; Haffele *et al*. 2013). Cover height appears to be most important when the primary predators are avian but also can be important when mammalian predators such as Striped Skunks *Mephitis mephitis* are the primary predator (Hines & Mitchell 1983; Crabtree *et al*. 1989), whereas cover density is most important when the primary predators are mammalian (Schrank 1972; Bowman & Harris 1980; Livezey 1981; Hines & Mitchell 1983; Rangen *et al*. 2000). Height and density of cover may directly impact success by obstructing movement of predators, increasing abundance of alternative prey, providing visual obstruction, or obstructing distribution of scent. While a large number of studies have found waterfowl select specific structural characteristics of vegetation for nesting, studies testing for the adaptive benefit have been inconsistent with only a few studies finding a positive relationship with nest success (Glover 1956; Crabtree et al. 1989; Clark & Shutler 1999; Durham & Afton 2003). The inconsistency of results among studies likely is due to variation in methodology and scale, the complex relationship between life history characteristics, cover characteristics, types of predators, availability of alternative prey or a combination of these other factors (Clark & Nudds 1991; Horn et al. 2005; Haffele 2012).

Some authors have proposed a tradeoff between concealment of the eggs from predation (current reproductive investment) and escaping from predation (future reproductive investment) leading to selection of cover of an intermediate height and density (Götmark et al.1995; Traylor et al. 2004; Miller et al. 2007; McRoberts et al. 2012); however, evidence supporting the selection of cover height or density as a stabilising selective trait is equivocal (Keith 1961; Duncan 1986; Clark & Shutler 1999; Haffele et al. 2013). Additionally, for some species, likely because cover > 45 cm tends to shade out shorter vegetation and thereby reduces density at ground level, nest success appears to be greatest when cover is at an intermediate height (16–45 cm; Crabtree et al. 1989; Haffele et al. 2013); thus, the perceived tradeoff between current and future reproductive success is not necessary for selection of intermediate nest cover height to occur in ground nesting birds (Hines & Mitchell 1983; Crabtree et al. 1989; Durham & Afton 2003; Haffele et al. 2013).

Level of reproductive investment varies among species based on the likelihood of future productivity (Fontaine & Martin 2006; Dassow et al. 2012). Theoretically, factors that influence 4th order nest site selection, such as vegetation height, should vary among bird species based on life-history traits leading to interspecific variation (Grant & Shaffer 2012). This theoretical relationship is supported among closely related species of ducks; shorter lived species that invest more into current reproduction tend to nest in more dense cover (Keith 1961; Duncan 1986; Greenwood et al. 1995; Haffele 2012). This interspecific variation in nesting cover requirements should be recognised when planning and implementing conservation and restoration of waterfowl nesting habitats and not establish only uniform dense nesting covers (Keith 1961; Livezey 1981; Lokemoen et al. 1990; Greenwood et al. 1995).

Specific composition of cover

Species composition of vegetation can influence nest site selection and success, and preferred vegetative composition varies among closely related species (Keith 1961; Duncan 1986; Crabtree et al. 1989; Lokemoen et al. 1990). Crabtree et al. (1989) found nest success to be greatest in vegetation composed of grasses and forbs and surmised this combination provided greater concealment by grasses at lower levels and forbs at higher levels. Although physical characteristics of vegetative cover likely have the greatest influence on nest site selection by birds (Schrank 1972; Gilbert et al. 1996), factors influenced by vegetative species composition itself also may directly or indirectly influence nest site selection. For example, although measures of height and density did not discern differences between forbs and grasses, Blue winged Teal and Northern Shoveler tended to select sites with more grasses while Mallard and Gadwall selected sites with more forbs (Livezey 1981; Lokemoen et al. 1990; Clark & Shutler 1999). These researchers could not discern if these differences in selection were due to smaller bodied teal being constrained by robust forbs, a difference in unmeasured characteristics such as overhead cover, or a direct selection for preferred species of vegetation. For example, certain species of vegetation being preferred or avoided as nest sites due to their ability to disrupt scent plumes of birds and nests (Aylor et al. 1993; Conover 2007). Regardless of the cause, because of the continuous loss of nesting cover in temperate regions of North America and replacement of native vegetation with exotic species, understanding the degree of specificity or level of plasticity in this trait may be important for determining the necessary composition of nesting cover that maximises benefits to waterfowl. Choosing a specific plant species to provide cover is a process that also has a component of phenology and between-year variation in the advancement of spring. This is especially true for early nesting species such as Mallard and Northern Pintail, which often start nest-building and egg-laying before the present year's plants provide cover adequate for nesting. Accordingly, in early nesting species, evergreens and last year's vegetation may act as cues for hens prospecting for nest sites. An issue related to this is the potential mismatch between nesting timetable of waterfowl and vegetation development that may arise as a consequence of global climate change (Drever & Clark 2007). If these processes become uncoupled, decreased nest success may result, which may ultimately affect population trajectories (Drever et al. 2012).

Vegetative litter and remnant down

Aldo Leopold (1933) first proposed leaf litter was an important component of nest site selection and success for ground nesting birds. The overall importance of litter depth in selection of nest-sites has been documented for certain grassland songbirds (Winter 1999; Davis 2005; Fisher & Davis 2010), but the relative importance for waterfowl is largely unknown. Leaf litter and remnant down may be important for providing the appropriate thermal microenvironment and concealing eggs from predation, especially for early nesting species prior to emergence of new vegetation (Bue et al. 1952; Duebbert 1969; Fast et al. 2010; Haffele 2012). Alternatively, depth of remnant leaf litter may be used as a predictor for the amount of vegetation that can be expected to grow in that location later in the nesting season (Haffele 2012). Height and amount of leaf litter were important factors for nest site selection by early (Mallard, Northern Pintail), intermediate (Blue-winged Teal) and late nesting ducks (Gadwall and Northern Shoveler; Haffele 2012). Haffele (2012) found strong selection by ducks for nest sites with deeper leaf litter, and found nest sites with more litter had lower nest mortality. Furthermore, Common

Eiders initiated nesting earlier in nest bowls with experimentally increased amount of down suggesting preference for nest bowls with remnant down from the previous year's nesting attempt. However, during heavy snow years, both Cackling Geese *Branta hutchinsii* and Emperor Geese *Chen canagica* selected sites with more short dead vegetation (Petersen 1990; Fast *et al.* 2010).

Microclimate

In addition to shielding the nest and hen from predators, nest site selection influences physical conditions of incubation such as shelter from wind, relative humidity, precipitation and excessive solar radiation diurnally and loss nocturnally (Walsberg 1981). Because the vast majority of egg mortality comes from predation with only a minor component of embryonic mortality associated with nest microclimate, studies associating nest site selection with microclimate are few relative to those associating nest site selection with predation. For example, Gloutney & Clark (1997) investigated the influence of nest site selection of Mallard and Blue-winged Teal on nest microclimate and concluded selective pressure of optimising physical conditions of incubation is secondary to the selective pressures of egg and hen survival, based on a combined measure of temperature and relative humidity. More recently, however, evidence is accumulating indicating the impact of the nest microclimate is not limited to the immediate impact of embryonic mortality but can influence both short- and long-term fitness of the adults and young, thus benefits associated with favourable microclimates may not be immediately recognised (Wilson & Verbeek 1995; Zicus *et al.* 2004; Hepp *et al.* 2006; Fast *et al.* 2007; DuRant *et al.* 2011). Furthermore, the cost of incubating eggs with less than optimal microclimates can affect the adult's ability to care for young and adult survive post hatch (Erikstad & Tveraa 1995; Öst *et al.* 2003). Selecting nest sites that minimise energetic costs of incubation could be especially important for smaller bodied species nesting in temperate areas or larger bodied species in arctic and sub-arctic regions (Piersma *et al.* 2003; Hilton *et al.* 2004). Selecting nest sites that provide appropriate insulation could come at a cost to nest concealment, leading to a tradeoff between appropriate microclimate and concealment (Shutler *et al.* 1998). Finally, nest site micro-climate could indirectly influence risk of predation by influencing incubation behaviour of adults. Less insulated clutches could be more energetically expensive to incubate forcing hens to leave the nest more to feed, thus increasing susceptibility to egg predation (Thompson & Raveling 1987; Afton & Paulus 1992; Durrant *et al.* 2011). Because of the limited amount of interest this area has received and the potential demographic impact of reduced reproduction, more studies simultaneously considering the impact of nest site selection on microclimate, predator avoidance and the long-term ramifications on fitness of both adults and young seem warranted.

Date of ice or snow melt

To breed successfully, the ice- and snow-free period must be sufficient for waterfowl to lay and incubate eggs, for young to grow and for young and moulting adults to attain flight and acquire adequate nutrient reserves to sustain them for the first stage of the fall migration. Because the amount of time to achieve these stages of the annual cycle is so limited, there is strong selective pressure for females to initiate nesting as soon as possible at latitudes where the time window for successful breeding may be limiting. Earlier nest initiation for arctic and subarctic breeding species has been found to influence nest survival, growth rates of young, adult body size, year of first breeding and first year survival; which are vital rates that influence individual fitness (Lindholm *et al.* 1994; Sedinger *et al.* 1995; Cooch 2002; Blums *et al.* 2005; Pilotte *et al.* 2014). This strong selective pressure for early nest initiation has caused the date that potential nesting sites become snow free to be an important component of nest site selection in species breeding in arctic habitats (Ely & Raveling 1984; Petersen 1990; Chaulk *et al.* 2007; Lecomte *et al.* 2008). Because winter winds often sweep snow from the highest potential nest sites, the date a nest site becomes ice or snow free is often strongly correlated to the height of the nest site. Thus, when selecting the first snow-free sites, waterfowl are often selecting for sites with higher elevation in the area (Mickelson 1975; Eisenhauer & Kirkpatrick 1977; Peterson 1990). Selection of higher nesting sites also has been reported in more temperate areas, likely decreasing the potential for flooding and increasing probability for early snow melt from tallest sites (Jarvis & Harris 1967; Ely & Raveling 1984; O'Neil 1988).

Kinship

For some semi- and colonial nesting species or for clumps of nests of dispersed nesting species, related individuals have been found to nest in closer proximity than would be expected under a random distribution (van der Jeugd *et al.* 2002; Fowler *et al.* 2004; Sonsthagen *et al.* 2010). Clusters of related individuals can arise due to adult and natal breeding philopatry, phenotype matching, and kinship associations (Ely & Scribner 1994; van der Jeugd *et al.* 2002; Fowler *et al.* 2004; Sonsthagen *et al.* 2010). However, evidence is consistent with the idea that individuals actively seek nest sites near closely related individuals (van der Jeugd *et al.* 2002; Sonsthagen *et al.* 2010; Fishman *et al.* 2011). For example, Barnacle Geese have been observed nesting near their siblings from the same brood on islands different from their natal island, indicating nesting proximity was not due to natal philopatry (van der Jeugd *et al.* 2002). A similar observation was not made regarding sisters from different broods (*i.e.* different years), a result most consistent with the hypothesis females are actively selecting nest sites near known kin (van der Jeugd *et al.* 2002). At this point, benefits of selecting nest sites near kin are speculative but may include increased willingness for cooperation in predator defence, joint defence of high quality food patches and relatives with which they share heredity, or decreased aggression between related neighbours (van der Jeugd *et al.* 2002; Sonsthagen *et al.* 2010; Fishman *et al.* 2011).

Fidelity, experience and information sharing

The costs and benefits of nest site fidelity and complementary behaviour of dispersal are thought to be an influential component of nest site selection (Hinde 1956; Greenwood 1982). There are a number of proposed advantages to exhibiting fidelity to a nesting site, including familiarity with food and other resources and with neighbours decreasing aggressive interactions (Greenwood 1982), all of which may ramify into to greater productivity (Harvey *et al.* 1979; Newton & Marquiss 1982; Gratto *et al.* 1985; Korpimäki 1988). One particularly important factor influencing an individual's decision to maintain breeding site fidelity or disperse appears to be past reproductive success. Making nest site selection decisions based on past experiences would require some level of consistency in the success of specific nest site. Although there often appears to be substantial annual variability in nest success within a nesting location (Haffele *et al.* 2013; Ringelman 2014), other studies have documented consistency in security of nest sites, such as nest cavities used by Common Goldeneyes *Bucephala clangula* (Elmberg & Pöysä 2011). Therefore, a number of studies have found migratory birds disperse farther when they fail in a reproductive attempt (Weatherhead & Boak 1986; Paton & Edwards 1996). This behaviour also has been observed in waterfowl, with a positive relationship between degree of fidelity and fecundity (MacInnes & Dunn 1988; Hepp & Kennamer 1992; Lindberg & Sedinger 1997; Öst *et al.* 2011).

In addition to the individual's past reproductive success, a number of studies have provided evidence that birds use success of neighbours (public information) to make initial nest site selection or determine whether to disperse or exhibit spatial fidelity (Boulinier *et al.* 1996; Doligez *et al.* 2003; Valone 2007). For waterfowl, there is some evidence of hetero- and conspecific attraction (Elmberg *et al.* 1997; Pöysä *et al.* 1998); however, use of public information for nest sites selection is only speculative for ground nesting birds and has received limited support for cavity nesting species (Pöysä 2006; Ringelman *et al.* 2012; *cf.* Roy *et al.* 2009).

Obstacles and structures

In addition to vegetation, waterfowl nesting in open landscapes, such as tundra, often use obstacles such as rocks or drift wood to conceal nests from predators and increase quality of the nesting microclimate (Ryder 1972; Noel *et al.* 2005; Fast *et al.* 2007; Öst & Steele 2010). Waterfowl nesting in association with such obstacles have been reported to have higher nest success and reduced weight loss during incubation (Kilpi & Lindström 1997; Öst *et al.* 2008b; Öst & Steele 2010). Similar to waterfowl nesting in vegetation, however, there appears to be a tradeoff between level of concealment and the ability of incubating hens to detect and quickly escape from predators (Öst *et al.* 2008a; Öst & Steele 2010).

Nesting cavities

A somewhat unique adaptation evolved by a number of birds including waterfowl is nesting in tree and artificial cavities. Birds likely began nesting in cavities in an effort to avoid nest predation (Lack 1954; Nilsson 1986). Ducks are secondary cavity nesters that rely on tree holes excavated by other wildlife, natural formation through injury to trees and subsequent decay, or nest boxes erected by humans (Bellrose & Holms 1994;

Nielsen *et al.* 2007). Although cavities were once thought to be limited in North America for Wood Duck *Aix sponsa*; this may no longer be the pervasive reality in most areas (Soulliere 1988; Nielsen *et al.* 2007; Denton *et al.* 2012). Lowney & Hill (1989), however, reported that densities of cavities suitable for Wood Duck nesting (*i.e.* adequate dimensions and internal surface for eggs) in Mississippi hardwood bottomlands were among the lowest reported for mature forests in North America. Additionally, in more northern regions of Europe and North America, areas where cavities are used by Smew *Mergellus albellus* and *Bucephala* species, cavities are still thought to limit some populations (Savard 1988; Pöysä & Pöysä 2002; Vaillancourt *et al.* 2009; Robert *et al.* 2010), mainly because natural forests contain many more old and hollow trees than do modern managed forests. Thus, some suggest nest box programmes are important to maintain breeding populations of cavity nesting ducks in some regions (Lowney & Hill 1989).

Although over-water cavities would intuitively appear to be more secure and nest success has been found to be higher during periods of flood (Nielsen & Gates 2007), there appears to be no clear selection for over water cavities for Wood Ducks. However, Wood Duck duckling survival was greatest for individuals hatched in predator-protected nest structures located amidst flooded scrub-shrub wetlands in Mississippi, which may have concealed nest structures from avian egg predators during egg laying and incubation or provided near cover for hens and broods after exodus from the nest (Davis *et al.* 2007, 2009). Moreover, regarding natural cavities, cavity characteristics appear to be more important than the cavity tree characteristics when ducks are selecting nest sites (Robb & Brookhout 1995; Yetter *et al.* 1999). Cavities used by waterfowl have been found to be higher from the ground or water, nearer forest opening or wetlands, and have entrances with smaller widths and heights (Prince 1968; Dow & Fredga 1985; Yetter *et al.* 1999; Robert *et al.* 2010). All of these characteristics have been found to reduce the potential for predation. Orientation of cavity entrance could also be important for nest microclimate (Gilmer *et al.* 1978), and while distance to wetlands was important for Common Goldeneyes (Pöysä *et al.* 1999), it did not appear to have a strong influence on cavity use by Wood Ducks (Robb & Bookhout 1995).

Conclusions and implications

Throughout the 20th and 21st centuries, both environmental impacts and conservation-management actions to remediate those impacts for breeding waterfowl have occurred on characteristics that affect nest site selection at the local or patch and microhabitat scales (*i.e.* 3rd and 4th order selection), those likely based almost entirely on more fine-grained interactions (Hutto 1985; Kaspari *et al.* 2010). The outcomes of habitat selection decisions at 3rd and 4th order scales influence species compositions of waterfowl communities in breeding areas, which may influence predator-prey relationships and competition for resources on an ecological time scale and behavioural and morphological characteristics on an evolutionary scale (Morris 2003). Outcomes from these intrinsic interactions likely provided selective forces acting upon individuals which may have influenced their 1st and 2nd order selection. Additionally, whether hierarchical habitat selection by migratory waterfowl is driven by bottom-up, top-down, or both processes in different time and space scales on individual survival and reproductive success remains intriguing and worthy of research (Kaminski & Elmberg 2014). Moreover, 1st order scales of selection of migratory waterfowl operate at a continental context and generally determine distribution or range limits of species, while 2nd order scales of selection often determine habitat(s) used within a biome(s) in which species and individuals distribute themselves. Decisions at the 1st and 2nd level seem based on large-grain abiotic (*e.g.* wetland system characteristics, climate, and landscape configurations) and biotic characteristics (*e.g.* terrestrial and aquatic communities) that may be perceived from a long distance but likely have developed through novel or philopatric experiences at smaller scales (Hutto 1985). Hutto (1985) termed these large-grained characteristics "extrinsic characteristics" because they are external to local habitats or patches; thus, these characteristics may not lend themselves to manipulation by management and conservation actions. Observed patterns from 1st and 2nd order selection are results of processes in evolutionary (ultimate factors) more than in ecological time frames (proximate factors) and are likely maintained through adult and natal philopatry (Klomp 1953; Hutto 1985). Even in migrants, like most waterfowl, glaciation and concurrent biome and climate shifts are background long-term influences of species pools and hence of present regional community composition. Nevertheless, the ecological time frame lets us study range expansions and retractions and abundance shifts relative to abiotic and biotic conditions. Both are important agents shaping waterfowl and other wildlife distributions and communities in the past and present (*e.g.* Schummer *et al.* 2010; Pearse *et al.* 2010). Furthermore, the scale has changed at which ecologists now recognise anthropogenic activities are modifying the environment. Climate change and the additive effects of natural resource exploitation and agricultural development by humans are now recognised to influence the environment at the scale of the biome or continent; thus, understanding ecological questions at a larger scale through macroecological studies will become more relevant for the management and conservation of waterfowl. As this review indicates, however, studies of 1st and 2nd order waterfowl nest site selection are relatively few and often indirect, forcing us to speculate based on results of 3rd and 4th order selection. Thus, an increased emphasis on studies addressing 1st and 2nd orders of habitat selection appears warranted. Yet, we must recognise that some characteristics are changing at the granular level of 1st and 2nd order selection and may be manifested or detected in 3rd or 4th orders of selection studies through outcomes of individual distribution, survival and reproductive success. Thus, even studies addressing 1st and 2nd order selection should associate the outcome of those selection decisions at the level of the individuals to understand processes and patterns promoting fitness.

We also recognise the process of nest site

selection interacts strongly with the process of community ecological development, yet most studies of this process are limited to very few components of the community (*e.g.* vegetation characteristics and nest success or predator abundance and nest success) and often only one or two small and often ambiguously defined landscape scales. We suggest the simplistic approach that has dominated past studies has limited our ability to understand the process of nest selection, limiting management and conservation actions; thus, a multi-scale community approach of study is warranted. This approach might best be achieved by following the advice of Bloom *et al.* (2013) and incorporating both habitat selection and demographic variables in the modelling process. We also recognise questions of causality are best addressed

Table 1. Non-ranked priority recommendations for future research on nest site (habitat) selection in waterfowl, derived but adapted from 16 suggestions proposed by Kaminski and Weller (1992). We recognise that a number of the recommendations address recruitment, whereas this review was limited to nest site selection. Thus, our recommendations are limited to the nest site selection component of recruitment.

Priority	Recommendation
1	Relate habitat selection to waterfowl survival and recruitment rate.
2	Test models and determine the effects on recruitment of possible inter and intra-specific density-dependent habitat selection.
3	Determine effects of different densities and communities of predators and prey on waterfowl habitat selection and how that process interacts with scale and community of vegetation.
4	Invoke hierarchical approaches in studies of habitat selection by individuals to obtain data on individuals' habitat use throughout their annual cycle and range for incorporation into population models and to guide habitat conservation planning and implementation.
5	Relate long-term changes in wetland and upland composition to corresponding changes in the variety of interacting factors (*e.g.* vegetation, food, predators and competitors) that influence waterfowl recruitment at multiple scales.
6	Determine effects of variation and loss in availability of intermittently, temporarily, and seasonally flooded wetlands on waterfowl habitat selection, dispersal and recruitment.
7	Determine the effect of habitat fragmentation on waterfowl habitat selection and community organisation.

through manipulative experiments; however, experimental studies are inherently difficult and costly to carry out at larger scales. This dilemma is a true challenge for waterfowl and wildlife research, conservation and management, but it deserves resolution and implementation through landscape-scale cooperatives of ecological and conservation partners.

Kaminski & Weller (1992) produced a thorough review of breeding habitats of Nearctic waterfowl. At the conclusion of their review, they identified 16 issues which they believed needed further study. Here, we suggest that a slightly modified list of seven of those 16 issues warrant priority consideration (Table 1). We conclude that only the first of the 16 recommendations has been explored with sufficient replication using species of adequately diverse life histories to allow for general inference and management actions, although some unique species of conservation concern are exceptions. This observation corroborates our assertion that the advancement in understanding the process of nest site selection has been limited by approach.

Finally, the time required to complete the task of reviewing thoroughly the substantial literature on multi-scale nest site selection by Holarctic waterfowl has prevented us from addressing habitat selection by broods, a second but no less important component of breeding habitat selection. In fact, a recent Mallard study suggests potential tradeoffs between nest site and brood habitat selection, suggesting simultaneous integration of these may be most appropriate for future studies (Bloom *et al.* 2013). We recommend undertaking the task of updating and advancing reviews of waterfowl pre-fledging ecology and summarising the current status of information regarding 3rd and 4th order brood site selection (*e.g.* Sedinger 1992), given that 1st and 2nd order selection occur at the time of nest site selection and were covered in this review.

Acknowledgements

We would like to thank Stuart Slattery, Dave Howerter, and Tony Fox for their input in the oral presentation of this manuscript at the 6th North American Duck Symposium, Ecology and Conservation of North American Waterfowl. We also thank Rick Kaminski and Aaron Pearse for constructive comments and edits, both of which greatly benefitted the quality of this manuscript, and Lauren Hill for transcribing the numerous citations for this manuscript.

References

Ackerman, J.T. 2002. Of mice and mallards: positive indirect effects of coexisting prey on waterfowl nest success. *Oikos* 99: 469–480.

Ackerman, J.T., Blackmer, A.L. & Eadie, J.M. 2004. Is predation on waterfowl nests density dependent? – Tests at three spatial scales. *Oikos* 107: 128–140.

Afton, A.D. & Paulus S.L. 1992. Incubation and brood care. *In* B.D.J. Batt, A.D. Afton, M.G. Anderson, C.D. Ankney, D.H. Johnson, J.A. Kadlec & G.L. Krapu (eds.), *Ecology and Management of Breeding Waterfowl*, pp. 62–108. University of Minnesota Press, Minneapolis, USA.

Alerstam, T. & Högstedt, G. 1982. Bird migration and reproduction in relation to habitats for survival and breeding. *Ornis Scandinavica* 13: 25–37.

Alerstam, T., Hedenström, A. & Åkesson, S. 2003. Long-distance migration: evolution and determinants. *Oikos* 103: 247–260.

Alexander, R.D. 1974. The evolution of social behavior. *Annual Review of Ecology and Systematics* 5: 325–383.

Allard, K. & Gilchrist, H.G. 2002. Kleptoparasitism of Herring Gulls taking eider eggs by Canada Geese. *Waterbirds* 25: 235–238.

Anderson, M.G. & Titman R.D. 1992. Spacing patterns. *In* B.D.J. Batt, A.D. Afton, M.G. Anderson, C.D. Ankney, D.H. Johnson, J.A. Kadlec & G.L. Krapu (eds.), *Ecology and Management of Breeding Waterfowl*, pp. 251–289. University of Minnesota Press, Minneapolis, USA.

Andrén, H. 1991. Predation: an overrated factor for over-dispersion of birds' nests? *Animal Behaviour* 41: 1063–1069.

Ankney, C.D. & MacInnes, C.D. 1978. Nutrient reserves and reproductive performance of female Lesser Snow Geese. *The Auk* 95: 459–471.

Arnold, T.W., Craig-Moore, L.M. & Armstrong, L.M. 2007. Waterfowl use of dense nesting cover in the Canadian Parklands. *The Journal of Wildlife Management* 71: 2542–2549.

Arrhenius, O. 1921. Species and area. *Journal of Ecology* 9: 95–99.

Arzel, C., Elmberg, J., Guillemain, M., Lepley, M., Bosca, F., Legagneux, P. & Nogues, J-B. 2009. A flyway perspective on food resource abundance in a long-distance migrant, the Eurasian Teal (*Anas crecca*). *Journal of Ornithology* 150: 61–73.

Ashmole, N. 1963. The regulation of numbers of tropical oceanic birds. *Ibis* 116: 217–219.

Aubin, A.E., Dzubin, A., Dunn, E.H. & MacInnes, C.D. 1993. Effects of summer feeding area on gosling growth in Snow Geese. *Ornis Scandinvica* 24: 255–260.

Aylor, D.E., Wang, Y. & Miller, D.R. 1993. Intermittent wind close to the ground within a grassy canopy. *Boundary-Layer Meteorology* 66: 427–448.

Baldwin, F.B., Alisauskas, R.T. & Leafloor, J.O. 2011. Nest survival and density of cackling geese inside and outside a Ross's goose colony. *The Auk* 128: 404–414.

Ball, I.J., Eng, R.L. & Ball, S.K. 1995. Population density and productivity of ducks on large grassland tracts in Northcentral Montana. *Wildlife Society Bulletin* 23: 767–773.

Bayard, T.S. & Elphick, C. 2010. How area sensitivity in birds is studied. *Conservation Biology* 24: 938–947.

Bellrose, F.C. 1980. *Ducks, Geese and Swans of North America*. Stackpole Books, Harrisburg, Pennsylvania, USA.

Bellrose, F.C. & Holms D.J. 1994. *The Ecology and Management of Wood Ducks*. Stackpole Books, Mechanicsburg, Pennsylvania, USA.

Bender, D.J., Contreras, T.A. & Fahrig, L. 1998. Habitat loss and population decline: a meta-analysis of the patch size effect. *Ecology* 79: 517–533.

Bennett, L.J. 1938. Redheads and ruddy ducks nesting in Iowa. *Transactions of the North American Wildlife and Natural Resources Conference* 3: 647–650.

Bent, A.C. 1923. *Life Histories of North American Wildfowl*. United States Natural History Museum Bulletin No. 126. U.S. Natural History Museum, Washington D.C., USA.

Bentzen, R.L., Powell, A.N. & Suydam, R.S. 2009. Strategies for nest-site selection by King Eiders. *Journal of Wildlife Management* 73: 932–938.

Bergin, T.M., Best, L.B. & Freemark, K.E. 2000. Effects of landscape structure on nest predation in roadsides of a Midwestern agroecosystem: a multiscale analysis. *Landscape Ecology* 15: 131–143.

Bêty, J., Gauthier, G., Giroux, J.-F. & Korpimäki, E. 2001. Are goose nesting success and lemming cycles linked? Interplay between nest density and predators. *Oikos* 93: 388–400.

Bêty, J., Gauthier, G. Korpimäki, E. & Giroux, J-F. 2002. Shared predators and indirect trophic interactions: lemming cycles and arctic-nesting geese. *Journal of Animal Ecology* 71: 88–98.

Bird, D.M. & Donehower, C.E. 2008. Gull predation and breeding success of Common Eiders on Stratton Island, Maine. *Waterbirds* 31: 454–462.

Block, W.M. & Brennan, L.A. 1993. The habitat concept in ornithology: theory and applications. *Current Ornithology* 11: 35–91.

Bloom, P.M., Clark, R.G., Howerter, D.W. & Armstrong, L.M. 2013. Multi-scale habitat selection affects offspring survival in a precocial species. *Oecologia* 173: 1249–1259.

Blums P., Nichols, J.D., Lindberg, M.S., Hines, J.E. & Mednis, A. 2003. Factors affecting breeding dispersal of European ducks on Engure Marsh, Latvia. *Journal of Animal Ecology* 72: 292–307.

Blums, P., Nichols, J.D. & Hines, J.E. 2005. Individual quality, survival variation and patterns of phenotypic selection on body condition and timing of nesting in birds. *Oecologia* 143: 365–376.

Bock, C.E. & Jones, Z.F. 2004. Avian habitat evaluation: should counting birds count? *Frontiers in Ecology and the Environment* 2: 403–410.

Boulinier, T., Danchin, E. & Monnat, J.Y. 1996. Timing of prospecting and the value of information in a colonial breeding bird. *Journal of Avian Biology* 27: 252–256.

Bousfield, M.A. & Syroechkovskiy, Y.V. 1985. A review of Soviet research on the Lesser Snow Goose on Wrangel Island, U.S.S.R. *Wildfowl* 36: 13–20.

Bowman, G.B. & Harris, L.D. 1980. Effect of spatial heterogeneity on ground-nest depredation. *Journal of Wildlife Management* 44: 806–813.

Braun, C.E., Harmon, K.W., Jackson, J.A. & Littlefield, C.D. 1978. Management of National Wildlife Refuges in the United States: its impacts on birds. *Wilson Bulletin* 90: 309–332.

Brook, R.W., Pasitschniak-Arts, M. & Howerter, D.W. 2008. Influence of rodent abundance on nesting success of prairie waterfowl. *Canadian Journal of Zoology* 86: 497–506.

Brown, J.H. 2014. Why are there so many species in the tropics? *Journal of Biogeography* 41: 8–22.

Brown, J.H. & Maurer, B.A. 1989. Macroecology: The division of food and space among species on continents. *Science* 243: 1145–1150.

Bue, I.B., Blankenship, L.H. & Marshall, W.H. 1952. The relationship of grazing practices to waterfowl breeding populations and production on stock ponds in western South Dakota. *Transactions of the North American Wildlife and Natural Resources Conference* 17: 396–414.

Chalfoun, A.D., Thompson, F.R. & Ratnaswamy, M.J. 2002. Nest predators and fragmentation: a review and meta-analysis. *Conservation Biology* 16: 306–318.

Chalfoun, A.D. & Martin, T.E. 2009. Habitat structure mediates predation risk for sedentary prey: experimental tests of alternative hypotheses. *Journal of Animal Ecology* 78: 497–503.

Chaulk, K.G., Robertson, G.J. & Montevecchi, W.A. 2007. Landscape feature and sea ice influence nesting common eider abundance and dispersion. *Canadian Journal of Zoology* 85: 301–309.

Chesser, T. & Levey, D.J. 1998. Austral migrants and the evolution of migration in new world birds: diet, habitat and migration revisited. *The American Naturalist* 152: 311–319.

Clark, R.G. & Nudds, T.D. 1991. Habitat patch size and duck nesting success: the crucial experiments have not been performed. *Wildlife Society Bulletin* 19: 534–543.

Clark, R.G. & Shutler, D. 1999. Avian habitat selection: pattern from process in nest-site use by ducks? *Ecology* 80: 272–287.

Cody, M.L. 1985. An introduction to habitat selection in birds. *In* M.L. Cody (ed.), *Habitat Selection in Birds*, pp. 3–56. Academic Press, Orlando, Florida, USA.

Coley, P.D., Bryant, J.P. & Chapin, F.S. 1985. Resource availability and plant antiherbivore defense. *Science* 230: 895–899.

Coluccy, J.M., Yerkes, T., Simpson, R., Simpson, J.W., Armstrong, L. & Davis, J. 2008. Population dynamics of breeding mallards in the great lakes states. *Journal of Wildlife Management* 72: 1181–1187.

Conner, E.F., Courtney, A.C. & Yoder, J.M. 2000. Individuals-area relationships: the relationship between animal population density and area. *Ecology* 81: 734–748.

Conover, M.R. 2007. *Predator–prey dynamics: the role of olfaction*. Taylor and Francis, Boca Raton, Florida, USA.

Cooch, E.G. 2002. Fledging size and survival in snow geese: timing is everything (or is it?). *Journal of Applied Statistics.* 29: 143–162.

Cooch, E.G., Lank, D.B., Dzubin, A., Rockwell, R.F. & Cooke, F. 1991. Body size variation in Lesser Snow Geese: environmental plasticity in gosling growth rates. *Ecology* 72: 503–512.

Cooch, E.G., Jefferies, R.L., Rockwell, R.F. & Cooke, F. 1993. Environmental change and the cost of philopatry: An example in the Lesser Snow Goose. *Oecologia* 93: 128–138.

Cooch, F., Rockwell, R.F. & Brault, S. 2001. Retrospective analysis of demographic responses to environmental change: a Lesser Snow Goose example. *Ecological Monographs* 71: 377–400.

Coulson, J.C. & Dixon, F. 1979. Colonial breeding in sea-birds. *In* G. Larwood. & B.T. Rosen (eds.), *Biology and Systematics of Colonial Organisms,* pp. 445–458. Academic Press, London, UK.

Coulton, D.W., Clark, R.G. & Wassenaar, L.L. 2011. Social and habitat correlates of immigrant recruitment of yearling female Mallards to breeding locations. *Journal of Ornithology* 152: 781–791.

Cox, G.W. 1968. The Role of Competition in the Evolution of Migration. *Evolution* 22: 180–192.

Cox, G.W. 1985. The evolution of avian migration systems between temperate and tropical regions of the New World. *The American Naturalist* 126: 451–474.

Crabtree R.L. & Wolfe, M.L. 1988. Effects of alternate prey on skunk predation of waterfowl nests. *Wildlife Society Bulletin* 16: 163–169.

Crabtree, R.L., Broome, L.S. & Wolfe M.L. 1989. Effects of habitat characteristics on gadwall nest predation and nest-site selection. *Journal of Wildlife Management* 53: 129–137.

Dahl, T.E. 2011. Status and trends of wetlands in the conterminous United States 2004 to 2009. U.S. Department of the Interior, Fish and Wildlife Service, Washington D.C., USA.

Dalby, L., McGill, B.J. & Fox, A.D. 2014. Seasonality drives global-scale diversity patterns in waterfowl (*Anseriformes*) via temporal niche exploitation. *Global Ecology and Biogeography* 23: 550–562.

Danell, K. & Sjöberg, K. 1977. Seasonal emergence of chironomids in relation to egglaying and hatching of ducks in a restored lake (northern Sweden). *Wildfowl* 28: 129–135.

Darwin, C. 1972. The Voyage of the Beagle. Bantam Books, New York, New York, USA.

Dassow, J., Eichholz, M.W., Weatherhead, P.J. & Stafford, J.D. 2012. Risk taking by upland nesting ducks relative to the probability of nest predation. *Journal of Avian Biology* 43: 61–67.

Davis, J.B., Cox, R.R., Kaminski, R.M. & Leopold, B.D. 2007. Survival of Wood duck ducklings and broods in Mississippi and Alabama. *The Journal of Wildlife Management* 71: 507–517.

Davis, J.B., Leopold, B.D., Kaminski, R.M. & Cox, R.R. 2009. Wood duck duckling mortality and habitat implications in floodplain systems. *Wetlands* 29: 607–614.

Davis, J.B., Guillemain, M., Kaminski, R.M. Arzel, C., Eadie, J.M. & Rees, E.C. 2014. Habitat and resource use by waterfowl in the northern hemisphere in autumn and winter. *Wildfowl* (Special Issue No. 4): 17–69.

Davis, S.K. 2005. Nest-site selection patterns and the influence of vegetation on nest survival of mixed-grass prairie passerines. *The Condor* 107: 605–616.

Denton, J.C., Roy, C.L., Soulliere, G.J. & Potter, B.A. 2012. Change in density of duck nest cavities at forests in the north central United States. *Journal of Fish and Wildlife Management* 3: 76–88.

Derksen, D.V. & Eldridge, W.D. 1980 Drought-displacement of Pintails to the Arctic Coastal Plain, Alaska. *Journal of Wildlife Management* 44: 224–229.

Dessborn, L., Elmberg, J. & Englund, G. 2011. Pike predation affects breeding success and habitat selection of ducks. *Freshwater Biology* 56: 579–589.

Dion, N., Hobson, K.A. & Larivière, S. 2000. Interactive effects of vegetation and predators on the success of natural and simulated nests of grassland songbirds. *The Condor* 102: 629–634.

Doligez, B., Cadet, C. & Danchin, E. 2003. When to use public information for breeding habitat selection? The role of environmental predictability and density dependence. *Animal Behaviour* 66: 973–988.

Donovan, T.M. & Lamberson, R.H. 2001. Area-sensitive distributions counteract negative effects of habitat fragmentation on breeding birds. *Ecology* 82: 1170–1179.

Dow, H & Fredga, S. 1985. Selection of nest sites by a hole-nesting duck, the Goldeneye *Bucephala clangula*. *Ibis* 127: 16–30.

Drever, M.C. & Clark, R.G. 2007. Spring temperature, clutch initiation date and duck nest success: a test of the mismatch hypothesis. *Journal of Animal Ecology* 76: 139–148.

Drever, M.C., Clark, R.G. & Derksen, C. 2012. Population vulnerability to climate change linked to timing of breeding in boreal ducks. *Global Change Biology* 18: 480–492.

Duebbert, H.F. 1966. Island nesting of the Gadwall in North Dakota. *The Wilson Bulletin* 78: 12–25.

Duebbert, H.F. 1969. High nest density and hatching success of ducks on South Dakota CAP land. *Transactions of the North American Wildlife and Natural Resources Conference* 34: 218–228.

Duebbert, H.F. & Kantrud, H.A. 1974. Upland duck nesting related to land use and predator reduction. *Journal of Wildlife Management* 38: 257–265.

Duebbert, H.F. & Lokemoen, J.T. 1976. Duck nesting in fields of undisturbed grass-legume cover. *Journal of Wildlife Management* 40: 39–49.

Duebbert, H.F. & Lokemoen, J.T. 1980. High duck nesting success in a predator-reduced environment. *Journal of Wildlife Management* 44: 428–437.

Duebbert, H.F., Lokemoen, J.T. & Sharp, D.E. 1983. Concentrated nesting of mallards and gadwalls on Miller Lake Island, North Dakota. *Journal Wildlife Management* 47: 729–740.

Durham, R.S. & Afton A.D. 2003. Nest-site selection and success of mottled ducks on agricultural lands in southwest Louisiana. *Wildlife Society Bulletin* 31: 433–442.

Duncan, D.C. 1986. Influence of vegetation on composition and density of island-nesting ducks. *Wildlife Society Bulletin* 14: 158–160.

Duncan, D.C. 1987. Nest-site distribution and overland brood movements of northern pintails in Alberta. *Journal of Wildlife Management* 51: 716–723.

DuRant, S.E., Hopkins, W.A., Hawley, D.M. & Hepp, G.R. 2011. Incubation temperature affects multiple measure of immunocompetence in young wood ducks (*Aix Sponsa*). *Biology Letters* 10: 1–6.

Dwernychuk, L.W. & Boag, D.A. 1972. Ducks nesting in association with gulls – an ecological trap? *Canadian Journal of Zoology* 50: 559–563.

Dyrcz A., Okulewicz J. & Witkowski J. 1981. Nesting of "timid" waders in the vicinity of "bold" ones as an antipredator adaptation. *Ibis* 123: 542–545.

Dzubin, A. & Gallop, J.B. 1972. Aspects of breeding Mallard ecology in Canadian parkland and grassland. *In* R.I. Smith, J.R. Palmer & T.S. Baskett (eds.), *Population Ecology of Migratory Birds – a Symposium*, pp. 113–152. United States Fish and Wildlife Service Wildlife Report No. 2. USFWS, Washington D.C., USA.

Dzus, E.H. & Clark, R.G. 1997. Overland travel, food abundance, and wetland use by Mallards: relationships with offspring survival. *Wilson Bulletin* 109: 504–515.

Earl, J.P. 1950. Production of mallards on irrigated land in the Sacramento Valley, California. *Journal of Wildlife Management* 14: 332–342.

Eichholz, M.W. & Sedinger, J.S. 1999. Regulation of incubation behavior in Black Brant. *Canadian Journal of Zoology* 77: 249–257.

Eichholz, M.W., Dassow, J.A., Weatherhead, P.J. & Stafford, J.D. 2012. Experimental evidence that nesting ducks use mammalian urine to assess predator abundance. *The Auk* 129: 638–664.

Eisenhauer, D.I. & Kirkpatrick, C.M. 1977. Ecology of the Emperor goose in Alaska. *Wildlife Monographs* 57: 3–62.

Elmberg, J. & Pöysä, H. 2011. Is the risk of nest predation heterospecifically density-dependent in precocial species belonging to different nesting guilds? *Canadian Journal of Zoology* 89: 1164–1171.

Elmberg, J., Nummi, P. & Pöysä, H. 1993. Factors affecting species number and density of dabbling duck guilds in North Europe. *Ecography* 16: 251–260.

Elmberg, J., Nummi, P. & Pöysä, H. 1994. Relationships between species number, lake size and resource diversity in assemblages of breeding waterfowl. *Journal of Biogeography* 21: 75–84.

Elmberg, J., Pöysä, H. & Sjöberg, K. 1997. Interspecific interactions and co-existence in dabbling ducks: observations and an experiment. *Oecologia* 111: 129–136.

Elmberg, J., Folkesson, K., Guillemain, M. & Gunnarsson, G. 2009. Putting density dependence in perspective: nest density, nesting phenology, and biome, all matter to survival of simulated mallard *Anas platyrhynchos* nests. *Journal of Avian Biology* 40: 317–326.

Ely, C.R. & Raveling, D.G. 1984. Breeding biology of Pacific White-Fronted Geese. *Journal of Wildlife Management* 48: 823–837.

Ely, C.R. & Scribner, K.T. 1994. Genetic diversity in Arctic-nesting geese: implications for management and conservation. *Transactions of the North American Wildlife and Natural Resources Conference* 59: 91–110.

Erikstad, K.E. & Tveraa, T. 1995. Does the cost of incubation set limits to clutch size in Common Eiders *Somateria mollissima*? *Oecologia* 103: 270–274.

Esler, D. & Grand, J.B. 1993. Factors influencing depredation of artificial duck nests. *Journal of Wildlife Management* 57: 244–248.

Evans, K.L., Warren, P.H. & Gaston, K.J. 2005. Species-energy relationships at the macroecological scale: a review of the mechanisms. *Biological Review* 80: 1–25.

Fabricius, E. & Norgren, H. 1987. *Lär känna Kanadagåsen*. Svenska Jägareförbundet [Swedish Sportsmens' Association], Stockholm, Sweden.

Fast, P.L., Gilchrist, G. & Clark, R.G. 2007. Experimental evaluation of nest shelter effects on weight loss in incubating common eiders *Somateria mollissima*. *Journal of Avian Biology* 38: 205–213.

Fast, P.L.F., Grant G.H. & Clark, R.G. 2010. Nest-site materials affect nest-bowl use by Common Eiders *Somateria mollissima*. *Canadian Journal of Zoology* 88: 214–218.

Fisher, R.J. & Davis, S.H. 2010. From Wiens to Robel: A review of procedures and patterns describing grassland bird habitat selection. *Journal of Wildlife Management* 74: 265–273.

Fishman, D.J., Craik, S.R., Zadworny, D. & Titman, R.D. 2011. Spatial-genetic structuring in a red-breasted merganser colony in the Canadian Maritimes. *Ecology and Evolution* 1: 107–118.

Flint, P.L., Grand, J.B. & Rockwell, R.F. 1998. A model of northern pintail productivity and population growth rate. *Journal of Wildlife Management* 62: 1110–1118.

Flint, P.L., Grand, J.B. Fondell, T.F. & Morse, J.A. 2006. Population dynamics of Greater Scaup breeding on the Yukon-Kuskokwim Delta, Alaska. *Wildlife Monographs* 162: 1–21.

Fontaine, J. J. & Martin, T.E. 2006. Parent birds assess nest predation risk and adjust their reproductive strategies. *Ecology Letters* 9: 428–434.

Forbes, S.L. & Kaiser, G.W. 1994. Habitat choice in breeding seabirds: when to cross the information barrier. *Oikos* 70: 377–384.

Forsman, J.T, Mönkkönen, M., Korpimäki, E. & Thomson, R.L. 2012. Mammalian nest predator feces as a cue in avian habitat selection decisions. *Behavioural Ecology* 10: 1–5.

Fournier, M. A. & Hines, J. E. 2001. Breeding ecology of sympatric Greater and Lesser Scaup (*Aythya marila* and *Aythya affinis*) in the Subarctic Northwest Territories. *Arctic* 54: 444–456.

Fowler, A.C. 2005. Fine-scale spatial structuring in cackling Canada geese related to reproductive performance and breeding philopatry. *Animal Behaviour* 69: 973–981.

Fowler, A.C., Eadie, J.M. & Ely, C.R. 2004. Relatedness and nesting dispersion within breeding populations of greater white-fronted geese. *The Condor* 106: 600–607.

Fox, T., Eide, N.E. & Bergersen, E. 2009. Resource partitioning in sympatric arctic-breeding geese: summer habitat use, spatial and dietary overlap of Barnacle and Pink-footed Geese in Svalbard. *Ibis* 151: 122–133.

Fretwell, S.D. 1972. *Populations in a Seasonal Environment*. Princeton University Press, Princeton, New Jersey, USA.

Fretwell, S.D. 1980. Evolution of migration in relation to factors regulating bird numbers. *In* A. Keast & E.S. Morton (eds.), *Migrant Birds in the Neotropics: Ecology, Behavior, Distribution, and Conservation,* pp. 517–527. Smithsonian Institute Press, Washington D.C., USA.

Fretwell, S.D. & Lucas, H.L. 1969. On territorial behavior and other factors influencing distribution in birds. I. Theoretical development. *Acta Biotheoretica* 19: 1–20.

Fritzell, E.K. 1975. Effects of agricultural burning on nesting waterfowl. *Canadian Field-Naturalist* 89: 21–27.

Garrettson, P.R. & Rohwer, F.C. 2001. Effects of mammalian predator removal on production of upland-nesting ducks in North Dakota. *Journal of Wildlife Management* 65: 398–405.

Gates, J.E. & Gysel, L.W. 1978. Avian nest dispersion and fledging success in field-forest ecotones. *Ecology* 59: 871–883.

Gauthier, G., Bety, J., Giroux, J.F. & Rochefort, L. 2004. Trophic interactions in a high arctic Snow Goose Colony. *Integrative and Comparative Biology* 44: 119–129.

Gilbert, D.W., Anderson, D.R., Ringelman, J.K. & Szymczak, M.R. 1996. Response of nesting ducks to habitat and management on the Monte Vista National Wildlife Refuge, Colorado. *Wildlife Monographs* 131: 3–44.

Gilmer, D.S., Ball, I.J. & Mathisen, J.E. 1978. Natural cavities used by Wood Ducks in North-Central Minnesota. *Journal of Wildlife Management* 42: 288–298.

Gloutney, M.L. & Clark R.G. 1997. Nest-site selection by mallards and blue-winged teal in relation to microclimate. *The Auk* 114: 381–395.

Glover, F.A. 1956. Nesting and production of the blue-winged teal (*Anas discors* Linnaeus) in northwest Iowa. *Journal of Wildlife Management* 20: 28–46.

Gosser, A.L. & Conover, M.R. 1999. Will the availability of Insular nesting sites limit reproduction in urban Canada goose populations? *Journal of Wildlife Management* 63: 369–373.

Götmark, F. & Åhlund, M. 1988. Nest predation and nest site selection among Eiders *Somateria mollissima*: the influence of gulls. *Ibis* 130: 111–123.

Grand, J.B. & Flint, P.L. 1996. Survival of Northern Pintail Ducklings on the Yukon-Kuskokwim Delta, Alaska. *The Condor* 98: 48–53.

Grand, J.B. & Flint, P.L. 1997. Productivity of nesting spectacle Eiders on the lower Kashunuk River, Alaska. *The Condor* 99: 926–932.

Grant, T.A. & Shaffer, T.L. 2012. Time-specific patterns of nest survival for ducks and passerines breeding in North Dakota. *The Auk* 129: 319–328.

Gratto, C.L., Morrison, R.I.G. & Cooke, F. 1985. Philopatry, site tenacity and mate fidelity in the semipalmated sandpiper. *The Auk*. 102: 16–24.

Greenwood, P.J. 1982. The natal and breeding dispersal of birds. *Annual Review of Ecology and Systematics* 13: 1–21.

Greenwood, R.J., Sargeant A.B., Johnson D.H., Cowardin L.M. & Shaffer, T.L. 1995. Factors associated with duck nest success in the prairie pothole region of Canada. *Wildlife Monographs* 128: 3–57.

Greenwood, R.J., Pietruszewski, D.G. & Crawford, R.D. 1998. Effects of food supplementation on depredation of duck nests in upland habitat. *Wildlife Society Bulletin* 26: 219–226.

Grinnell, J. 1914. Barriers to distribution as regards birds and mammals. *The American Naturalist* 48: 248–254.

Gross, A.O. 1945. The black duck nesting on the outer coastal islands of Maine. *The Auk* 62: 620–622.

Gunnarsson, G. & Elmberg, J. 2008. Density-dependent nest predation – an experiment with simulated Mallard nests in contrasting landscapes. *Ibis* 150: 259–269.

Haffele, R.D. 2012. Nesting ecology of ducks in dense nesting cover and restored native plantings in northeastern North Dakota. MSc. thesis. Southern Illinois University Carbondale, Carbondale, USA.

Haffele, R.D., Eichholz, M.W. & Dixon, C.S. 2013. Duck productivity in restored species – rich native and species-poor non-native plantings. *PLoS ONE* 8: e68603. doi:10.1371/journal.pone.0068603.

Hammond, M.C. & Mann, G.E. 1956. Waterfowl nesting islands. *Journal of Wildlife Management* 20: 345–352.

Hanski, I., Hansson, L. & Henttonen, H. 1991. Specialist predators, generalist predators and the microtine rodent cycle. *Journal of Animal Ecology* 60: 353–367.

Harrison, S. & Bruna, E. 1999. Habitat fragmentation and large-scale conservation: what do we know for sure? *Ecography* 22: 225–232.

Harvey, P.H., Greenwood, P.J. & Perrins, C.M. 1979. Breeding area fidelity of great tits (*Parus major*). *Journal of Animal Ecology* 48: 305–313.

Haszard, S. & Clark, R.G. 2007. Wetland use by White-winged Scoters (*Melanitta fusca*) in the Mackenzie Delta Region. *Wetlands* 27: 855–863.

Hawkins, B.A., Field, R. & Cornell, H.V. 2003. Energy, water, and broad-scale geographic patterns of species richness. *Ecology* 84: 3105–3117.

Hawkins, B.A. & Porter, E.E. 2003. Relative influence of current and historical factors on mammal and bird diversity patterns in deglaciated North America. *Global Biology & Biogeography*. 12: 475–481.

Hepp, G.R. & Kennamer, R.A. 1992. Characteristics and consequences of nest-site fidelity in wood ducks. *The Auk* 109: 812–818.

Hepp, G.R, Kennamer, R.A & Johnson, M.H. 2006. Maternal effects in wood ducks: incubation temperature influences incubation period and neonate phenotype. *Functional Ecology* 20: 307–314.

Higgins, K.F. 1977. Duck nesting in intensively farmed areas of North Dakota. *Journal of Wildlife Management* 41: 232–242.

Hildén, O. 1965. Habitat selection in birds: a review. *Annales Zoologici Fennici* 2: 53–75.

Hinde, R.A. 1956. The biological significance of the territories of birds. *Ibis* 98: 340–369.

Hilton, G.M., Hansell, M.H., Ruxton, G.D., Reid, J.M. & Monaghan, P. 2004. Using artificial nests to test importance of nesting material and nest shelter for incubation energetics. *The Auk* 121: 777–787.

Hines, J.E. & Mitchell, G.J. 1983. Gadwall nest-site selection and nesting success. *Journal of Wildlife Management* 47: 1063–1071.

Hoekman, S.T., Mills, L.S., Howerter, D.W., Devries, J.H. & Ball, I.J. 2002. Sensitivity analysis of the life cycle of midcontinent mallards. *Journal of Wildlife Management* 66: 883–900.

Holling, C.S. 1965. The functional response of predators to prey density and its role in mimicry and population regulation. *Entomological Society of Canada Memoirs* S45: 5–60.

Holt, R.D. 1977. Predation, apparent competition, and the structure of prey communities. *Theoretical and Population Biology* 12: 197–229.

Holt, R.D. 2008. Theoretical perspectives on resource pulses. *Ecology* 89: 671–681.

Holt, R.D. & Lawton, J.H. 1994. The ecological consequences of shared natural enemies. *Annual Review of Ecology and Systematics* 25: 495–520.

Horn, D.J., Phillips, M.L. & Korford, R.P. 2005. Landscape composition, patch size and distance to edges: interactions affecting duck reproductive success. *Ecological Applications* 15: 1367–1376.

Hortal, J., Diniz-Filho, J.A., Bini, L.M., Rodríguez, M.Á., Baselga, A., Nogués-Bravo, D., Rangel, T.F., Hawkins, B.A. & Lobo, J.M. 2011. Ice age climate, evolutionary constraints and diversity patterns of European dung beetles. *Ecological Letters* 14: 741–748.

Hurlbert, A.H. & Stegen, J.C. 2014. When should species richness be energy limited, and how would we know? *Ecology Letters* 17: 401–413.

Hurlbert, A.H. & Haskell, J.P. 2003. The effect of energy and seasonality on avian species richness and community composition. *The American Naturalist* 161: 83–97.

Hutto, R.L. 1985. Habitat selection by nonbreeding, migratory land birds. *In* M.L. Cody (ed.), *Habitat Selection in Birds*, pp. 455–476. Academic Press, Orlando, Florida, USA.

Iles, D.T., Rockwell, R.F. & Matulonis, P. 2013. Predators, alternative prey and climate influence annual breeding success of a long-lived sea duck. *Journal of Animal Ecology* 82: 683–693.

Jane, S.D. & Bowmaker, J.K. 1988. Tetrachromatic colour vision in the duck (*Anas platyrynchos* L.): Microspectophotometry of visual pigments and oil droplets. *Journal of Comparative Physiology A* 162: 225–235.

Jarvis, R.L. & Harris, S.W. 1967. Canada goose nest success and habitat use at Malheur Refuge. *Murrelet* 48: 46–51.

Johnson, D.H. 1980. The comparison of usage and availability measurements for evaluating resource preference. *Ecology* 61: 65–71.

Johnson, D.H. & Shaffer. T.L. 1987. Are mallards declining in North America? *Wildlife Society Bulletin* 15: 340–345.

Johnston, D.W. & Odum, E.P. 1956. Breeding bird populations in relation to plant succession on the piedmont of Georgia. *Ecology* 37: 50–62.

Kaminski, R.M. & Elmberg, J. 2014. An introduction to habitat use and selection by waterfowl in the northern hemisphere. *Wildfowl* (Special Issue No. 4): 9–16.

Kaminski, R.M. & Prince, H.H. 1984. Dabbling duck-habitat associations during spring in Delta Marsh, Manitoba. *Journal of Wildlife Management* 48: 37–50.

Kaminski, R.M. & Weller, M.W. 1992. Breeding habitats of Nearctic waterfowl. *In* B.D.J. Batt, A.D. Afton, M.G. Anderson, C.D. Ankney, D.H. Johnson, J.A. Kadlec & G.L. Krapu (eds.), *Ecology and Management of Breeding Waterfowl*, pp. 568–589. University of Minnesota Press, Minneapolis, USA.

Kaspari, M., Stevelson, B.S. & Shik, J. 2010. Scaling community structure: how bacteria, fungi, and ant taxocenes differentiate along a tropical forest floor. *Ecology* 91: 2221–2226.

Keith, L.B. 1961. A study of waterfowl ecology on small impoundments in southeastern Alberta. *Wildlife Monographs* 6: 3–88.

Kellett, D.K. & Alisauskas, R.T. 1997. Breeding biology of King Eiders nesting on Karrak Lake, Northwest Territories. *Arctic* 50: 47–54.

Klomp, H. 1953. De Terreinkeus van de Kievit, *Vanellus vanellus* (L.). *Ardea* 41: 1–139.

Klug, P.E., Wolfenbarger, L.L. & McCarty, J.P. 2009. The nest predator community of grassland birds responds to agroecosystem habitat at multiple scales. *Ecography* 32: 973–982.

Koivula, M. & Viitala, J. 1999. Rough-legged Buzzards use of vole scent marks to assess hunting areas. *Journal of Avian Biology* 30: 329–332.

Korshgen, C.E. 1977. Breeding stress of female eiders in Maine. *Journal of Wildlife Management* 41: 360–373.

Korpimäki, E. 1988. Effects of territory quality on occupancy, breeding performance and breeding dispersal in Tengmalm's Owl. *Journal of Animal Ecology* 57: 97–108.

Kossack, C.W. 1950. Breeding habits of Canada geese under refuge conditions. *American Midland Naturalist* 43: 627–649.

Krapu, G.L. & Reinecke, K.J. 1992. Foraging ecology and nutrition. *In* B.D.J. Batt, A.D. Afton, M.G. Anderson, C.D. Ankney, D.H. Johnson, J.A. Kadlec & G.L. Krapu (eds.), *Ecology and Management of Breeding Waterfowl*, pp. 1–29. University of Minnesota Press, Minneapolis, USA.

Krapu, G.L., Klett. A.T. & Jorde, G.D. 1983. The effect of variable spring water conditions on Mallard reproduction. *The Auk* 100: 689–698.

Krapu, G.L., Pietz, P.J., Brandt, D.A. & Cox, R.R., Jr. 2006. Mallard brood movements, wetland use, and duckling survival during and following a prairie drought. *Journal of Wildlife Management* 70: 1436–1444.

Krasowski, T. P. & Nudds, T.D. 1986. Microhabitat structure of nest sites and nesting success of diving ducks. *Journal of Wildlife Management* 50: 203–208.

Kristan, W.B. 2006. Sources and expectations for hierarchical structure in bird-habitat associations. *The Condor* 108: 5–12.

Kuehl, A.K. & Clark, W.R. 2002. Predator activity related to landscape features in northern Iowa. *Journal of Wildlife Management* 66: 1224–1234.

Lack, D. 1933. Habitat selection in birds with special reference to the effects of afforestation on the breckland avifauna. *Journal of Animal Ecology* 2: 239–262.

Lack, D. 1954. *The Natural Regulation of Animal Numbers*. Clarendon Press, London, UK.

Lack, D. 1965. Evolutionary ecology. *Journal of Animal Ecology* 34: 223–231.

Lack, D. 1971. *Ecological Isolation in Birds*. Blackwell Scientific Publishing, Oxford, UK.

Larivière S. & Messier, F. 1998. Effect of density and nearest neighbours on simulated waterfowl nests: can predators recognize high density nesting patches. *Oikos* 83: 12–20.

Larivière, S. & Messier. F. 2000. Habitat selection and use of edges by striped skunks in the Canadian prairies. *Canadian Journal of Zoology* 78: 366–372.

Larivière, S. & Messier, F. 2001. Space-use patterns by female striped skunks exposed to aggregations of simulated duck nests. *Canadian Journal of Zoology* 79: 1604–1608.

Larsson, K. & Forslund P. 1991. Environmental induced morphological variation in the Barnacle Goose, *Branta leucopsis*. *Journal of Environmental Biology* 4: 619–636.

Lecomte, N., Gauthier, G. & Giroux, J.F. 2008. Breeding dispersal in a heterogeneous landscape: the influence of habitat and nesting success in greater snow geese. *Oecologia* 155: 33–41.

Leitch, W.G. 1951. Saving, maintaining and developing waterfowl habitat in western Canada. *Transactions of the North American Wildlife and Natural Resources Conference* 16: 94–99.

Leopold, A. 1933. *Game Management*. University of Wisconsin Press, Madison, Wisconsin, USA.

Lima, S.L. 2009. Predators and the breeding bird: behavioral and reproductive flexibility under the risk of predation. *Biological Reviews* 84: 485–513.

Lindberg, M.S. & Sedinger, J.S. 1997. Ecological consequences of nest site fidelity in Black Brant. *The Condor* 99: 25–38.

Lindholm, A., Gauthier, G. & Desrochers, A. 1994. Effects of hatch date and food supply on gosling growth in Arctic-nesting greater snow geese. *The Condor* 96: 898–908.

Livezey, B.C. 1981. Location and success of duck nests evaluated through discriminant analysis. *Wildfowl* 32: 23–27.

Lokemoen, J.T., Duebbert, H.F. & Sharp, D.E. 1990. Homing and reproductive habits of mallards, gadwalls, and blue-winged teals. *Wildlife Monographs* 106: 1–28.

Lowney, M.S. & Hill, E.P. 1989. Wood Duck nest sites in bottomland hardwood forests of Mississippi. *Journal of Wildlife Management* 53: 378–382.

Low, J.B. 1941. Nesting of the ruddy duck in Iowa. *The Auk* 58: 506–517.

MacArthur, R.H., Diamond, J.M. & Karr, J.R. 1972. Density compensation in island faunas. *Ecology* 53: 330–342.

MacInnes, C.D. & Dunn, E.H. 1988. Components of clutch size variation in arctic-nesting Canada Geese. *The Condor* 90: 83–89.

Mack, G.G., Clark, R.G. & Howerter, D.W. 2003. Size and habitat composition of female mallard home ranges in the prairie-parkland region of Canada. *Canadian Journal of Zoology* 81: 1454–1461.

Magee, P.A. 1993. Detrital accumulation and processing in wetlands. *In* D.H. Cross & P. Vohs (eds.), *Waterfowl Management Handbook*, pp. 3–14. Fish and Wildlife Leaflet No. 13. U.S. Department of the Interior, Fish and Wildlife Service, Washington D.C., USA.

Major, R.E. & Kendal, C.E. 1996. The contribution of artificial nest experiments to understanding avian reproductive success: a review of methods and conclusions. *Ibis* 138: 298–307.

Martin, T.E. 1998. Are microhabitat preferences of coexisting species under selection and adaptive? *Ecology* 79: 656–670.

McLandress, M.R. 1983. Temporal changes in habitat selection and nest spacing in a colony of Ross' and lesser snow geese. *The Auk* 100: 335–343.

McRoberts, J.T., Quintana, N.T. & Smith, A.W. 2012. Greater Scaup, *Aythya marila*, nest site characteristics on grassy islands, New Brunswick. *Canadian Field-Naturalist* 126: 15–19.

Mickelson, P.G. 1975. Breeding biology of cackling geese and associated species on the Yukon-Kuskokwin Delta, Alaska. *Wildlife Monographs* 45: 3–35.

Miller, D.A., Grand, J.B., Fondell, T.F. & Anthony R.M. 2007. Optimizing nest survival and female survival: consequences of nest site selection for Canada geese. *Condor* 109: 769–780.

Milne, H. 1976. Body weights and carcass composition of the Common Eider. *Wildfowl* 27: 115–122.

Moles, A.T., Bonser, S.P. & Poore, A.G.B. 2011. Assessing the evidence for latitudinal gradients in plant defence and herbivory. *Functional Ecology* 25: 380–388.

Morris, D.W. 2003. Toward an ecological synthesis: a case for habitat selection. *Oecologia* 136: 1–13.

Morrison, W.E. & Hay, M.E. 2012. Are lower-latitude plants better defended? Palatability of freshwater macrophytes. *Ecology* 91: 65–74.

Mulhern, J.H., Nudds, T.D. & Neal, B.R. 1985. Wetland selection by Mallards and Blue-winged teal. *Wilson Bulletin* 97: 473–485.

Murphy, G.E.P. & Romanuk, T.N. 2014. A meta-analysis of declines in local species richness from human disturbances. *Ecology and Evolution* 4: 91–103.

Newton, I. & Marquiss, M. 1982. Fidelity to breeding area and mate in Sparrowhawks *Accipiter nisus*. *Journal of Animal Ecology* 51: 327–341.

Nicolai, C.A., Flint, P.L. & Wege, M.L. 2005. Annual survival and site fidelity of Northern pintail banded on the Yukon-Kuskokwim Delta, Alaska. *Journal of Wildlife Management* 69: 1202–1210.

Nicolai, C.A., Sedinger, J.S. & Wege, M.L. 2008. Differences in growth of Black Brant goslings between a major breeding colony and outlying breeding aggregations. *The Wilson Journal of Ornithology* 120: 755–766.

Nichols, J.D., Conley, W., Batt, B. & Tipton, A.R. 1976. Temporally dynamic reproductive strategies and the concept of R- and K-Selection. *The American Naturalist* 110: 995–1005.

Nielsen, C.L. & Gates, R.J. 2007. Reduced nest predation of cavity-nesting wood ducks during flooding in a bottomland hardwood forest. *The Condor* 109: 210–215.

Nielsen, C.L., Gates, R.J. & Zwicker, E.H. 2007. Projected availability of natural cavities for Wood Ducks in Southern Illinois. *Journal of Wildlife Management*. 71: 875–883.

Nilsson, S.G. 1986. Evolution of hole-nesting in birds: on balancing selection pressures. *The Auk* 103: 432–435.

North American Waterfowl Management Plan, Plan Committee. 2012. North American Waterfowl Management Plan 2012. People conserving waterfowl and wetlands. Canadian Wildlife Service, U.S. Fish and Wildlife Service, Secretaria de Medio Ambiente y Recursos Naturales.

Nudds, T.D. 1983 Niche dynamics and organization of waterfowl guilds in variable environments. *Ecology* 64: 319–330.

Nudds, T.D. 1992. Patterns in breeding duck communities. *In* B.D.J. Batt, A.D. Afton, M.G. Anderson, C.D. Ankney, D.H. Johnson, J.A. Kadlec & G.L. Krapu (eds.), *Ecology and Management of Breeding Waterfowl*, pp. 540–567. University of Minnesota Press, Minneapolis, USA.

Oberholser, H.C. & McAtee, W.L. 1920. Waterfowl and their food plants in the sandhill region of Nebraska. United States Department of Agriculture Bulletin No. 794. U.S. Department of Agriculture, Washington D.C., USA.

Oetting, R.B. & Dixon, C.D. 1975. Waterfowl nest densities and success at Oak Hammock Marsh, Manitoba. *Wildlife Society Bulletin* 3: 166–171.

O'Neill, R.V., D.L. Deangelis, D.L., Wade, J.B. & Allen, T.F.H. 1986. *A Hierarchical Concept of Ecosystems*. Princeton University Press, Princeton, New Jersey, USA.

O'Neil, T.A. 1988. Controlled pool elevation and its effects on Canada goose productivity and nest location. *The Condor* 90: 228–232.

Öst, M. & Steele, B.B. 2010. Age-specific nest-site preference and success in eiders. *Oecologia* 162: 59–69.

Öst, M., Ydenberg, R., Kilpi, M. & Lindström, K. 2003. Condition and coalition formation by brood-rearing common eider females. *Behavioural Ecology* 14: 311–317.

Öst, M., Wickman, M., Matulionis, E. & Steele, B. 2008a. Habitat-specific clutch size and cost of incubation in eiders reconsidered. *Oecologia* 158: 205–216.

Öst, M., Smith, B.D. & Kilpi, M. 2008b. Social and maternal factors affecting duckling survival in eiders *Smateria mollissima*. *Journal of Animal Ecology* 77: 315–325.

Öst, M., Lehikoinen, A., Jaatinen, K. & Kilpi, M. 2011. Causes and consequences of fine-scale breeding dispersal in a female-philopatric species. *Oecologia* 166: 327–336.

Pasitschniak-Arts, M. & Messier, F. 1996. Predation on artificial duck nests in a fragmented prairie landscape. *Ecoscience* 3: 436–441.

Pasitischniak-Arts, M., Clark, R.G. & Messier, F. 1998. Duck nesting success in a fragmented prairie landscape: is edge effect important? *Biological Conservation* 85: 55–62.

Paasitschniak-Arts, N. & Messier F. 1998. Effects of edges and habitats on small mammals in a prairie ecosystem. *Canadian Journal of Zoology* 76: 2020–2025.

Parker, H. & Holm, H. 1990. Patterns of nutrient and energy expenditure in female common eiders nesting in the high Arctic. *The Auk* 107: 660–668.

Paton, P.W.C. & Edwards, T.C. 1996. Factors affecting interannual movements of snowy plovers. *The Auk* 113: 534–543.

Pearse, A.T., Krapu, G.L., Brandt, D.A. & Kinzel, P.J. 2010. Changes in agriculture and abundance of snow geese affect carrying capacity of sandhill cranes in Nebraska. *Journal of Wildlife Management* 74: 479–488.

Pearse, A.T. & Ratti, J.T. 2004. Effects of predator removal on Mallard duckling survival. *Journal of Wildlife Management* 68: 342–350.

Pehrsson, O. 1986. Duckling production of the Oldsquaw in relation to spring weather and small rodent fluctuations. *Canadian Journal of Zoology* 64: 1835–1841.

Pennings, S.C., Siska, E.L. & Bertness, M.D. 2001. Latitudinal differences in plant palatability in Atlantic Coast salt marshes. *Ecology* 82: 1344–1359.

Pennings, S.C., Zimmer, M. & Dias, N., Sprung, M., Davé, N., Ho, C.K., Kunza, A., McFarlin, C., Mews, M., Pfauder, A. & Salgado, C. 2007. Latitudinal variation in plant-herbivore interactions in European salt marshes. *Oikos* 116: 543–549.

Person, B.T., Herzog, M.P. & Ruess, R.W. 2003. Feedback dynamics of grazing lawns: coupling vegetation change with animal growth. *Oecologia* 135: 583–592.

Petersen, M.R. 1990. Nest-site selection by emperor geese and cackling Canada geese. *Wilson Bulletin* 102: 413–426.

Phillips, M.L., Clark, W.R. & Sovada, M.A. 2003. Predator selection of prairie landscape features and its relation to duck nest success. *Journal of Wildlife Management* 67: 104–114.

Phillips, M.L., Clark, W.R. & Nusser, S.M. 2004. Analysis of predator movement in prairie landscapes with contrasting grassland composition. *Journal of Mammalogy* 85: 187–195.

Piersma, T. 1997. Do global patterns of habitat use and migration strategies co-evolve with relative investments in immunocompetence due to spatial variation in parasite pressure? *Oikos* 80: 623–631.

Piersma, T., Lindstrom, A., Drent, R.H., Tulp, I., Jukema, J., Morrison, T.I.G, Reneerkens, J., Schekkerman, H. & Visser G.H. 2003. High daily energy expenditure of incubating shorebirds on high arctic tundra: a circumpolar study. *Functional Ecology* 17: 356–362.

Pirnie, M.D. 1935. *Michigan Waterfowl Management*. Franklin DeKleine Company, Lansing, Michigan, USA.

Pilotte, C., Reed, E.T., Rodrigue, J. & Giroux, J.F. 2014. Factors influencing survival of Canada geese breeding in southern Quebec. *The Journal of Wildlife Management* 78: 231–239.

Pöysä, H. 2006. Public information and conspecific nest parasitism in goldeneyes: targeting safe nests by parasites. *Behavioural Ecology* 10: 459–460.

Pöysä, H. & Paasivaara, A. 2006. Movements and mortality of common goldeneye *Bucephala clangula* broods in a patchy environment. *Oikos* 115: 33–42.

Pöysä H. & Pöysä, P. 2002. Nest-site limitation and density dependence of reproductive output in the common goldeneye *Bucephala clangula*: implications for the management of cavity-nesting birds. *Journal of Applied Ecology* 39: 502–510.

Pöysä, H., Elmberg, J. & Sjöberg, K. 1998. Habitat selection rules in breeding mallards (*Anas platyrhynchos*): a test of two competing hypotheses. *Oecologia* 114: 283–287.

Pöysä, H., Milonoff, M., Ruusila, V. & Virtanen, J. 1999. Nest site selection in relation to habitat edge: experiments in the common goldeneye. *Journal of Avian Biology* 30: 79–84.

Pöysä, H., Elmberg, J., Sjöberg, K. & Nummi, P. 2000. Nesting mallards (*Anas platyrhynchos*) forecast brood-stage food limitation when selecting habitat: experimental evidence. *Oecologia* 122: 582–586.

Prince, H.H. 1968. Nest sites used by Wood Ducks and Common Goldeneyes in New Brunswick. *Journal of Wildlife Management* 32: 489–500.

Probst, R., Pavlicev, M. & Viitala. J. 2002. UV reflecting vole scent marks attract a passerine, the Great Grey Shrike *Lanius excubitor*. *Journal of Avian Biology* 33: 437–440.

Quakenbush, L., Suydam, R. & Deering, M. 2004. Breeding biology of Steller's Eiders near Barrow, Alaska, 1991–99. *Arctic* 57: 166–182.

Rangen, S.A., Clark, R.G. & Hobson. K.A. 2000. Visual and olfactory attributes of artificial nests. *The Auk* 117: 136–146.

Rappole, J.H. & Jones, P. 2002. Evolution of old and new world migration systems. *Ardea* 90: 925–937.

Raupp, M.J. & Denno, R.F. 1979. The influence of patch size on a guild of Sapfeeding insects that inhabit the Salt Marsh Grass *Spartina patens*. *Entomological Society of America* 8: 412–417.

Raveling, D.G. 1989. Nest-predation rates in relation to colony size of Black Brant. *Journal of Wildlife Management* 53: 87–90.

Reynolds, R.E., Shaffer, T.L., Renner, R.W. 2001. Impact of the conservation reserve program on duck recruitment in the U.S. prairie pothole region. *Journal of Wildlife Management* 65: 765–780.

Ribic, C.A., Koford, R.R., Herkert, J.R., Johnson, D.H., Niemuth, N.D., Naugle, D.E, Bakker, K.K., Sample, D.W. & Renfrew, R.B. 2009. Area sensitivity in North American grassland birds: patterns and processes. *The Auk* 126: 233–244.

Ricklefs, R.E. 1969. An analysis of nesting mortality in birds. *Smithsonian Contributions to Zoology* 9: 1–48.

Ringelman, K.M. 2014. Predator foraging behavior and patterns of avian nest success: What can we learn from an agent-based model? *Ecological Modeling* 272: 141–149.

Ringelman, K.M., Eadie, J.M. & Ackerman, J.T. 2012. Density-dependent nest predation in waterfowl: the relative importance of nest density versus nest dispersion. *Oecologia* 169: 695–702.

Robb, J.R. & Bookhout, T.A. 1995. Factors influencing Wood Duck use of natural cavities. *Journal of Wildlife Management* 59: 372–383.

Robbins, C.S., Dawson, D.K. & Dowell, B.A. 1989. Habitat area requirements of breeding forest birds of the middle Atlantic states. *Wildlife Monographs* 103: 3–34.

Robert, M., Vaillancourt, M. & Drapeau, P. 2010. Characteristics of nest cavities of Barrow's Goldeneyes in eastern Canada. *Journal of Field Ornithology* 81: 287–293.

Rolland, C., Danchin, E. & de Fraipoint, M. 1998. The evolution of coloniality in birds in relation to food, habitat, predation, and life-history traits: a comparative analysis. *The American Naturalist* 151: 514–529.

Romesburg, H.C. 1981. Wildlife science: gaining reliable knowledge. *Journal of Wildlife Management* 45: 293–313.

Root, R.B. 1973. Organization of a plant-arthropod association in simple and diverse habitats: the fauna of collards (*Brassica oleracea*). *Ecological Monographs* 43: 95–124.

Rotella J.J. & Ratti, J.T. 1992. Mallard brood movements and wetland selection in southwestern Manitoba. *Journal of Wildlife Management* 56: 508–515.

Roth, R.R. 1976. Spatial heterogeneity and bird species diversity. *Ecology* 57: 773–782.

Roy, C., Eadie, J.M. & Schauber, E.M. 2009. Public information and conspecific nest parasitism in wood ducks: does nest density influence quality of information? *Animal Behaviour* 77: 1367–1373.

Ryder, J.P. 1972. Biology of nesting Ross' Geese. *Ardea* 60: 185–215.

Sachs, J.L., Hughes, C.R., Neuchterlein, G.L. & Buitron, D. 2007. Evolution of coloniality in birds: a test of hypotheses with the Red-necked Grebe (*Podiceps grisegena*). *The Auk* 124: 628–642.

Safine, D.E. & Lindberg, M.S. 2008. Nest habitat selection of White-winged scoters on Yukon Flats, Alaska. *The Wilson Journal of Ornithology* 120: 582–593.

Sargeant A.B. & Raveling D.G. 1992. Mortality during the breeding season. *In* B.D.J. Batt, A.D. Afton, M.G. Anderson, C.D. Ankney, D.H. Johnson, J.A. Kadlec & G.L. Krapu (eds.), *Ecology and Management of Breeding Waterfowl*, pp. 396–422. University of Minnesota Press, Minneapolis, USA.

Sargeant, A.B., Allen, S.H. & Eberhardt, R.T. 1984. Red fox predation on breeding ducks in midcontinent North America. *Wildlife Monographs* 89: 3–41.

Savard, J.P. 1988. Use of nest boxes by Barrow's Goldeneyes: nesting success and effect on the breeding population. *Wildlife Society Bulletin* 16: 125–132.

Schamber, J.L., Flint, P.L. & Grand, J.B. 2009. Population dynamics of Long-tailed Ducks breeding on the Yukon-Kuskokwin Delta, Alaska. *Arctic* 62: 190–200.

Schmidt, J.H., Lindberg, M.S., Johnson, D.S. & Verbyla, D.L. 2011. Season length influences breeding range dynamics of trumpeter swans *Cygnus buccinator*. *Wildlife Biology* 17: 364–372.

Schmutz, J.K., Robertson, R.J. & Cooke, F. 1983. Colonial nesting of the Hudson Bay eider duck. *Canadian Journal of Zoology* 61: 2424–2433.

Schlaepfer, M.A., Runge, M.C. & Sherman, P.W. 2002. Ecological and evolutionary traps. *Trends in Ecology and Evolution* 17: 474–480.

Schrank, B.W. 1972. Waterfowl nest cover and some predation relationships. *Journal of Wildlife Management* 36: 182–186.

Schummer, M.L., Kaminski, R.M., Raedeke, A.H. & Graber, D.A. 2010. Weather-related indices for autumn-winter dabbling duck abundance in middle North America. *Journal of Wildlife Management* 94: 94–101.

Sedinger, J.S. 1992. Ecology of prefledging waterfowl. *In* B.D.J. Batt, A.D. Afton, M.G. Anderson, C.D. Ankney, D.H. Johnson, J.A. Kadlec & G.L. Krapu (eds.), *Ecology and Management of Breeding Waterfowl*, pp. 109–127. University of Minnesota Press, Minneapolis, USA.

Sedinger, J.S. & Flint, P.L. 1991. Growth rate is negatively correlated with hatch date in black brant. *Ecology* 72: 496–502.

Sedinger, J.S., Flint, P.L. & Lindberg, M.S. 1995. Environmental influence on life-history traits: growth, survival, and fecundity in Black Brant (*Branta bernicla*). *Ecology*: 76: 2404–2414.

Sedinger, J.S., Lindberg, M.S., Person, B.P., Eichholz, M.W., Herzog, M.P. & Flint, P.L. 1998. Density dependent effects on growth, body size, and clutch size in Black Brant. *The Auk* 115: 613–620.

Sedinger, J.S., Herzog, M.P. & Ward, D.H. 2004. Early environment and recruitment of Black Brant (*Branta bernicla nigricans*) into the breeding population. *The Auk* 121: 68–73.

Shutler, D., Gloutney, M.L. & Clark, R.G. 1998. Body mass, energetic constraints, and duck nesting ecology. *Canadian Journal of Zoology* 76: 1805–1814.

Siegel-Causey, D. & Kharitonov, S.P. 1990. The evolution of coloniality. *Current Ornithology* 7: 285–330.

Sittler, B., Olivier, G. & Berg, T.B. 2000. Low abundance of king eider nests during low lemming years in Northeast Greenland. *Arctic* 53: 53–60.

Simpson, J.W., Yerkes, T.J., Smith, B.D. & Nudds, T.D. 2005. Mallard duckling survival in the Great Lakes region. *The Condor* 107: 898–909.

Sliwinski, M.S. & Koper, N. 2012. Grassland bird responses to three edge type in a fragmented mixed grass prairie. *Avian Conservation and Ecology* 7: 6.

Sonsthagen, S.A., Talbot, S.L., Lanctot, R.B. & Mccracken, K.G. 2010. Do common eiders nest in kin groups? Microgeographic genetic structure in a philopatric sea duck. *Molecular Ecology* 19: 647–657.

Sovada, M.A., Zicus, M.C. & Greenwood, R.J. 2000. Relationships of habitat patch size to predator community and survival of duck nests. *Journal of Wildlife Management* 64: 820–831.

Soulliere, G.J. 1988. Density of suitable wood duck nest cavities in a Northern Hardwood Forest. *Journal of Wildlife Management*. 52: 86–89.

Southwood, T.R.E. 1977. Habitat, the templet for ecological strategies? *Journal of Animal Ecology* 46: 377–365.

Stephens, D.W. & Krebs, J.R. 1986. *Foraging Theory*. Princeton University Press, New Jersey, USA.

Stephens, S., Koons, D.N. & Rotella, J.J. 2004. Effects of habitat fragmentation on avian nesting success: a review of the evidence at multiple spatial scales. *Biological Conservation* 115: 101–110.

Stephens, S.E., Rotella, J.J., Lindberg, M.S., Taper, M.L. & Ringelman, J.K. 2005. Duck nest survival in the Missouri Coteau of North Dakota: landscape effects at multiple spatial scales. *Ecological Applications* 15: 2137–2149.

Summers, R.W., Underhill, L.G., Syroechkovski, E.E., Lappo, H.G., Prŷs-Jones, R.P. & Karpov, V. 1994. The breeding biology of Dark-bellied Brent Geese *Branta b. bernicla* and King Eiders *Somateria spectabilis* on the northeastern Taimyr Peninsula, especially in relation to Snowy Owl *Nyctea scandiaca* nests. *Wildfowl* 45: 110–118.

Swanson, G.A., Meyer, M.I. & Serie, J.R. 1974. Feeding ecology of breeding blue-winged teal. *Journal of Wildlife Management* 38: 396–407.

Talent, L.G., Jarvis, R.L. & Krapu, G.L. 1983. Survival of Mallard broods in south-central North Dakota. *The Condor* 85: 74–78.

Thompson, S.C. & Raveling, D.G. 1987. Incubation behavior of emperor geese compared with other geese: interactions of predation, body size, and energetics. *The Auk* 104: 707–716.

Tinbergen, N., Impekoven, M. & Franck, D. 1967. An experiment on spacing-out as a defence against predation. *Behavior* 38: 307–321.

Traylor, J.T. Alisauskas, R.T. & Kehoe, F.P. 2004. Nesting ecology of White-winged scoters, at Redberry Lake, Saskatchewan. *The Auk* 121: 950–962.

Underhill, L.G., Prŷs-Jones, R.P. & Syroechkovski, E.E. 1993. Breeding of waders and Brent Geese *Branta bernicla bernicla* at Pronchishcheva Lake, northeastern Taimyr, Russia, in a peak and a decreasing lemming year. *Ibis* 135: 277–292.

Vaillancourt, M., Drapeau, P., Gauthier, S. & Robert, M. 2008. Availability of standing trees for large cavity-nesting birds in the eastern boreal forest of Quebec, Canada. *Forest Ecology and Management* 255: 2272–2285.

Valone, T.J. 2007. From eavesdropping on performance to copying the behavior of others: a review of public information use. *Behavioral Ecology and Sociobiology* 62: 1–14.

van der Jeugd, H.P., van der Veen, I.T. & Larsson, K. 2002. Kin clustering in barnacle geese: familiarity or phenotype matching? *Behavioural Ecology* 13: 786–790.

Van Horne, B. 1983. Density as a misleading indicator of habitat quality. *Journal of Wildlife Management* 47: 893–901.

van Kleef, H.H., Willems, F. & Volkov, A.E. 2007. Dark-bellied brent geese breeding near snowy owl nest lay more and larger eggs. *Journal of Avian Biology* 38: 1–6.

van Riper, C., III. 1984. The influence of nectar resources on nesting success and movement patterns of the Common Amakihi. *The Auk* 101: 38–46.

Vermeer, K. 1968. Ecological aspects of ducks nesting in high densities among Larids. *The Wilson Bulletin* 80: 78–83.

Vickery P.D., Hunter, M.L. & Wells, J.V. 1992. Evidence of incidental nest predation and its effects on nests of threatened grassland birds. *Oikos* 63: 281–288.

Viitala, J., Korpimäki, E., Palokangas, P. & Koivula, M. 1995. Attraction of kestrels to vole scent marks visible in ultraviolet light. *Nature* 373: 425–427.

Wagner, R.H. 1993. The pursuit of extrapair copulations by female birds: A new hypothesis of colony formation. *Journal of Theoretical Biology* 163: 333–346.

Wagner, R.H. & Danchin, E. 2003. Conspecific copying: A general mechanism of social aggregation. *Animal Behaviour* 65: 405–408.

Walker, J., Lindberg, M.S. & MacCluskie, M.C. 2005. Nest survival of Scaup and other ducks in the Boreal Forest of Alaska. *Journal of Wildlife Management* 69: 582–591.

Walsberg, G. E. 1981. Nest-site selection and the radiative environment of the warbling vireo. *The Condor* 83: 86–88.

Walters, C.J. 1986. *Adaptive Management of Renewable Resources*. McGraw-Hill, New York, USA.

Weatherhead P.J. & Boak, K.A. 1986. Site infidelity in song sparrows. *Animal Behaviour* 34: 1299–1310.

Webster, J.R. & Benfield, E.F. 1986. Vascular plant breakdown in freshwater ecosystems. *Annual Review of Ecology and Systematics* 17: 567–594.

Weller, M.W. 1979. Density and habitat relationships of blue-winged teal nesting in northwestern Iowa. *Journal of Wildlife Management* 43: 367–374.

Whitcomb, R.F., Robbins, C.S., Lynch, J.F., Whitcomb, B.L., Klimkiewicz, K. & Bystrak, D. 1981. Effects of forest fragmentation on avifauna of the eastern deciduous forest. *In* R.L. Burgess & D.M. Sharpe (eds.), *Forest Island Dynamics in Man-dominated Landscapes*, pp. 125–205. Springer-Verlag, New York, USA.

Wiens, J.A. 1989. Spatial scaling in ecology. *Functional Ecology* 3: 385–397.

Willms, M.A. & Crawford, R.D. 1989. Use of earthen islands by nesting ducks in North Dakota. *Journal of Wildlife Management* 53: 411–417.

Willson, M.F. 1976. The breeding distribution of North American migrant birds: a critique of MacArthur (1959). *Wilson Bulletin* 88: 582–587.

Wilson, S.E. & Verbeek, N.A.M. 1995. Patterns of Wood Duck nest temperatures during egg-laying and incubation. *The Condor* 97: 963–969.

Winter, M., Johnson, D.H. & Shaffer, J.A. 2006. Patch size and landscape effects on density and nesting success of grassland birds. *The Journal of Wildlife Management* 70: 158–172.

Wittenberg, J.F. & Hunt, G.L. 1985. The adaptive significance of coloniality in birds. *In* D.S. Farner, J.R. King, & K.C. Parkes (eds.), *Avian biology. Vol. VIII*, pp. 1–78. Academic Press, New York, New York.

Wright, D.H. 1983. Species-energy theory: an extension of species-area theory. *Oikos* 41: 495–506.

Yetter, A.P., Havera, S.P. & Hine, C.S. 1999. Natural-cavity use by nesting Wood Ducks in Illinois. *Journal of Wildlife Management* 63: 630–638.

Young, A.D. & Titman, R.D. 1986. Cost and benefits of Red-breasted Mergansers nesting in tern and gull colonies. *Canadian Journal Zoology* 64: 2339–2343.

Zicus, M.C., Rave, D.P. & Riggs, M.R. 2004. Factors influencing incubation egg-mass loss for three species of waterfowl. *The Condor* 106: 506–516.

Zoellick, B.W., Ulmschneider, H.M. & Cade, B.S. 2004. Isolation of Snake River islands and mammalian predation of waterfowl nests. *Journal of Wildlife Management* 68: 650–662.

Waterfowl habitat use and selection during the remigial moult period in the northern hemisphere

ANTHONY D. FOX[1]*, PAUL L. FLINT[2], WILLIAM L. HOHMAN[3] & JEAN-PIERRE L. SAVARD[4]

[1]Department of Bioscience, Aarhus University, Kalø, Grenåvej 14, DK-8410 Rønde, Denmark.
[2]USGS, 4210 University Drive, Anchorage, AK 99508-4626, USA.
[3]USDA/NRCS, National Wildlife Team, 501 W. Felix St. Bldg. 23, Fort Worth, Texas 76115, USA.
[4]Scientist Emeritus, Environment Canada, 801-1550 Avenue d'Estimauville, Québec, G1J 0C3, Canada.
*Correspondence author. E-mail: tfo@dmu.dk

Abstract

This paper reviews factors affecting site selection amongst waterfowl (Anatidae) during the flightless remigial moult, emphasising the roles of predation and food supply (especially protein and energy). The current literature suggests survival during flightless moult is at least as high as at other times of the annual cycle, but documented cases of predation of flightless waterfowl under particular conditions lead us to infer that habitat selection is generally highly effective in mitigating or avoiding predation. High energetic costs of feather replacement and specific amino-acid requirements for their construction imply adoption of special energetic and nutritional strategies at a time when flightlessness limits movements. Some waterfowl meet their energy needs from endogenous stores accumulated prior to remigial moult, others rely on exogenous supply, but this varies with species, age, reproductive status and site. Limited evidence suggests feather proteins are derived from endogenous and exogenous sources which may affect site selection. Remigial moult does not occur independently of other annual cycle events and is affected by reproductive investment and success. Hence, moult strategies are affected by age, sex and reproductive history, and may be influenced by the need to attain a certain internal state for the next stage in the annual cycle (*e.g.* autumn migration). We know little about habitat selection during moult and urge more research of this poorly known part of the annual cycle, with particular emphasis on identifying key concentrations and habitats for specific flyway populations and the effects of disturbance upon these. This knowledge will better inform conservation actions and management actions concerning waterfowl during moult and the habitats that they exploit.

Key words: Anatidae, energy balance, feather synthesis, moult, predation, protein, survival.

Normally annual replacement of avian remiges (hereafter flight feathers) is essential, because damage, abrasion and exposure to ultra-violet light degrade such tissues (Stresemann & Stresemann 1966; Bergman 1982). Allometric relationships underlying the production of feather tissues proportionally prolong replacement of flight feathers in larger (> 300 g) birds (Rohwer *et al.* 2009). For this reason, large bird species extend flight feather moult over two or more seasons (*e.g.* Bridge 2007) or, as is the case amongst birds that do not rely on the powers of flight for feeding, undergo a simultaneous moult of all flight feathers that renders them flightless temporarily, including species of the Alcidae, Anatidae, Anhingidae, Bucerotidae, Gavidae, Gruidae, Heliornithidae, Jacanidae, Pelecanoididae, Phoenicopteridae, Podicipedidae, Rallidae and Scolopacidae (Stresemann & Stresemann 1966; Marks 1993). With the notable exception of the Bucerotidae, where breeding females undergo simultaneous moult whilst sealed inside nest cavities by males (Stonor 1937; Moreau 1937), the common feature of all these avian families is their occupancy of aquatic or marine habitats. Indeed, of the avian families that forage primarily on or under water throughout the annual cycle (*i.e.* excluding seabirds that forage on land or on the wing; Fregatidae, Laridae, Procellariiformes and Sternidae), members of the Anatidae, Alcidae, Gavidae, Pelecanoididae and Podicipedidae replace their flight feathers synchronously whilst in wetlands, most often on water. To minimise the omnipresent mortality risk posed by aquatic and aerial predators throughout the annual cycle, these species exploit aquatic systems that expose them to relatively low predation risk within and outside the flightless remigial moult period. Furthermore, the moult of males of several species of waterfowl from a bright breeding plumage (the alternate plumage according to Palmer 1972 and Weller 1980, basic according to Pyle 2005) to a cryptic eclipse plumage prior to the wing feather moult (the basic plumage according to Palmer 1972 and Weller 1980, the alternate according to Pyle 2005) is likely, at least in part, an adaptation to reduce conspicuousness during this highly vulnerable period. Loss of the powers of flight also reduces foraging opportunities, so a likely determinant of habitat selection during remigial moult is the need to derive sufficient energy and nutrients to satisfy maintenance requirements and supplementary needs of feather replacement (Hanson 1962; Hohman *et al.* 1992).

For example, the massive aggregations of Eared Grebes *Podiceps nigricollis* that moult on the Great Salt Lake, Utah, USA encapsulate elements of wetland selection that likely characterise a moulting site for waterbirds during remigial moult, namely the flat trophic structure of a hyper-saline lake with little emergent or submersed vegetation means aquatic or aerial predators are rare or non-existent, and the protein content of the super-abundant and highly accessible prey (in this case the Brine Shrimp *Artemia franciscana*) in the lake provides optimal conditions for obtaining energy and nutrients for feather growth (Jehl 1990; Wunder *et al.* 2012). Flightless waterbirds constrained to a wetland minimise depredation by diving in the case of ducks or, as in the case of dabbling ducks, geese and swans, escaping terrestrial mammals by swimming into vegetation or open water distant from shore. Prior to remigial moult, birds must select habitats that fulfil their nutritional needs, specifically those for energy and protein with specific amino acids (Murphy & King 1984a,b). Given the need to meet normal energy and protein requirements, plus the extra needs of wing moult, and avoidance of predation whilst unable to fly, how do moulting waterfowl select their moulting habitats? Specifically, with regard to improving conservation and management options to protect and enhance such habitats, what are the key features of these habitats? Answering these questions is difficult, because few studies have ever specifically examined habitat use and selection by moulting waterfowl, in the sense that a particular feature or features are selected over others. Early examples attempted to describe the physical, floristic, and invertebrate features within sites that attracted, for example, moulting Green-winged Teal *Anas crecca* (Kortegaard 1974), but rarely have researchers investigated habitat use before, during and following the remigial moult to enlighten factors affecting habitat choice at that critical period. Thus, while we can describe habitats used by moulting waterfowl, there is insufficient literature available to relate use to available habitats to infer preferences or selection (*sensu* Kaminski & Weller 1992; Kaminski & Elmberg 2014).

In this review, we adopt a comparative approach among waterfowl taxa to examine habitat selection in relation to strategies that fulfil the needs for energy and protein during the remigial moult, while minimising risks to survival posed by flightlessness. Those few studies that exist suggest that safety from predators is paramount in the selection of habitats for moult but food resources and availability are also likely to be critical. Nonetheless, numerous questions remain. What general features can we distinguish about moulting sites that are similar or different from habitats used at other times of the annual cycle? How can identification of these features help us understand the potential process of habitat selection by moulting waterfowl? Can existing variation within and among species be used to address survival and possible fitness consequences of selecting specific habitats? Lastly, how do answers to these questions provide insight into how the conservation and management of moulting habitats might be improved for indigenous waterfowl in the northern hemisphere? Intuitively, moult is not independent of other annual cycle events and is highly dependent on reproductive investment and outcome (*i.e.* whether an individual is paired or unpaired, a breeder or non-breeder, a successful or unsuccessful breeder). Thus, we also attempt to assess how this may influence habitat use during remigial moult and the manner in which moult is completed. Finally, we consider the future key policy, conservation, and research needs in this arena. We start by considering the evidence for selection of habitats based on safety from predation and assess the specific energetic and nutritional needs of remigial moult. The focus is on remigial moult rather than body plumage moult, because both occur simultaneously during the summer moult into the eclipse plumage in the northern hemisphere (Weller 1980) and the period of flightlessness while remiges are lost and regrown and associated habitat and resource use by waterfowl are most crucial to individual survival and future fitness prospects.

Role of predation during remigial moult

The limited evidence available suggests that survival during the post-breeding and remigial moult is the same, if not greater, than during other periods in the annual cycle for such species as Mallard *Anas platyrhynchos* (Kirby & Cowardin 1986), Mottled Duck *A. fulvigula* (Bielefeld & Cox 2006), Black Duck *A. rubripes* (Bowman 1987), Wood Duck *Aix sponsa* (Thompson & Baldassarre 1988; Davis *et al.* 2001), Harlequin Duck *Histrionicus histrionicus* (Iverson & Esler 2007), Barrow's Goldeneye *Bucephala islandica* (Hogan *et al.* 2013a) and scoters *Melanitta* sp. (Anderson *et al.* 2012). However, survival probability by itself does not imply that predation risk has no role in shaping habitat use by post breeding moulting waterfowl as it may simply indicate that waterfowl have mitigated this risk through adaptations.

Indeed, there are reports of flightless Pink-footed Geese *Anser brachyrhynchus* being depredated by Walrus *Odobenus rosmarus rosmarus* in Svalbard (Fox *et al.* 2010) and flightless moulting Common Eider *Somateria mollissima* and Greylag Geese *Anser anser* being pursued by Killer Whales *Orcinus orca* in the Shetland Islands, Scotland (D.

Gifford, *in litt.*). Interestingly, both situations relate to expanding Palearctic populations of geese moulting in newly colonised areas, where local terrestrial moulting habitats have become saturated as a result of increasing local density. These examples may therefore be atypical in the sense that these populations have yet to reach fitness equilibrium with regard to colonisation of formerly unoccupied territory and exposure to novel predators. In the case of Mottled Duck, drying wetland conditions in peninsular Florida in late summer concentrate wildlife, including alligators *Alligator mississippiensis*, in and around remaining wetlands, resulting in alligator depredation of post-breeding moulting Mottled Duck (Bielefeld & Cox 2006).

An exception to "normal" survival rates during moult is the periodic occurrence of mass mortality due to diseases frequently occurring during moult and summer drought periods (Bellrose 1980). Dabbling ducks exposed to botulism during the post-breeding period may suffer severe mortality (*e.g.* > 350,000 Northern Pintail *Anas acuta* in prairie Canada, Miller & Duncan 1999 and < 12,000 Mallard, Fleskes *et al.* 2010; Evelsizer *et al.* 2010), as have Redhead *Aythya americana* (> 3,000 at one site, Wobeser & Leighton 1988) and sea ducks have suffered mortality related to virus exposure (Hollmén *et al.* 2003), although other die-off events may have been the result of contaminant exposures (Henny *et al.* 1995). The flocking behaviour of moulting waterfowl may facilitate disease spread within groups, and large mass mortalities are likely to have population level effects (Reed & Rocke 1992), but these events are rare and are unlikely to represent a major factor in moult ecology and site selection.

Factors that mitigate or reduce mortality during moult

Moult migration

Ecologists have hypothesised that an advantage of moult migration is selection of habitats with lower predation pressure or less diverse predator communities (Salomonsen 1968). In many species of ducks and geese, moult migrations are to higher latitude locations where predator populations may be numerically, or more seasonally, constrained (Yarris *et al.* 1994). For sea duck species, moult migrations may or may not represent an increase in latitude, but almost always represent a movement to marine areas where predators are fewer. Thus, moult migration itself may represent an adaptation to minimise mortality during the flightless period.

Flocking behaviour

Factors affecting abundance of moulting birds at a given site have not been explored. Most, if not all, waterfowl species moult in flocks that vary in size from a few birds to thousands of individuals. Solitary moulting does not seem to have evolved in any waterfowl species, although northern breeding female ducks completing brood rearing late will moult alone. In general, birds in groups are less susceptible to predation than solitary birds and safety is, to some degree, proportional to group size and to position within the group (Petit & Bilstein 1987; Elgar 1989; Tamisier & Dehorter 1999; *cf.* Davis *et al.* 2007). Leafloor *et al.* (1996) reported social interactions tend to synchronise moult timing in captive female Mallard and that this may have important survival advantages. The group response of moulting waterfowl to disturbance events suggests that one of the advantages to moulting in flocks is the increased detection of potential predators (Kahlert 2003). Thus, flocking behaviour may have survival advantages in spite of the potential for increased competition for food and more rapid pathogen transmission.

Behavioural activity also affects group size. Foraging flocks of moulting waterfowl are usually much smaller than roosting flocks which can aggregate thousands of individuals (Reed 1971; Jepsen 1973; Joensen 1973). Roosting in dense flocks likely has important bioenergetic and predator protection advantages. Moulting Surf Scoter *Melanitta perspicillata* (O'Connor 2008) and King Eider *Somateria spectabilis* (Frimer 1994) stop feeding when disturbed and regroup in large flocks offshore, a behaviour common to most moulting sea ducks and divers and some dabbling ducks (Oring 1964). Similarly, resting flocks of flightless Red-breasted Merganser *Mergus serrator* are larger than foraging flocks (JPLS unpublished data). The tendency of flocked birds (diving and sea ducks) to dive synchronously to feed may be an anti-predator or a foraging efficiency strategy (Schenkeveld & Ydenberg 1985). Certainly, simultaneous diving is a typical response to avian predators (PLF pers. obs.). Some species, especially dabbling ducks, rely on cover for protection and disperse in vegetation when disturbed. In some species of sea ducks, females often moult in smaller groups than males and may use different habitats (Gauthier & Bédard 1976; Jepsen 1973; Diéval 2006; Diéval *et al.* 2011). Dabbling ducks seem to rely more on cover during the remigial moult, and unlike some sea ducks, tend to form smaller flocks during remigial moult than before or after this period (Kortegaard 1974).

Selection for escape habitat

Moulting birds may show stronger selection for escape mechanisms as opposed to foraging habitat. For example, most northern geese moult in treeless areas and moulting concentrations can reach thousands of birds, where terrestrially feeding birds rely on adjacent open-water rivers, lakes, and other wetlands for escape from predators (Derksen *et al.* 1982; Madsen & Mortensen 1987). For these arctic geese, the primary predator influencing forage behaviour was likely Arctic Fox *Alopex lagopus*. Fox & Kahlert (2000) and Kahlert (2003) found that moulting Greylag Geese only foraged in close proximity to water (*i.e.* escape habitat) even though abundant unexploited forage of equal quality was available in other locations. Thus, these geese seem to be selecting for open habitats with good visibility but restrict their habitat use based on access to escape habitat from the local predators, although predators actually were absent during the study (Kahlert *et al.* 1996). In contrast, moulting ducks spend most of their feeding and roosting times on water where predators may harass and attack them from the air or under water. Dabbling ducks rely on emergent vegetation for concealment and escape and select lakes and marshes with presence of emergent vegetation (Kortegaard 1974; Fleskes *et al.* 2010). Moulting diving ducks escape danger by diving and dispersing and select wetlands of sufficient depth to avoid aerial predation. Thus, most dabbling ducks select habitats with emergent vegetation to moult whereas moulting diving and sea ducks generally avoid them and favour open areas (Oring 1964). However, predation on flightless Common Eider by Killer Whales has been documented (Smith 2006; Booth & Ellis 2006) indicating bird vulnerability to marine predators. Most sea ducks moulting in coastal waters forage at shallower water depths when flightless, perhaps linked to their impaired diving capacities as most use their wings underwater (Comeau 1923). Selection of shallow water may reduce the amount of time spent foraging and possibly minimise heat loss during submergence in cold waters. Further, use of shallow waters may be a strategy to minimise predation as water depth may limit exposure to some marine mammals (*e.g.* Killer Whales). In the extreme, some moulting Kerguelen Pintail *Anas eatoni* apparently moult in a cave on the Kerguelen archipelago to escape predation (Buffard 1995).

Behavioural modifications

Numerous studies have also documented general behavioural shifts associated with moult that might influence predation risk (Hohman *et al.* 1992). In general, moulting waterfowl reduce active behaviours and spend more time roosting (*e.g.* Döpfner *et al.* 2009; Portugal *et al.* 2011). The degree to which this behavioural modification is adopted is likely linked to energy balance, body mass dynamics and ability of birds to store reserves prior to moult. Nonetheless, by minimising time in potentially less safe habitats (*i.e.* foraging habitats) and selecting relatively safe roosting habitats, moulting birds may be minimising predation risk.

Minimisation of flightless period

Several studies have suggested that body mass dynamics of moulting waterfowl represent an adaptive mechanism to minimise length of the flightless period (Douthwaite 1976; Brown & Saunders 1998; Owen & Ogilvie 1979). The logic is that wing loading ultimately determines ability to fly. Therefore, mass loss during moult reduces wing loading and allows birds to regain flight before primaries are fully grown. For many species, mass loss may allow birds to gain flight when primaries are about 70% of ultimate length (Taylor 1995; Howell 2002; Flint *et al.* 2003; Dickson 2011). If flightlessness itself increases mortality risk, then this adaptation to minimise the flightless period would be an adaptation to reduce mortality. Interestingly, there are some species (*e.g.* scoters) which have protracted flightless periods and corresponding high survival (Anderson *et al.* 2012). These species have slow rates of feather growth and no mass loss while flightless (Dickson *et al.* 2012). As such, there appears to be little selective pressure to minimise the flightless period for scoters because they use habitats with adequate food and encounter little apparent risk of predation, although we cannot ignore the alternative hypothesis that these ducks need full wing length to fly.

Role of food

Is there a high energetic cost of remigial moult?

Replacing feathers is energetically costly, so waterfowl face potentially increased energy demands to meet the costs of feather growth (Payne 1972; Thompson & Boag 1976; Dolnik & Gavrilov 1979; Qian & Xu 1986; Portugal *et al.* 2007), estimated at 1.3 times basal metabolic rate in Mallard (Prince 1979). However, most species moult at the warmest point in the annual cycle and other factors suggest that the costs of feather replacement are not necessarily difficult to meet from external sources. For example, many waterfowl engage in the moult of body feathers synchronously with remiges (Weller 1980; Taylor 1995; Howell *et al.* 2003). Further, many species restrict food intake during moult compared to other times of the year. Finally, while many species lose mass during moult which could indicate an inability to balance energy intake with demand, this is not the case for all species or sites (Lewis *et al.* 2011a,b; Dickson 2011). Accordingly, moult is potentially a time period of energetic constraint, yet it appears that waterfowl have adapted to mitigate this cost.

Mass loss, what does it mean?

There has been debate as to whether mass loss during remigial moult in waterfowl is:

(1) an adaptive trait, whereby fat stores provide an endogenous source of energy to regrow feathers as rapidly as possible whilst reducing reliance on external energy sources, access to which

potentially increases predation risk (as suggested for some geese; Fox & Kahlert 2005);

(2) a simple reflection of the elevated energetic costs of feather synthesis which birds meet by catabolism of body "reserves" (the energetic stress hypothesis; Hohman 1993) rather than body "stores" acquired exogenously prior to the moult (*sensu* van der Meer & Piersma 1994);

(3) due to predation risk, that imposes cryptic feeding behaviour and exploitation of habitats where foraging is less effective, such foraging constraints necessitate exploitation of fat (which does not necessarily preclude pre-moult accumulation of fat stores; Panek & Majewski 1990); or

(4) an adaptive trait to reduce the length of the flightless period because lighter body mass enables Anatidae to regain flight earlier on incompletely re-grown flight feathers earlier than if heavier (Douthwaite 1976; Owen & Ogilvie 1979; Brown & Saunders 1998).

For at least one population of Mallard, Fox *et al.* (2013) showed (using supplementary feeding) that there was no support for (2) and (3) above and that (4) alone was not the primary factor that shaped weight loss. Rather they considered that the accumulation and subsequent depletion of fat stores, together with reductions in energy expenditure, enables Mallard to re-grow feathers as rapidly as possible by exploiting habitats that offer safety from predators, but do not necessarily enable them to balance energy budgets during the flightless period of remigial feather re-growth. In other words, both sexes of Mallard showed prior mass gain (mostly fat stores) to fuel energy demands during wing moult, just as migratory populations accumulate such stores to fuel migration. Male and non-breeding female Mallard could meet up to 82% of all energy expenditure whilst flightless from energy stores alone, and Pochard *Aythya ferina* could derive up to 92% of such energy demands (Fox & King 2011). Fondell *et al.* (2013) also provided evidence against (2) as Black Brant *Branta bernicla nigricans* with access to the most nutritious forage lost the most mass. However, Lewis *et al.* (2011b) emphasised that the adaptive relationship described in (1) is not fixed, as the overall rates of mass loss declined across several decades for moulting Black Brant.

Many species show mass loss during flightlessness (see review in Hohman *et al.* 1992 and references therein), including temperate moulting Greylag Geese. On the Danish island of Saltholm, their mass loss equated to depletion of fat stores accumulated prior to moult and which again could support a large proportion of the energy expenditures during moult if the geese opted not to move between safe resting areas during daylight and their night-time feeding grounds (Kahlert 2006a). Through hyperphagia, male Northern Shoveler *Anas clypeata* accumulated reserves prior to moulting and used stored resources to grow feathers, an adaptation to declining cladoceran availability in mid-summer (DuBowy 1985). However, even within a species, not all populations lose mass at the same rate implying that, while overall mass loss may be adaptive, there is some influence of local feeding conditions (Fox *et al.* 1998; Fox & Kahlert 2005; Fondell *et al.* 2013).

Factors that mitigate or influence energetic costs and mass loss

Meeting protein needs during remigial moult

Flight feathers comprise *c.*0.7% of the total body mass of a Greylag Goose (A.D. Fox, unpubl. data) and 0.2% of a female Mallard (Heitermeyer 1988) and remiges some 25% of feather mass, so while a substantial part of overall plumage, the absolute mass of flight feathers is relatively not that great. However, the simultaneous replacement of the largest feathers of most waterfowl over a relatively short period necessitates access to food that provides the basic nutrients for their synthesis. This includes amino acids containing sulphur for β-keratin synthesis (Hohman *et al.* 1992), which are generally less common in avian tissues and diet than in feathers (Murphy & King 1982, 1984a,b). The extent to which protein invested in feather tissue is derived from exogenous versus endogenous sources in moulting waterfowl remains unclear, but the only study (Fox *et al.* 2009) suggests both sources are used. Recent studies suggest a progressive change in isotopic composition, shifting from largely endogenous sourced protein to protein derived from the diet along the length of the feather as moult migrant geese come into equilibrium with a new isoscape (S. Rohwer pers. comm.).

Studies of the diet of moulting Greylag Geese on Saltholm showed there was selection for the most highly digestible and protein-rich species available; as the quality of this forage declined, the diet became increasingly diverse (Fox *et al.* 1998). However, that same study showed that birds exploited the best quality forage that was closest to open water to which birds could escape when threatened; leaving food resources distant from the water's edge unexploited, strongly suggested that predation risk during flightlessness was more important than simply food supply (Fox & Kahlert 2000, even in this case, where predators are absent, geese still responded vehemently to predator-like stimuli, Kahlert 2003, 2006b). However, in the case of moulting Greylag Geese on Saltholm, stable isotope data evidenced that geese used protein accumulated on the mainland in Sweden for feather synthesis (Fox *et al.* 2009). Proteins were released from organs which change in size during moult (Fox & Kahlert 2005), and excretion of nitrogen in the form of urea and uric acid nearly ceased during the middle part of moult suggesting considerable physiological mechanisms that reduced reliance on external sources of nitrogen during wing moult (Fox & Kahlert 1999).

Meeting lipid needs during moult

Lipids accumulated prior to moult are primarily used to meet elevated energy demands during moult to offset the temporarily increased demands of feather synthesis (Young & Boag 1982; Fox & Kahlert 2005; Fox *et al.* 2008). Many dabbling ducks (*e.g.* DuBowy 1985; Sjöberg 1988; Panek & Majewski 1990; Moorman *et al.* 1993; King & Fox 2012) and diving ducks (*e.g.* Fox & King 2011) lose mass during the flightless moult (*cf.* Fox *et al.* 2008; Dickson 2011; Hogan 2012; Hogan *et al.* 2013b), likely due to consumption of body fat stores. As in the case of Saltholm moulting Greylag Geese (Fox & Kahlert 2005), Folk *et al.* (1966) reported mass loss in moulting Mallard, and Young & Boag (1982) documented a reduction in fat stores in Mallard through moult, although they asserted there was no overall change in total carcass lipids, total proteins, or total body mass. However, Panek & Majewski (1990) showed 12% declines in body mass amongst males and females through moult. Taylor (1993) reported that moulting Black Brant lost 71–88% of stored lipid reserves and ended moult with only structural lipids remaining (*i.e.* 2–4% of fresh body mass).

So why do some Anatids lose mass during moult, while others do not? Not only can using fat stores potentially free moulting waterfowl from feeding or at least as intensively as would otherwise be necessary (*e.g.* Fox & King 2011) but there is also evidence from the difference in energy stores between moulting individuals that energy stores may affect the rate of feather growth. In Barrow's Goldeneye, van der Wetering & Cooke (2000) found that remigial growth rate in recaptured individuals was positively correlated with size-adjusted body mass at initial capture and that the daily rate of body mass loss was greater amongst birds that started moult in better condition. Hence, the lipid status in which an individual starts moult may have considerable implications for rate of feather growth (and hence duration of flightlessness) and may also be a function of the environment in which it moults to meet needs of maintenance and feather replacement.

Other species, such as some arctic moulting geese, unconstrained by feeding restrictions because of diel light conditions during summer (Fox *et al.* 1999), or sea ducks, such as the Common Scoter *Melanitta nigra* that live on protein and energy-rich food and occupy habitat that subsequently is the source of its winter food supply (Fox *et al.* 2008), show no such accumulation of fat stores in advance of wing moult and appear entirely able to supply their energy expenditure from exogenous sources during moult. Indeed, Canvasback *Aythya valisineria* males can actually accumulate mass in the form of lipid stores from exogenous sources toward the end of remige growth in preparation for autumn moult into breeding plumage, presumably because they undertake remigial moult in habitats with high food quality (Thompson & Drobney 1996). Surf and White-winged Scoter *Melanitta deglandi* moulting in coastal British Columbia and Alaska also gained weight during their remigial moult. Barrow's Goldeneye moulting on arctic wetlands lost weight during remigial moult (van de Wetering & Cooke 2000) but those moulting in northern Alberta gained weight (Hogan 2013b) suggesting high variability in moult ecology within species, perhaps linked to the time constraints imposed on birds in these different biogeographic settings.

Furthermore, the evidence presented above suggest Mallard under certain circumstances do not lose mass during moult (*e.g.* Young & Boag 1982), and there is evidence that Greylag Geese in the north of their range in Iceland and northern Norway lose far less mass than do those on Saltholm (Arnor Sigfusson, Carl Mitchell and Arne Follestad, pers. comm.). Rates of mass loss in Black Brant have varied through time such that birds moulting on the same lakes now lose less mass compared to several decades ago, suggesting that this trait is highly adaptive and may depend on local circumstances associated with the specific moult site and individual status (Lewis *et al.* 2011b). Hence, moult mass dynamics appear to vary within species with site and potentially reproductive status.

Data from moulting Black Brant from three different areas (*i.e.* brood-rearing flocks, failed breeding birds on the Yukon Delta, and failed- or non-breeding birds from Teshekpuk Lake) show major differences in mass dynamics before and during moult (Fondell *et al.* 2013, see Table 1 for an overall summary). Subjectively, forage varied across these three groups, with little forage available near the nesting colony for brood rearing birds, intermediate levels of forage available to failed breeding birds on the Yukon Delta, and abundant food available near Teshekpuk Lake for the moult migrant functional non-breeders. However, adults rearing young did not lose mass during moult. Failed breeding birds on the Yukon Delta lost intermediate amounts of weight during moult and those at Teshekpuk Lake lost the most. In these cases, mass loss was negatively correlated with apparent forage availability, yet mass at the onset of moult also varied across these three areas and was linked with the forage available at each area. So successful breeding birds started moult at a relatively low mass and

Table 1. Summary of moult behaviour of discrete elements of the brant population from the Yukon-Kuskokwim Delta in Alaska, showing the location, duration of flightlessness, mass dynamics and food quality of the habitats used for brood-rearing parents, failed- and non-breeding birds (Singer *et al.* 2012; Fondell *et al.* 2013).

Population segment	Moult site location	Flightless period (days)	Feather growth rate (mm/day)	Starting body mass	Food quality	Mass loss during moult
Brood rearing parents	At colony (few km)	30	5	Low	Low	None
Failed breeders	Yukon Delta (many km)	21	7.5	Intermediate	Intermediate	Little
Non-breeders	Teshekpuk (950 km)	21	7.5	High	High	Most

maintained body mass, whereas moult migrants started at a higher mass and lost most of that during moult. Thus, all groups converged on a similar mass at the end of moult. This convergence would suggest that mass loss is adaptive in that there is some optimal mass that reduces wing loading and facilitates flight and dispersal. These data suggest that there would appear to be some selection pressure to minimise the length of the flightless period. It is also interesting that moult migrants start at the highest mass implying that they are more than able to make up for the energetic costs of migration and perhaps bring fat stores as a hedge against uncertain food resources at the ultimate moult destination.

Black Brant rearing broods delay the onset of moult until about 16 days after hatch, thus regaining flight at about the same time as goslings fledge (Singer *et al.* 2012). So for waterfowl that stay with their broods, there may be little selective pressure to reduce the length of the flightless period because there is little advantage to adults flying before their young. Failed- and non-breeding birds utilise available forage to gain mass prior to moult. They can use supplementary body fat stores potentially to invest in more rapid growth of flight feathers (compared to the brood-rearing adults) to allow an early return to flight. However, female ducks raising broods face a trade-off between protecting the brood and departing with sufficient time to complete remigial moult in appropriate habitats.

Forage resources

In previous sections on the energetic dynamics of moult, we described a range of species that showed fat store accumulation prior to moult, where these stores were depleted as feather regrowth proceeded. This strategy may be an adaptation toward energetic independence from exogenous energy-rich foods in situations where foraging brings accompanying predation risks. However, evidence indicated that the rate of depletion of fat stores varied with remigial growth rate and initial mass in some species, suggesting endogenous food stores could accelerate feather growth and reduce duration of flightlessness and predation risk. Hence, the presence of lipid body stores does not necessarily imply that moulting birds have reduced need for exogenous energy, because there are many examples of moulting birds selecting food-rich environments. For instance, at the Ismaninger Teichgebiet, a complex of fish ponds in Bavaria, Germany, where 3 out of 30 impoundments were left fishless, these ponds each attracted, on average, about 2,000 birds, mostly moulting Eared Grebes, Gadwall *Anas strepera*, Mallard, Pochard, Tufted Duck *Aythya fuligula*, Red-crested Pochard *Netta rufina* and Coot *Fulica atra*, compared to <100 moulting waterbirds on the remaining lakes stocked with Carp *Cyprinus carpio* (Köhler & Köhler 1998). The implication of this extraordinary difference in moulting waterbird density was that the abundance of macroinvertebrates and algae in the ponds without carp provided improved feeding conditions over those ponds with carp. Intriguingly, Northern Shoveler was the only species to show similar (but low) moulting densities on stocked and fishless lakes, likely because it is a pelagic dabbling duck specialising on filter feeding of macro- and micro-invertebrates (Ankney & Afton 1988). Amongst Gadwall at this site, both sexes lost relatively little body mass, which they recovered before the end of wing moult (Gehrold & Köhler 2013). This contrasts with the consistent loss of mass amongst both sexes of the same species at a moulting site in southeast England (King & Fox 2012). The unplanned Ismaninger Teichgebiet "experiment" strongly suggested that locally high densities of moulting waterfowl may react to a combination of factors in moulting habitats, but the high density of food in the fishless ponds may have overridden any anti-predator function.

The density of moulting Barrow's Goldeneye on 21 ponds of the Old Crow Flats, Canada, was positively correlated with total phosphorous levels (van de Wetering 1997) suggesting a positive relationship between goldeneye abundance and primary productivity of the wetlands. In northwest North America, large moulting concentrations of waterfowl are associated with highly productive large shallow boreal and sub-boreal lakes (Munro 1941; van de Wetering 1997; Hogan et al. 2011), large shallow boreal (Bailey 1983a,b) and arctic wetlands (King 1963, 1973) and coastal estuaries (Flint et al. 2008), while in Iceland, numbers of moulting Barrow's Goldeneye and Red-breasted Merganser were correlated positively with food abundance (Einarsson & Gardarsson 2004). Loss of the ability to fly not only affects the ability of birds to avoid predation, it also spatially limits the ability to gather food during the flightless period, which likely restricts the ability to exploit the best foraging conditions. Even where this is not the case, the physical constraint on movements likely enhances local depletion of food resources.

Feeding ecology may also impose constraints on the size of moulting flocks. In general, species feeding on shellfish moult in larger flocks than species feeding on mobile invertebrates, fish or vegetation. For shellfish feeding species, foraging in groups is not limited to the moulting season but occurs during non-breeding seasons when birds are gregarious. Common Eider moult in mussel rich habitats, such as in the Danish Wadden Sea and Kattegat where aggregations reach tens of thousands, while scoters moult in areas with large sandy subtidal areas rich in bivalve shellfish resources (Joensen 1973; Laursen et al. 1997). These numbers likely represent abundance at moulting locations as foraging flocks at moulting sites tend to be smaller than those later in the winter (Follestad et al. 1988; Rail & Savard 2003). Mergansers feed on fish and moulting flocks are usually smaller than those of scoters and eiders: Red-breasted Merganser flocks moulting in the coastal waters of Anticosti Island in the St Lawrence, Canada averaged 39 birds/flock with a range of 1–322 birds (Craik et al. 2009, 2011); Common Merganser M. merganser flocks moulting on fresh and salt water appear to be in a similar range of size (Kumari 1979; Pearce et al. 2009). Goldeneyes forage on invertebrates and also moult in smaller groups than scoters or eiders (Jepsen 1973; van de Wetering 1997). Freshwater diving and dabbling ducks that consume aquatic vascular plant material may rely on large beds of such plants to support them through remigial moult, where larger feeding resources attract dense flocks of moulting birds (Bailey 1983a,b).

Flock behaviour for moulting may actually enhance forage quality and quantity via positive feedback. This hypothesis is particularly true for goose grazing systems where regular foraging tends to maintain plant productivity at a higher biomass and prolong the peak of nitrogen content (Cargill & Jefferies 1984). Fox & Kahlert (2000, 2003) demonstrated that moulting Greylag Geese maintained both protein and biomass production at the greatest possible levels by frequent re-grazing of the sward. There are no similar documented positive feedback relationships for other forage systems (*e.g.* invertebrates) that would be relevant for other waterfowl; however, behavioural modifications like simultaneous diving may enhance feeding efficiency. Thus, for some species of waterfowl, flocking behaviour during the flightless period may actually increase forage quantity, quality, foraging efficiency, or a combination of these. Moulting waterfowl therefore seem to concentrate in areas of high forage abundance, hardly surprising as large aggregations of birds would require a concentrated food source, from which they may gain an additional advantage from the decrease in predation risk to individual birds.

Behavioural modifications

One means of rebalancing an energy budget burdened by the additional energy demands of major feather synthesis is to reduce other forms of energy and other nutrient expenditure. Birds replacing flight feathers conserve energy by not flying, which is the most expensive activity of all avian energy expenditure, usually estimated at 12–15 times basal metabolic rate in waterfowl (*e.g.* Prince 1979; Madsen 1985). In moulting Common Eider, Guillemette *et al.* (2007) showed that daily and resting metabolic rates increased by 9 and 12%, but also that flightlessness reduced daily and resting metabolic rate by 6 and 14%, respectively, helping to balance energy demands of feather synthesis. Indeed, many authors concur that moulting waterfowl are much more secretive, but also far more quiescent than in the period prior to and following flightlessness, reducing activity substantially (*e.g.* Adams *et al.* 2000; Döpfner *et al.* 2009) even in captive waterfowl fed *ad libitum* (Portugal *et al.* 2010), although these studies collected activity data only during daylight. Hogan *et al.* (2013a,b) reported that moulting Barrow's Goldeneye foraged primarily at night on one lake and diurnally on another, possibly in relation to prey behaviour and availability suggesting adaptability to local conditions but perhaps also due to greater vulnerability to predation on the smaller lake where they fed nocturnally, although they could not quantify this risk. Hence, by their lack of flight and generally reduced activities, moulting waterfowl can substantially reduce their energy expenditure during flightlessness.

In contrast to the general observations that flightless moulting waterfowl are less active whilst replacing feathers, an alternative strategy would be for birds to feed more or on more energy dense or nutritious foods to meet elevated needs of moult. This hypothesis has rarely been tested because of the paucity of studies in this field, but has certainly been advanced for Mallard by Hartman (1985) and seems to occur in moulting Steller's Eider which feed on more energetically rich prey during the remigial moult (Petersen 1981).

Strategies invoked during remigial moult may also differ between species (Döpfner *et al.* 2009): Mute Swans *Cygnus olor* reduced swimming activities but increased foraging during moult; Red-crested Pochard *Netta rufina* increased locomotion and feeding activities; and Gadwall and Tufted Duck spent less time foraging. However, nocturnal activities were not monitored during this study and it is known that Redhead and Red-crested Pochard feed primarily at night during remigial moult (Bailey 1981; van Impe 1985), which has been suggested to provide an exogenous source of energy to offset thermoregulatory costs during the coolest part of the 24 hour cycle. Future study of the diurnal and nocturnal activities and energy budgets of moulting waterfowl would be extremely valuable to enlightening our understanding of habitat use and selection by moulting waterfowl.

Physiological modifications

An alternative strategy would be for moulting waterfowl to reconstruct body parts to help meet the energetic needs of remigial moult, an aspect of phenotypic plasticity. In this way, organs or muscles that are costly to maintain are reduced in size to minimise energy consumption. Shorebirds are well-known for making radical and rapid adjustments to the digestive apparatus in response to food supply (Piersma *et al.* 1993) and during refuelling episodes on migration (Piersma *et al.* 1999a,b) and grebes adjust organ sizes to meet the energetic costs of migration (Jehl & Henry 2010, 2013) and muscle mass for those that moult (Piersma 1988). Maintenance costs consume the vast majority of normal energetic expenditures of an organism, and most internal organs (but particularly the liver and gastrointestinal tract) are energetically costly to maintain in a larger state than is functionally necessary (Ferrell 1988). Hence, the level of downsizing of organs that occurs in the Greylag Goose during moult on the Danish island of Saltholm is likely associated with energetic savings over and above the alternative explanation under the use-disuse hypothesis (Fox & Kahlert 2005). That study found reduction of 41% in intestine mass and 37% reductions in liver and heart mass during moult, although these were increasingly reconstructed as birds progressed toward completion of moult. In a study of high arctic Brant, Ankney (1984) found no such changes in liver or intestine mass through moult, but this phenomenon may be because arctic geese are more able to meet their energy demands from herbivory during moult than are Greylag Geese moulting farther south. Dramatic changes in digestive organ size have been described in other waterfowl undergoing remigial moult (*e.g.* DuBowy 1985; Thompson & Drobney 1996), but this aspect of energy conservation and the degree to which such plasticity in organs can contribute to reducing energy expenditure has been rarely studied.

There is also evidence of changes in

muscle size during moult. In Mallard (Young & Boag 1982), Long-tailed Duck *Clangula hyemalis* (Howell 2002), Black Brant (Taylor 1993), Northern Shoveler and Blue-winged Teal *Anas discors* (DuBowy 1985) flight muscle mass decreased and leg muscle mass increased consistently with the predictions arising from the hypothesis of phenotypic plasticity (see also Ankney 1979, 1984; Fox & Kahlert 2005). Despite major changes in muscle architecture, these tend not to contribute to major overall changes in body mass during flightless moult in most studied Anatidae (*e.g.* Ankney 1979, 1984; Thompson & Drobney 1996), so such changes are less easy to dissociate from the simple hypothesis of use/disuse.

Competition within and among species

Some moulting sites attract single species (*e.g.* Harlequin Duck or Steller's Eider; Boertmann & Mosbech 2002; P. Flint unpubl. data) whereas others are used by several species of waterfowl including dabbling, diving and even sea ducks. For example Ohtig Lake (5.5 x 2.5 km) in Alaska supports 20,000 moulting ducks of at least 10 species, as does Taksleslúk Lake (19 x 6 km) which attracts 10,000 moulters (King 1963, 1973). When disturbed, birds form mixed species flocks and swim toward natural sanctuary habitats. These lakes obviously fulfil the diverse needs of these species, combining vast areas of shallow open water with dense shoreline cover. Even smaller lakes may attract several species. Moulting Barrow's Goldeneye often form loose groups with moulting Canvasback and scaup *Aythya* sp. when resting and tighter groups if disturbed (van de Wetering 1997). These associations seem based on selection of similar habitat by moulting species rather than attraction to sites used by a given species. However, species likely benefit from each other from protection in numbers. In northern Greenland, Common Eider and King Eider moult at similar locations but use different habitats with King Eider foraging in deeper waters (Frimer 1995). In northern Alaska, four species of geese utilise the large thaw lakes north of Teshekpuk Lake (Flint *et al.* 2008). However, there is some spatial segregation with different species using somewhat different areas. Even in cases where multiple species are moulting within the same watershed, we do not know if they compete for the same forage (Lewis *et al.* 2011b).

Selection of moulting habitats by sea ducks is likely based on food resources and on the presence of congeners. Thousands of eiders, scoters and Long-tailed Duck may use a given moulting site. Approximately 30,000 Long-tailed Duck moult along 531 km of coastline in the Beaufort Sea, dispersed among 73 areas supporting from a few to 2,500 birds, depending on local habitat configuration (*e.g.* presence of islands, sand spits; river deltas; Gollop & Richardson 1974). However, within that site they form several foraging flocks of various sizes. Flint *et al.* (2004) studied radio-tagged Long-tailed Duck moulting in lagoon systems and demonstrated that while flocks were consistently observed in the same locations, there was considerable turnover of individuals within some flocks. As such, aggregations could simply be the result of numerous individuals sampling and selecting habitat rather than benefitting from flocking behaviour. Hagy and Kaminski (2012) report for wintering dabbling ducks that they continue to sample and forage in emergent wetlands in spite of food depletion in the wetlands. Because waterfowl may not be able to assess food abundance without sampling, they continue to forage in patches to assess resource abundance. Forage sampling may be an adaptive strategy for waterfowl as a consequence of temporal and spatial dynamics of wetlands used by Anatids. Finally, almost nothing is known about social intra- and inter-specific interactions during moult, especially of interaction between males and females and between adults and sub-adults.

Habitat selection

Moult migration: selection of moult location at macro-scales

Given available food and safety from predation during the flightless moult period, the default setting for waterfowl ought to be to moult on summer areas, typically within the breeding range, to conserve energy expended in migrating. Yet despite this expectation, many waterfowl show a well-developed moult migration (Salomonsen 1968), and the review by Hohman *et al.* (1992) revealed a wide and bewildering range of moult migration strategies among and within species and populations. In North America, there is a general northward movement into boreal forest and tundra biomes and associated offshore areas, although notable exceptions to northward movements are resident species, female Anatids that successfully rear a brood, and some species such as scoter, King Eider, Northern Pintail, and some *Aythya* species that moult on or near the wintering grounds (Sheaffer *et al.* 2007; Luukkonen *et al.* 2008; Oppel *et al.* 2008). Ducks, geese and swans also aggregate in biologically productive areas in the arctic, such as major river deltas to regrow flight feathers, although such aggregations are also known from temperate regions (*e.g.* the Volga Delta for Northern Pintail; Dobrynina & Kharitonov 2006). Indeed, although great variability exists among Anatid moulting strategies throughout the northern hemisphere, habitat use and selection have evolved to promote individual survival and as rapid as possible growth of remiges to regain flight.

For many sexually immature, unpaired or otherwise non-breeding birds, the spring migration is functionally a moult migration, especially for males, because moulting is the only major annual-cycle physiological process experienced by these non-breeding individuals during spring and summer. In species with delayed maturation, sub-adult females often return to their natal grounds before moult (Eadie & Gauthier 1985; Pearce & Petersen 2009). North American Aythyini pair during spring, but all females probably return to their natal sites whereas males follow mates or use unfamiliar breeding areas hoping to invest in reproduction. Functionally non-breeding sub-adult geese may associate with parents before moving to other areas to moult. For the males of many dabbling and diving duck species, they are able to exploit the abundance of food in the northern spring and summer to undertake a moult migration away from the nesting areas early in the season after having acquired necessary body stores.

Reproductive success has a major influence on the timing of moult migration and moult, so while successful breeding geese moult with their offspring on natal grounds, male and female ducks rarely if ever do. Breeding failure is typically greatest during egg laying, early incubation, and early brood rearing in ducks, so the onset of the "post-breeding" period may be variable, depending on possible re-nesting effort and success and species-, sex-, and age-specific variation. Although individual waterbirds are flightless for 3–7 weeks (Dickson *et al.* 2012), not all individuals moult at the same time (Jepsen 1973; Austin & Fredrickson 1986; Hohman *et al.* 1992), so at some moulting locations flightless birds can be present over a period of 2–3 months (Dickson *et al.* 2012). Amongst ducks, sub-adults and unpaired adult males generally moult first, followed by paired males, unsuccessful breeding females and successful breeding females (Jepsen 1973; Joensen 1973; Savard *et al.* 2007; Hogan 2012). Breeding status also influences the timing and duration of moult as was shown in the example of moulting Black Brant in Alaska above (Table 1).

An early departure from nesting or brooding females, and an ability to acquire body stores and migrate to moulting habitats, enables male ducks to exploit productive habitats (not necessarily suitable for brood rearing) whilst avoiding intra-specific competition from brood rearing females and ducklings. Scoters breed on inland lakes but moult in coastal waters, frequenting habitats similar to those used during winter. Some populations even moult and winter in the same locations (Fox *et al.* 2008) but most do not (Bordage & Savard 1995; Savard *et al.* 1998). Even species that moult on or near breeding areas use different habitats. Moulting scaups and goldeneyes use larger lakes for moulting than during the breeding season. Colonial Common Eider that frequent coastal waters during their entire life cycle segregate brood rearing and moulting habitats (Dieval *et al.* 2011). Mergansers and Harlequin Duck that breed on rivers, typically avoid these habitats for moulting and regroup on larger lakes (Common Merganser) or in coastal waters (mergansers and Harlequins Duck; Pearce *et al.* 2009; Robertson & Goudie 1999; Mallory & Metz 1999). Adult female Common Merganser may moult with offspring in rivers in Alaska, but it is likely a rare event resulting from late nest initiation and success (J. Pearce, *in litt.*).

One of the earliest studies of waterfowl moult migration was by Sven Ekman (1922) who described movement of Lesser White-fronted Geese *Anser erythropus* uphill within the same area to moult. The same phenomenon was evident amongst Greenland White-fronted Geese *Anser albifrons flavirostris* in west Greenland, which exploited lowland wetlands during spring arrival and subsequent breeding period, but successively moved to plateau areas to moult (Fox & Stroud 1981). In both cases, this behaviour likely was a response to the successive delay in growth of plants at a higher altitude, because the Greenland

White-fronted Geese finally fed on the north facing slopes, the site of the most delayed plant growth around the lakes. In the west Greenland study area, adequate water bodies existed at all altitudes (*i.e.* offering safety from predation), but movement of foraging geese uphill suggested that food availability and quality were fundamental in choice of their moult habitat.

For most northern breeding geese, the period of moult occurs simultaneously with brood rearing period. Adequate moulting habitat exists in close proximity to the nesting grounds, yet non-breeding and failed breeding Greenland White-fronted Geese often moult in the same habitats but away from families (Fox & Stroud 1981). When such habitat is limiting within an area of the breeding grounds, food limitation may influence non- or failed-breeders to disperse to alternative moulting areas. Amongst Greenland White-fronted Geese, this circumstance involves the non-breeding or failed nesting birds moving uphill within their mountainous summer range to exploit the delayed thaw and plant growth at higher altitudes, thereby reducing direct competition from breeders and broods (Fox & Stroud 1981). A similar shift but on a far greater spatial scale is also evident amongst the Pink-footed Geese that breed in Iceland, where the non-breeding individuals undertake a moult migration to north eastern Greenland (Taylor 1953, Mitchell *et al.* 1999). In this case, travelling north to exploit the delayed arrival of spring growth, an area where the growing season is too short to support breeding birds. The same species shows an analogous shift in Svalbard, where the delay in thaw is along a west-east axis, with non- and failed breeders travelling to Edgeøya and moulting further east of the core nesting range (Glahder *et al.* 2007).

As migration often has some genetic basis (Berthold *et al.* 2003), the same likely is true for moult migration, albeit mediated by individual reproductive outcome and status. For example, translocated Canada Geese *Branta canadensis* established in Yorkshire, England undertook an apparently innate, northward moult migration to Scotland (Dennis 1964), perhaps as their conspecifics did on the North American continent. However, learning may be important as well, as demonstrated by the selection of new moulting locations. For example, the Yorkshire Canada Geese subsequently started to moult locally (Garnett 1980), as Canada Geese have started to moult in the salt marshes of the St. Lawrence Estuary in recent years (Canadian Wildlife Service unpublished data) and urban sites have been used by moulting Canada Geese for decades creating management problems (Breault & McKelvey 1991; Moser *et al.* 2004). Factors affecting the selection of a moulting location appear complex. Birds often fly thousands of kilometres to distant moulting areas, although they could moult in suitable nearby areas (Brodeur *et al.* 2002; Robert *et al.* 2008, Chubbs *et al.* 2008; Savard & Robert 2013). Some female Common Eider breeding in the Gulf of St. Lawrence moult along Anticosti Island, an important moulting location for about 30,000 eiders, located about 100 km from the colony but others from the same colony moult in Maine about 800 km away. Such discrepancies may reflect reproductive status of the females, or the time at which they abandoned their attempt to breed, not least because the timing of arrival to moult sites may affect the degree to which local density affects the decision to settle or not or could have some genetic basis. Nevertheless, how can these two vastly different strategies have equal fitness benefits, especially given the differential energetic investment needed to move between these different areas?

Most waterfowl exhibit site fidelity to their moulting location and even their moulting site within a location (Szymczak & Rexstad 1991; Bollinger & Derksen 1996; Bowman & Brown 1992; Breault & Savard 1999; Flint *et al.* 2000; Phillips & Powell 2006; Knoche *et al.* 2007), suggesting local knowledge about conditions during flightlessness may be an advantage and therefore also a factor in habitat selection. Equally, the reliability of appropriate conditions, such as food resources and lack of predation, is likely to favour return to specific areas (Salomonsen 1968). However, some individuals are known to change moult location between years. Female breeding philopatry, combined with winter pairing, means that drake waterfowl are likely to find themselves in very different locations annually at the end of egg laying (Peters *et al.* 2012). For example, a drake Northern Pintail might pair with a hen that nests in the prairies in one year and with one that nests in Alaska the next, confronting that individual with radically different moult migration conditions (*e.g.* geographical and nutritional) among years even when reproductive investment does not vary. This possibility is borne out by observation as well, for example, a moulting adult male White-winged Scoter, captured in the St. Lawrence Estuary, moulted on the Labrador coast the following year and in Hudson Bay during the third year, whereas five others returned to their previous moulting area in the St. Lawrence (JPLS, unpubl. data). Likely, age, sex, pairing status, reproductive success and body condition may all interact and affect selection of a moulting location (Jepsen 1973; Petersen 1981). In eastern North America, Barrow's Goldeneye moult in a variety of habitats from inland lakes to estuarine and even marine wetlands (Robert *et al.* 2002; Savard & Robert 2013). An adult female Barrow's Goldeneye moulted one year on an inland lake near James Bay, 930 km from her nesting area and the following year in the St. Lawrence Estuary, only 132 km from her breeding area (Savard & Robert 2013) indicating plasticity in choice of moulting location and habitat.

In many cases, moulting birds occupy habitats that are unsuitable for breeding birds because some critical component is missing (*e.g.* nesting habitat nearby) or season length is inadequate. However, in other cases, the habitat seems suitable, but breeding birds are geographically separated from moulting non-breeders. Generally family groups are behaviourally dominant over non-breeders and arctic nesting ganders, for instance, will aggressively displace non-breeding geese, despite being numerically outnumbered. But growing goslings require an abundance of high quality forage and broods cannot defend their entire home range from competitive foragers. Given that moulting locations are traditionally used and individuals show high site fidelity, non-breeding birds may have displaced breeders from these "optimal" habitats. This hypothesis avoids the "group selection" argument that is required to explain why moulting birds voluntarily abandon breeding habitats, if those breeding habitats are the best available.

Selection of the moult location at the intermediate scale

Given that birds have selected habitat at the large scale (*i.e.* via moult migration or not), they next select habitat at an intermediate scale in terms of actual moulting location. Functionally, this level of choice represents selection of the watershed or wetland complex and is likely to be the unit scale to which an individual shows high levels of inter-annual site fidelity. The scale of this selection is ultimately determined by wetland size, complexity, and continuity (Lewis *et al.* 2011a,b). Multiple studies have shown that individuals have generally high rates of fidelity to specific wetland complexes, implying generally consistent conditions across years; nevertheless, there are several examples where moulting birds have shifted distributions or colonised new areas over time (Flint *et al.* 2008).

Lewis *et al.* (2010a, b) examined pre-moult patterns of movements for Black Brant that had undergone a short distance moult migration. In that study, individual brant used a range of wetlands over a broad area before ultimately selecting a specific moult location. They concluded that patterns of movements were consistent with birds functionally prospecting for moult locations. In most cases, birds visited a range of potential moult locations, before returning to a previously used lake to moult. As such, site selection can have a fidelity component where birds show a preference for locations across years, but by prospecting multiple sites each year, they are able to detect potentially new high quality sites allowing them to adapt to habitat change (Flint *et al.* 2014).

Selection of moult location at the very fine scale

When birds have selected a wetland complex, how they utilise habitat within these areas during the flightless period represents their balance between nutrient acquisition for maintenance and moult and survival. Settlement at a local scale is more likely to reflect annual habitat and ambient conditions. For example hydrology, air and water temperatures, wind exposure, extent of escape cover, and food availability affect habitat selection at the fine scale, which as a result may show lesser levels of inter-annual site fidelity. Fox & Kahlert (2000) showed that food may be broadly distributed within moulting sites, but birds only utilised forage in close proximity to escape habitat. Lewis *et al.* (2011a) did not measure forage availability, but showed that moulting Black Brant used a home range that was a functional strip of foraging habitat along shorelines. Further, Lewis *et al.* (2011a) found a relationship between initial body mass and home range size such that birds with increased body mass had decreased home ranges. This pattern fits the notion that stored reserves are primarily used to minimise activity during the moult. Thus, moulting locations can range from a restricted locality of a few hectares to over hundreds of square kilometres, depending on body condition of moulting birds, the size and dispersion of suitable habitats (Joensen 1973; Gilliland *et al.* 2002), and the scale of resolution (*i.e.* Ungava Bay *vs.* Labrador coast; southern *vs.* eastern coast of Ungava Bay; "inlet A" *vs.* "inlet B"; *etc.*). Harlequin Duck moult along rocky coastlines and feed within a few metres of the tide line (Robertson & Goudie 1999). The configuration of their foraging habitat is basically linear imposing a limit on the sizes of home range and moulting flocks (usually < 50 birds). In contrast, moulting scoters and eiders forage mostly in subtidal zones at depths < 10 m, which vary greatly in area; thus, their home range size is likely determined by bathymetry and moulting flocks can reach thousands of birds.

Effects of disturbance

It has been noted that most species of waterfowl tend to moult in relatively undisturbed locations. While flocks were observed to respond strongly to disturbance stimuli, Lacroix *et al.* (2003) found no clear effect of a localised seismic survey on displacement of moulting Long-tailed Duck. In cases of persistent harassment, such disturbance may lead to drowning, considering flightless birds with incomplete plumage are less efficient divers, have lower thermal efficiency, and probably less buoyant than fully feathered birds. Comeau (1923) reported fishermen harassing flightless birds until they drowned. Derksen *et al.* (1982) reported that moulting geese formed tight flocks and immediately ran to water when disturbed. Further, Madsen (1984) and Derksen *et al.* (1982) noted that relatively low levels of human disturbance could cause moulting geese to abandon a wetland, but uncertainty exists as to what extent such disturbance and displacement might have on rates of site fidelity among years and individual survival. However, intentional regular disturbance precluded birds from moulting on specific wetlands in an urban environment (Castelli & Sleggs 2000). Thus, disturbance may play a role in habitat selection by moulting waterfowl with demonstrated displacement within years and potential for displacement among years. The thresholds and stimuli involved should be the focus of research attention in the future to improve our ability to undertake impact assessments and provide management recommendations.

Adaptation to change

There is little evidence that moulting habitats may be limiting, but habitat loss could potentially result in limits on moulting habitat availability. We have mentioned the expanding populations of geese that have fully occupied the available freshwater lochs as potential moult sites and have commenced moulting in marine waters where they are exposed to novel predation by marine mammals (Glahder *et al.* 2007). This phenomenon suggests exposure to a new source of mortality as a result of intra-specific competition and density dependent processes on land. We know little evidence to support the idea that inter-specific competition could also impose limitations on moulting birds. The only evidence comes from two studies of interactions between goose species moulting in the same arctic areas. The first is the study of Madsen &

Mortensen (1987), which showed that allopatric Barnacle Geese and Pink-footed Geese in Greenland fed on the same graminoid plants, but more moss occurred in the diet under sympatry, especially amongst Barnacle Geese. Barnacle Geese spent more time feeding in sympatry than when feeding alone, while Pink-footed Geese showed no change. The other case is the study of the interactions between the endemic Greenland White-fronted Geese in west Greenland and recently colonised Canada Geese *Branta canadensis interior* to this region. In this case, both species fed on similar plants in allopatry, but where they moulted together, Canada Geese showed no diet shift, whilst White-fronted Geese fed on lower quality forage, such as moss, and tended to feed on the periphery of areas where they would feed in isolation even though little overt aggression was witnessed between the species (Kristiansen & Jarrett 2002). Although such a shift in diet is not evidence of competitive interactions, such a mechanism may have fitness consequences, given that feeding at the study site was restricted to 200 m of the edge of water where food may have become limited because geese were reluctant to forage away from water to which they would retreat when threatened by predators such as Arctic Fox. Other studies of the same species, in extensive wetlands where feeding limitations may not be so manifest, found little evidence for such shifts in diet and no evidence of changes in local abundance of White-fronted Geese in the face of increasing Canada Geese (Levermann & Raudrup undated; Boertmann & Egevang 2002).

Moulting waterfowl are also faced with broadscale ecological changes as dictated by global climate or more localised weather dynamics (*e.g.* Pacific Decadal Oscillation, North Atlantic Oscillation; Schummer *et al.* 2014). Such changes may allow for range expansion or invoke range contraction. For example, distributional shifts of moulting geese along the Arctic Coastal Plain may have been caused by climatic ecological changes (Flint *et al.* 2008; Lewis *et al.* 2011b; Tape *et al.* 2013). Further, recent surveys document a substantial range expansion for moulting Black Brant in this area (Flint *et al.* 2014). Ground observations document an increase in grazing lawns along the coast and experimental manipulations demonstrate that longer, warmer summers likely increase forage plant productivity. Thus, environmental changes appear to have influenced the distribution and abundance of forage plants, and moulting geese have expanded their range accordingly.

Habitat loss and degradation are likely to become an increasing challenge, for example, as a result of loss of boreal forest habitat, changes in precipitation, oil and gas exploration. Yet we know almost nothing about how waterfowl adapt when wetlands and food resources used for moult are no longer available.

Post-moult requirements

We have already discussed the influence of age, sex, body condition, breeding success, and environmental resources on habitat selection during the flightless period. Further, wing moult does not appear to be a period of particularly high mortality, nor does it necessarily expose birds to unusually high energetic constraints. Nonetheless, moulting physiological ecology may have cross-seasonal effects on subsequent annual-cycle events. Here we consider the potential effects of post-moult requirements on habitat selection and body mass dynamics during the moult. Following wing moult, the next critical period in the life cycle of waterfowl is completion of the body moult (Pyke 2005) and fall migration and associated staging. Fifty years ago, Harold Hanson (1962) speculated that "…there could be little doubt that stress of remigial moult is particularly heavy on females, following the energy demands of egg laying and care of young. It seems possible that the apparent differential stress of moult may be a primary reason for the preponderance of males in populations of adult waterfowl". Hanson's general point is likely highly relevant for many female ducks that invest heavily (especially somatic lipids) in a clutch, incubation and rearing offspring, followed by the energetic challenges of remigial moult and hyperphagia in preparation for autumn migration. This has recently been confirmed in Mallard, where radio tracking of brood rearing females showed "…reduced survival of females that raise broods presumably resulted from insufficient time to moult and prepare for fall migration" (Arnold & Howerter 2012). The key point here is likely the "time" element. For Black Brant, brood rearing and failed or non-breeding birds moult on about the same date (9 July) but brood rearing birds have slower feather growth resulting in a longer flightless period (Taylor 1995; Singer et al. 2012). Given that Taylor (1993) demonstrated that non-breeding moulting Black Brant functionally exhausted all available stored lipids, brood rearing birds may not finish moult in worse condition. Thus, brood rearing birds may finish moult in comparable conditions, but at a later date leaving less time to rebuild reserves for migration. Therefore, the primary cost of wing moult for breeding birds is likely related to time between completion of wing moult and onset of fall migration.

Given time constraints, some portion of habitat selection during moult may be dictated by habitat needs immediately following wing moult. Lewis *et al.* (2010a,b) showed that Black Brant departed inland moulting lakes and moved to nearby coastal estuaries upon resumption of flight. Importantly, this shift occurred before flight feathers were fully grown. Thus, apparent habitat suitability changes when birds regain flight. They then stage in these coastal estuaries for several weeks before initiating autumn migration. This phenomenon raises the possibility that habitat selection during the moult was influenced by proximity of suitable staging areas or individual ability to exploit moult habitats that facilitate foraging and body condition to enable flight to more distant post-moult staging areas without fitness costs. Several studies have also demonstrated that while birds lose mass during the moult they begin to gain mass as soon as they regain flight (Brown & Saunders 1998; Howell 2002). The behaviour of birds following wing feather moult seems variable with some moving out of the moulting area (Lewis *et al.* 2010a,b) and others remaining at the same location until autumn migration (Brodeur *et al.* 2002; Robert *et al.* 2002). Some waterfowl also complete body moult at or near their wing moult location taking advantages of productivity of these sites. We know very little of this period of the life cycle of waterfowl. We believe that cross-seasonal effects are important factors in determining habitat use by moulting waterfowl, but we suggest that it is the period immediately following moult that likely has the greatest effect.

Conclusions

Post-breeding waterfowl appear to select moult habitats on the basis of avoiding: (i) predation whilst unable to fly, and (ii) the nutritional stress associated with flight feather replacement (especially amongst brood rearing female ducks). Indeed, habitats will be selected to reduce risk of mortality to an absolute minimum; this outcome is borne out by studies that confirm high survival during moult. Despite the apparent wealth of references cited here, this part of the annual cycle remains poorly studied and even less well understood, particularly with respect to habitat selection. Waterfowl show a remarkable array of structural and metabolic changes associated with moult which must confirm this period as critical to completion of the life cycle, because waterfowl remiges and plumages must be replaced annually. Different waterfowl species adopt a wide range of potential mechanisms to acquire adequate nutrients and energy to survive flight feather moult, presumably as rapidly as possible, without compromise to the quality of the feather structures (although this is not known) that must support the bird throughout the coming year. Increasingly, observations suggest moulting Anatidae reduce their general level of activity, yet whilst many authors speculate on the fitness costs of various behaviours during moult, none have actually shown the magnitude of the relative costs and benefits of these different strategies to survival and fulfilling the life-cycle. If a moulting waterbird drastically reduces feeding activity, what is the cost (in terms of reduced energetic intake) versus the gain (in terms of reduced energy expenditure and elevated survival probability)? With the exception of Kahlert (2006a, which could not address survival probability) there have been few attempts to empirically demonstrate the effects of adopting such strategies.

We should also consider that perhaps a very important factor in choosing a moult site is not merely that the individual emerges from remigial moult with a new set of flight feathers but also attains improved body composition, organ status, and condition for onward movement to the next stage in the annual cycle. We also suggest that researchers reconsider the interpretation of mass loss during moult. Generally, we find support for the hypothesis that mass loss is an adaptation to minimise the length of the flightless period. However, this review also demonstrates that local conditions influence rates of mass loss. Mass loss is variable within and among species, among years, locations, ages, sexes and breeding success. In cases where forage is abundant, mass loss may be behaviourally driven, stimulating birds to forage and complete moult at an increased rate. Conversely, when forage is inadequate, mass loss may be compulsory, but the associated early return to flight allows birds to seek more favourable conditions. Accordingly, we suggest that mass loss is a mechanism used by moulting waterfowl or a consequence of moulting and environmental conditions to adapt to variable constraints. As such, mass loss cannot be used to infer habitat limitation. An especially exciting research prospect would be to use manipulative experiments to see how movement from selected habitats to unoccupied ones might affect mass loss during wing moult. Samples of flightless birds could be captured and moved to similar but unoccupied habitats to compare survival, body mass dynamics, and behaviour with those left to remain at selected habitats. Such approaches might also challenge the notion that social dynamics combined with site fidelity may have reduced the possibilities of colonization of potentially suitable moulting habitats not currently exploited.

Finally, our narrow focus on remigial moult in this review has perhaps had the effect of emphasising loss of flight (and the associated survival risks and foraging challenges) at the expense of nutritional aspects of body moult in the Anatidae generally. Completion of body moult (*i.e.* other than flight feathers) in other regions and other times of year also has important consequences for performance of other annual cycle events (not least for attracting a mate) and therefore fitness and deserves far greater research attention.

Conservation and management implications

What is clear from this synthesis is that there is no "one size fits all". We see different species with contrasting patterns in phenotypic plasticity and the ways they may go about reducing energy expenditure, whilst others (such as non-breeding arctic geese) have little difficulty in balancing energy budgets during the replacement of remigial feathers. Yet even within species, individuals may adopt different strategies at different stages of their own lives, dependent on breeding status, confirming that not all individuals or species have the same goals through to the end of moult. Faced with the challenges of moult, waterfowl may take multiple routes to achieve the same end. An individual selecting between these alternatives is likely to do so according to its internal state and the environmental factors that it encounters, so we need to better understand this part of the process before we can define concrete research/management options for specific species in specific situations.

Faced with such massive variation and uncertainty, but confronted by an urgent need to think about developing conservation policy to address the needs of waterfowl moult, we suggest some priorities for actions in the immediate future. The first is to define moult habitat for each population flyway and identify the larger or most sensitive concentrations (but beware the likelihood of rapid changes in feeding ecology and use of moulting resorts, *e.g.* Nilsson *et al.* 2001). Such a simple inventory exercise, to know "how many and where", would provide a framework for establishing site-safeguard networks and potentially identify critical habitat types to ensure these figure prominently in the design of site safeguard networks as well as in land-use planning and environmental impact assessment where these habitats are threatened. The second priority is to attempt to define the moult strategies for each population, for example determining if the various subsets of a species are moult migrants, depend upon fat stores for energy supply or on local resources. Particular focus should be placed upon breeding females, because of the nutritional and energetic stresses on them following incubation and brood rearing, and the specific implications of their moult patterns for habitat protection, protection from disturbance and hunting. This will affect whether site safeguard and/or protection from disturbance is more important to be implemented on the moulting grounds or on the post-breeding pre-moulting habitats where fat stores are accumulated. It is also important to: 1) assess the degree of wetland loss and global change (including climate) that affect moult behaviour and distribution, 2) determine the effects of internal state specificity on the strategies adopted by different individuals at different stages of the life cycle, 3) design site safeguard and habitat management programmes to optimise the availability and suitability of moult habitat along the flyways of the northern hemisphere but that can also accommodate change, and 4) define relationships between breeding, moult and wintering locations (pooling present and future knowledge from satellite telemetry studies for example).

Management implications and policy needs

Moulting waterfowl are sensitive to disturbance and therefore show enhanced susceptibility to human activities that increasingly encroach on moulting habitats. However, we still know very little about threshold levels that can effect change, such as shifts in moulting location or strategies between seasons as a result of disturbance. We know that this happens, because there are documented cases of Greylag Goose moulting grounds being completely abandoned from one year to another (*e.g.* Nilsson *et al.* 2001). These aspects need to be the subject of much greater study to establish management prescriptions for more effective management of moulting sites where human disturbance is a factor in affecting carrying capacity. There are some indications that flightless sea ducks are extremely vulnerable to disturbance, especially because of their reduced diving capacities (Comeau 1923), thus moulting sites may need special consideration. The exploitation of moulting flocks for subsistence occurred in the north (King 1973) but its current level is unknown as is its impact at the population level. Little work has been done on the carrying capacity of moulting locations and its annual variability, yet such knowledge is essential if we are to be able to provide management advice in generic or specific case of conflict. Gill net fisheries at important moult sites should be managed to avoid by-catch casualties. Also contingency plans are needed at all important moulting locations to minimise the impact of possible oil spills. Northern moulting locations both inland and coastal have not all been identified. As northern development is likely to increase as a result of climate warming it is urgent to identify all sites of particular significance to flyway populations. In particular, there are increasing pressures on boreal forest, near-shore, and arctic biomes posed by mining, oil and gas exploration and exploitation, agriculture, timber harvest, direct and indirect (*e.g.* transport, contaminants) impacts of oil, disease and contaminants, and the general increase in accessibility to these areas that brings associated levels of disturbance and recreational pressures.

Acknowledgements

We are deeply grateful to a great many colleagues and contacts for their contributions to discussions about Anatidae moult throughout our careers that have contributed to this review and we especially acknowledge the contributions of Hugh Boyd, Fred Cooke, Johnny Kahlert, Roy King, Myrfyn Owen, Steve Portugal and Sievert Rohwer. Thanks to Rick Kaminski for the invitation to contribute to the ECNAW conference in Memphis, for engineering the production of proceedings from that meeting, encouraging our contribution and contributing significantly to the quality of this review. Our grateful thanks go to John Pearce and Jane Austin as well as the editor and referees Johnny Kahlert and Matthieu Guillemain for improving earlier drafts.

References

Adams, P.A., Robertson, G.J. & Jones, I.L. 2000. Time-activity budgets of Harlequin Ducks molting in the Gannet Islands, Labrador. *Condor* 102: 703–708.

Anderson, E.M., Esler, D., Boyd, W.S., Evenson, J.R., Nysewander, D.R., Ward, D.H., Dickson, R.D., Uher-koch, B.D., van Stratt, C.S. & Hupp, J.W. 2012. Predation rates, timing, and predator composition for scoters (*Melanitta* spp.) in marine habitats. *Canadian Journal of Zoology* 90: 42–50.

Ankney, C.D. 1979. Does the wing moult cause nutritional stress in lesser snow geese? *Auk* 96: 68–72.

Ankney, C.D. 1984. Nutrient reserve dynamics of breeding and moulting brant. *Auk* 101: 361–370.

Ankney, C.D. & Afton, A.D. 1988. Bioenergetics of breeding Northern Shovelers: diet, nutrient reserves, clutch size and incubation. *Condor* 90: 459–472.

Arnold, T.W. & Howerter, D.W. 2012. Effects of radio transmitters and breeding effort on harvest and survival rates of female mallards. *Wildlife Society Bulletin* 36: 286–290.

Austin, J.E. & Fredrickson, L.H. 1986. Molt of female Lesser Scaup immediately following breeding. *Auk* 103: 293–298.

Bailey, R.O. 1981. *The postbreeding ecology of the Redhead Duck (Aythya americana) on Long Island Bay, Lake Winnipegosis, Manitoba*. Ph.D. thesis, McGill University, Montréal, Canada.

Bailey, R.O. 1983a. Distribution of postbreeding diving ducks (Aythyini and Mergini) on southern boreal lakes in Manitoba. *Canadian Wildlife Service Progress Notes* 136: 1–8.

Bailey, R.O. 1983b. Use of southern boreal lakes by moulting and staging diving ducks. *In* H. Boyd (ed.), *Proceeding of the First Western Hemisphere Waterfowl and Waterbird Symposium*, pp. 54–59. International Wetlands and Waterbird Research Bureau, Slimbridge, UK.

Bellrose, F.C. 1980. *Ducks, Geese and Swans of North America*. Stackpole Books, Harrisburg, USA.

Bergman, G. 1982. Why are the wings of *Larus f. fuscus* so dark? *Ornis Fennica* 59: 77–83.

Berthold, P., Gwinner, E. & Sonnenschein, E. (eds.). 2003. *Avian Migration*. Springer-Verlag, Berlin, Germany.

Bielefeld, R.R. & Cox, R.R. 2006. Survival and cause-specific mortality of adult female Mottled Ducks in East-central Florida. *Wilson Society Bulletin* 34: 388–394.

Boertmann, D. & Egevang, C. 2002. Canada Geese *Branta canadensis* in West Greenland: in conflict with Greenland White-fronted Geese *Anser albifrons flavirostris*? *Ardea* 90: 335–336.

Boertmann, D. & Mosbech, A. 2002. Moulting Harlequin Ducks in Greenland. *Waterbirds* 25: 326– 332.

Bollinger, K.S. & Derksen, D.V. 1996. Demographic characteristics of moulting Black Brant near Teshekpuk Lake, Alaska. *Journal of Field Ornithology* 67: 141–158.

Booth, C. J. & Ellis, P. 2006. Common Eiders and Common Guillemots taken by killer whales. *British Birds* 99: 533–535.

Bordage, D. & J.-P.L. Savard. 1995. Black scoter (*Melanitta nigra*). *In* A. Poole & F. Gill (eds.), *The Birds of North America*, No. 177. The Academy of Natural Sciences, Philadelphia and the American Ornithologists Union, Washington DC, USA.

Bowman, T.D. 1987. *Ecology of male Black Ducks moulting in Labrador*. M.Sc. thesis, University of Maine, Orono, Maine, USA.

Bowman, T.D. & Brown, P.W. 1992. Site fidelity of Black Ducks to a moulting area in Labrador. *Journal of Field Ornithology* 63: 32–34.

Breault, A. & McKelvey, R.W. 1991. *Canada Geese in the Fraser Valley: A Problem Analysis*. Technical Report Series No. 133. Canadian Wildlife Service, Pacific and Yukon Region, Vancouver, Canada.

Breault, A.M. & Savard, J.-P.L. 1999. Philopatry of Harlequin Ducks moulting in southern British Columbia. *In* R.I. Goudie, M.R. Petersen & G.J. Robertson (eds.), *Behaviour and Ecology of Sea Ducks*, pp. 41–44. Canadian Wildlife Service Occasional Paper No. 100. Canadian Wildlife Service, Ottawa, Canada.

Bridge, E.S. 2007. Influence of morphology and behavior on wing-moult strategies in seabirds. *Marine Ornithology* 34: 7–19.

Brodeur, S., Savard, J.-P.L., Robert, M., Laporte, P. Lamothe, P., Titman, R.D., Marchand, S., Gilliland, S. & Fitzgerald, G. 2002. Harlequin Duck *Histrionicus histrionicus* population structure in eastern Nearctic. *Journal of Avian Biology* 33: 127–137.

Brown, R.E. & Saunders, D.K. 1998. Regulated changes in body mass and muscle mass in moulting Blue-winged Teal for an early return to moult. *Canadian Journal of Zoology* 76: 26–32.

Buffard, E. 1995. Anti-predator behaviour of flightless Kerguelen Pintail *Anas eatoni* moulting in a cave on the Kerguelen archipelago. *Wildfowl* 46: 66–68.

Cargill, S.M. & Jefferies, R.L. 1984. The effects of grazing by Lesser Snow Geese on the vegetation of a sub-arctic salt marsh. *Journal of Applied Ecology* 21: 669–686.

Castelli, P.M., & Sleggs, S.E. 2000. Efficacy of border collies to control nuisance Canada Geese. *Wildlife Society Bulletin* 28: 385–392.

Chubbs, T.E., Trimper, P.G., Humphries, G.W., Thomas, P.W., Elson, L.T. & Laing, D.K. 2008. Tracking seasonal movements of adult male Harlequin Ducks from central Labrador using satellite telemetry. *Waterbirds* 31(Special Publication 2): 173–182.

Comeau, N.A. 1923. Notes on the diving of loons and ducks. *Auk* 40: 525.

Craik, S.R., Savard, J. P.L. & Titman, R.D. 2009. Wing and body molts of male Red-breasted Mergansers in the Gulf of St. Lawrence, Canada. *Condor* 111: 71–80.

Craik, S.R., Savard, J.-P.L. Richardson, M.J. & Titman, R.D. 2011. Foraging ecology of flightless male red-breasted Mergansers in the Gulf of St. Lawrence, Canada. *Waterbirds* 34: 280–288.

Davis, J.B., Kaminski, R.M., Leopold, B.D. & Cox, R.R., Jr. 2001. Survival of female Wood

Ducks during brood rearing in Alabama and Mississippi. *Journal of Wildlife Management* 65: 738–744.

Davis, J.B., Cox, R.R., Jr., Kaminski, R.M. & Leopold, B.D. 2007. Survival of Wood Duck ducklings and broods in Mississippi and Alabama. *Journal of Wildlife Management* 71: 507–517.

Dennis, R. 1964. Capture of moulting Canada Geese in the Beauly Firth. *Wildfowl Trust Annual Report* 15: 71–74.

Derksen, D.V., Eldridge, W.D. & Weller, M.V. 1982. Habitat ecology of Pacific Black Brant and other geese moulting near Teshekpuk Lake, Alaska. *Wildfowl* 33: 39–57.

Dickson, R.D. 2011. Postbreeding ecology of White-winged Scoters (*Melanitta fusca*) and Surf Scoters (*M. perspicillata*) in western North America: wing moult phenology, body mass dynamics and foraging behaviour. M.Sc. thesis, Simon Fraser University, Vancouver, Canada.

Dickson, R.D., Esler, D., Hupp, J.W., Anderson, E.M., Evenson, J.R. & Barrett. J. 2012. Phenology and duration of remigial moult in Surf Scoters (*Melanitta perspicillata*) and White-winged Scoters (*Melanitta fusca*) on the Pacific coast of North America. *Canadian Journal of Zoology* 90: 932–944.

Diéval, H. 2006. Répartition de l'Eider à duvet pendant les périodes d'élevage des jeunes et de mue des adultes le long du fleuve Saint-Laurent. M.Sc. thesis, University of Québec in Montréal, Montréal, Canada.

Diéval, H., Giroux, J.-F. & Savard, J.-P.L. 2011. Distribution of common eiders *Somateria mollissima* during the brood-rearing and moulting periods in the St. Lawrence Estuary. *Wildlife Biology* 17: 124–134.

Dobrynina, I.N. & Kharitonov, S.P. 2006. The Russian waterbird migration atlas: temporal variation in migration routes, pp. 582–589. *In* G.C. Boere, C.A. Galbraith & D.A. Stroud (eds.), *Waterbirds Around the World*. The Stationery Office, Edinburgh, UK.

Dolnik, V.R. & Gavrilov, V.M. 1979. Bioenergetics of moult in the chaffinch (*Fringilla coelebs*). *Auk* 96: 253–264.

Döpfner, M., Quillfeldt, P. & Bauer, H.-G. 2009. Changes in behavioral time allocation of waterbirds in wing-moult at Lake Constance. *Waterbirds* 32: 559–571.

Douthwaite, R.J. 1976. Weight changes and wing moult in the Red-billed Teal. *Wildfowl* 27: 123–27.

DuBowy, P.J. 1985. Seasonal organ dynamics in post-breeding male Blue-winged Teal and Northern Shovelers. *Comparative Biochemistry and Physiology* 82A: 899–906.

Eadie, J.M. & Gauthier, G. 1985. Prospecting for nest sites by cavity-nesting ducks of the Genus *Bucephala*. *Condor* 87: 528–534.

Einarsson, A. & Gardarsson, A. 2004. Moulting diving ducks and their food supply. *Aquatic Ecology* 38: 297–307.

Ekman, S. 1922. *Djürvärldens Utbredningshistoria på den Skandinaviska halvön*. Bonnier, Stockholm, Sweden.

Elgar, M.A. 1989. Predation, vigilance and group size in mammals and birds: a critical review of the empirical evidence. *Biological Reviews* 64: 13–33.

Evelsizer, D.D., Clark, R.G. & Bollinger, T.K. 2010. Relationships between local carcass density and risk of mortality in molting mallards during avian botulism outbreaks. *Journal of Wildlife Diseases* 46: 507–513.

Ferrell, C.L. 1988. Contribution of visceral organs to animal energy expenditures. *Journal of Animal Science* 66: 23–34.

Fleskes, J.P., Mauser, D.M., Yee, J.L., Blehert, D.S. & Yarris, G.S. 2010. Flightless and post-moult survival and movements of female Mallards moulting in Klamath Basin. *Waterbirds* 33: 208–220.

Flint, P.L., Petersen, M.R., Dau, C.P., Hines, J.E. & Nichols, J.D. 2000. Annual survival and site fidelity of Steller's eiders molting along the Alaska Peninsula. *Journal of Wildlife Management* 64: 261–268.

Flint, P.L., Reed, J.A., Franson, J.C., Hollmén, T.E., Grand, J.B., Howell, M.D., Lanctot, R.B., Lacroix, D.L. & Dau, C.P. 2003. *Monitoring Beaufort Sea Waterfowl and Marine Birds*. U.S. Geological Survey Report, OCS Study MMS 2003–037. USGS, Alaska Science Center, Anchorage, Alaska, USA.

Flint, P.L., Lacroix, D.L., Reed, J.A. & Lanctot, R.B. 2004. Movements of flightless Long-tailed Ducks during wing molt. *Waterbirds* 27: 35–40.

Flint, P.L., Mallek, E., King, R.J., Schmutz, J.A., Bollinger, K.S. & Derksen, D.V. 2008. Changes in abundance and spatial distribution of geese molting near Teshekpuk Lake, Alaska: interspecific competition or ecological change? *Polar Biology* 31: 549–556.

Flint, P.L., Meixell, B.W. & Mallek, E.J. 2014. High fidelity does not preclude colonization: range expansion of molting Black Brant on the Arctic coast of Alaska. *Journal of Field Ornithology* 85: 75–83.

Folk, C., Hudec, K. & Toufar, J. 1966. The weight of the mallard, *Anas platyrhynchos*, and its changes in the course of the year. *Zoologicke Listy* 15: 249–250.

Follestad, A., Larsen, B.H., Nygard, T. & Rov, N. 1988. Estimating numbers of moulting eiders *Somateria mollissima* with different flock size and flock structure. *Fauna Norvegica Series C, Cinclus* 11: 97–99.

Fondell, T.F., Flint, P.L., Schmutz, J.A., Schamber, J.L. & Nicolai, C.A. 2013. Variation in body mass dynamics among moult sites in Brant Geese *Branta bernicla nigricans* supports adaptivity of mass loss during moult. *Ibis* 155: 593–604.

Fox, A.D. & Kahlert, J. 1999. Adjustments to nitrogen metabolism during wing moult in greylag geese. *Functional Ecology* 13: 661–669.

Fox, A.D. & Kahlert, J. 2000. Do moulting Greylag Geese *Anser anser* forage in proximity to water in response to food availability and/or quality? *Bird Study* 47: 266–274.

Fox, A.D. & Kahlert, J. 2003. Repeated grazing of a salt marsh grass by moulting greylag geese *Anser anser* – does sequential harvesting optimise biomass or protein gain? *Journal of Avian Biology* 34: 89–96.

Fox, A.D. & Kahlert, J. 2005. Changes in body organ size during wing moult in non-breeding Greylag Geese on Saltholm, Denmark. *Journal of Avian Biology* 36: 538–548.

Fox, A.D. & King, R. 2011. Body mass loss amongst moulting Pochard *Aythya ferina* and Tufted Duck *A. fuligula* at Abberton Reservoir, South-east England. *Journal of Ornithology* 152: 727–732.

Fox, A.D. & Stroud, D.A. 1981. *Report of the 1979 Greenland White-fronted Goose Study Expedition to Eqalungmiut Nunaat, West Greenland*. Greenland White-fronted Goose Study (GWGS), Aberystwyth, UK.

Fox, A.D., Kahlert, J. & Ettrup, H. 1998. Diet and habitat use of moulting Greylag Geese *Anser anser* on the Danish island of Saltholm. *Ibis* 140: 676–683.

Fox, A.D., Kahlert, J., Walsh, A.J., Stroud, D.A., Mitchell, C., Kristiansen, J.N. & Hansen, E.B. 1999. Patterns of body mass change during moult in three different goose populations. *Wildfowl* 49: 45–56.

Fox, A.D., Hartmann, P. & Petersen, I.K. 2008. Body mass and organ size change during wing moult in common scoter. *Journal of Avian Biology* 39: 35–40.

Fox, A.D., Hobson, K.E. & Kahlert, J. 2009. Isotopic evidence for endogenous protein contributions to Greylag Goose *Anser anser* flight feathers. *Journal of Avian Biology.* 40: 108–112.

Fox, A.D., Fox, G.F., Liaklev, A. & Gerhardsson, N. 2010. Predation of flightless pink-footed geese *Anser brachyrhynchus* by Atlantic walruses *Odobenus rosmarus rosmarus* in southern Edgeøya, Svalbard. *Polar Research* 29: 455–457.

Fox, A.D., King, R. & Owen, M. 2013. Wing moult and mass change in free-living Mallard *Anas platyrhynchos. Journal of Avian Biology* 44: 1–8.

Frimer, O. 1994. The behaviour of moulting King Eiders *Somateria spectabilis. Wildfowl* 45: 176–187.

Frimer, O. 1995. Comparative behaviour of sympatric moulting populations of Common Eider *Somateria mollissima* and King Eider *Somateria spectabilis* in central West Greenland. *Wildfowl* 46: 129–139.

Garnett, M.G.H. 1980. Moorland breeding and moulting of Canada Geese in Yorkshire. *Bird Study* 27: 219–226.

Gauthier, J. & Bédard, J. 1976. Les déplacements de l'Eider commun (*Somateria mollissima*) dans l'estuaire du Saint-Laurent. *Canadian Naturalist* 103: 261–283.

Gehrold, A. & Köhler, P. 2013. Wing-moulting waterbirds maintain body condition under good environmental conditions: a case study of Gadwalls (*Anas strepera*). *Journal of Ornithology* 154: 783–793.

Gilliland, S.G., Robertson, G.J., Robert, M., Savard, J.-P.L., Amirault, D., Laporte, P. & Lamothe, P. 2002. Abundance and distribution of Harlequin Ducks moulting in eastern Canada. *Waterbirds* 25: 333–339.

Glahder, C.M., Fox, A.D., O'Connell, M., Jespersen, M. & Madsen, J. 2007. Eastward moult migration on non breeding Svalbard Pink-footed Geese (*Anser brachyrhynchus*) in Svalbard. *Polar Research* 26: 31–36.

Gollop, M.A. & Richardson, W.J. 1974. Inventory and habitat evaluation of bird breeding and moulting areas along the Beaufort Sea coast from Prudhoe Bay, Alaska to Shingle Point, Yukon Territory, July, 1973. *Arctic Gas Biological Report*, Serial 26(1): 1–61.

Guillemette, M., Pelletier, D., Grandbois, J.M. & Butler, P.J. 2007. Flightlessness and the energetic cost of wing moult in a large sea duck. *Ecology* 88: 2936–2945.

Hagy, H.M. & Kaminski, R.M. 2012. Winter waterbird and food dynamics in autumn-managed moist-soil wetlands in the Mississippi Alluvial Valley. *Wildlife Society Bulletin* 36: 512–523.

Hanson, H.C. 1962. The dynamics of condition factors in Canada geese and their relation to seasonal stresses. *Arctic Institute of North America Technical Paper* 12: 1–68.

Hartman, G. 1985. Foods of male Mallard before and during moult, as determined by faecal analysis. *Wildfowl* 36: 65–71.

Heitermeyer, M.E. 1988. Protein costs of the prebasic moult of female Mallards. *Condor* 90: 263–266.

Henny, C.J., Rudis, D.D., Roffe, T.J. & Robinson-Wilson, E. 1995. Contaminants and sea ducks in Alaska and the circumpolar region. *Environmental Health Perspectives* 103 (Suppl. No. 4): 41–49.

Hogan, D. 2012. *Postbreeding ecology of Barrow's goldeneyes in northwestern Alberta.* M.Sc. thesis, Simon Fraser University, Burnaby, Canada.

Hogan, D., Tompson, J.E., Esler, D. & Boyd, W.S. 2011. Discovery of important postbreeding sites for Barrow's Goldeneye in the boreal transition zone of Alberta. *Waterbirds* 34: 261–388.

Hogan, D., Tompson, J.E. & Esler, D. 2013a. Survival of Barrow's Goldeneyes during remigial molt and fall staging. *Journal of Wildlife Management* 77: 701–706.

Hogan, D., Tompson, J.E. & Esler, D. 2013b. Variation in body mass and foraging effort of

Barrow's Goldeneyes (*Bucephala islandica*) during remigial molt. *Auk* 130: 313–322.

Hohman, W.L. 1993. Body composition dynamics of ruddy ducks during wing moult. *Canadian Journal of Zoology* 71: 2224–2228.

Hohman, W.L., Ankney, C.D. & Gordon, D.H. 1992. Ecology and management of postbreeding waterfowl. *In* B.D.J. Batt, A.D. Afton, M.G. Anderson, C.D. Ankney, D.H. Johnson, J.A. Kadlec & G.L. Krapu (eds.), *Ecology and Management of Breeding Waterfowl*, pp. 128–189. University of Minnesota Press, Minneapolis, USA.

Hollmén, T.E., Franson, J.C., Flint, P.L., Grand, J.B., Lanctot, R.B., Docherty, D.E. & Wilson, H.M. 2003. An adenovirus linked to mortality and disease in Long-tailed Ducks (*Clangula hyemalis*) in Alaska. *Avian Diseases* 47: 1434–1440.

Howell, M.D. 2002. *Molt dynamics of male long-tailed ducks on the Beaufort Sea*. M.Sc. thesis, Auburn University, Alabama, USA.

Howell, M.D., Grand, J.B. & Flint, P.L. 2003. Body molt of male Long-tailed Ducks in the near-shore waters of the North Slope, Alaska. *Wilson Bulletin* 115: 170–175.

Iverson, S.A. & Esler, D. 2007. Survival of female Harlequin Ducks during wing moult. *Journal of Wildlife Management* 71: 1220–1224.

Jehl, J.R. 1990. Aspects of the moult migration. *In* E. Gwinner (ed.), *Bird Migration: Physiology and Ecophysiology*, pp. 102–113. Springer, Berlin, Germany.

Jehl, J.R. & Henry, A.E. 2010. The postbreeding migration of Eared Grebes. *Wilson Journal of Ornithology* 122: 217–227.

Jehl, J.R. & Henry, A.E. 2013. Intra-organ flexibility in the eared grebe *Podiceps nigricollis* stomach: A spandrel in the belly. *Journal of Avian Biology* 44: 97–101.

Jepsen, P.U. 1973. Studies of the moult migration and wing-feather moult of the Goldeneye (*Bucephala clangula*) in Denmark. *Danish Review of Game Biology* 8(6): 1–23.

Joensen, A.H. 1973. Moult migration and wing-feather moult of seaducks in Denmark. *Danish review of Game Biology* 8(4): 1–42.

Kahlert, J. 2003. The constraint on habitat use in wing-moulting Greylag Geese *Anser anser* caused by anti-predator displacements. *Ibis* 145: E45–E52.

Kahlert, J. 2006a. Effects of feeding patterns on body mass loss in moulting Greylag Geese *Anser anser*. *Bird Study* 53: 20–31.

Kahlert, J. 2006b. Factors affecting escape behaviour in moulting Greylag Geese *Anser anser*. *Journal of Ornithology* 149: 567–577.

Kahlert, J., Fox, A.D. & Ettrup, H. 1996. Nocturnal feeding in moulting Greylag Geese *Anser anser* – an anti-predator response? *Ardea* 84: 15–22.

Kaminski, R.M. & Weller, M.W. 1992. Breeding habitats of Nearctic waterfowl. *In* B.D.J. Batt, A.D. Afton, M.G. Anderson, C.D. Ankney, D.H. Johnson, J.A. Kadlec & G.L. Krapu (eds.), *Ecology and Management of Breeding Waterfowl*, pp. 568–589. University of Minnesota Press, Minneapolis, USA.

Kaminski, R.M. & Elmberg, J. 2014. An introduction to habitat use and selection by waterfowl in the northern hemisphere. *Wildfowl* (Special Issue No. 4): 9–16.

King, J.G. 1963. Duck banding in arctic Alaska. *Journal of Wildlife Management* 27: 356–362.

King, J.G. 1973. A cosmopolitan duck moulting resort; Taksleluk Lake Alaska. *Wildfowl* 24: 103–109.

King, R. & Fox, A.D. 2012. Moulting mass dynamics of female Gadwall *Anas strepera* and male Wigeon *A. penelope* at Abberton Reservoir, South East England. *Bird Study* 59: 252–254.

Kirby, R.E. & Cowardin, L.M. 1986. Spring and summer survival of female mallards from north central Minnesota. *Journal of Wildlife Management* 50: 38–43.

Knoche, M.J., Powell, A.N., Quakenbush, L.T., Wooller, M.J. & Phillips, L.M. 2007. Further evidence for site fidelity to wing molt locations by King Eiders: Intergrating stable isotope analyses and satellite telemetry. *Waterbirds* 30: 52–57.

Köhler, P. &. Köhler, U. 1998. Considerable increase of moulting waterfowl in fishponds without carp in the Ismaninger Teichgebiet (Bavaria, Germany). *Sylvia* 34: 27–32.

Kortegaard, L. 1974. An ecological outline of a moulting area of teal, Vejlerne, Denmark. *Wildfowl* 25: 134–142.

Kristiansen, J.N. & Jarrett, N.S. 2002. Inter-specific competition between Greenland White-fronted Geese *Anser albifrons flavirostris* and Canada Geese *Branta canadensis interior* moulting in West Greenland. *Ardea* 90: 1–13.

Kumari, E. 1979. Moult and moult migration of waterfowl in Estonia. *Wildfowl* 30: 90–98.

Lacroix, D.L., Lanctot, R.B., Reed, J.A. & McDonald, T.L. 2003. Effect of underwater seismic surveys on molting male Long-tailed Ducks in the Beaufort Sea, Alaska. *Canadian Journal of Zoology* 81: 1862–1875.

Laursen, K., Pihl, S., Durinck, J., Hanse, M., Skov, H., Frikke, J. & Danielsen, F. 1997. Numbers and distribution of waterbirds in Denmark. *Danish Review of Game Biology* 15(1): 1–181.

Leafloor, J.O., Ankney, C.D. & Risi, K.W. 1996. Social enhancement of wing moult in females Mallards. *Canadian Journal of Zoology* 74: 1376–1378.

Levermann, N. & Raundrup, K. undated. *Do the Greenland white-fronted geese stand a chance against the invasive Canadians?* Unpublished report, Biologisk Institut, Copenhagen University, Copenhagen, Denmark. [In Danish.]

Lewis, T.L., Flint, P.L., Schmutz, J.A. & Derksen, D.V. 2010a. Temporal and spatial shifts in habitat use by Black Brant immediately following flightless molt. *Wilson Journal of Ornithology.* 122: 484–493.

Lewis, T.L., Flint, P.L., Schmutz, J.A. & Derksen, D.V. 2010b. Pre-moult patterns of habitat use and moult site selection by Brent Geese *Branta bernicla nigricans*: Individuals prospect for moult sites. *Ibis* 152: 556–568.

Lewis, T.L., Flint, P.L., Derksen, D.V. & Schmutz, J.A. 2011a. Fine scale movements and habitat use of Black Brant during the flightless wing molt in Arctic Alaska. *Waterbirds* 34: 177–185.

Lewis, T.L., Flint, P.L., Derksen, D.V., Schmutz, J.A., Taylor, E.J. & Bollinger, K.S. 2011b. Using body mass dynamics to explain long-term shifts in habitat use of arctic molting geese: evidence for ecological change. *Polar Biology* 34: 1751–1762.

Luukkonen, D.R., Prince, H.H. & Mykut, R.C. 2008. Movements and survival of molt migrant Canada Geese from southern Michigan. *Journal of Wildlife Management* 72: 449–462.

Madsen, J. 1984. Study of the possible impact of oil exploration on goose populations in Jameson Land, East Greenland. A progress report. *Norsk Polarinstitutt Skrifter* 181: 141–151.

Madsen, J. 1985. Relations between change in spring habitat selection and daily energetics of Pink-footed Geese *Anser brachyrhynchus*. *Ornis Scandinavica* 16: 222–228.

Madsen, J. & Mortensen, C.E. 1987. Habitat exploitation and interspecific competition in east Greenland. *Ibis* 129: 25–44.

Mallory, M. & Metz, K. 1999. Common Merganser (*Mergus merganser*). *In* A. Poole & F. Gill (eds.), *The Birds of North America*, No. 442. The Birds of North America, Inc., Philadelphia, USA.

Marks, J.S. 1993. Molt of bristle-thighed curlews in the northwestern Hawaiian islands. *Auk* 110: 573–587.

Miller, M.R. & Duncan, D.C. 1999. The Northern Pintail in North America: Status and conservation needs of a struggling population. *Wildlife Society Bulletin* 27: 788–800.

Mitchell, C., Fox, A.D., Boyd, H., Sigfusson, A. & Boertmann, D. 1999. Pink-footed Goose *Anser brachyrhynchus*: Iceland/Greenland. *In* J. Madsen, G. Cracknell & A.D. Fox (eds.), *Goose Populations of the Western Palearctic. A Review of Status and Distribution,* pp. 68–81. Wetlands International Publ. 48, Wetlands International, Wageningen, the Netherlands. National Environmental Research Institute, Rønde, Denmark.

Moorman, T.E., Baldassarre, G.A. & Hess Jr, T.R. 1993. Carcass mass and nutrient dynamics of Mottled Ducks during remigial molt. *Journal of Wildlife Management* 57: 224–228.

Moreau, R.E. 1937. The comparative breeding biology of the African Hornbills (Bucerotidae). *Proceedings of the Zoological Society of London, Series a – General and Experimental* 107: 331–346.

Moser, T.J., Lien, R.D., VerCauteren, K.C., Abraham, K.F., Andersen, D.E., Bruggink, J.G., Coluccy, J.M., Graber, D.A., Leafloor, J.O., Luukkonen, D.R. & Trost, R.E. (eds.) 2004. *Proceedings of the 2003 International Canada Goose Symposium, 19–21 March 2003.* Madison, Wisconsin, USA.

Munro, J.A. 1941. Studies of waterfowl in British Columbia Greater Scaup Duck, Lesser Scaup Duck. *Canadian Journal of Research D* 19: 113–138.

Murphy, M.E. & King, J.R. 1982. Amino acid composition of the plumage of the White-crowned Sparrow. *Condor* 84: 435–438.

Murphy, M.E. & King, J.R. 1984a. Sulfur amino acid nutrition during moult in the White-crowned Sparrow. 2. Nitrogen and sulfur balance in birds fed graded levels of the sulfur-containing amino acids. *Condor* 86: 324–332.

Murphy, M.E. & King, J.R. 1984b. Dietary sulfur amino acid availability and moult dynamics in White-crowned Sparrows. *Auk* 101: 164–167.

Nilsson, L., Kahlert, J. & Persson, H. 2001. Moult and moult migration of Greylag Geese *Anser anser* from a population in Scania, south Sweden, *Bird Study* 48: 129–138.

O'Connor, M. 2008. Surf Scoter (*Melanitta perspicillata*) ecology on spring staging grounds and during the flightless period. M.Sc. thesis, McGill University, Montréal, Canada.

Oppel, S., Powell, A.N. & Dickson, D.L. 2008. Timing and distance of king eider migration and winter movements. *Condor* 110: 296–305.

Oring, L. 1964. Behavior and ecology of certain ducks during the postbreeding period. *Journal of Wildlife Management* 28: 223–233.

Owen, M. & Ogilvie, M.A. 1979. Wing moult and weights of Barnacle Geese in Spitzbergen. *Condor* 81: 42–52.

Palmer, R.S. 1972. Patterns of molting. *In* D.S. Farner & J.R. King (eds), *Avian Biology. Volume 2,* pp. 65–102. Academic Press, New York, USA.

Panek, M. & Majewski, D.P. 1990. Remex growth and body mass of Mallards during wing moult. *Auk* 107: 255–259.

Payne, R.B. 1972. Mechanisms and control of moult. *In* D.S. Farner & J.R. King (eds.), *Avian Biology. Volume 2,* pp. 103–155. Academic Press, New York, USA.

Pearce, J.M. & Petersen, M.R. 2009. Post-fledging movements of juvenile Common Mergansers (*Mergus merganser*) in Alaska as inferred by satellite telemetry. *Waterbirds* 32: 133–137.

Pearce, J.M., Zwiefelhofer, D. & Maryanski, N. 2009. Mechanisms of population heterogeneity among moulting Common Mergansers on Kodiak Island, Alaska: implications for genetic assessments of migratory connectivity. *Condor* 111: 283–293.

Peters, J.L, Bolender, K.A. & Pearce, J.M. 2012. Behavioural *vs*. Molecular sources of conflict between nuclear and mitochondrial DNA: the role of male-biases dispersal in a Holarctic sea duck. *Molecular Ecology* 21: 3562–3575.

Petersen, M.R. 1981. Populations, feeding ecology and moult of Steller's Eiders. *Condor* 83: 256–262.

Petit, D.R. & Bildstein, K.L. 1987. Effect of group size and location within the group on the foraging behaviour of white ibises. *Condor* 99: 602–609.

Phillips, L.M. & Powell, A.N. 2006. Evidence for wing moult and breeding site fidelity in King Eiders. *Waterbirds* 29: 148–153.

Piersma, T. 1988. Breast muscle atrophy and constraints on foraging during the flightless period of wing moulting Great Crested Grebes. *Ardea* 76: 96–106.

Piersma, T., Dietz, M.W., Dekinga, A., Nebel, S., van Gils, J., Battley, P.F. & Spaans, B. 1999a. Reversible size-changes in stomachs of shorebirds: when, to what extent, and why? *Acta Ornithologica* 34: 175–181.

Piersma, T., Gudmundsson, G.A. & Lilliendahl, K. 1999b. Rapid changes in the size of different functional organ and muscle groups during refuelling in a long-distance migrating shorebird. *Physiological and Biochemical Zoology* 72: 405–415.

Piersma, T., Koolhaas, A. & Dekinga, A. 1993. Interactions between stomach structure and diet choice in shorebirds. *Auk* 110: 552–564.

Portugal, S.J., Green, J.A. & Butler, P.J. 2007. Annual changes in body mass and resting metabolism in captive barnacle geese (*Branta leucopsis*): the importance of wing moult. *Journal of Experimental Biology* 210: 1391–1397.

Portugal, S.J., Isaac, R., Quinton, K.L. & Reynolds, S.J. 2010. Do captive waterfowl alter their behaviour patterns during their flightless moult? *Journal of Ornithology* 151: 443–448.

Portugal, S. J., Green, J.A., Piersma, T., Eichhorn, G. & Butler, P.J. 2011. Greater energy stores enable flightless moulting geese to increase resting behaviour. *Ibis* 153: 868–874.

Prince, H.H. 1979. Bioenergetics of post-breeding dabbling ducks. *In* T.A. Bookhout (ed.), *Waterfowl and wetlands: an integrated review*, pp. 103–117. Proceedings of the 1977 Symposium of the North Central Section of The Wildlife Society. Madison, Wisconsin, USA.

Pyle, P. 2005. Molts and plumages of ducks (Anatinae). *Waterbirds* 28: 208–219.

Qian, G.H. & Xu, H.F. 1986. Moult and resting metabolic rate in the common teal *Anas crecca* and the shoveler *A. clypeata*. *Acta Zoologica Sinica* 32: 68–73.

Rail, J.-F. & Savard J.-P.L. 2003. *Identification des Aires de Mue et de Repos au Printemps des Macreuses* (Melanitta *sp.*) *et de l'Eider à Duvet* (Somateria mollissima) *dans l'Estuaire et le Golfe du Saint-Laurent*. Technical Report Series No. 408. Canadian Wildlife Service, Quebec Region, Quebec, Canada.

Reed, A. 1971. Pre-dusk rafting flights of wintering goldeneyes and other diving ducks in the Province of Quebec. *Wildfowl* 22: 61–62.

Reed, T.M. & Rocke, T.E. 1992. The role of avian carcasses in botulism epizootics. *Wildlife Society Bulletin* 20: 175–182.

Robert, M., Benoit, R. & Savard, J.-P.L. 2002. Relationship among breeding, moulting and wintering areas of male Barrow's Goldeneyes (*Bucephala islandica*) in eastern North America. *Auk* 119: 676–684.

Robert, M., Mittelhauser, G.H., Jobin, B., Fitzgerald, G. & Lamothe, P. 2008. New insights on Harlequin Duck population structure in eastern North America as revealed by satellite telemetry. *Waterbirds* 31 (Special Publication 2): 159–172.

Robertson, G.J. & Goudie, R.I. 1999. Harlequin Duck (*Histrionicus histrionicus*). In A. Poole & F. Gill (eds.), *The Birds of North America*, No. 547. The Birds of North America, Inc., Philadelphia, USA.

Robertson, G. J., Cooke, F., Goudie, R.I. & Boyd, W.S. 1998. Moult speed predicts pairing success in male Harlequin Ducks. *Animal Behaviour* 55: 1677–1684.

Rohwer, S., Ricklefs, R.E., Rohwer, V.G. & Copple, M.M. 2009. Allometry of the duration of flight feather moult in birds. *PLoS Biol* 7(6): e1000132.

Salomonsen, F. 1968. The moult migration. *Wildfowl* 19: 5–24.

Savard, J.-P.L. & Robert, M. 2013. Relationships among Breeding, Moulting and Wintering Areas of Adult Female Barrow's Goldeneyes (*Bucephala islandica*) in Eastern North America. *Waterbirds* 36: 34–42.

Savard, J.-P.L., Bordage, D. & Reed, A. 1998. Surf Scoter (*Melanitta perspicillata*). In A. Poole & F. Gill (eds.), *The Birds of North America*, No. 363. The Birds of North America, Inc. The Academy of Natural Sciences, Philadelphia, USA.

Savard, J.-P.L., Reed, A. & Lesage, L. 2007. Chronology of breeding and moult migration in Surf Scoters (*Melanitta perspicillata*). *Waterbirds* 30: 223–229.

Schenkeveld, L.E. & Ydenberg, R.C. 1985. Synchronous diving by Surf Scoter flocks. *Canadian Journal of Zoology* 63: 2516–2519.

Schummer, M.L., Cohen, J., Kaminski, R.M., Brown, M.E. & Wax, C.L. 2014. Atmospheric teleconnections and Eurasian snow cover as predictors of a weather severity index in relation to Mallard *Anas platyrhynchos* autumn–winter migration. *Wildfowl* (Special Issue No. 4): 000–000.

Sheaffer, S.F., Malecki, R.A., Swift, B.L., Dunn, J. & Scribner, K. 2007. Management implications of molt migration by the Atlantic flyway resident population of Canada Geese, *Branta canadensis*. *Canadian Field Naturalist* 121: 313–320.

Singer, H.V., Sedinger, J.S., Nicolai, C.A., van Dellen, A.W. & Person, B.T. 2012. Timing of adult remigial wing molt in female Black Brant (*Branta bernicla nigricans*). *Auk* 129: 239–246.

Sjöberg, K. 1988. The flightless period of free-living male Teal *Anas crecca* in northern Sweden. *Ibis* 130: 164–171.

Smith, W.E. 2006. Moulting Common Eiders devoured by killer whales. *British Birds* 99: 264.

Stonor, C.R. 1937. On the attempted breeding of a pair of Trumpeter Hornbills (*Bycanistes buccinator*) in the gardens in 1936; together with some remarks on the physiology of the moult in the female. *Proceedings of the Zoological Society of London, Series A – General and Experimental* 107: 89–94.

Stresemann, E. & Stresemann, V. 1966. Die Mauser der Vögel. *Journal of Ornithology* 107: 3–448.

Szymczak, M.R. & Rexstad, E.A. 1991. Harvest distribution and survival of a gadwall population. *Journal of Wildlife Management* 61: 191–201.

Tamisier, A. & Dehorter, O. 1999. *Camargue Canards et Foulques*. Centre Ornithologique du Gard. Nimes, France.

Tape, K.D., Flint, P.L., Meixell, B.W. & Gaglioti, B.J. 2013. Inundation, sedimentation and subsidence creates goose habitat along the Arctic coast of Alaska. *Environmental Research Letters* 8: 045031.

Taylor, J. 1953. A possible moult-migration of Pink-footed Geese. *Ibis* 95: 638–641.

Taylor, E.J. 1993. Molt and bioenergetics of Pacific Black Brant (*Branta bernicla nigricans*) on the Arctic Coastal Plain, Alaska. Ph.D. thesis, Texas A&M University, College Station, Texas, USA.

Taylor, E.J. 1995. Molt of Black Brant (*Branta bernicla nigricans*) on the arctic coastal plain, Alaska. *Auk* 112: 904–919.

Thompson, D.C. & Boag, D.A. 1976. Effect of moulting on the energy requirements of Japanese quail. *Condor* 78: 249–252.

Thompson, J.D. & Baldassarre, G.A. 1988. Postbreeding habitat preference of wood ducks in northern Alabama. *Journal of Wildlife Management* 52: 80–85.

Thompson, J.E. & Drobney, R.D. 1996. Nutritional implications of moult in male Canvasbacks: variation in nutrient reserves and digestive tract morphology. *Condor* 98: 512–526.

Van der Meer, J. & Piersma, T. 1994. Physiologically inspired regression models for estimating and predicting nutrient stores and their composition in birds. *Physiological Zoology* 67: 305–329.

Van de Wetering, D. 1997. *Moult characteristics and habitat selection of postbreeding male Barrow's Goldeneye* (Bucephala islandica) *in northern Yukon*. Technical Report Series Number 296. Canadian Wildlife Service, Pacific and Yukon Region, British Columbia.

Van de Wetering, D. & Cooke, F. 2000. Body weight and feather growth of male Barrow's Goldeneye during wing molt. *Condor* 102: 228–231.

Van Impe, J. 1985. Mues des remiges chez la Nette a huppe rousse *Netta rufina* (Pallas) en Espagne du Nord. *Alauda* 53: 2–10.

Weller, M.W. 1980. 3. Molts and Plumages of Waterfowl. *In* F.C. Bellrose *Ducks, Geese and Swans of North America,* pp. 34–38. Stackpole Books, Harrisburg, USA.

Wobeser, G. & Leighton, T. 1988. Avian Cholera epizootic in wild duck. *Canadian Veterinary Journal* 29: 1015–1016.

Wunder, M.B., Jehl, J.R. & Stricker, C.A. 2012. The early bird gets the shrimp: confronting assumptions of isotopic equilibrium and homogeneity in a wild bird population. *Journal of Animal Ecology* 81: 1223–1232.

Yarris, G.S., McLandress, R. & Perkins, A.E.H. 1994. Molt migration of postbreeding female mallards from Suisun Marsh, California. *Condor* 96: 36–45.

Young, D.A. & Boag, D.A. 1982. Changes in the physical condition of male mallards (*Anas platyrhynchos*) during moult. *Canadian Journal of Zoology* 60: 3220–3226.

Photograph: Moulting Brent Geese, by Gerrit Vyn.

Drivers of waterfowl population dynamics: from teal to swans

DAVID N. KOONS[1]*, GUNNAR GUNNARSSON[2], JOEL A. SCHMUTZ[3] & JAY J. ROTELLA[4]

[1]Department of Wildland Resources and the Ecology Center, Utah State University, Logan, Utah 84322, USA.
[2]Division of Natural Sciences, Kristianstad University, SE-291 88 Kristianstad, Sweden.
[3]U.S. Geological Survey, Alaska Science Center, 4210 University Drive, Anchorage, Alaska 99508, USA.
[4]Department of Ecology, Montana State University, Bozeman, Montana 59717, USA.
*Correspondence author. E-mail: david.koons@usu.edu

Abstract

Waterfowl are among the best studied and most extensively monitored species in the world. Given their global importance for sport and subsistence hunting, viewing and ecosystem functioning, great effort has been devoted since the middle part of the 20th century to understanding both the environmental and demographic mechanisms that influence waterfowl population and community dynamics. Here we use comparative approaches to summarise and contrast our understanding of waterfowl population dynamics across species as short-lived as the teal *Anas discors* and *A.crecca* to those such as the swans *Cygnus* sp. which have long life-spans. Specifically, we focus on population responses to vital rate perturbations across life history strategies, discuss bottom-up and top-down responses of waterfowl populations to global change, and summarise our current understanding of density dependence across waterfowl species. We close by identifying research needs and highlight ways to overcome the challenges of sustainably managing waterfowl populations in the 21st century.

Key words: climate change, demographic buffering and lability, density dependence, ducks, elasticity, environmental stochasticity, geese, population dynamics, swans.

Competition for and selection of available habitats throughout the annual cycle, trophic interactions and associated life history trade-offs can all affect individual fitness (Dawkins & Krebs 1979; Stearns 1992; Manly *et al.* 2002). Advances in our understanding of these topics were reviewed by the first three plenary sessions of the 6th North American Duck Symposium and Workshop, "Ecology and Conservation of North American Waterfowl," (ECNAW) in Memphis, Tennessee in January 2013. Ultimately, factors that have an impact on individual fitness can scale up to affect population dynamics via a change in mean fitness and also through the assembly of co-occurring species in a waterfowl community (Ricklefs 2008). The population remains the primary biological unit on which waterfowl management objectives are based (*e.g.* the North American Waterfowl Management Plan; Canadian Wildlife Service & U.S. Fish and Wildlife Service 2012), but a greater focus on the community of waterfowl species is starting to emerge (Péron & Koons 2013).

To set the stage for papers that review the roles of harvest (Cooch *et al.* 2014) as well as the combination of habitat and harvest (Osnas *et al.* 2014) when managing waterfowl populations, we begin by focusing on the basic demographic mechanisms governing waterfowl population dynamics, and how they interact with environmental conditions. We take a comparative approach by considering the full suite of waterfowl species that range from those as small as teal to those as large as swans, and assess the state of knowledge of waterfowl population ecology relative to relevant theories and to what is known for other animal taxa. First, we compare the functional response of population dynamics to life-cycle perturbations across waterfowl life history strategies. Second, we discuss how resistance to perturbation in one part of the life-cycle might in turn have a mechanistic connection to other stages in the life-cycle that are "more free" to vary over time and thereby can potentially make important contributions to observed changes in abundance. Third, we discuss areas that require additional research to provide a better understanding of the interplay between dynamic environmental conditions and demography in an era of global change that, fourth, can be augmented or tempered by density-regulating mechanisms.

Perturbation analyses across the slow–fast continuum in waterfowl

Waterfowl have evolved a diverse array of life histories: teal live a life almost as short as many passerines, whereas many swans and geese have annual survival rates of ~0.90 that can lead to average life spans of 10 years, and lifetimes in excess of 30 years for the longest-lived individuals. Although some waterfowl are quite long-lived, "penguins and albatrosses they are not" (*sensu* C.D. Ankney), as typical life spans in those groups are 20 years and some individuals live past 60 (U.S.G.S. Bird Banding Laboratory longevity records). Given their slow pace of life, long-lived species can afford to delay reproduction and invest in offspring "slowly" over their lifespan while balancing life history trade-offs. In contrast, in short-lived species a "fast" start to reproduction early in life is favoured to help ensure that individuals pass on genes to the next generation. This inter-specific pattern of life history strategies in birds and mammals has aptly become known as the "slow–fast continuum" (*sensu* Harvey & Zammuto 1985; Sæther 1988; Gaillard *et al.* 1989; Promislow & Harvey 1990; for an analogous life history theory in plants, see Grime 1977; Silvertown *et al.* 1992).

With greater longevity comes greater complexity in the age structure of a population. In turn, immediate population growth rates and long-term abundances are more sensitive to perturbations of the stable age distribution in long-lived species with slow life histories (*e.g.* geese, swans and eiders) than they are in short-lived species with fast life histories (*e.g.* teal; Koons *et al.* 2005, 2006a, 2007). These management-relevant effects of perturbed age structure on immediate and future abundance arise through demographic processes known as "transient dynamics" and "population momentum" (*cf.* Koons *et al.* 2006b). The importance of age structure in models for guiding harvest management is emphasised by Cooch *et al.* (2014) in this special issue (see also Hauser *et al.* 2006), but the general topics of age structure and its impact on transient dynamics and population momentum across waterfowl life history strategies are ripe areas for future research.

To the contrary, the impact of changes in vital rates (*i.e.* survival and reproductive success) on the long-term population growth rate is better studied for many waterfowl species. One popular metric for measuring the relative impact of vital rate perturbations is the "elasticity", which measures the effect of equal proportionate changes in vital rates on the focal metric of population dynamics (most often λ, the deterministic finite rate of growth; Caswell 2001). Elasticity analyses have been published for an array of waterfowl species including Mallard *Anas platyrhynchos* (Hoekman *et al.* 2002, 2006), Northern Pintail *Anas acuta* (Flint *et al.* 1998), Lesser Scaup *Aythya affinis* (Koons *et al.* 2006c), Greater Scaup *A. marila* (Flint *et al.* 2006), Long-tailed Duck *Clangula hyemalis* (Schamber *et al.* 2009), King Eider *Somateria spectabilis* (Bentzen & Powell 2012), Common Eider *Somateria mollissima* (Gilliland *et al.* 2009; Wilson *et al.* 2012), as well as Barnacle Goose *Branta leucopsis* (Tombre *et al.* 1998), Emperor Goose *Chen canagica* (Schmutz *et al.* 1997) and Snow Goose *Chen caerulescens* (Cooch *et al.* 2001; Aubry *et al.* 2010). We used studies like these and other published demographic data to construct simple matrix population models (following Oli & Zinner 2001) to compare the elasticities of λ to proportional changes in annual fertility (*i.e.* the rate of recruiting females to 1 year of age per adult female) and survival after the hatch year (AHY) across waterfowl life histories.

From dabbling ducks to pochard, shelduck, sea ducks, geese and swans, the elasticity of the population growth rate to changes in AHY survival increases with generation time (*i.e.* the average difference in age between parents and newborn offspring) whereas that to changes in fertility decreases with generation time (Fig. 1), a pattern common to all birds (Sæther & Bakke 2000; Stahl & Oli 2006). Given current data, only the Blue-winged Teal *Anas discors* has a higher elasticity for fertility than for AHY survival (Fig. 1). This implies that, all else being equal, the population growth rate will respond more readily to changes in AHY survival than it will to changes in fertility for the majority of waterfowl species that have moderate or long generation times. For species with very fast life histories like teal, however, the reverse is true.

Demographic buffering and lability in waterfowl life-cycles

Despite the greater sensitivity of λ to proportional changes in survival for

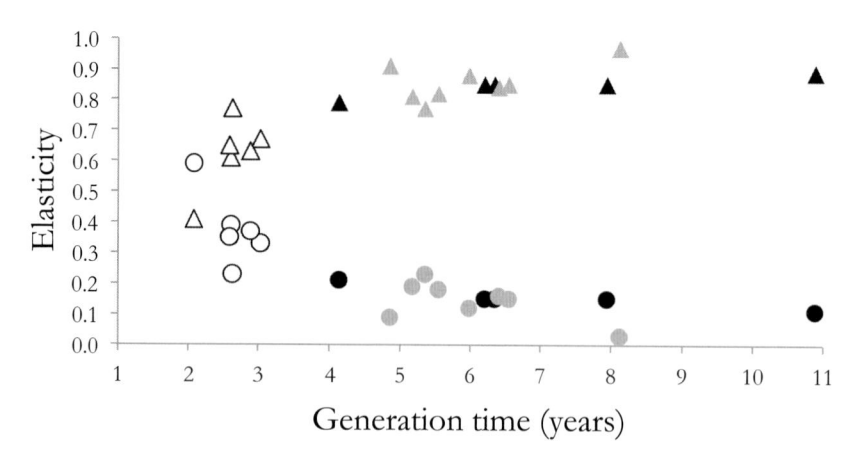

Figure 1. The elasticity of population growth rate (λ) to changes in annual fertility (circles) and adult survival (triangles) for 19 species of Nearctic and Palearctic ducks (white), shelduck and sea ducks (grey), and geese and swans (black), in relation to generation time (*i.e.* the average difference in age between parents and newborn offspring). On the far left are results for Blue-winged Teal and on the far right are results for Tundra Swan *Cygnus c. columbianus*. Elasticity results or demographic data used to conduct an elasticity analysis were extracted from: Patterson 1982; Rohwer 1985; Kennamer & Hepp 1987, 2000; Hepp *et al.* 1989; Hepp & Kennamer 1993; Flint *et al.* 1998; Cooch *et al.* 2001; Hoekman *et al.* 2002; Oli *et al.* 2002; Rohwer *et al.* 2002 and references therein; Flint *et al.* 2006; Grand *et al.* 2006 and references therein; Koons *et al.* 2006c; Gilliland *et al.* 2009.

many species, changes in the underlying components of fertility can still have management-relevant effects on λ. It is often the case that management actions cannot elicit the desired response in a vital rate (Baxter *et al.* 2006; Koons *et al.* 2014). When managers have difficulty influencing the vital rate(s) with the greatest elasticity, they might look to those that are most prone to being affected by changes in environmental conditions or management actions (*i.e.* the most labile). Actual changes in population abundance are not only a function of the vital rates with the greatest elasticities, but also density-dependent feedbacks (see section below) and the degree to which each vital rate changes over time (with fluctuations in age structure also contributing to changes in abundance; Davison *et al.* 2010). In fact, the contribution of each vital rate to past variation in population growth rate can be calculated analytically by using Life Table Response Experiments (LTRE), which comprise multiple approaches for best addressing the study hypothesis and design (Caswell 2001, 2010; see Cooch *et al.* 2001 for a waterfowl example). Over time, the contribution of a vital rate to past variation in population growth rate is proportional to the product of its squared elasticity and the square of its coefficient of process variation (when estimates are available, vital rate correlations should also be incorporated into calculations). Thus, a vital rate can make important contributions to changes in population growth by having a large elasticity, a large process variance, or both (eqn. 7 in Heppell *et al.* 1998).

In waterfowl, vital rate contributions to retrospective variation in λ have only been computed for Mallard (Hoekman *et al.* 2002; Amundson *et al.* 2012), King Eider (Bentzen & Powell 2012), Common Eider (Wilson *et al.* 2012), Barnacle Goose (Tombre *et al.* 1998) and Snow Goose (Cooch *et al.* 2001). A summary of these studies indicates that although adult survival may have the greatest elasticity, nesting success, pre-fledging survival and juvenile survival are often more labile to environmental conditions. As a result, variation in these fertility components can sometimes contribute more to observed variation in population dynamics than do changes in adult survival (see Gaillard *et al.* 1998 for similar results in ungulates). Yet, a larger contribution of fertility components to observed changes in λ does not necessarily imply that changing these vital rates will have a greater impact on population growth than would similar-sized changes in adult survival. Rather, fertility components could fluctuate enough over time to make important contributions to population dynamics (Caswell 2000). Managers might thus use LTRE and other variance decomposition methods (*e.g.* life-stage simulation analysis; Wisdom *et al.* 2000) as a platform for identifying the vital rates that can make important contributions to population growth through their natural lability to changing environmental conditions or management actions.

To gain a broader insight into which vital rates are most labile to environmental conditions across waterfowl life histories, it is necessary to understand the basic theory of population growth in stochastic environments. From first principles, the rate of growth for any population experiencing environmental stochasticity can be approximated as:

$$\ln \lambda_s \cong \ln \bar{\lambda} - \frac{\sigma^2}{2}, \quad (1)$$

where $\ln \bar{\lambda}$ denotes the mean and σ^2 the variance of annual population growth rate (N_t/N_{t-1}) across environmental conditions on the log scale (Lewontin & Cohen 1969; Tuljapurkar 1982). Of key importance, increased variance in annual population growth rate decrements the long-term rate of growth in a stochastic environment ($\ln \lambda_s$).

Recognising this relationship, Gillespie (1977), and later Pfister (1998), noted that the vital rates with the greatest potential to affect mean fitness (*i.e.* $\ln \lambda_s$) should exhibit the least amount of temporal variance because organisms should presumably be selected to avoid fluctuations in vital rates that would have severe negative impacts. Over the long-term, natural selection should have favoured such life history properties; a concept that has become known as the Demographic Buffering hypothesis (DB; Gaillard *et al.* 2000; Boyce *et al.* 2006). There is general support for the DB hypothesis in plants and diverse animal species (*e.g.* Morris *et al.* 2008; Dalgleish *et al.* 2010; Rotella *et al.* 2012) including birds (Schmutz 2009). Avian species with high adult survival elasticities tend to exhibit less variation in adult survival over time than do species with low adult survival elasticities. In all comparative studies, however, empirical fits to DB predictions are not perfect; there is often a great deal of deviation from the predicted negative relationship between vital rate elasticities and temporal variation in vital rates (Jäkäläniemi et al. 2013). Among waterfowl, currently available data for temporal variation in adult survival do not support the DB hypothesis (Fig. 2).

Mixed support for the DB hypothesis could occur for a variety of reasons. For example, few studies are able to study the entire life-cycle of a population over a long enough period of time to attain proper estimates of temporal "process variation" in each vital rate (see Rotella et al. 2012 for an example of the benefits provided by such estimates). Unfortunately, ignoring the issue of vital rate variance decomposition in tests of the DB hypothesis will inevitably inflate type II errors (Morris & Doak 2002). In addition, rapid anthropogenic alterations to the environment might have exceeded the capacity of organisms to buffer or respond to environmental fluctuations (Schmutz 2009). At a fundamental level, the DB concept is also restricted and cannot always capture the full effects of environmental stochasticity on $\ln \lambda_s$. Optimising fitness in a stochastic world is a balancing act of increasing mean vital rates, which affects the first term on the right-hand side of eqn. 1, but also decreasing variance in vital rates, which minimises the negative effect of the second term. It has only recently been recognised that temporal variation in vital rates can have a positive impact on $\ln \lambda_s$ when the variation induces an increase in $\ln \bar{\lambda}$ that is sufficiently large to outweigh the negative effect of σ^2 (see eqn. 1; Drake 2005; Koons et al. 2009). This can occur

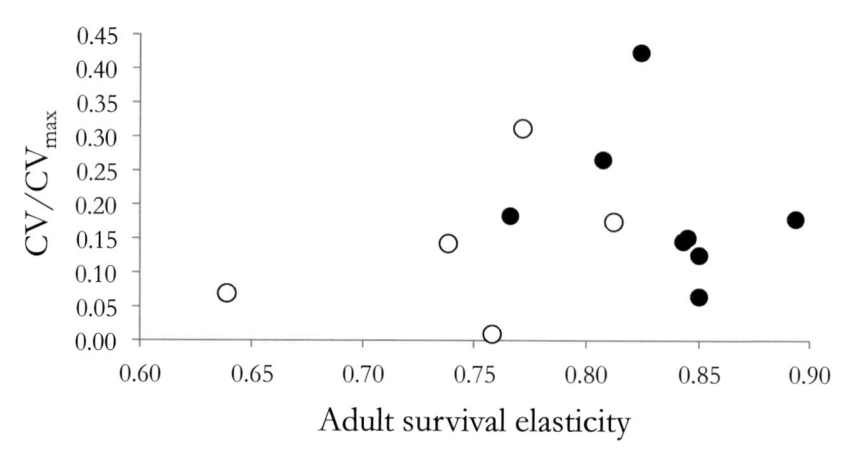

Figure 2. The relationship ($P > 0.10$, n.s.) between adult survival elasticities and the relative temporal "process variation" in adult survival across 13 waterfowl populations (white for ducks and black for geese; no sea duck data were available for these analyses). The coefficient of process variation (CV) was divided by the theoretically maximum possible value of CV for a given mean survival probability (see Morris & Doak 2004). Similar results were achieved using raw values of CV. The waterfowl data were extracted from Schmutz (2009) and references therein, as was the approach to developing matrix population models for the calculation of elasticities.

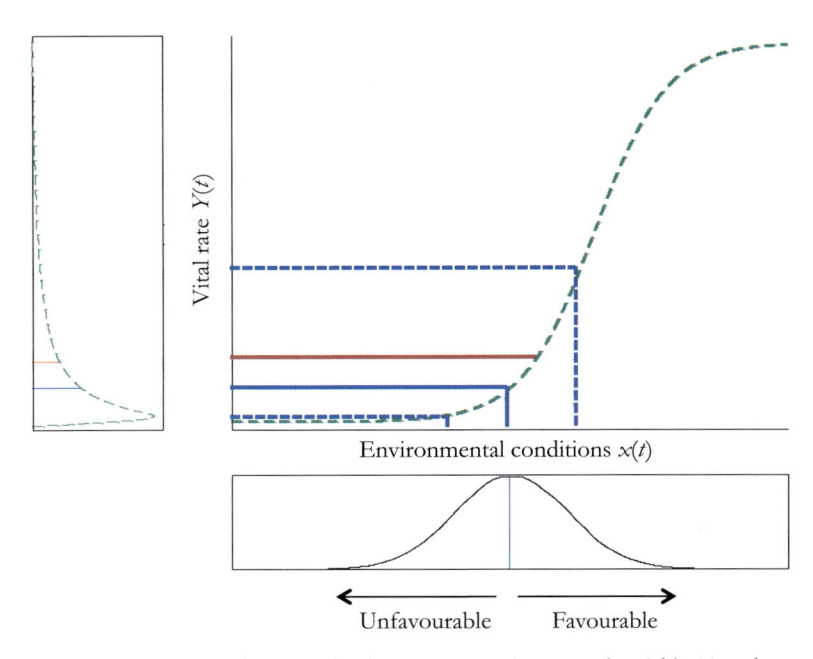

Figure 3. A hypothetical sigmoid relationship between an environmental variable (*x*) and any vital rate *Y*. The lower box illustrates the mean (blue bar) and temporal distribution of environmental conditions *x(t)*. The left box indicates the resulting temporal distribution of vital rate *Y(t)*. The solid blue bars in the main figure illustrate vital rate performance in the average environment. The dashed blue bars show vital rate performance 1 s.d. below and above the average environment. As conditions vary over time, favourable environments produce larger increases in vital rate performance than decreases experienced in equally unfavourable environments. In turn, this raises the mean of *Y* and the long-term stochastic growth rate (red lines). The Demographic Buffering hypothesis implicitly assumes a linear relationship between *x(t)* and *Y(t)*, in which case variation in *x(t)* has no effect on the mean of *Y*, only its statistical variance.

when a vital rate has a convex relationship with prevailing environmental conditions (Fig. 3). Although the effect of "variance" in the strictest sense will still be negative, variation across convex relationships with environmental variables can "skew" the distribution of a vital rate and increase its mean value (Fig. 3), thereby increasing $\ln \bar{\lambda}$ (the interested reader should see Rice *et al.* 2011 for a decomposition of $\ln \bar{\lambda}$ that is more complete than the commonly used approximation shown in eqn. 1).

For vital rates that are on average low, such as nest success (*i.e.* the proportion of clutches where at least one egg hatches) in many duck populations, or offspring survival in some goose and swan populations, the potential for convex relationships with niche axes is strong. If such relationships occur, enhanced environmental stochasticity could actually be beneficial because favourable conditions could induce booms in vital rates that enhance the long-term stochastic population growth rate (Koons *et al.* 2009;

Walker *et al.* 2013). Further research on DB and lability in waterfowl is nevertheless needed to provide a better understanding of how vital rates will respond to global change across species, and how such responses will affect population growth rates.

Demographic drivers in an era of global change

For waterfowl management to keep pace with climate, land-use and water-use changes in the 21st century, a greater depth of knowledge is needed of the mechanisms affecting demography and population dynamics. Habitat and predation continue to be central topics of waterfowl research and management (*e.g.* Duebbert & Kantrud 1974) but we must also prepare for changes in the dynamic interplay between abiotic conditions and trophic interactions. Some regions of the Nearctic and Palearctic are becoming warmer while others are not; some may receive more precipitation and others less (IPCC 2013). Many landscapes are being converted to produce more ethanol, wind or solar power (Northrup & Wittemyer 2013; Wright & Wimberly 2013), and greater human demands for land and water will continue to affect the amounts and quality of habitat available to wildlife (Fischer & Heilig 1997; Lemly *et al.* 2000; Pringle 2000). The aforementioned changes describe shifts in the "mean" environmental conditions that often come to mind when we speak of global change in the 21st century. Some outcomes of these changes are predictable; less water or breeding habitat would likely lead to fewer ducks (Reynolds *et al.* 2001; Stephens *et al.* 2005), whereas more corn planted for ethanol production could compound the ongoing problem of over-abundant goose populations (Ankney 1996; Abraham *et al.* 2005).

In addition to predicted changes in mean climatic conditions, increases in the variability of climate could also occur in many parts of the Nearctic and Palearctic (*e.g.* Wetherald 2009), including the prairie pothole region of North America (Johnson *et al.* 2010). This could imply greater extremity in climate from year to year; for instance, very dry years followed by heavy precipitation that leads to flooding, or balmy winters followed by bitter cold ones. Other than the direct effects of exposure, climate tends to affect waterfowl populations through changes in bottom-up food resources, top-down predation, and density-dependent interactions (Nudds 1992).

Trophic ecologists in Scandinavia have shown that the shape of a predator-prey functional response can determine fundamentally whether increased temporal variation in a resource has a positive or negative effect on the mean vital rates of a consumer (*e.g.* Henden *et al.* 2008). In other words, trophic interactions can dictate the relative advantages of DB *versus* lability in a population. In other systems, climate-driven pulses in primary productivity can result in counter-acting direct and time-lagged effects on primary consumers because of complex trophic interactions such as apparent competition (Schmidt & Ostfeld 2008). For example, in the prairie pothole region, Walker *et al.* (2013) found that duck nest success in year t was positively associated with pond density and primary productivity in the same year (t), but negatively related to these variables in previous years (t–1 and *t*–2). Findings like these could possibly be attributed to the positive numerical response of alternate prey to climate-induced change in primary productivity, which could swamp out generalist predators in the current year, but then lead to numerical responses in predator populations that later have a deleterious effect on waterfowl reproductive success (Ackerman 2002; Brook *et al.* 2008; Schmidt & Ostfeld 2008; Iles *et al.* 2013a; Walker *et al.* 2013).

Changing mean and variance of climatic conditions can also affect waterfowl populations through differential changes in phenology across trophic levels. For example, the match-mismatch hypothesis predicts that a consumer should try to "match" its life-cycle events with the timing of maximal resource availability or quality because failure to do so results in a "mismatch" between resource and consumer phenologies that reduces fitness (Visser & Both 2005). Research on the match-mismatch hypothesis in ducks is advancing (*e.g.* Drever & Clark 2007; Sjöberg et al. 2011), and species with less flexible nesting phenology (*e.g.* scaup and scoters; Gurney *et al.* 2011) might suffer more from climate-induced shifts in the timing of resource availability compared to those that are more flexible (*e.g.* Mallard; Drever *et al.* 2012).

Research on the match-mismatch hypothesis in geese indicates that warming of the Arctic has led to, on average, earlier greening of graminoid plants that can readily take advantage of early growing degree days (van der Jeugd *et al.* 2009; Doiron *et al.* 2013; Gauthier *et al.* 2013). Some arctic geese have, however, evolved to balance their timing of migration with photoperiod, plant phenology and resource availability along migration paths, as well as plant phenology and snow cover on the breeding grounds (Strong & Trost 1994; Gwinner 1996; Bauer *et al.* 2008, Tombre *et al.* 2008). Thus, they cannot always "match" their nesting phenology with the greening of graminoid forage on the breeding grounds (Gauthier *et al.* 2013). A year of late nesting relative to the phenology of graminoid greening can result in reduced food quality for goslings that in turn inhibits their growth, body condition and ultimately survival (Dickey *et al.* 2008; Aubry *et al.* 2013). However, the trend toward earlier greening is not consistent from year to year because the variability in growing degree days in spring seems to be getting larger. Although the trend toward earlier greening may force a mismatch between arctic geese and their preferred forage, stochasticity in the near-term will still provide some years where they can match their nesting to plant phenology (Aubry *et al.* 2013).

Climate-driven changes in phenology can also affect the intensity of top-down predation and even result in completely novel predator-prey interactions. For example, climate-driven declines in the extent and duration of sea ice each year have been linked to earlier onshore arrival of Polar Bears *Ursus maritimus* in many parts of the Arctic and reduced opportunities for the Polar Bears to hunt seals (Family: Phocidae and Otariidae; Stirling & Derocher 1993; Stirling & Parkinson 2006). This has resulted in a mismatch for Polar Bears with their preferred seal prey. Ironically, this now exposes Polar Bears to a novel overlap with the breeding seasons of many ground-nesting waterfowl populations (Rockwell & Gormezano 2009), from which they are readily consuming eggs and offspring (Madsen *et al.* 1998; Noel *et al.* 2005; Drent & Prop 2008; Smith *et al.* 2010; Iles *et al.* 2013b; Gormezano & Rockwell 2013; Iverson *et al.* 2014). Novel predator-prey interactions such as these have strong potential to affect waterfowl population dynamics (Rockwell *et al.* 2011). Climate change could thus affect waterfowl populations through changes in both bottom-up and top-down interactions. Yet the strength of these interactions, as well as those among conspecifics, may be moderated by population density and the ability of individual species to respond to the novel selection pressures induced by climate change.

Density dependence: the ever-elusive regulator of populations

Population density affects population dynamics through both intra- and inter-specific interactions. For example, density can influence competition for territories and mates, competition for and depletion of limited foods, rates of pathogen transmission, and functional responses of predators. Density can also act in positive ways. In arctic geese, for example, colonial nesting density and intermediate levels of grazing have both been shown to enhance vital rates (*e.g.* Raveling 1989; Hik & Jefferies 1990; Aubry *et al.* 2013). Both positive and negative density-dependent interactions scale up to affect vital rates in ways that shape the pace of population growth and, ultimately, negative density dependence at some point in the life-cycle places an upper bound on the capacity of a population to grow (Turchin 2003). Ever since Malthus (1798), density dependence has therefore been central to our way of thinking about population dynamics. Given its role in affecting population growth, harvest management has often been used to manipulate population density in an attempt to optimise long-term yield from populations (*e.g.* maximum sustainable yield theory; Walters 1986). On the other hand, habitat management attempts to manipulate the carrying capacity by providing more per capita resources, and thereby increasing the ceiling on abundance where reproductive success and mortality balance each other out (Smith *et al.* 1989). Only recently has waterfowl management begun a formal integration of these concepts, which are tied both to density dependence and to environmental change (Runge *et al.* 2006; Mattsson *et al.* 2012), a topic specifically addressed by Osnas *et al.* (2014) in this volume. For these reasons, and others, it is critical to improve our understanding of how density dependence operates over the annual life-cycle, and across the diverse array of waterfowl life histories (Gunnarsson *et al.* 2013).

Measuring the influence of density dependence on population dynamics in the wild, however, has been said to be "like a search for the holy grail" (Krebs 1995). Two demographic approaches have nevertheless been employed to make progress toward understanding density dependence: surveys of population abundance, and studies of life history traits (Lebreton & Gimenez 2013). Time series analyses of how surveyed abundance responds to levels of population abundance in previous years are *sensu stricto* phenomenological (Krebs 2002) but can, with care, be used to test for the presence of density dependence, measure its impact relative to environmental variables, and even examine the interactive effects of variation in population density and environmental variables on population dynamics (*e.g.* Stenseth *et al.* 2003; Rotella *et al.* 2009). The long-standing problem, however, has been the lack of independence between explanatory and response variables and the related issue of shared sampling variation (*i.e.* uncertainty in abundance) between the axes being analysed, both of which bias estimation towards greater strength (and presence) of density dependence than actually exists (Freckleton *et al.* 2006; Lebreton & Gimenez 2013). Unfortunately, many older published studies did not account for analytical problems that induce these false positives. The results of those studies should probably be disregarded and, where available data allow, the information should be re-analysed with modern methods to improve our understanding of density dependence. Modern state-space statistical models for time-series data can account for these issues (*e.g.* Knape & de Valpine 2012; Delean *et al.* 2013) and help make use of widely available monitoring data to gain insight into the influence of density dependence on waterfowl population dynamics over time (Murray *et al.* 2010), space (Viljugrein *et al.* 2005), and across life histories (Jamieson & Brooks 2004; Sæther *et al.* 2008; Murray *et al.* 2010).

The approach of studying the effects of population density on life history traits nicely avoids the issue of dependence between explanatory and response variables. That said, uncertainty in estimates of population abundance (the explanatory variable) can result in bias toward false absence of density dependence (McArdle 2003); a conservative outcome that is often more favoured in science than a false positive (Lebreton & Gimenez 2013). When conducted with explicit attention to biotic interactions, the life history trait approach can provide a deeper understanding of density dependence than can analyses of abundance time series (Krebs 2002). Such approaches have been used nicely in both observational and experimental waterfowl studies to elucidate the mechanistic effects of population density on fidelity and adult breeding probability (Sedinger *et al.* 2008), clutch size (Cooch *et al.* 1989; Sedinger *et al.* 1998), nesting success (Raveling 1989 for geese; Gunnarsson & Elmberg 2008; Ringleman *et al.* 2012 for ducks; see Gunnarsson *et al.* 2013 for a review), offspring growth (Lindholm *et al.* 1994; Schmutz & Laing 2002; Person *et al.* 2003), offspring survival (Williams *et al.* 1993; Nicolai & Sedinger 2012 for geese; Gunnarsson *et al.* 2006; Amundson *et al.* 2011 for ducks; see Gunnarsson *et al.* 2013 for review), subsequent post-fledging survival (*e.g.* Schmutz 1993; Sedinger & Chelgren 2007; Aubry *et al.* 2013) and even the effects of nutrient limitation during development on eventual adult body size (*e.g.* Cooch *et al.* 1991a,b; Sedinger *et al.* 1995; Loonen *et al.* 1997). Moreover, density dependence at one stage of the life-cycle (*e.g.* nesting) can affect population density and its impact later in the life-cycle (*e.g.* offspring rearing and then post-fledging) through a sequential cohort process (Elmberg *et al*. 2005).

Although its effects are complex, great progress has been made in recent years toward understanding the role of density dependence in waterfowl (Gunnarsson *et al*. 2013). Lack of long-term and experimental studies for many taxa and the aforementioned issues with estimation currently prevent us from being able to make robust conclusions about patterns in density dependence across waterfowl life histories. When resources are limited, density dependence can have strong effects on reproductive success of the most fecund (*e.g.* Mallard; Kaminski & Gluesing 1987) and even the most long-lived of waterfowl species (*e.g.* arctic geese and swans; Williams *et al*. 1993; Nummi & Saari 2003). Rarely, however, has population density been shown to have an effect on adult survival (Dugger *et al*. 1994; Ludwichowski *et al*. 2002; Menu *et al*. 2002; Sedinger *et al*. 2007), which we might expect because adult survival has high elasticity values and should thus be highly buffered against environmental changes induced by population density (see sections above).

Given existing evidence, and following the Eberhardt hypothesis that ungulate ecologists have used to focus their study of density dependence (Eberhardt 1977, 2002; Bonenfant *et al*. 2009), we hypothesise that waterfowl become more robust to the effects of density dependence as they develop into the prime ages of adulthood, but might again become susceptible at older ages; for instance, because of immunosenescence and density-related pathogen transmission (Palacios *et al*. 2011).

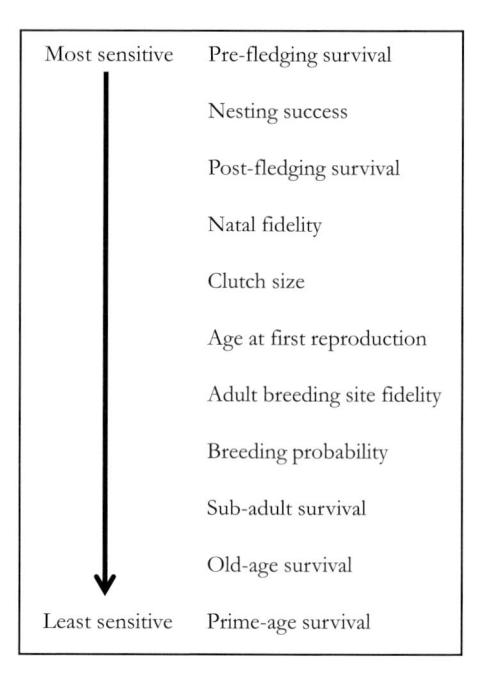

Figure 4. A hypothesised sequence of vital rate sensitivity to density dependence over the age-structured life-cycle of a hypothetical waterfowl species.

Most sensitive → Pre-fledging survival, Nesting success, Post-fledging survival, Natal fidelity, Clutch size, Age at first reproduction, Adult breeding site fidelity, Breeding probability, Sub-adult survival, Old-age survival, Least sensitive → Prime-age survival

For a given waterfowl taxonomic group, the rank-order of vital rate sensitivity to density dependence can be organised as a list (see Fig. 4), but the organisation and presence of density dependence across the life-cycle could shift with life history strategy (*e.g. r–K* selection; MacArthur & Wilson 1967; Pianka 1970) or perhaps more so with "lifestyle" (*e.g.* diet, nest-site preference, mating strategy, *etc.*; Dobson 2007; Sibly & Brown 2007) and prevailing environmental conditions. At this point, these ideas are nascent and are presented here for future waterfowl ecologists to advance or repudiate as science progresses. Knowing where density dependence operates in the life-cycle, and how it varies over time and space will eventually help managers apply actions in the most appropriate seasons, habitats and environmental conditions for achieving desired population responses relative to resource investments (du Toit 2010).

Conclusions

Our understanding of waterfowl population dynamics has come a long way since the middle part of the 20th century. A strong focus on studying vital rates has helped develop and improve population models for several species, and new tools have allowed researchers to identify differences in vital rate contributions to population dynamics across life histories. Progress has been made in research on bottom-up and top-down mechanisms affecting populations, and light has even been shed on the once elusive mechanism of density dependence.

There are nevertheless key gaps that need to be filled in order to sustain healthy waterfowl populations amidst the challenges presented by global change. For example, significant portions of Palearctic waterfowl populations occupy regions where research and monitoring are scarce (*e.g.* Russia and the boreal forest of North America), which makes it difficult to develop scientifically robust studies and management of populations that do not necessarily recognise geographic borders and survey boundaries. We outlined briefly our current understanding of density dependence in waterfowl, and presented a framework for organising an understanding of the key life-cycle stages where density dependence has a significant impact on population dynamics (Fig. 4). Although often assumed in energetic models, explicit studies of density dependence during the staging and wintering periods of the life-cycle are scarce. The cross-seasonal approach to studying life-cycle dynamics may be the best way to fulfil these informational needs (see Sedinger & Alisauskas 2014).

In addition to adaptive management approaches, formal experiments are needed to separate density-dependent from density-independent processes across gradients of resource availability. Long-term observational studies additionally offer the test of time, and are perhaps the best way to understand the complex effects of environmental change and stochasticity on populations (Clutton-Brock & Sheldon 2010). Where possible, experimental and observational approaches should be combined to enhance learning within the adaptive management framework.

The relatively new Integrated Population Model (IPM, not to be confused with "integral" population models) offers an innovative way to combine data from different research approaches to test hypotheses and, importantly, to link information from detailed field studies with large-scale monitoring data to guide adaptive management (Besbeas *et al.* 2002). In addition, by using the constraint that only birth, immigration, death and emigration can affect population abundance, IPMs can combine abundance data with vital rate data in a way that can reduce bias and improve precision in demographic estimates (Abadi *et al.* 2010). Utilising these features, IPMs are already being used to provide a synthetic view of the mechanisms that shape waterfowl population dynamics (Péron *et al.*

2012), and are even being used to study the demographic effects of species interactions at large scales (Péron & Koons 2012). IPMs thus offer a way to model the dynamic mechanisms that affect waterfowl populations and communities at scales that are relevant to managing migratory species. Waterfowl are among the most extensively studied vertebrates on the planet, and we predict that this rich tradition will contribute to great advancements in population ecology, evolution and applied natural resource management in the 21st century.

Acknowledgements

We thank the ECNAW scientific committee for inviting us to give our plenary talk at the 2013 symposium in Memphis, Tennessee, as well as R.G. Clark and J.S. Sedinger for helpful comments on an earlier version of the manuscript. DNK thanks R.F. Rockwell for discussions about waterfowl population ecology over the years and NSF (DEB 1019613) for financial support.

References

Abadi, F., Gimenez, O., Arlettaz, R. & Schaub, M. 2010. An assessment of integrated population models: bias, accuracy, and violation of the assumption of independence. *Ecology* 91: 7–14.

Abraham, K.F., Jefferies, R.L. & Alisauskas, R.T. 2005. The dynamics of landscape change and Snow Geese in mid-continent North America. *Global Change Biology* 11: 841–855.

Ackerman, J.T. 2002. Of mice and Mallards: positive indirect effects of coexisting prey on waterfowl nest success. *Oikos* 99: 469–480.

Amundson, C.L., & Arnold, T.W. 2011. The role of predator removal, density-dependence, and environmental factors on Mallard duckling survival in North Dakota. *Journal of Wildlife Management* 75: 1330–1339.

Amundson, C.L., Pieron, M.R., Arnold, T.W. & Beaudoin, L.A. 2013. The effects of predator removal on Mallard production and population change in northeastern North Dakota. *Journal of Wildlife Management* 77: 143–152.

Ankney, C.D. 1996. An embarrassment of riches: too many geese. *Journal of Wildlife Management* 60: 217–223.

Aubry, L.M., Rockwell, R.F. & Koons, D.N. 2010. Metapopulation dynamics of mid-continent Lesser Snow Geese: implications for management. *Human-Wildlife Interactions* 4: 11–32.

Aubry, L.M., Rockwell, R.F., Cooch, E.G., Brook, R.W., Mulder, C.P. & Koons, D.N. 2013. Climate change, phenology, and habitat degradation: drivers of gosling body condition and juvenile survival in Lesser Snow Geese. *Global Change Biology* 19: 149–160.

Bauer, S., Van Dinther, M., Høgda, K.-A., Klaassen, M. & Madsen, J. 2008. The consequences of climate-driven stop-over sites changes on migration schedules and fitness of Arctic geese. *Journal of Animal Ecology* 77: 654–660.

Baxter, P.W.J., McCarthy, M.A., Possingham, H.P., Menkhorst, P.W. & McLean, N. 2006. Accounting for management costs in sensitivity analyses of matrix population models. *Conservation Biology* 20: 893–905.

Bentzen, R.L. & Powell, A.N. 2012. Population dynamics of King Eiders breeding in northern Alaska. *Journal of Wildlife Management* 76: 1011–1020.

Besbeas, P., Freeman, S.N., Morgan, B.J.T. & Catchpole, E.A. 2002. Integrating mark-recapture-recovery and census data to estimate animal abundance and demographic parameters. *Biometrics* 58: 540–547.

Bonenfant, C., Gaillard, J.-M., Coulson, T., Festa-Bianchet, M., Loison, A., Garel, M., Loe, L.E., Blanchard, P., Pettorelli, N., Owen-Smith, N., du Toit, J. & Duncan, P. 2009. Empirical evidence of density-dependence in populations of large herbivores. *Advances in Ecological Research* 41: 314–352.

Boyce, M.S., Haridas, C.V., Lee, C.T., & the NCEAS Stochastic Demography Working Group. 2006. Demography in an increasingly variable world. *Trends in Ecology & Evolution* 21: 141–148.

Brook, R.W., Pasitschniak-Arts, M., Howerter, D.W. & Messier, F. 2008. Influence of rodent abundance on nesting success of prairie waterfowl. *Canadian Journal of Zoology* 86: 497–506.

Canadian Wildlife Service & U.S. Fish and Wildlife Service. 2012. North American waterfowl management plan. Canadian Wildlife Service and U.S. Fish and Wildlife Service Report, U.S. Fish and Wildlife Service, Washington D.C., USA.

Caswell, H. 2000. Prospective and retrospective perturbation analyses: their roles in conservation biology. *Ecology* 81: 619–627.

Caswell, H. 2001. *Matrix Population Models: Construction, Analysis and Interpretation, 2nd edition*. Sinauer Associates Inc., Sunderland, Massachusetts, USA.

Caswell, H. 2010. Life table response experiment analysis of the stochastic growth rate. *Journal of Ecology* 98: 324–333.

Clutton-Brock, T. & Sheldon, B.C. 2010. Individuals and populations: the role of long-term, individual-based studies of animals in ecology and evolutionary biology. *Trends in Ecology and Evolution* 25: 562–573.

Cooch, E.G., Lank, D.B., Rockwell, R.F. & Cooke, F. 1989. Long-term decline in fecundity in a Snow Goose population: evidence for density dependence. *Journal of Animal Ecology* 58: 711–726.

Cooch, E.G., Lank, D.B., Rockwell, R.F., Dzubin, A. & Cooke, F. 1991a. Body size variation in Lesser Snow Geese: environmental plasticity in gosling growth rates. *Ecology* 72: 503–512.

Cooch, E.G., Lank, D.B., Rockwell, R.F. & Cooke, F. 1991b. Long-term decline in body size in a Snow Goose population: evidence of environmental degradation? *Journal of Animal Ecology* 60: 483–496.

Cooch, E.G., Rockwell, R.F. & Brault, S. 2001. Retrospective analysis of demographic responses to environmental change: an example in the Lesser Snow Goose. *Ecological Monographs* 71: 377–400.

Cooch, E.G., Guillemain, M., Boomer, G.S., Lebreton, J.-D. & Nichols, J.D. 2014. The effects of harvest on waterfowl populations. *Wildfowl* (Special Issue No. 4): 220–276.

Dalgleish, H.J., Koons, D.N. & Adler, P.B. 2010. Can life-history traits predict the response of forb populations to changes in climate variability? *Journal of Ecology* 98: 209–217.

Davison, R., Jacquemyn, H., Adriaens, D., Honnay, O., de Kroon, H. & Tuljapurkar, S. 2010. Demographic effects of extreme weather events on a short-lived calcareous grassland species: stochastic life table response experiments. *Journal of Ecology* 98: 255–267.

Dawkins, R. & Krebs, J.R. 1979. Arms races between and within species. *Proceedings of the Royal Society of London, B* 205: 489–511.

Delean, S., Brook, B.W. & Bradshaw, C.J.A. 2013. Ecologically realistic estimates of maximum population growth using informed Bayesian priors. *Methods in Ecology and Evolution* 4: 34–44.

Dickey, M.–H., Gauthier, G. & Cadieux, M. 2008. Climatic effects on the breeding phenology and reproductive success of an Arctic-nesting goose species. *Global Change Biology* 14: 1973–1985.

Doiron, M., Legagneux, P., Gauthier, G. & Lévesque, E. 2013. Broad-scale satellite normalized difference vegetation index data predict plant biomass and peak date of nitrogen concentration in Arctic tundra vegetation. *Applied Vegetation Science* 16: 343–351.

Dobson, F.S. 2007. A lifestyle view of life-history evolution. *Proceedings of the National Academy of Sciences, USA* 104: 17565–17566.

Drake, J.M. 2005. Population effects of increased climate variation. *Proceedings of the Royal Society of London, B* 272: 1823–1827.

Drent, R., & Prop, J. 2008. Barnacle Goose *Branta leucopsis* survey on Nordenskiöldkysten, west Spitsbergen 1975–2007: breeding in relation to carrying capacity and predator impact. *Circumpolar Studies* 4: 59–83.

Drever, M.C. & Clark, R.G. 2007. Spring temperature, clutch initiation date and duck nest success: a test of the mismatch hypothesis. *Journal of Animal Ecology* 76: 139–148.

Drever, M.C., Clark, R.G., Derksen, C., Slattery, S.M., Toose, P. & Nudds, T.D. 2012. Population vulnerability to climate change linked to timing of breeding in boreal ducks. *Global Change Biology* 18: 480–492.

Duebbert, H.F. & Kantrud, H.A. 1974. Upland duck nesting related to land use and predator reduction. *Journal of Wildlife Management* 38: 257–265.

Dugger, B.D., Reinecke, K.J. & Fredrickson, L.H. 1994. Late winter survival of female Mallards in Arkansas. *Journal of Wildlife Management* 58: 94–99.

du Toit, J.T. 2010. Considerations of scale in biodiversity conservation. *Animal Conservation* 13: 229–236.

Eberhardt, L.L. 1977. Optimal policies for conservation of large mammals, with special references to marine ecosystems. *Environmental Conservation* 4: 205–212.

Eberhardt, L.L. 2002. A paradigm for population analysis of long-lived vertebrates. *Ecology* 83: 2841–2854.

Elmberg, J., Gunnarsson, G., Pöysä, H., Sjöberg, K. & Nummi, P. 2005. Within-season sequential density dependence regulates breeding success in Mallards *Anas platyrhynchos*. *Oikos* 108: 582–590.

Fischer, G. & Heilig, G.K. 1997. Population momentum and the demand on land and water resources. *Philosophical Transactions of the Royal Society, B* 352: 869–889.

Flint, P.L., Grand, J.B. & Rockwell, R.F. 1998. A model of Northern Pintail productivity and population growth rate. *Journal of Wildlife Management* 62: 1110–1118.

Flint, P.L., Grand, J.B., Fondell, T.F. & Morse, J.A. 2006. Population dynamics of Greater Scaup breeding on the Yukon-Kuskokwim Delta, Alaska. *Wildlife Monographs* 162: 1–22.

Freckleton, R.P., Watkinson, A.R., Green, R.E. & Sutherland, W.J. 2006. Census error and the detection of density dependence. *Journal of Animal Ecology* 75: 837–851.

Gaillard, J.-M., Pontier, D., Allainé, D., Lebreton, J.-D. & Trouvilliez, J. 1989. An analysis of demographic tactics in birds and mammals. *Oikos* 56: 59–76.

Gaillard, J.-M., Festa-Bianchet, M. & Yoccoz, N.G. 1998. Population dynamics of large herbivores: variable recruitment with constant adult survival. *Trends in Ecology & Evolution* 13: 58–63.

Gaillard, J.-M., Festa-Bianchet, M., Yoccoz, N.G., Loison, A. & Toigo, C. 2000. Temporal variation in fitness components and population dynamics of large herbivores. *Annual Review of Ecology and Systematics* 31: 367–393.

Gauthier, G., Bêty, J., Cadieux, M.-C., Legagneux, P., Doiron, M., Chevallier, C., Lai, S., Tarroux, A. & Berteaux, D. 2013. Long-term monitoring at multiple trophic levels suggests heterogeneity in responses to climate change in the Canadian Arctic tundra. *Philosophical Transactions of the Royal*

Society, B 368: 20120482. http://dx.doi.org/10.1098/rstb.2012.0482.

Gillespie, J.H. 1977. Natural selection for variance in offspring numbers: a new evolutionary principle. *American Naturalist* 111: 1010–1014.

Gilliland, S.G., Gilchrist, H.G., Rockwell, R.F., Robertson, G.J., Savard, J.-P.L., Merkel, F. & Mosbech, A. 2009. Evaluating the sustainability of harvest among northern Common Eiders *Somateria mollissima borealis* in Greenland and Canada. *Wildlife Biology* 15: 24–36.

Gormezano, L.J. & Rockwell, R.F. 2013. What to eat now? Shifts in Polar Bear diet during the ice-free season in western Hudson Bay. *Ecology and Evolution* 3: 3509–3523.

Grand, J.B., Koons, D.N., Arnold, J.M. & Derksen, D.V. 2006. *Modeling the Recovery of Avian Populations*. USGS Report to the Alaska Science Center, Anchorage, USA.

Grime, J.P. 1977. Evidence for the existence of three primary strategies in plants and its relevance to ecological and evolutionary theory. *American Naturalist* 111: 1169–1194.

Gunnarsson, G., Elmberg, J., Sjöberg, K., Pöysä, H. & Nummi, P. 2006. Experimental evidence for density-dependent survival in Mallard (*Anas platyrhynchos*) ducklings. *Oecologia* 149: 203–213.

Gunnarsson, G. & Elmberg, J. 2008. Density-dependent nest predation – an experiment with simulated Mallard nests in contrasting landscapes. *Ibis* 150: 259–269.

Gunnarsson, G., Elmberg, J., Pöysä, H. Nummi, P., Sjöberg, K., Dessborn, L. & Arzel, C. 2013. Density dependence in ducks: a review of the evidence. *European Journal of Wildlife Research* 59: 305–321.

Gurney, K.E.B., Clark, R.G., Slattery, S.M., Smith-Downey, N.V., Walker, J., Armstrong, L.M., Stephens, S.E., Petrula, M., Corcoran, R.M., Martin, K.H., DeGroot, K.A., Brook, R.W., Afton, A.D., Cutting, K., Warren, J.M., Fournier, M. & Koons, D.N. 2011. Time constraints in temperate-breeding species: influence of growing season length on reproductive strategies. *Ecography* 34: 628–636.

Gwinner, E. 1996. Circannual clocks in avian reproduction and migration. *Ibis* 138: 47–63.

Harvey, P.H. & Zammuto, R.M. 1985. Patterns of mortality and age at first reproduction in natural populations of mammals. *Nature* 315: 319–320.

Hauser, C.E., Cooch, E.G. & Lebreton, J.-D. 2006. Control of structured populations by harvest. *Ecological Modelling* 196: 462–470.

Henden, J.-A., Bårdsen, B.-J., Yoccoz, N.G. & Imset, R.A. 2008. Impacts of differential prey dynamics on the potential recovery of endangered Arctic Fox populations. *Journal of Applied Ecology* 45: 1086–1093.

Hepp, G.R., Kennamer, R.A. & Harvey, W.F. 1989. Recruitment and natal philopatry of Wood Ducks. *Ecology* 70: 897–903.

Hepp, G.R. & Kennamer, R.A. 1993. Effects of age and experience on reproductive performance of Wood Ducks. *Ecology* 74: 2027–2036.

Heppell, S.S. 1998. Application of life-history theory and population model analysis to turtle conservation. *Copeia* 1998: 367–375.

Hik, D.S. & Jefferies, R.L. 1990. Increases in the net above-ground primary production of a saltmarsh forage grass: a test of the predictions of the herbivore-optimization model. *Journal of Ecology* 78: 180–195.

Hoekman, S.T., Mills, L.S., Howerter, D.W., Devries, J.H. & Ball, I.J. 2002. Sensitivity analyses of the life cycle of midcontinent Mallards. *Journal of Wildlife Management* 66: 883–900.

Hoekman, S.T., Gabor, T.S., Petrie, M.J., Maher, R., Murkin, H.R. & Lindberg, M.S. 2006. Population dynamics of Mallards breeding in agricultural environments in eastern Canada. *Journal of Wildlife Management* 70: 121–128.

Iles, D.T., Rockwell, R.F., Matulonis, P., Robertson, G.J., Abraham, K.F., Davies, C. & Koons, D.N. 2013a. Predators, alternative prey and climate influence annual breeding success of a long-lived sea duck. *Journal of Animal Ecology* 82: 683–693.

Iles, D.T. Peterson, S.L., Gormezano, L.J., Koons, D.N. & Rockwell, R.F. 2013b. Terrestrial predation by Polar Bears: not just a wild goose chase. *Polar Biology* 36: 1373–1379.

Intergovernmental Panel on Climate Change (IPCC). 2013. Climate change 2013: the physical science basis. *In* T.F. Stocker, D. Qin, G.-K. Plattner, M. Tignor, S.K. Allen, J. Boschung, A. Nauels, Y. Xia, V. Bex & P.M. Midgley (eds.), *Contribution of Working Group I to the Fifth Assessment Report of the Intergovernmental Panel on Climate Change*, 1535 pp. Cambridge University Press, Cambridge, UK.

Iverson, S.A., Gilchrist, H.G., Smith, P.A., Gaston, A.J. & Forbes, M.R. 2014. Longer ice-free seasons increase the risk of nest depredation by Polar Bears for colonial breeding birds in the Canadian Arctic. *Proceedings of the Royal Society, B* 281: 20133128. http://dx.doi.org/10.1098/rstb.2013.3128.

Jamieson, L.E. & Brooks, S.P. 2004. Density dependence in North American ducks. *Animal Biodiversity and Conservation* 27: 113–128.

Jäkäläniemi, A., Ramula, S. & Tuomi, J. 2013. Variability of important vital rates challenges the demographic buffering hypothesis. *Evolutionary Ecology* 27: 533–545.

Johnson, W.C., Werner, B., Guntenspergen, G.R., Voldseth, R.A., Millett, B., Naugle, D.E., Tulbure, M., Carroll, R.W.H., Tracy, J. & Olawsky, C. 2010. Prairie wetland complexes as landscape: functional units in a changing climate. *BioScience* 60: 128–140.

Kaminski, R.M. & Gluesing, E.A. 1987. Density- and habitat-related recruitment in Mallards. *Journal of Wildlife Management* 51: 141–148.

Kennamer, R.A. & Hepp, G.R. 1987. Frequency and timing of second broods in Wood Ducks. *Wilson Bulletin* 99: 655–662.

Kennamer, R.A. & Hepp, G.R. 2000. Integration of long-term research with monitoring goals: breeding Wood Ducks on the Savannah River Site. *Studies in Avian Biology* 21: 39–49.

Knape, J. & De Valpine, P. 2012. Are patterns of density dependence in the global population dynamics database driven by observation errors? *Ecology Letters* 15: 17–23.

Koons, D.N., Grand, J.B., Zinner, B. & Rockwell, R.F. 2005. Transient population dynamics: relations to life history and initial population state. *Ecological Modelling* 185: 283–297.

Koons, D.N., Grand, J.B. & Arnold, J.M. 2006a. Population momentum across vertebrate life histories. *Ecological Modelling* 197: 418–430.

Koons, D.N., Rockwell, R.F. & Grand, J.B. 2006b. Population momentum: implications for wildlife management. *Journal of Wildlife Management* 70: 19–26.

Koons, D.N., Rotella, J.J., Willey, D.W., Taper, M.L., Clark R.G., Slattery, S., Brook, R.W., Corcoran, R.M. & Lovvorn, J.R. 2006c. Lesser Scaup population dynamics: what can be learned from available data? *Avian Conservation and Ecology* 1(3): 6. [online] URL: http://www.ace-eco.org/vol1/iss3/art6/.

Koons, D.N., Holmes, R.R. & Grand, J.B. 2007. Population inertia and its sensitivity to changes in vital rates and population structure. *Ecology* 88: 2857–2867.

Koons, D.N., Pavard, S., Baudisch, A. & Metcalf, C.J.E. 2009. Is life-history buffering or lability adaptive in stochastic environments? *Oikos* 118: 972–980.

Koons, D.N., Rockwell, R.F. & Aubry, L.M. 2014. Effects of exploitation on an overabundant species: the Lesser Snow Goose predicament. *Journal of Animal Ecology* 83: 365–374.

Krebs, C.J. 1995. Two paradigms of population regulation. *Wildlife Research* 22: 1–10.

Krebs, C.J. 2002. Two complementary paradigms for analysing population dynamics. *Philosophical Transactions of the Royal Society, B* 357: 1211–1219.

Lebreton, J.-D. & Gimenez, O. 2013. Detecting and estimating density dependence in wildlife populations. *Journal of Wildlife Management* 77: 12–23.

Lemly, A.D., Kingsford, R.T. & Thompson, J.R. 2000. Irrigated agriculture and wildlife conservation: conflict on a global scale. *Environmental Management* 25: 485–512.

Lewontin, R.C. & Cohen, D. 1969. On population growth in a randomly varying environment. *Proceedings of the National Academy of Sciences, USA* 62: 1056–1060.

Lindholm, A., Gauthier, G. & Desrochers, A. 1994. Effects of hatch date and food-supply on gosling growth in Arctic nesting Greater Snow Geese. *Condor* 96: 898–908.

Loonen, M.J.J.E., Oosterbeek, K. & Drent, R.H. 1997. Variation in growth of young and adult size in Barnacle Geese *Branta leucopsis*: evidence for density dependence. *Ardea* 85: 177–192.

Ludwichowski, I., Barker, R. & Bräger, S. 2002. Nesting area fidelity and survival of female Common Goldeneyes *Bucephala clangula*: are they density-dependent? *Ibis* 144: 452–460.

Madsen, J., Bregnballe, T., Frikke, J. & Kristensen, I.B. 1998. Correlates of predator abundance with snow and ice conditions and their role in determining timing of nesting and breeding success in Svalbard Light-bellied Brent geese *Branta bernicla hrota*. *Norsk Polarinstitutt Skrifter* 200: 221–234.

MacArthur, R.H. & Wilson, E.O. 1967. *The Theory of Island Biogeography*. Princeton University Press, Princeton, New Jersey, USA.

Malthus. T.R. 1798. An essay on the principle of population. *In* G. Hardin (ed.), *Population, Evolution, and Birth Control: A Collage of Controversial Ideas* (1964), pp. 4–16. Freeman, San Francisco, California, USA.

Manly, B.F., McDonald, L., Thomas, D.L., McDonald, T.L. & Erickson, W.P. 2002. *Resource Selection by Animals: Statistical Design and Analysis for Field Studies, 2nd edition*. Kluwer, Boston, Massachusetts, USA.

Mattsson, B.J., Runge, M.C., Devries, J.H., Boomer, G.S., Eadie, J.M., Haukos, D.A., Fleskes, J.P., Koons, D.N., Thogmartin, W.E. & Clark R.G. 2012. A modeling framework for integrated harvest and habitat management of North American waterfowl: Case-study of Northern Pintail metapopulation dynamics. *Ecological Modelling* 225: 146–158.

McArdle, B.H. 2003. Lines, models, and errors: regression in the field. *Limnology and Oceanography* 48: 1363–1366.

Menu, S., Gauthier, G. & Reed, A. 2002. Changes in survival rates and population dynamics of Greater Snow Geese over a 30-year period: implications for hunting regulations. *Journal of Applied Ecology* 39: 91–102.

Morris, W.F. & Doak, D.F. 2002. *Quantitative Conservation Biology: Theory and Practice of Population Viability Analysis*. Sinauer, Sunderland, Massachusetts, USA.

Morris, W.F. & Doak, D.F. 2004. Buffering of life histories against environmental stochasticity: accounting for a spurious correlation between the variabilities of vital rates and their contributions to fitness. *American Naturalist* 163: 579–590.

Morris, W.F., Pfister, C.A., Tuljapurkar, S., Haridas, C.V., Boggs, C.L., Boyce, M.S., Bruna, E.M., Church, D.R., Coulson, T., Doak, D.F., Forsyth, S., Gaillard, J.-M., Horvitz, C.C., Kalisz, S., Kendall, B.E., Knight, T.M., Lee, C.T. & Menges, E.S. 2008. Longevity can buffer plant and animal populations against changing climatic variability. *Ecology* 89: 19–25.

Murray, D.L., Anderson, M.G. & Steury, T.D. 2010. Temporal shift in density dependence among North American breeding duck populations. *Ecology* 91: 571–581.

Nicolai, C.A. & Sedinger, J.S. 2012. Are there trade-offs between pre- and post-fledging survival in Black Brent geese? *Journal of Animal Ecology* 81: 788–797.

Noel, L.E., Johnson, S.R., O'Doherty, G.M. & Butcher, M.K. 2005. Common Eider (*Somateria mollissima v-nigrum*) nest cover and depredation on central Alaskan Beaufort Sea barrier islands. *Arctic* 58: 129–136.

Northrup, J.M. & Wittemyer, G. 2013. Characterising the impacts of emerging energy development on wildlife, with an eye towards mitigation. *Ecology Letters* 16: 112–125.

Nudds, T.D. 1992. Patterns in breeding duck communities. *In* B.D.J. Batt, A.D. Afton, M.G. Anderson, C.D. Ankney, D.H. Johnson, J.A. Kadlec & G.L. Krapu (eds.), *Ecology and Management of Breeding Waterfowl*, pp. 540–567. University of Minnesota Press, Minneapolis, USA.

Nummi, P. & Saari, L. 2003. Density-dependent decline of breeding success in an introduced, increasing Mute Swan population. *Journal of Avian Biology* 34: 105–111.

Oli, M.K. & Zinner, B. 2001. Partial life-cycle analysis: a model for pre-breeding census data. *Oikos* 93: 376–387.

Oli, M.K., Hepp, G.R. & Kennamer, R.A. 2002. Fitness consequences of delayed maturity in female Wood Ducks. *Evolutionary Ecology Research* 4: 563–576.

Osnas, E.E., Runge, M.C., Mattsson B.J., Austin, J., Boomer, G.S., Clark, R.G., Devers, P., Eadie, J. M., Lonsdorf, E.V. & Tavernia, B.G. 2014. Managing harvest and habitat as integrated components. *Wildfowl* (Special Issue No. 4): 305–328.

Palacios, M.G., Winkler, D.W., Klasing, K.C., Hasselquist, D. & Vleck, C.M. 2011. Consequences of immune system aging in nature: a study of immunosenescence costs in free-living Tree Swallows. *Ecology* 92: 952–966.

Patterson, I.J. 1982. *The Shelduck – A Study in Behavioural Ecology*. Cambridge University Press, Cambridge, UK.

Péron, G., Nicolai, C.A. & Koons, D.N. 2012. Demographic response to perturbations: the role of compensatory density dependence in a North American duck under variable harvest regulations and changing habitat. *Journal of Animal Ecology* 81: 960–969.

Péron, G., & Koons, D.N. 2012. Integrated modeling of communities: parasitism, competition, and demographic synchrony in sympatric ducks. *Ecology* 93: 2456–2464.

Péron, G., & Koons, D.N. 2013. Intra-guild interactions and projected impact of climate and land use changes on North American pochard ducks. *Oecologia* 172: 1159–1165.

Person, B.T., Herzog, M.P., Ruess, R.W., Sedinger, J.S., Anthony, R.M. & Babcock, C.A. 2003. Feedback dynamics of grazing lawns: coupling vegetation change with animal growth. *Oecologia* 135: 583–592.

Pianka, E.R. 1970. On r- and K-Selection. *American Naturalist* 104: 592–597.

Pfister, C.A. 1998. Patterns of variance in stage-structured populations: evolutionary predictions and ecological implications. *Proceedings of the National Academy of Sciences, USA* 95: 213–218.

Pringle, C.M. 2000. Threats to U.S. public lands from cumulative hydrologic alterations outside of their boundaries. *Ecological Applications* 10: 971–989.

Promislow, D.E.L. & Harvey, P.H. 1990. Living fast and dying young: a comparative analysis of life-history variation among mammals. *Journal of Zoology* 220: 417–437.

Raveling, D.G. 1989. Nest-predation rates in relation to colony size of Black Brant. *Journal of Wildlife Management* 53: 87–90.

Reynolds, R.E., Shaffer, T.L., Renner, R.W., Newton, W.E., & Batt, B.D.J. 2001. Impact of the Conservation Reserve Program on duck recruitment in the US Prairie Pothole Region. *Journal of Wildlife Management* 65: 765–780.

Rice, S.H., Papadopoulos, A. & Harting, J. 2011. Stochastic processes driving directional evolution. *In* P. Pontarotti (ed.), *Evolutionary Biology – Concepts, Biodiversity, Macroevolution and Genome Evolution*, pp. 21–33. Springer-Verlag, Berlin, Germany.

Ricklefs, R.E. 2008. Disintegration of the ecological community. *American Naturalist* 172: 741–750.

Ringelman, K.M., Eadie, J.M. & Ackerman, J.T. 2012. Density-dependent nest predation in waterfowl: the relative importance of nest density versus nest dispersion. *Oecologia* 169: 695–702.

Rockwell, R.F. & Gormezano, L.J. 2009. The early bear gets the goose: climate change, Polar Bears and Lesser Snow Geese in western Hudson Bay. *Polar Biology* 32: 539–547.

Rockwell, R.F., Gormezano, L.J. & Koons, D.N. 2011. Trophic matches and mismatches: can Polar Bears reduce the abundance of nesting Snow Geese in western Hudson Bay? *Oikos* 120: 696–709.

Rohwer, F.C. 1985. The adaptive significance of clutch size in prairie ducks. *Auk* 102: 354–361.

Rohwer, F.C., Johnson, W.P., and Loos, E.R. 2002. Blue-winged Teal (*Anas discors*). *In* A. Poole (ed.), *The Birds of North America Online*. Cornell Laboratory of Ornithology, Ithaca, USA. Available at http://bna.birds.cornell/bna/species/625doi:10.2173/bna.625 (last accessed 14 March 2014).

Rotella, J.J., Link, W.A., Nichols, J.D., Hadley, G.L., Garrott, R.A., and Proffitt, K.M. 2009. An evaluation of density-dependent and density-independent influences on population growth rates in Weddell Seals. *Ecology* 90: 975–984.

Rotella, J.J., Link, W.A., Chambert, T., Stauffer, G.E. & Garrott, R.A. 2012. Evaluating the demographic buffering hypothesis with vital rates estimated for Weddell Seals from 30 years of mark–recapture data. *Journal of Animal Ecology* 81: 162–173.

Runge, M.C., Johnson, F.A., Anderson, M.G., Koneff, M.D., Reed, E.T. & Mott, S.E. 2006. The need for coherence between waterfowl harvest and habitat management. *Wildlife Society Bulletin* 34: 1231–1237.

Sæther, B.-E. 1988. Pattern of covariation between life history traits of European birds. *Nature* 331: 616–627.

Sæther, B.-E. & Bakke, O. 2000. Avian life history variation and contribution of demographic traits to the population growth rate. *Ecology* 81: 642–653.

Sæther, B.-E., Lillegård, M., Grøtan, V., Drever, M.C., Engen, S., Nudds, T.D. & Podruzny, K.M. 2008. Geographical gradients in the population dynamics of North American prairie ducks. *Journal of Animal Ecology* 77: 869–882.

Schamber, J.L., Flint, P.L., Grand, J.B., Wilson, H.M. & Morse, J.A. 2009. Population dynamics of Long-tailed Ducks breeding on the Yukon-Kuskokwim Delta, Alaska. *Arctic* 62: 190–200.

Schmidt, K.A. & Ostfeld, R. S. 2008. Numerical and behavioral effects within a pulse-driven system: consequences for shared prey. *Ecology* 89: 635–646.

Schmutz, J.A. 1993. Survival and pre-fledging body mass in juvenile Emperor Geese. *Condor* 95: 222–225.

Schmutz, J.A. 2009. Stochastic variation in avian survival rates: life-history predictions, population consequences, and the potential responses to human perturbations and climate change. *Environmental and Ecological Statistics* 3: 441–461.

Schmutz, J.A. & Laing, K.K. 2002. Variation in foraging behavior and body mass in broods of Emperor Geese (*Chen canagica*): evidence for interspecific density dependence. *Auk* 119: 996–1009.

Schmutz, J.A., Rockwell, R.F. & Petersen, M.R. 1997. Relative effects of survival and reproduction on the population dynamics of Emperor Geese. *Journal of Wildlife Management* 61: 191–201.

Sedinger, J.S. & Alisauskas R.T. 2014. Cross-seasonal effects and the dynamics of waterfowl populations. *Wildfowl* (Special Issue No. 4): 277–304.

Sedinger, J.S. & Chelgren, N.D. 2007. Survival and breeding advantages of larger Black Brant goslings: within and among cohort variation. *Auk* 124: 1281–1293.

Sedinger, J.S., Flint, P.L. & Lindberg, M.S. 1995. Environmental influence on life-history traits: growth, survival, and fecundity in Black Brant (*Branta bernicla*). *Ecology* 76: 2404–2414.

Sedinger, J.S., Lindberg, M.S., Person, B.T., Eichholz, M.W., Herzog, M.P. & Flint, P.L. 1998. Density dependent effects on growth, body size and clutch size in Black Brant. *Auk* 115: 613–620.

Sedinger, J.S., Nicolai, C.A., Lensink, C.J., Wentworth, C. & Conant, B. 2007. Black Brant harvest, density dependence, and survival: a record of population dynamics. *Journal of Wildlife Management* 71: 496–506.

Sedinger, J.S., Chelgren, N.D., Lindberg, M.S. & Ward, D.H. 2008. Fidelity and breeding probability related to population density and individual quality in Black Brent geese (*Branta bernicla nigricans*). *Journal of Animal Ecology* 77: 702–712.

Sibly, R.M. & Brown, J.H. 2007. Effects of body size and lifestyle on evolution of mammal life histories. *Proceedings of the National Academy of Sciences, USA* 104: 17707–17712.

Silvertown, J., Franco, M. & McConway, K. 1992. A demographic interpretation of Grime's triangle. *Functional Ecology* 6: 130–136.

Sjöberg, K., Gunnarsson, G., Pöysä, H., Elmberg, J. & Nummi, P. 2011. Born to cope with climate change? Experimentally manipulated hatching time does not affect duckling survival in the Mallard *Anas platyrhynchos*. *European Journal of Wildlife Research* 57: 505–516.

Smith, L.M., Pederson, R.L. & Kaminski, R.M. (eds.) 1989. *Habitat Management for Migrating and Wintering Waterfowl in North America*. Texas Tech University Press, Lubbock, Texas, USA.

Smith, P.A., Elliott, K.H., Gaston, A.J. & Gilchrist, H.G. 2010. Has early ice clearance increased predation on breeding birds by Polar Bears? *Polar Biology* 33: 1149–1153.

Stahl, J.T. & Oli, M.K. 2006. Relative importance of avian life-history variables to population growth rate. *Ecological Modelling* 198: 23–39.

Stearns, S.C. 1992. *The Evolution of Life Histories*. Oxford University Press, New York, USA.

Stenseth, N.C., Viljugrein, H., Saitoh, T., Hansen, T.F., Kittilsen, M.O., Bølviken, E. & Glöckner, F. 2003. Seasonality, density dependence, and population cycles in Hokkaido Voles. *Proceedings of the National Academy of Sciences, USA* 100: 11478–11483.

Stephens, S.E., Rotella, J.J., Lindberg, M.S., Taper, M.L. & Ringelman, J.K. 2005. Duck nest survival in the Missouri Coteau of North Dakota: landscape effects at multiple spatial scales. *Ecological Applications* 15: 2137–2149.

Stirling, I. & Derocher, A.E. 1993. Possible impacts of climatic warming on Polar Bears. *Arctic* 46: 240–245.

Stirling, I. & Parkinson, C.L. 2006. Possible effects of climate warming on selected populations of Polar Bears (*Ursus maritimus*) in the Canadian Arctic. *Arctic* 59: 261–275.

Strong, L.L. & Trost, R.E. 1994. Forecasting production of Arctic nesting geese by monitoring snow cover with advanced very high resolution radiometer (AVHRR) data. *Proceedings of the Pecora Symposium* 12: 425–430.

Tombre, I.M., Black, J.M. & Loonen, M.J.J.E. 1998. Critical components in the dynamics of a Barnacle Goose colony: a sensitivity analysis. *In* F. Mehlum, J.M. Black & J. Madsen (eds.), *Research on Arctic Geese*, pp. 81–89. Proceedings of the Svalbard Goose Symposium, Oslo, Norway, Norsk Polarinstitutt Skrifter 200.

Tombre, I.M., Høgda, K.A., Madsen, J., Griffin, L.R., Kuijken, E., Shimmings, P., Rees, E. & Verscheure, C. 2008. The onset of spring and timing of migration in two arctic nesting goose populations: the pink-footed goose *Anser brachyrhynchus* and the barnacle goose *Branta leucopsis*. *Journal of Avian Biology* 39: 691–703.

Tuljapurkar, S.D. 1982. Population dynamics in variable environments. II. Correlated environments, sensitivity analysis and dynamics. *Theoretical Population Biology* 21: 114–140.

Turchin, P. 2003. *Complex Population Dynamics*. Princeton University Press, Princeton, New Jersey, USA.

van der Jeugd, H.P., Eichhorn, G., Litvins, K.E. Stahl, J., Larsson, K., van der Graaf, A.J. & Drent, R.H. 2009. Keeping up with early springs: rapid range expansion in an avian herbivore incurs a mismatch between reproductive timing and food supply. *Global Change Biology* 15: 1057–1071.

Viljugrein, H., Stenseth, N.C., Smith, G.W. & Steinbakk, G.H. 2005. Density dependence in North American ducks. *Ecology* 86: 245–254.

Visser, M.E. & Both, C. 2005. Shifts in phenology due to global climate change: the need for a yardstick. *Proceedings of the Royal Society, B* 272: 2561–2569.

Walker, J., Rotella, J.J., Stephens, S.E., Lindberg, M.S., Ringelman, J.K., Hunter, C. & Smith, A.J. 2013. Time-lagged variation in pond density and primary productivity affects duck nest survival in the Prairie Pothole Region. *Ecological Applications* 23: 1061–1074.

Walters, C. 1986. *Adaptive Management of Renewable Resources*. MacMillan, New York, USA.

Wetherald, R. 2009. Changes of variability in response to increasing greenhouse gases. Part II: hydrology. *Journal of Climate* 22: 6089–6103.

Williams, T.D., Cooch, E.G., Jefferies, R.L. & Cooke, F. 1993. Environmental degradation, food limitation and reproductive output: juvenile survival in Lesser Snow Geese. *Journal of Animal Ecology* 62: 766–777.

Wilson, H.M., Flint, P.L., Powell, A.N., Grand, J.B. & Moran, C.L. 2012. Population ecology of breeding Pacific Common Eiders on the Yukon-Kuskokwim Delta, Alaska. *Wildlife Monographs* 182: 1–28.

Wisdom, M.J., Mills, L.S. & Doak, D.F. 2000. Life-stage simulation analysis: estimating vital-rate effects on population growth for conservation. *Ecology* 81: 628–641.

Wright, C.K. & Wimberly, M.C. 2013. Recent land use change in the western corn belt threatens grasslands and wetlands. *Proceedings of the National Academy of Sciences, USA* 110: 4134–4139.

Conspecific brood parasitism in waterfowl and cues parasites use

HANNU PÖYSÄ[1]*, JOHN M. EADIE[2] & BRUCE E. LYON[3]

[1]Finnish Game and Fisheries Research Institute, Joensuu Game and Fisheries Research, Yliopistokatu 6, FI-80010 Joensuu, Finland.
[2]Department of Wildlife, Fish and Conservation Biology, University of California, One Shields Avenue, Davis, California 95616, USA.
[3]Department of Ecology and Evolutionary Biology, University of California, Santa Cruz, California 95064, USA.
*Corresponding author. E-mail: hannu.poysa@rktl.fi

Abstract

Conspecific brood parasitism (CBP) occurs in various insects, fishes and birds, but it is disproportionately common in waterfowl (Anatidae). Studies of CBP in Anatids therefore have helped to develop a fundamental conceptual framework with which to explain this intriguing behaviour. Yom-Tov (1980) first drew attention to CBP, and Andersson and Eriksson (1982) also hinted at the fascinating behavioural, ecological and evolutionary aspects of CBP in waterfowl. Several reviews followed these early papers, but much has been learned more recently about CBP in waterfowl. Here we aim to review the traditional conceptual framework of CBP in waterfowl and to consider empirical studies that have attempted to test related hypotheses. The survey provided support for the hypotheses that CBP allows some females to reproduce when not otherwise possible, whereas other females use parasitic egg-laying as a way to enhance their fecundity. A recently developed framework that considers CBP as part of a flexible life-history strategy could provide a useful direction for future studies of CBP. A second aim of this review is to consider the use of cues by conspecific brood parasites seeking suitable places to lay eggs parasitically. Recent studies have revealed remarkable cognitive abilities in parasitic females, but the actual mechanisms remain unknown. Clearly, breeding females are sensitive to cues such as nest site security, patterns of previous nest use or success, clutch size, and perhaps even the degree of kinship between hosts and other parasites. Indeed, additional investigations of CBP are needed to provide a better understanding of the processes and patterns of this avian reproductive strategy.

Key words: Anseriformes, brood parasitism, information use, life-history, nest predation risk.

Conspecific brood parasitism (CBP) is an alternative reproductive tactic in which a female lays eggs in the nests of other conspecific individuals and leaves the subsequent care of the eggs and young to the host female. CBP has been documented in at least 234 species of birds and is particularly prevalent in Anseriformes where it has been reported in 76 of the 161 species (Yom-Tov 2001).

Several comprehensive reviews have been published on the hypotheses for the occurrence and evolution of CBP in birds (Eadie *et al*. 1988; Rohwer & Freeman 1989; Sayler 1992; Lyon & Eadie 2008) and this paper does not intend to provide another broad-ranging overview of the CBP breeding strategy. However, empirical and theoretical research on CBP has grown since the original paper by Yom-Tov (1980) and much of this work has focused on waterfowl (Fig. 1). Development of theory to explain CBP grew through the 1990s peaking in 2001–2005. Empirical studies

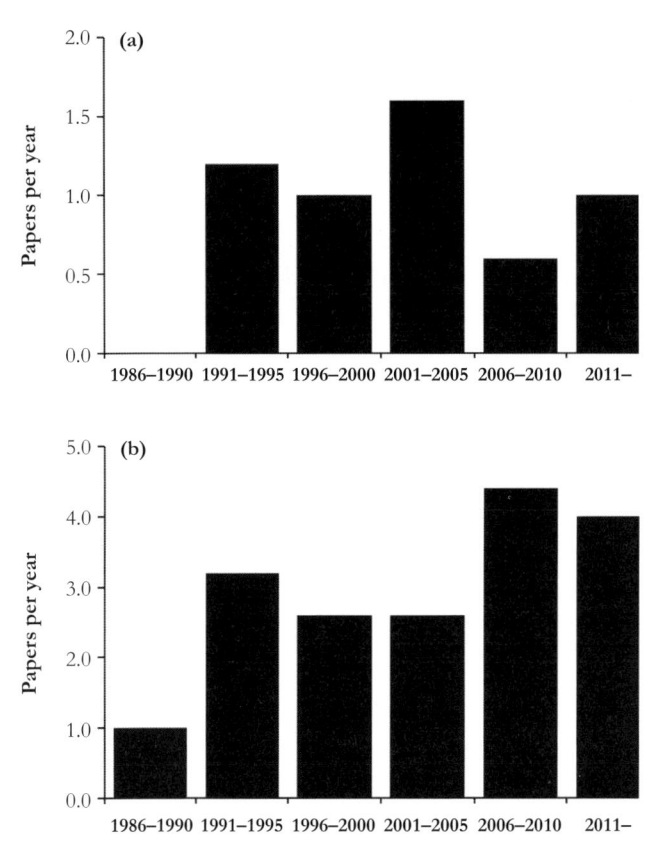

Figure 1. Number of (a) theoretical papers published on CBP in general (top panel; $n = 24$ papers in total), and (b) empirical papers published on CBP in waterfowl (bottom panel; $n = 77$) over the past three decades.

lagged and reached their highest frequency in the last 5–10 years (Fig. 1). Many advances have been made, but there is a surprising amount that is not yet known. Indeed, for many species, it is still not clear which females within a population pursue this behaviour, nor do we fully understand the fitness consequences to parasites or their hosts. Much of the work to date has focussed on ecological factors that correlate with the occurrence of CBP, but longitudinal studies of females are still rare. Similarly, most studies are observational, albeit with an expanded toolkit of molecular genetic techniques which help to ascertain maternity. Experimental studies are uncommon with a few notable exceptions (Eadie 1989; Pöysä 2003a,b; Pöysä *et al.* 2010; Odell & Eadie 2010). Despite these gaps, the field is now at a point where some retrospection would be valuable. The initial goal of this paper therefore is to review briefly the traditional set of hypotheses posed to account for CBP in waterfowl and to evaluate how existing empirical work meets those expectations. We then offer an alternative conceptual framework proposed by Lyon & Eadie (2008) that could advance our understanding of this behaviour more effectively.

Secondly, considerable growth in this field involves the information that might be available to parasites and hosts to modulate their behaviour in an adaptive manner. How do females choose a nest or host to parasitize? What information might be available to females to shape their behavioural decisions? Brood parasites gain fitness by having other females provide parental care for their offspring. Even though parasitism represents a relatively cheap way to gain fitness at the expense of other individuals, this does not mean that parasites should lay their eggs indiscriminately. To the extent that fitness from parasitism can be enhanced by decisions that parasites make regarding where to lay their eggs, or how many eggs to lay in a given nest, natural selection should favour those decisions or tactics. However, parasites must be able to gather useful information about potential host nests that they can use to inform their laying decisions. Do they gather this information, and if so, what cues do they use? A second goal of this paper therefore is to examine the growing body of work that is beginning to explore the cues used by conspecific brood parasites. The focus here is on specific cues that the parasites may use to select host nests into which to lay eggs. This is not meant to imply that cues and decisions used by hosts are not important. For example, hosts may desert nests in response to CBP (Eadie 1989; Jaatinen *et al.* 2009), resulting in the direct loss of parasitic eggs. Hosts could also influence which parasites gain access to their nests (Åhlund 2005).

Conspecific brood parasitism in waterfowl

Here we provide a brief overview of the set of traditional hypotheses that have been suggested to explain CBP, outlined in earlier more comprehensive reviews (including Andersson 1984; Eadie *et al.* 1988; Sayler 1992; Lyon & Eadie 2008), and examine the evidence from the literature in support of these hypotheses for waterfowl. We then present a revision of the framework of hypotheses proposed by Lyon and Eadie (2008) to help guide further research in this field.

Is it parasitism or inadvertent competition?

Some early researchers considered parasitic egg-laying (also called "egg-dumping" in the older literature) in waterfowl to be non-adaptive, either because it reflects abnormal behaviour ("loss of maternal instinct") or because it is a side-effect of competition for suitable nest sites in hole-nesting waterfowl (Erskine 1990). Semel and Sherman (2001) resurrected this idea and proposed a similar mechanism to account for "apparent" CBP in Wood Duck *Aix sponsa*. They proposed that some nest sites are preferred, perhaps because of their quality or because young females return to their natal nest on breeding for the first time. Contests for these nests ensue with more than one female laying eggs in the nest, but ultimately only a single female incubates the clutch. The usurped females become *de facto* parasites. Parasitism was not the focus of their behaviour and arises only as an inadvertent consequence of laying eggs and failure to establish final ownership of the nest (*i.e.* accidental parasitism).

However, several recent lines of evidence argue against the accidental parasite hypothesis as an explanation for CBP in waterfowl, and in other birds as well. First, several researchers (Eadie 1989; Pöysä 2003b; Odell & Eadie 2010) observed frequent parasitic egg-laying in nests in which eggs were added experimentally to empty nests without a host female being present. Hence, parasitic laying in these instances cannot be explained as a result of the laying female being "ousted" from the nest by the host female who incubates the eggs, since there was no host with whom to compete. Second, Eadie (1989) conducted removal experiments with Barrow's Goldeneye *Bucephala islandica* and Common Goldeneye *B. clangula* and found that, when the host female was removed, putative parasites continued to lay eggs but did not incubate the eggs, despite the fact that there was no female to prevent them from doing so. Conversely, when the parasitic females were removed, the hosts continued to lay and ultimately incubate the clutch, demonstrating that the different response by parasites was not simply due to an effect of experimental disturbance. This suggests that parasites and hosts behave very differently right from the outset.

Finally, recent work by Åhlund (2005) has demonstrated striking differences in the behavioural tactics of parasitic and host Common Goldeneye females at the nest: hosts and parasites differed in the timing of egg-laying, deposition of down, covering eggs on departure, and time spent on the nest as the egg-laying sequence progressed. These observations suggest that CBP is a genuine reproductive tactic and not just a consequence of nest site competition (Åhlund 2005). Similarly, recent experimental studies have revealed sophisticated responses of parasitically laying Common Goldeneye females to variation in nest (egg) predation risk (Pöysä 2003a; Pöysä *et al.* 2010), while other studies have demonstrated clear fitness advantages of CBP for parasitically laying Common Goldeneye females (Åhlund & Andersson 2001). The fact that CBP occurs in such a large number of waterfowl species, many of which do not nest in tree cavities and do not compete for specific nests sites, argues against the accidental parasite hypothesis as a general explanation of this behaviour. These observations, together with the discovery that brood parasites often make fine-tuned, adaptive egg-laying decisions in waterfowl and other species (Brown & Brown 1991; Lyon & Everding 1996; Pöysä 1999; Lyon & Eadie 2008 and below), confirm that CBP is generally an adaptive alternative reproductive strategy in both waterfowl and other birds.

Traditional hypotheses

Adaptive hypotheses about the CBP have traditionally been classified into four types (summarised from Lyon & Eadie 2008) as follows:

Best-of-a-bad-job (BOBJ). According to this hypothesis, females lay eggs parasitically when they are unable to breed otherwise (constraint), or when environmental conditions are unfavourable such that the prospects for successful reproduction by nesting are low (restraint). A variety of ecological and physiological factors have been proposed to influence a female's ability to nest on her own, including nest site or territory limitation, body condition, age and experience.

Nest loss. A variant of the BOBJ hypothesis focuses on nest loss as the causative factor. Females that lose their nest to predation during egg-laying or early incubation may be able to lay some additional eggs (or may have eggs already developing in the ovary) but are not able (or it is not worthwhile) to establish a new nest. This hypothesis could be classified as a form of constraint (BOBJ), but many researchers have discussed it as a separate mechanism (and so we list it here in that form for comparison).

"Professional" or life-long specialist parasites. Under this hypothesis, females never raise their own young and only lay in the nests of other females. It is argued that these females have higher lifetime fitness (when rare in the population), because they are emancipated from the costs of parental care and so are able to invest in additional production of eggs. Under a game theoretic version of this hypothesis, negative frequency-dependent selection works to stabilise the frequencies of nesting and parasite females in the population (a mixed evolutionarily stable strategy; ESS).

Fecundity enhancement. This hypothesis posits that nesting females also lay some additional eggs parasitically and, by doing so, are able to increase fitness beyond that possible through nesting alone, presumably by bypassing some of the constraints or costs of raising the additional eggs/young on their own.

We were able to locate 17 studies that have attempted to test at least some of these hypotheses for waterfowl (Table 1). Of the four traditional hypotheses, no support has been found for either the nest loss hypothesis (0 of 5 studies that examined this hypothesis) or the life-long "professional"

Table 1. Summary of empirical studies in which the applicability of traditional hypotheses of conspecific brood parasitism (CBP) have been assessed. Tests of a specific best-of-a-bad-job hypothesis are given in parentheses.

Species	Hypotheses				Source
	Best-of-a-bad-job	Nest loss	Life-long parasite	Fecundity enhancement	
Bar-headed Goose	Yes			No	Weigmann & Lamprecht 1991
Snow Goose	Yes	No			Lank et al. 1989
Barnacle Goose	Yes				Forslund & Larsson 1995
Barnacle Goose	Yes		No	No	Anderholm et al. 2009
Maned Duck	Yes				Briggs 1991
Canvasback	Yes				Sorenson 1993
Redhead	Yes		No	Yes	Sorenson 1991
Wood Duck	Yes	No	No		Semel & Sherman 2001
Common Eider	No				Robertson et al. 1992
Common Eider	Yes/No			No	Robertson 1998
Common Eider	No (nest site limitation)			No	Waldeck et al. 2004
Common Eider	No (nest site limitation)	No			Waldeck et al. 2011
Common and Barrow's Goldeneye	Yes (nest site limitation)		No	Yes	Eadie 1989, 1991
Common Goldeneye	No (nest site limitation)	No			Pöysä 1999
Common Goldeneye			No	Yes	Åhlund & Andersson 2001
Common Goldeneye	No (nest site limitation)				Åhlund 2005
Ruddy Duck	No	No	No	Yes	Reichart et al. 2010

parasite hypothesis (0 of 6 studies) although the later may be difficult to detect given that it would be beyond the scope of most studies to follow a "pure" parasite throughout her entire lifetime (requires detailed observational data, recaptures, genotyping of all eggs in a population, *etc*.). Nonetheless, in most cases where females have been followed through time, they have been observed to switch between nesting and parasitism (or to use both strategies in the same year – dual nesters) suggesting that pure parasites, if they occur, are rare (Eadie 1989; Sorenson 1991; Åhlund & Andersson 2001; Reichart *et al*. 2010). To date, the hypotheses that are best supported are versions of the BOBJ hypothesis (10 of 16 studies) and, to a lesser extent, the fecundity enhancement hypothesis (4 of 8 studies; Table 1). Thus, at least for most waterfowl, there is support for the idea that some females pursue CBP due to constraint or restraint, whereas other females appear to do so to enhance total reproductive output. Few of these studies were able to determine which females did what. Perhaps the most interesting are the results of Åhlund and Andersson (2001) who showed that parasitic Common Goldeneyes comprised both females that only laid parasitically in a given season (pure parasites) and others that laid parasitically and also had a nest of their own (dual nesters). The reproductive "payoffs" varied considerably – dual nesters produced 1.5 times more offspring than non-parasitic (nesting) females and 2 times that of pure parasites. By combining parasitism with normal nesting, some females were able to double their reproduction. Similar patterns occur in Common and Barrow's Goldeneyes in British Columbia (Eadie 1989; Jaatinen *et al*. 2009, 2011). Clearly, more than one hypothesis can apply to the same population. How best to make sense of the range of outcomes summarised in Table 1 is considered below.

A revised framework for future research

Lyon and Eadie (2008) pointed out that the traditional set of hypotheses are potentially confounded at several levels, conflating what a female does (nest, parasitize, or both) with fitness benefits of doing so, with ecological factors influencing her decision (nest loss, nest limitation, host availability, *etc*.), and finally with the evolutionary dynamics that maintain some frequency of CBP in the population (frequency-dependent ESS). Lyon and Eadie (2008) proposed a revision to the traditional set of hypotheses and they based this revision partly on a conceptual framework derived from Sorenson's (1991) reproductive decision model. Under this model, the ability to lay some eggs parasitically allows females to fine-tune reproductive investment because without the possibility of CBP, females are faced with an all-or-none decision to nest or not to reproduce at all. Key to this framework is the fundamental difference between two contexts of brood parasitism – parasitism by non-nesting females and parasitism by nesting females. For non-nesting females, parasitism allows for an intermediate investment between no reproduction and nesting. Thus, females prevented from nesting can gain some fitness through parasitism where otherwise none would be possible. Parasitism by nesting females, in contrast, allows females to increase their reproductive effort when conditions are very good without entailing a full second nest effort. Parasitic egg-laying allows these females to adjust reproductive effort upwards in smaller increments to match expected returns.

This conceptual framework is a useful advance in two ways. First, it unifies all four possible nesting options (not breeding, parasitize only, nest only, parasitize and nest) as part of a single continuum that varies from low-to-high reproductive investment, and low-to-high expected fitness benefits. This captures the variation found both within and among species of waterfowl (Table 1). Second, parasitism can be combined with nesting in various ways over a female's lifetime to provide a flexible life-history, whereby females are able to modify their reproductive investment and options to variable ecological and social conditions. This framework moves the field forward from considering a large number of single independent hypotheses for each type of parasitism that intermix ecological factors, proximal influences, fitness benefits and evolutionary dynamics (the traditional framework; Table 1) into a more general life-history context with hypotheses that focus on the specific life-history trajectories of females and the expected fitness returns from pursuing those alternatives.

With this new framework, Lyon and Eadie (2008) proposed a modified categorisation of hypotheses for CBP, focusing on the three key fitness components: (a) current fecundity, (b) offspring survival, and (c) adult survival (*i.e.* future fecundity). Of particular importance is the distinction between parasites with and without their own nests because these two contexts likely involve different constraints, and different hypotheses may apply. Accordingly, there are three questions that must be addressed to understand the benefit of CBP: (a) does the female have a nest or not, (b) what fitness components and life-history trade-offs play a role in leading to increased fitness benefits via parasitism, and (c) what ecological, social, or physiological factors influence these trade-offs? Table 2 summarises Lyon and Eadies' (2008) revised hypothesis framework. Data are not available to test these hypotheses (few studies have followed the life-histories of individual parasites) and so we cannot place current studies in this new context. However, this is a new, more integrated framework for future studies of CBP in waterfowl and other birds. Perhaps the biggest requirement to improve understanding of this intriguing reproductive system is to determine more clearly what females are doing, both within a breeding season (nest, parasitize, or both) and among breeding seasons. Once a female's nesting status is determined, the context and suite of relevant hypotheses can be analysed more carefully and thus thoroughly evaluated (Table 2). This opens a wide range of new and intriguing questions about information use by the females which pursue these alternative pathways, and it is perhaps here where some of our newest insights on CBP have emerged. The second half of this review focuses on these new developments.

Table 2. A framework of hypotheses on the adaptive benefits of conspecific brood parasitism, modified from Table 1 in Lyon & Eadie 2008. This framework emphasises the distinction between hypotheses that apply to females without nests (strategy a) *versus* parasitizing females that also nested (strategy b).

Strategy	Mechanism	Fitness component enhanced	Traditional hypotheses
(a) Non-nesting parasite	**Egg production** (bypass costs of nesting and allocate more effort into egg production)	Current fecundity	Not emphasised; could explain life-long parasites
	Nest/territory limitation (unable to obtain a nest site or territory)	Adult survival and/or current fecundity	BOBJ (constraint, salvage strategy, nest limitation)
	Energy/condition/ experience (females in poor condition, young)	Adult survival	BOBJ (restraint, salvage strategy, energy limitation)
	Quality of brood rearing (parasites lay in high quality nests; good hosts/ safe sites)	Offspring survival	Nest predation could apply
(b) Nesting parasite	**Nest loss** (loss during egg-laying)	Current fecundity	BOBJ (constraint, salvage strategy, nest loss)
	Clutch size/brood size constraints (high quality females in excellent condition increase egg production, bypass brood size constraints)	Current fecundity and/or offspring survival	Fecundity enhancement Side-payment Dual nesting
	Cost of reproduction (reduce cost of care in own nest/brood to enhance future reproduction)	Adult survival	Not emphasised

Cues used by parasitic breeders

There is a long history of discussion about the distinction between signals and cues (see Danchin *et al.* 2008; Wagner & Danchin 2010). Signals are traits that have been designed by selection to convey information, whereas cues incidentally contain information but were not selected for that purpose (Maynard Smith & Harper 2003; Danchin *et al.* 2008). Information about the attributes of potential hosts or their nests are most likely cues, and signals might be expected in species where kin selection plays a role in facilitating the occurrence of CBP. Studies of obligate brood parasites nicely illustrate the importance of information use to parasitic tactics. Reproductive fitness for obligate parasites depends entirely on the success of their parasitic eggs, so there is strong selection to employ tactics that enhance the survival of the eggs and young. For example, obligate parasites often need to find nests of the right host species and collect information that enables them to time their laying to match that of the host's breeding cycle (Davies 2000). Although total female fitness in species with conspecific parasitism does not depend nearly as heavily on gains from parasitic eggs as obligate parasites, fitness from parasitism can nonetheless be substantial (Åhlund & Andersson 2001). Thus, whenever parasitism is a well-developed component of reproduction, we expect that the use of cues to select hosts will be important as well. The question then is whether conspecific brood parasites show tactics similar to those of obligate brood parasites and, if so, what types of information are used when making choices about parasitism.

Cues can reflect characteristics of the physical environment or the social environment. Types of information of particular interest in the context of CBP are "personal information" and "social information". The former is the information obtained by an individual's own interaction with the environment (*e.g.* its experience or history, such as the failure of a previous breeding attempt), whereas the latter refers to information obtained by observing other individuals (*e.g.* their location, phenotypic condition and reproductive performance). Danchin *et al.* (2008) term social information and non-private personal information (*i.e.* the information accessible to other individuals) as "public information". Social information may also be based on signals, *i.e.* traits that evolve and are involved in true communication between individuals (see Danchin *et al.* 2008). However, the role of signals in CBP has not been addressed (for interspecific brood parasitism, see Parejo & Avilés 2007).

How might conspecific brood parasites use information to modify parasitic egg-laying behaviour adaptively? Three issues must be considered: 1) how do researchers test for evidence of cue use; 2) what behaviours might enable potential parasites to acquire information; and 3) what specific cues might be used? The last question further entails a nested series of choices that parasites might make, each involving a distinct cue or set of cues, such as: (a) what general area to use (habitat cues), (b) which nests or females to parasitize, and (c) when to parasitize a host nest. In addition to these choices, parasites need to decide how many eggs to lay in a given nest and, if parasites also have their own nest, how many eggs to allocate to nesting *versus* parasitism. Information gained by parasites could influence any of these decisions and so it is important to clarify which aspect is being addressed. This in turn requires that researchers are thoughtful about their methodology and careful in interpreting the patterns observed.

Methods used to detect cue use

Three methods can be used to elucidate the tactics and cues used by brood parasites. Simplest is a comparison showing that parasitized and non-parasitized nests differ with respect to some attribute likely to be important to the success of the parasite, such as the quality of the territory, nest or host. However, this method provides somewhat weak evidence for parasitic tactics because the patterns may reflect the outcome of host defenses and not parasitic tactics. It also can be possible to obtain false positive evidence for non-random patterns of parasitism. For example, spatiotemporal clustering in attributes of host nests can result in patterns when data are analysed for the entire population but not at spatial and temporal scales that are relevant to the choices that individual parasites face (*e.g.* Lyon 1993; McRae & Burke 1996). To assess parasitic tactics properly, it may be necessary to understand the spatial and temporal patterns of parasitism and then assess the choices parasites make with respect to the pool of hosts that are actually available given the spatial and temporal constraints on parasitism (Brown & Brown 1991; Andersson & Åhlund 2000). Monte Carlo randomisations provide one powerful method for assessing patterns of brood parasitism in this context (Emlen & Wrege 1986; Lyon 1993). McRae and Burkes' (1996) study of Moorhens *Gallinula choropus* highlights the value of controlling for the spatial pool of available hosts. Host-parasite relatedness was higher than expected at the population level but was not different from random expectation given the pool of hosts actually available to brood parasites. Controlling for the pool of potential hosts revealed that parasites were not specifically targeting relatives (McRae & Burke 1996). However, neither population comparisons nor contrasts that control for spatiotemporal patterns of potential hosts provide definitive evidence for which cues parasites actually use to select nests. The problem is that factors that correlate with parasitism may not be the actual cues that parasites use when choosing nests to parasitize. Only experiments manipulating putative cues provide fully convincing evidence for cue use. These experiments are rarely done, but they have been conducted in a few waterfowl species (see below).

How do females obtain information?

Studies addressing behavioural aspects of nest site selection in Barrow's Goldeneye and Common Goldeneye have revealed that females gather information by prospecting for potential nest sites prior to the next breeding season (Eadie & Gauthier 1985; Zicus & Hennes 1989; Pöysä *et al.* 1999). Pöysä (2006) found for the latter species that this behaviour is associated with CBP: nest sites that were visited more frequently by prospecting females in year t had a higher probability of being parasitized in year $t + 1$, suggesting that parasites gather information through nest-site prospecting to target parasitic laying in particular nests.

Prospecting activity peaks after most nests have hatched (and ducklings have left nests), matching the time when cues of a successful nest (eggshell membranes and fragments; Fig. 2) are highly visible (J. Eadie, unpubl. data; H. Pöysä, unpubl. data). It is not

A. Wood Duck

(a) Not used (shavings, undisturbed)

(b) Abandoned or laying (no down)

(c) Active incubation (down, eggs covered)

(d) Active incubation/laying (female, eggs, down)

(e) Successful hatch (shell membranes, egg caps)

(f) Failed/depredated (rotten, broken eggs, shells)

B. Common Goldeneye

(a) Successful hatch (7 ducklings hatched and left the nest)

(b) Successful hatch (9 ducklings hatched of 11 eggs and left the nest)

Figure 2. Examples of cues available in (A) Wood Duck nests and (B) Common Goldeneye nests after nesting. **A:** Wood Duck nests showing various stages of nesting, from not used (a) through to failed (f). **B:** Common Goldeneye nests showing two different examples of successful nests.

known if prospecting behaviour is associated with CBP in the Barrow's Goldeneye but, interestingly, prospecting activity seems to be higher at nest sites that had been parasitized earlier in the season (see Fig. 2 in Eadie & Gauthier 1985).

Prospecting behaviour is not restricted to hole-nesting species. Schamel (1977) mentions that preferred nest-sites are visited (prospected) regularly by non-breeding females throughout the summer (*i.e.* after hatching) in the ground-nesting Common Eider *Somateria mollissima*, another duck in which CBP is common (*e.g.* Robertson 1998; Waldeck *et al.* 2008). Fast *et al.* (2010) showed experimentally that nest-site materials left from the previous year influence the use of nest bowls by Common Eider females: nest bowls containing down were occupied earlier than control nest bowls with no down. As discussed by the authors, one explanation could be that nest down may indicate previous nest success and nest-site safety to females prospecting for nests. It would be interesting to study whether these aspects, *i.e.* prospecting behaviour and cues indicating previous nest success, are associated with the occurrence of CBP in the species.

Cues used by parasites to locate nests

Cues used by parasitically laying females to find and select suitable host nests have been studied extensively in the context of interspecific (obligate) avian brood parasitism. The main hypotheses can be classified as those dealing with nest placement (*e.g.* nest exposure, characteristics of the surrounding habitat) and those dealing with host behaviour (*e.g.* conspicuous host behaviour, host activity; see Patten *et al.* 2011). While characteristics associated with nest placement in waterfowl may not be as diverse as they are in passerines, the most important group of host species for interspecific brood parasites, some general patterns emerge. First, CBP in waterfowl is more frequent in cavity-nesting species than in species that nest in emergent vegetation or upland (Rohwer & Freeman 1989; Sayler 1992; Eadie *et al.* 1998), implying that the ease in locating nest sites could play a role in CBP. Support for this idea also comes from cavity-nesting Wood Ducks where highly visible nest boxes are more frequently parasitized than less visible nest boxes (Semel *et al.* 1988; Roy Nielsen *et al.* 2006a; but see Jansen & Bollinger 1998 for a less clear effect). On the other hand, nest box visibility does not seem to affect the frequency of parasitism in another cavity-nesting duck, the Barrow's Goldeneye (Eadie *et al.* 1998). Similarly, Åhlund (2005, p. 434) mentions that parasitism rate does not differ between nests near the shore and nests further inland in the Common Goldeneye population he studied; visibility presumably differed considerably between the nest site types. In line with this, an experiment addressing nest site selection in Common Goldeneye revealed that females (potential parasites) prospect shore and forest boxes equally, irrespective of differences in the visibility of the nest boxes (Pöysä *et al.* 1999), suggesting that females are very capable at finding nest sites. Hence, while highly visible nest sites may be easier to locate, there must be other cues parasites use to locate and select nests.

Early observations suggested that parasitic females use the activity of other females (potential hosts) to locate nests. For example Weller (1959) describes several cases in which Redhead *Aythya americana* females apparently observed the nest-building and egg-laying activities of other females, leading the author to suggest that parasitic females used this behaviour to find nests. Similarly, several authors have suggested that Wood Duck females have a "decoying effect" on one another, leading often to heavily parasitized nests (Heusmann *et al.* 1980; Semel & Sherman 1986; see also Roy Nielsen *et al.* 2006a). Inspired by these observations Wilson (1993) carried out an experiment using Wood Duck decoys and found evidence for the hypothesis that parasitic Wood Duck females use the presence of conspecifics as a cue in the selection of nests. On the other hand, decoy nests with experimental eggs but no host were parasitized at the same rate as real nests that did have a host in Common Goldeneyes (Pöysä 2003b), Barrow's Goldeneyes (Eadie 1989) and Wood Ducks (Odell & Eadie 2010), suggesting that the presence of a conspecific host is not a necessary cue for parasites.

Cues used by parasites to select a nest

Location of potential nests for egg laying is the first step in the process of nest selection of parasites. However, not all of the located nests will eventually be parasitized (H. Pöysä, unpubl. data) suggesting that parasites actively select among potential nest sites. It is not always easy to make a clear distinction between these two steps in the process of nest selection, and the design of some studies does not allow a clear separation (see text on methodology above). In this section we consider only those studies that deal with the final step of the process, *i.e.* actual selection of nests by parasites, and review a variety of cues that have been identified.

Nest site quality or state

Empirical studies have explored several possible cues parasites may use to select a nest (Table 3). Several studies have addressed nest site characteristics while characteristics of the host female have received less attention. We are aware of only one study for waterfowl, on nest-box-breeding Common Goldeneye, in which both nest and host traits were considered, and nest site characteristics turned out to be more important than those of the host female (Paasivaara *et al.* 2010). Parasitism in relation to nest site quality has been well-studied in a non-waterfowl species, the Cliff Swallow *Hirundo pyrrhonota* (Brown & Brown 1991). Parasitic Cliff Swallows show remarkable sophistication in their ability to target host nests that are more likely to be successful than average, in part due to lower infestation by blood sucking nest parasites. Nest age was also identified as one cue used by parasitic females (Brown & Brown 1991): nest age may be a reliable indicator of the safety of a particular nest site.

Nest success and nest site safety, traits that do not necessarily mean the same thing, have been found to be associated with the occurrence of CBP in some waterfowl species but not in others (Table 3). For example, in the Common Goldeneye parasitism in a given year occurred more frequently in nest sites that were not depredated (at least one duckling hatched and left the nest) during the previous nesting attempt than in nest sites that were either depredated or control nest sites (Pöysä 1999). A later study revealed a mechanism by which parasitically laying females identify safe nest sites, *i.e.* by nest site prospecting during the previous year (Pöysä 2006). Remnants of successful hatching of a clutch (see Fig. 2) thus seem to be important cues by which parasitic Common Goldeneye females select target nests. A critical prerequisite in this hypothesis is that nest success is predictable between successive breeding seasons, as found for Common Goldeneyes in which nest depredation is the main determinant of variation in breeding success (Pöysä 2006). Predictability of nest success, coupled with the ability of parasites to assess it and lay accordingly, make parasitic laying an advantageous evolutionary strategy (Pöysä & Pesonen 2007).

Roy *et al.* (2009) tested the nest success hypothesis for Wood Ducks and found, contrary to the prediction, that previously unsuccessful nest sites were more likely to be parasitized in the following year. These authors also found that previous success did not consistently predict future success. An important difference between this study and Pöysä's studies is that in the Wood Duck the main cause of failing was nest desertion, probably caused by a high rate of parasitic laying (Roy *et al.* 2009). If females are simply evaluating cues related to nest predation (*e.g.* broken eggs) a nest with a large number of deserted eggs may still indicate a safe nest with respect to predation risk, and be targeted by parasites. This would present an interesting situation of conflicting cues (predation risk or nest abandonment) and females might be predisposed to one source of information (a sensory trap). Alternatively, this pattern might result if parasitic egg-laying in deserted nests was more frequent because of a lack of host defence (*i.e.* at tended nests, hosts may prevent access whereas this would not be the case at untended nests). This highlights the difficulty of inferring cue use from patterns of nest use and only experimental studies are likely to tease these apart. At any rate, patterns of parasitic egg-laying in Wood Duck females did not correspond to patterns of nest success. The authors suggested that high nest density may have confounded the quality of information and caused parasites to make poor decisions (see Roy *et al.* 2009).

The nest success hypothesis also has been tested for Common Eider, and results suggest that parasitic females did not use nest-site safety as a cue for egg laying (Lusignan *et al.* 2010; Table 3). Specifically, Lusignan *et al.* (2010) found that nests in dense woody vegetation had the highest probability of survival but the lowest frequency of CBP. On the other hand, nests in highly visible artificial wooden nest shelters had the highest rate of parasitism and ranked second in terms of nest survival. This finding suggests that nest visibility had a greater effect on parasitism rate than nest site safety (see Lusignan *et al.* 2010).

Another way that parasitically laying Common Eider females could use cues to choose high quality nests was suggested by Ruxton (1999). He was inspired by the

Table 3. Empirical studies that have tested for evidence of the use of cues in the selection of nests by parasitically laying females. Type of cue and study (observational or experimental) are given, together with main findings. Interpretation of results and further comments according to original articles.

Species	Type of cue	Type of study	Support for cue use	Main finding and comments	Source
Common Goldeneye	Nest predation risk	Experimental	Yes	Parasitism more frequent in previously non-depredated nests	Pöysä 1999
Common Goldeneye	Nest predation risk	Experimental (simulation before egg laying)	No	Parasites did not respond to simulated depredation when laying in nests of which they did not have previous experience of success	Pöysä 2003a
Common Goldeneye	Nest predation risk	Experimental	Yes	Parasitism more frequent on lakes with low nest depredation risk	Pöysä 2003a
Common Goldeneye	Nest predation risk	Experimental (simulation during egg laying)	Yes/No	Parasites that experienced simulated partial clutch depredation stopped egg laying in the nest, whereas parasites that had not experienced it laid in the nest later in the season	Pöysä et al. 2010
Common Goldeneye	Nest success	Observational	Yes	Parasitism more frequent in previously successful nests	Pöysä 2006
Wood Duck	Nest success	Observational	Yes/No	Parasitism more frequent in previously unsuccessful nests	Roy et al. 2009
Common Eider	Nest-site safety/visibility	Observational	No	Nests with highest probability of survival had lowest frequency of parasitism (nest-site visibility more influential)	Lusignan et al. 2010

Table 3 (*continued*).

Species	Type of cue	Type of study	Support for cue use	Main finding and comments	Source
Common Goldeneye	Nest site traits	Observational (number of nesting attempts, occupation rate)	Yes	Probability of parasitism increased with the number of previous nesting attempts and occupation rate	Paasivaara *et al.* 2010
Common and Barrow's Goldeneye	Host presence	Experimental	No	Dummy nests that did not have a host were used and parasitized	Eadie 1989
Common Goldeneye	Host presence	Experimental	No	Dummy nests that did not have a host were equally parasitized as real nests with a host	Pöysä 2003b
Common Goldeneye	Host traits	Observational (nesting experience, condition)	Yes/No	Probability of parasitism increased with host nesting experience (and less clearly with host body condition) but, in general, host traits less important than nest site traits	Paasivaara *et al.* 2010
Wood Duck	Clutch size	Experimental	Yes	Parasitism more frequent in simulated nests with smaller clutch size (hatch success probably higher in smaller clutches)	Odell & Eadie 2010
Black Brant	Egg size	Observational	Yes	Parasitic eggs differed less in size from eggs in their host's nests than did random eggs placed in random nests	Lemons & Sedinger 2011
Common Eider	Host trait	Observational (condition)	No	Parasitized females were of same body mass as non-parasitized females	Waldeck *et al.* 2011

observation that predation rates were lower for parasitized Common Eider nests in one population (Robertson 1998). This finding, coupled with observations that nest depredation rates were correlated with female attentiveness (Swennen *et al.* 1993), led Ruxton (1999) to suggest that parasitically laying females could use variation in the female attentiveness at the nest during the egg-laying period as a cue of nest depredation risk and select target nests accordingly. The hypothesis predicts that there should be a positive relationship between an individual female's attentiveness at the nest during early egg-laying and her risk of parasitism (Ruxton 1999). To our knowledge this prediction has not been tested. It is important to note that nest attentiveness could also indicate female quality, making it difficult to distinguish between female quality and nest site quality because of the correlation between these two variables. Thus although it may be possible to correlate apparent cues (*e.g.* pattern of female nest attentiveness) with parasitic behaviour, we need to be careful to consider the underlying information that these patterns might represent (*i.e.* female quality). The "cues" that we measure may not necessarily be the cues that parasites perceive or respond to.

There are only a few studies in which cues affecting nest site selection and the laying decisions of parasites have been addressed experimentally (Table 3). In addition to Pöysä (1999, see above), two experimental studies have addressed the role of nest depredation risk and actual (partial) nest depredation in affecting the laying decision of parasites in the Common Goldeneye (Pöysä 2003a; Pöysä *et al.* 2010). Those experiments revealed that parasitically laying females respond to varying degrees of nest depredation risk (*i.e.* they prefer laying in simulated nests that are in safe environments) but their response to simulated nest depredation varied depending on whether females experienced simulated partial clutch depredation (Table 3). These experimental findings suggest, first, that nest depredation and nest depredation risk are important cues, and second, that both personal information and social information are used in the selection of target nests by parasitically laying Common Goldeneye females. Other experimental studies have found that host presence was not an important cue in nest selection by parasitically laying Common Goldeneye (Pöysä 2003b) and Barrow's Goldeneye females (Eadie 1989).

Parasitic Wood Duck females appear to respond to variation in the number of eggs in a nest (Odell & Eadie 2010). In this study a choice of nests containing clutches of 5, 10, 15 or 20 experimental eggs was offered to Wood Duck females, and the number of eggs laid in the simulated nests declined in direct relation to the number of experimental eggs in the nest. This finding is of particular interest because it suggests the possibility that parasitically laying females are able to assess the number of eggs in a nest, a cue associated with important fitness consequences because large clutches often have low hatching success (Roy Nielsen *et al.* 2006b; Odell & Eadie 2010). Lemons and Sedinger (2011) report a remarkable pattern for the Black Brant *Branta bernicla nigricans* in which parasitic eggs match the size of host eggs, suggesting that parasitically laying females recognise host egg size and lay accordingly, probably to improve hatching success. How they might do so is completely unknown.

Territory or host quality

The importance of host female quality (*i.e.* body condition) has been addressed in Common Goldeneyes and Common Eiders, and appeared not to be an important cue for parasitically laying females (Paasivaara *et al.* 2010; Waldeck *et al.* 2011). It should be noted that nesting schedule, a feature that also may reflect host female quality, has been found to be associated with the occurrence of CBP in many waterfowl species (*i.e.* clutches laid early in the season are more frequently parasitized than late clutches: Dow & Fredga 1984; Sorenson 1991; Robertson *et al.* 1992); however, other factors were not controlled in these studies (see Paasivaara *et al.* 2010). Older females often nest earlier and if CBP occurs more frequently early in the season, then a pattern would emerge of older (and perhaps more experienced or higher quality) females being parasitized disproportionately. The causative arrow could however be in the opposite direction: parasites might target older experienced females and, if older females nest earlier, then CBP would be more frequent in early nests. This would require careful experiments or statistical controls to decouple this pattern. Monte Carlo randomisation analyses could be used to determine if parasites selected non-randomly from among the host nests available, as noted above. Very few such tests have been conducted for waterfowl and this remains an interesting and important direction for future work.

Timing of host laying cycle

Synchronising the timing of egg-laying with the host's laying cycle is thought to be important for interspecific brood parasites (Davies 2000) and for conspecific brood parasites as well. A role for host cues that reveal the timing of their cycle (but which have yet to be confirmed) has been found to be important in the few studies of non-waterfowl species that have examined these patterns while also taking random expectations into account (Emlen & Wrege 1986; Lyon & Everding 1996). Brown and Brown (1988) found that parasitic Cliff Swallows, which parasitize hosts by transferring eggs physically in their beaks, were remarkably good at synchronising timing with the host's laying and incubation period.

Matching the timing of egg-laying to a host's own clutch is particularly important in precocial birds such as waterfowl because the young hatch synchronously and leave the nest simultaneously within 24–48 h after hatching. Mismatched timing of egg-laying by the parasite can result in eggs failing to hatch, or young hatching after the host female has already left with her brood (see Bellrose & Holm 1994; review in Sayler 1992). Nonetheless, parasitic eggs are laid after the onset of incubation in several species (Jones & Leopold 1967; Clawson *et al.* 1979; Heusmann *et al.* 1980; Eriksson & Andersson 1982; Eadie 1989; Bellrose & Holm 1994; Šťovíček *et al.* 2013; review in Sayler 1992). Sayler (1992) describes a case of interspecific brood parasitism in which a parasitic Redhead female laid in a

Canvasback *Aythya valisineria* nest while it contained hatched ducklings. In contrast, Wood Ducks in some populations laid up to 80% of parasitic eggs prior to host incubation (Clawson *et al*. 1979).

Matching the timing of egg-laying with hosts has been documented in several non-waterfowl species, but more research would be useful to explore the cues and mechanisms used by parasites to fine-tune the timing of parasitic egg-laying and to understand better the constraints of doing so. To add further complexity, exciting new work by Hepp and colleagues has shown that even slight differences in incubation temperature can have significant impacts on the post-natal development and survival of young (Hepp *et al*. 1990, 2006; Kennamer *et al*. 1990; DuRant *et al*. 2010, 2011, 2012a,b). Thus, timing of egg-laying and incubation efficiency could have a large impact on parasite (and host) fitness. Odell (2008) found that eggs of parasitic Wood Ducks in California had higher levels of androgens than host eggs and this might accelerate the development of parasitic eggs laid at the end of the host laying period or after the initiation of incubation. Typically, host females spend large portions of the day on the nest during incubation, and yet parasitic females do not appear to use host presence as a cue to avoid these nests. Possibly, parasites cannot detect accurately the stage of incubation and simply the presence of another female or evidence of an active nest provides a sufficient incentive to induce egg-laying. Alternatively, cases of mismatched timing of laying with respect to the host's laying cycle might be influenced by host availability, for instance some females may have no potential hosts in the laying stage to parasitize when they have a parasitic egg ready for laying. A similar explanation may account for the fact that parasitic females will often lay eggs in deserted nests, occasionally leading to large accumulations of abandoned eggs (termed "dump nests" in the older literature). Sayler (1992) noted that parasitic Redhead females will often follow each other to nests and lay a series of eggs in those nests over several days. Similar behaviour has been observed in Wood Ducks (Semel & Sherman 2001) and Common Goldeneyes (Eadie 1989; Åhlund 2005). Sayler (1992) suggested that these nests appear active to parasitic females given the presence of other females and eggs being laid in the nest, even though the host has already abandoned the nest.

A curious (and opposite) pattern has been documented in Common Eiders. In several populations, researchers have found that parasitic eggs are often laid in nests *before* the host female begins to lay her own eggs. The host female thus lays her eggs in a nest in which another egg is already present, and then subsequently completes her clutch and incubates the nest. Common Eiders often reuse nest bowls in successive years and this pattern could be explained if females are simply competing for certain nest sites, with the first female being ousted (the accidental parasite). However, Robertson (1998) suggested that this phenomenon was due to nest takeover and adoption, possibly in response to nest predation risk. He reported that in nests where females took over nests, predation on the first eggs was lower than in comparable nests with only a single female (*i.e.* no takeover). Robertson (1998) argued that the presence of an egg in a nest would indicate that the site was a safe nest location (since the egg had not been depredated), and therefore the nest may be attractive to another female. The benefit of obtaining a safe nest site could outweigh the potential cost of caring for the additional eggs. Ruxton (1999) further suggested that females can detect variation in the predation risk associated with different nests, and use this information to target nests with low predation risk as sites for laying parasitically, a hypothesis similar to Pöysä's (1999, 2003a, 2006; Pöysä et al. 2010) hypothesis to explain CBP in Common Goldeneyes. Waldeck and Andersson (2006), using protein fingerprinting techniques, found that another female laid before the host started laying in 41% of mixed clutches. Similarly, Hario et al. (2012) found that 58% of parasitic eggs in a population in Finland were the first or second eggs laid. Waldeck and Andersson (2006) reported that nests that were taken over have higher early survival than other nests, consistent with the hypothesis that CBP in Common Eiders is driven by selection for safe nest sites. It is still unclear whether "nest takeover" is a form of CBP at all, although it is often presented in that context (see discussion in Roberston 1998). Clearly, much more work remains to better understand the cues that parasitic females use to fine-tune the timing of egg-laying.

What determines the patterns of parasitism?

We have, to this point, focused on the cues that parasites use. This is not intended to imply that cues that hosts might use, and the role of hosts in determining the patterns or outcomes of parasitic interactions, are not important. Indeed, there are a number of decisions hosts might make to influence the occurrence of parasitism, each involving different cues. An analysis of host decisions is beyond the scope of our paper, and limited information is available. However, it will be important in future studies to integrate cue use and decision-making by hosts for several reasons: 1) it links cognition and decision-making to parasitism broadly; 2) decisions are linked in a game theoretic way that include both parasite and host responses (*e.g.* Andersson & Eriksson's 1982 clutch size model); and 3) for some decisions (*e.g.* who lays in a nest), it can be very difficult to determine whether the host or the parasite determines the outcome (and hence who is using what cues).

This last question is particularly germane in our efforts to understanding the relative roles that hosts and parasites play in determining the patterns of parasitism. While some authors emphasise the active role of hosts (*e.g.* Andersson & Åhlund 2000), others suggest that hosts do not play an important role (*e.g.* Pöysä 2004). Evidence that host-mediated facilitation does not play a central role in the laying decisions of parasites comes from experiments in which parasitic laying has been induced in simulated nests that do not have a host present (Pöysä 2003a,b; Odell & Eadie 2010; Pöysä et al. 2010). It is noteworthy that laying in simulated nests is frequent in Common Goldeneyes even when active real nests are available (H. Pöysä, unpubl. data), indicating that this behaviour is not simply due to a shortage of host nests. Pöysä (2003b) specifically addressed the importance of host recognition by parasites in the selection of nests and found it not to be important. These experiments suggest that host females do not play an active role as facilitators of parasitism. However, the role of hosts in CBP remains uncertain. One issue that has recently been of growing interest, and is particularly relevant to the question of the role of hosts in facilitating CBP, is the potential influence of kinship amongst hosts and parasites.

A role for kinship?

The idea that hosts and parasites might be related, and hence that CBP is not a form of parasitism *per se*, but rather a cooperative behaviour facilitated by kin selection, was suggested over 30 years ago by Andersson (1984). In waterfowl, females are the philopatric sex so the premise of the kinship hypothesis is that females might return to their natal area and lay eggs in the nests of close kin. A central feature of the mechanism is that hosts are in the driver's seat. By allowing kin to lay eggs in the host nest, the host may be facilitating reproduction by a relative where otherwise none would have been possible, thereby increasing the hosts' own inclusive fitness. The idea was often cited but rarely tested until a number of new theoretical models revisited this idea (Zink 2000; Andersson 2001; Lopez-Sepulcre & Kokko 2002; Jaatinen *et al.* 2011a). The current consensus of these models is that kinship can facilitate CBP, provided that costs to the host are low and some degree of kin recognition exists (Lyon & Eadie 2000; Eadie & Lyon 2011). If costs to hosts are high, then parasites should avoid laying eggs in a relative's nest so as not to reduce the host's fitness, and thereby lower the parasite's inclusive fitness. Testing the kin selection hypothesis requires detailed information not only on host-parasite relatedness, but also on the costs to the hosts, the degree (or existence) of kin recognition and the extent to which parasites *versus* hosts control or facilitate CBP.

A number of empirical studies using molecular genetic techniques (DNA microsatellites and isoelectric focusing of egg albumin proteins) have now documented high host-parasite relatedness in waterfowl, including the Wood Duck, Common Eider, Barrow's Goldeneye and Common Goldeneye (review in Eadie & Lyon 2011). However, the mechanisms leading to high host-parasite relatedness remain unknown, although kin recognition and discrimination against unrelated parasites by hosts have been suggested in the Common Goldeneye (Andersson & Åhlund 2000; but see Pöysä 2004). The finding that hosts and parasites are often related opens the possibility that parasites or hosts could recognise kin and that kinship could provide a cue in nest/host selection. Interestingly, Jaatinen *et al.* (2011b) found that the response of parasitic Barrow's Goldeneye females to relatedness depends on their nesting status: parasitic females that had a nest of their own ("nesting parasites") responded to relatedness by laying more eggs with increasing relatedness to the host, while non-nesting parasites did not respond to relatedness. The authors discuss several possible reasons why nesting and non-nesting parasites differed in their response to relatedness but the mechanisms underlying this finding are currently unknown. Finally, Pöysä *et al.* (2014) provided experimental evidence that nest predation risk and interaction between related parasites are associated with kin-biased co-parasitism (related parasites non-randomly laying in the same nest) in the Common Goldeneye. In reviewing the evidence to date, the initial results suggest that kinship could play a role as a cue at least for parasites that have a nest of their own, but it remains to be determined whether kinship is a central driver in the evolution of CBP, or instead just one of several factors that influence the frequency and occurrence of this intriguing behaviour (Eadie & Lyon 2011).

Conclusions and future issues

Our understanding of CBP has advanced considerably in the 35 years since Yom-Tov (1980) first brought it to our attention, and research on waterfowl (Anseriformes) has contributed disproportionately to this knowledge. We now have a much better understanding of the ecological and social conditions under which CBP occurs, and over two dozen studies of waterfowl have tested the existing set of hypotheses proposed to account for this behaviour. Collectively, these studies support the hypotheses that CBP allows some females to reproduce when they otherwise could not, while other females use parasitic egg-laying as a way to enhance total fecundity. A life-history approach offers a new framework by which to integrate all of these possibilities into a theory of flexible life-history, and the set of traditional hypotheses for CBP can be readily integrated into this new framework. We suggest that this new framework will provide a useful direction and impetus for the next generation of studies of CBP, fuelled by an increasing battery of molecular genetic techniques and a growing array of technological tools to track females and their reproductive trajectories throughout their lifespan.

We have also focused on an emerging, exciting area for future investigations of CBP – namely the use of cues and information by conspecific brood parasites as they seek suitable places to lay their parasitic eggs. Recent empirical studies of CBP in waterfowl have revealed remarkable cognitive abilities in parasitic females, although the actual mechanisms remain unknown in most cases. In particular, the use of public information by parasites in locating and selecting nests that have high prospects of success is a promising avenue worth exploring to gain insight into the evolution of nest/host selection and egg laying decisions of parasites. Interestingly, the importance of public information has also been stressed recently in the context of interspecific brood parasitism; Parejo and Avilés (2007) suggested that parasites might eavesdrop on the sexual signals of their hosts to find high quality foster parents for their own offspring. The role of parental quality as a cue in CBP has received little support in waterfowl but more research on this aspect is needed. The ability of parasites to evaluate the number of eggs in a nest and to modify their own laying behaviour accordingly, as demonstrated with Wood Ducks, is intriguing and worth further exploration in other species.

The role that kinship plays as a cue in the laying decisions of parasites (and hosts) remains a challenging task for future studies. High host-parasite relatedness has been reported for several species, but it is unclear whether direct assessment of relatedness is involved or if some unmeasured correlate leads related females to select the same nest site. For example, as suggested by Pöysä (2004), high natal and nest site philopatry and preference of both hosts and parasites to lay in safe nest sites will also generate high host-parasite relatedness. Experimental studies and examination of the cost of parasitism to hosts and the ability of females to recognise or interact differentially with kin are required to disentangle these effects.

Acknowledgements

We thank the scientific committee of the Ecology and Conservation of North American Waterfowl symposium for the opportunity to give the plenary talk from which this paper is extracted. Gary Hepp served as the lead editor of this manuscript and we appreciate his enduring patience and useful comments on the manuscript. We also thank Rick Kaminski and Eileen Rees for useful editorial comments. Support for the development of the ideas and research reported in this paper was provided by funding from Finnish Game and Fisheries Research Institute (to HP), NSF (IOS-1355208 to JME; IOS- 1354894 to BEL), and the Dennis G. Raveling Waterfowl Professorship at U.C. Davis (to JME).

References

Åhlund, M. 2005. Behavioural tactics at nest visits differ between parasites and hosts in a brood-parasitic duck. *Animal Behaviour* 70: 433–440.

Åhlund, M. & Andersson, M. 2001. Female ducks can double their reproduction. *Nature* 414: 600–601.

Anderholm, S., Waldeck, P., van der Jeugd, H.P., Marshall, R.C., Larsson, K. & Andersson, M. 2009. Colony kin structure and host-parasite relatedness in the Barnacle Goose. *Molecular Ecology* 18: 4955–4963.

Andersson, M. 1984. Brood parasitism within species. *In* C.J. Barnard (ed.), *Producers and Scroungers. Strategies of Exploitation and Parasitism*, pp. 195–228. Chapman & Hall, New York, USA.

Andersson, M. 2001. Relatedness and the evolution of conspecific brood parasitism. *American Naturalist* 158: 599–614.

Andersson, M. & Åhlund, M. 2000. Host-parasite relatedness shown by protein fingerprinting in a brood parasitic bird. *Proceedings of the National Academy of Sciences, USA* 97: 13188–13193.

Andersson, M. & Åhlund, M. 2012. Don't put all your eggs in one nest: spread them and cut time at risk. *American Naturalist* 180: 354–363.

Andersson, M. & Eriksson, M.O.G. 1982. Nest parasitism in goldeneye *Bucephala clangula*: some evolutionary aspects. *American Naturalist* 120: 1–16.

Bellrose F.C. & Holm D.J. 1994. *Ecology and Management of the Wood Duck*. Stackpole Books, Mechanicsburg, Pennsylvania, USA.

Briggs, S.V. 1991. Intraspecific nest parasitism in Maned Ducks *Chenonetta jubata*. *Emu* 91: 230–35.

Brown, C.R. & Brown, M.B. 1988. A new form of reproductive parasitism in Cliff Swallows. *Nature* 331: 66–68.

Brown, C.R. & Brown, M.B. 1991. Selection of high-quality hosts by parasitic Cliff Swallows. *Animal Behaviour* 41: 457–465.

Clawson, E.L., Hartman, G.W. & Fredrickson, L.H. 1979. Dump nesting in a Missouri Wood Duck population. *Journal of Wildlife Management* 43: 347–355.

Danchin, É., Giraldeau, L.-A. & Wagner, R.H. 2008. An information-driven approach to behaviour. In É. Danchin, L.-A. Giraldeau & F. Cézilly (eds.), *Behavioural Ecology*, pp. 97–130. Oxford University Press, Oxford, UK.

Davies, N.B. 2000. *Cuckoos, Cowbirds and Other Cheats*. T. & A.D. Poyser, London, UK.

Dow, H. & Fredga, S. 1984. Factors affecting reproductive output of the goldeneye duck *Bucephala clangula*. *Journal of Animal Ecology* 53: 679–692.

DuRant S.E., Hepp G.R., Moore I.T., Hopkins B.C. & Hopkins W.A. 2010. Slight differences in incubation temperature affect early growth and stress endocrinology of wood duck (*Aix sponsa*) ducklings. *Journal of Experimental Biology* 213: 45–51.

DuRant, S.E., Hopkins, W.A. & Hepp, G.R. 2011. Embryonic developmental patterns and energy expenditure are affected by incubation temperature in Wood Ducks (*Aix sponsa*). *Physiological and Biochemical Zoology* 84: 451–457.

DuRant, S.E., Hopkins, W.A., Hawley, D.M. & Hepp, G.R. 2012a. Incubation temperature affects multiple measures of immunocompetence in young Wood Ducks (*Aix sponsa*). *Biology Letters* 8: 108–111.

DuRant, S.E., Hopkins, W.A., Wilson, A. & Hepp, G.R. 2012b. Incubation temperature affects the metabolic cost of thermoregulation in a young precocial bird. *Functional Ecology* 26: 416–422.

Eadie, J.M. 1989. Alternative reproductive tactics in a precocial bird: the ecology and evolution of brood parasitism in goldeneyes. Ph.D. thesis, University of British Columbia, Vancouver, Canada.

Eadie, J.M. 1991. Constraint and opportunity in the evolution of brood parasitism in waterfowl. In B.D. Bell, R.O. Cossee, J.E.C. Flux, B.D. Heather, R.A. Hitchmough, C.J.R. Robertson & M.J. Williams (eds.), *Proceedings of the Twentieth International Ornithological Conference*, pp. 1031–1040. Hutcheson, Bowman & Stewart Ltd., Wellington, New Zealand.

Eadie, J.M. & Gauthier, G. 1985. Prospecting for nest sites by cavity-nesting ducks of the genus *Bucephala*. *Condor* 87: 528–534.

Eadie, J.M. & Lyon, B. 2011. The relative role of relatives in conspecific brood parasitism. *Molecular Ecology* 20: 5114–5118.

Eadie, J.M., Kehoe, F.P. & Nudds, T.D. 1988. Pre-hatch and post-hatch brood amalgamation in North American Anatidae: a review. *Canadian Journal of Zoology* 66: 1709–1721.

Eadie, J.M, Sherman, P. & Semel, B. 1998. Conspecific brood parasitism, population dynamics, and the conservation of cavity-nesting birds. In T. Caro (ed.), *Behavioral Ecology and Conservation Biology*, pp. 306–340. Oxford University Press, Oxford, UK.

Emlen, S.T. & Wrege, P.H. 1986. Forced copulation and intra-specific parasitism: two costs of social living in the White-fronted Bee-eater. *Ethology* 71: 2–29.

Eriksson, M.O.G. & Andersson, M. 1982. Nest parasitism and hatching success in a population of goldeneyes *Bucephala clangula*. *Bird Study* 29: 49–54.

Erskine, A.J. 1990. Joint laying in *Bucephala* ducks – "parasitism" or nest-site competition? *Ornis Scandinavica* 21: 52–56.

Fast, P.L.F., Gilchrist, H.G. & Clark, R.G. 2010. Nest-site materials affect nest-bowl use by Common Eiders (*Somateria mollissima*). *Canadian Journal of Zoology* 88: 214–218.

Forslund, P. & Larsson, K. 1995. Intraspecific nest parasitism in the Barnacle Goose: behavioural tactics of parasites and hosts. *Animal Behaviour* 50: 509–517.

Hario, M., Koljonen, M.-L. & Rintala, J. 2012. Kin structure and choice of brood care in a

Common Eider (*Somateria m. mollissima*) population. *Journal of Ornithology* 153: 963–973.

Hepp, G.R., Kennamer, R.A. & Harvey W.F. 1990. Incubation as a reproductive cost in female Wood Ducks. *Auk* 107: 756–764.

Hepp, G.R., Kennamer, R.A. & Johnson M.H. 2006. Maternal effects in Wood Ducks: incubation temperature influences incubation period and neonate phenotype. *Functional Ecology* 20: 308–314.

Heusmann, H.W., Bellville, R. & Burrell, R.G. 1980. Further observations on dump nesting by Wood Ducks. *Journal of Wildlife Management* 44: 908–915.

Jaatinen, K., Öst, M., Waldeck, P. & Andersson, M. 2009. Clutch desertion in Barrow's Goldeneyes (*Bucephala islandica*) – effects on non-natal eggs, the environment and host female characteristics. *Annales Zoologici Fennici* 46: 350–360.

Jaatinen, K., Lehtonen, J. & Kokko, H. 2011a. Strategy selection under conspecific brood parasitism: an integrative modeling approach. *Behavioral Ecology* 22: 144–155.

Jaatinen, K., Öst, M., Gienapp, P. & Merilä, J. 2011b. Differential responses to related hosts by nesting and non-nesting parasites in a brood-parasitic duck. *Molecular Ecology* 20: 5328–5336.

Jansen, R.W. & Bollinger, E.K. 1998. Effects of nest-box visibility and clustering on Wood Duck brood parasitism in Illinois. *Transactions of the Illinois State Academy of Science* 91: 161–166.

Jones, R.E. & Leopold, A.S. 1967. Nesting interference in a dense population of Wood Ducks. *Journal of Wildlife Management* 31: 221–228.

Kennamer, R.A., Harvey, W.F. & Hepp, G.R. 1990. Embryonic development and nest attentiveness of Wood Ducks during egg-laying. *Condor* 92: 587–592.

Lank, D.B., Cooch, E.G., Rockwell, R.F. & Cooke, F. 1989. Environmental and demographic correlates of intraspecific nest parasitism in Lesser Snow Geese *Chen caerulescens caerulescens*. *Journal of Animal Ecology* 58: 29–45.

Lemons, P.R. & Sedinger, J.S. 2011. Egg size matching by an intraspecific brood parasite. *Behavioral Ecology* 22: 696–700.

Lopez-Sepulcre, A. & Kokko, H. 2002. The role of kin recognition in the evolution of conspecific brood parasitism. *Animal Behaviour* 64: 215–222.

Lusignan, A.P., Mehl, K.R., Jones, I.L. & Cloutney, M.L. 2010. Conspecific brood parasitism in Common Eiders (*Somateria mollissima*): do brood parasites target safe nest sites? *Auk* 127: 765–772.

Lyon, B.E. 1993. Tactics of parasitic American coots: host choice and the pattern of egg dispersion among host nests. *Behavioral Ecology and Sociobiology* 33: 87–100.

Lyon, B.E. & Eadie, J.M. 2000. Family matters: kin selection and the evolution of conspecific brood parasitism. *Proceedings of the National Academy of Sciences USA* 97: 1942–1944.

Lyon, B.E. & Eadie, J.M. 2008. Conspecific brood parasitism in birds: a life-history perspective. *Annual Review of Ecology, Evolution and Systematics* 39: 343–363.

Lyon, B.E. & Everding, S. 1996. A high frequency of conspecific brood parasitism in a colonial waterbird, the Eared Grebe (*Podiceps nigricollis*). *Journal of Avian Biology* 27: 238–244.

Maynard Smith, J. & Harper, D. 2003. *Animal Signals*. Oxford University Press, Oxford, UK.

McRae, S.B. & Burke, T. 1996. Intraspecific brood parasitism in the Moorhen: parentage and parasite-host relationships determined by DNA fingerprinting. *Behavioral Ecology and Sociobiology* 38: 115–129.

Odell, N.S. 2008. Female reproductive investment in a conspecific brood parasite. Ph.D. thesis. University of California, Davis, USA.

Odell, N.S. & Eadie, J.M. 2010. Do Wood Ducks use the quantity of eggs in a nest as a cue to the nest's value? *Behavioral Ecology* 21: 794–801.

Paasivaara, A., Rutila, J., Pöysä, H. & Runko, P. 2010. Do parasitic Common Goldeneye *Bucephala clangula* females choose nests on the basis of host traits or nest site traits? *Journal of Avian Biology* 41: 662–671.

Parejo, D. & Avilés, J.M. 2007. Do avian brood parasites eavesdrop on heterospecific sexual signals revealing host quality? A review of the evidence. *Animal Cognition* 10: 81–88.

Patten, M.A., Reinking, D.L. & Wolfe, D.H. 2011. Hierarchical cues in brood parasite nest selection. *Journal of Ornithology* 152: 521–532.

Pöysä, H. 1999. Conspecific nest parasitism is associated with inequality in nest predation risk in the Common Goldeneye (*Bucephala clangula*). *Behavioral Ecology* 10: 533–540.

Pöysä, H. 2003a. Parasitic Common Goldeneye (*Bucephala clangula*) females lay preferentially in safe neighbourhoods. *Behavioral Ecology and Sociobiology* 54: 30–35.

Pöysä, H. 2003b. Low host recognition tendency revealed by experimentally induced parasitic egg laying in the Common Goldeneye (*Bucephala clangula*). *Canadian Journal of Zoology* 81: 1561–1565.

Pöysä, H. 2004. Relatedness and the evolution of conspecific brood parasitism: parameterizing a model with data for a precocial species. *Animal Behaviour* 67: 673–679.

Pöysä, H. 2006. Public information and conspecific nest parasitism in goldeneyes: targeting safe nests by parasites. *Behavioral Ecology* 17: 459–465.

Pöysä, H. & Pesonen, M. 2007. Nest predation and the evolution of conspecific brood parasitism: from risk spreading to risk assessment. *American Naturalist* 169: 94–104.

Pöysä, H., Milonoff, M., Ruusila, V. & Virtanen, J. 1999. Nest-site selection in relation to habitat edge: experiments in the Common Goldeneye. *Journal of Avian Biology* 30: 79–84.

Pöysä, H., Lindblom, K., Rutila, J. & Sorjonen, J. 2010. Response of parasitically laying goldeneyes to experimental nest predation. *Animal Behaviour* 80: 881–886.

Pöysä, H., Paasivaara, A., Lindblom, K., Rutila, J. & Sorjonen, J. 2014. Co-parasites preferentially lay with kin and in safe neighbourhoods: experimental evidence from goldeneye ducks. *Animal Behaviour* 91: 111–118.

Reichart, L.M., Anderholm, S., Munoz-Fuentes, V. & Webster, M.S. 2010. Molecular identification of brood-parasitic females reveals an opportunistic reproductive tactic in Ruddy Ducks. *Molecular Ecology* 19: 401–413.

Robertson, G.J. 1998. Egg adoption can explain joint egg-laying in Common Eiders. *Behavioral Ecology and Sociobiology* 43: 289–296.

Robertson, G.J., Watson, M.D. & Cooke, F. 1992. Frequency, timing and costs of intraspecific nest parasitism in the Common Eider. *Condor* 94: 871–879.

Rohwer, F.C. & Freeman, S. 1989. The distribution of conspecific nest parasitism in birds. *Canadian Journal of Zoology* 67: 239–253.

Roy, C., Eadie, J.M., Schauber, E.M., Odell, N.S., Berg, E.C. & Moore, T. 2009. Public information and conspecific nest parasitism in Wood Ducks: does nest density influence quality of information? *Animal Behaviour* 77: 1367–1373.

Roy Nielsen, C.L., Gates, R.J. & Parker, P.G. 2006a. Intraspecific nest parasitism of Wood Ducks in natural cavities: comparisons with nest boxes. *Journal of Wildlife Management* 70: 835–843.

Roy Nielsen, C.R., Parker, P.G. and Gates, R.J. 2006b. Intraspecific nest parasitism of cavity-nesting wood ducks: costs and benefits to hosts and parasites. *Animal Behaviour* 72: 917–926.

Ruxton, G.D. 1999. Are attentive mothers preferentially parasitised? *Behavioral Ecology and Sociobiology* 46: 71–72.

Sayler, R.D. 1992. Ecology and evolution of brood parasitism in waterfowl. *In* B.D.J. Batt, A.D. Afton, M.G. Anderson, C.D. Ankney, D.H. Johnson, J.A. Kadlec & G.L. Krapu (eds.), *Ecology and Management of Breeding Waterfowl*, pp. 290–322. University of Minnesota Press, Minneapolis, USA.

Schamel, D. 1977. Breeding of the Common Eider (*Somateria mollissima*) on the Beaufort Sea coast of Alaska. *Condor* 79: 478–485.

Semel, B. & Sherman, P.W. 1986. Dynamics of nest parasitism in Wood Ducks. *Auk* 103: 813–816.

Semel, B. & Sherman, P.W. 2001. Intraspecific parasitism and nest-site competition in Wood Ducks. *Animal Behaviour* 61:787–803.

Semel, B., Sherman, P.W. & Byers, S.M. 1988. Effects of brood parasitism and nest-box placement on Wood Duck breeding ecology. *Condor* 90: 920–930.

Sorenson, M.D. 1991. The functional significance of parasitic egg laying and typical nesting in Redhead Ducks: an analysis of individual behaviour. *Animal Behaviour* 42: 771–796.

Sorenson, M.D. 1993. Parasitic egg-laying in Canvasbacks – frequency, success and individual behavior. *Auk* 110: 57–69.

Šťovíček, O., Kreisinger, J., Javůrková, V. & Albrecht, T. 2013. High rates of conspecific brood parasitism revealed by microsatellite analysis in a diving duck, the Common Pochard *Aythya ferina*. *Journal of Avian Biology* 44: 369–375.

Swennen, C., Ursem, J.C.H. & Duiven, P. 1993. Determinate laying and egg attendence in Common Eiders. *Ornis Scandinavica* 24: 48–52.

Wagner, R.H. & Danchin, É. 2010. A taxonomy of biological information. *Oikos* 119: 203–209.

Waldeck, P. & Andersson, M. 2006. Brood parasitism and nest takeover in Common Eiders. *Ethology* 112: 616–624.

Waldeck, P., Kilpi, M., Öst, M. & Andersson, M. 2004. Brood parasitism in a population of Common Eider (*Somateria mollissima*). *Behaviour* 141: 725–739.

Waldeck, P., Andersson, M., Kilpi, M. & Öst, M. 2008. Spatial relatedness and brood parasitism in a female-philopatric bird population. *Behavioral Ecology* 19: 67–73.

Waldeck, P., Hagen, J.I., Hanssen, S.A. & Andersson, M. 2011. Brood parasitism, female condition and clutch reduction in the Common Eider *Somateria mollissima*. *Journal of Avian Biology* 42: 231–238.

Weigmann, C. & Lamprecht, J. 1991. Intraspecific nest parasitism in Bar-headed Geese, *Anser indicus*. *Animal Behaviour* 41: 677–688.

Weller, M.W. 1959. Parasitic egg laying in the Redhead (*Aythya americana*) and other North American Anatidae. *Ecological Monographs* 29: 333–365.

Wilson, S.F. 1993. Use of Wood Duck decoys in a study of brood parasitism. *Journal of Field Ornithology* 63: 337–340.

Yom-Tov, Y. 1980. Intraspecific nest parasitism in birds. *Biological Reviews* 55: 93–108.

Yom-Tov, Y. 2001. An updated list and some comments on the occurrence of intraspecific nest parasitism in birds. *Ibis* 143: 133–143.

Zicus, M.C. & Hennes, S.K. 1989. Nest prospecting by Common Goldeneyes. *Condor* 91: 807–812.

Zink, A.G. 2000. The evolution of intraspecific brood parasitism in birds and insects. *American Naturalist* 155: 395–405.

The effects of harvest on waterfowl populations

EVAN G. COOCH[1]*, MATTHIEU GUILLEMAIN[2], G. SCOTT BOOMER[3], JEAN-DOMINIQUE LEBRETON[4] & JAMES D. NICHOLS[5]

[1]Department of Natural Resources, Cornell University, Ithaca, New York 14853, USA.
[2]Office National de la Chasse et de la Faune Sauvage, CNERA Avifaune Migratrice, La Tour du Valat, Le Sambuc, 13200 Arles, France.
[3]Division of Migratory Bird Management, U.S. Fish & Wildlife Service, Laurel, Maryland 20708, USA.
[4]Centre d'Écologie Fonctionnelle et Évolutive, UMR 5175, CNRS, 34293 Montpellier Cedex 5, France.
[5]Patuxent Wildlife Research Center, U.S. Geological Survey, Laurel, Maryland 20708, USA.
*Correspondence author. E-mail: evan.cooch@cornell.edu

Abstract

Change in the size of populations over space and time is, arguably, the motivation for much of pure and applied ecological research. The fundamental model for the dynamics of any population is straightforward: the net change in the abundance is the simple difference between the number of individuals entering the population and the number leaving the population, either or both of which may change in response to factors intrinsic and extrinsic to the population. While harvest of individuals from a population constitutes a clear extrinsic source of removal of individuals, the response of populations to harvest is frequently complex, reflecting an interaction of harvest with one or more population processes. Here we consider the role of these interactions, and factors influencing them, on the effective harvest management of waterfowl populations. We review historical ideas concerning harvest and discuss the relationship(s) between waterfowl life histories and the development and application of population models to inform harvest management. The influence of population structure (age, spatial) on derivation of optimal harvest strategies (with and without explicit consideration of various sources of uncertainty) is considered. In addition to population structure, we discuss how the optimal harvest strategy may be influenced by: 1) patterns of density-dependence in one or more vital rates, and 2) heterogeneity in vital rates among individuals within an age-sex-size class. Although derivation of the optimal harvest strategy for simple population models (with or without structure) is generally straightforward, there are several potential difficulties in application. In particular, uncertainty concerning the population structure at the time of harvest, and the ability to regulate the structure of the harvest itself, are significant complications. We therefore review the evidence of effects of harvest on waterfowl populations. Some of this evidence has focussed on correspondence of data with more phenomenological models and other evidence relates to specific mechanisms, including density-dependence and heterogeneity. An important part of this evidence is found in the evolution of model weights under various adaptive harvest management programmes of the U.S. Fish and Wildlife Service for North American waterfowl.

Overall, there is substantial uncertainty about system dynamics, about the impacts of potential management and conservation decisions on those dynamics, and how to optimise management decisions in the presence of such uncertainties. Such relationships are unlikely to be stationary over space or time, and selective harvest of some individuals can potentially alter life history allocation of resources over time – both of which will potentially influence optimal harvest strategies. These sources of variation and uncertainty argue for the use of adaptive approaches to waterfowl harvest management.

Key words: additive mortality, compensatory mortality, harvest, population structure.

Annual migrations of waterfowl have long provided human populations with a regular source of protein and outdoor recreation. Their use as a valuable quarry and food has been facilitated by the great concentrations of these birds at some wintering or migration stopover sites, the ease with which their eggs could be collected, and the flightless moulting period making adults particularly vulnerable to trapping during late summer. Not surprisingly, waterfowl remains are very common at prehistoric human settlement sites (*e.g.* Ericson & Tyrberg 2004), and antique artefacts of waterfowl hunting are numerous (*e.g.* Egyptian paintings or Roman mosaics; see Arnott 2007). Subsistence hunting of waterfowl is still a traditional activity, especially in the Arctic where some duck and goose species breed (Padding *et al.* 2006). Commercial harvest of waterfowl is also legal and heavily practiced in some parts of species wintering ranges (Balmaki & Barati 2006). However, much of current waterfowl harvest is now through recreational hunting. Because of their cosmopolitan distribution over both the northern and the southern hemisphere, there is virtually no area around the globe where there are wetlands and no waterfowl harvest of any type.

Throughout much of the world, and certainly in North America and Europe, there is widespread recognition that waterfowl hunting requires some form of regulation. This recognition reflects the assumption that unregulated harvest has the potential to reduce waterfowl populations to dangerously low levels. As a result, government organisations worldwide have imposed various restrictions on the hunting of waterfowl, including, for example, establishment of seasons of the year and times of day when hunting is not permitted, areas within which hunting is not permitted, daily limits to the number of birds that can be harvested, restrictions on types of baits and other attractants (*e.g.* types of decoys) that can be used, and restrictions on the number and types of shells permitted. These various restrictions represent management actions designed to bring about desired outcomes with respect to harvested waterfowl populations. Implementation of such restrictions thus pre-supposes knowledge of the relationships between these regulations and harvest rates, and between harvest rates and waterfowl population change. However, both sets of relationships are characterised by uncertainty. Some of this uncertainty is likely not resolvable. For example, the exact harvest rates that result from a specific set of hunting regulations are always likely to be viewed as random variables arising from a distribution that characterises this source of partial controllability (Johnson *et al.* 1993, 1997). However, the relationships between harvest rates and both waterfowl survival rates and population change are represented by competing hypotheses and thus by uncertainty that is potentially resolvable by evidence. These hypotheses are the focus of this review, as we consider available evidence and ways to provide further resolution as a means of improving future management of waterfowl resources.

Despite the relatively narrow focus of our review, we remind the reader that harvest regulation is one of a relatively large number of potential actions that can be used to manage waterfowl populations. For example, a variety of management actions has been developed to improve habitat on waterfowl breeding grounds. Some of these actions are very specific and local, such as erecting nesting structures for cavity nesting species (Hawkins & Bellrose 1940; Bellrose 1990), constructing islands as potential nesting sites with reduced access to predators (Hammond & Mann 1956; Giroux 1981), and planting dense nesting cover for prairie nesting species (Duebbert *et al.* 1981; McKinnon & Duncan 1999). Other actions, such as the Conservation Reserve Program (Reynolds *et al.* 1994, 2001), are much less specific, and are designed to influence habitat across broad geographic areas. Active control of nesting predators can be viewed as a form of breeding habitat improvement. Predator reduction has been successfully applied throughout the world (Garrettson & Rohwer 2001; Kauhala 2004; Whitehead *et al.* 2008; Pieron & Rohwer 2010); however, in many parts of the world (*e.g.* North America) controversy remains about whether this action should be considered. Management actions affecting migration and wintering habitat have also been identified and implemented for waterfowl (Gilmer *et al.* 1982; Smith *et al.* 1989). In summary, a variety of potential management actions exists, and integrated programs of waterfowl management should include consideration of multiple actions (including harvest regulations) in order to achieve programme objectives (Runge *et al.* 2006).

History of waterfowl harvest management

In North America prior to the mid-1800s, waterfowl were viewed as extremely abundant and accordingly were hunted for sale and recreation throughout the year (Phillips & Lincoln 1930; Day 1949). Population declines in the late 1800s and early 1900s led to concerns about effects of harvest and to the beginning of government intervention. The United States government was granted authority to regulate waterfowl harvest, and the Migratory Bird Treaty Act of 1918 specified that hunting would be permitted only when deemed compatible with protection and maintenance of populations. During the period 1930–1950, the perception of declines and low populations led to restrictions in hunting regulations (U.S. Fish and Wildlife Service 1988) and to the initiation of monitoring programmes designed to assess waterfowl population status (Martin *et al.* 1979; Smith *et al.* 1989; Nichols 1991a). Over the next 25 years, these monitoring programmes were expanded and improved, and resulting data were employed to develop population models for use in establishing harvest regulations for key species (Crissey 1957; Geis *et al.* 1969). During this period (1951–1975), these models and conventional wisdom led to restriction of hunting regulations during years when breeding grounds were dry and population sizes low, producing disagreements about the effectiveness of such restrictions and the perceived lack of consideration of the desires of the hunting public (Nichols 2000). However, these political disagreements were not well-grounded in science, and the management of waterfowl hunting in North America was generally viewed as a good example of the scientific management of animal populations (Nichols *et al.* 1995).

In the early 1970s, analyses of Mallard *Anas platyrynchos* ringing and recovery data, using newly developed inference methods, led Anderson & Burnham (1976) to the conclusion that historical data did not provide strong support for the premise that had guided Mallard harvest for the prior 50 years, that changes in Mallard harvest rates had produced corresponding changes in Mallard survival and population size. This landmark study introduced structural uncertainty to North American waterfowl harvest management; that is, uncertainty in hypotheses about how changes in waterfowl harvest translate into changes in population dynamics. Subsequent efforts to resolve this uncertainty and manage waterfowl harvest in the face of it include a period (1979–1985) of stabilised hunting regulations (McCabe 1987) and a subsequent period (1985–1990) of risk-aversive conservatism (Sparrowe & Patterson 1987; U.S. Fish and Wildlife Service 1988). However, neither of these approaches led to resolution of the uncertainty, nor to a widely accepted approach for dealing with it.

In the early 1990s, members of the Office of Migratory Bird Management, U.S. Fish and Wildlife Service (USFWS), began to give serious consideration to implementing an adaptive approach to harvest management. Although the central ideas underlying adaptive management had been described and developed by Walters (1986), the approach had never been fully implemented on even a small scale. In 1992, Fred Johnson of the USFWS assembled an *ad hoc* working group of state and federal waterfowl biologists to discuss alternative approaches for waterfowl harvest management. The ideas of adaptive harvest management (AHM) were discussed, and the group decided to develop this approach, becoming the interagency working group for AHM. The proposed approach to AHM was outlined in Johnson *et al.* (1993) and formally adopted by the USFWS for mid-continent Mallard in 1995 (Nichols *et al.* 1995; Williams & Johnson 1995; Johnson *et al.* 1997). AHM was especially attractive because it provided a means of simultaneously reducing uncertainty while managing in the face of it. The AHM programme for mid-continent Mallard is still used each year to establish recommended hunting regulations. Its success has led to the development of AHM programmes for other Mallard populations and other waterfowl species in North America (*e.g.* Atlantic Flyway Canada Geese *Branta canadensis*; Hauser *et al.* 2007). Inferences reviewed in this paper about the relationship of hunting regulations and harvest rates to waterfowl populations are based both on specific analyses and on the results of AHM programmes.

The general pattern of increasing protection of waterfowl and regulation of harvest has occurred in Europe as well, where waterfowl were also considered as an almost infinite resource until the end of the 19th Century, and were commercially exploited as such. Duck decoys, in particular, were used to trap birds at their wintering and migration stopover sites, sometimes in an industrial manner (33,000 teal were caught in a single season on one island in the North Sea, leading to the building of a duck canning factory; Phillips 1923). Such commercial harvests gradually lost popularity and were abandoned throughout Europe during the 20th century, although trade of legally-harvested waterfowl by recreational hunters is still legal in some countries. Because of the number of different countries in Europe, it is more difficult to reach international agreements, and national waterfowl management policies have sometimes developed towards different systems and at different paces. The main current legal framework is the 'Bird Directive' (adopted by the European Commission in 1979), which limits in particular the periods of the year during which birds can be harvested anywhere along their flyways, and the African-Eurasian Waterbird Agreement (AEWA), which aims to coordinate research, monitoring and policy at the Palearctic flyway scale (including beyond the European Union). The European waterfowl management policy is therefore far less developed than the American system, although the eventual set-up of a proper international adaptive harvest management scheme is a goal for the future (Elmberg *et al.* 2006). In fact, an adaptive management programme for the Svalbard Pink-footed Goose *Anser brachyrynchus* population is under current development (*e.g.* Johnson *et al.* 2014).

Life history characteristics of waterfowl

Despite the general similarities in morphology and behavioural habits, waterfowl form a very diverse family of birds when it comes to body size, with up to a 32-fold difference in body mass between a 330g Green-winged Teal *Anas carolinensis* and a 10.5 kg Trumpeter Swan *Cygnus buccinator*. Such differences in body mass have obvious consequences in terms of, for instance, energy needs (Miller & Eadie

2006), which translate into very different life history strategies. These differences have long been recognised, as exemplified in the following statement from the mid-20th century: "Another inference from the general observation that geese are bigger than ducks is that large waterfowl survive better, and produce fewer offspring, than small ones" (Boyd 1962).

Waterfowl therefore can be broadly organised along a "fast-slow" gradient, with faster duck species having short life expectancies but a high annual reproductive output, as opposed to slower geese and swans surviving much longer but producing fewer offspring per breeding attempt (Gaillard *et al.* 1989). In line with ducks producing more offspring which survive more poorly, density-dependent feedback on individual survival is thought to be more common in ducks, at least during some stages of their life cycle (Gunnarsson *et al.* 2013), while this is not so much the case in geese and swans. Life history variation exists even within ducks (subfamily *Anatinae*). For example, among North American ducks, Patterson (1979) characterised Mallards, Blue-winged Teal *Anas discors*, and Northern Pintail *Anas acuta* as relative "*r*-strategists" (faster life histories) and Redhead *Aythya americana*, Canvasback *Aythya valisineria* and Scaup *Aythya marila* as "*K*-strategists" (slower life histories).

The constraint imposed by their smaller size prevents ducks from carrying substantial body reserves along long migratory flights (Klaassen 2002), leading to their characterisation as "income breeders"; *sensu* Drent & Daan (1980), in that they mostly rely on the energy available at or near their breeding grounds to fuel their reproduction. In contrast, the larger size of geese and swans permits the storage of lipid and protein reserves well before reproduction, often as early as on the wintering grounds, leading to their characterisation as "capital breeders". Such differences in life history strategies have profound consequences for population structure and, hence, the modelling of population dynamics. Duck populations are generally considered as being relatively simple in structure, with little need to incorporate age structure beyond the first-year/adult dichotomy (Devineau *et al.* 2010). Conversely, the low reproductive rate and long survival of geese, swans and many sea ducks lead to more complex populations, with delayed age of first breeding, extended age-specificity, *etc*. The greater heterogeneity among individuals within such populations often requires the use of more structured models.

Modelling considerations: structure & heterogeneity

Models are used in harvest management to allow us to predict the numerical response of a population subjected to a certain level of harvest. However, all population models, regardless of their application (*e.g.* harvest management), represent approximations to reality which can never be fully specified. Utility of the model in the context of harvest management is primarily determined by the degree to which the model correctly represents the functional form relating the management control option (say, varying harvest pressure through legislative action), and the response of the population to harvest. Much literature has focussed on the question of whether harvest mortality in waterfowl is "additive" or "compensatory" to natural mortality (see below), and the impacts of that distinction on optimal harvest management. Second, the ability of a population model to reflect accurately the dynamics of a population will be strongly influenced by the degree to which the model structure adequately accounts for important differences among individuals, both in terms of underlying vital rates (survival, fertility), but also potentially in the functional response of those vital rates to perturbation (*i.e.* harvest). At one extreme, simple scalar models assume that all individuals have identical latent probabilities of survival and reproduction – we refer to such models as *scalar projection models*. As noted in the preceding section, for many duck populations, such simple scalar or near-scalar models are often sufficient. At the other extreme, we imagine a model containing sufficient structure to model the dynamics of each individual in the population. We refer to such models as *individually-based projection models*. It is clear that this latter class of models represents the closest approximation to full reality. Such a model would completely account for heterogeneity among individuals in the population (the role of individual heterogeneity will be re-visited later).

However, a fully individually-based population model is generally intractable, both in terms of construction, analysis and application in a management context. For example, estimation of a time-specific survival probability for individual *i* over time step t to $t + 1$ requires inference about a binomial parameter based on a single Bernoulli trial. Based on whether the individual is alive or dead at time $t + 1$, we must somehow estimate the underlying probability of survival, and there is simply not enough information in this single observation to allow us to do this well (see Cohen 1986). As such, we are generally left constructing a model which represents a compromise between a simple scalar model, and a fully individually-based model. Such "intermediate" models are based on the reasonable idea that much (if not all) of the variation among individuals can be explained by one or more factors (demographic, genotypic, spatial, developmental), which can be used to structure a population into (generally) discrete classes of individuals grouped together by sharing one or more of these factors (we note that in some cases, discretization is a mathematically convenient approximation to the continuous state-space). For many waterfowl species, particularly longer-lived swans and geese, and many sea ducks, there is significant variation in both survival and fertility as a function of the age of the individual. A model which differentiates among individuals based on differences in such factors is known generally as a *structured model*. Such models are parameterised not only in terms of potential differences in survival and fertility among classes of individual, but also in terms of the probability of making transitions among classes (due to aging, growth or movement).

This section addresses the impact of population structure on the projected impacts of that harvest on waterfowl population dynamics; we defer consideration of the role of different functional forms relating harvest to the numerical response to the next section. Here, we describe the conditions under which harvest would lead to change in population abundance, and the structural factors influencing the magnitude and time course of such change.

Equilibrium harvest for scalar populations

We introduce some of the basic considerations in model-based harvest management through a numerical example. We consider first a simple deterministic scalar population in discrete time, without density-dependence, where the population size N at time $t + 1$ is given as the product of the current population size at time t and a scalar multiplier, λ.

$$N_{t+1} = \lambda N_t. \qquad (1)$$

As long as $\lambda > 1$, then $N_{t+1} > N_t$ (i.e. the population will grow). In the absence of harvest or some other "control" measure, the population will grow without bound. In such cases, we focus on calculating the equilibrium harvest rate, E, which represents the maximum harvest which does not lead to the increase or decline of the population over time (i.e. the harvest condition under which $N_{t+1}/N_t = 1$):

$$\frac{N_{t+1}}{N_t} = (\lambda - E). \qquad (2)$$

Thus, for example, a population with $\lambda = 1.05$ is projected to grow at 5% per time step. The equilibrium harvest rate then is simply $E = (1.05–1) = 5\%$. In terms of absolute numbers, if $N_t = 1,000$, then the equilibrium harvest would be 50 individuals at each time step.

Equilibrium harvest for structured populations

Now we consider a structured population. While there are a number of different classes of structured models, we will focus on the use of matrix-based models, which are canonical models for discrete-time population dynamics, where individuals are classified (grouped) into discrete "(st)ages" (Caswell 2001; Lebreton 2005). To simplify the presentation, we will consider deterministic models, with no density-dependence. We'll assume the minimal structured model with 2 age classes (juveniles and adults, where the adult class consists of all individuals ≥ 1 years of age). Fertility and survival transitions for our example population are given in the life-cycle graph shown in Fig. 1 (constructed assuming a post-breeding census; Caswell 2001). Assuming $S_A = 0.65$, $S_J = 0.5$, and $F = 0.8$, then the projection matrix model $\mathbf{A}$ can be constructed directly from the life-cycle graph as

$$\mathbf{A} = \begin{bmatrix} S_J F & S_J F \\ S_J & S_A \end{bmatrix} = \begin{bmatrix} 0.40 & 0.52 \\ 0.50 & 0.65 \end{bmatrix}, \quad (3)$$

from which we derive the following standard metrics for projected growth (λ), stable age proportions (w_i), and age-specific reproductive values (v_i; for details see Caswell 2001):

$$\lambda = 1.05, \mathbf{w} = \begin{bmatrix} 0.3478 \\ 0.4348 \end{bmatrix}, \mathbf{v} = \begin{bmatrix} 1.0 \\ 1.5 \end{bmatrix}.$$

Thus, in the absence of harvest and assuming time-invariance and no density-

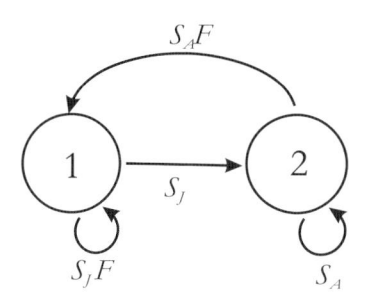

Figure 1. Life-cycle graph and structure of the underlying life history for the 2-age class (adults, juveniles) example. The life-cycle graph is based on a post-breeding census. Node 1 is the number of juveniles (offspring) in the population, and node 2 is the number of adults (age ≥ 1 year). The arcs connecting the nodes reflect survival (left-to-right) and fertility (right-to-left). S_A and S_J are the survival probabilities for adults and juveniles, respectively. F is the reproductive rate, and is assumed to be invariant with age for age > 1 year.

dependence, the population is projected to grow without bound at 5% per time step, eventually stabilising at a juvenile:adult ratio of 4:5. As indicated by the reproductive value vector **v** (Fisher 1930), each adult in the population is worth 1.5 juveniles to future population growth, suggesting that harvest of an adult individual is potentially of greater impact than harvest of a juvenile individual. (Here, we have normalised **v**, and **w** so that **v'w** = 1. It is customary to express **v** such that v_1 = 1, so that the reproductive value of each stage is compared to that of the first stage, and **w** such that the elements n_i sum to 1, so they represent the proportion of the population in each stage class).

The difference in reproductive value between adults and juveniles represents a key consideration which differentiates modelling the harvest of a structured population, and harvest of a simple scalar population. Reproductive value is a well-known concept in evolutionary biology (Stearns 1992 provides a general review) and has been identified as affecting the optimal age- or stage-specific harvest of a population (MacArthur 1960; Grey & Law 1987; Brooks & Lebreton 2001; Kokko 2001; Lebreton 2005; Hauser *et al.* 2006). Harvest of individuals of higher reproductive value will, generally, have a greater proportional impact on population dynamics than harvest of individuals with lower reproductive value (although the relative value of individuals may change following a perturbation (Caswell 2001; Cameron & Benton 2004) and is a function of whether or not the population is increasing or decreasing at the time of harvest (Mertz 1971)). Thus, the inclusion of structure adds extra dimensions of uncertainty, but also additional flexibility and opportunities, to harvest management. Most obviously, a structured model may require a structured harvest to reach an optimal harvest objective.

Constant harvest

We can demonstrate the role of structure and age-specific reproductive value by means of a simple numerical example. Suppose at time *t* the population consists of ~1,000 individuals. Based on a simple scalar model, a projected growth rate of λ = 1.05 (as per the preceding example) implies that we could harvest at most 5% of the population each time step. Given, say, 1,000 individuals in the population at the time of harvest, this would correspond to a constant harvest of 50 individuals.

But, which 50 individuals? Juveniles, adults, or some of each? Consider the 2-age-class structured model introduced earlier (eqn. 3), and the consequences of harvesting 50 juveniles, *versus* harvesting 50 adults from a starting population of ~1,000 individuals (assuming for the moment that we could selectively harvest a particular age class – the implications of violating this assumption are considered later). If the population structure at the time of the harvest is proportional to the equilibrium age structure (*i.e.* consisted of 444 juveniles, and 555 adults), then the population – and thus each age class – is projected to increase at 5% per time step. Under these conditions, it might seem reasonable to assume that harvesting 50 adults, or 50 juveniles, or any vector summing to 50 total individuals (*e.g.* 25 adults and 25 juveniles), would have the same effect on long-term dynamics (namely, no change in population size between now and the next time step following harvest). However, this is not the case – in fact, the direction and magnitude of the change in the population is determined by the relative proportions of each age class in the harvest.

Since such a result might seem counter-intuitive, it is useful to evaluate the correct equilibrium harvest conditions for a structured population. Let the dynamics of a structured population subjected to a constant harvest be given by:

$$\mathbf{N}_{t+1} = \mathbf{A}\mathbf{N}_t - \mathbf{E}, \qquad (4)$$

where $\mathbf{A}$ is the matrix projection model, and $\mathbf{E}$ is the harvest vector where the ith element represents the number of individuals in stage i that is harvested during each time period (here, we assume a specific, constant number of individuals harvested for each age class. We consider proportional harvest later). Under equilibrium harvest for a discrete-time projection model, $\mathbf{N}_{t+1} = \mathbf{N}_t = \mathbf{N}^*$. Thus, eqn. (4) can be rearranged to show that:

$$\mathbf{N}^* = (\mathbf{A} - \mathbf{I})^{-1}. \qquad (5)$$

If $\mathbf{A}$ is primitive (which is generally the case for population projection models, which are generally positive and square), then the unharvested population will eventually grow as:

$$\mathbf{N}_t \sim (\mathbf{v}'\mathbf{N}_0)\lambda^t\mathbf{w}.$$

Following Hauser *et al.* (2006), if the difference between the harvested population and the equilibrium state after some time t is:

$$\mathbf{N}_t - \mathbf{N}^* \sim [\mathbf{v}'(\mathbf{N}_0 - \mathbf{N}^*)]\lambda^t\mathbf{w}, \qquad (6)$$

then the equilibrium harvest vector $\mathbf{E}$ is given as:

$$\mathbf{v}'\mathbf{E} = \mathbf{v}'\big[\mathbf{N}_0(\lambda - 1)\big]. \qquad (7)$$

Now (λ–1) is the long-term proportional increase of the unharvested population (*e.g.* if $\lambda = 1.05$, then the population will increase by (1.05–1) = 5% per year in the long term). The reproductive value of this 'excess' proportion of the initial population $\mathbf{N}_0$ (the right-hand side of eqn. (7) must be equal to the reproductive value of the harvest (the left-hand side of eqn. (7). This ensures that the harvest is sustainable and that the population will approach a steady state over time.

Returning to our numerical example, if the initial population $\mathbf{N}_0$ is known, then we find harvest vectors $\mathbf{E}$ that satisfy the equilibrium condition (eqn. 7). The initial population may have the stable stage distribution of the unharvested system, e.g. $\mathbf{N}_0 = (444,555)'$. Then:

$$\begin{bmatrix} 1.0 & 1.5 \end{bmatrix} \begin{bmatrix} E_J \\ E_A \end{bmatrix}$$

$$= (1.05-1)\begin{bmatrix} 1.0 & 1.5 \end{bmatrix} \begin{bmatrix} 444 \\ 555 \end{bmatrix},$$

leading to:

$$E_J + 1.5 E_A = 63.825,$$

where E_J and E_A are the numbers of juveniles and adults harvested per time step, respectively. The equilibrium condition (eqn. 7) is that a particular linear combination of the harvest taken from each class is held at a constant. It is important to note that the coefficients for harvest from a class (i.e. E_J, E_A) are the reproductive values of individuals in those classes (i.e. $v_1 = 1.0$ for E_J, $v_2 = 1.5$ for E_A). The constant (in this case 63.825) is dependent on the initial population size and structure $\mathbf{N}_0$. The particular solution to the equilibrium harvest equation for our present example, where $\mathbf{N}_0 = (444,555)'$, is shown in Fig. (2). If the harvest $\mathbf{E} = (E_J, E_A)'$ that is actually taken falls below this line, then the population will eventually increase. If the harvest falls above this line then the population will eventually decline. Harvest that falls on this line (i.e. satisfying the equilibrium condition) will cause the population to stabilise over time to a population size and structure given by equation (7). We introduced this example by claiming that a total harvest of 50 individuals would cause the population to increase or decrease over the long term. We see clearly from this figure that a harvest of 50 adults only, and no juveniles, is above the line, leading to a population decrease. In contrast, a harvest of 50 juveniles only, and no adults, is below the line, leading to a population increase.

Note that the structure and size of this steady state population (i.e. at equilibrium) are dependent on the particular harvest vector $\mathbf{E}$ that is used. For example, if we choose to harvest only adults then the equilibrium population size is calculated to be > 1,000 (i.e. above the starting population size):

$$\mathbf{E} = \begin{bmatrix} 0.00 \\ 42.55 \end{bmatrix} \text{ and } \mathbf{N}^* = \begin{bmatrix} 464.2 \\ 541.5 \end{bmatrix} = 1005.7.$$

If instead we choose to harvest only juveniles, the equilibrium population size is calculated to be < 1,000 (i.e. below the starting population size), then:

$$\mathbf{E} = \begin{bmatrix} 63.83 \\ 0.00 \end{bmatrix} \text{ and } \mathbf{N}^* = \begin{bmatrix} 406.2 \\ 580.2 \end{bmatrix} = 986.4.$$

Three important points should be noted here. First, as mentioned earlier, the coefficients for harvest from a given age-class (i.e. E_J, E_A) are the reproductive values of individuals in those classes (i.e. $v_1 = 1.0$ for E_J, $v_2 = 1.5$ for E_A). In other words, the equilibrium harvest of juveniles only (63.825) would need to be 1.5 times larger than the equilibrium harvest of adults only (42.550; 63.825/42.550 = 1.5). This is because the harvest of a single adult from the population is demographically equivalent to the harvest of 1.5 juveniles. This linear relationship between the reproductive value vector and the equilibrium harvest vector is not limited to simple 2-age class models – for any number

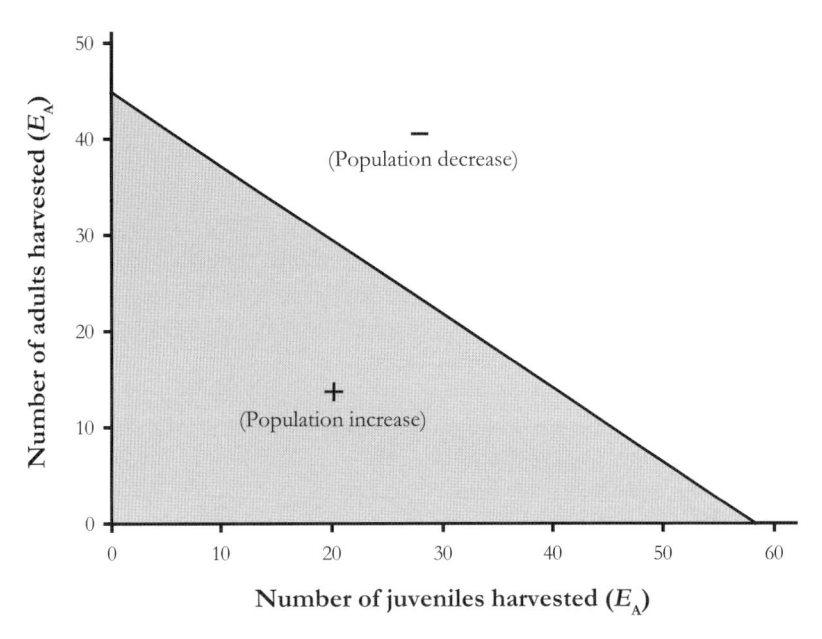

Figure 2. Equilibrium harvest for constant harvest of a fixed number of adults (vertical axis) and juveniles (horizontal axis), for the population projection model described by eqn. (3). The equilibrium harvest is specific to the initial population size and structure, which here we assume to be 1,000 individuals in the stable age proportions (*i.e.* $\mathbf{N}_0$ = (444,555)′). Harvest at any point below the equilibrium (shaded area) will cause the population to increase, whereas harvest at any point above the equilibrium will cause the population to decrease. Adapted from Hauser *et al.* (2006).

of age classes (≥ 2), the equilibrium expression is a k-dimensional plane, for k age classes in the model (Hauser *et al.* 2006; for most long-lived waterfowl species, $k ≥ 5$). The coefficients of equilibrium solution are the reproductive values of the corresponding age classes.

The second point is that the population size at equilibrium is smaller for a juvenile-only harvest (986.4), and more skewed towards adults, compared to an adult-only harvest (1,005.7, 46% juvenile). This dependence of the final size and structure of the population on the structure of the population at the time of the harvest, and the structure of the harvest itself, is an important consideration addressed later (see also Koons *et al.* 2014a).

Finally, it is possible that the population does not stabilise at equilibrium abundance, but instead grows unbounded, even if the harvest satisfies eqn. (7). This can occur if in some time steps the number of individuals that needs to be removed for a given age class is larger than the number of individuals existing in that age class at that time. In such cases, the full "equilibrium harvest" cannot be taken, and the population grows unbounded (Hauser *et al.* 2006). (This issue does not occur if we implement a proportional harvest, as developed in the next section).

Proportional harvest

In the preceding, we considered a *constant* harvest, where a constant number of individuals is harvested (from a given age class) at each harvest decision. Such a scenario is arguably unrealistic for waterfowl harvest, where harvest regulations are based on an assumed relationship between various regulatory options and the proportional probabilities of mortality due to harvest (*i.e.* kill rate, K_i, the probability of being harvested during time interval t). Under *proportional* harvest, the same proportion of individuals is removed from each age class each time period (although the proportion may differ among age classes). The harvest model becomes:

$$\mathbf{N}_{t+1} = (\mathbf{I} - \mathbf{K})\mathbf{A}\mathbf{N}_t,$$

where $\mathbf{K} = \text{diag}(K_1, K_2, \ldots, K_k)$, and $0 \leq K_i \leq 1$ is the proportion of age class i to be harvested. If harvest occurs immediately after reproduction (generally the case for waterfowl), then:

$$\mathbf{N}_{t+1} = \mathbf{A}(\mathbf{I} - \mathbf{K})\mathbf{N}_t. \quad (8)$$

To find the equilibrium condition under proportional harvest, we again set $\mathbf{N}_{t+1} = \mathbf{N}_t = \mathbf{N}^*$, and solve for $\mathbf{N}^*$:

$$\mathbf{N}^* = \mathbf{A}(\mathbf{I} - \mathbf{K})\mathbf{N}^*. \quad (9)$$

That is, we choose harvest vector $\mathbf{K}$ so that 1 is the dominant eigenvalue of $(\mathbf{I} - \mathbf{K})\mathbf{A}$.

We denote the corresponding right and left eigenvectors as $\mathbf{v}_K$ and $\mathbf{w}_K$, respectively. Then the population under harvest $\mathbf{K}$ will approach the equilibrium:

$$\mathbf{N}^* = (\mathbf{v}'_K \mathbf{N}_0)\mathbf{w}_K. \quad (10)$$

We solve for $\mathbf{K}$ in a straightforward way. Using the matrix $\mathbf{A}$ from our 2-age-class model, then from eqn. (9):

$$\begin{vmatrix} -0.7 - 0.3 K_J & 0.6 - 0.6 K_J \\ 0.5 - 0.5 K_A & -0.35 - 0.65 K_A \end{vmatrix} = 0,$$

which can be simplified to:

$$K_A = \frac{11 - 81 K_J}{151 - 21 K_J},$$

which is a non-linear expression in K_A and K_J (although over the limited range of plausible values of $0 \leq K_i \leq 1$ for this problem, the equilibrium solution appears linear; Fig. 3). If the harvest chosen falls on this curve then the population will approach a steady state over time. If the harvest chosen falls below the curve then the population will grow geometrically over time. If the harvest chosen falls above the curve then the population will decline geometrically.

Two important differences should be noted compared to the constant harvest scenario introduced earlier. First, the equilibrium condition (eqn. 9) is not dependent on $\mathbf{N}_0$ (although the equilibrium population vector $\mathbf{N}^*$ is; eqn. 10), and is not a simple function of reproductive values (although it is clear from the equilibrium condition that a higher harvest rate for juveniles is required in order to achieve equivalence with a given harvest rate for adults, consistent with the interpretation of the relative value of juveniles and adults under a constant harvest model). Second, because we are dealing here with a proportional harvest, K_i, where $0 \leq K_i \leq 1$, then the situation where a full equilibrium harvest cannot be taken – as was the case

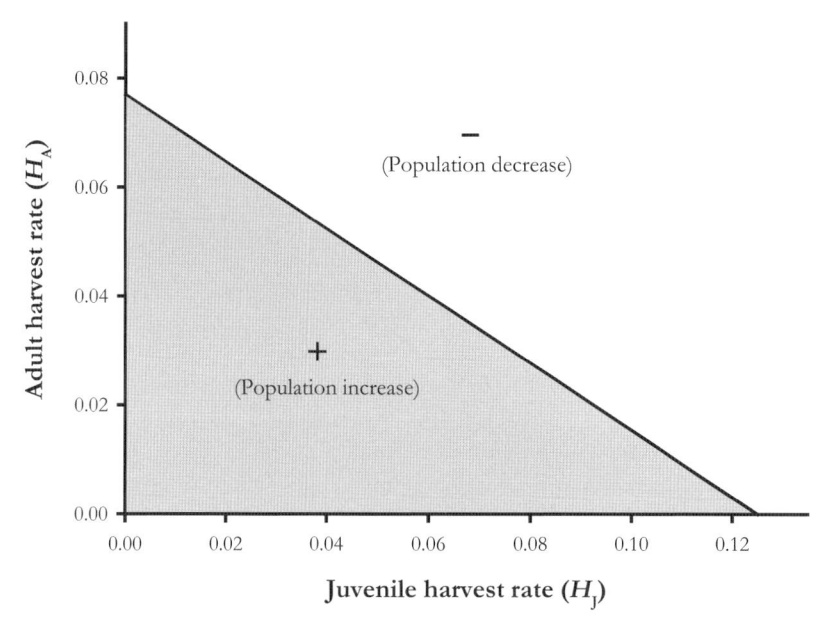

Figure 3. Equilibrium harvest for harvest of a fixed *proportion* of adults (vertical axis) and juveniles (horizontal axis), for the population projection model described by eqn. (3). The equilibrium harvest is specific to the initial population size and structure, which here we assume to be 1,000 individuals in the stable age proportions (*i.e.* $\mathbf{N}_0 = (444,555)'$). Harvest at any point below the equilibrium (shaded area) will cause the population to increase, whereas harvest at any point above the equilibrium will cause the population to decrease. Adapted from Hauser *et al.* (2006).

for a constant harvest if the equilibrium harvest for a particular age class was larger than the number of individuals in that age class – cannot occur. As such, under a proportional harvest, there is always a bounded equilibrium. However, as was the case for a constant harvest, the size and structure of the equilibrium population varies depending on both the structure of the harvest vector $\mathbf{K}$, and the structure of the population at the time of harvest, $\mathbf{N}_0$ (Hauser *et al.* 2006). Since waterfowl harvest management is almost universally based on proportional harvest, we do not discuss constant harvest models further.

"Structure", by any other name…

In the preceding, we focussed exclusively on age structure (*i.e.* where the age of the individual was the only determinant of variation in survival or fertility among individuals specified in the model). For some species of waterfowl, especially longer-lived swans, geese and many sea ducks, age structure is clearly an important consideration.

However, while age structure may generally be a less important consideration for many short-lived duck species, there are other forms of structure which may be important considerations in model construction, some especially for ducks, and others for waterfowl generally (*e.g.* spatial structure). Brooks & Lebreton (2001) describe a simple application of the methods described in the preceding to a metapopulation, where the value of the individual is conditioned on both age and location. Further, the models presented above considered individuals of only one sex. While this may be appropriate in some cases, it may not always be the case, especially for species where the dynamics of the population are influenced by the form of the mating system (*e.g.* polygamy), or more commonly, where survival rates differ significantly between the sexes. In such cases, models where the dynamics of the two sexes are linked by the pair bond and maternity function are more appropriate. Such sex-linked models can generate rather complex dynamics (Caswell & Weeks 1986; Lindström & Kokko 1998).

Finally, all waterfowl populations (and wild populations in general) are characterised by differences among individuals that extend beyond the sources of variation already discussed. Even when models structurally separate males and females, or young and old, or one location *versus* another, within a given "node" (say, females of age 2, that are in location X), there are remaining differences among individuals. These differences are commonly referred to as reflecting "individual heterogeneity". While it is quite likely that these differences vary continuously among individuals, as a first approximation, we can consider modelling this additional heterogeneity (*i.e.* differences beyond those explained by structural elements such as age,

or sex, or location) based on a finite set of "classes" of individuals (this is strictly analogous to the practice of using finite mixture models to approximate heterogeneity in analysis of mark-encounter data; *e.g.* Pledger *et al.* 2003). For example, consider a population where individuals can be characterised as either "high quality" or "low quality", based on their total latent survival probability (where a "low quality" individual is one with a lower probability of survival). In theory, we could reconfigure a matrix model which assumes that all individuals within a given structural class have identical latent vital rates (*e.g.* Fig. 1), to account for two discrete "quality" classes within that structural class. For example, if we assume that "quality" differences influence only juvenile and adult survival, but not fertility, then we might restructure the 2 age-class matrix model we considered earlier (Fig. 1) to reflect the unequal contributions of individuals of different quality to population growth (Fig. 4).

The projection matrix model **A** can be constructed directly from this life-cycle graph as:

$$\mathbf{A} = \begin{bmatrix} \pi S_{J,h} F + (1-\pi) S_{J,l} & S_{A,h} F & S_{A,l} F \\ \pi S_{J,h} & S_{A,h} & 0 \\ (1-\pi) S_{J,l} & 0 & S_{A,l} \end{bmatrix},$$

where $S_{x,q}$ is the latent survival probability of individuals of age class x, and quality class q (where $q = high$ or low), F is the fertility rate, and π is the probability that a juvenile is (or will become) a "high" quality individual.

In this sense, heterogeneity models are (at least in simple, discrete form) structurally equivalent to models structured based on

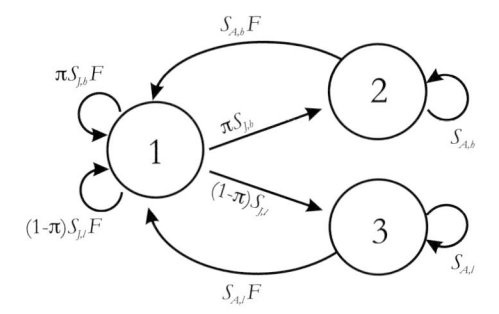

Figure 4. Life-cycle graph and structure of the underlying life history for a 2-age class (adults, juveniles) model, with discrete (finite mixture) heterogeneity in juvenile and adult survival. The life-cycle graph is based on a post-breeding census. Node 1 is the number of juveniles (offspring) in the population, and represents the combined fertility contributions of high quality (node 2) and low quality (node 3) adult (age ≥ 1 year) individuals, where differences in quality are characterised by lower juvenile and adult survival rates among lower quality individuals. The arcs connecting the nodes reflect survival (left-to-right) and fertility (right-to-left). $S_{A,q}$ and $S_{J,q}$ are the survival probabilities for adults and juveniles, respectively, for quality class q. The parameter π determines the probability that a new juvenile (node 1) becomes a high quality adult. F is the reproductive rate, and is assumed to be invariant with age and quality for age > 1 year.

age, location, or gender, and we can apply the same methods discussed earlier to evaluate the dynamics of the population described by this "heterogeneity" model to evaluate the relative value of harvest of individuals of low or high quality to those dynamics (*i.e.* reproductive value conditioned on both age and quality). The possible role of unspecified heterogeneity on numerical response to harvest is considered in a later section.

For some taxa (either sessile organisms, organisms that are temporarily sessile at the time of harvest – *e.g.* fish in the net in a commercial fishery, or where the mechanism of harvest is state-selective – *e.g.* use of nets of specific mesh size or other gear to capture only certain size-classes of fish), it is possible to harvest an individual selectively on the basis of its individual state (*e.g.* you keep the big ones, and toss back the small ones). For most waterfowl, however, there will often be considerable uncertainty in establishing the "state" of an individual at the time of harvest, and optimisation of harvest based on state structure of the population will be only partially controllable. Even in cases where the extent of uncertainty about system state is reduced (for example, if harvest occurs at locations where the targeted individuals represent specific age- or sex-classes), waterfowl harvest in many cases is simply a random selection of individuals with differential vulnerability to the harvest.

Consequences of hunting for waterfowl populations

As noted above, a key to making wise management decisions is to be able to make predictions about system response to potential management actions. For hunted populations, predictions will usually include the hunting mortality rate that will result from any prescribed set of regulations. Studies contrasting ring recovery rates in years of differing hunting regulations have provided evidence supporting the inference of higher harvest rates in years of more liberal hunting regulations (see reviews of

Nichols & Johnson 1989; Nichols 1991; Johnson & Moore 1996, and more recent species-specific evidence of Johnson et al. 1997; Francis et al. 1998; Calvert & Gauthier 2005; Alisauskas et al. 2011; Peron et al. 2012; Iverson et al. 2014). Of course numerous variables (e.g. environmental conditions and resulting migration timing, and regional hunter activity) in addition to hunting regulations are expected to influence hunting mortality rates. As a result, hunting mortality rates predicted to correspond to a specific set of hunting regulations are best characterised as a probability distribution. Examples of such distributions estimated for mid-continent Mallard and Black Duck are provided by Johnson et al. (1997) and USFWS (2013), respectively.

In addition to predicting the hunting mortality rate, predictions are required for the changes in survival rates (probability of surviving all mortality sources), reproductive rates, and the rate of movement in and out of the focal population expected to accompany this level of hunting mortality. Much of the uncertainty in waterfowl management involves these relationships, and most of the North American waterfowl programmes in adaptive harvest management include multiple models (and corresponding hypotheses) as a means of dealing with this uncertainty. We consider these relationships below.

Survival

The most direct influence of hunting mortality should be on the total (all sources) mortality rate. A general form for this relationship can be expressed as (eqn. 11):

$$E(S_t) = S_0(1 - \beta K_t), \quad (11)$$

where E denotes expected value, S_t is the probability that a bird alive at the beginning of the hunting season in year t survives and is alive at this same time the next year ($t + 1$), S_0 is the probability that a bird alive at the beginning of the hunting season in year t would survive to that same time the following year $t + 1$ in the complete absence of any hunting mortality, K_t is the probability that a bird alive at the beginning of the hunting season in year t would die from hunting causes before that same time the following year $t + 1$ in the complete absence of any non-hunting mortality, and β is the slope parameter relating S_t and K_t.

Equation (11) is very general and can be used to model a variety of relationships between hunting and survival depending on the value of β, with $\beta = 1$ and $\beta = 0$ indicating plausible models with maximal and minimal effects of hunting on survival, respectively. S_0 and K_t are each defined as applying when the other mortality source is not operating. As such, they are referred to as *net rates* in the literature of competing mortality risks (e.g. Berkson & Elveback 1960; Chiang 1968; see below). S_0 is not defined as time-specific (it is not subscripted by t), for consistency with historical development, but time-specificity is certainly possible conceptually.

Because S_0 and K_t are net rates, they cannot usually be estimated directly. Instead, without extra information about the timing of the different mortality sources, we are usually restricted to estimation of so-called 'crude' rates (*sensu* Chiang 1968). Specifically, a crude, source-specific mortality rate is the probability of dying from that source in the presence of all other mortality sources. In the case of modelling and inference about duck populations, K'_t is defined as the crude hunting mortality rate, or the probability that a bird alive at the beginning of the hunting season in year t would die from hunting causes during the hunting season of year t in the presence of non-hunting mortality that occurs during this period. Even K'_t cannot typically be estimated directly, but instead requires information from multiple sources. Ring recovery data are the common source of information about hunting mortality, and corresponding models permit direct estimation of ring recovery rates, f_t, the probability that a ringed bird alive at the beginning of the hunting season of year t is shot and retrieved by a hunter during the hunting season of year t and its ring correctly reported (see Brownie et al. 1985; Williams et al. 2002). Define c_t as the probability that a bird shot by a hunter during the hunting season of year t is retrieved by the hunter ($1 - c_t$ denoting 'crippling loss'), and λ_t as the probability that a retrieved bird is reported. Then:

$$f_t = K'_t c_t \lambda_t. \qquad (12)$$

If probabilities associated with ring reporting and retrieval are constant over time, then K'_t is related to f_t by a proportionality constant, making f_t a reasonable index of K'_t (see Anderson & Burnham 1976; Burnham & Anderson 1984; Burnham et al. 1984). And because non-hunting mortality during the hunting season is often thought to be small relative to hunting mortality, K'_t, and thus f_t, are thought to be reasonable indices to K_t for the common situation (for North American waterfowl) of ringing occurring just before the hunting season.

Additive mortality hypothesis

Anderson & Burnham (1976) used equation (11) to define two endpoint hypotheses designed to bracket the possible relationships between total survival and hunting mortality. They used the term "additive" for the situation where $\beta = 1$, corresponding to a model in which hunting and non-hunting mortality sources are viewed as independent competing risks (Berkson & Elveback 1960; Chiang 1968). The term "additive" is applicable, as the instantaneous risks associated with the two mortality sources are added in order to obtain the total (both sources) probability of dying during an interval in which both sources apply. This additive mortality model is commonly used in fisheries management (Beverton & Holt 1957; Ricker 1958; Hilborn & Walters 1992) and is intuitively reasonable.

Although equation (11) applies to net rates of hunting and non-hunting mortality, regardless of their temporal patterns of occurrence, the intuition underlying this expression is perhaps most apparent when the two mortality sources are completely separated in time. So assume that only hunting mortality occurs during the hunting season and that all non-hunting mortality is restricted to the period following the hunting season. Then, equation (11) simply states that the probability of surviving the year is the product of first surviving hunting mortality during the hunting season and then surviving non-hunting mortality sources during the rest of the year. Because survival is a multiplicative process in time, with the survival probability for one period being applied to the survivors of the previous period, equation (11) should correspond to intuition. North American waterfowl management was based on the additive mortality hypothesis prior to 1976 (see Geis *et al.* 1969; Nichols 2000). Finally, note that equation (11) specifies a linear decrease in total annual survival, S_t, as net hunting mortality, K_t, increases (Anderson & Burnham 1976; Nichols *et al.* 1984; Fig. 5a).

Compensatory mortality hypothesis

Anderson & Burnham (1976) used equation (11) to specify a compensatory mortality hypothesis, under which $\beta = 0$ for some range of values of K_t, specifically for $K_t \leq C$ where C is a threshold subject to the inequality, $C \leq 1 - S_0$. So $S_t = S_0$ for $K_t \leq C$, that is changes in kill rate below the threshold induce no variation in total survival, which remains equal to net survival from non-hunting sources only. Because of this complete lack of influence of hunting mortality on total survival, at least for a range of hunting mortality rates, this basic model is sometimes referred to as depicting "complete compensation" (Conroy & Krementz 1990; Fig. 5b.). Kill rates greater than the threshold necessarily result in declines in total annual survival (linear decline under Anderson & Burnham 1976).

Conroy & Krementz (1990) noted that the diversity of life history characteristics among waterfowl (*e.g.* with respect to fast-slow and *r–K* variation) should lead to predictions about the degree to which particular species would be expected to exhibit more additive *versus* more compensatory mortality. Species with relatively "slow" life histories have higher survival rates and are expected to exhibit less ability to compensate for hunting losses than species with faster life histories characterised by much lower survival rates. At a minimum, total annual survival rates and net survival rates from non-hunting sources impose a constraint on the maximum value of a compensation threshold, $C \leq 1 - S_0$. Conroy & Krementz (1990) thus noted that a variety of hypotheses about hunting-survival relationships exists between the two endpoint hypotheses of additivity and complete compensation. They referred to these hypotheses as "partial compensation" and noted that they are characterised by $0 < \beta < 1$ in equation (11) below some threshold C (Fig. 5c).

Possible mechanisms for compensation. The family of hypotheses defined by equation (11) thus covers the full range of possible relationships between hunting mortality and total survival, ranging from complete additivity, to partial compensation to complete compensation (Fig. 5). However, a cost of this flexibility is that the model is phenomenological, in the sense that it provides no hint of plausible mechanisms that might underlie most of the possible hypotheses. Additivity, with $\beta = 1$, is consistent with intuition about how different mortality sources might interact. An individual can only die of one source and each death translates to fewer survivors at the end of any time period (*e.g.* 1 year). However, hypotheses reflecting some degree of compensation, $\beta < 1$, are not underlying compensatory mortality (Anderson & Burnham 1976; Nichols 1991). The usual explanation under density-dependence is that population size at the end of the hunting season is a determinant of subsequent survival during the portion of the year without hunting. In years where hunting mortality is large, abundance at the end of the hunting season is reduced, and each individual alive at this time has an increased probability of surviving the rest of the year. In years of low hunting mortality, abundance at the end of the hunting season is increased, and each individual has a lowered chance of surviving the rest of the year. Although we can fit equation (11) to data that derive from this mechanism, the actual mechanism is that the probability of surviving the hunting season is $1 - K_t$, and the magnitude of $S_{0,t}$ (the year-specific probability of surviving non-hunting mortality following the hunting season) now depends on abundance at the end of the hunting season, and thus on K_t.

Johnson et al. (1993) suggested a more mechanistic model designed to incorporate the above thinking about density-dependent non-hunting survival. The model consists of the following 2 expressions:

$$E(S_t) = S_{0,t}(1 - K_t), \qquad (13)$$

and

$$S_{0,t} = \frac{e^{a+bN_t(1-K_t)}}{1+e^{a+bN_t(1-K_t)}}, \qquad (14)$$

where N_t is the population size at the beginning of the hunting season in year t, and a and b are parameters specifying the exact nature of the relationship between the probability of surviving non-hunting mortality and population size. Thus,

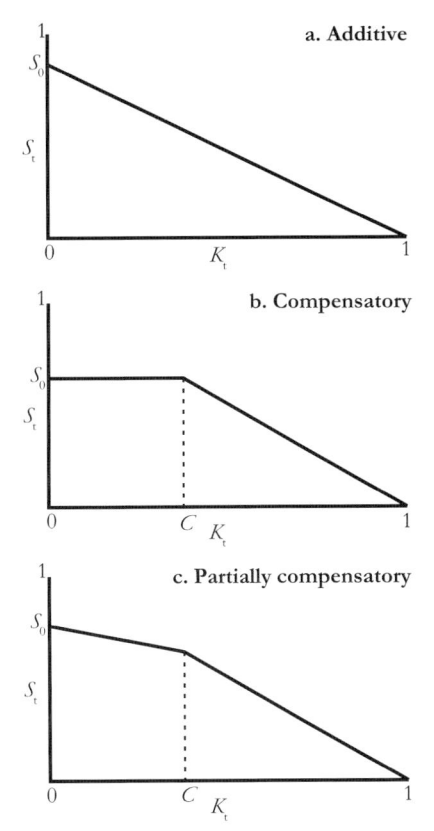

Figure 5. Additive (a), compensatory (b), and partially compensatory (c) mortality hypotheses. S_0 represents the net probability of surviving non-hunting mortality sources, which is also the theoretical survival rate in the absence of harvest. C is the threshold beyond which kill rate K affects S most strongly (where $C \leq 1 - S_0$). Adapted from Conroy & Krementz (1990).

necessarily intuitive and require some sort of underlying mechanism. Most of the discussions about compensatory mortality have involved one of two mechanisms, density-dependence and individual heterogeneity.

Density-dependence is the most frequently cited mechanistic hypothesis equation (13) simply expresses the standard competing risks relationship that total survival is the product of the probability of surviving two risks, the probability of surviving hunting mortality (in the complete absence of non-hunting mortality) and the probability of surviving non-hunting mortality (in the complete absence of hunting mortality). Equation (14) then specifies that the probability of surviving non-hunting mortality risks is a function of the expected abundance at the end of the hunting season, $N_t (1 - K_t)$. We note that strict application of this model with the above definitions assumes that only hunting mortality (no non-hunting mortality) occurs during the hunting season. However, if most of the mortality occurring during the hunting season results directly from hunting, then we can use K'_t as an index of K_t in the above expression. Of course there are other functional forms than the linear-logistic relationship of equation (14), and any such relationship would be more mechanistic than equation (11).

Characterisation of a model as phenomenological or mechanistic is of course subjective, and in reality these terms apply to regions along a continuum of mechanistic detail. For example, our use of population size at the end of the hunting season as the determinant of the probability of surviving the rest of the year represents a simplification (see Lebreton 2005). Density-dependence in ecological relationships typically involves some resource that is potentially in short supply, such that a more mechanistic depiction for equation (14) would be to substitute for N_t the number of animals *per unit of limiting resource* at the end of the hunting season. Despite substantive research on food resources during winter, clear linkages with subsequent survival implications are difficult to discern.

A second mechanism that can underlie compensatory responses (*i.e.* responses in which $\beta < 1$ in equation 11) involves heterogeneity among individual birds in underlying probabilities of surviving both hunting and non-hunting mortality sources (Johnson *et al.* 1986, 1988; Nichols 1991; Lebreton 2005; Sedinger & Herzog 2012; Lindberg *et al.* 2013). Arguments about the relevance of individual heterogeneity to animal population dynamics can be traced back at least as far as Errington (1943, 1967), who wrote about predation (one mortality source) and the fact that predated individuals would likely not have survived other sources had they survived predation. Writing about Muskrats *Ondatra zibethicus* that suffered predation, he wrote, "…they usually represented wastage, and, from the standpoint of the population biology of the species, it did not matter much what befell them" (Errington 1967: 155), and "the predation is centred upon….what is identifiable as the more biologically expendable parts of the population" (Errington 1967: 225). Errington's arguments about predation could be relevant to human hunting of waterfowl as well, if the segment of the population that experiences the higher hunting mortality rate also experiences the higher probability of dying from non-hunting mortality sources (Johnson *et al.* 1986, 1988; Lebreton 2005; Lindberg *et al.* 2013). With respect to the above modelling discussion of structured populations, hunting mortality is largest for the individuals of low quality with the lowest reproductive values.

For simplicity and ease of presentation assume that a population of ducks is characterised by heterogeneity of survival probabilities that can be approximated as a 2-point (finite) mixture (see "structure" section above, page 233), with the groups labelled as g = 1, 2. For ease of interpretation, also assume that the anniversary date each year is the beginning of the hunting season. Only hunting mortality occurs during the hunting season, and only non-hunting mortality occurs following the hunting season. This temporal partitioning of mortality sources is perhaps not a bad approximation for waterfowl and leads to views of seasonal hunting and non-hunting survival and mortality rates as net rates (rates that occur when only the focal mortality source is operating).

Define the following parameters:

π_t = probability, at the beginning of year t, that a randomly selected individual in the population is a member of group 1,

$1 - K_t^g$ = probability that a bird in group g (g = 1 or 2) survives exposure to hunting mortality during the hunting season of year t,

$1 - \bar{K}_t = \pi_t(1 - K_t^1) + (1 - \pi_t)(1 - K_t^2)$ = probability that a randomly selected member of the population survives exposure to hunting mortality during the hunting season of year t (mean net hunting survival), and

$S_{0,t}^g$ = probability that a bird in group g (g = 1 or 2) survives exposure to non-hunting mortality sources in year t, given that it is alive at the end of the hunting season.

The average probability of surviving non-hunting mortality sources for a randomly selected individual alive at the end of the hunting season is:

$$\bar{S}_{0,t} = S_{0,t}^1 \left(\frac{\pi_t(1-K_t^1)}{\pi_t(1-K_t^1)+(1-\pi_t)(1-K_t^2)} \right) + S_{0,t}^2 \left(\frac{(1-\pi_t)(1-K_t^2)}{\pi_t(1-K_t^1)+(1-\pi_t)(1-K_t^2)} \right). \quad (15)$$

Unlike the average probability of surviving hunting mortality, this population level average includes not only the initial probabilities of group membership, π_t and $1 - \pi_t$, but also the relative probabilities of surviving the hunting season. Thus, the terms in large parentheses reflect the expected proportions of the population at the end of the hunting season in each of the groups 1 and 2, respectively. The key concept in considering the influence of heterogeneity on population level effects of hunting is that the composition of the heterogeneous population changes (as reflected in these proportions) over time. If probabilities of surviving both hunting and non-hunting mortality are greater for one group than another, then this high survival group will increase in representation during the hunting season. This will lead to a greater average probability of surviving non-hunting sources than if both groups had experienced similar hunting mortality rates.

The population in the above example consists of two groups of birds, and each group is characterised by its own probabilities of surviving hunting and non-hunting mortality sources. If we assume additive mortality within each group of birds, then the probability that a bird in group g survives exposure to all mortality sources in year t can be written as:

$$S_t^g = S_{0,t}^g (1 - K_t^g). \quad (16)$$

Define $\overline{S}_t$ as the probability that an individual randomly selected from the population at the beginning of the year survives exposure to all mortality sources in year t (mean total survival at the population level). For the 2-group population considered above, we can write this total survival as:

$$\overline{S}_t = \pi_t S_{0,t}^1 (1 - K_t^1) + (1 - \pi_t) S_{0,t}^2 (1 - K_t^2). \quad (17)$$

Consider two hypothetical groups of individuals in a heterogeneous population with corresponding probabilities of surviving hunting and non-hunting mortality sources given in Table 1. Individuals of group 1 experience higher probabilities of surviving both hunting and non-hunting mortality sources than individuals of group 2. Indeed annual probabilities of survival, computed using equation (17), are over twice as high for individuals of group 1, so the heterogeneity is substantial (Table 1). Average annual survival at the population level is computed using equation (16) as 0.53. The final row of Table 1 assumes a homogeneous population in which each individual is characterised by probabilities of surviving hunting and non-hunting mortality sources computed as weighted averages of the group-specific vital rates, with weights π_t and $1 - \pi_t$. The probability of surviving non-hunting mortality sources differs from the average for a heterogeneous population, because the homogeneous population value ignores the change in composition that occurs during the hunting season. The lower probability of surviving non-hunting mortality sources produces a lower annual survival rate for the homogeneous population (Table 1). These differences between heterogeneous and homogenous populations are attributable to the change in composition of the heterogeneous population (also see Vaupel & Yashin 1985; Johnson et al. 1986), and are consistent with the basic mechanism underlying Errington's (1967) ideas of the "doomed surplus".

The emphasis in this section has been on heterogeneity in survival probabilities, and the example in Table 1 suggests that substantial differences in survival among individuals do not necessarily produce large differences in total annual survival. Thus, both density-dependence and heterogeneous survival can mediate the effects of hunting on populations, but both processes are limited in their ability to compensate for hunting losses. Heterogeneous vital rates can also include reproduction. If the individuals that are better able to survive hunting and non-hunting mortality sources are also the better reproducers, then heterogeneity offers even greater potential for compensatory effects. Indeed, Lindberg et al. (2013) provided evidence that female Pacific Black Brant *Branta bernicla nigricans* exhibit heterogeneous survival and recruitment probabilities that lead to increased population growth rates, relative to growth of hypothetical homogeneous populations.

Evidence

Anderson & Burnham (1976) specified the two extreme hypotheses, additive mortality

Table 1. Survival rates in a heterogeneous population comprised of two groups of individuals with different survival rates. Mortality sources are restricted to seasons, with hunting mortality occurring first, followed by non-hunting mortality. Rates include net probabilities of surviving hunting (1–K) and non-hunting (S_0) mortality sources, as well as total annual survival (S). Rates are presented for individuals in group 1 and group 2. Average rates are based on the proportions of the populations in each group to which each rate applies. The homogeneous rates correspond to a population in which each individual experiences source-specific survival rates that are simple averages of those for the two groups.

Group (g)	Proportion of population	1–K	S_0	S
1	0.5	0.95	0.75	0.71
2	0.5	0.75	0.45	0.34
Average		0.85	0.62[a]	0.53
Homogenous		0.85	0.60	0.51

[a] Conditional on the expected population composition at the beginning of the season during which non-hunting mortality applies (computed using equation 15).

and compensatory mortality, and then analysed extensive ringing and recovery data for Mallard in North America to draw inferences about which hypothesis seemed to correspond most closely to these birds. They took advantage of the ring recovery models that had just been developed (Seber 1970; Brownie & Robson 1976; Brownie *et al.* 1978) to estimate annual survival rates and hunting mortality (kill) rates, and then to use these estimates with various analytic approaches for inference about hunting effects. They concluded that the Mallard data largely supported the compensatory mortality hypothesis, and that Mallard experienced hunting mortality rates that were typically below threshold levels.

A variety of improvements in analytic methods for testing these hypotheses followed the publication of Anderson & Burnham (1976), as did efforts to apply these various methods to other waterfowl species. These methods and results constitute a substantial literature that has been reviewed periodically (Nichols *et al.* 1984; USFWS 1988; Nichols 1991; Nichols & Johnson 1996). The most recent reviews (Nichols 1991; Nichols & Johnson 1996) show a mixed bag of results, with a number of studies providing evidence favouring the compensatory mortality hypothesis, some favouring the additive mortality hypothesis, and many providing equivocal results. This uncertainty led Nichols & Johnson (1996) to conclude that an adaptive approach to harvest management would be useful for the purposes of both managing harvest and learning about harvest effects (see the adaptive harvest management section, page 255 below).

Pöysä et al. (2004) reviewed previous studies on Mallard in North America, contrasting earlier work with the more recent efforts by Smith & Reynolds (1992, also see Sedinger & Rexstad 1994; Smith & Reynolds 1994). Based primarily on the Smith & Reynolds (1992) inferences from a more recent period (1979–1989), Pöysä et al. (2004) suggested that Mallard populations may have experienced a change in response to harvest over time, where hunting effects became additive, at least to some degree. Sedinger & Herzog (2012) argued that the conclusions of Pöysä et al. (2004) were unwarranted, but Pöysä et al. (2013) noted that despite some relevant points, the criticisms of Sedinger & Herzog (2012) did not cause them to change their conclusions.

Studies of compensatory *versus* additive mortality appearing after the review by Nichols & Johnson (1996) include four papers on ducks and several papers on goose species. The life history differences between ducks and geese (see above) lead to the observation that geese tend to have higher annual survival probabilities than ducks, and thus less potential to compensate for hunting losses. This observation leads to the expectation that many goose species will exhibit additive mortality, whereas ducks are more likely to exhibit possible compensatory mechanisms. Francis et al. (1998) analysed ringing and recovery data for American Black Duck *Anas rubripes* over three groupings of years characterised by increasingly restrictive hunting regulations. They found evidence of increases in survival rates, some consistent with the additive mortality hypothesis and some smaller than expected under this hypothesis

(Francis et al. 1998). Conroy et al. (2002) developed various models to assess habitat and density-dependent effects on Black Duck survival. Model weights indicated support for models that reflected the additive mortality hypothesis. Rice et al. (2010) found no evidence that Pintail survival rates varied among groups of years characterised by different hunting regulations, but concluded that serious evaluation of effects of hunting was beyond their scope of investigation. Peron et al. (2012) developed an integrated population model for Redhead that used ringing and recovery data as well as information about abundance from the Waterfowl Breeding Population and Habitat Survey and about harvest age and sex ratios from the USFWS Harvest Survey. They found no evidence that Redhead survival varied in response to either daily bag limit or recovery rate, providing some support for the compensatory mortality hypothesis.

Alisauskas et al. (2011) conducted an extensive analysis of Lesser Snow Goose *Chen caerulescens caerulescens* population responses to increased harvest associated with special conservation measures designed to reduce abundances. They found some evidence of decreased annual survival rates associated with the additional harvest pressure for southern nesting populations, but no such evidence from the much larger northern population segments. However, Alisauskas et al. (2011) estimated much smaller increases in harvest rates associated with the conservation measures than had been hoped. An analysis of the southern La Pérouse Bay population of Lesser Snow Geese led Koons et al. (2014b) to conclude that young females exhibited evidence of compensation for an early time period, but evidence favoured additivity for young females in later years and adult females for the entire period of study. Two analyses of ringing and recovery data from G. Gauthier's long-term study of Greater Snow geese *Chen caerulescens atlantica* have focussed on effects of hunting on survival. Gauthier *et al.* (2001) provided evidence of additivity for the decade preceding a spring conservation harvest designed to reduce abundance. Calvert & Gauthier (2005) analysed ring recovery data for the initial years of the spring conservation harvest and found evidence of decreased survival (consistent with additive mortality) for adult Greater Snow Geese but not juveniles. Iverson *et al.* (2014) analysed ring recovery data for a population of Canada Geese in Ontario. Their analysis stratified individuals by reproductive status, and they found evidence of additivity for breeding adults, but not for non-reproductive birds. Sedinger *et al.* (2007) investigated variation in survival and recovery rates of Black Brant over the period 1950–2003 and found evidence of a decrease in recovery rate estimates over time, from early to more recent decades. These decreases in recovery rates were accompanied by an increase in annual survival rates, until recent decades when recovery rates became very small.

Results of studies summarised in previous reviews and the more recent work cited above provide mixed results. The majority of studies of effects of hunting on geese have provided at least some support for the additive mortality hypothesis, as predicted based on their typically slow life history and associated high survival rates. Studies of ducks have yielded period- and species-specific results, with some studies supporting the additive mortality hypothesis and others the partial and completely compensatory mortality hypotheses. Many of the papers reporting these analytic results ended with caveats and recommendations. The caveats virtually all involved the typically correlative nature of the efforts to study effects of hunting, and the acknowledgement that weak inferences are a likely result of this restriction. The recommendations were for either experimentation or adoption of an adaptive approach to harvest management as potential approaches to yielding stronger inferences. Adaptive approaches have been adopted for some species and do permit additional inferences about effects of hunting (see below).

Some methodological challenges

A variety of approaches exists for drawing inferences about effects of hunting. With the development of ring recovery models (Seber 1970; Brownie & Robson 1976; Brownie *et al.* 1978, 1985), waterfowl ringing programmes now permit estimation of ring recovery rates (indices to both crude and – for pre-season ringing – net hunting mortality rates, see page 236) and total annual survival rates. One straightforward approach to inference about hunting effects is to contrast recovery rates and annual survival rates for years of differing hunting regulations. If recovery rates indeed differ as predicted by the regulations changes, then an expectation under the additive mortality hypothesis is that annual survival rates are reduced in years with more liberal hunting regulations (*e.g.* larger daily bag limits or longer seasons).

Another seemingly straightforward approach is to correlate ring recovery rates and annual survival rates estimated from ring recovery data. However, as noted by Anderson & Burnham (1976) when these rate parameters are estimated from the same set of ring recovery data, the estimates are characterised by non-negligible sampling covariances. Because of these sampling covariances, simple correlation analyses using, for example, point estimates of time- or area-specific survival and recovery rates will yield inferences that confound true process covariation and sampling covariation, yielding correlation statistics that cannot be interpreted as pertaining strictly to the underlying mortality processes.

As a means of dealing with sampling variation, Burnham *et al*. (1984) proposed the direct fitting of the equation (11) (and a related power function model) to ring recovery data using a deterministic ultrastructural model. This approach to inference has been used in several waterfowl analyses (Barker *et al*. 1991; Smith & Reynolds 1992; Rexstad 1992). Otis & White (2004) developed a random effects approach to this kind of modelling by considering recovery rate as a random effect that covaries to some degree with annual survival because of possible effects of harvest on survival. This approach permits direct estimation of the process correlation (not confounded with sampling correlation) between recovery and survival rates, with a negative correlation expected under additivity and no correlation predicted under complete compensation. This approach was used for Canada Geese in Ontario by Iverson *et al*. (2014).

This basic approach of drawing inferences about effects of hunting by directly estimating the β of equation (11) or by estimating covariation of recovery and annual survival rates is appropriate for situations in which recovery rates are reasonable indices to net hunting mortality rates. We noted above that estimates of retrieval rates for hunter-killed birds and ring reporting rates are needed to translate ring recovery rates to crude hunting mortality rates. Retrieval rates are typically assumed to be approximately time-invariant, but these rates have received little study. Ring reporting rates have been studied, and if these vary over time and/or space, then they can and should be incorporated directly into inferences about hunting mortality rates. Reporting rate estimates can be incorporated into ring recovery analysis as constants (with or without sampling variation), or the raw data (*e.g.* recoveries from reward-ringed birds) used to estimate reporting rate can be incorporated into the analyses via joint likelihoods that include, for example, both standard and reward rings. We also noted that crude hunting mortality rates (rates obtained in the presence of other (non-hunting) mortality) are most useful for inferences about hunting when they are estimated from ringing that occurs just before the hunting season. When ringing occurs at other times of the year (*e.g.* post-season only), resulting data are not as useful for drawing inferences about hunting (*e.g.* see Nichols & Hines 1987) absent additional assumptions, as non-hunting survival becomes a potentially important source of variation in estimates of recovery rates and crude hunting mortality rates.

A variant (*e.g.* Sedinger *et al.* 2010) on this approach of investigating the correlation between ring recovery rates and annual survival substitutes Seber's (1970) reporting parameter, r_t, for ring recovery rate, f_t, where $f_t = r_t (1 - S_t)$. Use of a random effects approach, similar to that of Otis & White (2004), permits direct estimation of the process correlation absent any confounding with sampling variation. However, process correlations between S_t and r_t predicted under the compensatory and additive mortality hypotheses are not as straightforward as those expected for S_t and f_t. In analyses where ring recoveries are restricted to hunting recoveries, we can write Seber's reporting rate parameter as:

$$r_t = \left(\frac{K'_t}{1-S_t}\right) c_t \lambda_t$$

$$= \left(\frac{K'_t}{K'_t + (1 - S'_{0,t})}\right) c_t \lambda_t . \quad (18)$$

The term in brackets is the probability that a bird that died during year *t* died as a result of hunting. The remaining terms are the probabilities of retrieval and ring reporting defined for equation (12). We have already indicated that rates of retrieval and ring reporting can either be estimated or else are frequently assumed to be constant over time. So the question for selecting f_t versus r_t is whether we prefer K'_t (see eqn. 12), or $K'_t/[K'_t + (1 - S'_{0,t})]$ (eqn. 18), as an index of net hunting mortality rate, K_t.

The difficulty in using r_t is that it is potentially influenced by non-hunting mortality in a manner that leads to a positive correlation with S_t. For example, if K'_t is constant, but $(1 - S'_{0,t})$ varies, then r_t will be larger when non-hunting mortality is smaller and total survival larger. Consider the numerical example of Table 2 in which we model additive mortality and again make the simplifying assumption that only hunting mortality occurs during the hunting season (thus $K'_t = K_t$) and only non-hunting mortality occurs following the hunting season (thus $1 - S'_{0,t} = (1 - K_t)(1 - S_{0,t})$; *i.e.* a bird must survive hunting mortality in order to be exposed to non-hunting mortality). We consider two years in Table 2, with constant rates of kill, retrieval, and ring reporting for both years, and differences only in the net probabilities of dying from non-hunting causes, $1 - S_{0,t}$. As a result of the variation in non-hunting mortality, both annual survival and the Seber (1970) reporting parameter are larger for the year of lower non-hunting mortality (year 2). So the process correlation between r_t and annual survival would be positive and possibly interpreted as evidence against additive mortality, whereas in reality additive mortality governed the survival process for both years. Note that there is no such indication of a positive correlation between ring recovery rate, f_t, and annual survival. In addition, note that Seber's (1970) reporting parameter, r_t, while incorporating what we have termed ring reporting rate, λ_t, is a very different quantity (eqn. 18) that attains very different values (Table 2).

Another basic approach to inference about additive and compensatory mortality is based on the relationship between net probabilities of experiencing hunting and non-hunting mortality, where "net"

Table 2. Numerical example illustrating the process correlation between Seber's (1970) reporting parameter, r_t, and total annual survival, S_t, under the additive mortality hypothesis. In the absence of variation in net hunting mortality, K_t, higher net probabilities of surviving non-hunting mortality, $S_{0,t}$, lead to higher r_t and S_t. Other definitions: c_t = probability that a bird shot in year t is retrieved, λ_t = probability that the ring of a bird shot and retrieved in the hunting season of year t is reported, f_t = ring recovery rate, $1 - S'_{0,t}$ = crude non-hunting mortality rate, and θ_t = probability that a bird which died during year t died as a result of hunting.

Year (t)	$K_t = K'_t$	c_t	λ_t	f_t [a]	$S_{0,t}$	$1 - S'_{0,t}$ [b]	θ_t [c]	S_t [d]	r_t [e]
1	0.200	0.800	0.750	0.120	0.600	0.320	0.385	0.480	0.231
2	0.200	0.800	0.750	0.120	0.800	0.160	0.556	0.640	0.333

[a] $f_t = K'_t c_t \lambda_t$; [b] $1 - S'_{0,t} = (1 - K'_t)(1 - S_{0,t})$; [c] $\theta_t = \left(\dfrac{K'_t}{K'_t + (1 - S'_{0,t})} \right)$; [d] $S_t = (1 - K'_t) S_{0,t}$;

[e] $r_t = \left(\dfrac{K'_t}{K'_t + (1 - S'_{0,t})} \right) c_t \lambda_t = \theta_t c_t \lambda_t$

indicates a mortality probability that would apply in the absence of any other mortality source (Chiang 1968). Both proposed mechanisms underlying compensatory mortality hypotheses suggest a change in one net mortality probability (non-hunting) as a function of variation in another net mortality probability (hunting). Density-dependence results in lower net non-hunting survival in years where hunting mortality is low, as many birds are alive at the end of the hunting season (other things being equal). Competition for resources is then hypothesised to result in lower net non-hunting survival of the survivors, than if hunting mortality had been larger. Under the heterogeneity hypothesis, the larger the hunting mortality rate the greater the change in composition of the heterogeneous population, leaving relatively more high survival group individuals, and thus greater average non-hunting survival. In summary, both density-dependence and heterogeneity lead to the prediction of a negative correlation between the net survival probabilities associated with the two mortality sources. Specifically, lower hunting mortality is associated with higher non-hunting mortality and *vice versa*.

If only hunting mortality occurs during the waterfowl hunting season and only non-hunting mortality occurs after the hunting season, then this temporal separation permits direct estimation of both net mortality rates via a ringing programme that includes ringing at two times of the year, the beginning and end of the hunting season. Inference based on a single ringing period each year is more difficult, however. Schaub & Pradel (2004) and Schaub & Lebreton (2004) used multistate models with recovery data based on annual ringing to estimate separate survival rates associated with two different mortality sources. However, it is important to recognise that these are "crude" rates (*sensu* Chiang 1968), as they are conditional on the deaths occurring as a result of the other mortality source. Above the crude hunting mortality rate, K'_t, is defined as the probability that a bird alive at the beginning of the hunting season in year t would die from hunting causes during the hunting season of year t in the presence of the non-hunting mortality that occurs during this period. Similarly, the crude mortality rate $(1 - S'_{0,t})$ could be defined as the probability that a bird alive at the beginning of the hunting season in year t would die from non-hunting causes throughout the year t in the presence of the hunting mortality that occurred earlier in the year.

There are two important difficulties to consider when using estimates of these crude rates, K'_t and $1 - S'_{0,t}$, to draw inferences about effects of hunting. The first difficulty is similar to that noted above for direct inferences about the relationship between annual survival and recovery rates, and involves the sampling covariance between time-specific estimates of these crude rates. These can be dealt with using either frequentist or Bayesian approaches (see Schaub & Lebreton 2004). The other difficulty is that these crude rates of hunting and non-hunting mortality are expected to exhibit negative process covariation for reasons that have nothing to do with changing net rates. This was noted by Schaub & Lebreton (2004) and Servanty *et al.* (2010) who refer to this component of process correlation as an "intrinsic bias" (Peron 2013 labelled it "competition bias"). To demonstrate this natural negative correlation between the two crude rates under the additive mortality hypothesis (*i.e.,* in the absence of any correlation between net rates), consider the following example of temporally separated hunting and non-hunting mortality. Assume that hunting mortality occurs first, and that this mortality is larger in year t_1 than in year t_2, $K'_{t1} = K_{t1} > K'_{t2} = K_{t2}$ (note that because of the timing of mortality, these are both crude and net rates). Assume that net non-hunting survival is the same in the 2 years; that is, survivors of the hunting season have equal chances of surviving non-hunting mortality in the two years, $S_{0,t1} = S_{0,t2}$. Thus there is no association between the net rates associated with non-hunting and hunting mortality in the two years. However, because of the greater hunting mortality in year 1, the crude non-hunting mortality rate will be smaller in year 1 than in year 2:

$$(1 - S'_{0,t1}) = (1 - S_{0,t1})(1 - K_{t1}) < (1 - S'_{0,t2})$$
$$= (1 - S_{0,t2})(1 - K_{t2}).$$

Smaller numbers of birds are available to be exposed to non-hunting mortality sources in year 1; hence fewer die from these sources, despite equal net non-hunting rates in the 2 years. So the crude rates covary negatively, but this covariance is induced by variation in only one net rate and has nothing to do with the covariance between net rates that underlies compensatory mortality.

Although mortality sources were separated in time for our example, for ease of presentation, we note that the issue remains regardless of the timing of source-specific rates. Even when non-hunting and hunting mortality occur simultaneously, it is still true that more birds removed by one source (*e.g.* hunting) will leave fewer birds available to die of the competing source (non-hunting). Schaub & Lebreton (2004: 83) noted this problem when attempting to draw inferences about compensation or additivity using source-specific mortality data, but their important caveat does not seem to be appreciated by all others who have used this approach. Servanty *et al.* (2010) dealt with this issue by eliciting expert opinion about the timing and magnitude of non-hunting mortality, and incorporating these opinions using a Bayesian inference framework. In summary, we emphasise that attempts to use estimates of crude source-specific mortality rates (based on Schaub & Pradel 2004; Schaub & Lebreton 2004) to draw inferences about hunting effects require careful analysis and interpretation.

It is interesting that in work that is closely related in some ways to that of Servanty *et al.* (2010), but carried out 35 years previous, Brownie (1974) developed an approach to estimate the instantaneous risks (these translate directly into net mortality rates) associated with hunting and non-hunting using the extra information about the date of recovery of each hunting season ring recovery. She was able to estimate these risks directly, but had to assume that each was time-constant; the hunting risk throughout the hunting season and the non-hunting risk throughout the year. Even though she was able to separately estimate these annual risks without having to guess at their magnitude, she noted both the restrictive nature of her assumptions about temporal constancy of risks and the high sampling covariance between the estimates of risk, and abandoned this approach to inference about effects of hunting.

Finally, we note that radio-telemetry data are well suited to estimate net source-specific mortality rates (the rates from which strong inferences about compensation can be most readily obtained) directly. Time-specific deaths from sources other than a focal source can be immediately censored, providing direct inference about the net mortality rate of the focal source (Heisey & Fuller 1985; Heisey & Patterson 2006). Sandercock *et al.* (2011) provide a nice example of this approach to inference about compensation using a non-waterfowl species.

Reproduction

While assessment of the impact of harvest on waterfowl dynamics has generally (and intuitively) focussed on the direct relationship between harvest and survival (preceding section), harvest can potentially drive population dynamics in other ways, through the indirect influence of removal of individuals on other components of fitness. In this and the following section, we briefly consider the effects of harvest on components of reproduction, and on migration and movements.

Harvest can potentially influence reproduction in several ways. First, and perhaps most obviously, harvest clearly removes both the immediate and future (residual) contributions of a harvested individual to population growth. However, the typical accounting of the reproductive value of an individual, reflecting current and future reproductive contributions lost to harvest (*sensu* MacArthur 1960), assumes that such individual contributions to population growth represent independent events (*i.e.* that removal of one individual does not change the reproductive value of any other individual). However, in many cases per capita reproduction is influenced by the number (abundance or density) and structure (age, sex, spatial) of conspecifics. Density-dependence in population growth has been demonstrated at large spatial and temporal scales (Viljugrein *et al.* 2005; Sæther *et al.* 2008; Murray *et al.* 2010), and for many species of waterfowl (especially shorter lived ducks), variation in reproductive output is a dominant driver of annual population dynamics. Thus, it is reasonable to assume that any activity which reduces "density" (*e.g.* harvest) has the potential to increase population growth, by increasing reproductive productivity.

However, this conclusion is arguably overly simplistic, in at least a couple of respects. First, as noted by Lebreton (2009; also see above), estimates of density or population size are only a proxy for what affects demographic performance – a correlation between density or population size and reproductive performance is phenomenological, and does not generally indicate the important mechanisms underlying the observed relationship. This increases the uncertainty in projecting the impacts of harvest on population production, which is an important factor in decisions involving annual harvest regulations. For example, a reduction in total number of breeding individuals may not necessarily result in increased production at the population level, if the harvested individuals are lower quality birds which do not contribute significantly to annual production. Uncertainty about the functional form relating production to changes in abundance or density can be particularly important (Kokko 2001; Runge & Johnson 2002). Similarly, the influence of breeding population size on various components of post-laying fitness (*i.e.* events that might occur following the primary production of the clutch) may be more difficult to predict. For example, increased density of nesting birds may increase nest survival (Ringleman *et al.* 2014), but might lead to increased competition among juveniles for limiting resources, leading to reduced juvenile growth and survival (Cooch *et al.* 1991; Sedinger *et al.* 1995). It is probably true that for many species, the impact of population abundance (density) on both pre- and post-laying components of reproductive fitness will reflect a complex interaction of both frequency- and density-dependent effects, strongly influenced by various environmental effects which can significantly modify the relationships (Lebreton 2009).

Second, harvest not only has the potential to change the size of the population, but its structure (age, sex, spatial, heterogeneity) as well. Any harvest which changes the structure of a population will influence reproductive output, if those structural elements themselves influence one or more components of reproductive fitness, independently or in addition to potential density-dependent effects. For example, for species with polygynous mating systems, differential harvest of males and females might result in a skew to the sex-ratio in the breeding population, which could potentially influence the probability of any individual female laying a clutch, or the probability of that clutch surviving to fledging (in species where males and females play different roles in nest guarding and brood rearing), or both. Or, for species where breeding propensity may be based on relative proportions of individuals of different age or breeding experience at the start of breeding, differences in harvest vulnerability between younger and older birds could potentially be strong drivers for annual variation in the proportion of individuals breeding, which in turn would have a strong impact on population production. Changes to the structure of the population by harvest are not unexpected, since harvest is generally not random with respect to individual contributions to population growth. This non-randomness can be intentional (say, for example, due to a male only season or greater bag limit for males than females, or a region- or season-specific regulatory programme which allows for different harvest as a function of location and time of year), or an artefact of the interaction of non-specific (*i.e.* presumed random) harvest with structural differences in susceptibility to harvest. Such differences could reflect differences in vulnerability due to heterogeneity in individual reproductive performance (*e.g.* as a function of differences in timing of breeding, whether an individual bred at all that year, the timing and pattern of migration, number of offspring produced, or physiological condition following breeding).

Evidence

Much of the preceding is couched in terms of "potential impacts" of harvest on reproductive output, at either the individual or population level. In general, there seems to us to be a lack of consideration of the processes underlying the relationship between harvest and reproduction in waterfowl, beyond the obvious and logically trivial observation that a harvested individual has no future reproductive potential. Even seemingly simple (yet quite important) relationships between (say) harvest vulnerability and reproductive output are poorly quantified. For example, despite the long-held belief amongst many goose biologists that adults with young will have different vulnerability to harvest than adults without young, there are few rigorous attempts to quantify this relationship.

The relative paucity of empirical studies to date on the role of harvest on reproduction likely reflects several factors. First, there has been a general tendency to focus on impacts of harvest on survival, since: (i) the relationship of harvest and survival is potentially (and presumed to be) more accessible to management action, and (ii) the impacts of survival on overall population growth are often higher than possible impacts on reproduction, and thus are arguably more important to quantify. This is especially true for longer-lived species, where it has been shown repeatedly (to the point of redundancy, perhaps) that projected population growth is more sensitive to changes in adult survival than any other single demographic parameter (in fact, this result is a logical necessity for any species with a generation length greater than ~4–5 years; Lebreton & Clobert 1991; Caswell 2001; Niel & Lebreton 2005).

Second, evaluating the impacts of harvest on components of reproductive output (*e.g.* egg laying, nesting success, breeding proportions, recruitment) or population structure (*e.g.* sex ratio, spatial distribution) often requires intensive data from breeding ground studies, with adequate samples of marked, known-aged individuals for estimating some parameters. At present, there are a number of such studies involving breeding populations of geese, where high density nesting and strong natal philopatry lend themselves to collection of extensive, detailed demographic data. For example, Sedinger & Nicolai (2011) and Lindberg *et al.* (2013) have provided compelling evidence from their long-term study of Pacific Black Brant that harvest has both direct and indirect impacts on reproduction. Several near-replicate studies of a number of different goose species have clearly shown the negative impacts of density on clutch size (Cooch *et al.* 1989; Sedinger *et al.* 1998) and post-hatch growth and survival and recruitment of goslings (Cooch *et al.* 1991; Sedinger *et al.* 1995, 1998; Lepage *et al.* 1999).

In contrast, there are relatively few comparable studies involving breeding ducks, where natal dispersal and difficulty in capturing and marking broods makes analysis of variation of many reproductive parameters much more difficult. Exceptions to this general difficulty in studying nesting ducks are provided by cavity-nesting species, which can attain high nesting densities and exhibit high philopatry relative to most ground-nesting ducks. In Wood Ducks *Aix sponsa*, for example, high nesting densities have been associated with reductions in reproductive parameters such as breeding probability, nesting and hatching success, as well as with increased nest abandonment (Haramis & Thompson 1985; review in Nichols & Johnson 1990). With this exception of cavity-nesters, most of the empirical tests of the effect of density on components of reproductive rate for ducks relate to nest survival (*e.g.* Prop & Quinn 2003; Ringelman *et al.* 2014 and references therein). However, large-scale aggregate measures of reproductive rate (*e.g.* age ratios in autumn) have been related to population size and density in prairie-nesting Mallard of North America, providing some evidence of negative density-dependence (see Anderson 1975; Brown *et al.* 1976; Kaminski & Gluesing 1987; Johnson *et al.* 1997; also see page 260).

Even when a relationship between harvest and one or more components of reproductive performance has been established, the larger consequences on the dynamics (and management) of the population have not generally been considered. For example, while there have been a number of studies of the role of mate loss on reproductive success for several waterfowl species (Cooke *et al.* 1981; Martin *et al.* 1985; Forslund & Larsson 1991; Manlove *et al.*1998; Lercel *et al.* 1999; Hario *et al.* 2002), to our knowledge, there has been no rigorous assessment of the role of sex-specific differences in harvest vulnerability on population dynamics in waterfowl, although such studies are quite common for other taxa (see Milner *et al.* 2007). Similarly, a number of studies have suggested that lower quality individuals (based on various criteria, such as body condition) are likely more vulnerable to harvest (Hepp *et al.* 1986; Dufour *et al.* 1993; Heitmeyer *et al.* 1993; Pace & Afton 1999). Such non-random selection clearly has the potential to alter the structure and production of the breeding population. However, there has been little consideration of such selective harvesting and resulting changes in population structure on overall dynamics and management – the recent paper by Lindberg *et al.* (2013) is an important exception.

Migration/movement

Rates of movement are often ignored in discussions of population dynamics, yet they are important vital rates that bring about changes in population size at various scales. As with the other components of their population dynamics, there is convincing evidence that hunting can affect the movement rates of waterfowl.

Escape flights of waterfowl facing hunters are a common field observation and lead to local redistribution of the birds towards hunting-free refuges (Béchet *et al.* 2004). Ample demonstrations have been published of local increases in waterfowl numbers after reserve creation (Bellrose 1954; Madsen & Fox 1995; Fox & Madsen 1997; Madsen 1998a). Some authors demonstrated that such local movements were due to greater emigration rates from areas with more hunting pressure (*e.g.* hand-reared Mallard in Legagneux *et al.* 2009). However, protected areas may be attractive to birds not only because they are free from hunting, but also because they receive specific waterfowl-friendly habitat management. Several studies nevertheless have demonstrated that local redistribution of waterfowl towards protected areas was genuinely linked with hunting. For example, Madsen (1998b) established experimental reserves within a Danish fjord, which were moved from year to year. He documented concomitant changes in the annual distribution of hunted waterfowl species, which matched the movement of the reserves over time. Protected species conversely did not adjust their distribution to that of the reserves. Within a given year, Cox & Afton (1997) also recorded redistribution of female Northern Pintail depending on whether hunting was or was not occurring: the ducks increased their diurnal use of protected areas during two successive hunting periods, while this use decreased during the periods pre-hunting, post-hunting, and during the time between the two split hunting periods. It should be noted, though, that Link *et al.* (2011) did not obtain the same result in Mallard, which may indicate differential susceptibility of the different species to hunting disturbance and/or differences in habitat selection processes related with differential food availability.

Beyond such local scale effects, hunting has also been shown to affect waterfowl movement at the regional level (review in Madsen & Fox 1995). Ebbinge (1991) documented changes in the number of geese at the national scale in some countries after hunting was banned or, conversely, reintroduced in neighbouring countries. Madsen & Jepsen (1992) also reported that increased shooting pressure (together with changes in farming practices) led to an earlier departure of the Pink-footed Goose population from Denmark to the Netherlands during autumn migration. In ducks, Väänänen (2001) showed that Finnish *Anas* species move towards protected areas at the onset of the hunting season, and that their numbers also decline at the regional scale after this date, which may indicate that hunting precipitates fall migration in these species. This is consistent with Cox & Afton (2000) recording greater emigration rates of female Northern Pintail from Louisiana (largely towards other regions further north in the alluvial valley of the Mississippi River) during hunting periods than before, between or after these periods.

The potential effect of hunting on waterfowl movement rates is therefore clearly established. However, whether this later translates into lower survival or breeding output at the population level is not so clear. Gill & Sutherland (2000) and Gill *et al.* (2001) have already commented on the difficulty to predict the consequences of such disturbance at the population level based on observations of local responses of individuals (either through change in local time-activity budgets or in movement rates). Indeed, the local behaviour of individuals is dependent upon their respective abilities to respond locally (*i.e.* based on individual body condition, and/or the availability of alternative sites to the disturbed area), and the consequent effect at the population level may largely depend upon density-dependent processes.

To our knowledge, the only species in which the consequences of hunting for both individual movement and population dynamics have been demonstrated is the Greater Snow Goose. Spring hunting of this species at a migration area was introduced as a means to reduce the population and the problems caused to northern breeding habitats by these superabundant birds. The implementation of spring hunting changed the regional migration movements of these birds, with fewer unidirectional movements and more westward and reverse movements (Béchet *et al.* 2003). These changed movement patterns, a greater proportion of time spent alert and in flight, and greater use of less profitable habitats, resulted in poorer body condition of the geese after the initiation of the spring hunts (Béchet *et al.* 2004). Lower breeding output was in turn eventually recorded (Mainguy *et al.* 2002). Such an effect of hunting for population dynamics *via* changes in movement rates is also possible in other waterfowl species, but has yet to be demonstrated.

Adaptive harvest management

Overview

As noted in the above historical review, AHM was implemented by the USFWS for mid-continent Mallard in 1995 as a means of simultaneously managing in the face of uncertain harvest–population relationships and reducing this uncertainty. AHM programmes have now been implemented for other Mallard populations as well as for other waterfowl species (USFWS 2013). Here we briefly describe AHM components and the AHM process and then review what has been learned through its application to different Mallard populations.

"Adaptive management" has taken on many different meanings (Williams *et al.* 2009; Williams & Brown 2012), and our focus here will be on the passive process used by the USFWS in their various AHM programmes. This process requires some key elements: objectives, potential actions, model(s), a monitoring programme and a decision algorithm. Harvest management objectives are expressed as an objective function which typically is to maximise average annual harvest over a long time horizon. The time horizon places value on harvests from future waterfowl populations and thus serves to promote conservation. The potential actions are typically sets of regulations packages that specify season lengths and daily bag limits. At least one package is relatively restrictive (short seasons and small bag limits), at least one is relatively liberal (long seasons and large bags), and at least one package is moderate, where ideas about liberal and restrictive are typically population- and species-specific. An alternative approach to discrete regulations packages would be to specify a target harvest rate, treating this rate as a continuous control variable in the optimisation. However, because the regulations themselves are the actions selected by the USFWS, and because there is substantial uncertainty about the relationship between regulations and harvest rate (partial controllability), the USFWS has focussed on these discrete sets of regulations. In an effort to integrate the various potential actions available to manage waterfowl populations (including habitat management, for example), current deliberations are considering the modification of existing AHM programmes to incorporate additional objectives and additional kinds of actions (Runge *et al.* 2006; Osnas *et al.* 2014).

Models are required in order to predict population responses to management actions. Optimisation essentially entails a comparison of these predictions in order to select the action that is expected to do the best job of achieving management objectives. Models used in AHM are not simply models of waterfowl population processes, but are tailored to the specific purpose of predicting the consequences of actions to population change. Frequently, there is substantial uncertainty about population responses and associated models, and we refer to this as structural uncertainty. We attempt to incorporate this uncertainty into the models in one of two ways. The most commonly used approach to date has been the use of a discrete model set containing multiple models of system responses. Relative degrees of belief, or model weights, are associated with each model. These weights determine the influence of each model in the optimisation and are updated through time based on a comparison of model-based predictions and observations from monitoring programmes. The evolution of model weights over time reflects our learning. The other approach to dealing with model uncertainty is to employ a very general model and express the uncertainty as the variance of a key model parameter. In this case also, monitoring data are used to update estimates of the parameter and hopefully reduce its variance.

Monitoring programmes provide data that play several roles in the management process. They provide estimates of system state (*e.g.* population size) in order to make state-dependent decisions. State-dependence simply refers to the fact that we will typically take very different management actions when our focal population far exceeds a desired level as when our population is below the desired level. Estimates of system state are also used to see how well we are doing at meeting management objectives. Estimates of system state are essential to the process of learning, in providing an estimate of truth against which to compare our model-based predictions, thus providing a way to update model weights (discrete model set) or better estimate key model parameters that characterise our uncertainty.

Finally, some sort of decision algorithm must use these various components to determine an optimal action for each possible system state. Harvest management programmes are typically recurrent decision problems. For example, we usually make decisions about hunting regulations annually. Because optimal actions are state-specific and because an action taken at time t influences the state of the system, and thus the optimal decision, at time $t + 1$, the optimal decision at time t depends on projections of future system states and decisions. Dynamic optimisation algorithms are thus needed, with stochastic dynamic programming (Bellman 1957; Williams *et al.* 2002) being the method of choice for problems for which dimension is not too large. Although optimisation is recommended for decision making, we note that adaptive management can still be carried out, and learning can still occur, even when sub-optimal actions are taken.

Programmes of adaptive management typically require a deliberative or set-up phase during which the above components are all developed (*e.g.* Williams *et al.* 2007). At the first decision point, the optimisation algorithm is used to produce optimal state-dependent actions, based on the objectives, the available actions, and the model(s). These recommended actions are sometimes referred to as a policy matrix or decision matrix. The current state of the system is estimated via the monitoring programme and these estimates are used with the decision matrix to select the appropriate action for that decision point. The action is taken and drives the system to a new state, which is estimated by the monitoring programme. In the case of multiple models, the new estimate of system state is compared with the predictions of the different models, and the degrees of belief or weights associated with the models are modified using Bayes formula (see Williams *et al.* 2002). In the case of a single model characterised by uncertainty in one or more key parameters, these parameter estimates are updated with the new data. This completes the first step in the adaptive management process.

At this point, the process enters the iterative phase, and the next step is to consider the action to take at the next decision point. In the case of "passive" adaptive management (Nichols & Williams 2013), which has been used for most North American waterfowl management, the optimisation algorithm is run with the new model weights or updated parameter estimates. This produces a new decision matrix characterising the updated state of knowledge of system dynamics. This matrix is used with the estimate of current system state to select the optimal action for the upcoming decision point. This action is then taken, the system is driven to a new state identified by monitoring, information about that new state is used to update model weights and/or parameter estimates, and the process proceeds iteratively in this manner. In the case of waterfowl harvest, the decision points occur annually. After some experience (*e.g.* several years) with the process, there may be cause to reconsider some of its components. This has been referred to as double-loop learning (Williams *et al.* 2007) and entails moving back into the deliberative phase to revisit any of the process components, from objectives through models and monitoring. Indeed, this process of revisiting components is occurring now for some of the processes developed for adaptive harvest management of North American waterfowl, with special attention focussed on objectives and actions (*e.g.* incorporation of habitat management into decisions; Runge *et al.* 2006).

This basic process has been used to establish annual hunting regulations for mid-continent Mallard for nearly 19 years and for eastern Mallard for 14 years. As these are the longest-running programmes for North American waterfowl harvest management, they provide the best opportunities to learn about effects of harvest and to begin to discriminate among competing models of such effects. In the following two sections, we briefly review the competing models used in these two programmes and show the evolution of support (or lack of it) for these models.

Mid-continent and eastern Mallard

The mid-continent Mallard stock includes birds breeding in the traditional survey area of the waterfowl breeding population and habitat survey (WBPHS) strata 13–18, 20–50, and 75–77, plus birds observed in the states of Michigan, Minnesota, and Wisconsin (Fig. 6; USFWS 2013). Harvest policies derived for the Mississippi and Central Flyways are based on this population. The eastern Mallard stock includes birds breeding in the WBPHS strata 51–54 and 56 in the provinces of Ontario and Quebec plus birds breeding in the eastern states of Virginia northward into New Hampshire which are monitored through the Atlantic Flyway Breeding Waterfowl Survey. Harvest policies derived for the Atlantic Flyway are based on this population.

For each Mallard stock, a set of four models representing different combinations of recruitment and survival relationships are used to represent structural uncertainty and predict Mallard population responses to environmental changes and harvest regulations. For mid-continent Mallard, strong and weak density-dependent relationships are used to predict recruitment as a function of breeding population size and the number of Canadian ponds observed in the WBPHS (Fig. 7); while the eastern Mallard strong and weak recruitment models are based on a non-linear relationship that predicts annual recruitment as a function of breeding

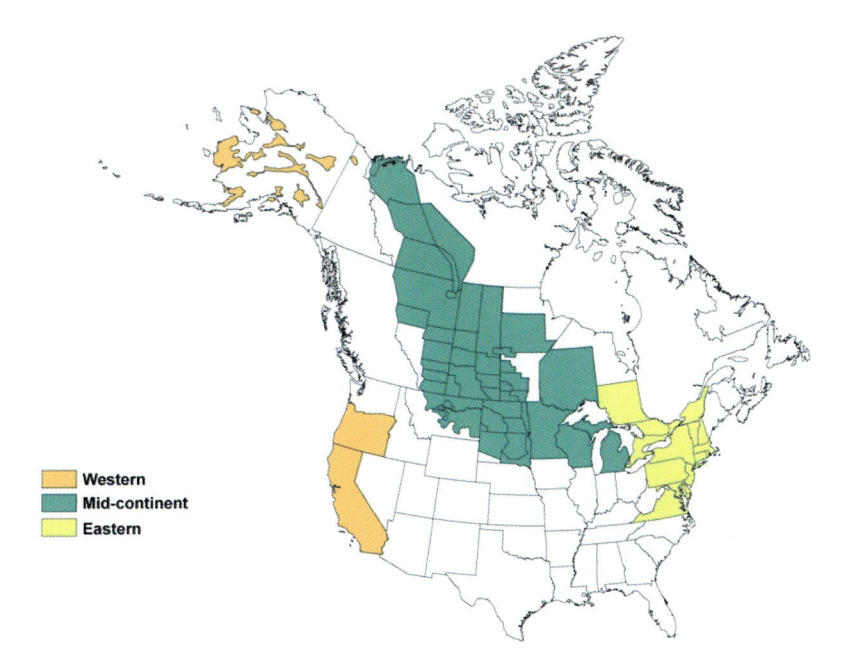

Figure 6. Distribution of eastern, mid-continent, and western Mallard stocks in North America, as described in the USFWS's adaptive harvest management programme.

population size only. Each model set also includes two survival sub-models to represent uncertainty in the relationship between harvest mortality and survival. These models are based on an ultra-structural formulation (see eqn. 11) that predicts annual survival as a function of the survival rate in the absence of harvest and the kill rate expected under a particular set of harvest regulations. Under the additive model, survival rates decline linearly with increasing harvest rates, while under the compensatory model, survival rates remain unchanged until a threshold harvest rate ($C \leq 1 - S_0$) has been exceeded and then decline linearly with increasing kill rates (see Fig. 7). Combining the recruitment and survival sub-models results in four models: additive hunting mortality and strong density-dependent recruitment (SaRs); compensatory hunting mortality and weak density-dependent recruitment (ScRw); additive hunting mortality and weak density-dependent recruitment (SaRw); and compensatory hunting mortality and strong density-dependent recruitment (ScRs). The mid-continent and eastern Mallard model sets were last updated in 2002 (Runge *et al.* 2002) and 2012 (USFWS 2013), respectively.

Under current AHM protocols, the relative belief we have in each model is quantified with an individual model weight. Because these weights sum to 1, they serve as individual measures of relative credibility. When AHM programmes were first implemented for these two populations, individual prior model weights were set equal (0.25), reflecting equal confidence in the

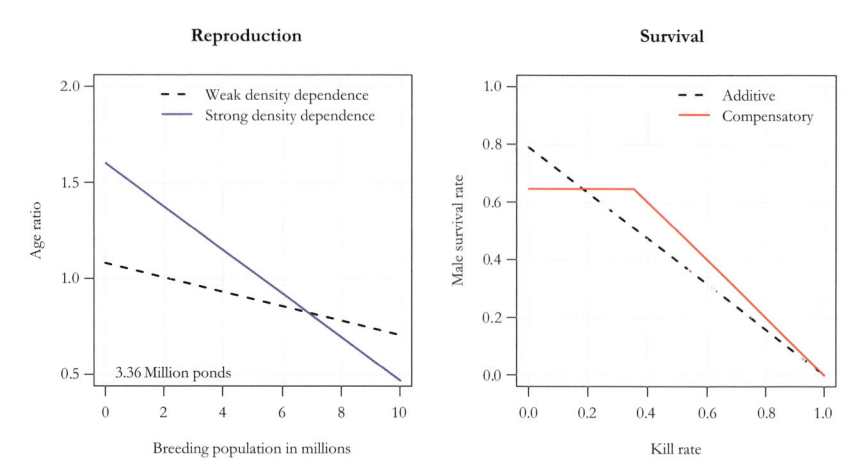

Figure 7. Reproductive and survival sub-models used to represent different relationships for recruitment and male survival in the mid-continent Mallard model set. Predicted levels of recruitment assume 3.36 million Canadian ponds.

ability of each model to predict Mallard population sizes. As observed population estimates were compared to individual model predictions, model weights and our relative beliefs in each model have been updated with Bayes theorem. For mid-continent Mallard, model weights remained essentially unchanged until 1999, when each model predicted a population decline in the face of a significant population increase (Fig. 8). Because the weak density-dependent model predictions were closer to the observed population estimates, they gained more weight and credibility. Since 1999, models SaRw and ScRw (weak density-dependent recruitment) have continued to gain credibility as weights for models SaRs and ScRs (strong density-dependent recruitment) have declined. For eastern Mallard, changes in model weights have been slower, with the additive mortality weak density-dependent model (SaRw) gradually accumulating more weight over time (Fig. 8).

The evolution of model weights over time resolves structural uncertainty and represents learning in adaptive management (Williams *et al.* 2002). Current model weights suggest strong evidence for weak density-dependent reproduction (96% and 68%) *versus* strong density-dependent reproduction (4% and 32%) in the mid-continent and eastern Mallard populations, respectively. For both the mid-continent and eastern Mallard stocks, model weights favor the additive mortality model (67% and 70%) compared to the compensatory mortality model (33% and 30%), suggesting support for the additive harvest mortality hypothesis. As the number of comparisons of model predictions to observed population estimates has increased over time, evidence has accumulated indicating that predictions from model SaRw are more reliable compared to other predictions from the model set.

However, it must be recognised that conclusions based on an interpretation of

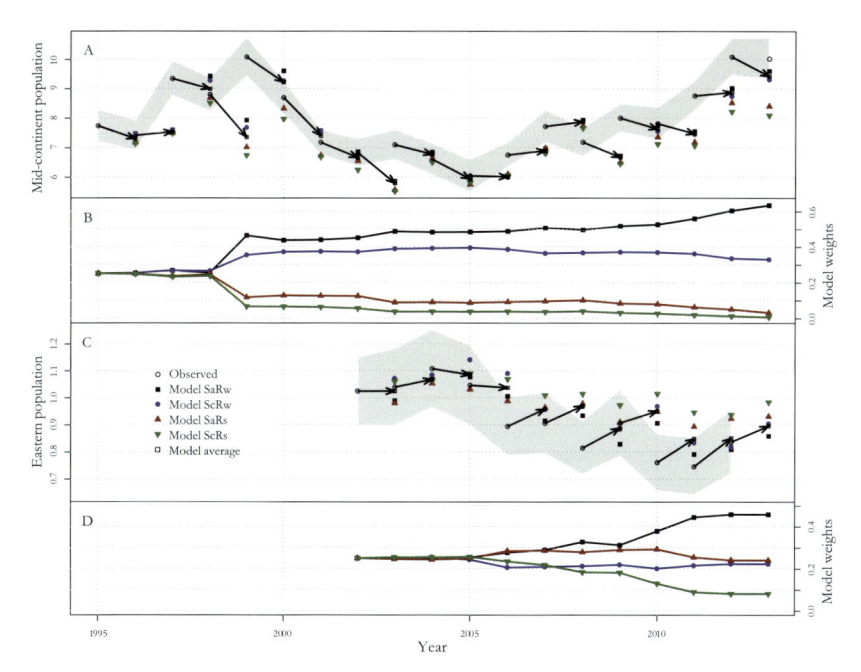

Figure 8. Panels A and C: population estimates of mid-continent and eastern Mallards (in millions) compared to predictions of each member of the mid-continent and eastern Mallard model set (SaRw = additive mortality and weakly density-dependent reproduction, ScRw = compensatory mortality and weakly density-dependent reproduction, SaRs = additive mortality and strongly density-dependent reproduction, ScRs = compensatory mortality and strongly density-dependent reproduction). The grey shading represents 95% confidence intervals for observed population estimates. For each model set, the arrow represents a weighted mean annual prediction. Panels B and D: annual changes in model weights for each member of the mid-continent and eastern Mallard model sets; weights were assumed to be equal in 1995 and 2002, respectively. For the eastern Mallard population, model weights were not updated in 2013 because breeding population estimates were not available.

relative credibility measures (model weights) are conditional on the set of hypotheses represented in the entire model set and that ecological interpretations are limited by the dependency of regulations on population status (Johnson *et al.* 2002; Sedinger & Herzog 2012). Recently, Sedinger & Herzog (2012) questioned whether current AHM model weights provide support for the additive mortality hypothesis, suggesting that this outcome may result from a spurious relationship between harvest and annual survival rates because density-dependent mortality is not explicitly considered in the AHM model set. They base part of their argument on results from Conn & Kendall (2004), who demonstrated through simulation, that the model weight updating procedures used in AHM may result in model weights that support the additive mortality hypothesis in cases where the true underlying dynamics were generated with a density-dependent, compensatory survival model. We note that the mid-continent Mallard model set and the prediction variance estimates that governed the model weight updating procedures simulated by Conn & Kendall (2004) were completely revamped during the 2002 revisions (for details see Runge *et al.* 2002). Simulation experiments using the current AHM protocols have verified that model weights converge on the correct model when the true generating model is included in the AHM model set (G.S. Boomer, unpubl. data).

We acknowledge that the true relationship between harvest and survival may not be represented well by the current model set and also note that model predictions may be consistent with observations without accurately representing the demographic relationships that determine population dynamics (Johnson *et al.* 2002). The current formulations of survival sub-models were included in the AHM model set to account for uncertainty in the relationship between harvest mortality and annual survival, specifying endpoints on a range of possible responses to harvest mortality. Certainly, an alternative harvest-survival model may perform better than each survival model in the current model set. The consideration of alternative beliefs in decision making is a hallmark of adaptive management, which provides a rigorous process to evaluate the ability of alternative models to predict the consequences of management actions. The current implementation of AHM provides an ideal framework to consider alternative models describing population responses to harvest management. We note that current efforts are underway to evaluate the sub-model performance of the eastern and mid-continent Mallard AHM protocols through the double-loop learning process. The continued successful application of adaptive harvest management requires that model sets are updated with the most recent information to ensure that key demographic relationships are properly represented and a full suite of population responses to management actions is considered.

Summary and conclusions

A defensible argument could be made that there has been and continues to be more interest (and pages published) related to harvest of biotic resources than perhaps any other subject in biology (studies related to disease dynamics being the only likely competitor). Our intent in this review was to provide a reasonably complete review of some of the current and historical interest in harvest, and harvest management, as pertains to waterfowl. However, beyond the usual difficulties of reviewing such a large literature, we were faced with the additional challenge presented by the fact that "ducks aren't geese, and geese aren't ducks…" (C.D. Ankney, pers. comm.). We have proposed that differences in life histories between most duck and goose species (reviewed briefly at the start of this paper) result in important differences in estimation and management of the impacts of harvest on their respective population dynamics.

Perhaps the most obvious and important difference between modelling goose and duck dynamics (with or without harvest) is the presence of significant "age structure" for goose species, and several sea ducks.

Specifically, for geese, we need to adequately account for significant age-specific differences in survival and fertility in the model. While the presence of greater degrees of age-structure for geese (and probably most sea ducks) is important, all species of waterfowl are likely "structured" to some degree, independent of possible age differences. For example, differences in survival or reproductive output as a function of location, or amongst individuals of differing "quality", are both forms of "structure" which are likely common to most waterfowl populations. We note that recent work by F.A. Johnson (pers. comm.) has extended this idea to multispecies management, where the structural component involves relative proportions of different species, and how one or more species can be managed when harvest might be in part a function of the structure and dynamics of the focal species. When populations are structured, the harvest required to achieve a particular management objective can be described as a vector (specifying the number or proportion of each structural class in the harvest), the elements of which are determined by: (i) the number of structural classes, and (ii) the reproductive value vector of individuals in each of those structural classes at the time of harvest. The inclusion of structure adds extra dimensions of uncertainty to harvest management. Most obvious is the need for estimation of an increased number of demographic rates and functional forms relating one or more vital rates to various intrinsic (population density, population structure, *etc.*) and extrinsic (*e.g.* climate variables) factors. While population size is often estimated annually, modelling and management of structured models also ideally requires an estimate of the structural composition of the population at each time step. In cases where harvest can be selective of individuals in different structural classes, an optimal harvest strategy may prescribe a structured harvest (*e.g.* so many big ones, so many small ones, *etc.*), with the actual structure of the harvest depending on harvest objectives.

However, for waterfowl species, we are usually limited in both our ability to characterise structure at the time of harvest (*i.e.* structure is often only partially observable), and to select the age or stage of the individuals that we harvest (*i.e.* harvest of specific classes of individuals is often only under partial control). More often than not, actual harvest is a function of an interaction of: (i) the harvest regulatory option(s), and (ii) the relative vulnerability to harvest of different classes of individuals in the population (young, old, male, female, *etc.*). In principle, adaptive harvest management approaches can be applied here, since they can explicitly account for such uncertainties. The larger technical challenge with applying AHM to structured populations is that the state of the population at the time of the annual harvest decision is only partially observable in many cases, and may need to be reconstructed in some form. Alternatively, you might accept that you are harvesting an unknown mixture, and that your expected returns will likely be reduced because of this uncertainty. To date, there is limited experience with application of AHM to such partially observable structured systems.

The differences between ducks and geese have important ramifications beyond the structure of population models. Of particular interest in the management context is that the management objectives can and occasionally do differ significantly between ducks and geese. For most ducks, the management objective can be stated simply as "we want enough birds in the population to keep it viable, in the presence of sustained mortality impacts from sport harvest". While this is generally also true for geese, for some populations of some goose species, there is the unique additional objective relating to limiting or reducing the size of the population. For example, light geese (primarily Lesser Snow Geese and Ross's Geese *Chen rossii*) in several parts of North America are demonstrably overabundant (Ankney 1996), and the current management objective for these populations is to reduce numbers to a point where habitat damage is mitigated, and (importantly) where normal harvest pressure can serve to keep the population in check (Batt 1997; Rockwell *et al.* 1997). What is interesting in this situation is that the "biocontrol" objective for light geese represents a clear paradigm shift for waterfowl management, in a couple of respects. First, it is entirely counter to the usual management perspective that 'more is better'. Among waterfowl managers, this view is probably heavily influenced by the perspective that there is "no such thing as too many ducks". The prevailing view is that because ducks are strongly limited by specific habitat requirements (*e.g.* ponds in breeding wetland areas), there is little potential for duck populations to exceed carrying capacity (which typically is one of several criteria by which a species might be considered as "over abundant"). Moreover, even if duck populations increased significantly relative to historical numbers, it is difficult for most biologists and managers to imagine duck populations causing major negative impacts to the environment (save, perhaps, for the increased likelihood of disease with increased density). In contrast, many goose species, primarily because of the significant differences they exhibit in foraging behaviour and bill morphology, and little evidence they have approached their overall carrying capacity, have been increasingly observed to have strongly negative impacts on their habitat (both winter and breeding), lead to detrimental collateral impacts on other wildlife species, and in some cases significant economic liability for agricultural crops (Abraham & Jefferies 1997; Batt 1997, 1998; Moser 2001).

Second, the "biocontrol" problem forces consideration of the underlying assumption that harvest, and therefore harvest management, is an efficient and effective tool to facilitate change in abundance. Traditionally, waterfowl management was premised on the belief that small changes in bag limits would lead to desired and detectable numerical responses in the target populations. However, despite special legislation passed to encourage very large increases in light goose harvest, there is strong empirical evidence to suggest that goose populations have not been controlled, and continue to increase (Alisauskas *et al.* 2011). There are at least two proposed explanations for the failure of harvest to significantly reduce goose numbers and limit further population growth. One is that the number of geese was simply too large for the current number of hunters to possibly control. While this is undoubtedly true to some degree, this problem is likely compounded by: 1) a functional response by geese becoming more wary and thus less vulnerable, or 2) some aspect of hunter behaviour, such as an attenuation in effort/hunter or general willingness to pursue Snow Geese after an initial interest for doing so during the beginning of the conservation order. Both a functional behavioural response by geese and changes in hunter behaviour were observed following implementation of liberalisations in Greater Snow Goose harvest regulations (Béchet *et al.* 2003; Calvert *et al.* 2007). Additional focus on understanding these aspects of partial controllability and hunter behaviour would be useful.

Regardless of whether the underlying mechanisms involve hunter behaviour or bird behaviour, the result for management is a need to modify model components relating hunting regulations to harvest rate. Rather than model harvest rate resulting from any fixed set of hunting regulations as characterised by a single probability distribution (Johnson *et al.* 1997; USFWS 2013), it will be necessary to consider a family of distributions associated with different population sizes. Above some threshold population size, average harvest rates for any fixed set of hunting regulations are hypothesised to decrease as a function of abundance. Incorporation of such density-dependent harvest rates into our management models highlights the fact that there are real and practical limits on the degree to which harvest can influence waterfowl population dynamics. Under an objective of population control, optimal management would attempt to maintain population size below the threshold at which population growth exceeds the capacity for control by harvest (see Hauser *et al.* 2007 for an example applied to the Atlantic Flyway population of Canada Geese). Thus, for fixed (frequently at maximally liberal levels for control objectives) hunting regulations, goose populations appear to respond in a *positive* density-dependent manner. This is quite different than typical duck management, where optimal strategies are conditioned by the expectation of some level of negative density-dependent feedback over the range of population sizes typically encountered.

Much of our presentation has focussed on "models" – specifically, population projection models which allow us to make predictions about the numerical trajectory of populations over time and space, and the degree to which those trajectories might be influenced by perturbation, natural or anthropogenic (in the context of this review, meaning "harvest"). The consideration of models focuses on two key points. Models represent canonical and (relatively) transparent representations about what we think we know concerning the factors which determine the dynamics of populations. Second, we can, and often do, use these models in applications. Here, we have focussed on models where dynamics are potentially impacted by harvest. These harvest models are used to project the consequences to waterfowl population dynamics of specified changes in rates of survival, fecundity, and in some cases movement. The major sources of uncertainty in waterfowl harvest management involve the translation of hunting regulations (the actions that we take) into various hunting mortality rates, and the subsequent translation of specific hunting mortality rates into changes in total rates of survival and fecundity.

With respect to the first source of uncertainty, hunting mortality rates are stochastically predictable for sets of hunting regulations with which we have experience. Uncertainty about translation of hunting mortality rates into population level effects is greater, but potentially reducible. The additive mortality hypothesis and the compensatory mortality hypothesis represent endpoints in which effects of hunting mortality on total survival are maximal and minimal, respectively. The additive mortality hypothesis corresponds to the idea of independent competing risks, whereas the compensatory mortality hypothesis is thought to require mechanisms such as density-dependent mortality or individual heterogeneity in survival probabilities.

A summary of evidence based on analyses for various waterfowl species leads to the general inference that the additive hunting mortality hypothesis represents a reasonable approximation to reality for most goose species. Results are quite varied for duck species to the point that generalisations are not really possible. It is not clear whether this inability to generalise is attributable to underlying processes varying across species, locations and even time periods, or instead to the difficulties of drawing inferences from observational studies. These latter difficulties led to our description of some commonly used techniques for such inferences, with warnings about possible pitfalls. Effects of changes in hunting mortality on reproductive output generally entail density-dependent responses, and there is evidence of negative density-dependent reproduction in a number of waterfowl species. Finally, there is also good evidence that hunting can influence waterfowl movement rates at spatial scales ranging from local to regional, but the population dynamic consequences of these movements are not well understood.

Traditionally, the use of such models (say, in harvest management) has been presented as something distinct from efforts to use models as a basis for discriminating among associated hypotheses. However, we submit that this dichotomy is a false one, and that use of models within a structured, adaptive management framework not only serves as a transparent, defensible mechanism to evaluate harvest objectives, but the process of harvest in turn provides a powerful experimental framework in which the various hypotheses about system dynamics inherent in the population models can be tested. Adaptive management was developed to aid decision making for recurrent decision problems characterised by potentially reducible structural uncertainty. The establishment of annual hunting regulations is certainly a recurrent decision problem, and the various hypotheses about population dynamic effects of hunting represent reducible structural uncertainty. A programme of adaptive harvest management (AHM) was established for mid-continent Mallard in 1995 and has been used to by the USFWS to develop hunting regulations for these birds since that time. AHM for eastern Mallard was initiated by the USFWS in 2000. AHM programmes have since been developed for other species and populations as well, but mid-continent and eastern Mallard provide our longest running examples. Evolution of model weights (measures of model credibility) shows that predictions of the models including weakly density-dependent reproduction and additive hunting mortality have performed best over the respective periods of operation of these two AHM programmes. In addition to this reduction in uncertainty (learning), AHM provides a basis for making decisions in the face of such uncertainty about processes governing population responses to management actions.

While there was probably no serious expectation that application of an AHM framework to harvest management of mid-continent Mallard would unequivocally come down as favouring one hypothesis or the other, results to date have formed the basis for much new thinking about not only the underlying mechanisms of compensation (*e.g.* the possible role of individual heterogeneity or alternative models to represent density-dependent survival), but also the policy elements of AHM. Perhaps one of the greatest benefits of AHM is that it provides a structured decision making approach that allows one to agree to a formal process while explicitly considering alternative beliefs (and disagreements) about management outcomes (Johnson & Case 2000). As the United States waterfowl harvest management community has entered the double-loop learning phase of AHM, much emphasis has been placed on the consideration of alternative harvest management objectives and possible linkages to habitat management programmes with formal connections to the human dimensions of waterfowl management (Runge *et al.* 2006; NAMWP 2014; Osnas *et al.* 2014). In addition to updating AHM model sets, the double-loop learning process of AHM has also offered an opportunity to think critically about large scale system change and how decision frameworks will need to be adjusted to cope with this new form of uncertainty (Nichols *et al.* 2011). As the interest in AHM has expanded, so has the variety of populations and systems for which it has been proposed (*e.g.* Hauser *et al.* 2007; Johnson *et al.* 2014). In developing these new applications, managers and waterfowl biologists are forced to identify important differences in the systems in question (*e.g.* system dimension and alternative harvest management objectives), and how well or directly experiences accumulated to date with (primarily) single species near-scalar duck models, apply to other systems, including (in particular) geese.

Acknowledgements

We thank the organisers of the North America Duck Symposium for inviting this submission. In particular, we thank Dave Koons and Rick Kaminski for keeping us on track for what was intended to be a "brief review" of the subject. They are not to be blamed, however, for what is a clear failure on our part to attain anything approaching brevity. The paper benefitted from comments by Dave Koons, Mark Lindberg, Eric Osnas, Eileen Rees and Gary White.

References

Abraham, K.F. & Jefferies, R.L. 1997. High goose populations: causes, impacts and implications. In B.D.J. Batt, (ed.), *Arctic Ecosystems in Peril: Report of the Arctic Goose Habitat Working Group*, pp. 7–72. Arctic Goose Joint Venture Special Publication. U.S. Fish and Wildlife Service, Washington D.C., USA, and Canadian Wildlife Service, Ottawa, Canada.

Alisauskas, R.T., Rockwell, R.F., Dufour, K.W., Cooch, E.G., Zimmerman, G., Drake, K.L., Leafloor, J.O., Moser, T.J. & Reed, E.T. 2011. Effect of population reduction efforts on harvest, survival, and population growth of mid-continent lesser snow geese. *Wildlife Monographs* 179: 1–42.

Anderson, D.R. 1975. Population ecology of the mallard: V. Temporal and geographic estimates of survival, recovery and harvest rates. USFWS Resource Publication No.125. U.S. Fish and Wildlife Service, Washington D.C., USA.

Anderson, D.R. & Burnham, K.P. 1976. Population ecology of the mallard: VI. The effect of exploitation on survival. USFWS Resource Publication No. 128. U.S. Fish and Wildlife Service, Washington D.C., USA.

Ankney, C.D. 1996. An embarrassment of riches: too many geese. *Journal of Wildlife Management* 60: 217–223.

Arnott, W.G. 2007. *Birds in the Ancient World from A to Z*. Routledge, Oxford, UK & New York, USA.

Balmaki, B. & Barati, A. 2006. Harvesting status of migratory waterfowl in northern Iran: a case study from Gilan Province. In G.C. Boere, C.A. Galbraith & D.A. Stroud (eds), *Waterbirds around the World*, pp. 868–869. The Stationery Office, Edinburgh, UK.

Batt, B.D.J. (ed.). 1997. *Arctic Ecosystems in Peril: Report of the Arctic Goose Habitat Working Group*. Arctic Goose Joint Venture Special Publication. U.S. Fish and Wildlife Service, Washington D.C., USA, and Canadian Wildlife Service, Ottawa, Canada.

Batt, B.D.J. (ed.). 1998. *The Greater Snow Goose: Report of the Arctic Goose Habitat Working Group*. Arctic Goose Joint Venture Special Publication. U.S. Fish and Wildlife Service, Washington D.C., USA, and Canadian Wildlife Service, Ottawa, Canada.

Béchet, A., Giroux, J-F., Gauthier, G., Nichols, J.D. & Hines, J.E. 2003. Spring hunting changes the regional movements of migrating greater snow geese. *Journal of Applied Ecology* 40: 553–564.

Béchet, A., Giroux, J-F. & Gauthier, G. 2004. The effects of disturbance on behaviour, habitat use and energy of spring staging snow geese. *Journal of Applied Ecology* 41: 689–700.

Bellrose, F.C. 1954. The value of waterfowl refuges in Illinois. *Journal of Wildlife Management* 18: 160–169.

Bellrose, F.C. 1990. History of wood duck management in North America. In L. Fredrickson, G.V. Burger, S.P. Havera, D.A. Gruber, R.E. Kirby & T.S. Taylor (eds.), *Proceedings of the 1988 North American Wood Duck Symposium*, pp. 13–20. Gaylord Memorial Laboratory, St. Louis, USA.

Bellman, R.E. 1957. *Dynamic Programming*. Princeton University Press, Princeton, USA.

Berkson, J. & Elveback, L. 1960. Competing exponential risks, with particular reference to the study of smoking and lung cancer. *Journal of the American Statistical Association* 55: 415–428.

Beverton, R.J.H., & Holt, S.J. 1957. *On the Dynamics of Exploited Fish Populations*. Chapman and Hall, London, UK.

Bischof, R., Mysterud, A. & Swenson, J.E. 2008. Should hunting mortality mimic the patterns of natural mortality? *Biology Letters* 4: 307–310.

Black, J.M. 1996. *Partnerships in Birds – the Study of Monogamy*. Oxford University Press, Oxford, UK.

Boyd, H. 1962. Population dynamics and the exploitation of ducks and geese. In E.D. Le Cren & M.W. Holdgate (eds.), *The Exploitation of Natural Animal Populations. A Symposium of the British Ecological Society, Durham 28th–31st March 1960*, pp. 85–95. John Wiley & Sons, New York, USA.

Brooks, E.N. & Lebreton, J-D. 2001. Optimizing removals to control a metapopulation: application to the yellow legged herring gull (*Larus cachinnans*). *Ecological Modelling* 136: 269–284.

Brown, G.M., Jr., Hammack, J. & Tillman, M.F. 1976. Mallard population dynamics and management models. *Journal of Wildlife Management* 40: 542–555.

Brownie, C. 1974. *Estimation of Instantaneous Mortality Rates from Band Return Data When 'Within-season Date of Kill' is Also Available*. Biometrics Unit Report No. BU-536-M. Biometrics Unit, Cornell University, Ithaca, New York, USA.

Brownie, C. & Pollock, K.H. 1985. Analysis of multiple capture-recapture data using band-recovery methods. *Biometrics* 41: 411–420.

Buebbert, H.F. 1981. Breeding birds on waterfowl production areas in northeastern North Dakota. *Prairie Naturalist* 13: 19–22.

Bunnefeld, N., Reuman, D.C., Baines, D. & Milner-Gulland, E.J. 2011. Impact of unintentional selective harvesting on the population dynamics of red grouse. *Journal of Animal Ecology* 80: 1258–1268.

Burnham, K.P. & Anderson, D.R. 1984. Tests of compensatory vs additive hypotheses of mortality in mallards. *Ecology* 65: 105–112.

Burnham, K.P., White, G.C. & Anderson, D.R. 1984. Estimating the effect of hunting on annual survival rates of adult mallards. *Journal of Wildlife Management* 48: 350–361.

Cameron, T.C. & Benton, T.G. 2004. Stage-structured harvesting and its effects: an empirical investigation using soil mites. *Journal of Animal Ecology* 73: 996–1006.

Caswell, H. 2001. *Matrix Population Models: Construction, Analysis and Interpretation, 2nd Edition*. Sinauer Associates, Sunderland, USA.

Caswell, H. & Weeks, D.E. 1986. 2-sex models – chaos, extinction, and other dynamic consequences of sex. *American Naturalist* 128: 707–735.

Charnov, E.L. 1993. *Life History Invariants*. Oxford University Press, Oxford, UK.

Chiang, C.L. 1968. *Introduction to Stochastic Processes in Biostatistics*. John Wiley, New York, USA.

Cohen, J.E. An uncertainty principle in demography and the unisex issue. *The American Statistician* 40: 32–39.

Conn, P.B. & Kendall, W.L. 2004. Evaluating mallard adaptive management models with time series. *Journal of Wildlife Management* 68: 1065–1081.

Conroy, M.J. & Krementz, D.G. 1990. A review of the evidence for the effects of hunting on American black duck populations. *Transactions of the North American Wildlife and Natural Resources Conference* 55: 501–517.

Conroy, M.J., Miller, M.W. & Hines, J.E. 2002. Identification and synthetic modeling of factors affecting American black duck populations. *Wildlife Monographs* 150.

Cooch, E.G., Lank, D.B., Rockwell, R.F. & Cooke, F. 1989. Long-term decline in fecundity in a snow goose population – evidence for density dependence. *Journal of Animal Ecology* 58: 711–726.

Cooch, E.G., Lank, D.B., Rockwell, R.F. & Cooke, F. 1991. Long-term decline in body size in a snow goose population – evidence of environmental degradation. *Journal of Animal Ecology* 60: 483–496.

Cooke, F., Bousfield, M.A. & Sadura, A. 1981. Mate change and reproductive success in the lesser snow goose. *Condor* 83: 322–327.

Cox, R.R. & Afton, A.D. 1997. Use of habitats by female northern pintails wintering in southwestern Louisiana. *Journal of Wildlife Management* 61: 435–443.

Crissey, W.F. 1957. Forecasting waterfowl harvest by flyways. *Transactions of the North American Wildlife and Natural Resources Conference* 22: 256–268.

Day, A. 1949. *North American Waterfowl*. Stackpole Books, Harrisburg, USA.

Devineau, O., Guillemain, M., Johnson, A.R. & Lebreton, J-D. 2010. A comparison of green-winged teal *Anas crecca* survival and harvest between Europe and North America. *Wildlife Biology* 16: 12–24.

Drent, R.H. & Daan, S. 1980. The prudent parent: energetic adjustments in avian breeding? *Ardea* 68: 225–252.

Duebbert, H.F., Jacobson, E.T., Higgins, K.F. & Podoll, E.B. 1981. *Establishment of Sseeded Grasslands for Wildlife Habitat in the Prairie Pothole Region*. U.S. Fish and Wildlife Service Special Scientific Report: Wildlife No. 234. U.S. Fish and Wildlife Service, Washington D.C., USA.

Dufour, K.W., Ankney, C.D. & Weatherhead, P.J. 1993. Condition and vulnerability to hunting among mallards staging at Lake St. Clair, Ontario. *Journal of Wildlife Management* 57: 209–215.

Ebbinge, B.S. 1991. The impact of hunting on mortality-rates and spatial-distribution of geese wintering in the western Palearctic. *Ardea* 79: 197–209.

Elmberg, J., Nummi, P., Pöysä, H., Sjöberg, K., Gunnarsson, G., Clausen, P., Guillemain, M., Rodrigues, D. & Väänänen, V-M. 2006. The scientific basis for a new and sustainable management of migratory European ducks. *Wildlife Biology* 12: 121–128.

Ericson, P.G.P. & Tyrberg, T. 2004. The early history of the Swedish avifauna. A review of the subfossil record and early written sources. Kungl. *Vitterhets Historie och Antikvitets Akademiens Handlingar, Antikvariska serien* 45. Almqvist & Wiksell International, Stockholm, Sweden.

Errington, P.L. 1943. An analysis of mink predation upon muskrats in north-central United States. *Iowa Agricultural Research Bulletin* 320: 728–924.

Errington, P.L. 1967. *Of Predation and Life*. Iowa State University Press, Ames, USA.

Fox, A.D. & Madsen, J. 1997. Behavioural and distributional effects of hunting disturbance on waterbirds in Europe: Implications for refuge design. *Journal of Applied Ecology* 34: 1–13.

Gaillard, J.M., Pontier, D., Allaine, D., Lebreton, J.D., Trouvilliez, J. & Clobert, J. 1989. An analysis of demographic tactics in birds and mammals. *Oikos* 56: 59–76.

Garrettson, P.R., & F.C. Rohwer. 2001. Effects of mammalian predator removal on upland-nesting duck production in North Dakota. *Journal of Wildlife Management* 65: 398–405.

Gauthier, G. & Lebreton J-D. 2004. Population models for Greater Snow Geese: a comparison of different approaches to assess potential impacts of harvest. *Animal Biodiversity and Conservation* 27: 503–514.

Geis, A.D., Martinson, R.K & Anderson, D.R. 1969. Establishing hunting regulations and allowable harvest of mallards in United States. *Journal of Wildlife Management* 33: 848–859.

Gill, J.A. & Sutherland, W.J. 2000. Predicting the consequences of human disturbance from behavioural decisions. *In* L.M. Gosling & W.J. Sutherland (eds.), *Behaviour and Conservation*, pp. 51–64. Cambridge University Press, Cambridge, UK.

Gill, J.A., Norris, K. & Sutherland, W.J. 2001. Why behavioural responses may not reflect the population consequences of human disturbance. *Biological Conservation* 97: 265–268.

Gilmer, D.S., Miller, M.R., Bauer, R.D. & Ledonne, J.R. 1982. California's Central

Valley Wintering Waterfowl – Concerns and Challenges. *Transactions of the North American Wildlife and Natural Resources Conference* 47: 441–452.

Giroux, J-F. 1981. Ducks nesting on artificial islands during drought. *Journal of Wildlife Management* 45: 783–786.

Grey, D.R. & Law, R. 1987. Reproductive values and maximum yields. *Functional Ecology* 1: 327–330.

Gunnarsson, G., Elmberg, J., Pöysä, H., Nummi, P., Sjöberg, K., Dessborn, L. & Arzel, C. 2013. Density-dependence in ducks: a review of the evidence. *European Journal of Wildlife Research* 59: 305–321.

Hammond, M.C. & Mann, G.E. 1956. Waterfowl nesting islands. *Journal of Wildlife Management* 20: 345–352.

Hario, M., Hollmén, T.E., Morelli, T.L. & Scribner, K.T. 2002. Effects of mate removal on the fecundity of common eider *Somateria mollissima* females. *Wildlife Biology* 8: 161–168.

Haramis, G.M. & Thompson, D.Q. 1985. Density-production characteristics of box-nesting wood ducks in a northern green-tree impoundment. *Journal of Wildlife Management* 49: 429–436.

Hauser, C.E., Cooch, E.G. & Lebreton, J-D. 2006. Control of structured populations by harvest. *Ecological Modelling* 196: 462–470.

Hauser, C.E., Runge, M.C., Cooch, E.G., Johnson, F.A. & Harvey, W.F. 2007. Optimal control of Atlantic population Canada geese. *Ecological Modelling* 201: 27–36.

Hawkins, A.S. & Bellrose, F.C. 1940. Wood duck habitat management in Illinois. *North American Wildlife Conference Transactions* 5: 392–395.

Heisey, D.M. & Fuller, T.K. 1985. Evaluation of survival and cause-specific mortality rates using telemetry data. *Journal of Wildlife Management* 49: 68–674.

Heisey, D.M. & Patterson, B.R. 2006. A review of methods to estimate cause-specific mortality in presence of competing risks. *Journal of Wildlife Management* 70: 1544–1555.

Heitmeyer, M.E., Fredrickson, L.H. & Humburg, D.H. 1993. Further evidence of biases associated with hunter-killed mallards. *Journal of Wildlife Management* 57: 733–740.

Hepp, G.R., Blohm, R.J., Reynolds, R.E., Hines, J.E. & Nichols, J.D. 1986. Physiological condition of autumn-banded mallards and its relationship to hunting vulnerability. *Journal of Wildlife Management* 50: 177–183.

Hilborn, R. & Walters, C.J. 1992. *Quantitative Fisheries Stock Assessment: Choice, Dynamics and Uncertainty.* Chapman and Hall, New York, USA.

Hutchings, J.A. & Baum, J.K. 2005. Measuring marine fish biodiversity: temporal changes in abundance, life history and demography. *Philosophical Transactions of the Royal Society of London. Series B, Biological Sciences* 360: 315–338.

Johnson, D.H., Burnham, K.P. & Nichols, J.D. 1984. The role of heterogeneity in animal population dynamics. *Proceedings of the International Biometrics Conference* 13: 1–15.

Johnson, D.H., Nichols, J.D., Conroy, M.J. & Cowardin, L.M. 1988. Some considerations in modeling the mallard life cycle. Pp. 9–20 in M. W. Weller, ed. *Waterfowl in Winter.* University of Minnesota Press, Minneapolis, USA.

Johnson, F.A. & Case, D.J. 2000. Adaptive regulation of waterfowl harvests: lessons learned and prospects for the future. *Transactions of the North American Wildlife and Natural Resources Conference* 65: 94–108.

Johnson, F.A., Williams, B.K., Nichols, J.D., Hines, J.E, Kendall, W.L., Smith, G.W. & Caithamer, D.F. 1993. Developing an adaptive management strategy for harvesting waterfowl in North America. *Transactions of the North American Wildlife and Natural Resources Conference* 58: 565–583.

Johnson, F.A., Moore, C.T., Kendall, W.L., Dubovsky, J.A., Caithamer, D.F., Kelley, J.T. & Williams, B.K. 1997. Uncertainty and the management of mallard harvests. *Journal of Wildlife Management* 61: 203–217.

Johnson, F.A., Kendall, W.L. & Dubovsky, J.A. 2002. Conditions and limitations on learning in the adaptive management of mallard harvests. *Wildlife Society Bulletin* 30: 176–185.

Johnson, F.A., Jensen, G.H., Madsen, J. & Williams, B.K. 2014. Uncertainty, robustness and the value of information in managing an expanding Arctic goose population. *Ecological Modelling* 273: 186–199.

Kaminski, R.M. & Gluesing, E.A. 1987. Density- and habitat-related recruitment in mallards. *Journal of Wildlife Management* 51:141–148.

Kauhala, K. 2004. Removal of medium-sized predators and the breeding success of ducks in Finland. *Folia Zoologica* 53: 367–378.

Keyfitz, N. 1971. On the momentum of population growth. *Demography* 8: 71–80.

Klaassen, M. 2002. Relationships between migration and breeding strategies in arctic breeding birds. *In* P. Berthold, E. Gwinner & E. Sonnenschein (eds.), *Avian Migration*, pp. 237–249. Springer-Verlag, Berlin, Germany.

Kokko, H. 2001. Optimal and suboptimal use of compensatory responses to harvesting: timing of hunting as an example. *Wildlife Biology* 7: 141–152.

Koons, D.N., Rockwell, R.F. & Grand, J.B. 2006. Population momentum: Implications for wildlife management. *Journal of Wildlife Management* 70: 19–26.

Koons, D.N., Gunnarsson, G., Schmutz, J.A. & Rotella, J.J. 2014a. Drivers of waterfowl population dynamics: from teal to swans. *Wildfowl* (Special Issue No. 4): 169–191.

Koons, D.N., Rockwell, R.F. & Aubry, L.M. 2014b. Effects of exploitation on an overabundant species: the lesser snow goose predicament. *Journal of Animal Ecology* 83: 365–374.

Lebreton, J.-D. 2005. Dynamical and statistical models for exploited populations. *Australian & New Zealand Journal of Statistics* 47: 49–63.

Lebreton, J.-D. 2009. Assessing density-dependence: where are we left? *In* D.L. Thomson, E.G. Cooch & M.J. Conroy (eds.), *Modeling Demographic Processes in Marked Populations*, pp. 19–42. Environmental and Ecological Statistics, Springer, New York, USA.

Lebreton, J.-D. & Clobert, J. 1991. Bird population dynamics, management, and conservation: the role of mathematical modelling. *In* C.M. Perrins, J. Lebreton & G.J. Hirons (eds.), *Bird Population Studies*, pp. 105–125. Oxford University Press, Oxford, UK.

Legagneux, P., Inchausti, P., Bourguemestre, F., Latraube, F. & Bretagnolle, V. 2009. Effect of predation risk, body size and habitat characteristics on emigration decisions in mallards. *Behavioral Ecology* 20: 186–194.

Lercel, B.A., Kaminski, R.M. & Cox, R.R. 1999. Mate loss in winter affects reproduction of mallards. *Journal of Wildlife Management* 63: 621–629.

Lepage, D., Desrochers, A. & Gauthier, G. 1999. Seasonal decline of growth and fledging success in snow geese *Anser caerulescens*: an effect of date or parental quality? *Journal of Avian Biology* 30: 72–78.

LeSchack, C.R., Afton, A.D. & Alisauskas, R.T. 1998. Effects of male removal on female reproductive biology in Ross' and lesser snow geese. *Wilson Bulletin* 110: 56–64.

Lindberg, M.S., Sedinger, J.S. & Lebreton, J.-D. 2014. Individual heterogeneity in black brant survival and recruitment with implications for harvest dynamics. *Ecology and Evolution* 3: 4045–56.

Lindström, J. & Kokko, H. 1998. Sexual reproduction and population dynamics: the role of polygyny and demographic sex differences. *Proceedings of the Royal Society B – Biological Sciences* 265: 483–488.

Link, P.T., Afton, A.D., Cox, R.R. & Davis, B.E. 2011. Daily movements of female mallards wintering in southwestern Louisiana. *Waterbirds* 34: 422–428.

MacArthur, R.H. 1960. On the relation between reproductive value and optimal predation. *Proceedings of the National Academy of Sciences of the United States of America* 46: 143–145.

Madsen, J. 1998a. Experimental refuges for migratory waterfowl in Danish wetlands. I. Baseline assessment of the disturbance effects of recreational activities. *Journal of Applied Ecology* 35: 386–397.

Madsen, J. 1998b. Experimental refuges for migratory waterfowl in Danish wetlands. II. Tests of hunting disturbance effects. *Journal of Applied Ecology* 35: 398–417.

Madsen, J. & Jepsen, P.U. 1992. Passing the buck. Need for a flyway management plan for the Svalbard Pink-footed Goose. *In* M. van Roomen & J. Madsen (eds.), *Waterfowl and Agriculture: Review and Future Perspective of the Crop Damage Conflict in Europe*, pp. 109–110. IWRB Special Publications No. 21. IWRB, Slimbridge, UK.

Madsen, J. & Fox, A.D. 1995. Impacts of hunting disturbance on waterbirds – a review. *Wildlife Biology* 1: 193–207.

Mainguy, J., Béty, J., Gauthier, G. & Giroux, J.F. 2002. Are body condition and reproductive effort of laying Greater Snow Geese affected by the spring hunt? *Condor* 104: 156–161.

Manlove, C.A. & Hepp, G.R. 1998. Effects of mate removal on incubation behavior and reproductive success of female wood ducks. *Condor* 100: 688–693.

Martin, F.W., Pospahala, R.S. & Nichols, J.D. 1979. Assessment and population management of North American migratory birds. *In* J. Cairns, Jr., G.P. Patil & W.E. Waters (eds.), *Environmental Biomonitoring, Assessment, Prediction, and Management – Certain Case Studies and Related Quantitative Issues*. Statistical Ecology, Vol. 11. pp. 187–239, International Cooperative. Publishing House, Fairland, USA.

Martin, K., Cooch, F.G. & Rockwell, R.F. 1985. Reproductive performance in lesser snow geese: are two parents essential? *Behavioral and Ecological Sociobiology* 17: 257–263.

McCabe, R.E. 1987. Results of stabilized duck hunting regulations. *Transactions of the North American Wildlife and Natural Resources Conference* 52: 177–326.

McKinnon, D.T. & Duncan, D.C. 1999. Effectiveness of dense nesting cover for increasing duck production in Saskatchewan. *Journal of Wildlife Management* 63: 382–389.

Mertz, D.B. 1971. Life history phenomena in increasing and decreasing populations. *In* E.C. Pielou & W.E. Waters (eds.), *Statistical ecology*, Vol. II. *Sampling and Modeling Biological Populations and Population Dynamics*, pp. 361–399. Pennsylvania State University Press, USA.

Miller, M.R. & Eadie, J. McA. 2006. The allometric relationship between resting metabolic rate and body mass in wild waterfowl (*Anatidae*) and an application to estimation of winter habitat requirements. *Condor* 108: 166–177.

Milner, J.M., Nilsen, E.B. & Andreassen, H.P. 2007. Demographic side effects of selective hunting in ungulates and carnivores. *Conservation Biology* 21: 36–47.

Moser T.J. (ed.). 2001. *The Status of Ross's Geese*. Arctic Goose Joint Venture Special Publication. U.S. Fish and Wildlife Service, Washington D.C., USA, and Canadian Wildlife Service, Ottawa, Canada.

Murray, D.L., Anderson, M.G. & Steury, T.D. 2010. Temporal shift in density-dependence among North American breeding duck populations. *Ecology* 91: 571–581.

Nichols, J.D. 1991a. Extensive monitoring programmes viewed as long-term population studies: the case of North American waterfowl. *Ibis* 133 suppl. 1: 89–98.

Nichols, J.D. 1991b. Science, population ecology, and the management of the American black duck. *Journal of Wildlife Management* 55: 790–799.

Nichols, J.D. 2000. Evolution of harvest management for North American waterfowl: Selective pressures and preadaptations for adaptive harvest management. *Transactions of the North American Wildlife and Natural Resources Conference* 65: 65–77.

Nichols, J.D. & Hines, J.E. 1987. *Population Ecology of the Mallard: VIII. Winter Distribution Patterns and Survival Rates of Winter-banded Mallards.* U.S. Geological Survey Resource Publication No. 162, USGS, Washington D.C., USA.

Nichols, J.D. & Johnson, F.A. 1990. Wood duck population dynamics: a review. *In* L.H. Fredrickson, G.V. Burger, S.P. Havera, D.A. Graber, R.E. Kirby & T.S. Taylor (eds.), *Proceedings of the 1988 North American Wood Duck Symposium*, pp. 83–105. St. Louis, USA.

Nichols, J.D. & Williams, B.K. 2013. Adaptive management. *In* A.H. El-Shaarawi & W.W. Piegorsch (eds.), *Encyclopedia of Environmetrics, 2nd edition*, pp. 1–6. John Wiley, New York, USA.

Nichols, J.D., Conroy, M.J., Anderson, D.R. & Burnham, K.P. 1984. Compensatory mortality in waterfowl populations – a review of the evidence and implications for research and management. *Transactions of the North American Wildlife and Natural Resources Conference* 49: 535–554.

Nichols, J.D., Johnson, F.A. & Williams, B.K. 1995. Managing North American waterfowl in the face of uncertainty. *Annual Review of Ecology and Systematics* 26: 177–199.

Nichols, J.D., Koneff, M.D., Heglund, P.J., Knutson, M.G., Seamans, M.E., Lyons, J.E., Morton, J.M., Jones, M.T., Boomer, G.S. & Williams, B.K. 2011. Climate change, uncertainty and natural resource management. *Journal of Wildlife Management* 75: 6–18.

Niel, C. & Lebreton, J.-D. 2005. Using demographic invariants to detect overharvested bird populations from incomplete data. *Conservation Biology* 19: 826–835.

Osnas, E.E., Runge, M.C., Mattsson, B.J., Austin, J., Boomer, G.S., Clark, R.G., Devers, P. Eadie, J.M., Lonsdorf, E.V. & Tavernia, B.G. 2014. Managing harvest and habitat as integrated components. *Wildfowl* (Special Issue No. 4): 305–328.

Pace, R.M. & Afton, A.D. 1999. Direct recovery rates of lesser scaup banded in the Northwest Minnesota: sources of heterogeneity. *Journal of Wildlife Management* 63: 389–395.

Padding, P.I., Gobeil, J.F. & Wentworth, C. 2006. Estimating waterfowl harvest in North America. *In* G.C. Boere, C.A. Galbraith & D.A. Stroud (eds.), *Waterbirds Around the World*, pp. 849–852. The Stationery Office, Edinburgh, UK.

Patterson, J.H. 1979. Can ducks be managed by regulation? Experience in Canada. *Transactions of the North American Wildlife and Natural Resources Conferences* 44: 130–139.

Peron, G., Nicolai, C.A. & Koons, D.N. 2012. Demographic response to perturbations: the role of compensatory density-dependence in a North American duck under variable harvest regulations and changing habitat. *Journal of Animal Ecology* 81: 960–969.

Peron, G. 2013. Compensation and additivity of anthropogenic mortality: life-history effects and review of methods. *Journal of Animal Ecology* 82: 408–417.

Phillips, J.C. 1923. *A Natural History of the Ducks. Volume II: The Genus Anas.* Houghton Mifflin Co., Boston, USA.

Phillips, J., and Lincoln, F.C. 1930. *American Waterfowl: Their Present Situation and Outlook for the Future.* Houghton Mifflin, Boston, USA.

Pieron, M.R. & Rohwer, F.C. 2010. Effects of large-scale predator reduction on nest success of upland nesting ducks. *Journal of Wildlife Management* 74: 124–132.

Pledger, S., Pollock, K.H. & Norris, J.L. 2003. Open capture-recapture models with heterogeneity: I. Cormack-Jolly-Seber model. *Biometrics* 59: 786–794.

Pöysä, H., Elmberg, J., Gunnarsson, G., Nummi, P. & Sjöberg, K. 2004. Ecological basis of sustainable harvesting: is the prevailing paradigm of compensatory mortality still valid? *Oikos* 104: 612–615.

Pöysä, H., Dessborn, L., Elmberg, J., Gunnarsson, G., Nummi, P., Sjöberg, K., Suhonen, S. & Söderquist, P. 2013. Harvest mortality in North American mallards: A reply to Sedinger and Herzog. *Journal of Wildlife Management* 77: 653–654.

Prop, J. & Quinn, J.L. 2003. Constrained by available raptor hosts and islands: density-dependent reproductive success in red-breasted geese. *Oikos* 102: 571–580.

Pulliam, H.R. 1988. Sources, sinks, and population regulation. *American Naturalist* 132: 652–661.

Reynolds, R.E., Shaffer, T.L., Sauer, J.R. & Peterjohn, B.G. 1994. Conservation Reserve Program: benefit for grassland birds in the northern plains. *Transactions of the North American Wildlife and Natural Resources Conference* 59: 328–336.

Reynolds, R.E., Shaffer, T.L., Renner, R.W., Newton, W.E. & Batt, B.D.J. 2001. Impact of the Conservation Reserve Program on duck recruitment in the U.S. prairie pothole region. *Journal of Wildlife Management* 65: 765–780.

Ricker, W.E. 1954. Stock and recruitment. *Journal of Fish Research Board of Canada* 11: 559–623.

Rice, M.B., Haukos, D.A., Dubovsky, J.A. & Runge, M.C. 2010. Continental survival and recovery rates of northern pintails using band-recovery data. *Journal of Wildlife Management* 74: 778–787.

Ringelman, K.M., Eadie, J.M. & Ackerman, J.T. 2014. Adaptive nest clustering and density-dependent nest survival in dabbling ducks. *Oikos* 123: 239–247.

Rockwell, R. F., Cooch, E.G. & Brault, S. 1997. Dynamics of the midcontinent population of lesser snow geese: projected impacts of reduction in survival and fertility on population growth rates. In B.D.J. Batt (ed.), *Arctic Ecosystems in Peril: Report of the Arctic Goose Joint Venture Habitat Working Group*, pp. 73–100. Arctic Goose Joint Venture Special Publication. U.S. Fish and Wildlife Service, Washington D.C., USA and Canadian Wildlife Service, Ottawa, Canada.

Runge, M.C. & Johnson, F.A. 2002. The importance of functional form in optimal control solutions of problems in population dynamics. *Ecology* 83: 1357–1371.

Runge, M.C., Johnson, F.A., Dubovsky, J.A., Kendall, W.L., Lawrence, J. & Gammonley, J. 2002. *A Revised Protocol for the Adaptive Harvest Management of Mid-continent Mallards*. U.S. Fish and Wildlife Service, U.S. Department of the Interior Technical report, Washington D.C., USA.

Runge, M.C., Johnson, F.A., Anderson, M.G., Koneff, M.D., Reed, E.T. & Mott, S.E. 2006. The need for coherence between waterfowl harvest and habitat management. *Wildlife Society Bulletin* 34: 1231–1237.

Sæther, B-E., Engen, S., Grøtan, V., Bregnballe, T., Both, C., Tryjanowski, P., Leivits, A., Wright, J., Møller, A.P., Visser, M.E. & Winkel, W. 2008. Forms of density regulation and (quasi-) stationary distributions of population sizes in birds. *Oikos* 117: 1197–1208.

Sandercock, B.K., Nilsen, E.B., Broseth, H. & Pedersen, H.C. 2011. Is hunting mortality additive or compensatory to natural mortality? Effects of experimental harvest on the survival and cause-specific mortality of willow ptarmigan. *Journal of Animal Ecology* 80: 244–258.

Schaub, M. & Lebreton, J.-D. 2004. Testing the additive versus compensatory hypothesis of mortality from ring recovery data using a random effects model. *Animal Biodiversity and Conservation* 27: 73–85.

Schaub, M. & Pradel, R. 2004. Assessing the relative importance of different sources of mortality from recoveries of marked animals. *Ecology* 85: 930–938.

Sedinger, J.S. & Herzog, M.P. 2012. Harvest and dynamics of duck populations. *Journal of Wildlife Management* 76: 1108–1116.

Sedinger, J.S. & Nicolai, C.A. 2011. Recent trends in first-year survival for Black Brant breeding in southwestern Alaska. *Condor* 113: 511–517.

Sedinger, J.S., Flint, P.L. & Lindberg, M.S. 1995. Environmental influence on life-history traits – growth, survival and fecundity in Black Brant (*Branta Bernicla*). *Ecology* 76: 2404–2414.

Sedinger, J.S., Lindberg, M.S., Person, B.T., Eichholz, M.W., Herzog, M.P. & Flint, P.L. 1998. Density-dependent effects on growth, body size and clutch size in Black Brant. *Auk* 115: 613–620.

Sedinger, J.S., White, G.C., Espinosa, S., Partee, E.T. & Braun, C.E. 2010. Assessing compensatory versus additive harvest mortality: an example using greater sage-grouse. *Journal of Wildlife Management* 74: 326–332.

Servanty, S., Choquet, R., Baubet, E., Brandt, S., Gaillard, J.-M., Schaub, M., Toïgo, C., Lebreton, J.-D., Buoro, M. & Gimenez, O. 2010. Assessing whether mortality is additive using marked animals: a Bayesian state-space modeling approach. *Ecology* 91: 1916–1923.

Smith, L.M., Pederson, R.L. & Kaminski, R.M. (eds.). 1989. *Habitat Management for Migrating and Wintering Waterfowl in North America*. Texas Tech University Press, Lubbock, USA.

Stearns, S. 1992. *The Evolution of Life Histories*. Oxford University Press, Oxford, UK.

Sparrow, R.D. & Patterson, J.H. 1987. Conclusions and recommendations from studies under stabilized duck hunting regulations, management implications and future directions. *Transactions of the North American Wildlife and Natural Resources Conference* 52: 320.

U.S. Fish and Wildlife Service. 2013. *Adaptive Harvest Management: 2013 Duck Hunting Season*. U.S. Department of Interior, Washington D.C., USA.

Väänänen, V.-M. 2001. Hunting disturbance and the timing of autumn migration in Anas species. *Wildlife Biology* 7: 3–9.

Vaupel, J.W. & Yashin, A.I. 1985. Heterogeneity ruses – some surprising effects of selection on population-dynamics. *American Statistician* 39: 176–185.

Viljugrein, H., Stenseth, N.C., Smith, G.W. & Steinbakk, G.H. 2005. Density-dependence in North American ducks. *Ecology* 86: 245–254.

Walters, C.J. 1986. *Adaptive Management of Renewable Resources*. McGraw Hill, New York, USA.

Whitehead, A.L., Edge, K.A., Smart, A.F., Hill, G.S. & Willans, M.J. 2008. Large scale predator control improves the productivity of a rare New Zealand riverine duck. *Biological Conservation* 141: 2784–2794.

Williams, B.K. & Johnson, F.A. 1995. Adaptive management and the regulation of waterfowl harvests. *Wildlife Society Bulletin* 23: 430–436.

Williams, B.K., Nichols, J.D & Conroy, M.J. 2002. *Analysis and Management of Animal Populations*. Academic Press, San Diego, USA.

Williams, B.K., Szaro, R.C. & Shapiro, C.D. 2007. *Adaptive Management: The U.S. Department of the Interior Technical Guide*. Adaptive Management Working Group, U.S. Department of the Interior, Washington D.C., USA.

Williams, B.K. & Brown, E.D. 2012. *Adaptive Management: The U.S. Department of the Interior Applications Guide*. Adaptive Management Working Group, U.S. Department of the Interior, Washington D.C., USA.

Cross-seasonal effects and the dynamics of waterfowl populations

JAMES S. SEDINGER[1]* & RAY T. ALISAUSKAS[2]

[1]Department of Natural Resources and Environmental Science,
University of Nevada Reno, 1664 N. Virginia St., Reno, Nevada 89557, USA.
[2]Environment Canada, 115 Perimeter Road, Saskatoon, Saskatchewan S7N 0X4, Canada.
*Correspondence author. E-mail: jsedinger@cabnr.unr.edu

Abstract

Cross-seasonal effects (CSEs) on waterfowl populations link together events and habitats that individuals experience as carry-over effects (COEs) throughout the annual cycle. The importance of CSEs has been recognised since at least the 1950s. Studies of nutrient dynamics beginning in the 1970s, followed by regression analyses that linked production of young to winter habitat conditions, confirmed the importance of CSEs. CSEs have been most apparent in large-bodied waterfowl, but evidence for CSEs in much smaller passerines suggests the potential for CSEs in all waterfowl. Numerous studies have established effects of winter weather on body condition and reproduction in both ducks and geese. Additionally, the ubiquitous use (during laying and incubation) of nutrients stored previously during spring migration suggests that such nutrients commonly influence reproductive success in waterfowl. Carry-over effects from the breeding season to autumn and winter are less well understood, although nutrition during the growth period in geese has been widely demonstrated to influence subsequent survival and reproduction. Only a few studies have examined effects of breeding on reproduction in later years. Because pathogens and parasites can be carried between seasonal habitats, disease represents an important potential mechanism underlying CSEs; so far, however, this role for diseases and parasitism remains poorly understood. CSEs were originally of interest because of their implications for management of seasonal habitats and CSEs represent a fundamental rationale for the habitat joint ventures in North America. Substantial research examining the role of COEs in individual fitness and of CSEs on population dynamics has now been conducted. New techniques (*e.g.* stable isotopes, geolocators) developed over the last decade, combined with more traditional marking programmes have created opportunities to understand CSEs more fully and to inform the management of seasonal habitats for waterfowl.

Key words: carry-over effect, climate, cross-seasonal effect, fitness, population, reproduction, survival.

The concept that variation in habitat quality can influence population dynamics of waterfowl has been established for some time (Lynch 1952). For example, reduced snow accumulation and spring rains in mid-continent prairie-parkland ecosystems result in fewer ponds, which reduces annual production of young ducks (Pospahala *et al.* 1974). Kaminski & Gluesing (1987) extended this notion to include cross-seasonal effects (CSEs) when they showed that the influence of breeding habitat availability on autumn age-ratios (an index of the breeding success of the population) was modified by population density in Mallard *Anas platyrhynchos*. Such density-dependence in the reproductive process is incorporated into models of population dynamics used to manage harvest rates (Johnson *et al.* 1997). For ducks, the number of breeding individuals in the population is influenced by the production of young from the previous breeding season and the number of individuals surviving through winter, so that density-dependent effects on reproduction represent a CSE. Newton (2006) showed how density-dependent mortality operating at different segments of the annual cycle interacted with density-dependent reproduction to regulate population size.

Before the 1970s, attempts to understand waterfowl population dynamics in North America focused largely on habitat conditions in the mid-continent prairie-parkland breeding areas and on the role of harvest as the major drivers of population trends (Johnson *et al.* 1992). The interests of a number of prominent waterfowl scientists in the late 1970s to early 1980s coincided, stimulating discussion and thought on the role of habitats used by waterfowl outside the breeding season in governing their subsequent breeding activity and thus their population dynamics. Ankney & MacInnes (1978) and Raveling (1979) demonstrated that female geese stored nutrients acquired before nesting and used these nutrient reserves during egg laying and incubation. Ankney & MacInnes (1978) further showed that clutch size of Lesser Snow Geese *Chen caerulescens caerulescens* was directly related to their levels of endogenous nutrients. Shortly thereafter, Heitmeyer & Fredrickson (1981) used regression approaches to illustrate that winter wetland indices and winter precipitation predicted age-ratios in Mallard harvested the next autumn in the Mississippi Flyway, suggesting that winter conditions influenced production of young at the population level. Raveling & Heitmeyer (1989) used similar approaches to show that winter habitat conditions in California were important predictors of productivity in Northern Pintail *Anas acuta*, but that importance of winter habitat also depended on population size and breeding habitat conditions. In 1982, 40 waterfowl scientists met in Puxico, Missouri (Anderson & Batt 1983) to identify key aspects of the winter ecology of waterfowl that were still largely unstudied and not well understood. Specifically, this group focused on the potential role of CSEs in governing dynamics of waterfowl populations and proposed fundamental questions and approaches toward an improved understanding of CSEs. There was strong support for hypothesis testing, as well as comparative and experimental approaches to research combined with long-term studies. Participants identified the need to understand a number of basic aspects of waterfowl biology including the extent and basis for philopatry, potential of winter habitats to limit waterfowl populations, the ecology of winter habitats, the bioenergetics of winter moult, the role of flocking behaviour in winter ecology and other biological processes. A number of these questions have been addressed successfully, but more importantly, the questions themselves have evolved as our understanding of CSEs has increased.

Our goals in this paper are to: 1) provide a functional definition of CSEs; 2) discuss how body size and life-histories affect the potential for CSEs; 3) describe mechanisms underlying CSEs; 4) discuss the role of CSEs in management of waterfowl; and 5) propose approaches for detecting, understanding and measuring the magnitude of CSEs.

Cross-seasonal effects

We largely subscribe to the definitions of CSEs developed by Norris (2005) and Harrison *et al.* (2011). These definitions attribute lagged population processes in one season to conditions during a previous season. For example, breeding propensity by Mallard in prairie-parkland ecosystems might depend on habitat conditions, feeding opportunities and climate experienced by individuals during the previous winter in the Mississippi Alluvial Valley. We distinguish between CSEs at the population level and carry-over effects (COEs) that operate at the individual level, but the two are linked as population CSEs can result from COEs experienced by individuals. The influence of COEs and CSEs may be modified further by population size and density-dependence (Harrison *et al.* 2011). We exclude from CSEs those environmental or maternal effects experienced early in life on traits such as body size, even though they may influence subsequent survival and reproduction by individuals (*e.g.* Sedinger *et al.* 1995; Cooch 2002), because such effects are relatively permanent and not necessarily subject to modification of individual fitness by events occurring later in life. We allow for COEs on both survival and reproduction with potential lagged effects over multiple seasons, although we exclude direct effects of behaviour or habitat on fitness during the same season. For example, we do not view reduced survival of female ducks associated with nesting (Arnold *et al.* 2010) as a COE but we do consider reduced survival of breeding individuals over the next winter (*e.g.* Daan *et al.* 1996). Similarly, we do not consider reduced within-winter survival associated with occupancy of a particular habitat (Fleskes *et al.* 2007) as a COE, but do include the potential lagged effect of lower survival during spring migration associated with use of a particular winter habitat as a COE.

Generally, we are interested in COEs that influence breeding success as a result of previous events during the same annual cycle (*e.g.* Alisauskas 2002; Bêty *et al.* 2003). Presumably, a COE may extend from one breeding season to the next, although effects extending over a full annual cycle may be mediated by events during intervening seasons (Fig. 1). We are also concerned with factors that affect survival in seasons subsequent to use of particular habitats or

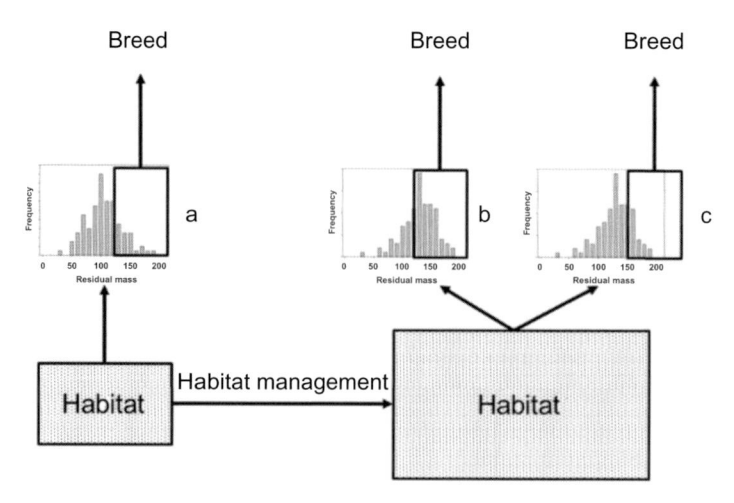

Figure 1. Hypothetical links between habitat availability, nutritional status (measured as body mass) and the probability of breeding. We show increases in mean mass in response to increases in wintering habitat but two different responses to increased mass. In scenario (b), threshold mass required for breeding does not change in response to increased mass, while in scenario (c) threshold mass increases in response to habitat availability, resulting in a similar proportion of individuals breeding before and after habitat enhancement. Under scenario (c), individuals benefit nutritionally from increased food availability but other attributes (potentially latent) prevent many individuals from translating improved nutritional status into increased reproductive success. We do not advocate scenario (c) as the expected response to improved habitat conditions. In fact, numerous examples suggest that individuals respond positively to increased food availability. The intention, here, is to indicate that complexities, including latent attributes of individuals, may influence population level responses to habitat management.

investments in reproduction. In this paper, we consider COEs that have the potential to influence both reproduction and survival. We also explore how COEs that operate at the individual level as precursors may scale up to CSEs at the population level. Such scaling up brings population level processes such as density-dependence into play.

Body size, life-histories and the potential for cross-seasonal effects

Body size is clearly related to the ability to store nutrients in one season for use in another (Alisauskas & Ankney 1992), which results in part from lower mass-specific metabolic rate in larger species (Kleiber 1975). Use of previously stored (endogenous) nutrients for egg formation and incubation generally increases with body size (Alisauskas & Ankney 1992; Meijer & Drent 1999; Alisauskas & DeVink 2015), although details about use of endogenous nutrients for breeding may vary as a function of the ecology or life-histories of individual species (*e.g.* Alisauskas & Ankney 1992; Alisauskas & DeVink 2015). Because greater reliance on endogenous nutrients should increase the potential for

COEs, one might expect that COEs are more common in larger-bodied waterfowl species (but see below). In fact, the largest numbers of examples of CSEs exist for larger bodied waterfowl, especially geese (Table 1), although this may reflect to some extent the greater ease with which individuals of larger-bodied species can be monitored (*e.g.* Ebbinge & Spaans 1995).

Other aspects of life-histories covary with body size and also might be predicted to influence the potential for COEs. For example, duration of maintenance of social bonds within families increases with body size; most geese and swans normally show lifelong monogamy and their young remain in family units through the first winter (with their parents) and sometimes through a second winter (either with parents or solely in sibling groups, *e.g.* Prevett & MacInnes 1980; Scott 1980). Because size of family groups influences social status (Boyd 1953; Raveling 1970; Poisbleau *et al.* 2006), breeding success in one year affects social status in the following winter. Social status related to family size affects dominance and aggressive defence of food resources. Presumably, enhanced access to food improves daily efficiency of nutrient storage during spring hyperphagia as geese travel north to breeding areas. Thus, a series of sequential COEs from the arctic summer, through the following winter and spring migration may influence reproductive investment by individual females in a subsequent breeding season.

Timing of pair formation in ducks is a function of body size. For example, larger bodied Mallard form pair bonds in early autumn, while small-bodied Blue-winged Teal *Anas discors* and Green-winged Teal *A. crecca* may not be paired until later in the year (MacKinney 1986). Additional energy demands of males from larger species associated with maintenance of pair bonds throughout autumn and winter might increase the potential that males of larger species are more influenced by habitat quality in winter than is the case for smaller bodied species. Alternatively, costs of delayed pairing by females of larger bodied species, and associated reduced access to food resources from intra- and interspecific competition, might interact disadvantageously with habitat quality to create greater potential for COEs in females of larger-bodied species.

Despite the clear logic underlying the linkage between body size and potential for COEs, there may be species-specific deviations from this general pattern in waterfowl. Examples from non-waterfowl species indicate that some caution is warranted when considering such relationships. Variation in quality of wintering habitat, and its influences on reproductive performance by American Redstarts *Setophaga ruticilla* is a very clear illustration of COEs (Marra *et al.* 1998), and there exist numerous other examples from small passerines and shorebirds (Norris & Marra 2007). Body size is probably too small in these passerine and shorebird examples for substantial mass to be carried between sequentially used seasonal habitats. So, the existence of COEs in such small birds suggests that mechanisms underlying COEs may involve complexities beyond nutritional dynamics. We address some of these potential mechanisms below.

Cross-seasonal effects in waterfowl

Evidence of CSEs transmitted from wintering areas to breeding is extensive (Table 1). Studies based on regression of population age ratios in post-breeding censuses against precipitation in the previous winter, demonstrate ecological links between landscapes separated sometimes by lengthy time lags of several months or great distances of thousands of kilometres. For example, winter precipitation can influence habitat, and thus feeding conditions for wintering waterfowl that in turn affect the production of young at the population level as indexed by age ratios during the following hunting season (Heitmeyer & Fredrickson 1981; Raveling & Heitmeyer 1989). Boyd *et al.* (1982) found that winter precipitation in the U.S. along the Gulf of Mexico affected the number of adult Lesser Snow Geese and their breeding success to a greater extent than spring temperatures and precipitation in the Dakotas and southern Manitoba, which are used during spring migration. More recently, winter climate has been related to age-ratios in Barnacle Geese *Branta leucopsis* (Trinder *et al.* 2009), pre-breeding body mass of Common Eiders *Somateria mollissima* (Descamps *et al.* 2010), as well as their breeding propensity (Jónsson *et al.* 2009) and breeding success (Lehikoinen *et al.* 2006). Correlations between expansion of cereals and other agricultural crops in both North America and Europe, and the growth of goose populations on both continents, support inferences from earlier regression-based studies about the importance of food supply to population growth. Finally, capture-mark-recapture (CMR) studies which demonstrate that quality of wintering locations influences breeding probability (*e.g.* Sedinger *et al.* 2011) provide additional evidence for the impact of winter habitats on production of offspring. Evidence for CSEs currently appears stronger for geese than for ducks (Table 1), but we suspect that this in part represents the greater ability to monitor individual geese throughout the annual cycle than is the case for ducks.

Evidence for CSEs linking spring migration and breeding are especially strong (Table 1). Studies using stable isotopes demonstrate that nutrients acquired away from breeding areas contribute to egg production in numerous species of ducks and geese (Table 1). In some cases contributions of endogenous nutrients to eggs were relatively modest (< 30% of the total). Such modest contributions to egg formation may, however, provide essential supplements to dietary nutrients if the latter are insufficient to meet the daily needs of females for egg production. Such needs can be substantial during early clutch formation, when yolk and albumin formation overlap (Alisauskas & Ankney 1992). Drent & Daan (1980) considered variation in reliance on nutrient reserves for egg formation along a capital-income continuum. Some studies have characterised egg formation as income-based if more than half of egg nutrients supplied were exogenous. However, the pervasiveness of nutrient reserve use (particularly of fat) in waterfowl (Ankney & Alisauskas 1991; Alisauskas & Ankney 1992) suggests to us that usage of nutrient reserves might be largely obligatory, or at least highly adaptive, and so *any* usage

Table 1. Examples of cross-seasonal and carry-over effects in waterfowl. These examples are restricted to cases where one of two conditions was met: 1) investigators demonstrated a direct link between individual state in one season and individual demographic rates in a subsequent season; or 2) investigators demonstrated a relationship between habitat conditions or average state at the population level and population demographic rates in a subsequent season.

Carry-over effect	Species	Reference
Individual level effects		
Breeding propensity and clutch size lower in springs with hunting on staging area	Greater Snow Goose	Mainguy et al. 2002
Females with greater fat before spring migration laid earlier and larger clutches	Greater Snow Goose	Béty et al. 2003
Endogenous reserves provided > 20% of egg nutrients based on stable isotopes	Greater Snow Goose	Gauthier et al. 2003
Increased lipid stores on spring staging area and population increase associated with increased corn production	Greater Snow Goose	Gauthier et al. 2005
Positive correlation between endogenous lipid and protein on arrival and clutch size	Lesser Snow Goose	Ankney & MacInnes 1978
Endogenous nutrients provided 55% of nutrients in eggs based on stable isotopes	Lesser Snow Goose	Hobson et al. 2011
Endogenous nutrients provided 25–36% of nutrients in eggs based on stable isotopes	Lesser Snow Goose	Sharp et al. 2013
Earlier hatching goslings are larger and are recruited at a higher rate	Lesser Snow Goose	Cooke et al. 1984; Cooch et al. 1991a; Aubry et al. 2013
Migration strategy and condition during spring related to production of offspring	Pink-footed Goose	Madsen 2001

Table 1 (*continued*).

Carry-over effect	Species	Reference
More than 50% of nutrients in eggs from endogenous sources based on stable isotopes	Canada Goose	Sharp *et al.* 2013
Larger goslings survive autumn migration at a higher rate	Barnacle Goose	Owen & Black 1989
Greater nutrient reserves and earlier migration positively related to reproductive success	Barnacle Goose	Prop *et al.* 2003
Heavier females on arrival to breeding area laid more eggs and were more attentive to nests	Barnacle Goose	Tombre *et al.* 2012
Females with greater nutrient reserves in spring produced more offspring	Dark-bellied Brent	Ebbinge & Spaans 1995
Endogenous nutrients used during egg laying and incubation	Dark-bellied Brent	Spaans *et al.* 2007
Earlier hatching goslings survive better and are recruited at a higher rate	Black Brant	Sedinger & Flint 1991; Sedinger *et al.* 1995; Sedinger & Chelgren 2007
Increased sea surface temperatures on wintering area reduced breeding propensity	Black Brant	Sedinger *et al.* 2006
Winter location affected breeding propensity and timing of nesting	Black Brant	Sedinger *et al.* 2011; Schamber *et al.* 2012
Breeding affected quality of wintering area	Black Brant	Sedinger *et al.* 2011
Use of particular brood rearing areas affects breeding probability the next year	Black Brant	Nicolai & Sedinger 2012
Females that hatch earlier are recruited at a higher rate	Mallard	Dzus & Clark 1998

Table 1 (*continued*).

Carry-over effect	Species	Reference
Females with greater nutrient reserves on arrival had greater breeding propensity, nested earlier and laid larger clutches	Mallard	Devries *et al.* 2008
Females fasted in winter were heavier but laid fewer eggs	Black Duck	Barboza and Jorde 2002
>30% of protein in eggs from endogenous source in one of three years	Lesser Scaup	Cutting *et al.* 2011
Endogenous nutrients used for maintenance during egg laying	Redhead	Hobson *et al.* 2004
Immunosuppression by low body mass females. Females that abandoned broods and had low lymphocyte levels had lower return rates the next year	Common Eider	Hanssen *et al.* 2003
Increased incubation demand reduced immune function and clutch size the next year	Common Eider	Hanssen *et al.* 2005
Wintering location related to timing of nesting and weakly to clutch size	King Eider	Mehl *et al.* 2004
Endogenous nutrients contributed significantly to egg production based on stable isotopes	King Eider	Oppel *et al.* 2010
Population level effects		
Increased commodity acreage correlated with increased population size	Lesser Snow Goose	Abraham *et al.* 2005
Autumn age-ratios positively related to mean body mass and fat in spring	Lesser Snow Goose	Alisauskas 2002
Smaller clutches fewer young breeders in years with drought on spring staging areas	Lesser Snow Goose	Davies & Cooke 1983
Estimated number of adults and young correlated with previous winter weather	Lesser Snow Goose	Boyd *et al.* 1982

Table 1 (*continued*).

Carry-over effect	Species	Reference
Increased disturbance reduced nutrient reserves during spring staging and age-ratios following breeding	Pink-footed Goose	Drent *et al.* 2003
Increased age-ratios with increased availability of agricultural foods on wintering areas	Greenland White-fronted Goose	Fox *et al.* 2005
Age ratios positively correlated with winter temperature	Barnacle Goose	Trinder *et al.* 2009
Winter precipitation positively related to age ratios the next autumn	Mallard	Heitmeyer & Fredrickson 1981
Winter precipitation positively related to age ratios the next autumn	Northern Pintail	Raveling & Heitmeyer 1989
Plumage quality during pair formation varied among moulting areas	Green-winged Teal	Legagneux *et al.* 2012
Winter index of North Atlantic Oscillation positively correlated with body mass of adult females at hatch, and subsequent age ratios	Common Eider	Lehikoinen *et al.* 2006
More females nested following warm-wet winters	Common Eider	Jónsson *et al.* 2009
NAO positively related to prenesting body condition in two separate nesting populations	Common Eider	Descamps *et al.* 2010
More females nested following warm-wet winters	Common Eider	Jónsson *et al.* 2009

might be properly considered as a capital breeding strategy.

Numerous studies have shown that nutritional state in spring affects success in producing offspring (Table 1). Similarly, earlier migrants typically are more likely to produce offspring (e.g. Prop et al. 2003). Because these studies were without experimental manipulation and so primarily observational, we cannot rule out the potential for spurious correlation between condition, or timing of migration, and reproductive success; that is, other unmeasured variables may have been associated with both nutritional status and production of offspring, but without the former variate directly influencing the latter. Two kinds of experiments, however, suggest that nutritional status can have an important influence on reproductive success in waterfowl. First, implementation of a spring hunt on a major spring staging area for Greater Snow Geese C. caerulescens atlanticus, reduced nutritional status of staging geese, associated with increased disturbance (Féret et al. 2003), and reduced the proportion of females that attempted to nest and reproductive investment by those females that did nest (Mainguy et al. 2002; Bêty et al. 2003). Second, reduced availability of agricultural habitats for Pink-footed Geese Anser brachyrhynchus during spring migration in Norway reduced both nutritional status of migrating geese and their subsequent reproductive success (Drent et al. 2003).

CSEs on adults from breeding to autumn or winter are generally less well understood than those linking other seasons. Limitations in our understanding reflect the difficulty of monitoring highly mobile individuals outside the breeding season and the difficulty of monitoring non-breeding individuals that typically migrate away from breeding areas to moult. Nevertheless, there are *a priori* reasons to expect that such CSEs exist and empirical evidence exists for such CSEs.

Female waterfowl invest substantial nutrients into both egg formation and incubation (Alisauskas & Ankney 1992; Afton & Paulus 1992; Alisauskas & DeVink 2015) and these nutrients must be replenished before autumn migration (Sedinger & Bollinger 1987). Post-breeding storage of nutrients occurs simultaneously with brood rearing and moult. Increased vigilance and attentiveness of adults toward their offspring may be at the expense of reduced foraging effort (Schindler & Lamprecht 1987; Sedinger & Raveling 1990; Sedinger et al. 1995), potentially reducing nutrient intake. The simultaneous need to grow feathers and restore nutrient reserves has the potential to reduce feather quality, rate of restoration of nutrient reserves or both. Schmutz & Ely (1999) and Eichholz (2001) each found that autumn survival was related to nutritional status, so the ability of adults to restore nutrients after hatch has implications for fitness later in the year.

Breeding success in geese can also influence events during the non-breeding season because family cohesion persists for a full year after hatch and family groups are dominant to pairs without young or singles (Raveling 1970; Prevett & MacInnes 1980; Lamprecht 1986; Poisbleau et al. 2006). Social status is related to position in foraging flocks and food intake (Boyd 1953; Gregoire & Ankney 1990; Black et al. 1992); thus, successful reproduction has the potential to influence nutrient dynamics the following winter through its effect on social status. Consistent with this hypothesis, Sedinger *et al.* (2011) showed that Black Brant *B. bernicla nigricans* that bred tended to shift to higher quality wintering areas the following winter. Similar mechanisms could occur in ducks if more "attractive" individuals pair earlier in autumn and experience enhanced social status as a result. Gregoire & Ankney (1990) also noted that family structure varied by winter habitat, so that any COEs of nutrition related to previous breeding success and resulting family size during winter on subsequent breeding may interact with the types of winter habitats occupied.

Finally, numerous studies have demonstrated COEs based on experience during their first summer for young waterfowl, especially geese. Growth conditions during the first summer influence survival during the next autumn (Owen & Black 1989; Hill *et al.* 2003; Sedinger & Chelgren 2007; Aubry *et al.* 2013) and ultimately, recruitment into the breeding population (Cooke *et al.* 1984; Sedinger *et al.* 2004). An important determinant of survival probability for goslings over the year after they fledge is phenology of hatch, both within years and between years (Cooch 2002; Slattery & Alisauskas 2002; Aubry *et al.* 2013). Slattery & Alisauskas (2002) found that dispersal distance by goslings after hatching influenced their subsequent survival probability over the next year. Dzus & Clark (1998) showed that hatch date influenced recruitment in Mallard, and Alisauskas & Kellett (2014) found a negative effect of relative initiation date of nests that produced King Eider *S. spectabilis* ducklings, with equivocal support for an additive negative influence of spring thaw date on the age that ducklings were recruited as breeders. Both studies support the notion that events early in life can influence recruitment and life-time reproductive success.

Investments in breeding by female waterfowl may result in lagged trade-offs with subsequent breeding or with other life-history traits. Viallefont *et al.* (1995) demonstrated that Lesser Snow Geese breeding for the first time were less likely to breed the following year, suggesting the existence of costs associated with these first breeding attempts. Specific brood-rearing areas used by female Black Brant influenced their likelihood of breeding the next year (Nicolai & Sedinger 2012). Assessment of breeding costs is complicated by heterogeneity in individual quality (van Noordwijk & de Jong 1986) and variation in the probability of breeding among individuals (Sedinger *et al.* 2008; Hoye *et al.* 2012). Additionally, as noted above, successful breeding by geese enhances their social status (Poisbleau *et al.* 2006), which may compensate for some of the costs of breeding (*e.g.* Sedinger *et al.* 2011). Leach & Sedinger (unpubl. data) manipulated brood size and found that removing broods can negatively affect probability of breeding the following year, but increasing brood size can also reduce breeding the following year for larger initial brood sizes. These results are generally consistent both with the social advantages of breeding in geese, and the idea that successful breeding can induce costs for future breeding.

Drake (2006) experimentally manipulated nest success of Ross's Geese *C. rossii* and found that survival of successful nesters was consistently lower than those with destroyed nests, and that the strength of this effect varied among years, consistent with COEs of successful breeding on subsequent survival. However, we note that there is a general paucity of studies of CSEs on the survival of breeding adults, and suggest that the lack of such results reflects logistical or technical difficulties in assessing the survival of adults in breeding *versus* non-breeding states. Because non-breeding waterfowl typically undergo moult migrations to areas remote from breeding habitats (Hohman *et al.* 1992), it is uncommon that samples of both breeders and non-breeders are monitored simultaneously. As illustrated by Drake (2006), this can be remedied to a certain extent by experimentation. The limited evidence for absence of evidence for CSEs on survival should, thus, be viewed as an absence of appropriate data, rather than an absence of such CSEs.

Mechanisms for cross-seasonal effects

The most direct mechanism underlying CSEs in waterfowl, especially those linking winter and spring migration areas to breeding, relates directly to nutritional state, and is, therefore, related to individual abilities to store more nutrients earlier than others (Prop *et al.* 2003; Drent *et al.* 2003). Individual variation in schedules of nutrient storage induces and governs variation in departure phenology. For example, Alisauskas (1988) found that Lesser Snow Geese constituting the northernmost vanguard of the mid-continent Lesser Snow Goose population during spring migration were fatter than conspecifics farther south at the same calendar date. Earlier arrival during migration increases the potential for protracted residency at each staging area, which may permit greater nutrient acquisition (Drent *et al.* 2003). The result is that individuals that store high levels of nutrients early at any staging area tend to maintain this advantage throughout spring migration, resulting in higher probability of early arrival onto nesting areas, earlier breeding and larger clutches (Prop *et al.* 2003; Drent *et al.* 2003; Bêty *et al.* 2003).

It is unlikely that the same nutrients stored in winter or early in migration actually contribute directly to reproduction because the substantial energy costs of long-distance migratory flight result in a major turnover of nutrients throughout spring migration. Although most assessments of the contribution of endogenous nutrients to egg production distinguish between body reserves acquired by the female before arrival on the breeding grounds (*i.e.* "capital" breeders) and food acquired locally on or near the breeding territories (for "income" breeders: Drent & Daan 1980; Gauthier *et al.* 2003; Hobson *et al.* 2004; Cutting *et al.* 2011), they do not assign a geographic origin to endogenous nutrients. As suggested by Klaassen *et al.* (2006), it is important to distinguish between distant *versus* local capital breeding (*i.e.* whether intensive feeding in preparation for breeding is a few kilometres from the breeding areas or at more distant sites), in addition to determining whether the birds use capital or income breeding strategies.

The substantial investments required for egg-laying by waterfowl create the potential for COEs, yet few examples of such COEs exist. We exclude costs of breeding, such as reduced survival, that occur during the breeding season (Madsen *et al.* 2002; Arnold *et al.* 2010) because such trade-offs, while real, do not represent COEs. Clearly, the ability of breeding adults to restore nutrients depleted during breeding may influence fitness during seasons following breeding. Several studies document a relatively slow replenishment of nutrients before the post-breeding moult (*e.g.* Ankney & MacInnes 1978; Ankney 1984; Fox *et al.* 2013). Non- and failed-breeding Black Brant gained more mass before moult than did adults with broods (Fondell *et al.* 2013), suggesting the possibility that mass gain following nesting might be constrained by competing demands of tending broods. Virtually all species lose mass during moult (Hohman *et al.* 1992), and while there is debate about whether such mass loss is adaptive (*e.g.* Fox & Kahlert 2005), adult waterfowl do not restore nutrient reserves at this time of year. Thus, investments in reproduction, combined with constraints on the ability to replenish these nutrients following breeding, have the potential to influence fitness in seasons following breeding.

Waterfowl undergo complete remigial moult and become flightless at the same time that they are restoring nutrients depleted during breeding, which has the potential to influence feather quality as well as somatic nutrient levels. Both moult location (Norris *et al.* 2007) and competing nutritional demands during moult (Norris *et al.* 2004) influence feather colour in passerines. Legagneaux *et al.* (2012) demonstrated that the iridescent colouration in Green-winged Teal was influenced by the location where individuals moulted, consistent with the hypothesis that nutrient availability influenced feather quality and perhaps attractiveness to mates. Feathers of yearling Black Brant during their second summer are highly degraded relative to those of adults (J. Sedinger pers. obs.). Because development of these feathers occurred during the growth period, when nutrient availability is typically limiting tissue production (*e.g.* Sedinger 1984; Sedinger *et al.* 2001), the poor quality of feathers on yearling geese is consistent with the notion that nutrient availability could influence feather quality, as is the case in other birds (Butler *et al.* 2008). Absence of data on feather quality and the relationship between feather quality and fitness in waterfowl have hindered assessment of linkages between feather quality and fitness (COE) or population dynamics (CSE) of waterfowl. We suggest that such research is warranted.

Immune function and associated disease status both provide mechanisms that could facilitate CSEs. Long-distance migrants appear to be more exposed, or more susceptible, to parasitic infection (Figuerola & Green 2000). Two studies demonstrate that parasite load during brood-rearing has a negative influence on the survival of young after fledging (Slattery & Alisauskas 2002; Souchay *et al.* 2013). Nematode levels were negatively associated with lipid levels for Lesser Snow Geese during spring migration (Shutler *et al.* 2012). The authors could not differentiate between the hypotheses that:

1) nematode infestation inhibited lipid deposition; 2) some individuals were chronically infected, which reduced the size of nutrient stores; and 3) individuals that invested more in lipid storage and less in immune function were more susceptible to nematode infection. The first two hypotheses would be consistent with COEs associated with disease.

Cholera *Pasturella multicida* has been known for some time to cause substantial mortality on both breeding and wintering areas (Rosen & Bischoff 1949; Wobeser *et al.* 1983). The dynamics of Cholera transmission are not well understood but evidence suggests that individuals serve as carriers (Samuel *et al.* 1999, 2004, 2005), creating the potential that exposure of individuals in one seasonal habitat can lead to disease transmission and mortality in another season, with possible population level consequences. Sub-lethal roles of disease beyond immediate mortality effects are also not well understood. Infection by low-pathogenic avian influenza has been associated with less efficient feeding and delayed migration in Bewick's Swans *Cygnus columbianus bewickii* (van Gils *et al.* 2007). Latorre-Margalef *et al.* (2009) found that Mallard infected with avian influenza had lower body mass during migration, which could influence timing of migration and investment in reproduction. The association between the timing of migration and reproductive success (see above) suggests that infection could cause lower productivity. Such sub-lethal effects are likely to be of broader importance in waterfowl, especially when individuals are concentrated because wintering or migration habitat is limited (Galsworthy *et al.* 2011; Gaidet *et al.* 2012).

Investment in reproduction can reduce immune function in waterfowl (Hanssen *et al.* 2003, 2005) and other birds (Deerenberg *et al.* 1997; Schmidt-Hempel 2003), with the potential for lower survival or reduced reproductive success in the future. Immune challenges in Mallard (post-fledging) influences colour preference and response to novel environments (Butler *et al.* 2012), as well as the ability to mount an immune response as adults (Butler & McGraw 2012). Although Norris & Evans (2000) demonstrated that investment in breeding affected immune function, which in turn influenced infection rate and reduced fitness, such studies are rare. The presence of complex sequential causative events relating individual variation in immune function to variation in fitness could translate to population level effects if pervasive. We believe studies that address such mechanisms underlying CSEs in waterfowl may be relevant to a full understanding of population dynamics.

Implications of cross-seasonal effects for management

The potential for CSEs was implicit in the development of winter habitat joint ventures in North America during the 1980s, as suggested by references to specific links between habitat and waterfowl population objectives in joint venture documents (*e.g.* Central Valley Habitat Joint Venture 1990). In some cases, specific mechanisms behind assumed CSEs were not made explicit until sometime later (Koneff

2006 unpublished letter to the North American Waterfowl Management Plan evaluation team). Establishment of the joint ventures stimulated substantial new research into dynamics of food availability in both wetlands (Naylor 2002; Hagy & Kaminski 2012a,b; Leach *et al.* 2012) and agricultural habitats (Stafford *et al.* 2010; Foster *et al.* 2010), which typically demonstrated the potential for food to become limiting by December and January (Naylor 2002; Foster *et al.* 2010). These modern studies built on earlier estimates of food abundance and carrying capacity of winter habitats for waterfowl in Europe (Ebbinge *et al.* 1975; Owen 1977) and North America (Givens *et al.* 1964). Demonstration of winter food depletion, combined with increases in many goose populations associated with increased availability of agricultural foods (Abraham *et al.* 2005; Fox *et al.* 2005; Gauthier *et al.* 2005), indicate that agroecosystems or direct management of winter habitats for waterfowl has great potential to influence dynamics of their populations. Refining winter habitat management will require improved understanding of the interplay among winter food abundance, winter population density, COEs, availability of breeding habitat and breeding population density (*e.g.* Runge & Marra 2005; Norris & Marra 2007). We contend that methods now exist to improve our understanding of all of these questions (see below).

Disease transmission increases at higher densities (Rosen & Bischoff 1949; Gaidet *et al.* 2012), so the availability of winter habitat can potentially influence rates of infection at the population level. Infection may influence survival rates (*e.g.* Slattery & Alisauskas 2002) and migration behaviour (van Gils *et al.* 2007; Latorre-Margalef *et al.* 2009). Thus, availability of winter habitat not only has the potential to influence CSEs, but habitat management on the wintering grounds could influence disease processes and so affect the dynamics of waterfowl populations.

Harvest pressure influences behaviour (Fox & Madsen 1997; Webb *et al.* 2011) and habitat use (Béchet *et al.* 2003; Moore & Black 2006), which in turn can affect nutritional status and investment in breeding, at least in geese (Mainguy *et al.* 2002). Direct effects of harvest on mortality rates and population dynamics vary as a function of body size (Rexstad 1992; Sedinger *et al.* 2007; Sedinger & Herzog 2012; Péron *et al.* 2012). Indirect effects of harvest acting through nutrient dynamics are poorly understood for ducks and most populations of geese (but see Pearse *et al.* 2012). Nevertheless, examples cited here indicate that the effects of harvest are likely to be more complex in all waterfowl than are currently assumed.

Density-dependent effects on recruitment are well established in both ducks (Kaminski & Gleussing 1987) and geese (Loonen *et al.* 1997; Lake *et al.* 2008) although mechanisms are less well understood for ducks. In geese, a substantial proportion of density-dependence occurs through food limitation during growth (Cooch *et al.* 1991b; Sedinger *et al.* 1998) and reduced survival during the first year associated with small body size (Sedinger & Chelgren 2007; Sedinger & Nicolai 2011). Thus, density-dependent effects on recruitment in geese reflect a CSE/COE effect on each year's cohort of young. A number of years of predator removal in the North American prairies has resulted in substantially increased nest success (Pieron & Rohwer 2010). Increased production of ducklings, however, has not produced any increase in local densities (Amundson *et al.* 2013). A reasonable hypothesis for the lack of a population response to increased duckling production is that increased brood densities, resulting from higher nest success, increased competition for food, and slower growth and lower post-fledging survival of ducklings, similar to mechanisms reported for geese.

Remaining issues and approaches to estimating cross-seasonal effects

Much of our understanding of CSEs is based on individual responses to an interplay between physiological states and environmental conditions (Table 1). An understanding of how individual effects translate into population level responses will require an appreciation of how population density feeds back on individual responses (Runge & Marra 2005; Norris & Marra 2007). Higher density could, for instance, induce shifts in the mean and variance of individual states (Fig. 1). An important determinant of whether individual response has relevance to population level effects is the degree to which heterogeneity exists in response to population density. For example, Lindberg *et al.* (2013) did not detect a relationship between the proportion of "low quality" individuals in Black Brant cohorts and *per capita* food abundance. Thus, it seems that populations may include a relatively fixed proportion of "low quality" individuals, incapable of response to density-dependent habitat conditions.

Assessment of most habitat joint ventures has been based on waterfowl response in the season for which a specific habitat is managed. For winter habitat joint ventures, effectiveness is often measured in predicted "use-days", based on measures of food abundance and models of nutritional requirements (*e.g.* Stafford *et al.* 2006). Alternatively, nest success is used to assess habitat programmes on breeding areas (*e.g.* Stephens *et al.* 2005). These approaches cannot assess the true value of habitat in the context of the complete annual cycle of waterfowl, however, because they do not account for CSEs, which may modulate the response of individual birds to concurrent habitat conditions. Moreover, an understanding of the role that CSEs play in the dynamics of waterfowl populations requires knowledge about proportional habitat use by specific populations at any point in the year, and how this varies over the annual cycle. Although challenging, such information would permit an assessment of the reproductive success of individuals in relation to management actions on previously used wintering areas. The situation may be reinforced by individuals that fledged in a particular breeding habitat migrating to, and also wintering in, the same habitats that had a role in their production. It is unclear how such COEs may have influenced the evolution of both breeding and winter philopatry, but the adaptiveness of occupying high quality habitats which enables pre-breeding ducks to improve their condition probably also has relevance for the survival of the new recruits to the population and the likelihood of them joining the breeding cohort in due course. Such a scenario may have played a role in the evolution of persistent family bonds in geese.

As mentioned above, spring phenology of arctic snow melt governs the timing of nesting, hatch and gosling growth for all arctic-nesting geese and also for many of the sea duck species. High plasticity in growth rates is an apparent adaptation to high variability in nutritional supply to goslings both within and among years, permitting nutritional flexibility in years of nutritional stress, although at the cost of reduced survival and adult body size when nutrients are limited during the growth period. Such intraspecific variation in body size and morphology influences not only subsequent survival, but can influence habitats and broad-scale landscapes occupied during the following winter. Alisauskas (1998) found that Lesser Snow Geese in their traditionally used coastal marsh habitats during winter were larger than conspecifics from inland agricultural landscapes associated with rice agriculture, both of which, were larger than Lesser Snow Geese wintering farther north near the Missouri River Valley of Iowa and Nebraska. Alisauskas (1998) suggested that perhaps large scale expansion of winter range associated with the correlation between body size and bill morphology may have been coupled with density-dependent morphological change on breeding areas. Thus winter range dynamics appear to result from an interplay of winter food distribution and density-dependent effects expressed on the lifelong morphology of geese from events experienced as growing goslings in the arctic.

Studies of geese have been especially successful in identifying cross-seasonal and cross-ecosystem linkages because of the relative ease of marking and observing them at multiple sites throughout the year. Ross's Geese and Snow Geese are highly gregarious, very abundant and occur at very high densities throughout the year, which poses great challenges in marking sufficient numbers so that individuals can be detected and studied. Ring recoveries provide a mechanism for linking together breeding, migration and wintering habitats (*e.g.* Alisauskas *et al*. 2011). We believe that many waterfowl scientists assume similar approaches are not possible for diving or dabbling ducks (genera *Aythya* and *Anas*) for a number of reasons, including insufficient site fidelity and the difficulty of using markers that can be detected at a distance. In our view, this is too pessimistic. Several studies of marked ducks on breeding areas demonstrated excellent success at re-encountering breeding females marked in earlier years (Sowls 1955; Arnold & Clark 1996; Anderson *et al*. 2001; Blums *et al*. 2002), indicating substantial fidelity to relatively small, well-defined breeding areas. We acknowledge it may be difficult or impossible to encounter individuals from these kinds of studies on migration or at wintering sites, as has been undertaken for geese. The emergence of new technologies should, however, improve the ability to identify links between breeding, staging and wintering areas. Stable isotopes in tissues grown during known time periods provide one mechanism for assigning individuals to habitats remote from the breeding range (Hobson *et al.* 2004; Clark *et al.* 2006), or winter areas (Mehl *et al.* 2005). This approach can take advantage of specific feather tracts moulted outside the breeding season (Smith & Sheeley 1993; Combs & Fredrickson 1996) or toe nails which are grown continuously (Clark *et al.* 2006). Geolocators, which record the timing of sunrise and sunset, to allow reconstruction of latitude and longitude, provide an alternative approach (*e.g.* Eichhorn *et al.* 2006). Both geolocators and stable isotopes require that individuals be captured but the studies cited demonstrate that large numbers of breeding females can be captured during nesting. The addition of either geolocators or stable isotope methods to traditional marking and recapture on long-term study sites could enhance substantially knowledge of CSEs for species other than geese.

Our understanding of the role of CSEs in both the dynamics of populations and fitness of individuals has increased dramatically over the past three decades. This information and new technologies are now sufficiently developed to allow more detailed assessments of the role of seasonal habitats in the dynamics of specific populations. We believe it is now possible to use modern methods in the context of CSEs to improve and prioritise management of seasonal habitats. We also believe it is now possible to incorporate events and decisions throughout the complete annual cycle into our understanding of life-history strategies. We are beyond the demonstration phase; we suggest that it is now time to move waterfowl biology forward.

Acknowledgements

Sedinger was supported by the College of Agriculture, Biotechnology and Natural Resources. His long-term studies of Black Brant have been supported by the National Science Foundation (grants OPP 92 14970, DEB 98 15383, OPP 99 85931, OPP 01 96406, DEB 0743152, DEB 12 52656), Alaska Science Center, U.S. Geological Survey, Ducks Unlimited, the Black Brant Group (Morro Bay, California), Phil Jebbia (in memory of Marnie Shepherd), and Migratory Bird Management (Region 7) of the U.S. Fish and Wildlife Service. Alisauskas' work has been supported by the California Department of Fish & Game, the Polar Continental Shelf Project, Delta Waterfowl, Ducks Unlimited Canada, the University of Saskatchewan, the Central and Mississippi Flyway Councils, Environment Canada and the U.S. Fish and Wildlife Service. The manuscript benefitted from comments by D. Koons and C. Nicolai.

References

Abraham, K.F., Jefferies, R.L. & Alisauskas, R.T. 2005. The dynamics of landscape change and snow geese in mid-continent North America. *Global Change Biology* 11: 841–855.

Afton, A.D. & Paulus, S.L. 1992. Incubation and brood care. *In* B.D.J. Batt, A.D. Afton, M.G. Anderson, C.D. Ankney, D.H. Johnson, J.A. Kadlec & G.L. Krapu (eds.), *Ecology and Management of Breeding Waterfowl*, pp. 62–108. University of Minnesota Press, Minneapolis, Minnesota, USA.

Alisauskas, R.T. 1988. Nutrient reserves of Lesser Snow Geese during winter and spring migration. Ph.D. thesis, University of Western Ontario, London, Canada.

Alisauskas, R.T. 1998. Winter range expansion and relationships between landscapes and morphometrics of midcontinent Lesser Snow Geese. *Auk* 115: 851–862.

Alisauskas, R.T. 2002. Arctic climate, spring nutrition, and recruitment in midcontinent lesser snow geese. *Journal of Wildlife Management* 66: 181–193.

Alisauskas, R.T. & Ankney, C.D. 1992. The cost of egg laying and its relationship to nutrient reserves in waterfowl. *In* B.D.J. Batt, A.D. Afton, M.G. Anderson, C.D. Ankney, D.H. Johnson, J.A. Kadlec, & G.L. Krapu (eds.), *Ecology and Management of Breeding Waterfowl*, pp. 30–61. University of Minnesota Press, Minneapolis, Minnesota, USA.

Alisauskas, R.T. & DeVink, J.-M. 2015. Dealing with deficits: breeding costs, nutrient reserves and cross-seasonal effects in seaducks. *In* D.V. Derksen, J.-P. Savard, J. Eadie & D. Esler (eds.), *Ecology and Conservation of North American Sea Ducks*, pp. 000–000. *Studies in Avian Biology* No. 00, Cooper Ornithological Society, c/o Ornithological Societies of North America, Waco, Texas, USA.

Alisauskas, R.T. & Kellett, D.K. 2014. Age-specific *in situ* recruitment of female King Eiders estimated with mark-recapture. *Auk* 131: 129–140.

Alisauskas, R.T., Rockwell, R.F., Dufour, K.W., Cooch, E.G., Zimmerman, G., Drake, K.L., Leafloor, J.O., Moser, T.J. & Reed, E.T. 2011. Harvest, survival and abundance of midcontinent lesser snow geese relative to population reduction efforts. *Wildlife Monographs* 179: 1–42.

Amundson, C.T., Pieron, M.R., Arnold, T.W. & Beaudoin, L.A. 2013. The effects of predator removal on mallard production and population change in northeastern North Dakota. *Journal of Wildlife Management* 77: 143–152.

Anderson, M.G. & Batt, B.D.J. 1983. Workshop on the ecology of wintering waterfowl. *Wildlife Society Bulletin* 11: 22–24.

Anderson, M.G., Lindberg, M.S. & Emery, R.B. 2001. Probability of survival and breeding for juvenile female Canvasbacks. *Journal of Wildlife Management* 65: 385–397.

Ankney, C.D. & Alisauskas R.T. 1991. Nutrient-reserve dynamics and diet of breeding female Gadwalls. *Condor* 93: 799–810.

Ankney, C.D. & MacInnes, C.D. 1978. Nutrient reserves and reproductive performance of female Lesser Snow Geese. *Auk* 95: 459–471.

Arnold, T.W. & Clark, R.G. 1996. Survival and philopatry of female dabbling ducks in southcentral Saskatchewan. *Journal of Wildlife Management* 65: 385–397.

Arnold, T.W., Roche, E.A., Devries J.H. & Howerter, D.W. 2010. Costs of reproduction in breeding female mallards: predation risk during incubation drives annual mortality. *Avian Conservation and Ecology* 7: 1.

Aubry, L.M., Rockwell, R.F., Cooch, E.G., Brooks, R.W., Mulder, C.P.H. & Koons, D.N. 2013. Climate change, phenology, and habitat degradation: drivers of gosling body condition and juvenile survival in lesser snow geese. *Global Change Biology* 19: 149–160.

Barboza, P.S. & Jorde, D.G. 2002. Intermittent fasting during winter and spring affects body composition and reproduction of a migratory duck. *Journal of Comparative Physiology B* 172: 419–434.

Béchet, A., Giroux, J.-F., Gauthier, G., Nichols, J.D. & Hines, J.E. 2003. Spring hunting changes the regional movements of migrating greater snow geese. *Journal of Applied Ecology* 40: 553–564.

Bêty, J., Gauthier, G. & Giroux, J.-F. 2003. Body condition, migration, and timing of reproduction in snow geese: a test of the condition-dependent model of optimal clutch size. *American Naturalist* 162: 110–121.

Black, J.M., Carbon, C., Wells, R.L. & Owen, M. 1992. Foraging dynamics in goose flocks: the cost of living on the edge. *Animal Behaviour* 44: 41–50.

Blums, P., Nichols, J.D., Hines, J.E. & Mednis, A. 2002. Sources of variation in survival and breeding site fidelity in three species of European ducks. *Journal of Animal Ecology* 71: 438–450.

Boyd, H. 1953. On encounters between wild white-fronted geese in winter flocks. *Behaviour* 5: 85–129.

Boyd, H., Smith, G.E.J. & Cooch, F.G. 1982. The Lesser Snow Geese of the eastern Canadian arctic. *Canadian Wildlife Service Occasional Paper* No. 46. Canadian Wildlife Service, Ottawa, Canada.

Butler, L.K., Rohwer, S. & Speidel, M.G. 2008. Quantifying structural variation in contour feathers to address functional variation and life history trade-offs. *Journal of Avian Biology* 39: 629–639.

Butler, M.W. & McGraw, K.J. 2012. Developmental immune history affects adult immune function but not carotenoid-based ornamentation in mallard ducks. *Functional Ecology* 26: 406–415.

Butler, M.W., Toomey, M.B., McGraw, K.J. & Rowe, M. 2012. Ontogenetic immune challenges shape adult personality in mallard ducks. *Proceedings of the Royal Society B* 279: 326–333.

Central Valley Habitat Joint Venture. 1990. Central Valley Habitat Joint Venture Implementation Plan. Sacramento, USA. Accessible at http://www.centralvalleyjointventure.org/assets/pdf/cvjv_implementation_plan.pdf (last accessed on 15 March 2014).

Clark, R.G., Hobson, K.A. & Wassenaar, L.I. 2006. Geographic variation in the isotopic (dD, $d^{13}C$, $d^{15}N$, $d^{34}S$) composition of feathers and claws from lesser scaup and northern pintail: implications for the studies of migratory connectivity. *Canadian Journal of Zoology* 84: 1395–1401.

Combs, D.L. & Fredrickson, L.H. 1995. Molt chronology of male Mallards wintering in Missouri. *Wilson Bulletin* 107: 359–365.

Cooch, E.G. 2002. Fledging size and survival in snow geese: timing is everything (or is it?). *Journal of Applied Statistics* 29: 143–162.

Cooch, E.G., Lank, D.B., Dzubin, A., Rockwell, R.F. & Cooke, F. 1991a. Body size variation in Lesser Snow Geese: seasonal variation in gosling growth rate. *Ecology* 72: 503–512.

Cooch, E.G., Lank, D.B., Rockwell, R.F. & Cooke, F. 1991b. Long-term decline in body size in a snow goose population: evidence of environmental degradation? *Journal of Animal Ecology* 60: 483–496.

Cooke, F., Findlay, C.S. & Rockwell, R.F. 1984. Recruitment and the timing of reproduction in Lesser Snow Geese (*Chen caerulescens caerulescens*). *Auk* 101: 451–458.

Cutting, K.A., Hobson, K.A., Rotella, J.J., Warren, J.M., Wainwright-de la Cruz, S.E. & Takekawa, J.Y. 2011. Endogenous contributions to egg protein formation in lesser scaup *Aythya affinis*. *Journal of Avian Biology* 42: 505–513.

Daan, S., Deerenberg, C. & Dijkstra, C. 1996. Increased daily work precipitates natural death in the kestrel. *Journal of Animal Ecology* 65: 539–544.

Davies, J.C. & Cooke, F. 1983. Annual nesting productivity in snow geese: prairie droughts and arctic springs. *Journal of Wildlife Management* 47: 291–296.

Deerenberg, C., Apanius, V., Daan, S. & Bos, N. 1997. Reproductive effort decreases antibody responsiveness. *Proceedings of the Royal Society B* 264: 1021–1029.

Descamps, S., Yoccoz, N.G., Gaillard, J.-M., Gilchrist, H.G., K. Erikstad, K., Hanssen,

S.A., Cazelles, B., Forbes, M.R. & Bêty, J. 2010. Detecting population heterogeneity in effects of North Atlantic Oscillations on seabird body condition: get into the rhythm. *Oikos* 119: 1526–1536.

Devries, J.H., Brook, R.W., Howerter, D.W. & Anderson, M.G. 2008. Effects of spring body condition and age on reproduction in Mallards (*Anas platyrhynchos*). *Auk* 125: 618–628.

Drake, K.L. 2006. The role of dispersal in population dynamics of breeding Ross's Geese. Ph.D. thesis, University of Saskatchewan, Saskatchewan, Canada.

Drent, R.H. & Daan, S. 1980. The prudent parent: energetic adjustments in avian breeding. *Ardea* 68: 225–252.

Drent, R., Both, C., Green, M., Madsen, J. & Piersma, T. 2003. Pay-offs and penalties of competing migratory schedules. *Oikos* 103: 274–292.

Dzus, E.H. & Clark, R.G. 1998. Brood survival and recruitment of Mallards in relation to wetland density and hatching date. *Auk* 115: 311–318.

Ebbinge, B.S., Canters, K. & Drent, R. 1975. Foraging routines and estimated daily food intake in Barnacle Geese wintering in the northern Netherlands. *Wildfowl* 26: 5–19.

Ebbinge, B.S. & Spaans, B. 1995. The importance of body reserves accumulated in spring staging areas in the temperate zone for breeding in dark bellied brent geese *Branta bernicla bernicla* in the high arctic. *Journal of Avian Biology* 26: 105–113.

Eichholz, M.W. 2001. The implications of agriculture in interior Alaska for population dynamics of Canada Geese. Ph.D. thesis, University of Alaska Fairbanks, Alaska, USA.

Eichhorn, G., Afanasyev, V., Drent, R.H. & van der Jeugd, H.P. 2006. Spring stopover routines in Russian Barnacle Geese *Branta leucopsis* tracked by resightings and geolocation. *Ardea* 94: 667–678.

Féret, M., Gauthier, G., Béchet, A., Giroux, J.-F. & Hobson, K.A. 2003. Effect of a spring hunt on nutrient storage by greater snow geese in southern Quebec. *Journal of Wildlife Management* 67: 796–807.

Figuerola, J. & Green, A.J. 2000. Haematozoan parasites and migratory behavior in waterfowl. *Evolutionary Ecology* 14: 143–153.

Fleskes, J.P., Yee, J.L., Yarris, G.S., Miller, M.R. & Casazza, M.L. 2007. Pintail and mallard survival in California relative to habitat, abundance, and hunting. *Journal of Wildlife Management* 71: 2238–2248.

Fondell, T.F., Flint, P.L., Schmutz, J.A., Schamber, J.L. & Nicolai, C.A. 2013. Variation in body mass dynamics among sites in Black Brant *Branta bernicla nigricans* supports adaptivity of mass loss during moult. *Ibis* 155: 593–604.

Foster, M.A., Gray, M.J. & Kaminski, R.M. 2010. Agricultural seed biomass for migrating and wintering waterfowl in the southeastern United States. *Journal of Wildlife Management* 74: 489–495.

Fox, A.D. & Kahlert, J. 2005. Changes in body mass and organ size during wing moult in non-breeding greylag geese *Anser anser*. *Journal of Avian Biology* 36: 538–548.

Fox, A.D. & Madsen, J. 1997. Behavioural and distributional effects of hunting disturbance on waterbirds in Europe: implications for refuge design. *Journal of Applied Ecology* 34: 1–13.

Fox, A.D., Madsen, J., Boyd, H., Kuijken, E., Norriss, D.W., Tombre, I.M. & Stroud, D.A. 2005. Effects of agricultural change on abundance, fitness components and distribution of two arctic-nesting goose populations. *Global Change Biology* 11: 881–893.

Fox, A.D., King, R. & Owen, M. 2013. Wing moult and mass change in free-living mallard *Anas platyrhynchos*. *Journal of Avian Biology* 44: 1–8.

Gaidet, N., Caron, A., Cappelle, J., Cumming, G.S., Balanç, G., Hammoumi, S., Cattoli, G.,

Abolnik, C., Servan de Almeida, R., Gil, P., Fereidouni, S.R., Grosbois, V., Tran, A., Mundava, J., Fofana, B., Ould El Mamy, A.B., Ndlovu, M., Mondain-Monval, J.-Y., Triplet, P., Hagemeijer, W., Karesh, W.B., Newman, S.H. & Dodman, T. 2012. Understanding the ecological drivers of avian influenza virus infection in wildfowl: a continental-scale study across Africa. *Proceedings of the Royal Society B* 279: 1131–1141.

Galsworthy, S.J., ten Bosch, Q.A., Hoye, B.J., Heesterbeek, J.A.P., Klaassen, M. & Klinkenberg, D. 2011. Effects of infection-induced migration delays on the epidemiology of avian influenza in wild mallard populations. *PLoS ONE* 6: e26118.

Gauthier, G., Bèty, J. & Hobson, K.A. 2003. Are Greater Snow Geese capital breeders? New evidence from a stable-isotope model. *Ecology* 84: 3250–3264.

Gauthier, G., Giroux, J.-F., Reed, A., Béchet, A. & Bélanger, L. 2005. Interactions between land use, habitat use and population increase in greater snow geese: what are the consequences for natural wetlands? *Global Change Biology* 11: 856–868.

Givens, L.S., Nelson, M.C. & Ekedahl, V. 1964. Farming for waterfowl. *In* J.P. Linduska (ed.), *Waterfowl tomorrow*, pp. 599–610. U.S. Department of Interior, Washington, USA.

Gregoire, P.E. & Ankney, C.D. 1990. Agonistic behavior and dominance relationships among Lesser Snow Geese during winter and spring migration. *Auk* 107: 550–560.

Hagy, H.M. & Kaminski, R.M. 2012a. Apparent seed use by ducks in moist-soil wetlands of the Mississippi Alluvial Valley. *Journal of Wildlife Management* 76: 1053–1061.

Hagy, H.M. & Kaminski, R.M. 2012b. Winter waterbird and food dynamics in autumn-managed moist-soil wetlands in the Mississippi alluvial valley. *Wildlife Society Bulletin* 36: 512–523.

Hanssen, S. A., Folstad, I. & Erikstad, K.E. 2003. Reduced immunocompetence and cost of reproduction in common eiders. *Oecologia* 136: 457–464.

Hanssen, S.A., Hasselquist, D., Folstad, I. & Erikstad, K.E. 2005. Cost of reproduction in a long-lived bird: incubation effort reduces immune function and future reproduction. *Proceedings of the Royal Society B* 272: 1039–1046.

Harrison, X.A., Blount, J.D., Inger, R., Norris, D.R. & Bearhop, S. 2011. Carry-over effects as drivers of fitness differences in animals. *Journal of Animal Ecology* 80: 4–18.

Heitmeyer, M. & Fredrickson, L.H. 1981. Do wetland conditions in the Mississippi Delta hardwoods influence mallard recruitment? *Transactions of the North American Wildlife and Natural Resources Conference* 46: 44–57.

Hill, M.R.J., Alisauskas, R.T., Ankney, C.D. & Leafloor, J.O. 2003. Influence of body size and condition on harvest and survival of juvenile Canada geese. *Journal of Wildlife Management* 67: 530–541.

Hobson, K.A., Atwell, L., Wassenaar, L.I. & Yerkes, T. 2004. Estimating endogenous nutrient allocations to reproduction in redhead ducks: a dual isotope approach using δD and $\delta^{13}C$ measurements of female and egg tissues. *Functional Ecology* 18: 737–745.

Hobson, K.A., Sharp, C.M., Jefferies, R.L., Rockwell, R.F. & Abraham, K.F. 2011. Nutrient allocation strategies to eggs by Lesser Snow Geese (*Chen caerulescens*) at a sub-arctic colony. *Auk* 128: 156–165.

Hohman, W.L., Ankney, C.D. & Gordon, D.H. 1992. Ecology and management of postbreeding waterfowl. *In* B.D.J. Batt, A.D. Afton, M.G. Anderson, C.D. Ankney, D.H. Johnson, J.A. Kadlec & G.L. Krapu (eds.), *Ecology and Management of Breeding Waterfowl*, pp. 128–189. University of Minnesota Press, Minneapolis, Minnesota, USA.

Hoye, B.J., Hahn, S., Nolet, B.A. & Klaassen, M. 2012. Habitat use throughout migration: linking individual consistency, prior breeding success and future breeding potential. *Journal of Animal Ecology* 81: 657–666.

Johnson, D.H., Nichols, J.D. & Schwartz, M.D. 1992. Population dynamics of breeding waterfowl. In B.D.J. Batt, A.D. Afton, M.G. Anderson, C.D. Ankney, D.H. Johnson, J.A. Kadlec & G.L. Krapu (eds.), *Ecology and Management of Breeding Waterfowl*, pp. 446–485. University of Minnesota Press, Minneapolis, Minnesota, USA.

Johnson, F.A., Moore, C.T., Kendall, W.L., Dubovsky, J.A., Caithamer, D.F., Kelley Jr., J.T. & Williams, B.K. 1997. Uncertainty and the management of mallard harvests. *Journal of Wildlife Management* 61: 203–217.

Jónsson, J.E., Gardarsson, A., Gill, J.A., Petersen, A. & Gunnarsson, T.G. 2009. Seasonal weather effects on the Common Eider, a subarctic capital breeder, in Iceland over 55 years. *Climate Research* 38: 237–248.

Kaminski, R.M. & Gluesing, E.A. 1987. Density- and habitat-related recruitment in mallards. *Journal of Wildlife Management* 51: 141–148.

Klaassen, M., Abraham, K.E., Jefferies, R.L. & Vrtiska, M. 2006. Factors affecting the site of investment, and the reliance on savings for arctic breeders: the capital–income dichotomy revisited. *Ardea* 94: 371–384.

Kleiber, M. 1975. *The Fire of Life an Introduction to Animal Energetics*. Robert E. Krieger Publishing Co., Huntington, New York, USA.

Lake, B.C., Schmutz, J.A., Lindberg, M.S., Ely, C.R., Eldridge, W.D. & Broerman, F.J. 2008. Body mass of prefledging Emperor Geese *Chen canagica*: large-scale effects of interspecific densities and food availability. *Ibis* 150: 527–540.

Lamprecht, J. 1986. Structure and causation of the dominance hierarchy in a flock of Bar-headed Geese (*Anser indicus*). *Behaviour* 96: 28–48.

Latorre-Margalef, N., Gunnarsson, G., Munster, V.J., Fouchier, R.A.M., Osterhaus, A.D.M.E., Elmberg, J., Olsen, B., Wallensten, A., Haemig, P.D., Fransson, T., Brudin, L. & Waldenström, J. 2009. Effects of influenza A virus infection on migrating mallard ducks. *Proceedings of the Royal Society B* 276: 1029–1036.

Leach, A.G., Straub, J.N., Kaminski, R.M., Ezell, A.W., Hawkins, T.S. & Leininger, T.D. 2012. Apparent seed use by ducks in moist-soil wetlands of the Mississippi Alluvial Valley. *Journal of Wildlife Management* 76: 1519–1522.

Legagneux, P., Clark, R.G., Guillemain, M., Eraud, C., Théry, M. & Bretagnolle, V. 2012. Large-scale geographic variation in iridescent structural ornaments of a long-distance migratory bird. *Journal of Avian Biology* 43: 355–361.

Lehikoinen, A., Kilpi, M. & Öst, M. 2006. Winter climate affects subsequent breeding success of common eiders. *Global Change Biology* 12: 1355–1365.

Loonen M.J.J.E., Oosterbeek, K. & Drent, R.H. 1997. Variation in growth of young and adult size in Barnacle Geese *Branta leucopsis*: evidence for density dependence. *Ardea* 85: 177–192.

Lynch, J.J. 1952. Escape from mediocrity. Unpublished memorandum, U. S. Fish and Wildlife Service, Abbeville Louisiana, USA.

MacKinney, F. 1986. Ecological factors influencing the social systems of migratory dabbling ducks. In D.I. Rubenstein & R.W. Wrangham (eds.), *Ecological Aspects of Social Evolution*, pp. 153–174, Princeton University Press, Princeton, New Jersey, USA.

Madsen, J. 2001. Spring migration strategies in Pink-footed Geese *Anser brachyrhynchus* and consequences for spring fattening and fecundity. *Ardea* 89: 43–55.

Madsen, J., Frederiksen, M. & Ganter, B. 2002. Trends in annual and seasonal survival of

Pink-footed Geese *Anser brachyrhynchus*. *Ibis* 144: 218–226.

Mainguy, J., Bêty, J., Gauthier, G. & Giroux, J.-F. 2002. Are body condition and reproductive effort of laying Greater Snow Geese affected by the spring hunt? *Condor* 104: 156–161.

Marra, P.P., Hobson, K.A. & Holmes, R.T. 1998. Linking winter and summer events in a migratory bird by using stable-carbon isotopes. *Science* 282: 1884–1886.

Mehl, K.R., Alisauskas, R.T., Hobson, K.A. & Kellett, D.K. 2004. To winter east or west? Heterogeneity in winter philopatry in a central-arctic population of King Eiders. *Condor* 106: 241–251.

Mehl, K.R., Alisauskas, R.T., Hobson, K.A. & Merkel, F.R. 2005. Linking breeding and wintering areas of king eiders: making use of polar isotopic gradients. *Journal of Wildlife Management* 63: 1297–1404.

Moore, J.E. & Black, J.M. 2006. Historical changes in black brant *Branta bernicla nigricans* use on Humboldt Bay, California. *Wildlife Biology* 12: 151–162.

Naylor, L.W. 2002. Evaluating moist-soil seed production and management in Central Valley wetlands to determine habitat needs for waterfowl. M.Sc. thesis, University of California Davis, Davis, California, USA.

Newton, I. 2006. Can conditions experienced during migration limit the population levels of birds? *Journal of Ornithology* 147: 146–166.

Nicolai, C.A. & Sedinger, J.S. 2012. Trade-offs between offspring fitness and future reproduction of adult female black brent. *Journal of Animal Ecology* 81: 798–805.

Norris, D.R. 2005. Carry-over effects and habitat quality in migratory populations. *Oikos* 109: 178–186.

Norris, D.R. & Marra, P.P. 2007. Seasonal interactions, habitat quality, and population dynamics in migratory birds. *Condor* 109: 535–547.

Norris, D.R., Marra, P.P., Kyser, T.K., Ratcliffe, L.M. & Montgomerie, R. 2007. Continent-wide variation in feather colour of a migratory songbird in relation to body condition and moulting locality. *Biology Letters* 3: 16–19.

Norris, D.R., Marra, P.P., Montgomerie, R., Kyser, T.K. & Ratcliffe, L.M. 2004. Reproductive effort, molting latitude and feather color in a migratory songbird. *Science* 306: 2249–2250.

Norris, K. & Evans, M.R. 2000. Ecological immunology: life history trade-offs and immune defense in birds. *Behavioral Ecology* 11: 19–26.

Oppel, S., Powell, A.N. & O'Brien, D.M. 2010. King eiders use an income strategy for egg production: a case study for incorporating individual dietary variation into nutrient allocation research. *Oecologia* 164: 1–12.

Owen, M. 1977. The role of wildlife refuges on agricultural land in lessening the conflict between farmers and geese in Britain. *Biological Conservation* 11: 209–222.

Owen, M. & Black, J.M. 1989. Factors affecting the survival of barnacle geese on migration from the breeding grounds. *Journal of Animal Ecology* 58: 603–617.

Pearse, A.T., Krapu, G. L. & Cox, R.R. Jr. 2012. Spring snow goose hunting influences body composition of waterfowl staging in Nebraska. *Journal of Wildlife Management* 76: 1393–1400.

Péron, G., Nicolai, C.A. & Koons, D.N. 2012. Demographic response to perturbations: the role of compensatory density dependence in a North American duck under variable harvest regulations and changing habitat. *Journal of Animal Ecology* 81: 960–969.

Pieron, M.R. & Rohwer, F.C. 2010. Effects of large-scale predator reduction on nest success of nesting ducks. *Journal of Wildlife Management* 74: 124–132.

Poisbleau, M., Fritz, H., Valeix, M., Perroi, P.-Y., Dalloyau, S. & Lambrechts, M.M. 2006. Social dominance correlates and family status in wintering dark-bellied brent geese, *Branta bernicla bernicla*. *Animal Behaviour* 71: 1351–1358.

Pospahala, R.S., Anderson, D.R. & Henny, C.J. 1974. *Population Ecology of the Mallard: Breeding Habitat Conditions, Size of the Breeding Populations, and Production Indices.* U.S. Department of the Interior, U.S. Fish & Wildlife Service, Washington DC, USA.

Prevett, J.P. & MacInnes, C.D. 1980. Family and other social groups in snow geese. *Wildlife Monographs* 71: 1–46.

Prop, J., Black, J.M. & Shimmings, P. 2003. Travel schedules to the high arctic: Barnacle Geese trade-off the timing of migration with accumulation of fat deposits. *Oikos* 103: 403–414.

Raveling, D.G. 1970. Dominance relationships and agonistic behavior of Canada geese in winter. *Behaviour* 37: 291–319.

Raveling, D.G. 1979. The annual cycle of body composition of Canada Geese with special reference to control of reproduction. *Auk* 96: 234–252.

Raveling, D.G. & Heitmeyer, M.E. 1989. Relationships of population size and recruitment of pintails to habitat conditions and harvest. *Journal of Wildlife Management* 53: 1088–1103.

Rexstad, E.A. 1992. Effect of hunting on annual survival of Canada geese in Utah. *Journal of Wildlife Management* 56: 297–305.

Rosen, M.N. & Bischoff, A.I. 1949. The 1948–49 outbreak of fowl cholera in birds in the San Francisco Bay area and surrounding counties. *California Fish and Game* 35: 185–192.

Runge, M.C. & Marra, P.P. 2005. Modeling seasonal interactions in the population dynamics of migratory birds. *In* R. Greenburg & P.P. Marra (eds.), *Birds of Two Worlds the Ecology and Evolution of Migration*, pp. 375–389, Johns Hopkins University Press, Baltimore, Maryland, USA.

Samuel, M.D., Takekawa, J.Y., Samelius, G. & Goldberg, D.R. 1999. Avian cholera mortality in lesser snow geese nesting on Banks Island, Northwest Territories. *Wildlife Society Bulletin* 27: 780–787.

Samuel, M.D., Shadduck, D.J. & Goldberg, D.R. 2004. Are wetlands the reservoir for avian cholera? *Journal of Wildlife Diseases* 40: 377–382.

Samuel, M.D., Shadduck, D.J., Goldberg, D.R. & Johnson, W.P. 2005. Avian cholera in waterfowl: the role of lesser snow and Ross's geese as disease carriers in the Playa Lakes region. *Journal of Wildlife Diseases* 41: 48–57.

Schindler, M. & Lamprecht, J. 1987. Increase of parental effort with brood size in a nidifigous bird. *Auk* 104: 688–693.

Schmidt-Hempel, P. 2003. Variation in immune defence as a question of evolutionary ecology. *Proceedings of the Royal Society B* 270: 357–366.

Schmutz, J.A. & Ely, C.R. 1999. Survival of greater white-fronted geese: effects of year, season, sex and body condition. *Journal of Wildlife Management* 63: 1239–1249.

Scott, D.K. 1980. Functional aspects of prolonged parental care in Bewick's Swans. *Animal Behaviour* 28: 938–952.

Sedinger, J.S. 1984. Protein and amino acid composition of tundra vegetation in relation to nutritional requirements of geese. *Journal of Wildlife Management* 48: 1128–1136.

Sedinger, J.S. & Bollinger, K.S. 1987. Autumn staging of Cackling Canada Geese on the Alaska Peninsula. *Wildfowl* 38: 13–18.

Sedinger, J.S. & Chelgren, N.D. 2007. Survival and breeding advantages of larger Black Brant goslings: within and among cohort variation. *Auk* 124: 1281–1293.

Sedinger, J.S. & Flint, P.L. 1991. Growth rate is negatively correlated with hatch date in Black Brant. *Ecology* 72: 496–502.

Sedinger, J.S. & Herzog, M.P. 2012. Harvest and dynamics of duck populations. *Journal of Wildlife Management* 76: 1108–1116.

Sedinger, J.S. & Raveling, D.G. 1990. Parental behavior of cackling Canada Geese during broodrearing: division of labor within pairs. *Condor* 92: 174–181.

Sedinger, J.S., Flint, P.L. & Lindberg, M.S. 1995. Environmental influence on life-history traits: growth, survival and fecundity in Black Brant (*Branta bernicla*). *Ecology* 76: 2404–2414.

Sedinger, J.S., Lindberg, M.S., Person, B.T., Eichholz, M.W., Herzog, M.P. & Flint, P.L. 1998. Density dependent effects on growth, body size and clutch size in Black Brant. *Auk* 115: 613–620.

Sedinger, J.S., Herzog, M.P., Person, B.T., Kirk, M.T., Obritchkewitch, T., Martin, P.P. & Stickney, A.A. 2001. Large-scale variation in growth of Black Brant goslings related to food availability. *Auk* 118: 1088–1095.

Sedinger, J.S., Herzog, M.P. & Ward, D.H. 2004. Early environment and recruitment of Black Brant into the breeding population. *Auk* 121: 68–73.

Sedinger, J.S., Ward, D.H., Schamber, J.L., Butler, W.I., Eldridge, W.D., Conant, B., Voelzer, J.F., Chelgren, N.D. & Herzog, M.P. 2006. Effects of El Niño on distribution and reproductive performance of Black Brant. *Ecology* 87: 151–159.

Sedinger, J.S., Chelgren, N.D., Lindberg, M.S. & Ward, D.H. 2008. Fidelity and breeding probability related to population density and individual quality in black brent geese (*Branta bernicla nigricans*). *Journal of Animal Ecology* 77: 702–712.

Sedinger, J.S., Schamber, J.L., Ward, D.H., Nicolai, C.A. & Conant, B. 2011. Carryover effects associated with winter location affect fitness, social status, and population dynamics in a long distance migrant. *American Naturalist* 178: E110–E123.

Sharp, C.M., Abraham, K.F., Hobson, K.A. & Burness, G. 2013. Allocation of nutrients to reproduction at high latitudes: insights from two species of sympatrically nesting geese. *Auk* 130: 171–179.

Shutler, D., Alisauskas, R.T. & McLaughlin, J.D. 2012. Associations between body composition and helminths of lesser snow geese during winter and spring migration. *International Journal for Parasitology* 42: 755–760.

Slattery, S.M. & Alisauskas, R.T. 2002. Use of the Barker model in an experiment examining covariate effects on first-year survival in Ross's geese (*Chen rossii*): a case study. *Journal of Applied Statistics* 29: 497–508.

Smith, L.M. & Sheeley, D.G. 1993. Molt patterns of wintering northern pintails in the southern high plains. *Journal of Wildlife Management* 57: 229–238.

Souchay, G., Gauthier, G. & Pradel, R. 2013. Temporal variation of juvenile survival in a long-lived species: the role of parasites and body condition. *Oecologia* 173: 151–160.

Sowls, L. K. 1955. *Prairie Ducks: a Study of their Behavior, Ecology, and Management*. University of Nebraska Press, Lincoln, Nebraska, USA.

Spaans, B., van't Hoff, C.A., van der Veer, W. & Ebbinge, B.S. 2007. The significance of female body stores for egg laying and incubation in Dark-bellied Brent Geese *Branta bernicla bernicla*. *Ardea* 95: 3–15.

Stafford, J.D., Kaminski, R.M., Reinecke, K.J. & Manley, S.W. 2006. Waste rice for waterfowl in the Mississippi Alluvial Valley. *Journal of Wildlife Management* 70: 61–69.

Stafford, J.D., Kaminski, R.M. & Reinecke, K.J. 2010. Avian foods, foraging and habitat conservation in world rice fields. *Waterbirds* 33: 133–150.

Stephens, S.E., Rotella, J.J., Lindberg, M.S., Taper, M.L. & Ringelman, J.K. 2005. Duck nest survival in the Missouri Coteau of North

Dakota: landscape effects at multiple spatial scales. *Ecological Applications* 15: 2137–2149.

Tombre, I.M., Erikstad, K.E. & Bunnes, V. 2012. State-dependent incubation behaviour in the high arctic barnacle geese. *Polar Biology* 35: 985–992.

Trinder, M.N., Hassell, D. & Votier, S. 2009. Reproductive performance in arctic-nesting geese is influenced by environmental conditions during the wintering, breeding and migration seasons. *Oikos* 118: 1093–1101.

van Gils, J.A., Munster, V.J., Radersma, R., Liefhebber, D., Uchier, R.A. & Klaassen, M. 2007. Hampered foraging and migratory performance in swans infected with low pathogenic avian influenza A virus. *PLoS ONE* 2: e184.

Van Noordwijk, A.J. & De Jong, G. 1986. Acquisition and allocation of resources: their influence on variation in life-history traits. *American Naturalist* 128: 137–142.

Viallefont, A., Cooke, F. & Lebreton, J.-D. 1995. Age-specific costs of first-time breeding. *Auk* 112: 67–76.

Webb, E.B., Smith, L.M., Vrtiska, M.P. & Lagrange, T.G. 2011. Factors influencing behavior of wetland birds in the Rainwater Basin during spring migration. *Waterbirds* 34: 457–467.

Wobeser, G., Kerbes, R. & Byersbergen, G.W. 1983. Avian cholera in Ross' and lesser snow geese in Canada. *Journal of Wildlife Diseases* 19: 12.

Photograph: Mallard benefitting from wetlands in Maryland, by Ron Nichols/USDA Natural Resources Conservation Service.

Managing harvest and habitat as integrated components

ERIK E. OSNAS[1]*, MICHAEL C. RUNGE[1], BRADY J. MATTSSON[2], JANE AUSTIN[3], G. SCOTT BOOMER[4], ROBERT G. CLARK[5], PATRICK DEVERS[4], JOHN M. EADIE[6], ERIC V. LONSDORF[7] & BRIAN G. TAVERNIA[1]

[1]U.S. Geological Survey, Patuxent Wildlife Research Center, Laurel, Maryland, USA.
[2]University of Natural Resources and Life Sciences, Vienna, Austria.
[3]U.S. Geological Survey, Northern Prairie Wildlife Research Center, Jamestown, North Dakota, USA.
[4]U.S. Fish and Wildlife Service, Population and Habitat Assessment Branch, Laurel, Maryland, USA.
[5]Environment Canada, Prairie and Northern Wildlife Research Center, Saskatoon, Saskatchewan, Canada.
[6]University of California, Davis, California, USA.
[7]Franklin and Marshall College, Lancaster, Pennsylvania, USA.
*Correspondence author. E–mail: erik.osnas@gmail.com

Abstract

In 2007, several important initiatives in the North American waterfowl management community called for an integrated approach to habitat and harvest management. The essence of the call for integration is that harvest and habitat management affect the same resources, yet exist as separate endeavours with very different regulatory contexts. A common modelling framework could help these management streams to better understand their mutual effects. Particularly, how does successful habitat management increase harvest potential? Also, how do regional habitat programmes and large-scale harvest strategies affect continental population sizes (a metric used to express habitat goals)? In the ensuing five years, several projects took on different aspects of these challenges. While all of these projects are still on-going, and are not yet sufficiently developed to produce guidance for management decisions, they have been influential in expanding the dialogue and producing some important emerging lessons. The first lesson has been that one of the more difficult aspects of integration is not the integration across decision contexts, but the integration across spatial and temporal scales. Habitat management occurs at local and regional scales. Harvest management decisions are made at a continental scale. How do these actions, taken at different scales, combine to influence waterfowl population dynamics at all scales? The second lesson has been that consideration of the interface of habitat and harvest management can generate important insights into the objectives underlying the decision context. Often the objectives are very complex and trade-off against one another. The third lesson follows from the second – if an understanding of the fundamental objectives is paramount, there is no escaping the need for a better understanding of human dimensions, specifically the desires of hunters and non-hunters and the role they play in conservation. In the end, the compelling question is how to better understand, guide and justify decisions about conservation investments in waterfowl management. Future efforts to integrate harvest and habitat management will include completion of the species-specific case-studies, initiation of policy discussions around how to integrate the decision contexts and governing institutions, and possible consideration of a new level of integration – integration of harvest and habitats management decisions across waterfowl stocks.

Key words: decision analysis, habitat management, harvest management, integration, objectives, population models, yield curve.

Waterfowl management in North America seeks the joint goals of providing hunting opportunity and conserving waterfowl populations by regulating harvest through the Migratory Bird Treaty Act of 1918 and of protecting and improving habitats through the North American Waterfowl Management Plan (NAWMP). Over time, waterfowl harvest management has evolved into one of the best applications of adaptive resource management in the world, where annual hunting regulations are set based on quantitative population models, data collected annually, and optimisation methods designed to make the best possible decision in the face of many sources of uncertainty to insure large sustainable harvests (Nichols *et al.* 1995, 2007). Habitat management is organised around a series of public private partnerships (Joint Ventures) that endeavour to protect and improve waterfowl habitat to meet specific population goals for each of several species. While both harvest and habitat management efforts are designed for the same populations, they are administered under different and independent regulatory contexts with goals that are possibly not consistent (Runge *et al.* 2006). The NAWMP population goals were developed in reference to waterfowl populations and habitat conditions of the 1970s, a decade of above-average habitat conditions, and without reference to a specific harvest policy. Harvest policy does not explicitly give a population goal, but instead is designed to achieve sustained harvests over a very long timeframe, which implicitly strives for an average population size to support that harvest. Because expected population size is linked to harvest rate and habitat conditions, interpreting the population goals stated in the NAWMP is impossible without reference to specific habitat conditions and harvest policy (Runge *et al.* 2006). In other words, population goals specified under harvest and habitat management plans are not currently coherent, and recognition of this problem has catalysed several efforts to reconcile these plans.

The first effort was the formation of a Joint Task Group (JTG) sanctioned by NAWMP and the International Association of Fish and Wildlife Agencies "to explore options and recommend preferred solutions to reconciling the use of [NAWMP] population objectives for harvest and habitat management" (Anderson *et al.* 2007). NAWMP partners needed a shared context for habitat and population goals; harvest managers needed to be able to translate NAWMP accomplishments into harvest opportunity. To accomplish this, a theoretical assessment framework was formulated around a population model that included density-dependent relationships in survival and reproduction, which could then be translated into ecological concepts of carrying capacity and a sustainable yield curve (Fig. 1, Appendix 1). The JTG

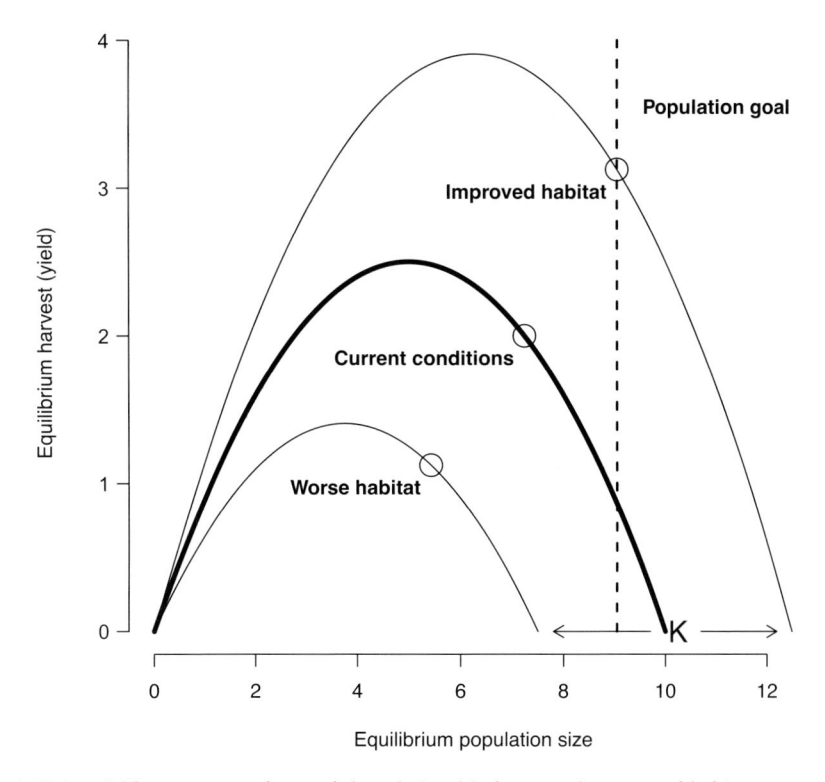

Figure 1. Using yield curves to understand the relationship between harvest and habitat management. The open circles represent a "right-shoulder strategy" – a harvest policy that maintains a yield less than the maximum sustained yield (the choice of the position on the right shoulder, here 80%, is a policy determination). Improved or worsened habitat is shown by an expanded or contracted yield curve (the curves shown are based on a 25% increase or decrease both in intrinsic growth rate, r, and in carrying capacity, K). A population goal (dashed line) is said to be "coherent" if it falls at the intersection of the desired harvest strategy and the desired habitat conditions.

explored various interpretations of NAWMP population goals in terms of sustainable harvest levels in light of the above-average habitat conditions of the 1970s. The main proposal was that population goals of the NAWMP are best interpreted as the expected population size found when a harvest strategy that seeks less than maximum sustained yield (a "shoulder strategy") is overlain with a yield curve that reflects the desired long-term average habitat conditions.

A sustainable yield curve (Fig. 1, Appendix 1) provides the conceptual framework needed to integrate harvest and habitat management (Runge *et al.* 2006; Anderson *et al.* 2007). A sustainable yield curve shows all the equilibrium points of a deterministic density-dependent population model, such as the familiar logistic growth model (Fig. 1). This model posits that birth rate declines and/or death rate increases with population density, for which there is good evidence in waterfowl (Vickery & Nudds 1984; Cooch *et al.* 1989; Sedinger *et al.* 1995; Johnson *et al.* 1997; Sedinger *et al.* 2001; Conroy *et al.* 2002; Runge & Boomer 2005; Viljugrein *et al.* 2005). If this is the case and both environmental conditions and harvest rate remain constant, then the population size and structure approaches a stable equilibrium, where birth and death rates are equal. In the absence of harvest, the population moves toward carrying capacity, the maximum equilibrium population size determined by the density-dependent effects. With harvest, the population will reach a stable equilibrium less than carrying capacity, and the yield curve describes this equilibrium population size for all harvests (Fig. 1). Of course the environment and harvest rates are not constant and populations do not behave deterministically. The environment can fluctuate on short time scales (annual variation in precipitation) or shift to new long-term average conditions (multi-year changes in agriculture policy, climate change, or permanent habitat loss or improvements). Short-term fluctuations in the environment or population size are not reflected in the yield curve; instead, the yield curve can be viewed as an average over these fluctuations for a large population. If long-term shifts change demographic rates, perhaps through changes in the strength of density-dependence, then the yield curve itself will expand or contract, so that a new equilibrium population size is realised for the same harvest rate, or a new harvest rate is required to realise the same population size (Fig. 1, Appendix 1). This is the crux of integrating habitat and harvest management – we can now ask to what extent demographic rates need to shift due to habitat management in order to obtain a desired population size and harvest rate. Conversely, we can ask what combination of harvest rate and population size (a harvest policy) we want to achieve under current habitat conditions.

The realisation among the waterfowl management community that population goals of the NAWMP only make sense in light of a particular harvest strategy has brought the human component of management to the forefront. Hunter satisfaction underlies both NAWMP goals and harvest policy but little is known about their relative importance to hunter satisfaction, which harvest managers need to know to set a desirable harvest and habitat policy. If we are below a population goal because of a shrinking habitat base, do we reduce harvest to maintain the population at goal? Or do we maintain harvest (by increasing the *per capita* harvest rate), thus driving the population even lower, in order to satisfy harvest desires of hunters? As long as populations remain large, relative to demographic and environmental stochasticity, so that extinction is not likely or ecological services are not jeopardised, these seem like fair policy alternatives. Yet we know little about how hunters and other stakeholders value population sizes and harvests. The JTG recognised this and called for an in-depth study of human dimensions underlying waterfowl management.

While the JTG was demonstrating conceptual and empirical approaches for providing a formal integration of harvest and habitat management objectives, a parallel process was underway to evaluate, for the first time, the NAWMP's effectiveness in achieving its biological goals (Assessment Steering Committee 2007). This assessment unearthed the many strengths and weaknesses inherent in planning and delivering large-scale conservation programmes, and produced wide-ranging recommendations related to NAWMP planning, adaptive processes, implementation strategies, institutional issues, integration among bird groups and funding. Most relevant here are recommendations that focused on developing improved ways of linking demographic and population responses to habitat management at scales ranging from Joint Ventures (JVs) to the continent, determining the impact of net landscape changes on waterfowl demography, and enhancing the ability to target financial investment in different regions of North America to advance NAWPM objectives. A comprehensive review of population and habitat objectives was also advocated. Addressing these recommendations alone creates significant challenges, both conceptually and operationally. For instance, habitat management typically occurs locally, with investments intended to return long-term benefits. On the other hand, harvest management decisions are made in short time steps, usually annually, with broad-scale impacts on birds. Thus, a central problem was to provide mechanisms allowing the integration of harvest and habitat objectives in ways that could reveal how management decisions could influence demographic rates within JVs or ecologically-related regions, and in turn scale up to affect continental population dynamics. Importantly, such an approach would provide new insights into optimal allocation of limited conservation resources to where they would have potential for greatest impacts.

In 2012 the NAWMP community reached consensus that a third goal, addressing the benefits of waterfowl populations and habitats to people, should be included explicitly in the NAWMP (NAWMP 2012a), and agreed that efforts to define objectives for hunters, conservation supporters and the general public should be continued. The conceptual and technical advances required to integrate these linked NAWMP goals (*i.e.* for habitat, harvest and people) have been explored in case studies focused on Black Duck *Anas rubripies*, Northern Pintail *Anas acuta* and scaup (*Aythia marila* and *A. affinis*) populations. These initiatives were expected to reveal new knowledge about relationships between population processes and management decisions, and also trade-offs at multiple scales. Furthermore, all three species are of high conservation concern due to their population status, trends in numbers, and their importance to hunters in North American flyways. Each initiative has pursued a different analytical approach in addressing unique problems encountered in integrating harvest and habitat management. Importantly, each initiative has also engaged the waterfowl community in consultation processes, to gain critical insights from managers and scientists alike on the provision of guidance about biological models that link demographic rates to regional habitat planning and harvest, which were used to inform JV management decisions. Future implementation of the NAWMP will be shaped in part by results of these case studies. Indeed, because there has been no clear route paving the way to successful integration of different components of the plan, there is high expectation that pilot projects will help to unveil how integration may be achieved (NAWMP 2012b).

Insights gained to date

In 2006, Runge *et al.* (2006) raised significant questions about the disconnected nature of waterfowl harvest and habitat management, suggesting the two programmes needed to be more coherent. The "coherence paper" (Runge *et al.* 2006) was received with scepticism by some who asked what the concern was about. The JV system is a highly successful model of collaborative landscape conservation, delivering waterfowl habitat management at a local and regional scale (Williams *et al.* 1999; Assessment Steering Committee 2007). Adaptive Harvest Management (AHM) has provided a stable and transparent way of setting harvest regulations (Nichols *et al.* 2007). Both habitat and harvest management appeared to be working well, so why make things more complicated? Was integration technically feasible? And even if it was, would it matter to the decisions being made? To some extent, we cannot yet fully answer these questions, as the various efforts to integrate habitat and harvest management are still works-in-progress, but the work of the last five years has changed much of the thinking about waterfowl management in North America. These advances include learning about potential consequences of habitat and harvest management decisions given inherent system dynamics (*i.e.* partial controllability), identifying critical information gaps (*i.e.* factors that seem most important for population dynamics but for which new information is required) and demonstrating that integration is entirely feasible and potentially useful. Thus, the case studies presented below give compelling reasons for integration, especially by providing a better understanding of population dynamics and in justifying decisions about investing scarce conservation resources in order to achieve the desired outcomes. Several important lessons about integration are emerging, as described below.

Integrating models

The first lesson concerns the efforts to build integrated models that can represent the effects of both harvest and habitat management, thus making underlying assumptions of habitat and harvest management explicit. One of the most difficult aspects of this task is not the integration of habitat and harvest in a theoretical sense (scc Appendix 1), but the integration across spatial scales and the estimation of relevant parameters. The key question is this: how do habitat actions at the local scale translate into population responses at the continental scale? Habitat management occurs at a local scale where land managers work to bring about long-term habitat change in a region, such as in a JV. Bird populations that are the target of habitat management are moving among habitats within a region and across regions. These movements act to average or dampen the effects of any one habitat effect, be it local or regional, and also make it very difficult to demonstrate empirically an effect of habitat improvements on population demographic rates. Harvest management, on the other hand, occurs at the continental scale, where season length and bag limit are set annually on the basis of population goals and an underlying model of the extent to which density-dependence influences population dynamics. Integration therefore requires linking dynamics across the local, regional and continental scales. In the Integrated Waterbird Management & Monitoring (IWMM) example given below, this integration requires an understanding of the spatial structure of the landscape that links local habitat conditions with the birds' migratory behaviour, and an estimation of the parameters (sometimes difficult to establish) that describe that process. In the pintail case study, for example, the challenge of parameter estimation includes drawing inference about intermediate-scale processes that have not been observed directly in the field.

Building mathematical models is a means to make assumptions explicit and to understand the consequences of actions. Once hypotheses have been explicitly stated, they should be rigorously tested when possible if the claim of science-based management is to have any merit. There is almost 20 years of doing just this under adaptive harvest management for waterfowl (Cooch *et al.* 2014) but a similar process for habitat management is less developed. Stating and testing assumptions about habitat management is an important recommendation of the NAWMP assessment efforts (Assessment Steering Committee 2007), and the efforts toward integration have made major advances in building models that represent assumptions about how habitat is linked to demographic rates. For example, the hypothesis that winter survival is driven by energy limitation underlies many decisions about habitat management, yet demonstrating a causal link between energy limitation and survival in wintering waterfowl has been elusive. Given that some studies show relatively high survival after the hunting season (*e.g.* Dugger *et al.* 1994; Fleskes *et al.* 2007) and that energy surpluses are evoked to explain increasing goose populations (Ankney 1996), testing this assumption and considering alternatives seems reasonable, including refinements to existing energy hypotheses and other mechanisms. Efforts of integration, discussed below, provide a framework to do just this.

Integrating objectives

The second lesson follows from the first. Habitat management is done at local or regional scales to meet local and regional objectives for populations, habitat and harvest, but ultimately habitat management is meant to scale up to affect continental population objectives. Because the entire initiative to integrate habitat management with harvest management has been framed as a decision, this has led the waterfowl management community to explore and articulate these objectives, to discover new objectives, and eventually realise that trade-offs between objectives will be necessary. Thus, the work to integrate habitat and harvest management has led the waterfowl management community to focus on and articulate objectives more clearly and to use a more structured approach (NAWMP 2012a), which in itself is a major accomplishment.

More complex objectives and trade-offs emerge as we consider the entire range of scales from local to continental. At the continental scale, the simplest of these trade-offs is that one cannot have higher average harvest rate without a subsequent decrease in expected population size if harvest mortality is not fully compensated by reduced natural mortality (Anderson & Burnham 1976; Cooch *et al.* 2014; Appendix 1). This has led to an examination and re-interpretation of the population objectives in the original NAWMP (NAWMP 2012a). At the regional and local scales, issues of population distribution and equity of hunter and non-hunter access arise. This can then lead to conflicts between objectives: should conservation resources be allocated to a region if it can be shown that habitat improvement in the region has little value for improving continental populations? Simply asking this question in the context of the decision reveals that other objectives, perhaps not yet fully articulated, exist.

Integrating human desires and governance

The third lesson follows directly from considering objectives in a decision context. When we realise that decisions are ultimately grounded in the objectives, we have to ask where those objectives come from. What are our hopes for waterfowl habitat and populations, and why? What do hunters want, and how does their conservation role affect the achievement of the other objectives? How do we structure governing institutions to best meet these objectives? This leads to focusing on human dimensions of waterfowl management and on the structure of governing institutions because the human component of waterfowl management is important, perhaps most important, for understanding objectives, making optimal decisions and designing institutions that can facilitate optimal decision making. This is especially the case when trade-offs between objectives or scales become necessary. The waterfowl management community must know how objectives rank in importance across stakeholder groups and this can only come from focused research on these groups to determine their values. For example, much of traditional waterfowl management and current harvest regulation is based on the assumption that hunter satisfaction increases with larger harvests. This may be true, but how is hunter satisfaction related to continental population size? If larger harvests are valued, how do hunters value season length and bag limit combinations that might lead to the same harvest? These questions can be extended to a broader set of stakeholders. For instance, how does non-hunter (or even anti-hunting) satisfaction depend on continental population size? How does it depend on local hunting activity? The waterfowl management community knows very little about the desires of hunters, and probably even less about the desires of non-hunters. If the general assumption is that hunter satisfaction increases with harvest and population size, is it reasonable to assume that non-hunting stakeholders' satisfaction is neutral to these metrics? Do these management choices affect the political and economic support hunters or non-hunters give to waterfowl and wetlands conservation? Finally, waterfowl hunters have historically had a strong tie to conservation of wetlands and waterfowl, but hunter numbers are declining to the extent that continued participation in waterfowl hunting is itself a major concern (Vrtiska *et al.* 2013). If decisions about harvest and habitat management affect hunter satisfaction can these decisions also be used to increase waterfowl hunter numbers through recruitment of new hunters and retention of current hunters? This is currently unknown, but by focusing on the decision context, human dimensions have moved to the forefront of the integration of habitat management and harvest. These issues are especially clear in the scaup example, given below, where the hunter population was included in the scaup population model, just as in models of classic predator-prey systems. However, all the integrated models in the world won't get us anywhere if different agencies, authorities, administrations or nations are not willing to integrate and coordinate policy and programmes that affect conservation and management at the continental level.

Expanding the range of objectives under consideration raises questions about the governance structures that support management. Government agencies have a public trust responsibility to include input from a broad array of stakeholders, the hunting and non-hunting public alike. But the strongest "stakeholder" input for government agencies comes from their statutory mandates. For waterfowl management, U.S. federal agencies must adhere to the Migratory Bird Treaty Act, which puts a primary emphasis on migratory bird populations and their habitats, with only secondary consideration given to consumptive and non-consumptive use. Thus, other entities (*e.g.* state agencies and non-government organisations (NGOs)) may be better enabled to pursue objectives like hunter satisfaction, but the current governance of habitat and harvest through the flyway system might not yet provide an effective structure for inclusion of these broader aims. These governance issues, along with the complicated governance issues associated with habitat protection and management by a diverse array of landholders, point to the need for thoughtful consideration of the institutional relationships that underlie waterfowl harvest and habitat management.

In short, the emerging efforts to integrate harvest and habitat management, which originated from a simple suggestion that the NAWMP population objectives might need revision (Runge *et al.* 2006), have precipitated a much deeper examination of the decision structures used for habitat and harvest management and the technical tools used to support them. It has been both a significant challenge, with many technical issues yet to address, as well as a unique opportunity to push the boundaries of our current approach. Success is not assured, but we will learn much even if our aspirations are not fully realised at the operational level. The following case studies describe these efforts in more detail.

Case studies

"Integrated Waterbird Management and Monitoring Initiative" (IWMM)

The U.S. Fish and Wildlife Service (USFWS), Joint Ventures (JVs), and the Flyway Councils work to conserve migratory waterbird populations by informing and implementing habitat and other management actions. The conservation of migratory waterbird populations and their habitats is an inherently challenging proposition given the geographic and temporal scope of species' life histories. Few biological models exist to address problems at such a large scale. Furthermore, the challenge is amplified by the fact that biological processes do not align with administrative programme boundaries and successful management for these species depends on linking management decisions at multiple spatial scales. At each scale, habitat management decisions involve allocating resources efficiently, in light of financial and personnel limitations and the need for public accountability, to maximise the benefits for waterbird populations. Decision analytic techniques hold promise for addressing these challenges in a structured and transparent manner (Wilson *et al.* 2007; McDonald-Madden *et al.* 2008; Thogmartin *et al.* 2009).

The IWMM seeks to provide decision support tools at multiple scales to aid waterbird habitat managers across the Atlantic and Mississippi Flyways. IWMM represents a joint initiative of conservation partners, including the USFWS, U.S. Geological Survey, state agencies and Ducks Unlimited. To date, IWMM has provided standardised waterbird and habitat monitoring protocols and a common database with reporting tools to participants across the two flyways, and coordinated pilot data collection has been underway for more than three years. Through the application of structured decision analytic techniques (Gregory *et al.* 2012), IWMM identified pressing waterbird management decisions at multiple spatial scales. At a flyway scale, decisions must be made about habitat acquisitions and restorations within the context of budgetary constraints. At a local scale, managers annually determine how to manage habitat within a wetland or collection of wetlands to maximise long-term benefits for waterbird populations. IWMM recognises that these habitat management decisions are naturally linked across scales.

IWMM's technical team is developing models for decision support at multiple scales. To address habitat acquisition and restoration decisions at a flyway scale, the team has developed a continental scale simulation model to couple waterbird survival during the migratory period to the amount and distribution of energy (in the form of appropriate habitat) across the flyways; thus, incorporating explicit hypotheses about energy limitation as a determinant of survival during the migratory period. The model uses geospatial land cover layers and land cover-specific roosting and caloric values to represent the quality of stopover sites. Portfolios of land acquisition and restoration decisions are evaluated by altering flyway food energy content and examining the change in survival that results from these decisions. The model identifies areas along the flyway that have a large benefit for the survival of individuals within a specific guild or species relative to the cost of management or acquisition. Insight can be provided to local managers about the importance of their general areas for non-breeding survival of specific guilds or species, linking management priorities at flyway and local scales.

For the eastern U.S., the flyway scale model is in the late stages of development (Eric Lonsdorf, unpubl. data). The model represents an important advance in guiding land acquisitions by explicitly linking hypotheses about waterbird biology to alternative acquisition and restoration decisions. The model can also help local managers understand the importance of their wetlands for specific waterbird guilds or species in a larger flyway context. For example, the flyway model can be incorporated into a structured decision making process to evaluate land acquisitions by the National Wildlife Refuge system. This framework can also allow for the inclusion of alternative models in an adaptive management programme to learn about population demographic rates while guiding refuge acquisition and management.

Although the IWMM initiative was not originally designed to integrate harvest and habitat, it has provided valuable tools and insights about the process of integration. The development of the IWMM predictive models has faced a deep challenge: integrating habitat management and waterbird demography across scales. This requires identifying the mechanisms that connect local-scale influences to broad-scale outcomes, in this case, physiological energetics and behavioural adaptations to the distribution of food resources (and thus energy) across the landscape, so that individual and local habitat mechanisms give rise to patterns of migration at the flyway scale. Processes that connect the local to the continental scales are often not observed directly, and yet they are the crux of understanding how habitat management translates into demographic change. There are significant challenges in estimating the parameters of these processes, but formal methods of expert judgment and modern Bayesian hierarchical methods are bearing fruit. The important lesson, perhaps, is that we cannot shy away from understanding the complex processes that link dynamics across scales; indeed, that is one of the most important aspects of developing integrated models.

Northern Pintail

In the integration efforts focused on pintail, the central focus was to build a formal mathematical framework to link habitat and harvest management across spatial scales. The integrated pintail model is a spatial version of that currently used in pintail harvest management (USFWS 2010; Mattsson *et al.* 2012). This is a spatial matrix-projection model with an annual time step partitioned into seasonal components to reflect the annual cycle of breeding, autumn migration and winter through spring migration. Breeding areas are separated into two spatial components to reflect regional differences – Alaska and the prairie potholes and parkland, with a third "breeding" class used to represent drought years when pintail are less likely to attempt breeding and are generally less observable (Runge & Boomer 2005). Wintering areas are divided into two regions – the California Central Valley and the Gulf Coast, to reflect potential differences in non-breeding season survival and density-dependence (Mattsson *et al.* 2012). The key aspects to the model are: 1) the migratory transitions that link winter grounds to breeding grounds, and 2) the density-dependent relationships for recruitment and survival in each breeding and wintering region, respectively (Fig. 2).

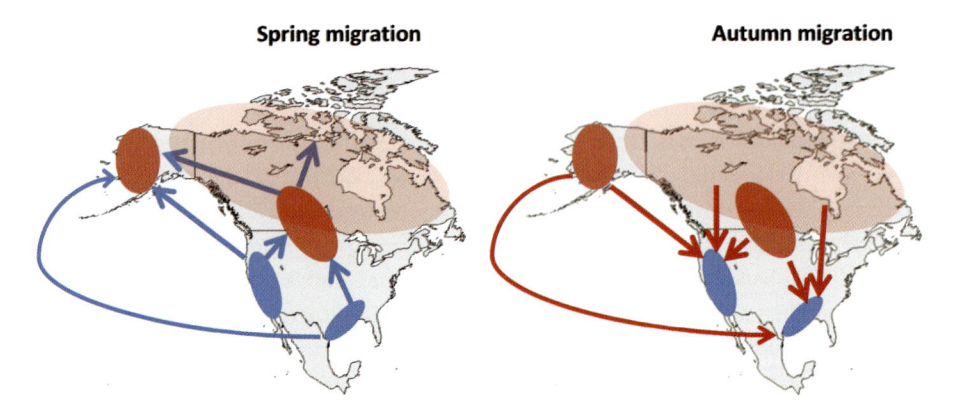

Figure 2. Pintail metapopulation model that allows analysis of the interaction of habitat and harvest management (after Mattsson *et al.* 2012). Two core breeding areas (Alaska and prairie-parklands, red) and two core wintering areas (primarily the California Central Valley of the Pacific flyway and the gulf coast of the Central and Mississippi flyways, blue) are linked through autumn (red arrows) and spring (blue arrows) migratory transitions. A third breeding season state is used to represent movements of pintail in drought years when breeding effort and observability of the population is low (large light red area). Arrows represent starting and ending locations of migration and not the geographic route of migrating pintail.

Using parameter values from published literature and expert judgment, the model can then be analysed to investigate the effects of habitat improvements on the yield curve. Efforts are now underway to use available long-term data sets on harvest, band recoveries and breeding population surveys to refine parameters for the model, including estimating the form and strength of the regional density-dependent relationships. Just as with IWMM, there are few direct data to inform intermediate processes. The ongoing approach to estimation is to use multiple data sources linked by the matrix projection model in a common hierarchical Bayesian statistical analysis, which can provide information on these hidden processes.

There have been two major accomplishments from work on the integration of pintail harvest and habitat data to date. The first is the demonstration that it is theoretically possible to link habitat and harvest across scales (Mattsson *et al.* 2012). The integrated pintail model shows how habitat improvements at the regional level might increase pintail demographic rates in that region, and hence can increase the continental yield curve. In the future, this model can be used to inform allocation of conservation resources. For example, preliminary analyses of the initial model suggests that proportional habitat-related improvements on prairie-parkland breeding areas could be more effective at increasing the yield curve than equal proportional improvements to wintering areas (Mattsson *et al.* 2012). Confirmation of this conclusion awaits formal parameter estimation and a better understanding of local processes through the development of mechanistic models, which are underway.

The second major accomplishment has been to motivate in-depth discussions about the assumptions and mechanisms of population regulation at the regional scale, as these are critical for translating local habitat management into continental demographic impacts. In several workshops, hypothesised mechanisms of density-dependence were elicited from local experts, resulting in a series of conceptual models that link changes in habitat to regional productivity or survival. In the prairie-parkland breeding region, competition for space is thought to be the leading driver of density-dependence. Annual variation in precipitation produces variation in pond numbers and distribution and this leads to variation in pintail breeding effort and distribution. However, the key element here is that, after controlling for precipitation-induced variation, pintail density has an additional effect. Thus in years of higher than average pintail numbers, more pintail nest in habitats of the prairie-parkland region where reproductive success is lower (J.H. Devries, pers. comm.). In the wintering regions, the focus is on the relationship between density and post-hunting season survival, through the effects of a limited food supply (and thus energy intake) on the birds' body mass in winter and spring. The assumption here is that habitat managers can increase pintail survival by providing more nutritious food resources; in years with higher post-hunting season population size and greater food depletion, survival is reduced compared to years with lower populations and identical habitat. These regional sub-models are intended to provide predictive tools that, when coupled with the continental demographic model, will provide an analytical framework for assessing the effects of local habitat changes on the continental yield curve.

Two insights emerge from the pintail work. First, one of the most challenging aspects of integrating habitat and harvest management is developing demographic models that integrate dynamics across spatial scales. Indeed, as with the IWMM project, the pintail project recognised that the crux of the modelling effort was the identification of intermediate-scale mechanisms of population regulation, such as the regional density-dependence relationships. The second insight arises from an ongoing challenge: these intermediate-scale mechanisms are difficult to observe directly, and so estimating the functions and parameters is likewise difficult. Fortunately, modern hierarchical statistical methods can potentially be used to make inference about such hidden processes, in this case by using observations at both the local and continental scales to gather insights about the mechanisms that link the dynamics across scales. Where possible, of course, efforts to measure demographic rates directly and to test hypotheses underlying habitat and harvest management should be encouraged, but modern inferential methods provide a promising alternative.

Scaup

Substantial declines in the continental scaup population in the 1980s and 1990s attracted concern from biologists and hunters alike. Biologists first approached the problem from the bottom up, examining long-term population and harvest data to develop hypotheses about factors that potentially were contributing to population decline (Austin *et al.* 2000; Afton & Anderson 2001; Austin *et al.* 2014). However, a model-based approach such as that used for the Northern Pintail was precluded because of broad uncertainties, particularly the absence of contemporary annual survival rates, sparse data on vital rates from breeding grounds in the boreal forest and taiga, and uncertainties about the population trends for each scaup species separately. It was clear that the research and monitoring necessary to fill these knowledge gaps, and to clarify the key factors affecting scaup, would require substantial resources, time and collaboration. At the same time, debate was growing about how adaptive harvest management affected the harvest of species other than Mallard *Anas platyrhynchus*, and was a particular concern for the scaup harvest which, like for other long-lived and slower-producing ducks, is more sensitive to the duration of the hunting season (Allen *et al.* 1999).

The waterfowl management community also was coming to recognise the importance of understanding the human components of waterfowl management, specifically, the hunter desires and factors that affect hunter participation (Case & Sanders 2008). Waterfowl hunters were expressing strong concerns about fewer scaup and the loss of hunting opportunities associated with restrictive regulations, and waterfowl managers were concerned about declining hunter numbers because of their important influence in conservation.

Moreover, waterfowl hunters and managers voiced concern about the decline of the diving duck hunting tradition – notably the use of large decoy sets in large open water bodies, often using low-profile layout boats – a practice that has changed little over the last century and are considered by many to be among the purest of all waterfowl hunting traditions. While this concern does not directly relate to conservation *per se*, it highlights the importance of human values in waterfowl management that is reflected in the NAWMP revision (NAWMP 2012a). Hence, waterfowl biologists and managers were challenged to develop approaches to decision-making in the face of deep sources of uncertainty in scaup biology as well as in the objectives of the scaup conservation and management community. As a result, efforts were initiated to address scaup conservation planning through the principles of structured decision making (Gregory *et al.* 2012), with a focus on how best to allocate scarce conservation resources among management actions on an annual basis.

The structured decision-making approach first required a clear, explicit statement of scaup conservation and management goals and objectives. Participants in the process quickly realised that there was one over-arching goal: to conserve scaup populations at levels that satisfy societal values. Under this goal, a resulting objectives hierarchy established linkages among three fundamental objectives (Fig. 3): 1) achieve continental habitat conditions capable of supporting a target scaup population; 2) maintain or increase the sustainable scaup harvest; and 3) sustain the diving duck hunting tradition. This last objective explicitly brings people into the objectives and recognises the important contribution of hunters to waterfowl conservation, through direct financial contributions, advocacy and economic activity. Thus, sustaining the diving duck hunting tradition through maintaining the number, participation and identity of diving duck hunters becomes an explicit conservation objective. Although participants thought this was fundamental, it is not to say that maintaining hunter tradition and numbers is equally important compared to the other objectives. Determining the relative importance of each objective remains to be determined and will require input from all societal stakeholders, hunters and non-hunters included.

To predict the consequences of conservation actions on each objective, a coupled scaup-hunter model (*i.e.* predator-prey or consumer-resource type model) was developed with explicit relationships between potential management actions and key scaup demographic processes as well as diving duck hunter dynamics. Hunter recruitment and retention were modelled as a function of scaup population levels, and scaup harvest rates were driven by the number of diving duck hunters. These linked models project scaup and hunter numbers forward through time, predicting the numbers of breeding scaup, numbers of diving duck hunters and scaup harvest. Initial functional relationships relating retention and recruitment to management actions were developed based largely on expert judgement and limited knowledge from other species. Using the model,

320 Integrating harvest and habitat management

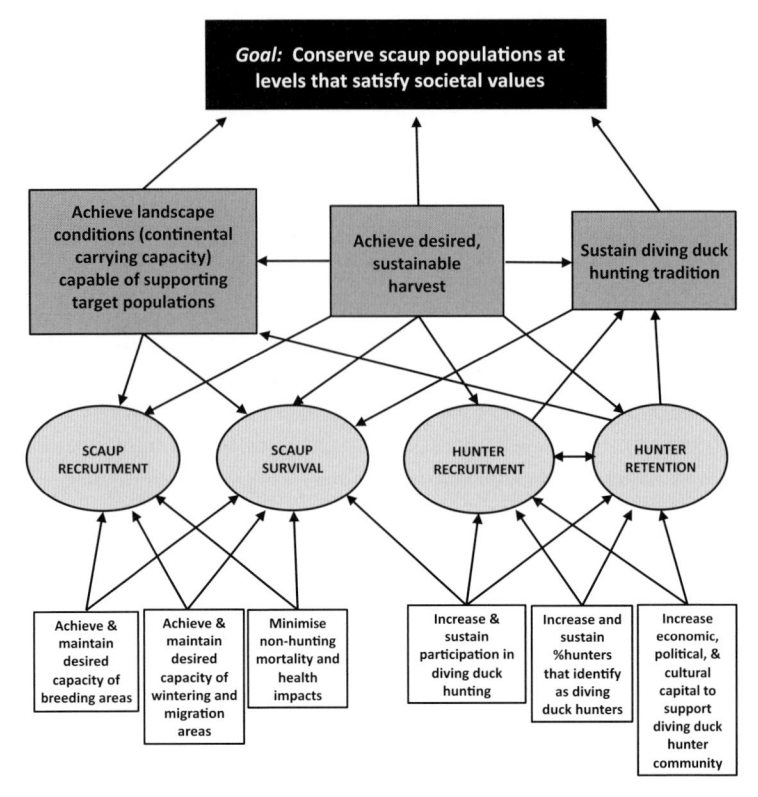

Figure 3. Objectives hierarchy for scaup conservation. After extensive discussions and revisions over three workshops, participants arrived at a set of three fundamental objectives under one overarching goal to conserve scaup populations at levels that satisfy societal values (black box). The fundamental objectives (medium grey boxes) identify issues of most concern. Each fundamental objective is linked to means objectives (white boxes) through simple functions that are hypothesised to affect survival and recruitment of scaup and hunters (light grey ovals).

waterfowl managers can ask: what affects each vital rate for both scaup and hunters? Also, what actions could be taken to alter or improve each vital rate? For each possible action, simple functional relationships were developed that linked how a management action affected a landscape variable (*e.g.* amount of nesting cover on the landscape), and how that in turn was related to a vital rate (*e.g.* probability of breeding success). Possible actions affecting hunters included harvest regulations (*e.g.* bag limit, season length), hunter access (*e.g.* alter amount of hunting habitat available through conservation or access programmes), or social networking (*e.g.* mentor programmes, web forums, community events). This modelling framework helps to identify portfolios of management actions (*e.g.* breeding or wintering habitat management) that have the greatest impact on scaup and hunter population change. Work is ongoing to finalise the modelling framework, establish a baseline parameterisation and conduct a sensitivity analysis (Austin *et al.* 2010). This will result in a robust platform to test assumptions about scaup and hunter population dynamics and to support decisions about the allocation of scarce conservation resources.

The key lessons from the work on scaup integration have been first, that there is a complex set of objectives involved when harvest and habitat management are considered together (see the "Integrating objectives" section above). Here participants thought that because encouraging and preserving hunter participation was fundamental, adding hunters explicitly into the objective was not only desirable but necessary. This ultimately took the form of a mathematical model to predict the consequence of management actions on hunter participation, a novel innovation in waterfowl management. Second, by admitting hunter objectives the question arises as to who is the decision maker and how can institutions be structured to best meet these hunter objectives (see "Integrating human desires and governance" above). These questions are largely unresolved for scaup, or for waterfowl management as a whole.

American Black Duck

At the time of the first NAWMP Continental Assessment (Assessment Steering Committee 2007) and release of the Joint Task Group report (Anderson *et al.* 2007), the Black Duck Joint Venture (BDJV) was finishing work on two priority issues: the development and implementation of the Eastern Waterfowl Breeding Survey and the completion of the technical framework of an international, adaptive harvest management strategy for Black Ducks (BDAHM). This confluence of events allowed the BDJV to re-evaluate priority information needs for the conservation of the Black Duck. The BDJV decided to focus greater effort on understanding Black Duck habitat ecology. The BDJV also agreed to focus on information needs required to determine where in the annual life-cycle limited financial resources should be allocated for habitat protection, restoration and enhancement to meet four fundamental objectives: 1) maintain Black Duck abundance at levels that meet legal and policy mandates; 2) maintain the relative distribution of breeding and non-breeding Black Ducks corresponding to the 1990–2012 period; 3) maintain carrying capacity to support the desired population and distribution; and 4) maintain consumptive and non-consumptive recreational opportunities commensurate with population sustainability and carrying capacity. Framing information needs in this context forced the community to address the issue of integrating habitat and harvest objectives (in "Integrating objectives" above). To address the trade-offs between objectives 1 and 4, above ("coherence", Runge *et al.* 2006), the desired NAWMP population goal was interpreted in reference to the BDAHM strategy, to harvest the population at 98% of maximum sustained yield (*i.e.* the 98% right-shoulder strategy). Therefore, objective (1) can be interpreted as achieving the NAWMP population goal for the Black Duck given a 98% right-shoulder harvest strategy. A future step is to re-evaluate, and potentially revise, the NAWMP population goal for Black Duck conditioned on the BDAHM strategy and capacity of the habitat JVs to increase continental carrying capacity.

The BDJV developed a conceptual annual life-cycle model relating Black Duck population dynamics to habitat limiting factors at the regional scale while accounting for annual harvest, much like that for Northern Pintail (Devers & Collins 2011; Mattsson *et al.* 2012). While progress on parameterising the life-cycle model and associated decision framework continues (see "Integrating models" above), several insights have emerged. The first insight concerns the decision context – "where" and "how much" habitat is needed; it became clear through this effort that there are multiple habitat decisions to be made at multiple scales. Habitat delivery on the breeding grounds of eastern Canada is independent of habitat delivery during the non-breeding season (*e.g.* the North American Wetland Conservation Canada programme *versus* the North American Wetland Conservation programme in the U.S.). In the case of each programme, no trade-off exists in terms of funds or resources between the breeding and non-breeding period; this is an important consideration as it establishes two separate decision processes. This is probably true for most waterfowl species. The decision context is also complex within countries because habitat decisions are made at the national, regional and local scales using a variety of funding mechanisms. Moreover, we cannot identify a single funding mechanism that is designed to fund projects based solely on waterfowl objectives. The vast majority of, if not all, habitat programmes are designed to achieve multiple objectives by providing habitat for threatened and endangered species, waterfowl and other species. The decentralised nature and multiple objectives of habitat conservation programmes create challenges regarding governance, not only across the waterfowl enterprise but more broadly within the wildlife conservation community. These insights about governance and sources of uncertainty have forced the BDJV community to think more broadly about the decision process, including the multi-objective nature of habitat programmes, and to consider a wide array of decision tools such as dynamic optimisation and robust decision-making (Lempert & Collins 2007).

The second insight is that, despite challenges related to integrated governance, a decision analytic approach based on an integrated population-habitat model allows the BDJV to make better decisions regarding the allocation of limited monitoring and research funds to address key uncertainties and assumptions ("Integrating models" above). For example, the Black Duck conceptual model assumes habitat restoration results in increased food availability (*i.e.* energetic carrying capacity) and post-hunting season survival. However, the BDJV lacks empirical data to parameterise this hypothetical relationship. To address this assumption, the BDJV has invested resources into a two-season banding programme to estimate post-hunting season survival and research to quantify the effect of restoration efforts on food availability. The results of these projects will be used to parameterise the life-cycle model and provide insight into the relationship between local scale habitat management and population response at the continental scale. In the long-term, the BDJV anticipates using the life-cycle model and analytical tools to guide the development of future research and monitoring projects.

The main lessons from the work on Black Duck have been, much like the scaup example, that the objectives become very complex and that governing institutions are not well-structured with respect to these complex objectives. The multi-species nature of habitat investment decisions allow for very little direct control to influence Black Duck conservation in particular. In these complex cases, the use of models becomes especially important for understanding the consequences of decisions.

Conclusions

Work on the integration of harvest and habitat management is ongoing. The species-specific initiatives must be completed, followed by a dedicated effort to implement these frameworks and use their guidance to inform management decisions. This will require commitment and buy-in by decision makers and local managers, and this can only come through continued engagement between the research and management communities, enhanced understanding of the underlying objectives of all stakeholders, and continued critical re-examination of the governance structures surrounding habitat and harvest management. Future work in this area should use the species-specific examples to build models that are capable of predicting the consequences of large-scale landscape change, such as those resulting from land use and climate changes. In addition, habitat and harvest management are not single-species endeavours. For harvest management, a common framework of hunting regulations affects many species at once, including species of significant conservation concern. In habitat management, decisions are rarely made in reference to a single species, or even just waterfowl. Thus, future efforts to integrate these management decisions must embrace a wider set of objectives, which undoubtedly will lead to more complex and difficult trade-offs.

Integration initiatives to date have shown that the management of harvest and habitat should not continue to be viewed as separate endeavours if the waterfowl management community desires to make optimal decisions with scarce resources. These two management regimes affect the same social-ecological system; thus, the question naturally arises as to whether the current governance system for waterfowl is in some sense sub-optimal (*cf*. Ostrom *et al*. 1999; Dietz *et al*. 2003; Ostrom 2009) and, if so, what parts need to change. The efforts to integrate harvest and habitat management have, if nothing else, raised this question and led to an examination of waterfowl management in a broader context that includes the waterfowl resource, habitat ecosystems and the objectives and desires of people interacting with those systems (NAWMP 2012a). Conservation decision-makers are challenged by increasingly complex decisions, fewer conservation and administrative resources, and changing social values. In addition, conservation planners are confronted with system change brought about by agricultural commodity markets in the short-term and climate in the long-term. Thus, waterfowl are only one component of a complex system, and the larger hope is that what started as a simple proposal to manage coherently hunter harvest and waterfowl habitat will lead to stronger and more adaptable technical, conceptual and institutional structures to address these larger challenges.

Acknowledgements

We thank the ECNAW scientific committee for inviting us to present this plenary and Dave Koons and Jim Nichols for constructive comments on the manuscript.

References

Afton, A.D. & Anderson, M.G. 2001. Declining scaup populations: a retrospective analysis of long-term population and harvest survey data. *Journal of Wildlife Management* 65: 781–796.

Allen, G.T., Caithamer, D.F. & Otto, M. 1999. A review of the status of greater and lesser scaup in North America. U.S. Fish and Wildlife Service, Office of Migratory Bird Management, Washington D.C., USA.

Anderson, M.G., Caswell, D., Eadie, J.M., Herbert, J.T., Huang, M., Humburg, D.D., Johnson, F.A., Koneff, M.D., Mott, S.E., Nudds, T.D. & others. 2007. Report from the Joint Task Group for clarifying North American Waterfowl Management Plan population objectives and their use in harvest management. U.S. Fish and Wildlife Service and U.S. Geological Survey, U.S. Department of the Interior, Washington D.C., USA.

Ankney, C.D. 1996. An embarrassment of riches: too many geese. *Journal of Wildlife Management* 60: 217–223.

Assessment Steering Committee. 2007. North American Waterfowl Management Plan: continental progress assessment final report. Plan Committee, North American Waterfowl Management Plan, Washington D.C., USA.

Austin, J.E., Afton, A.D., Anderson, M.G., Clark, R.G., Custer, C.M., Lawrence, J.S., Pollard, J.B. & Ringelman, J.K. 2000. Declining scaup populations: issues, hypotheses, and research needs. *Wildlife Society Bulletin* 28: 254–263.

Austin, J.E., Boomer, G.S., Brasher, M., Clark, R.G., Cordts, S.D., Eggeman, D., Eichholz, M.W., Hindman, L., Humburg, D.D., Kelley, J.R., Jr., Kraege, D., Lyons, J.E., Raedeke, A.H., Soulliere, G.J. & Slattery, S.M. 2010. Development of scaup conservation plan. Unpublished report from the Third Scaup Workshop – Migration, Winter, and Human Dimensions, 21–25 September 2009, Memphis Tennessee, USA. U.S. Geological Survey, Jamestown, North Dakota, USA.

Austin, J.E., Slattery, S. & Clark, R.G. 2014. Waterfowl populations of conservation concern: learning from diverse challenges, models and conservation strategies. *Wildfowl* (Special Issue No. 4): 470–497.

Beverton, R.J.H. & Holt, S.J. 1957. On the dynamics of exploited fish populations. Fisheries Investigation Series 2, Vol. 19,. Ministry of Agriculture Fisheries and Food (MAFF), London, UK.

Case, D.J. & Sanders, S. 2008. The future of waterfowl management: Framing future decisions for linking harvest, habitat, and human dimensions. Summary Report 10-9-08. D.J. Case and Associates, Mishawaka, Indiana, USA.

Conroy, M.J., Miller, M.W. & J.E. Hines. 2002. Identification and synthetic modeling of factors affecting American Black Duck populations. *Wildlife Monographs* 150: 1–64.

Cooch, E.G., Lank, D.B., Rockwell, R.F. & Cooke, F. 1989. Long-term decline in fecundity in a Snow Goose population: evidence for density dependence? *Journal of Animal Ecology* 58: 711–726.

Cooch, E.G., Guillemain, M., Boomer, G.S., Lebreton, J.-D. & Nichols, J.D. 2014. The effects of harvest on waterfowl populations. *Wildfowl* (Special Issue No. 4): 220–276.

Devers, P.K. & Collins, B. 2011. Conservation action plan for the American black duck, First Edition. U.S. Fish and Wildlife Service, Division of Migratory Bird Management, Washington D.C., USA.

Dietz, T., Ostrom, E. & Stern, P. C. 2003. The struggle to govern the commons. *Science* 302(5652): 1907–1912.

Dugger, B.D., Reinecke, K.J. & Fredrickson, L.H. 1994. Late winter survival of female mallards in Arkansas. *Journal of Wildlife Management* 58: 94–99.

Fleskes, J.P., Yee, J.L., Yarris, G.S., Miller, M.R. & Casazza, M.L. 2007. Pintail and mallard survival in California relative to habitat, abundance, and hunting. *Journal of Wildlife Management* 71: 2238–2248.

Fretwell, S.D. 1972. *Populations in a Seasonal Environment (No. 5)*. Princeton University Press, Princeton, New Jersey, USA.

Fretwell, S.D. & Lucas, H.L. 1969. On territorial behavior and other factors influencing habitat distribution in birds. *Acta Biotheoretica* 19: 16–36.

Getz, W.M. & Haight, R.G.. 1989. *Population Harvesting: Demographic Models of Fish, Forest, and Animal Resources. Vol. 27*. Princeton University Press, Princeton, New Jersey, USA.

Gregory, R., Failing, L., Harstone, M., Long, G., McDaniels, T. & Ohlson, D. 2012. *Structured Decision Making: a Practical Guide to Environmental Management Choices*. Wiley-Blackwell, West Sussex, UK.

Hilborn, R., Walters, C.J. & Ludwig, D. 1995. Sustainable exploitation of renewable resources. *Annual Review of Ecology and Systematics* 26: 45–67.

Johnson, F.A., Moore, C.T., Kendall, W.L., Dubovsky, J.A., Caithamer, D.F., Kelley Jr., J.R. & Williams, B.K. 1997. Uncertainty and the management of mallard harvests. *Journal of Wildlife Management* 61: 202–216.

Lempert, R.J. & Collins, M.T. 2007. Managing the risk of uncertain threshold responses: comparison of robust, optimum and precautionary approaches. *Risk Analysis* 27: 1009–1026.

Mattsson, B.J., Runge, M.C., Devries, J.H., Boomer, G.S., Eadie, J.M., Haukos, D.A., Fleskes, J.P., Koons, D.N., Thogmartin, W.E. & Clark, R.G. 2012. A modeling framework for integrated harvest and habitat management of North American waterfowl: Case-study of Northern Pintail metapopulation dynamics. *Ecological Modelling* 225: 146–158.

McDonald-Madden, E., Baxter, E.W.J. & Possingham, H.P. 2008. Subpopulation triage: how to allocate conservation effort among populations. *Conservation Biology* 22: 656–665.

North American Waterfowl Management Plan (NAWMP) Committee. 2012a. *North American Waterfowl Management Plan 2012: People Conserving Waterfowl and Wetlands*. Canadian Wildlife Service & U.S. Fish and Wildlife Service, Washington D.C., USA.

North American Waterfowl Management Plan (NAWMP) Committee. 2012b. *North American Waterfowl Management Plan: a Companion Document to the 2012 North American Waterfowl Management Plan*. Canadian Wildlife Service & U.S. Fish and Wildlife Service, Washington D.C., USA.

Nichols, J.D., Johnson, F.A. & Williams, B.K. 1995. Managing North American waterfowl in the face of uncertainty. *Annual Review of Ecology and Systematics* 26: 177–199.

Nichols, J.D., Runge, M.C., Johnson, F.A. & Williams, B.K. 2007. Adaptive harvest management of North American waterfowl populations: a brief history and future prospects. *Journal of Ornithology* 148(2): 343–349.

Ostrom, E. 2009. A general framework for analyzing sustainability of social-ecological systems. *Science* 325(5939): 419–422.

Ostrom, E., Burger, J., Field, C.B., Norgaard, R.B. & Policansky, D. 1999. Revisiting the commons: local lessons, global challenges. *Science* 284(5412): 278–282.

Runge, M.C. & Boomer, G.S. 2005. Population dynamics and harvest management of the continental Northern Pintail population. Final Report, 6 June 2005. U.S. Geological Survey, Patuxent Wildlife Research Center, Laurel, Maryland, USA.

Runge M.C., Johnson, F.A., Anderson, M.G., Koneff, M.D., Reed, E.T. & Mott, S.E. 2006. The need for coherence between waterfowl harvest and habitat management. *Wildlife Society Bulletin* 34: 1231–1237.

Sedinger, J.S., Flint, P.L. & Lindberg, M.S. 1995. Environmental influence on life-history traits: growth, survival, and fecundity in Black Brant (*Branta bernicla*). *Ecology* 76: 2404–2414.

Sedinger, J.S., Lindberg, M.S. & Chelgren, N.D. 2001. Age-specific breeding probability in Black Brant: effects of population density. *Journal of Animal Ecology* 70: 798–807.

Thogmartin, W.E., Fitzgerald, J.A. & Jones, M.T. 2009. Conservation design: where do we go from here? *Proceedings of the International Partners in Flight Conference* 4: 426–436.

U.S. Fish and Wildlife Service. 2010. Northern Pintail harvest strategy 2010. U.S. Department of the Interior, Fish and Wildlife Service, Division of Migratory Bird Management, Washington D.C., USA.

Vickery, W.L. & Nudds, T.D. 1984. Detection of density-dependent effects in annual duck censuses. *Ecology* 65: 96–104.

Viljugrein, H., Stenseth, N.C., Smith, G.W. & Steinbakk, G.H. 2005. Density dependence in North American ducks. *Ecology* 86: 245–254.

Vrtiska, M.P., Gammonley, J.H., Naylor, L.W. & Raedeke, A.H. 2013. Economic and conservation ramifications from the decline of waterfowl hunters. *Wildlife Society Bulletin* 37: 380–388.

Williams, B.K., Koneff, M.D. & Smith, D.A. 1999. Evaluation of waterfowl conservation under the North American Waterfowl Management Plan. *Journal of Wildlife Management* 63: 417–440.

Wilson, K.A., Underwood, E.C., Morrison, E.C., Klausmeyer, K.R., Murdoch, W.W., Reyers, B., Wardell-Johnson, G., Marquet, P.A., Rundel, P.W., McBride, M.F., Pressey, R.L., Bode, M., Hoekstra, J.M. Andelman, S.J., Looker, M., Rondinini, C., Kareiva, P., Shaw, M.R. & Possingham, H.P. 2007. Maximizing the conservation of the world's biodiversity: What to do, where and when. *PLoS Biology* 5: e223.

Appendix 1. A technical primer on integrating habitat and harvest: derivation of a sustainable yield curve based on density- and habitat-dependent demographic rates.

Harvest management has a long tradition of explicit quantitative demographic modelling as the basis for decisions (Beverton & Holt 1957; Getz & Haight 1989; Hilborn *et al.* 1995). Habitat management is implicitly based on an underlying population model (*e.g.* Fretwell & Lucas 1969; Fretwell 1972), but it is less common for the model to be made explicit in the context of habitat management decisions. For habitat and harvest management to be integrated in a meaningful and useful way, parameters of a population model (*i.e.* the demographic rates) must ultimately be functions of habitat characteristics and harvest rates. The simplest representation of a population model where density (N, numbers per unit area of space) varies over time (t), and where *per capita* birth (b) and death (d) rates are functions of N and habitat (H), is:

$$\frac{dN}{dt} = b(N,H) - d(N,H) \qquad \text{Equation (1)}$$

The birth and death functions, $b(N,H)$ and $d(N,H)$ respectively, are also functions of time, reflecting the seasonal nature of waterfowl reproduction and mortality, but this notation has been dropped for clarity. These functions might be very complex, for example N might be a vector that refers to densities at various locations (spatial complexity), times (delay effects), or to different species (interspecific competition), and H might be a vector that refers to the area of different habitat types or the state of resources within each habitat type. In addition, $b()$ or $d()$ might be non-linear, such that the effect of a density or habitat manipulation is not constant across all N or H. This might be the case for $d()$ with respect to N if hunting mortality is compensatory to other mortality sources (see Cooch *et al.* 2014, this volume, for a discussion of density-dependence and other mechanisms of compensation). While many biologists might envision very complex hypotheses about the form of these functions, practical limitations and parsimony will limit the form to fairly simple representations. The simplest form is a linear relationship in birth and death rates

$$b(N,H) = b_0 - b_1 N + b_2 H + b_3 NH \qquad \text{Equation (2a)}$$

$$b(N,H) = b_0 - b_1 N + b_2 H + b_3 NH \qquad \text{Equation (2a)}$$

where h is the harvest rate and b_i and d_i are parameters relating demographic rates to density and habitat. In addition, there is the constraint that $b(N,H) \geq 0$ and $d(N,H) \geq 0$. Even if reality is much more complex, this linear model can be thought of as a first approximation to a more complex model, especially if one is interested in small perturbations from a particular point of interest (*i.e.* current conditions).

With these relationships, equation (1) can be rewritten into the familiar logistic growth form with harvest,

$$\frac{dN}{dt} = rN\left(1 - \frac{N}{K}\right) - hN \qquad \text{Equation (3)}$$

with $r = b_0 - d_0 + (b_2 + d_2)H$ and $K = r/(b_1 + d_1 - (b_3 + d_3)H)$. Equation (3) has several interesting properties from the perspective of integrating harvest and habitat management. First, the "intrinsic growth rate", r – the growth rate as density approaches zero – is a function of the birth and death rate intercepts and the habitat coefficients. Thus, the "intrinsic growth rate" can vary with changes in habitat (see figures in Anderson *et al.* 2007). When *r* increases with *H*, this effect has been called "habitat quality" (Anderson *et al.* 2007; Mattsson *et al.* 2012, and see Fretwell & Lucas 1969 or Fretwell 1972 for an original formulation of essentially the same ideas) because this effect is independent of population density.

Second, "carrying capacity" (K) is a function of the "intrinsic growth rate", the birth and death rate density coefficients (b_1 and d_1), and the habitat-by-density interaction coefficients (b_3 and d_3). Therefore, habitat-related changes in "carrying capacity" may come from "quality" effects or through the habitat-density interaction coefficients, which have been called "habitat quantity effects" (Anderson *et al.* 2007; Mattsson *et al.* 2012) because the effect of habitat depends on the population size by changing the quantity of resources available per individual. Regardless of how the coefficients are named, it is important to realise that changes in habitat can have both "intercept" and "slope" effects on demographic rates. A major empirical challenge for integrating habitat and harvest is estimating these effects and determining the habitat dimension(s) along which they occur.

This second point is also important because the term "carrying capacity" is often used imprecisely in reference to habitat management, rather than as a specific rescaling of density- and habitat-specific demographic parameters. Without explicitly stating the functional form of birth and death rates, "carrying capacity" has little meaning other than as a dynamic equilibrium in the absence of harvest, which is never actually observed in exploited systems. Only with an explicit model (functional form) for demographic rates does "carrying capacity" have any practical utility.

Third, besides the undesirable state $N = 0$ when $h \geq r$, and the dynamic equilibrium $N = K$ when $h = 0$, there is a wide range of equilibrial N for $0 < h < r$ that satisfy:

$$rN\left(1 - \frac{N}{K}\right) - hN = 0 \qquad \text{Equation (4)}$$

If we let $Y = hN$ be the total sustainable harvest yield, a plot of Y versus N gives a "yield curve" (Fig. 1). Because we have explicitly written r and K as functions of habitat, we can show how the yield curve changes with habitat and how this depends on specific values of the coefficients (Fig. 1).

Some points about the yield curve deserve emphasis. First, the yield curve is not directly observable. Instead it is a representation of the dynamic equilibria of a population model only reached with constant parameter values and at infinite time. It can be thought of as a long-term attractor of population size if harvest and habitat remained constant. Yearly observations of population fluctuations or demographic rates are not changes in the yield curve but are instead, stochastic realisations around an average described by the yield curve. The yield curve then serves as a summary of the consequences of equations (2a,b); and the challenge for scientists and conservation planners is to propose a functional form of equations (2a,b), estimate the relevant parameters, and then make decisions based on the consequences as summarised by the yield curve. This is not a trivial task. Second, by "habitat change" we mean long term changes such as climate, agricultural policy, or actions of conservation planners that work to shift the equilibria of the model (*i.e.* the yield curve). Most conservation planners seek to affect long-term shifts in habitat that change population equilibria, not short-term fluctuations around existing conditions. Third, under AHM the USFWS derives harvest regulations through optimization of a stochastic version of a population model related to equation (1). The yield curve developed from the deterministic version of that model is extraordinarily helpful in understanding the results of the stochastic dynamic optimisation. Thus, the yield curve serves as a tool to communicate the relationship between expected harvest and expected population size.

Implementing the 2012 North American Waterfowl Management Plan: people conserving waterfowl and wetlands

DALE D. HUMBURG[1] & MICHAEL G. ANDERSON[2]

[1]Ducks Unlimited, Inc., One Waterfowl Way, Memphis, Tennessee 38120, USA.
[2]Ducks Unlimited Canada, Institute for Wetland and Waterfowl Research,
P.O. Box 1160, Stonewall, Manitoba R0C 2Z0, Canada.
*Correspondence author. E-mail: dhumburg@ducks.org

Abstract

The North American Waterfowl Management Plan (NAWMP) is a continental ecosystems model for wildlife conservation planning with worldwide implications. Since established in 1986, NAWMP has undergone continual evolution as challenges to waterfowl conservation have emerged and information available to support conservation decisions has become available. In the 2012 revision, the waterfowl management community revisited the fundamental basis for the Plan and placed greater emphasis on sustaining the Plan's conservation work and on integration across disciplines of harvest and habitat management. Most notably, traditional and non-traditional users (*i.e.* hunters and wildlife viewers) of the resource and other conservation supporters are integrated into waterfowl conservation planning. Challenges ahead for the waterfowl management enterprise include addressing tradeoffs that emerge when habitat for waterfowl populations versus habitat for humans are explicitly considered, how these objectives and decision problems can be linked at various spatial and temporal scales, and most fundamentally how to sustain NAWMP conservation work in the face of multi-faceted ecological and social change.

Key words: conservation planning, habitat, harvest, human dimensions, hunters.

Conservation planning for waterfowl in North America has, for nearly 30 years, emphasised continent-scale population objectives and associated goals for populations, habitat, and users at various geographical scales, such as administrative Flyways and Joint Ventures of the North American Waterfowl Management Plan (NAWMP). These linked features are not new to wildlife conservation and certainly not to waterfowl management. As modern waterfowl conservation was in its formative stages, Fredrick Lincoln, originator of the Flyways concept testified before the 75th Congress relating the key elements of populations, habitat and waterfowl hunting (U.S. Government Printing Office 1937):

Populations: *"It is my opinion at the present time that we have about a third of the number of ducks and geese that we had 10 or 15 years ago."*

Habitat: *"Furthermore, I am not satisfied that we can have the population we had 10 or 15 years ago, as I am not sure we could accommodate them all."*

Hunters: *"Nevertheless, I am satisfied that we are steadily progressing toward the time when we can enjoy reasonable sport."*

Efforts to develop a U.S. national waterfowl management plan during the late 1970s and early 1980s also included a focus on habitat, populations and recreational use of the resource. Richard Myshak, presenting a summary of the emerging national plan at the 1981 International Waterfowl Symposium in New Orleans (Myshak 1981), listed the goals for waterfowl management as: 1) preserve and manage the habitat needed to maintain and increase waterfowl numbers; 2) achieve optimum waterfowl population levels in relation to available habitat; and 3) provide optimum opportunity for people to use and enjoy waterfowl.

With concerns about deteriorating habitat, persistent drought in the northern plains during the 1980s, declining populations and controversy over the effects of hunting on waterfowl populations, the Canadian government at the same time initiated strategic planning for waterfowl conservation (Patterson 1985). Together, these U.S. and Canadian efforts formed the vanguard for negotiations that ultimately led to completion of the NAWMP in 1986. The NAWMP established explicit, continental scale, numeric objectives for waterfowl populations. In a summary statement, NAWMP's authors proposed:

"Meeting these goals would provide opportunity for 2.2 million hunters in Canada and the United States to harvest 20 million ducks annually. The harvest would include 6.9 million mallards, 1.5 million pintails and 675,000 black ducks. It would also provide benefits to millions of people interested in waterfowl for purposes other than hunting" (U.S. Department of the Interior and Environment Canada 1986, page 6).

Concerning specific habitat goals, the authors stated:

"The overall aim of this continental habitat program is to maintain and manage an appropriate distribution and diversity of high quality waterfowl habitat in North America that will (1) maintain current distributions of waterfowl populations and (2) under average environmental conditions, sustain an abundance of waterfowl consistent with [population] goals … (U.S. Department of the Interior and Environment Canada 1986, page 13).

Subsequent Plan updates continued evolution of the NAWMP by expanding the continental partnership to include Mexico, expanding habitat objectives to sustain growing waterfowl populations (NAWMP Committee 1994), broadening conservation strategy to regional landscapes, diversifying partnerships, and managing adaptively relative to environmental and human dynamics (NAWMP Committee 1998), and strengthening the biological foundation of waterfowl conservation planning (NAWMP

Committee 2004). Despite relatively specific goals and assumptions outlined in the 1986 NAWMP and continued updates to the Plan, ambiguity remained concerning the definition of "average environmental conditions," the extent to which harvest management should be used to achieve population goals, and lack of an explicit connection between habitat management and population goals (Runge *et al*. 2006).

Integration and efficiency were key themes as the waterfowl management community strived to develop coherence among habitat management, population management, and harvest (Runge *et al*. 2006; Anderson *et al*. 2007). Expanding dialogue about integrated management planning for waterfowl led to the Future of Waterfowl Management Workshop in August 2008 (Case & Sanders 2008) where participants agreed that work on human dimensions of waterfowl management should continue and that the next update of NAWMP should develop increasingly coherent goals for waterfowl harvest and habitat management.

Focus on integration and reassessment of fundamental goals for waterfowl management meant the 2012 NAWMP was viewed as a revision rather than as an update of the Plan (NAWMP Committee 2012a). An extensive series of stakeholder workshops during 2009–2011 was designed to break down administrative silos across waterfowl management public and private sectors. The workshops identified three strategic foci for NAWMP 2012: 1) relevance to contemporary society; 2) adaptable to changing ecological and social systems; and 3) effective and efficient with limited funding and staff resources. Ultimately, the consultation process yielded three fundamental goals for North American waterfowl management: 1) abundant and resilient waterfowl populations to support hunting and other uses without imperilling habitat; 2) wetlands and related habitats sufficient to sustain waterfowl populations at desired levels, while providing places to recreate and ecological services that benefit society; and 3) growing numbers of waterfowl hunters, other conservationists and citizens who enjoy and actively support waterfowl and wetlands conservation. These goals are important in two ways, firstly for the continued emphasis on healthy waterfowl populations and habitat to support them and secondly, in providing the new explicit goal for waterfowl supporters.

The context of the 2012 Plan was notably different than in the 1980s when a "duck crisis" was extant with record low numbers of breeding waterfowl and also deteriorating habitat conditions. In contrast, breeding waterfowl populations in the traditional survey areas in North America were at record levels during 2011–2013 (USFWS 2013) and with > 15 years of liberal hunting seasons and bag limits, the sense of urgency was less apparent. However, mid-continent breeding ground conditions aided by years of above average moisture masked the underlying deterioration of waterfowl habitat due to wetland drainage and the loss of grasslands. Additionally, growing impacts on the once pristine boreal forests in Canada, water challenges in the south and west United States, and Gulf Coast marsh loss due to sea-level rise and subsidence will likely soon have an effect on birds and in turn wildfowlers. Overall, waterfowl habitat in key North

American landscapes is being lost faster than it is being conserved, and threats to these landscapes are growing as human populations increase, water quality and quantity continue to erode, energy issues often dominate land use decisions, and a changing climate presents long-term pressures that exacerbate current threats. Moreover, numbers of waterfowl hunters have declined to half of 1970s levels (Vrtiska *et al.* 2013; Raftovich & Wilkins 2013) and conservation budgets are not keeping pace with challenges facing waterfowl. Indeed, the growing detachment of North Americans from nature (*e.g.* Louv 2006) is also a great concern for future conservation. Clearly, the need for continued focus on waterfowl conservation through NAWMP is paramount.

Priorities for implementation are found in > 30 key actions in the 2012 NAWMP Action Plan and in the following seven recommendations (NAWMP Committee 2012b):

1. **Develop, revise or reaffirm NAWMP objectives** so that all facets of North American waterfowl management share a common benchmark.

2. **Integrate waterfowl management** to ensure programs are complementary, inform resource investments and allow managers to understand and weigh tradeoffs among potential actions.

3. **Increase adaptive capacity** so structured learning expands as part of the culture of waterfowl management and programme effectiveness increases.

4. **Build support for waterfowl conservation** by reconnecting people with nature through waterfowl and by highlighting environmental benefits associated with waterfowl habitat conservation.

5. **Establish a Human Dimensions Working Group** to support development of objectives for people and ensure actions are informed by science.

6. **Focus resources on important landscapes** that have the greatest influence on waterfowl populations and those who hunt and view waterfowl.

7. **Adapt harvest management strategies** to support attainment of NAWMP objectives.

Here, we consider recommendations 1–3. Building support for waterfowl conservation (#4 above) has become primarily the responsibility of a new "Public Engagement Team" formed under the international NAWMP Committee. The Plan Committee and the National Flyway Council also have recently founded a new Human Dimensions Working Group (#5 above) for the purpose of providing social science technical support and advice to waterfowl conservation. Efforts to focus resources on the most important landscapes (#6 above) have been initiated by the NAWMP Science Support Team; and work to adapt harvest strategies relative to revised NAWMP goals (#7 above) is pursued by the existing Harvest Management Working Group chaired by the U.S. Fish & Wildlife Service.

Initial progress toward implementing the 2012 NAWMP Revision requires focus on recommendations 1–3 that will define actions by the waterfowl management community toward integration across populations, habitat, and waterfowl supporters. Chief among these is the need to revisit objectives established in the 1986 Plan. As an essential feature of structured decision-making and adaptive management (Williams *et al.* 2009), objective setting provides context for identifying management alternatives, monitoring and the future review of objectives. Thus, the focus for initial implementation will be on fundamental objectives and means to accomplish these objectives.

Objectives serve three primary purposes in conservation planning: 1) they operate as a communication and marketing tool to demonstrate the need for conservation; 2) they provide a biological basis and planning foundation; and 3) they function as a performance measure for assessing conservation accomplishments. Thus, managers must be clear about how best to craft and communicate revised objectives. Objectives should be linked at administrative and implementation scales whereby tradeoffs can be identified and efficiencies gained with available resources.

Population objectives

Objectives for waterfowl populations have remained largely unchanged since 1986. Benchmarks for several goose populations have been amended due to dramatic changes in abundance and distribution of geese; however, most duck objectives have not been revisited despite changes in bird numbers, breeding and non-breeding landscapes, and the hunter population. Experience gained since the mid-1980s provides perspectives on appropriateness of revisions in population objectives. Substantial land-use changes have occurred in some landscapes resulting in variation in the capacity of habitats to support waterfowl. Managers recognise the extent of variation in annual environmental conditions and question utility of striving for population averages. In addition, the degree of management influence on population dynamics remains uncertain. Finally, managers have increased their knowledge and experience of the responses of birds to habitat restoration and management and the impacts of harvest regimes.

Numeric population objectives have been particularly important for habitat managers who translated resource requirements of birds into objectives for protection, restoration and management of habitat. Population objectives, framed as averages, remain problematic as management targets because of variation in wetland conditions and other key environmental influences on waterfowl populations. Moreover, active, adaptive management requires sophisticated monitoring to track population vital rates and environmental conditions. Additionally, population objectives should also be consistent with goals for habitat and human use. Because these criteria frequently have not been met, a more rigorous conceptual perspective on population status, interaction of birds with their habitat and expectations for resource use is required.

As NAWMP population objectives are reassessed, legitimate alternatives will be considered. Among these are establishment of an objective at a relatively high level, a minimum level below which managers are concerned about sustaining populations, a "normal" operating range that reflects variation in population size and distribution attributable to uncontrolled environmental processes and the simultaneous management of multiple species and populations. Gains in management outcomes will be limited by the level of technical support required, data needed to inform decisions and the degree of complexity in the process. Although daunting, progress on these fronts has been made. For example, life-cycle modelling for Northern Pintail (*Anas acuta*: Mattson *et al.* 2012), scaup species (*Aythya affinis* and *A. marila*: Austin *et al.* 2014; Osnas *et al.* 2014) and American Black Duck (*Anas rubripes*: Devers & Collins 2011) has already seen considerable progress.

Objectives for waterfowl supporters

The 2012 NAWMP Revision explicitly acknowledged people as fundamental to the Plan. The decline in wildfowling is acknowledged and integrated into management planning. Considerable changes in social structure, an aging population and a shift to urban residence all contribute to this decline (Louv 2006; Wentz & Seng 2000). Most managers recognise the need to increase the relevance of waterfowl conservation to constituencies beyond hunters; however, this need is poorly understood and not accepted as important by the entire waterfowl management community. Three interest groups are specifically mentioned in the 2012 revision of the Plan – waterfowl hunters, bird-watchers and waterfowl conservation supporters. The particular weight placed on each in the management process is largely dependent on subjective values placed on numbers of birds, distribution, harvest opportunity, viewing and ecological services provided by landscapes that support birds and humans. There will not be a "right" answer with respect to objectives related to people. The emerging question is "Whose values matter and to what degree?" Values of waterfowl hunters, harvest managers, bird-watchers and landowners are different but all are legitimate, so tradeoffs inevitably will be necessary.

Objectives for waterfowl habitat

Protection, restoration and management of habitat are primary conservation tools affecting the capacity of North American landscapes to support waterfowl and waterfowl enthusiasts. Substantial gains over the period of NAWMP implementation, estimated at nearly 7 million ha (http://www.fws.gov/birdhabitat/NAWMP/index.shtm), have not necessarily kept pace with net changes in landscapes, but these are poorly quantified (NAWMP Assessment Steering Committee 2007). When developing habitat objectives, managers should take into account the association between waterbird numbers and the carrying capacity of the landscape, as well as the influence of variable environmental conditions on population demography and distribution.

Stepping-down continental objectives for habitat to regional or local scales is a logical process; however, it is largely dependent on selection of continental population objectives and an understanding of the influence of regional habitat on population processes. Thus, a key initial step for the revised NAWMP is to establish population objectives, despite considerable uncertainty about factors regulating populations of different species and the influence of habitats in different landscapes. Efficient allocation of conservation budgets also requires acknowledgment of the on-going status of habitats – whether secure or at risk in the near or long-term. Stepping-down revised population objectives will not be a trivial matter. Trends in land use and agricultural markets worldwide represent significant influences on waterfowl conservation efforts, and sustaining habitat carrying capacity for continental waterfowl populations will be challenging, especially with added complexity to satisfy objectives from all waterfowl enthusiasts. For successful waterfowl conservation, needs of human users of the resources must be considered and addressed using balanced strategies.

To date, most habitat management partnerships have considered waterfowl population objectives with only limited regard for human considerations except for addressing factors directly affecting habitat delivery (*e.g.* funding for conservation and for landowners' acceptance of programme options). Additionally, habitat for those other than traditional users (hunters) has been considered only rarely. Complexity in planning habitat management for the benefit of waterfowl will increase as managers acknowledge that landscapes valuable for waterfowl also have values beyond the interests of ducks and hunters. Habitat objectives that integrate goals for waterfowl, other wildlife, and humans present tradeoffs that may be quite different across landscapes. For instance, factors affecting waterfowl recruitment and survival versus those that determine engagement by users vary considerably among regions. Strictly from a waterfowl perspective, emphasis on breeding habitat is appropriate because the factors primarily affecting population growth rates occur during the breeding season (see Hoekman *et al.* 2002; Koons *et al.* 2006; Coluccy *et al.* 2008). Human populations, however, are distributed differently (*e.g.* most reside outside the breeding grounds), and habitat managed for users will present considerations beyond the traditional mission of habitat delivery for waterfowl alone. Waterfowl managers therefore will be challenged to formulate habitat objectives in the context of consumptive and non-consumptive human use plus continental waterfowl population objectives.

From individual objectives to an integrated system – challenges at multiple scales

The 2012 NAWMP Revision accepted that successful management of waterfowl populations, conservation of waterfowl habitat, and engagement of waterfowl users and supporters are inseparably linked components of waterfowl conservation. To manage the different components effectively and responsively, a management system that embraces these interrelationships should be employed. Such a coherent system will help focus on things that matter most for efficient achievement of all NAWMP goals.

An integrated management system should inform resource investment decisions by allowing managers to understand and weigh tradeoffs among potential actions. This approach will require increased adaptive capacity, and institutions and processes that enable united action. Features of an integrated management system should include quantifiable, coherent objectives; an overarching framework comprised of linked models; decision tools that help inform resource allocations at multiple spatial and temporal scales; coordination among multiple management authorities and decision nodes; and monitoring and assessment to track progress and enable adaptation.

As NAWMP planners proceed with development of an integrated system they face two immediate technical and process challenges: firstly, how will multiple objectives for waterfowl management be established? Can they rely on existing institutions and do they need the assistance of a new entity with overarching facilitation functions? Whatever the process, it will need to be iterative and adaptive. Secondly, how will managers monitor progress toward achieving expanded NAWMP objectives and adapt actions to results? For instance, what technical and human resources will be needed, and who will make the many adaptive decisions going forward? Indeed, no existing entity possesses clear responsibility for all interrelated decision-making that will emerge in an integrated system – not the Flyway Councils, not the Service Regulations Committee, not the Plan Committee, and not any single country.

During development of the 2012 Revision an *ad hoc* technical team tried but abandoned efforts to develop a singular formal structured decision making (SDM) framework for waterfowl management decisions. They recognised a daunting number of decision nodes, many decision makers and decision cycles operating at multiple spatial and temporal scales (Fig. 1), and noted that analytical challenges consistent with multiple objectives under the Plan were not independent. The team involved in preparing the Revision therefore advocated instead "linked decision processes" and a continuing commitment to adaptive management. However, how to link various nodes and scales is not readily apparent, and this need might vary greatly among individual management decision problems (NAWMP Action Plan 2012).

So what might comprise an integrated management system? Certainly, coherent quantifiable objectives would be one component, along with some concept of tradeoffs amid pursuit and fulfilment of multiple objectives. Multiple decision processes required for management of habitats, populations, harvest, users and supporters are likely to be diverse in nature, and we may be well-served by trying various approaches. Several candidate approaches have already been mentioned including elaboration of the Joint Task Group (JTG; Anderson *et al.* 2007) framework, SDM, scenario planning, decision-criteria matrices, resilience thinking and others (Appendix A in NAWMP Committee 2012). Each has advantages and limitations but can provide a basis for prediction, learning and improved decision making over time. In any case, a commitment to monitoring and assessment is critical for progress in understanding system dynamics and improving management performance.

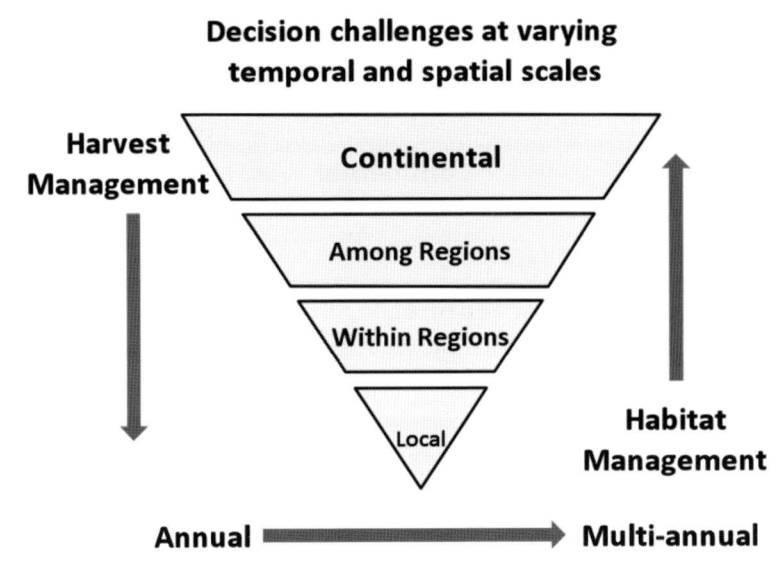

Figure 1. Schematic representation of waterfowl management decisions which are made by different managers and decision-making bodies at multiple spatial and temporal scales. Linking these decisions to bring coherence to the overall management of waterfowl populations is challenging. (Illustration by John M. Eadie, University of California-Davis).

Learning how to achieve multiple objectives simultaneously may be particularly challenging. Using a suite of different conservation projects, or at least some explicit tradeoffs in how individual parcels of habitat are managed, may be valuable. These kinds of tradeoffs need to be addressed in multiple places, as the nature of these tradeoffs will vary among environmental and social regions and over time.

When objectives are selected, an important next step will be to identify main sources of uncertainty that face attainment of objectives. These are likely to include matters of management control and weaknesses in our present knowledge of system dynamics. These uncertainties may also be expressed at multiple spatial and temporal scales and involve multiple human institutions.

Prioritizing among many monitoring and assessment efforts will be challenging, but we may find some approaches that inform multiple questions. Then managers must create the commitment to undertake this vital adaptive management work. A necessary related step will be to identify the main coordination challenges among existing administrative processes and institutions and ensure these are addressed in a manner that allows effective adaptive management for multiple, interrelated objectives.

Linking adaptive management cycles among spatial scales (Fig. 2) would be advantageous. Perhaps the easiest way to visualise this linkage is with a single suite of objectives for habitat conservation (Fig. 2). Adding objectives for waterfowl populations and users should work the same

Linking adaptive cycles across spatial, temporal and institutional scales

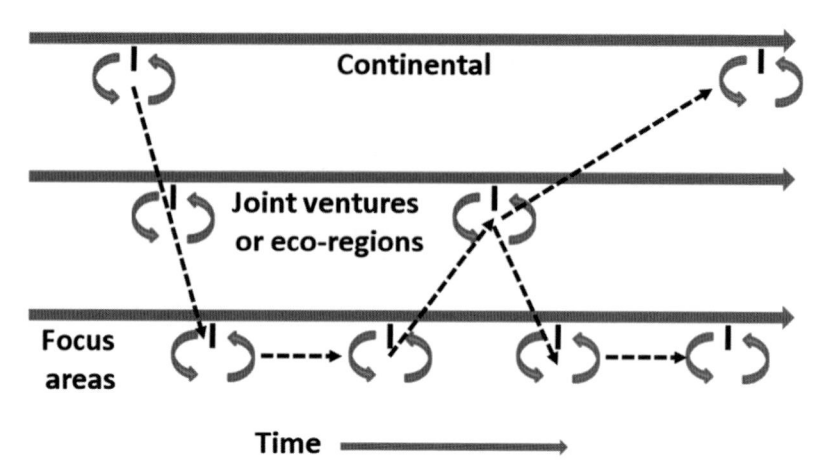

Figure 2. Links between adaptive management cycles at different spatial scales, required to ensure coherence and efficiency in waterfowl habitat management in North America. The left-hand set of links reflects the downward decision-making from continental to local scales; to the right, frequent decisions and feedback at the local scale contribute to to regional and ultimately continental decisions and outcomes.

in principle although with added complexity. Adaptive cycles should work most rapidly at the smaller spatial scales where scale-relevant responses should be detectable relatively quickly. Also, existence of many small focal areas presents opportunity for innovation and experimentation in ways that can accelerate learning about system dynamics and veracity of planning assumptions. At the continental and largest scale of interest for NAWMP, cycles of adaptation will happen more slowly but will have great impact when learning and change occur. Clearly, progress is made in well-connected learning organisations (Senge 1990, 2006; Bennis & Biederman 1997). Therefore, we must nurture strong linkages of information exchange between scales and among management units at equivalent scales, which should foster efficient and effective responses of the whole system to changes and acquisition of new knowledge.

Most of these linked system models are likely to be designed as decision-support tools for specific purposes and at various scales, and no single model is likely to serve the purpose for all decision-support needs. Linkage of decisions seems most important where true co-dependencies exist, such as between harvest potential and habitat carrying capacity or between demographic metrics such as winter survival rates and hunter access and success. Such linked system models should provide a means to predict consequences of management actions for attaining multiple objectives while resolving uncertainty. Some models will be empirically based and rigorous, relying on long-term data and well-documented demographic responses to management actions. Other models for poorly understood species or processes may be more qualitative or hypothetical.

Increasing adaptive capacity

Once objectives have been established, and key decisions identified and linked, the next logical step is to develop adaptive frameworks and actions that will allow waterfowl managers to learn from management efforts (North American Waterfowl Management Plan Committee 2012). The job of "increasing adaptive capacity," has at least two major components: 1) developing technical framework and plans to achieve increased capacity, and 2) mustering political and financial support and acquiring leadership to ensure implementation of the plan. Existing technical working groups should be able to address the technical framework, but new collective action seems necessary to garner resources and organise processes amongst institutions so that needed adaptive loops actually function.

With adoption of population, habitat and human goals in the new Plan, there is additional need for integration of goal-setting, modelling, monitoring activities and institutional support systems. The Plan Committee was adequately structured for its initial tasks of overseeing creation of the Joint Ventures, coordinating with the Flyway Councils, and generally guiding evolution of the 1986 Plan. However, changes began with the creation of the NAWMP Science Support Team (NSST) in 1999. The NSST, with an unfunded science-support mandate, struggled to generate deliverables requested by the Plan Committee. Appointments of JV science coordinators in the US and their part-time assignments to work on the NSST brought much-needed capacity to bear. Coupled with the work of temporary task groups like the NAWMP Continental Assessment team (NAWMP Assessment Steering Committee 2007), the NSST has made several advancements to guide habitat delivery of the Joint Ventures, but have proceeded well short of their plans and potential. Funding important research and planning activities that over-arch multiple JVs has remained particularly challenging.

Today, the broader vision of the 2012 NAWMP Revision has moved waterfowl management and the Plan Committee into a new realm. This new vision includes science support for social and ecological sciences and underscores the importance of the new Human Dimensions Working Group, the NSST and the Harvest Management Working Group. The time is rapidly approaching when increased, adaptive capacity under NAWMP will be mission-critical. When waterfowl managers have renewed explicit objectives to drive integrated decision frameworks, the adaptive capacity needed to support waterfowl management should become both more obvious and urgent.

In summary, by 2016 our collective high-priority waterfowl management goals should be to:

1. **Establish quantifiable objectives** for population and habitat conservation,

harvest opportunity and user participation at appropriate spatial scales and with acknowledged tradeoffs among them.

2. **Design an integrated framework** for making linked harvest, habitat and user-supporter management decisions where important dependencies exist among management objectives.

3. **Design and implement monitoring and evaluation programmes** to track progress toward objectives and inform each key decision problem.

4. **Seek ways to fund the process.**

In doing this we should recognise that we are unlikely to "get it right" from the outset, so we must plan to re-plan. We would be foolhardy to expect that a revised set of NAWMP objectives will serve our needs for the next 28 years as have the original 1986 objectives. This new endeavour will be challenging, technically and administratively – the valuing exercises, the modelling, the adaptive management frameworks, coordinated execution and finding fiscal support for the Plan will be needed to ensure its success. False starts or dead ends seem likely, so there may be advantages in exploring multiple options, especially at smaller scales where relatively rapid replication and learning may be most achievable. In this light, a commitment to managing adaptively may be more important than ever.

Acknowledgments

Co-authors of individual presentations at the 2012 NAWMP session at the Ecology and Conservation of North American Waterfowl Symposium included Ray Alisauskas, Mike Anderson, Tim Bowman, Bob Clark, John Eadie, Jody Enck, David Fulton, Dale Humburg, Kevin Hunt, Mark Koneff, Andrew Raedeke, Jim Ringelman, Greg Soulliere. Contributors to the 2012 NAWMP revision and to past and future waterfowl conservation include the countless scientists, managers and policy makers who will inform, apply and decide on the path forward. We thank Lisa Webb and Rick Kaminski for helpful comments on an earlier draft of this manuscript.

References

Anderson, M.G., Caswell, F.D., Eadie, J.M., Herbert, J.T., Huang, M. Humburg, D.D., Johnson, F.A., Koneff, M.D., Mott, S.E., Nudds, T.D., Reed, E.T., Ringelman, J.R., Runge, M.C. & Wilson, B.C. 2007. *Report from the Joint Task Group for Clarifying North American Waterfowl Management Plan Population Objectives and Their Use in Harvest Management.* NAWMP Joint Task Group unpublished report, U.S. Fish & Wildlife Service and U.S. Geological Survey, Washington D.C., USA. Accessible at http://nawmprevision.org/sites/default/files/jtg_final_report.pdf (last accessed 10 July 2014).

Austin, J.E., Slattery, S., & Clark, R.G. 2014. Waterfowl populations of conservation concern: learning from diverse challenges, models and conservation strategies. *Wildfowl* (Special Issue No. 4): 470–497.

Bennis, W. & Biederman P.W. 1997. *Organizing Genius: The Secret of Creative Collaboration.* Basic Books, New York, USA.

Case, D. & Sanders S. 2008. The future of waterfowl management workshop: framing future decisions for linking harvest, habitat and human dimensions. Report 10-9-08. D.J. Case and Associates,

Mishawaka, Indiana, USA. Accessible at http://www.nawmprevision.org/sites/default/files/future_of_waterfowl_mgt_workshop_final_report.pdf (last accessed 10.07.14).

Coluccy, J.M., Yerkes, T., Simpson, R., Simpson, J.W., Armstrong, L. & Davis, J. 2008. Population dynamics of breeding mallards in the Great Lakes states. *Journal of Wildlife Management* 72: 1181–1187.

Devers, P.K. & Collins, B. 2011. *Conservation Action Plan for the American Black Duck, First Edition*. U.S. Fish and Wildlife Service, Division of Migratory Bird Management, Laurel, Maryland, USA.

Hoekman, S.T., Mills, L.S., Howerter, D.W., DeVries, J.H. & Ball, I.J. 2002. Sensitivity analyses of the life cycle of mid-continent Mallards. *Journal of Wildlife Management* 66: 883–900.

Koons, D.N., Rotella, J.J., Willey, D.W., Taper, M., Clark, R.G., Slattery, S., Brook, R.W., Corcoran, R.M. & Lovvorn, J.R. 2006. Lesser scaup population dynamics: what can be learned from available data? *Avian Conservation and Ecology* 1(3): 6. Accessible at http://www.ace-eco.org/vol1/iss3/art6/ (last accessed 10 July 2014).

Louv, R. 2006. *Last Child in the Woods: Saving Our Children From Nature-Deficit Disorder*. Algonquin Books of Chapel Hill, Chapel Hill, North Carolina.

Mattsson, B.J., Runge, M.C., Devries, J.H., Boomer, G.S., Eadie, J.M., Haukos, D.A., Fleskes, J.P., Koons, D.N., Thogmartin, W.E. & Clark R.G. 2012. A modelling framework for integrated harvest and habitat management of North American waterfowl: Case-study of northern pintail metapopulation dynamics. *Ecological Modelling* 225: 146–158.

Myshak, R.J. 1981. National waterfowl management plans: United States viewpoint. *In* Ducks Unlimited (ed.), *Proceedings of the Fourth International Waterfowl Symposium, New Orleans, Louisiana*, pp. 50–52. Ducks Unlimited, Long Grove, Illinois, USA.

North American Waterfowl Management Plan Assessment Steering Committee. 2007. North American Waterfowl Management Plan: continental progress assessment. Final report. Canadian Wildlife Service, U.S. Fish & Wildlife Service, Secretaria de Medio Ambiente y Recursos Naturales and Secretaria de Desarrollo Social. Accessible at http://nawmprevision.org/sites/default/files/2007ContinentalAssessment.pdf (last accessed 10 July 2014).

North American Waterfowl Management Plan Committee. 1994. 1994 update, North American Waterfowl Management Plan: expanding the commitment. Canadian Wildlife Service, U.S. Fish & Wildlife Service, Secretaria de Medio Ambiente y Recursos Naturales and Secretaria de Desarrollo Social. Accessible at http://www.fws.gov/birdhabitat/NAWMP/Planstrategy.shtm (last accessed 10 July 2014).

North American Waterfowl Management Plan Committee. 1998. 1998 update, North American Waterfowl Management Plan: expanding the vision. Canadian Wildlife Service, U.S. Fish & Wildlife Service and Instituto Nacional de Ecologia. Accessible at http://www.fws.gov/birdhabitat/NAWMP/Planstrategy.shtm (last accessed 10 July 2014).

North American Waterfowl Management Plan Committee. 2004. North American Waterfowl Management Plan: 2004 strategic guidance strengthening the biological foundation. Canadian Wildlife Service, U.S. Fish & Wildlife Service and Secretaria de Medio Ambiente y Recursos Naturales. Accessible at http://www.fws.gov/birdhabitat/NAWMP/Planstrategy.shtm (last accessed 10 July 2014).

North American Waterfowl Management Plan Committee. 2012a. North American Waterfowl Management Plan: People conserving waterfowl and wetlands. Canadian Wildlife Service, U.S. Fish & Wildlife Service and Secretaria de Medio Ambiente y Recursos Naturales. Accessible at http://www.fws.gov/birdhabitat/NAWMP/Planstrategy.shtm (last accessed 10 July 2014).

North American Waterfowl Management Plan Committee. 2012b. NAWMP Action Plan: A Companion Document to the 2012 North American Waterfowl Management Plan. U.S. Fish and Wildlife Service, Washington D.C. Accessible at http://www.fws.gov/birdhabitat/NAWMP/Planstrategy.shtm (last accessed 10.07.14).

Osnas, E.E., Runge, M.C., Mattsson, B.J., Austin, J.E., Boomer, G.S., Clark, R.G., Devers, P., Eadie, J.M., Lonsdorf, E.V. & Tavernia, B.G. 2014. Managing harvest and habitat as integrated components. *Wildfowl* (Special Issue No. 4): 305–328.

Pahl-Wostl, C. 2009. A conceptual framework for analysing adaptive capacity and multi-level learning processes in resource governance regimes. *Global Environmental Change-Human and Policy Dimensions* 19: 354–365.

Patterson, J.H. 1985. The Canadian plan. *In* Ducks Unlimited (ed), *Proceedings of the Fifth International Waterfowl Symposium*, Kansas City, Missouri, pp. 15–17. Ducks Unlimited, Long Grove, Illinois, USA.

Raftovich, R.V. & Wilkins, K.A. 2013. Migratory bird hunting activity and harvest during the 2011–12 and 2012–13 hunting seasons. U.S. Fish and Wildlife Service, Laurel, Maryland, USA.

Runge, M.C., Johnson, F.A., Anderson, M.G., Koneff, M.D., Reed, E.T. & Mott, S.E. 2006. The need for coherence between waterfowl harvest and habitat management. *Wildlife Society Bulletin* 34: 1231–1237.

Senge, P.M. 1990. *The Fifth Discipline: The Art and Practice of the Learning Organization*. Doubleday, New York, USA.

Senge, P.M. 2006. *The Fifth Discipline: the Art and Practice of the Learning Organization. Revised Edition*. Doubleday, New York, USA.

U.S. Government Printing Office. 1937. Hearing before the Select Committee on Conservation of Wildlife Resources, House of Representatives. U.S. Government, Washington D.C., USA.

U.S. Department of the Interior & Environment Canada. 1986. *North American Waterfowl Management Plan*. U.S. Department of the Interior, Washington D.C., USA.

U.S. Fish and Wildlife Service. 2013. Waterfowl population status, 2013. U.S. Department of the Interior, Washington D.C., USA.

Vrtiska, M.P., Gammonley, J.H., Naylor, L.W. & Raedeke, A.H. 2013. Economic and conservation ramifications from the decline of waterfowl hunters. *Wildlife Society Bulletin*. 37: 380–388.

Wentz, J. & Seng, P. 2000. Meeting the challenge to increase participation in hunting and shooting. Final Report to the National Shooting Sports Foundation and International Hunter Education Association. Silvertip Productions, Ltd., Reynoldsburg, Ohio, USA.

Williams, B.K., Szaro, R.C. & Shapiro, C.D. 2009. *Adaptive Management: The U.S. Department of the Interior Technical Guide*. Adaptive Management Working Group, U.S. Department of the Interior, Washington D.C., USA.

Wetland issues affecting waterfowl conservation in North America

HEATH M. HAGY[1]*, SCOTT C. YAICH[2], JOHN W. SIMPSON[3], EDUARDO CARRERA[4], DAVID A. HAUKOS[5], W. CARTER JOHNSON[6], CHARLES R. LOESCH[7], FRITZ A. REID[8], SCOTT E. STEPHENS[9], RALPH W. TINER[10], BRETT A. WERNER[11] & GREG S. YARRIS[12]

[1]*Illinois Natural History Survey, Forbes Biological Station – Bellrose Waterfowl Research Center, University of Illinois at Urbana-Champaign, Havana, Illinois 62644, USA.
[2]Ducks Unlimited, One Waterfowl Way, Memphis, Tennessee 38120, USA.
[3]Winous Point Marsh Conservancy, 3500 S Lattimore Road, Port Clinton, Ohio 43452, USA.
[4]Ducks Unlimited Mexico, Ave. Vasconcelos 209 Ote. Residencial San Agustin, Garza Garcia, Nuevo León, Mexico.
[5]U.S. Fish and Wildlife Service, Texas Tech University, Lubbock, Texas 79409, USA.
[6]Department of Natural Resource Management, South Dakota State University, Brookings, South Dakota 57007, USA.
[7]U.S. Fish Wildlife Service, Habitat and Population Evaluation Team, Bismarck, North Dakota 58501, USA.
[8]Ducks Unlimited, 3074 Gold Canal Drive, Rancho Cordova, California 95670, USA.
[9]Ducks Unlimited Canada, P.O. Box 1160, Stonewall, Manitoba ROC 2ZO, Canada.
[10]U.S. Fish and Wildlife Service, 300 Westgate Center Drive, Hadley, Massachusetts 01035, USA.
[11]Program in Environmental Studies, Centre College, 600 West Walnut Street, Danville, Kentucky 40422, USA.
[12]Central Valley Joint Venture, 2800 Cottage Way, Sacramento, California 95825, USA.
* Correspondence author. E-mail: hhagy@illinois.edu

Abstract

This paper summarises discussions by invited speakers during a special session at the 6th North American Duck Symposium on wetland issues that affect waterfowl, highlighting current ecosystem challenges and opportunities for the conservation of waterfowl in North America. Climate change, invasive species, U.S. agricultural policy (which can encourage wetland drainage and the expansion of row-crop agriculture into grasslands), cost and competition for water rights, and wetland management for non-waterfowl species were all considered to pose significant threats to waterfowl populations in the near future. Waterfowl populations were found to be faced with significant threats in several regions, including: the Central Valley of California, the

Playa Lakes Region of the south-central U.S., the Prairie Pothole Region of the northern U.S. and western and central Canada, the boreal forest of northern Canada, the Great Lakes region and Latin America. Apart from direct and indirect threats to habitat, presenters identified that accurate and current data on the location, distribution and diversity of wetlands are needed by waterfowl managers, environmental planners and regulatory agencies to ensure focussed, targeted and cost-effective wetland conservation. Although populations of many waterfowl species are currently at or above long-term average numbers, these populations are thought to be at risk of decline in the near future because of ongoing and predicted nesting habitat loss and wetland destruction in many areas of North America.

Key words: agriculture, climate change, dabbling duck, national wetlands inventory, playa, policy, prairie pothole.

To the casual observer, it might seem that wetland-dependent wildlife face few conservation issues at present in North America. Dahl (2006) showed a 0.3% gain in deepwater and wetland area in the continental United States (*i.e.* excepting Alaska and Hawaii) between 1998 and 2004. During the early 21st century, numbers of breeding ducks have remained at or above their long-term average population estimates, and populations of several species (*e.g.* Blue-winged Teal *Anas discors* and Northern Shoveler *A. clypeata*) are at all-time highs (USFWS 2013). Even Lesser Scaup *Aythya affinis* and Northern Pintail *Anas acuta* populations have reversed historical declines and seem to be steady or increasing in number. The abundance of ponds and wetlands containing water in May (*i.e.* "May ponds") in breeding areas surveyed annually by the U.S. Fish and Wildlife Service and the Canadian Wildlife Service, which serves as an indicator of wetland habitat availability and waterfowl productivity, was 42% above the long-term average in summer 2013. Breeding populations and also the number of May ponds appear to be near or above levels observed in the early 1970s and late 1990s, both periods thought to be the "good old days" by waterfowl conservationists (Vrtiska *et al.* 2013). Moreover, waterfowl hunting regulations have remained liberal since the introduction of the Adaptive Harvest Management programme in 1995, allowing for maximum take (regulated by bag limits) of most species (Nichols *et al.* 2007; Vrtiska *et al.* 2013).

Despite currently large waterfowl population sizes, many threats loom that cause informed wetland and waterfowl conservationists to worry about the future. Dahl (2011) documented a loss in wetland area and only modest gains in the number of all wetlands and deepwater habitats combined during 2004–2009. Additionally, losses in vegetated wetlands have been largely offset by gains in agricultural and urban ponds and other non-vegetated wetlands, which likely are of less value to waterfowl, other waterbirds and other wildlife (Weller & Fredrickson 1974; Dahl

2011). Also of concern is that wetland losses have not been evenly distributed among regions and systems; for instance, losses in the Prairie Pothole Region and the lower Mississippi Alluvial Valley, which provide some of the most important habitat for breeding and wintering waterfowl in North America, have been more pronounced than in other regions (Dahl 2011; Johnston 2013). Unfortunately, we can expect that current May pond abundance and waterfowl breeding population size are facing probable declines in the future (Johnson *et al.* 2010; Johnston 2013). Agricultural policies that have long provided some protection for geographically-isolated wetlands through the "Swampbuster" provision in the U.S. Farm Bill now contain reduced or increasingly ineffective conservation provisions. Incentive-based wetland restoration, creation and protection programmes also face declining funding or elimination. Furthermore, mandates for ethanol production (*i.e.* the Renewable Fuels Standard) coupled with crop insurance policies have provided incentives for wetland drainage in the U.S. Great Plains (Reynolds *et al.* 2006; Johnston 2013). Reductions in federal spending and relatively high waterfowl populations may dissuade policy makers from prioritising wetland conservation policies in future Farm Bills. For these and many other reasons that will be highlighted subsequently, we deemed it necessary to convene a forum where scientists and conservation leaders could discuss current wetland policy and management issues that may affect waterfowl conservation efforts in the near future.

Recognising the ongoing and increasingly significant threats to wetlands and wetland wildlife, the Wetlands Working Group of the Wildlife Society held a special session at the 6th North American Duck Symposium – "Ecology and Conservation of North American Waterfowl", to describe and summarise issues affecting wetland conservation relating to waterfowl in North America. Here we present topics discussed at this session and provide an overview of current wetland issues affecting waterfowl conservation in North America. Our objectives are to: 1) outline the growing threats to wetlands and waterfowl in North America, 2) generally highlight current research and management that addresses these issues, and 3) provide recommendations for future actions that may benefit wetland and waterfowl conservation in North America.

Wetland policy

The United States

In the minds of biologists, hunters and the general public, waterfowl are stereotypically and appropriately linked to wetlands and other aquatic habitats. Yet, while the waterfowl management and scientific community has dedicated substantial resources to population and habitat management, there has been much less effort devoted to providing the scientific foundation for securing policies that maintain wetland habitats. The success or failure of these policies in maintaining the continent's wetland habitats will ultimately determine the level of success achievable by waterfowl conservationists.

The series of wetland status and trends reports produced by the United States Fish and Wildlife Service (USFWS) from the mid-1950s through to 2009 provides evidence of the impact of policies on wetlands in the U.S. The first report, examining the mid-1950s to the mid-1970s (Frayer et al. 1983), documented a loss of 113 million acres (c. 46 million ha) of wetlands with net losses approaching a half-million acres (c. 202,000 ha) annually. However, implementation of the Clean Water Act (CWA) in the mid-1970s provided some degree of federal protection to most wetlands, including the prairie potholes of the north-central United States and Canada, a key region for waterfowl production. The status and trends report for the mid-1970s to the mid-1980s (Dahl & Johnson 1991) documented a slowing of the national rate of net wetland loss to approximately one-third of pre-CWA rates. In 1985, the Swampbuster provision of the federal U.S. Farm Bill, which stopped agricultural subsidy payments to landowners who drained wetlands for farming (Dahl 2011; Johnston 2013), added another critical layer of protection to many wetlands at risk of being drained for agricultural uses.

To complement the regulatory protections of the Clean Water Act and disincentives of Swampbuster, voluntary incentive-based wetland conservation programmes such as the Wetland Reserve Program, the North American Wetlands Conservation Act, the Conservation Reserve Program and the USFWS Partners for the Fish and Wildlife Program were established in the late 1980s and 1990s. Concurrently, regulatory deceleration of wetland losses and the incentives towards maintaining and restoring wetlands were reflected in a net rate of loss 79% lower than that of the 1950s–1970s (Dahl 2000). The trend of increasing broad and protective wetland policies continued through the early 1990s, and by 2004 the net loss rate of wetlands most important to waterfowl and other wildlife had declined to approximately 80,000 acres (c. 32,000 ha) per year (Dahl 2006).

However, the tide of wetland conservation policy turned in 2001 with the U.S. Supreme Court in favour of the Solid Waste Agency of Northern Cook County's (SWANCC) appeal against the presence of migratory birds being used as the sole determinant for the U.S. Army Corps of Engineers' (USACE) jurisdiction over waters of the United States (SWANCC *versus* USACE). The Supreme Court's decision greatly narrowed the perceived jurisdiction of the Corps to regulate the drainage and infilling of wetlands not adjacent to open and clearly navigable waters (Dahl 2011). In response, the U.S. Environmental Protection Agency and USACE withdrew federal Clean Water Act protections from broad swaths of wetland categories, including so-called "geographically isolated wetlands" such as the prairie potholes, rainwater basins and playa wetlands of the Great Plains (Haukos & Smith 2003). At the same time, funding for many of the incentive-based conservation programmes peaked and has since declined.

The findings of the most recent assessment of wetland status and trends (Dahl 2011) mirrored this shift in conservation policy. For the first time in 50 years, wetland loss accelerated, increasing by 140% compared with 1998–2004. Policy-based funding for wetland conservation programmes has continued to decline, and changes to Farm Bill policy place *c*. 1.4 million wetlands in the Prairie Pothole Region of North and South Dakota at high risk of being drained and lost (Reynolds *et al.* 2006).

Canada

In the Prairie Pothole Region of Canada, wetlands represent a significant obstacle to production agriculture. As a result, wetland drainage continues to occur despite growing evidence of ecological goods and services that wetlands provide, including flood protection, carbon storage and groundwater recharge (Millar 1989). The jurisdiction for Canadian wetland policy resides at the provincial level (Rubec *et al.* 1998). As a result, effective policies that protect existing wetlands must be developed for each provincial jurisdiction if wetlands across Canadian landscapes important to waterfowl, such as the Prairie Pothole Region, are to be protected effectively.

High commodity prices and several years of above normal precipitation have resulted in high rates of wetland drainage to facilitate increased areas being put to agricultural production across the Prairie Pothole Region. For example, Ducks Unlimited Canada recently estimated that in Saskatchewan alone > 6,000 ha of wetlands were being drained on an annual basis (Ducks Unlimited Canada, unpubl. data). When contemporary cost estimates for wetland restoration are applied, the costs of restoring those drained wetlands would be > US$65 million. This rate of wetland loss makes maintaining an adequate wetland base to support healthy populations of breeding ducks impossible without wetland regulations that reduce loss rates.

In Alberta, implementation of a new wetland policy provides some wetland protection and requires mitigation at a ratio determined by the value of the affected wetland. However, although the new policy is largely enforced for developers and the energy sector, it is not applied consistently to agriculture (S. Stephens, pers. comm.). In Saskatchewan, policy prohibits the drainage of water from wetlands from an individual's property onto another landowner; however, these regulations have been poorly enforced, resulting in conflicts between neighbouring producers and significant unauthorised drainage across Saskatchewan. In Manitoba, existing policy protects semi-permanent and permanent ponds and lakes (Stewart & Kantrud 1971), but shallower and more ephemeral wetlands remain unprotected from drainage.

Given the different stages of progress on and viewpoints regarding wetland policy amongst the three provincial governments spanning prairie Canada, unique strategies for improving wetland policy and subsequent enforcement of regulations require diverse and nuanced approaches for each province. Currently, conservation advocates such as Ducks Unlimited Canada pursue strategies such as building a network of grassroots advocates and developing an understanding of how best to engage with those grassroots advocates in the process, providing support to affected landowners, building coalitions with agricultural industry groups around support for wetland policy, building stronger relationships with provincial staff and ministers in key ministries, developing new science to support the economic and ecological case for wetland regulation and developing a wetland monitoring system to facilitate measuring the impact of new wetland policies or lack thereof.

Latin America

Latin American countries have only relatively recently come to recognise the importance and value of their wetlands and begun to focus more attention on wetland conservation. In this region, earlier public policy efforts directed at natural resource conservation focused primarily on establishing systems of state and federal protected areas, but wetland protection was not usually a driving force behind site designations. As a result, past wetland conservation tended to be largely coincidental.

More recently, the Ramsar Convention's initiative to identify and protect Wetlands of International Importance ("Ramsar Sites") has become an important mechanism for promoting explicit recognition of the importance of wetlands and has focussed additional attention on wetland conservation in Latin America. In countries including Mexico, Colombia, Venezuela, Argentina, Chile and Brazil, government interest in the designation of Ramsar Sites has been responsible for spurring the development of national wetlands inventories and classification systems. For example, Mexico has made considerable progress in recording and classifying habitats across the entire country, with an explicit emphasis and priority being placed on regions with significant wetlands. Once in place, these inventories may prove useful as the foundation for promoting subsequent conservation activities by local, state and federal governmental entities, as well as non-governmental conservation organisations. Additionally, inventories provide guidance to outside funding institutions that can help target the allocation of resources to places and activities that can generate the greatest conservation return for their investment.

To optimise wetland and waterfowl conservation in the Latin American and Caribbean region, these nations and funding organisations should consider directing significant public policy effort toward the development of national conservation plans that include wetlands inventory data. These conservation plans should identify the most important habitats, provide information regarding the most significant site-specific conservation challenges, and propose pragmatic actions and policies that will need to be implemented to ensure long-term conservation and sustainable use of these wetlands and other wildlife habitats. The continued loss and degradation of many important wetland ecosystems, despite the existence of various international agreements and national policies, underscores the importance of developing realistic but effective conservation plans that involve and acknowledge the needs of all stakeholders in Latin America.

Important wetlands at-risk

Playa wetlands

Playas are dynamic, small, recharge wetlands located in the High Plains region of the western Great Plains in the central U.S. With ecological conditions that reflect the harsh, unpredictable environment of the High Plains, playas form a complex system providing numerous ecological functions and services, including habitat for migratory waterfowl (Haukos & Smith 1994). Essential to playa function is the erratic fluctuation between wet and dry states that creates a diversity of playa conditions or habitats throughout the entire High Plains (Smith et al. 2012). Inundation patterns and hydroperiods of playas vary annually with the average playa being inundated during January once every eleven years in Texas and New Mexico (Johnson et al. 2011a).

Playas provide habitat for migrating, wintering and breeding waterfowl (Ray et al. 2003; Baar et al. 2008; Haukos 2008). The number of inundated playas during winter determines the number of wintering waterfowl; Johnson et al. (2011a) reported that the percent of inundated playas varied from near zero in dry years to > 50% in wet years. During wet years, overwinter survival of Mallard *Anas platyrhynchos* and Northern Pintail in the High Plains is greater than for any other wintering area in North America (Bergan & Smith 1993; Moon & Haukos 2006). Estimated numbers of wintering ducks using southern playas during January ranges between 200,000 and 3 million depending on environmental conditions such as precipitation levels and winter temperatures (USFWS 1988; Haukos 2008).

The historical number of playas is unknown because of extensive landscape alteration in the High Plains during the past century (Smith et al. 2012). Recent estimates of playas vary greatly depending on the source and associated methodology used to identify playas, with published figures ranging from 30,000–80,000 playas (Smith et al. 2012; D. Haukos, pers. comm.). Although the large number of playas reported as present on the landscape gives the mistaken impression that there are sufficient functional playas capable of providing ecological services for waterfowl, Johnson et al. (2012) estimated that 17% of historical playas are no longer detectable on the southern Great Plains (Oklahoma, Texas and New Mexico). In addition, only 0.2% of existing playas have no wetland or watershed modification. Further, Johnson et al. (2012) estimated that 38.5% of historical playas had been lost from the landscape or experienced cultivation of the hydric soils, which can greatly reduce or eliminate natural forage for waterfowl. The greatest threat to playas is unsustainable sediment accumulation (Luo 1997; Smith 2003; Tsai 2007). Combining physical wetland loss, direct wetland cultivation and fill due to sediment accumulation results in an estimated 60% of historical playas that are no longer available to provide habitat for waterfowl (Johnson et al. 2012). Of the remaining playas on the southern Great Plains, none are fully functional (Johnson 2011b). These impacts to playa ecosystems likely contributed to the 32% decline in average body condition of Northern Pintail from the mid-1980s to early 2000s (Moon et al. 2007), with potential associated cross-seasonal effects on survival and reproductive capacity (Mattson et al. 2012).

Despite the acknowledged value of playas to waterfowl, conservation efforts have been stymied during the past three decades.

The vast number of playas, and the lack of perceptible physical differences in their characteristics (excepting inundation frequency) which provide value as wildlife habitat or contribute to ecological goods and services, have paralysed efforts to conserve these wetlands. The value of playas is greatest when they are considered in aggregate and regionally, although this approach is rarely used in conservation efforts (Smith *et al.* 2011; Johnson *et al.* 2012). Finally, there is lack of federal and state regulations or incentives to encourage the protection of playas, and no requirement to mitigate for any negative impacts on playa wetlands (Haukos & Smith 2003; Johnson *et al.* 2011b). The U.S. Department of Agriculture's Conservation Reserve Program, which has limited focus on wetlands compared with other habitats within the programme, is the main conservation initiative affecting playas on the High Plains. Unfortunately, playas in Conservation Reserve Program watersheds have altered hydrology characterised by reduced inundation frequency and hydroperiod possibly resulting from use of non-native vegetation in CRP plantings (Cariveau *et al.* 2011; Bartuszevige *et al.* 2012; O'Connell *et al.* 2012).

Conservation efforts should be coordinated at larger spatial and temporal scales to identify accurately the value of an individual playa. Moreover, conservation programmes need to be tailored specifically to playas as current efforts are not effective (Bartuszevige *et al.* 2012; O'Connell *et al.* 2012). Efforts to conserve playas will benefit from recognition that extreme environmental conditions are normal, and that these actually drive playa ecosystems. Relatively long temporal periods may exist between ecological states that provide high quality habitat for waterfowl. Finally, any conservation effort must consider the role and contribution of individual playas to the entire system when prioritising playas for conservation. Despite recognition of use of playas by waterfowl, the capacity of the playa system to support waterfowl is declining (Moon & Haukos 2006; Moon *et al.* 2007; Smith *et al.* 2011). Consequently, a multifaceted approach is needed to develop a playa conservation strategy that includes: 1) an educational effort to accumulate support for playa conservation, 2) modification of current conservation programmes so that playas are competitive for funding, and 3) greatly accelerating research efforts to accumulate knowledge relative to playa ecology, management and their status across the landscape.

Boreal forest wetlands

North America's boreal forest (hereafter, Boreal) is part of the largest terrestrial biome and unspoiled wetland and forest ecosystem in the world. This 600 million ha landscape stretches from western Alaska to Labrador and accounts for > 35% of the continent's forest-cover (Wells & Blancher 2011). Wetlands comprise 6% of the earth's land-cover, yet Canada alone has 25% of the world's wetlands (PEG 2011). Most of Canada's wetlands (> 85%) are in the Boreal, including bogs, fens, swamps, marshes and open water basins. Alaska's Boreal has > 2,000 rivers and streams that feed a water-rich wetland landscape. North America's Boreal holds 25% of the freshwater and

> 30% of soil carbon on the planet (PEG 2011). Despite their former isolation and vast coverage, North America's Boreal wetlands face increasing threats from climate change and expansion of industrial activities.

Prairie and boreal wetlands provide breeding habitat for the majority of duck pairs across North America (Slattery *et al.* 2011). Breeding season population estimates for the western Boreal region alone are 13–15 million birds, with many species having ≥ 50% of their breeding populations in the Boreal (Wells & Blancher 2011). The prairie and boreal biomes are arguably integrated ecologically as ducks may use the Boreal for nesting during prairie droughts and annual wing moult (Baldassarre & Bolen 2006). Consequently, extensive changes to boreal waterfowl habitat could have continental-level implications for waterfowl conservation objectives.

The perception of a pristine Boreal has changed rapidly because of the wide range of development activities occurring there, and development is predicted to increase substantially into the future (Bradshaw *et al.* 2009; Wells 2011). Seven distinct anthropogenic pressures threaten the North American Boreal, including agricultural expansion, petroleum exploration and development, forestry, hydroelectric development, mining, acid precipitation and climate change. Few regions have already and are expected to experience greater changes in mean temperatures than the Boreal (Soja *et al.* 2007; Bradshaw *et al.* 2009; Stocker *et al.* 2013), yet this biome has a great influence on global temperature and carbon storage (Bonan 2008). Impacts on Boreal wetlands may include loss of lakes and wetlands (> 40 ha in area) due to the melting of permafrost, increased evaporation and transpiration rates, and aggregation of floating emergent vegetation and associated inorganic sediments, resulting in regional decreases in surface water area (Smith *et al.* 2005; Riordan *et al.* 2006; Roach *et al.* 2011). The extent of these changes across the Boreal is currently unknown, but substantial increases are expected.

While increasing temperature may represent a threat beyond the control of classic waterfowl conservation mechanisms, other more direct anthropogenic landscape changes may be more amenable to sustainable development. Changes to hydrology can result in long-term drying (*e.g.* Bennett Dam on the Peace-Athabasca Delta) or flooding (*e.g.* Ramparts Dam proposed for the Yukon River and also several large operations in Quebec). Water pollution can potentially reach large blocks of watersheds because Boreal wetlands are often hydrologically connected through subsurface flow (Smerdon *et al.* 2005). Timber harvest may increase runoff and thus local flooding, and this can have a direct effect on the breeding success of cavity-nesting birds. Road construction can impound or drain water flowing to or from wetlands. We are only just beginning to understand the impact of these factors on waterfowl and their habitats, which challenges conservation efforts and necessitates a cautious approach to development and wildlife management in the region.

Protection of water quality, quantity and hydrologic patterns appears critical to conservation of waterfowl habitat within the Boreal. Because most Boreal wetlands recharge through lacustrine or riverine processes, those in the Alaskan boreal forest are protected under the U.S. Clean Water Act, but recent Supreme Court decisions (*e.g.* SWANCC *versus* USACE) have muddied the jurisdictional waters for many wetlands not immediately adjacent to navigable rivers or streams. In contrast there is almost no broad wetland protection in Boreal Canada, either at the federal or provincial/territorial levels, although recent legislation in Alberta may provide some level of protection. Widespread and enforceable legislative protections are critical to ensuring that the Boreal can support key North American waterfowl populations into the future.

Prairie wetlands

Wetlands potentially represent the most critical and limiting components of the landscape for breeding waterfowl (Kantrud & Stewart 1977). The Prairie Pothole Region of the north-central United States and south central Canada produces up to 75% of waterfowl in North America (Smith *et al.* 1964; Mitsch & Gosselink 2007). Wetland density in this region ranges from 4–38 potholes/km^2 (Baldassarre & Bolen 2006), but more than half of the original wetlands in the region have been lost or highly modified, principally for agriculture (Mitsch & Gosselink 2007). Moreover, conversion of native grassland and pastures to row-crop agriculture can have a dramatic effect on wetland integrity by increasing sediment and chemical runoff within the watershed (Zedler 2003). A myriad of factors including agricultural policy, changing wetland regulations, improved farming and land clearing technology, and climate change threaten wetland function and value for waterfowl in the Prairie Pothole Region (Johnston 2013; Wright & Wimberly 2013).

Prairie wetlands have been identified as particularly vulnerable to climate change. Evidence for this conclusion has come from an inter-institutional and multi-disciplinary team of investigators which has developed and used two simulation models, WETLAND SIMULATOR and WETLANDSCAPE, to project future consequences of climate change on prairie wetlands and waterfowl (*e.g.* Poiani & Johnson 1991; Poiani *et al.* 1995, 1996; Johnson *et al.* 2005, 2010; Werner *et al.* 2013). These researchers have reached four main conclusions after 20 years of research on the subject: 1) temperature matters, 2) geography matters, 3) impacts may have already occurred, and 4) threshold effects may yield future surprises. A representative simulation using weather data (1986–1989) from the Orchid Meadows field site demonstrated the effect of increasing air temperature on the length of time that water stands (hydroperiod) in a semi-permanent wetland basin (Johnson *et al.* 2004, 2010). Raising the temperature a modest 2°C shifted wetland permanence type from semi-permanent (not dry during the 4-year simulation) to seasonal (drying annually). A 4°C increase changed the wetland into one more typical of a temporary wetland that dried by late spring or mid-summer each year. This simulation, and hundreds more that have been completed across the Prairie Pothole Region (*e.g.* Poiani *et al.* 1996; Poiani & Johnson 2003), clearly illustrate how sensitive prairie wetland hydrology is to air temperature.

The Prairie Pothole Region is of modest geographic area, comprising *c.* 800,000 km² in the U.S. and Canada. Despite its size, a strong northwest to southeast climatic gradient exists within the region; mean annual temperature ranges from about 0–10°C and mean annual precipitation from about 35–90 cm (Millett *et al.* 2009). The intersection of these two climatic gradients produces different sub-regional climates, wetland functional dynamics and responsiveness to climatic change. Model simulations using data from regional weather stations with long-term records (≥ 100 years) show that the response of wetlands to climate change will be highly variable geographically (Johnson *et al.* 2010). The most favourable climate in the Prairie Pothole Region for wetland productivity during the 20th century is projected to shift eastward where there are fewer un-drained wetland basins and much less grassland available as nesting habitat for waterfowl. The naturally drier western edge of the Prairie Pothole Region, described as a "boom or bust" region for waterfowl production, may become largely a "bust" should the future climate be more arid as projected (Johnson *et al.* 2010). This possible future "mismatch" between the location of a productive wetland climate and functional wetland basins stands as a current challenge for wetland managers as they develop future plans to allocate resources for wetland conservation and management across the Prairie Pothole Region.

The northwest portion of the Prairie Pothole Region (west-Canadian prairies) warmed and dried late in the 20th Century (Millett *et al.* 2009). A hindcast simulation was conducted to determine if the change in climate between two 30-year periods (1946–1975 and 1976–2005) was sufficient to have affected wetland productivity. If so, the analysis would provide evidence that trends for warming and drying projected earlier for the mid 21st century (Johnson *et al.* 2005) may already have started in the late 20th century. The model indicated that climate changes were sufficient to have affected the wetland cover cycle, a major indicator of wetland productivity quantified by a cover cycle index (Werner *et al.* 2013). This analysis is the first to present evidence that climate change may already have affected wetland productivity in part of the Prairie Pothole Region.

Climate changes that exceed ecological thresholds can produce rapid and surprising changes in the functioning of natural ecosystems (*e.g.* Holling 1973; CCSP 2009). The most productive semi-permanent prairie wetlands pass through three stages during weather cycles: dry marsh, lake marsh and hemi-marsh (which includes both regenerating and degenerating sub-stages; van der Valk & Davis 1978). Climatic thresholds associated with drought must be reached and exceeded for habitats to enter the dry marsh stage, as must those associated with a precipitation deluge needed to enter the lake marsh stage. Between these two extremes, the most productive hemi-marsh stage is reached. Ratios that produce the highest indices for wetland productivity over decadal time intervals are approximately: 25:50:25 (dry, hemi and lake, respectively). Climate changes that cause wetlands to be "stuck" in either the lake or dry marsh extremes stop vegetation cycling and decrease productivity. Because the majority of ducks produced in North America are reared in Prairie Pothole Region wetlands, biologists are concerned that wetlands responsible for past high rates of waterfowl production could become drier and fail in the future by never or rarely reaching the lake marsh and hemi-marsh stages of the cycle (Sorenson *et al.* 1998; Johnson *et al.* 2010).

Twenty years of modelling and field research have found that prairie wetlands are highly sensitive to changes in climate and that they respond differently to wide-ranging sub-climates across the Prairie Pothole Region. Moreover, they may already have been negatively affected by climate warming in the Canadian prairies, and may not reach water level thresholds under a warmer climate needed to maintain historic dynamics and productivity. We suggest development of an early warning system to detect the onset of climate change across the Prairie Pothole Region by conducting simulation modelling and field monitoring in tandem to provide further understanding of changes to date and to improve accuracy in predicting for future changes in Prairie Pothole Region wetlands.

Prairie wetland conservation policy

Despite a changing climate and anthropogenic denudation of large areas of the landscape, all is not lost in the Prairie Pothole Region. To help protect critical habitat for waterfowl, visionary waterfowl biologists and managers recognised the importance of the region and initiated the USFWS's Small Wetlands Acquisition Program (SWAP) in 1958 with an amendment to the 1934 "Duck Stamp Act" (legislative documents:16 U.S.C. 718-718j, 48 Stat.452; P.L. 85-585; 72 Stat. 486) (Loesch *et al.* 2012). The SWAP amendment authorised that proceeds from the sale of duck stamps and the import duties on ammunition and firearms should be used for the acquisition of fee title (*i.e.* absolute ownership) or limited-interest title (restricted ownership) of Waterfowl Production Areas, and also for purchasing limited interest easements over Waterfowl Production Areas in Prairie Pothole Region states (USFWS 2013).

Over the past 50 years, the USFWS and its partners (*e.g.* sportsmen and women, private landowners and non-profit conservation organisations) have acquired ownership of nearly 0.7 million ha in National Wildlife Refuges and Waterfowl Production Areas and easements over an additional 1.1 million ha of Waterfowl Production Areas in the U.S. Prairie Pothole Region. The SWAP has thus acquired easements to conserve a network of privately owned wetlands and grasslands which provide nesting sites for breeding birds in proximity to larger Waterfowl Production Area wetland basins purchased for their importance as brood-rearing habitat. During the first 35 years, habitat was acquired by USFWS biologists who applied their knowledge of the area to prioritise acquisitions. More recently, spatially explicit habitat and biological data have been used to develop statistical models used by the USFWS to assess the Prairie Pothole Region landscape (Stephens *et al.* 2008). Habitat conservation efforts are then focused toward areas that produce the greatest benefits for migratory bird benefits, given the limited conservation funds (Reynolds *et al.* 2006; Niemuth *et al.* 2008).

In addition to traditional measures of conservation progress (*e.g.* money expended for land acquisitions and the number of acres protected), the success of the SWAP is assessed using measurable biological outcomes such as the abundance of waterfowl pairs and their breeding success. Through the purchase of wetland and grassland easements on private lands, the USFWS and its partners have secured breeding habitat for an estimated 1.1 million waterfowl pairs across 13 species of waterfowl. The resultant effort contributes approximately 708,000 recruits annually for Mallard, Northern Pintail, Gadwall *A. strepera*, Blue-winged Teal and Northern Shoveler annually (Cowardin *et al.* 1995; USFWS Habitat & Population Evaluation Team unpubl. data). While this large landscape-scale approach to waterfowl conservation has been highly successful, an additional 3.8 million ha of grassland (88% of the remaining grassland) and 0.7 million ha of wetlands (75% of the remaining wetlands) in the U.S. Prairie Pothole Region have been prioritised for protection (Ringleman 2005). In 2012, land values averaged across the northern plains states from North Dakota to Kansas were US$5,831/ha (USDA 2012) and, if applied to the 4.5 million-ha goals, would require > US$2.6 trillion in fee-title acquisition costs.

Through various partnership efforts including the Migratory Bird Conservation Fund, the Land and Water Conservation Fund and funding under the North American Wetlands Conservation Act, the USFWS and its partners continue to be active in pursuing conservation goals in order to maintain North America's waterfowl populations. Meeting existing wetland and grassland conservation goals is a daunting challenge (Doherty *et al.* 2013). Only through collaborative and complementary efforts, which incorporate a science-based approach to determining the best places in the landscape for directing conservation resources in the face of habitat loss, will effective conservation of wetlands in the Prairie Pothole Region be achieved.

Lower Great Lakes marshes

The lower Great Lakes coastal marshes are valuable areas for staging and wintering waterfowl and are among the most biologically significant wetlands within the Great Lakes region. These marshes have long been recognised for their importance in providing habitat for a wide variety of flora and fauna, and in particular for migratory birds. As an example, the coastal wetlands of northwest Ohio alone support *c.* 500,000 itinerant waterfowl during autumn migration (Ohio Division of Wildlife, unpubl. data). These marshes are also subject to a great number of anthropogenic stressors, including dredging, nutrient/pollutant loading, altered hydrological regimes and the introduction of non-native species. Today, a majority of the region's coastal marshes and wetlands have been drained or replaced by shoreline development or have been further degraded by altered hydrology and sediment deposition. Only 5% of the original 121,000 ha of Lake Erie marshes and swamps in northwest Ohio remain (Bookhout *et al.* 1989), and habitat loss continues to reduce the area available for diverse wetland plant communities capable of supporting waterfowl populations. Habitat loss of this magnitude underscores the importance of maintaining the remaining habitat at the highest level of quality possible.

A wide variety of invasive species now dominate wetland flora in many lower Great Lakes coastal marshes, having displaced native vegetation and in many cases important waterfowl resources (Mills *et al.* 1994; Zedler & Kercher 2004). In fact, invasive species are now considered the primary cause of wetland degradation in the region. The most abundant, widespread and harmful invasive plant species within these wetlands include Common Reed *Phragmites australis*, Reed Canary Grass *Phalaris arundinacea*, Curly Pondweed *Potamogeton crispus*, Eurasian Watermilfoil *Myriophyllum spicatum*, and non-native Cattail *Typha angustifolia* and *T. glauca*. Other less widespread but significant invasive species include European Frog-bit *Hydrocharis morsus-ranae*, Japanese Knotweed *Polygonum cuspidatum*, Yellow Flag Iris *Iris pseudacorus*, Purple Loosestrife *Lythrum salicaria* and Water Chestnut *Trapa natans*. Invasive species outbreaks continue to occur in this region and relative newcomers such as Flowering Rush *Butomus umbellatus* are quickly becoming established at nuisance levels.

In most cases, invasive plant species alter the biotic and abiotic environment of wetlands by excluding native plants, reducing plant diversity and modifying wetland processes (Drake *et al.* 1989; Davis *et al.* 1999; Meyerson *et al.* 1999; Windham & Lathrop 1999; Rooth *et al.* 2003). However, the indirect effects of invasive plants on wildlife are less well understood. For example, only a handful of studies have shown the effects of Common Reed on wildlife use and diversity, including studies on turtles (Bolton & Brooks 2010), toads (Greenberg & Green 2013), passerine birds (Meyer *et al.* 2010) and other wetland wildlife (Schummer *et al.* 2012). In contrast, a substantial research base of Common Reed biology, proliferation and management exists in the form of peer-reviewed articles, white papers and websites (*e.g.* http://www.greatlakesphragmites.net). Extensive research into chemical and biological control measures for Purple Loosestrife similarly led to the release of beetles *Galerucella* sp. as a highly successful biological control during the 1990s, despite a lack of data to show that the plant had negative impacts on the environment (Hager & McCoy 1998; Treberg & Husband 1999).

While no single management strategy can be employed to treat infestations of these diverse invasive plants, similarities do exist among species. Most often, managers employ an Integrated Pest Management (IPM) strategy that combines one or more techniques including mowing or harvesting, smothering, drowning, herbicide treatments, biological control agents, controlled burns and reseeding with native species (Radosevich 2007; Holt 2009). With the exception of Purple Loosestrife, where biological control proved successful, the most effective and widely used strategies typically include herbicide application within the IPM strategy. As an example, the most effective control of Common Reed includes a late-summer application of glyphosphate herbicide followed by a spring burn or other thatch removal method (J. Simpson, unpubl. data; MDEQ 2008). Variations of this method have also been applied effectively for many other emergent invasive plants. Other aquatic-approved herbicides, such as those formulated with imazypyr, are equally effective at removing invasive plants, but residual action limits subsequent regeneration of native species. Submerged or floating-leaved vegetation is typically managed with granular or similar broadcast herbicides in conjunction with mechanical mowing or harvesting. Ironically, some success has been also demonstrated by using non-native Common Carp *Cyprinus carpio* to reduce monocultures of submersed invasive plants (Kroll 2006), but carp often become established and can remove desirable native vegetation (Bajer *et al.* 2009).

In response to the logistical and financial hurdles associated with managing large non-native plant invasions, stakeholders in the Great Lakes Region of the U.S. and elsewhere have united to form cooperative weed management areas. These diverse groups now exist in most Great Lakes states and provinces and represent a cross section of government agencies, local units of government, non-profit conservation groups, community associations and individual landowners. In many cases, these associations form to address ecological, social and economic problems linked to vegetation management along developed shorelines and within recreational sites. Using private and government grant funds, these cooperative weed management areas have made progress by identifying and prioritising treatment sites, providing management tools, implementing post-treatment monitoring and research, and organising and educating landowners.

Invasive wetland plants are widespread and continually establishing across coastal and inland wetlands within the Great Lakes region. Management of invasive plants is unavoidable in order to continue providing quality wetland habitat for waterfowl and other wetland-dependent species. Management strategies continue to be refined, tested and researched, but research into the biological implications of these species should continue. Up-front research demonstrating the negative impacts of these species is essential for prioritising their management and focusing effort on species of greatest concern. Additionally, a greater understanding of the indirect effects of these plant species on waterfowl and other wetland-dependent wildlife is required to avoid expending exhaustive control measures on species whose ecological consequences are unproven.

Central Valley of California

The Central Valley of California supports an average of about 5.5 million wintering waterfowl annually, making it one of the most important regions for waterfowl in North America. However, the Central Valley has lost approximately 95% of its original wetlands due to flood control, urbanisation and conversion to agriculture (Fleskes 2012). During the past 20 years, conservation programmes such as the Wetlands Reserve Program, the North American Wetlands Conservation Act, the Migratory Bird Conservation Fund and the state's Inland Wetlands Conservation Program have provided a means to protect, enhance or restore former and existing wetlands throughout the Central Valley. Additionally, intensive management of remaining wetlands for food production, along with flooded grain (especially rice), has helped to mitigate for wetland loss and allowed continued support of large numbers of waterfowl.

While partners of the Central Valley Joint Venture have made considerable progress towards habitat goals of the North American Waterfowl Management Plan, changing policies and demand for limited resources, such as water, hinders management of existing wetlands and could impair farming practices that benefit nesting and wintering waterfowl. Water supply for certain National Wildlife Refuges and State Wildlife Areas, as well as other wetland complexes, was required under the provisions of the 1992 Central Valley Project Improvement Act. However, the full allocation of water required under the Act has been achieved only once in the past 20 years (G. Yarris, pers. comm.). In-stream flow requirements for fish species protected under various state and federal plans are competing pressures on the water available for wetland management, such as winter-flooding of rice fields, in the Central Valley. Moreover, the Clean Water Act, which protects wetland resources throughout the United States, increases management complexity in certain situations. Because of the altered hydrology of the Central Valley, most wetlands are managed with controlled flooding and drainage and thus are subject to the same regulations as other water diverters and dischargers. Current or proposed regulations will limit the discharge of contaminants and require expensive monitoring programmes to demonstrate compliance. Additionally, wetlands and flooded rice fields are ideal environments for methylation of mercury – the form of mercury which readily bioaccumulates and is toxic to humans and wildlife (Ackerman & Eagles-Smith 2010). Mercury is a legacy contaminant from the gold rush of the 1800s and is widespread throughout northern Central Valley watersheds. Regulations restricting methylmercury discharge into the San Joaquin-Sacramento Delta may inhibit wetland restoration and management and discourage flooding of rice fields during autumn and winter.

Ongoing conservation planning efforts in the Sacramento-San Joaquin Delta region of the Central Valley emphasise the restoration of anadromous fish runs (*e.g.* salmon *Salmo* and *Oncorhynchus* sp.) and other endangered fish (*e.g.* Delta Smelt *Hypomesus transpacificus*), possibly at the expense of waterfowl habitat. For example, proposed breaching of levees of some managed wetlands in the Suisun Marsh to restore tidal action and provide fish habitat will reduce managed wetlands in the region and require the restoration or creation of new managed wetlands elsewhere to compensate for this loss. The use of tidal wetlands by dabbling ducks is low compared to managed wetlands (Coates *et al.* 2012). Thus, tidal restoration may reduce the waterfowl carrying capacity of the Suisun Marsh, decreasing its importance for ducks in the Pacific Flyway.

Another recent constraint to wetland management is the mosquito abatement policies of vector control districts. Because many wetlands in California are near urban areas, summer irrigation for waterfowl food plant production, early autumn flooding for shorebird migration and other management activities that may produce mosquitoes are discouraged. Although alternative wetland management strategies are being developed in some cases (Washburn 2012), costs associated with mosquito control have created a disincentive to implement wetland management practices on both public and private wetlands (Olson 2010). For example, mosquito control costs have tripled on State Wildlife Areas since concerns of public exposure to West Nile Virus have come to the fore (B. Burkholder, pers. comm.).

Constraints, restrictions and regulations on wetlands and flooded agriculture in the Central Valley likely will continue into the future as the demand for water increases. Creative solutions to wetland restoration and management, and especially increased participation in policy development, will be critical for advancing the goals of the Central Valley Joint Venture and ensuring that sufficient habitat exists for all wetland-dependent species in the Pacific Flyway.

Looking ahead

Challenges

During our session, a number of key points and challenges to wetland conservation and waterfowl management became apparent. Firstly, unless there is an immediate and significant change in a) wetland protection measures, and b) agricultural policies that provide a disincentive to wetland drainage and conversion, the recent "good old days" of abundant wetlands for waterfowl are likely coming to a close. Secondly, the fate of large scale wetland conservation lies with private landowners – public land and areas protected by conservation easements will likely not sustain the current breeding populations of waterfowl in most of North America. Thirdly, wetland conservation policies and objectives must be robust to the wide variety of political, societal and environmental shifts or vagaries. One such environmental factor important to conservation priorities is changing climate, where simulations have shown potential changes in waterbird productivity and impacts on wetland availability when certain climate thresholds are exceeded. Fourthly, increasing demand for water due to urban and population growth, irrigated agriculture, and other commercial uses (*e.g.* hydraulic fracturing) combined with expected impacts of climate change will increase competition for and cost of water for managed wetlands and waterfowl habitats. Fifthly, increased wetland drainage for agriculture followed by increased crop irrigation increases water requirements while reducing the opportunities for aquifer recharge. Sixthly, updating and improving existing data on wetland distribution and quality for waterfowl is needed but will be difficult given declining government budgets and changes in agency priorities. Overall, managing waterfowl populations and their associated habitats in the face of climate change, invasive species and other biotic stressors will be challenging.

Opportunities

In spite of these challenges, there are also a number of opportunities in the near future that may directly or indirectly affect wetland and waterfowl conservation.

National wetlands inventory

Strategic conservation is critical to achieve significant progress towards wetland conservation goals (Stephens *et al.* 2008) and accurate information on the location, type and status and trends of wetlands is vital to this effort. In 1974, the USFWS established the National Wetlands Inventory Program (NWI) to provide information on the location, distribution and characteristics of U.S. wetlands. By late 2014, the NWI is expected to be complete for the lower 48 states, yet by that time much of the data will be > 25 years old. While NWI maps and geospatial data showing wetland types (Cowardin *et al.* 1979) have helped promote wetland conservation, continual updating and additional information (*e.g.* hydrogeomorphic properties) is needed to use NWI data for predicting wetland functions and determining more readily their value to organisms of interest (*e.g.* waterfowl). Recognising this need, the USFWS recently developed descriptors for landscape position, landform, water flow path and waterbody type (LLWW descriptors; Tiner 2003, 2011) to supplement NWI data on a case-by-case basis. When the Federal Geographic Data Committee established its wetland mapping standard (FGDCWS 2009) for the federal government, it suggested adding these attributes to increase the functionality of the NWI database.

When LLWW descriptors are added to existing NWI data, a "NWI+ database" is created. The NWI+ database is used to predict 11 functions of existing wetlands and, in some cases, potential function for wetland restoration sites. For each function, wetlands providing the function at high or moderate levels are predicted based on certain properties included in the database. Correlations between database features and functions were developed first by consulting the literature and then by peer review from regional scientists. For provision of waterfowl and waterbird habitat, in addition to the high and moderate categories, a third category for Wood Duck *Aix sponsa* habitat was created because this species frequents wooded swamps along rivers and streams as opposed to more open water wetlands (*e.g.* marshes) occupied by most other waterfowl and waterbirds. NWI+ data and the results of NWI+ analyses are displayed via an online map (NWI+ web mapper at http://aswm.org/wetland-science/wetlands-one-stop-mapping); NWI+ reports are also posted. This tool provides users with a first approximation of wetland functions across large geographic areas. To date, such data are available or will soon be posted for five entire states (CT, DE, MA, NJ and RI) while pilot or special projects are completed or are in progress for parts of other states (AK, CA, MD, MS, NH, NY, PA, SC, TX, VA, VT and WY).

The NWI+ data provide a better characterisation of wetlands, an expanded geospatial database and a preliminary landscape-level assessment of wetland functions. This information is valuable to fish and wildlife biologists, conservation planners, ecosystem modellers, regulatory personnel and the general public. Limited NWI funds do not allow these data to be produced nationwide, so NWI+ data are project area-focused. With further budget reductions imminent, such data, as well as updated traditional NWI data, will likely come mainly from user-funded initiatives. Other agencies/organisations have produced or are producing NWI+ data for parts of many states (MI, MN, MT, NM, OR, WI), while some states (CT, DE, NY, and PA) have funded NWI+ work in their state. NWI+ data will provide new opportunities for assessing and assigning functional values to wetlands at the time that they are mapped, and have the potential to increase the efficiency of conservation planning for target species or groups (*e.g.* dabbling ducks, wood ducks).

Influencing policy

Scientists, wetland managers and other conservationists should not simply react to policy shifts that influence wetland loss, but must also work to influence them. There are many opportunities to incorporate science into the policy debates that are shaping the future of waterfowl management. Waterfowl scientists and managers can, and must, focus increased efforts on providing information that can influence the future of wetland conservation policies, such as the Clean Water Act, that hold in the balance the future of tens of millions of acres of waterfowl habitat. Moreover, waterfowl conservationists should engage private landowners and convey to them the importance of wetlands for waterfowl as well as the myriad of other functions and benefits that these habitats provide for society. Although government restrictions on advocacy can limit the participation of many scientists in policy debates, experts should nonetheless have input to discussions regarding the anticipated effects of new and ongoing policies on wetlands.

Acknowledgements

We thank more than 100 wetland and waterfowl professionals for participating in our special session and the thousands of conservationists who work daily to restore and protect wetlands for waterfowl throughout the world. We thank our Associate Editor L. Webb and Editor E. Rees for helpful comments which improved this manuscript. Results, conclusions and opinions stated in this paper do not necessarily represent the views of the USFWS, The Wildlife Society or other state and federal agencies.

References

Ackerman, J.T. & Eagles-Smith, C.A. 2010. Agricultural wetlands as potential hotspots for mercury bioaccumulation: Experimental evidence using caged fish. *Environmental Science and Technology* 44: 1451–1457.

Baar, L., Matlack, R.S., Johnson, W.P. & Barron, R.B. 2008. Migration chronology of waterfowl in the Southern High Plains of Texas. *Waterbirds* 31: 394–401.

Baldassarre, G.A. & Bolen, E.G. 2006. *Waterfowl Ecology and Management, 2nd edition*. Kreiger Publishing, Malabar, Florida, USA.

Bajer, P.G., Sullivan, G. & Sorenson, P.W. 2009. Effects of rapidly increasing population of common carp on vegetative cover and waterfowl in a recently restored midwestern shallow lake. *Hydrobiologia* 632: 235–245.

Bartuszevige, A.M., Pavlacky, D.C., Jr., Burris, L. & Herbener, K. 2012. Inundation of playa wetlands in the western Great Plains relative to landcover context. *Wetlands* 32: 1103–1113.

Bergan, J.F. & Smith, L.M. 1993. Survival rates of female mallards wintering in the Playa Lakes Region. *Journal of Wildlife Management* 57: 570–577.

Bolton, R.M. & Brooks, R.J. 2010. Impact of the seasonal invasion of *Phragmites australis* (common reed) on turtle reproductive success. *Chelonian Conservation and Biology* 9: 238–243.

Bonan, G.B. 2008. Forests and climate change: forcings, feedbacks, and the climate benefits of forests. *Science* 13: 1444–1449.

Bookhout, T.A., Bednarik, K.E. & Kroll, R.W. 1989. The Great Lakes marshes. In L.M. Smith, R.L. Pederson & R.M. Kaminski (eds.), *Habitat Management for Migrating and Wintering Waterfowl in North America*, pp. 131–156. Texas Tech University Press, Lubbock, Texas, USA.

Bradshaw, C.J.A., Warkentin, I.G. & Sodhi, N.S. 2009. Urgent preservation of boreal carbon stocks and biodiversity. *Trends in Ecology and Evolution* 24: 541–548.

Cariveau, A.B., Pavlacky, D.C., Bishop, A.A. & LaGrange, T.G. 2011. Effects of surrounding land use on playa inundation following intense rainfall. *Wetlands* 31: 65–73.

Coates, P.S., Casazza, M.L., Halstead, B.J. & Fleskes, J.P. 2012. Relative value of managed wetlands and tidal marshlands for wintering northern pintails. *Journal of Fish and Wildlife Management* 3: 98–109.

Cowardin, L.M., Carter, V., Golet, F.C. & LaRoe, E.T. 1979. *Classification of Wetlands and Deepwater Habitats of the United States*. Report FWS/OBS-79/31. U.S. Fish and Wildlife Service, Washington D.C., USA.

Cowardin, L.M., Shaffer T.L. & Arnold, P.M. 1995. *Evaluations of Duck Habitat and Estimation of Duck Population Sizes with a Remote-Sensing-Based Approach*. Biological Science Report No. 2. U.S. Department of the Interior, Washington D.C., USA.

Dahl, T. E. 2000. *Status and Trends of Wetlands in the Conterminous United States 1986 to 1997*. U.S. Department of the Interior, U.S. Fish and Wildlife Service, Washington D.C., USA.

Dahl, T. E. 2006. *Status and Trends of Wetlands in the Conterminous United States 1998 to 2004*. U.S. Department of the Interior, Fish and Wildlife Service, Washington D.C., USA

Dahl, T. E. 2011. *Status and Trends of Wetlands in the Conterminous United States 2004 to 2009*. U.S. Department of the Interior, Fish and Wildlife Service, Washington D.C., USA.

Dahl, T. E. & Johnson, C.E. 1991. *Status and Trends of Wetlands in the Conterminous United States, mid-1970s to mid-1980s*. U.S. Department of the Interior, Fish and Wildlife Service, Washington D.C., USA.

Davis, M.A. 2009. *Invasion Biology*. Oxford University Press, Oxford, UK.

Doherty, K.E., Ryba, A.J., Stemler C.L., Niemuth, N.D. & Meeks, W.A. 2013. Conservation planning in an era of change: state of the U.S. Prairie Pothole Region. *Wildlife Society Bulletin* 37: 546–563.

Drake, J.A., Mooney, H.A., Di Castri, F., Groves, R.H., Kruger, F.J., Rejmánek, M. & Williamson, M. 1989. *Biological Invasions: a Global Perspective*. Wiley, Chichester, UK.

Federal Geographic Data Committee Wetlands Subcommittee (FGDCWS). 2009. *Wetlands Mapping Standard*. FGDC Document Number FGDC-STD-015-2009. U.S. Fish and Wildlife Service, Washington D.C., USA.

Fleskes, J.P. 2012. Wetlands of the Central Valley of California and Klamath Basin. In D. Batzer & A. Baldwin (eds.), *Wetland Habitats of North America: Ecology and Conservation Concerns*, pp. 357–370. University of California Press, Berkeley, California, USA.

Frayer, W.E., Monahan, T.J., Bowden, D.C. & Graybill, F.A. 1983. *Status and Trends of Wetlands and Deepwater Habitats in the Conterminous*

United States, 1950s to 1970s. Colorado State University, Fort Collins, Colorado, USA.

Greenberg, D.A., & Green, D.M. 2013. Effects of invasive plant on population dynamics in toads. *Conservation Biology* 27: 1049–1057.

Hager, H.A. & McCoy, K.D. 1998. The implications of accepting untested hypotheses: A review of the effects of purple loosestrife (*Lythrum salicaria*) in North America. *Biodiversity and Conservation* 7: 1069–1079.

Haukos, D.A. 2008. *Analyses of Selected Midwinter Waterfowl Data (1955–2008) in Region 2 (Central Flyway Portion).* U.S. Fish and Wildlife Service, Regional Migratory Bird Office, Albuquerque, New Mexico, USA.

Haukos, D.A. & Smith, L.M. 1994. Importance of playa wetlands to biodiversity of the Southern High Plains. *Landscape and Urban Planning* 28: 83–98.

Haukos, D.A. & Smith, L.M. 2003. Past and future impacts of wetland regulations on playas. *Wetlands* 23: 577–589.

Holling, C.S. 1973. Resilience and stability of ecological systems. *Annual Review of Ecology and Systematics* 4: 1–23.

Holt, J.S. 2009. Management of invasive terrestrial plants. *In* M.N. Clout & P.A. Williams (eds.), *Invasive Species Management: A Handbook of Principles and Techniques,* pp. 126–139. Oxford University Press, Oxford, UK.

Johnson, L. 2011. Occurrence, function and conservation of playa wetlands: the key to biodiversity of the Southern Great Plains. Ph.D. thesis, Texas Tech University, Lubbock, Texas, USA.

Johnson, W.P., Rice, M.B., Haukos, D.A. & Thorpe, P. 2011a. Factors influencing the occurrence of inundated playa wetlands during winter on the Texas High Plains. *Wetlands* 31: 1287–1296.

Johnson, L.A., Haukos, D.A., Smith, L.M. & McMurry, S.T. 2011b. Jurisdictional loss of playa wetlands caused by reclassification of hydric soils on the Southern High Plains. *Wetlands* 31: 483–492.

Johnson, L.A., Haukos, D.A., Smith, L.M. & McMurry, S.T. 2012. Physical loss and modification of Southern Great Plains playas. *Journal of Environmental Management* 112: 275–283.

Johnson, W.C., Millett, B.V., Gilmanov, T., Voldseth, R.A., Guntenspergen, G.R. & Naugle, D.E. 2005. Vulnerability of northern prairie wetlands to climate change. *Bioscience* 25: 863–872.

Johnson, W.C., Werner, B., Guntenspergen, G.R., Voldseth, R.A., Millett, B., Naugle, D.E., Tulbure, M., Carroll, R.W.H., Tracy, J. & Olawsky, C. 2010. Prairie wetland complexes as landscape functional units in a changing climate. *Bioscience* 60: 128–140.

Johnston, C.A. 2013. Wetland losses due to row crop expansion in the Dakota Prairie Pothole Region. *Wetlands* 33: 175–182.

Kantrud, H.A. & Stewart, R.E. 1977. Use of natural basin wetlands by breeding waterfowl in North Dakota. *Journal of Wildlife Management* 41: 243–253.

Kroll, R.W. 2006. Status of Winous Point marsh, fall 2006. Unpublished report to the Winous Point Marsh Conservancy, Port Clinton, Ohio, USA.

Loesch, C.R., Reynolds, R.E. & Hanson, L.T. 2012. An assessment of re-directing breeding waterfowl conservation relative to predictions of climate change. *Journal of Fish and Wildlife Management* 3: 1–22.

Luo, H.R., Smith, L.M., Allen, B.L. & Haukos, D.A. 1997. Effects of sedimentation on playa wetland volume. *Ecological Applications* 7: 247–252.

Mattson, B.J., Runge, M.C., Devries, J.H., Boomer, G.S., Eadie, J.M., Haukos, D.A., Fleskes, J.P., Koons, D.N., Thogmartin, W.E. & Clark, R.J. 2012. A prototypical modeling framework for integrated harvest and habitat management of

North American waterfowl: case-study of northern pintail metapopulation dynamics. *Ecological Modeling* 225: 146–158.

Meyer, S.W., Badzinski, S.S., Petrie, S.A. & Ankney, C.D. 2010. Seasonal abundance and species richness of birds in common reed habitats in Lake Erie. *Journal of Wildlife Management* 74: 1559–1567.

Meyerson, L.A., Chambers, R.M. & Vogt, K.A. 1999. The effects of Phragmites removal on nutrient pools in a freshwater tidal marsh ecosystem. *Biological Invasions* 1: 129–136.

Mitsch, W.J. & Gosselink, J.G. 2007. *Wetlands. 4th edition.* John Wiley & Sons, Inc., New York, USA.

Moon, J.A. & Haukos, D.A. 2006. Survival of female northern pintails wintering in the Playa Lakes Region of northwestern Texas. *Journal of Wildlife Management* 70: 777–783.

Moon, J.A., Haukos, D.A. & Smith, L.M. 2007. Changes in body condition of pintails wintering in the Playa Lakes Region. *Journal of Wildlife Management* 71: 218–221.

Michigan Department of Environmental Quality (MDEQ). 2008. *A Guide to the Control and Management of Invasive Phragmites, 2nd edition.* Lansing, Michigan, USA.

Millar, J.B. 1989. Perspectives on the status of Canadian prairie wetlands. *Freshwater Wetlands and Wildlife* 61: 821–852.

Millett, B.V., Johnson, W.C. & Guntenspergen, G.R. 2009. Climate trends of the North American Prairie Pothole Region. *Climatic Change* 93: 243–267.

Mills, E.L., Leach, L.H., Carlton, J.T. & Secor, C.L. 1994. Exotic species and the integrity of the Great Lakes. *Bioscience* 44: 666–676.

Nichols, J.D., Runge, M.C., Johnson, F.A. & Williams, B.K. 2007. Adaptive harvest management of North American populations: a brief history and future prospects. *Journal of Ornithology* 148: S343–S349.

Niemuth N.D., Reynolds, R.E., Granfors, D.A., Johnson, R.R., Wangler, B. & Estey, M.E. 2008. Landscape-level planning for conservation of wetland birds in the U.S. Prairie Pothole Region. *In* J.J. Millspaugh & F.R. Thompson (eds.), *Models for Planning Wildlife Conservation in Large Landscapes*, pp. 533–560. Elsevier Science, Burlington Massachusetts, USA.

O'Connell, J.L., Johnson, L.A. Smith, L.M., McMurry, S.T. & Haukos, D.A. 2012. Influence of land-use and conservation programs on wetland plant communities of the semi-arid United States Great Plains. *Biological Conservation* 146: 108–115.

Olson, B.W. 2010. An experimental evaluation of cost effective moist-soil management in the Sacramento Valley of California. M.Sc. thesis. University of California, Davis, California, USA.

Pew Environmental Group (PEG). 2011. *A Forest of Blue: Canada's Boreal.* Seattle, Washington, USA. PEG unpublished report. PEG, Seattle, Washington, USA. Accessible at http://borealscience.org/wp-content/uploads/2012/06/report-forestofblue.pdf (last accessed 8 February 2014).

Poiani, K.A. & Johnson, W.C. 1991. Global warming and prairie wetlands. *Bioscience* 41: 611–618.

Poiani, K.A., Johnson, W.C. & Kittel, T.G.F. 1995. Sensitivity of a prairie wetland to increased temperature and seasonal precipitation changes. *Water Resources Bulletin* 31: 283–294.

Poiani, K.A., Johnson, W.C., Swanson, G.A. & Winter, T.C. 1996. Climate change and northern prairie wetlands: simulations of long-term dynamics. *Limnology and Oceanography* 41: 871–881.

Poiani, K.A. & Johnson, W.C. 2003. Simulation of hydrology and vegetation dynamics of prairie wetlands in the Cottonwood Lake Area. *In*

T.C. Winter (ed.), *Hydrological, Chemical and Biological Characteristics of a Prairie Pothole Wetland Complex under Highly Variable Climatic Conditions – the Cottonwood Lake Area, East-Central North Dakota*, pp. 95–109 U.S. Geological Survey Professional Paper No. 1675. U.S. Geological Survey, Denver, Colorado, USA. Accessible at http://pubs.er.usgs.gov/publication/pp1675 (last accessed 4 August 2014).

Radosevich, S.R., Holt, J.S. & Ghersa, C.M. 2007. *Ecology of Weeds and Invasive Plants: Relationship to Agriculture and Natural Resource Management*. John Wiley and Sons, Inc. Hoboken, New Jersey, USA.

Ray, J.D., Sullivan, B.D. & Miller, H.W. 2003. Breeding ducks and their habitats in the High Plains of Texas. *Southwestern Naturalist* 48: 241–248.

Reynolds, R. E., Shaffer, T.L., Loesch, C.R. & Cox, R.R. 2006. The Farm Bill and duck production in the Prairie Pothole Region: increasing the benefits. *Wildlife Society Bulletin* 34: 963–974.

Ringelman, J.K. (ed.) 2005. Prairie Pothole Joint Venture 2005 implementation plan. U.S. Fish and Wildlife Service, Denver, Colorado, USA.

Riordan, B., Verbyla, D. & McGuive, A.A. 2006. Shrinking ponds in subarctic Alaska based on 1950–2002 remotely sensed images. *Journal of Geophysical Research* 111: 1–11.

Roach, J., Griffith, B., Verbyla, D. & Jones, J. 2011. Mechanisms influencing changes in lake area in Alaskan boreal forest. *Global Change Biology* 17: 2576–2583.

Rooth, J.E., Stevenson, J.C. & Cornwell, J.C. 2003. Increased sediment accretion rates following invasion by Phragmites australis: the role of litter. *Estuaries and Coasts* 26: 475–483.

Rubec, C. & Lynch-Stewart, P. 1998. Regulatory and non-regulatory approaches for wetland conservation in Canada. Technical Consultation on Designing Methodologies to Review Laws and Institutions Relevant to Wetlands. Environment Canada unpublished report. Environment Canada, Ottawa, Ontario, Canada. Accessible at http://www.ramsar.org/doc/wurc/wurchbk3cs2.doc (last accessed 8 April 2014).

Schummer, M.L., Palframan, J., McNaughton, E., Barney, T. & Petrie, S.A. 2012. Comparisons of bird, aquatic macroinvertebrate, and plant communities among dredged ponds and natural wetlands habitats at Long Point, Lake Erie, Ontario. *Wetlands* 32: 945–953.

Slattery, S.M., Morissette, J.L., Mack, G.G. & Butterworth, E.W. 2011. Waterfowl conservation planning: science needs and approaches. *In* J.V. Wells (ed.), *Boreal Birds of North America: a Hemispheric View of their Conservation Links and Significance*, pp. 23–40. University of California Press, Berkeley, USA.

Smerdon, B.D., Devito, K.J. & Mendoza, C.A. 2005. Interaction of ground water and shallow lakes on outwash sediments in the sub-humid Boreal Plains of Canada. *Journal of Hydrology* 314: 246–262.

Smith, A.G., Stoudt, J.H. & Gollop, J.B. 1964. Prairie potholes and marshes. *In* J. Linduska (ed.), *Waterfowl Tomorrow*, pp. 39–50. U.S. Government Printing Office, Washington D.C., USA.

Smith, L.C., Sheng, Y., MacDonald, G.M. & Hinzman, L.D. 2005. Disappearing arctic lakes. *Science* 308: 1429.

Smith, L.M. 2003. *Playas of the Great Plains*. University of Texas Press, Austin, USA.

Smith, L.M., Haukos, D.A., McMurry, S.T., LaGrange, T. & Willis, D. 2011. Ecosystem services provided by playa wetlands in the High Plains: potential influences of USDA conservation programs and practices. *Ecological Applications* 21: S82–S92.

Smith, L.M., Haukos, D.A. & McMurry, S. 2012. High Plains Playas. *In* D. Batzer & A. Baldwin (eds.), *Wetland Habitats of North America: Ecology and Conservation Concerns*, pp. 299–311. University of California Press, Berkeley, USA.

Soja, A.J., Tchebakova, N.M., French, N.H.F., Flannigan, M.D., Shugart, H.H., Stocks, B.J., Sukhinin, A.I., Parfenova, E.I., Chappin, F.S., III & Stackhouse, P.W., Jr. 2007. Climate-induced boreal forest change: Predictions verses current observation. *Global and Planetary Change* 56: 274–296.

Sorenson, L.G., Goldberg, R., Root, T.L. & Anderson, M.G. 1998. Potential effects of global warming on waterfowl populations breeding in the northern Great Plains. *Climatic Change* 40: 343–369.

Stephens, S.E., Walker, J.A., Blunck, D.R., Jayaraman, A., Naugle, D.E., Ringelman, J.K. & Smith, A.J. 2008. Predicting risk of habitat conversion in native temperate grasslands. *Conservation Biology* 22: 1320–1330.

Stewart, R.E. & Kantrud, H.A. 1971. *Classification of natural ponds and lakes in the glaciated prairie region*. U.S. Fish and Wildlife Service Resource Publication 92. U.S. Fish and Wildlife Service, Washington D.C., USA.

Stocker, T.F., Qin, D., Plattner, G.K., Tignor, M., Allen, S.K., Boschung, J., Nauels, A., Xia, Y., Bex, V. & Midgley, P.M. (eds.) 2013. *Climate Change 2013: The Physical Science Basis*. Contribution of Working Group I to the Fifth Assessment Report of the Intergovernmental Panel on Climate Change. Cambridge University Press, Cambridge, UK.

Tiner, R.W. 2003. *Correlating enhanced National Wetlands Inventory data with wetland functions for watershed assessments: a rationale for northeastern U.S. wetlands*. U.S. Fish and Wildlife Service, National Wetlands Inventory Program, Region 5, Hadley, Massachusetts, USA.

Tiner, R.W. 2011. *Dichotomous keys and mapping codes for wetland landscape position, landform, water flow path, and waterbody type descriptors: Version 2.0*. U.S. Fish and Wildlife Service, National Wetlands Inventory Program, Northeast Region, Hadley, Massachusetts, USA.

Treberg, M.A. & Husband, B.C. 1999. Relationship between the abundance of *Lythrum salicaria* (purple loosestrife) and plant species richness along the Bar River, Canada. *Wetlands* 19: 118–125.

Tsai, J.S., Venne, L.S., McMurry, S.T. & Smith, L.M. 2007. Influences of land use and wetland characteristics on water loss rates and hydroperiods of playas in the Southern High Plains, USA. *Wetlands* 27: 683–692.

van der Valk, A.G. & Davis, C.B. 1978. The role of seed banks in the vegetation dynamics of prairie glacial marshes. *Ecology* 59: 332–335.

Vrtiska, M.P., Gammonley, J.H., Naylor, L.W., & Raedeke, A.H. 2013. Economic and conservation ramifications from the decline of waterfowl hunters. *Wildlife Society Bulletin* 37: 380–388.

Washburn, N.B. 2012. Experimental evaluation of tradeoffs in mosquito production and waterfowl food production in moist-soil habitats of California's Central Valley. M.Sc. thesis. University of California, Davis, California, USA.

Weller, M.W. & Fredrickson, L.H. 1974. Avian ecology of a managed glacial marsh. *Living Bird* 12: 269–291.

Wells, J.V. 2011. Boreal forest threats and conservation status. *In* J.V. Wells (ed.), *Boreal birds of North America: a Hemispheric View of their Conservation Links and Significance*, pp. 1–6. University of California Press, Berkeley, USA.

Wells, J.V. & Blancher, P.J. 2011. Global role for sustaining bird populations. *In* J.V. Wells (ed.), *Boreal birds of North America: a*

Hemispheric View of their Conservation Links and Significance, pp. 7–22. University of California Press, Berkeley, USA.

Werner, B.A., Johnson, W.C. & Guntenspergen, G.R. 2013. Evidence for 20th century climate warming and wetland drying in the North American Prairie Pothole Region. *Ecology and Evolution* 3: 3471–3482.

Windham, L., & Lathrop, R.G. 1999. Effects of *Phragmites australis* (common reed) invasion on aboveground biomass and soil properties in brackish tidal marsh of the Mullica River, New Jersey. *Estuaries and Coasts* 22: 927–935.

United States Climate Change Science Program (CCSP). 2009. *Thresholds of Climate Change in Ecosystems*. U.S. Climate Science Program and the Subcommittee on Global Change Research Report, U.S. Geological Survey, Reston, Virginia, USA.

United States Department of Agriculture (USDA). 2012. Land values 2012 summary. National Agricultural Statistics Service Report 1949–1867. U.S. Fish and Wildlife Service, Washington D.C., USA.

United States Fish and Wildlife Service (USFWS). 1988. Playa lakes region waterfowl habitat concept plan, category 24 of the North American Waterfowl Management Plan. U.S. Fish and Wildlife Service, Albuquerque, New Mexico, USA.

United States Fish and Wildlife Service (USFWS). 2013. Annual report of lands under the control of the U.S. Fish and Wildlife Service. U.S. Fish and Wildlife Service, Division of Realty, Washington D.C., USA.

Zedler, J.B. 2003. Wetlands at your service: reducing impacts of agriculture at the watershed scale. *Frontiers in Ecology and the Environment* 1: 65–72.

Zedler, J.B. & Kercher S. 2004. Causes and consequences of invasive plants in wetlands: Opportunities, opportunists and outcomes. *Critical Reviews in Plant Sciences* 23: 431–452.

Photograph: Mallard and Northern Pintail in a seasonal emergent wetland in the Illinois River Valley, Illinois, USA, by Heath Hagy.

Opportunities and challenges to waterfowl habitat conservation on private land

WILLIAM L. HOHMAN[1*], ERIC B. LINDSTROM[2], BENJAMIN S. RASHFORD[3] & JAMES H. DEVRIES[4]

[1]U.S. Department of Agriculture/Natural Resources Conservation Service, National Wildlife Team, Fort Worth, Texas 76115, USA.
[2]Ducks Unlimited, Inc., Bismarck, North Dakota 58503, USA.
[3]Department of Agricultural and Applied Economics, University of Wyoming, Laramie, Wyoming 82071, USA.
[4]Ducks Unlimited Canada, P.O. Box 1160, Stonewall, Manitoba R0C2Z0, Canada.
*Correspondence author. E-mail: william.hohman@ftw.usda.gov

Abstract

The future of North American waterfowl populations is inseparably tied to management of private land in the United States (U.S.) and Canada. Private land ownership in major waterfowl habitat regions such as the Northern Great Plains, Lower Mississippi Alluvial Valley, Gulf Coast and California's Central Valley generally exceeds 90%, with agriculture being the dominant land-use in these regions. Planning and implementing avian conservation on private land in a strategic manner is complicated by a wide array of social, economic, political, administrative and scientific-technical issues. Prominent among these challenges are changing economic-drivers influencing land-use decisions, integration of bird conservation objectives at various scales, reconciling differences in wildlife habitat objectives between bird conservationists and land-users, administrative impediments to conservation planning and implementation, technology and scientific information gaps, and inadequate personnel capacity and financial constraints to effectively plan and deliver conservation. Given these unprecedented challenges to waterfowl habitat conservation, the need for effective public-private partnerships and collaboration has never been greater. With the goal of advancing collaborative waterfowl conservation on private land, the broad goals of this paper are to: (1) increase stakeholder awareness of opportunities and challenges to waterfowl habitat conservation on private land, and (2) showcase examples of collaborative efforts that have successfully addressed these challenges. To accomplish these goals this paper is organised into three sections: (1) importance of agricultural policy to private land conservation, (2) habitat potential on agricultural working land, and (3) strategic approaches to waterfowl habitat conservation. U.S. Department of Agriculture conservation programmes authorised through the Conservation Title of the 1985 Food Security

Act (hereafter, Farm Bill) and subsequent farm bills have provided unequalled potential for waterfowl habitat conservation on private land. Passage of the 2014 Farm Bill provides unique opportunities and alternative approaches to promote working land conservation strategies that are economically profitable and wildlife-friendly. However, reductions in private land conservation funding will require more effective targeting to maximise resource benefits. For example, in addition to conserving and restoring traditional habitats, we must work collaboratively to identify and promote working agricultural systems that are waterfowl-friendly and provide environmental services in addition to the production of food and fibre. Cultivation of rice *Oryza sativa* and winter cereals described below potentially represent two such situations. For over a quarter of a century the North American Waterfowl Management Plan (NAWMP) has served as a transformative model of partnership-based, landscape-scale conservation (DOI & EC 1986). Whereas the original plan and subsequent updates established abundant waterfowl populations as the plan's ultimate goal, the 2012 NAWMP revision seeks a formal integration of these objectives with societal needs and desires (DOI *et al.* 2012). The current plan recognises the critical importance of private working land; however, details are lacking, especially with respect to strategic targeting of conservation on private land. For example, the development of truly strategic plans to target waterfowl conservation on private land will require estimates of the benefits of various conservation alternatives, conservation costs, and the threat of habitat loss or conversion. We suggest development of spatially explicit models that inform landowners and managers at the field-level about the cost effectiveness of conservation and land-use options is critically needed.

Key words: agriculture, conservation, economics, environmental services, Farm Bill, habitat, private land, waterfowl.

The future of North American waterfowl populations is inseparably tied to the management of private land in the United States (U.S.) and Canada. Approximately 70% of the conterminous U.S. is held in private ownership, including > 90% of the land area in major waterfowl habitat regions such as the Northern Great Plains (NGP), Lower Mississippi Alluvial Valley (LMAV), Gulf Coast, Playa Lakes and California's Central Valley (Nickerson *et al.* 2011). Agriculture is the dominant land-use in these regions, with ~52% of the U.S. or 900 million acres (365 million ha) managed as cropland, pastureland or rangeland. Thus, the overwhelming majority of land-use decisions affecting waterfowl habitats are made by agricultural producers responding to a multitude of factors with various social and economic motivations.

The contemporary setting in which waterfowl managers operate is complex and continuously changing. Global factors associated with an increasing human population and natural resource exploitation and development have far-ranging impacts on land-use decisions, and ultimately on the availability and suitability of private land as waterfowl habitat. Competition in global commodity markets, water demands, current federal agricultural/energy policy and technological advancements in agriculture are fuelling agricultural intensification and expansion (Sohl et al. 2012). Moreover, the United Nation's Food and Agriculture Organization (FAO) projects that world food production will need to increase by 70% by 2050 to meet food demands for an estimated 9.1 billion humans (FAO 2009). The FAO anticipates 80% of production increases will come from increased yield and 20% from expansion of arable land; however, declines in the rate of growth in yields of major cereal crops from 1960 (3.2% per year) to 2000 (1.5% per year) suggest that FAO forecasts of production increases may be overly optimistic and additional land may need to be brought under cultivation.

Since passage of the 1985 Food Security Act (hereafter, Farm Bill), U.S. Department of Agriculture (USDA) conservation programmes authorised through the Conservation Title of the Farm Bill have provided unequalled potential for waterfowl habitat conservation on private land. This complex, multi-billion dollar legislation is typically reauthorised by Congress every five years and covers a broad range of programmes for commodities, crop insurance, farm credit, nutrition, forestry, energy and conservation. Recognised as the single largest private land conservation initiative in the U.S., the farm bill provides critical funding for important wildlife habitat, soil and water conservation programmes (Heard et al. 2000). Amendments to the original Farm Bill in 1990, 1996, 2002, 2008 and 2014 have retained and expanded conservation provisions such that there are now 13 agricultural conservation programmes with a combined funding level of $28.1 billion for 2014–2018 (CBO 2014). No other state or federal programme provides a comparable level of investment or impact for conservation initiatives on private land.

The consideration of fish and wildlife (hereafter, wildlife) in the delivery of conservation programmes was elevated in the 1996 Farm Bill. Wildlife currently is an explicit goal for the Conservation Reserve Program (CRP), the Environmental Quality Incentives Program (EQIP), and components of the Agricultural Conservation Easement Program (ACEP). Because of federal budget constraints and increased demand for agricultural commodities, the national cap on CRP acreage has been reduced from a peak of 45 million acres (18.21 million ha) in 1990 to 27.5 million acres (11.13 million ha) in 2014 and 24 million acres (9.71 million ha) in 2017 (Cuzio et al. 2013; Ducks Unlimited, unpubl. data). Declines in CRP acreage were only partially offset by increases in the size of the wetland easement programme (formerly the Wetland Reserve Program, WRP) from 2.125 to 3 million acres (0.86 to 1.23 million ha) through 2012. The attention of wildlife conservation groups has been focused on land retirement programmes in spite of the fact that farm bill's working lands programmes, such as EQIP, the

Agricultural Land Easement (formerly the Grassland Reserve Program, GRP), the Conservation Stewardship Program (CSP), and a "working lands" CRP concept, could receive greater funding and impact a far greater area. Consequently, the full potential for improving consideration of waterfowl and other wildlife in land-use decisions has yet to be realised.

Changes in climatic conditions (*e.g.* severe alterations in regional temperature and precipitation patterns), correlated to increasing levels of CO_2 and other greenhouse gases in the atmosphere, are already affecting the nation's natural resources, people, communities and economies that depend on healthy, functional ecosystems and the plants and animals that characterise them (NFWPCAS 2012). Uncertainties regarding the effects of climate change on ecosystems and associated biota, and on current land uses, pose significant challenges to both agricultural producers and waterfowl habitat managers.

Planning and implementing waterfowl habitat on private land is complicated by a wide array of social, economic, political, administrative and scientific/technical issues. Prominent among these challenges are how changing economic drivers influence land-use decisions, integration of bird conservation objectives at various scales, reconciliation of differences in wildlife objectives between bird conservationists and land-users, administrative impediments to conservation planning and implementation, technology and scientific information gaps, and constraints on the personnel and finances required to plan and deliver conservation effectively.

In the face of unprecedented challenges to waterfowl habitat conservation, the need for effective public-private partnerships and collaboration has never been greater. With the goal of advancing collaborative waterfowl conservation on private land, the broad aims of this paper are to: (1) increase stakeholder awareness of opportunities and challenges to waterfowl habitat conservation on private land, and (2) provide examples of collaborative efforts that have been successful in addressing these challenges. To accomplish these aims we have organised the paper into three sections: (1) importance of agricultural policy to private land conservation, (2) habitat potential on agricultural working land, and (3) strategic approaches to waterfowl habitat conservation.

Importance of agricultural policy to private land conservation

European settlement of North America beginning in the eighteenth century produced waves of change in land forms and vegetation (hereafter, landcover). Suitability of land for agriculture greatly influenced settlement patterns in North America (Maizel *et al.* 1998). As expansion rapidly proceeded westward during the 1800s and early 1900s, farms were created at the population frontier; areas too wet or too dry were farmed later when drainage or irrigation was possible. Other areas with poor climate, steep slopes, or soils unsuitable for use as cropland, grazed pasture or hay fields, were either farmed unsuccessfully or never farmed.

The influence of agriculture on pre-settlement landcover is especially evident in the fertile Great Plains region of North America. The vast grasslands, shrublands and savannas that characterise the region once represented the continent's largest ecosystem; however, conversion of grasslands to agricultural uses has been extensive, exceeding 99% in portions of the northern tallgrass prairie region of Iowa, Minnesota, eastern Dakotas and Manitoba (Samson & Knopf 1994; Noss *et al.* 1995). Associated with landcover change in the Great Plains came a concomitant change in communities of birds and other grassland-dependent wildlife. For example, dramatic declines in grassland bird species since the 1950s have been attributed to changes in the agricultural landscape of the region (Gerard 1995). Extensive loss and degradation of grasslands in the Great Plains resulted in its designation as one of the nation's most endangered ecosystems (Noss *et al.* 1995).

Wetlands in the Great Plains and other arable regions such as the LMAV and California's Central Valley have been similarly affected. Dahl (1990) reported that between the 1780s and 1980s, the U.S. (except Hawaii and Alaska) lost 53% of its original wetlands. In Canada, an estimated 40% of wetlands within the Prairie Pothole Region (PPR) have been lost to drainage since settlement (Millar 1989). Twenty-two U.S. states have lost > 50% of their wetlands, with California having the greatest wetland loss (> 90%, Dahl 1990). Long-term trends show freshwater emergent wetlands, especially forested wetlands, sustained the greatest loss of any freshwater wetland type (Dahl 2000). The rate of wetland conversion between the mid-1950s and 1970s was estimated at 458,000 acres/yr (185,400 ha/yr; Frayer *et al.* 1983). Extensive wetland losses occurred in the LMAV as bottomland hardwoods were cleared and drained for cultivation of agricultural crops. The rate of wetland losses slowed somewhat (to 290,000 acres/yr or 117,400 ha/yr) during the decade before the Emergency Wetlands Resources Act of 1986 was enacted to protect wetlands (Dahl 2000). The Act required the U.S. Fish and Wildlife Service (USFWS) to monitor the status and trends of wetlands and report details to the Congress at 10-year intervals.

The vast majority of inland wetland losses were due to agricultural conversion (Dahl 1990). The 1985 Farm Bill sought to stem further wetland losses by linking wetland conservation on agricultural land to the landowner's eligibility for USDA farm programme benefits, a provision commonly referred to as "Swampbuster". Similarly, Highly Erodible Land (HEL) provisions commonly referred to as "conservation compliance" and "Sodbuster" required producers who cultivated sensitive land to have fully implemented a USDA-approved conservation plan by 1985. Provisions for protection of highly erodible land and wetlands were retained in revisions to the farm bills through to 2008. While these provisions did not create wildlife habitat directly, they did, "… provide strong motivation for producers to apply conservation systems on their highly erodible land, to protect wetlands from conversion to croplands, and apply for enrolment in other USDA conservation programmes, especially the Conservation Reserve and Wetland Reserve Programs" (Brady 2005:5). Implementation of these provisions contributed to a reduction in soil erosion rates between 1981 and 2001 (Brady 2005). Under Swampbuster, during an era of declining wetland losses (*i.e.* 506,000 acres (205,000 ha) lost in 1992–1997 *vs.* 281,600 acres (114,000 ha) lost in 1997–2002), gross wetland losses due to agriculture declined from 26% during 1992–1997 to 18% during 1997–2002 (USDA NRCS 2000; 2013). Wetland restorations through other conservation programmes, especially CRP and WRP, resulted in net wetland gain on agricultural land in both 1997–2002 and 2002–2007, although the change during 2002–2007 was non-significant at the 95% confidence level (USDA NRCS 2013).

The contributions of farm bill programmes to waterfowl habitat conservation have been substantial (Heard *et al.* 2000). In the PPR, Reynolds (2000) estimated that between 1992 and 1997, the CRP contributed to a 30% improvement in duck production or 10.5 million additional ducks. Grassland birds likewise benefitted from the CRP in the NGP (Johnson 2000) and Midwest (Ryan 2000), as did early successional bird species (*e.g.* Northern Bobwhite *Colinus virginianus*) in the southeast U.S. (Burger 2000).

While farm bill provisions have helped discourage grassland and wetland conversion to cropland and provided incentives for the establishment of perennial cover on highly erodible land, some producers have continued to convert native grasslands to croplands. For example, Stephens *et al.* (2008) estimated that 90,300 acres (36,540 ha) of native grassland were converted to croplands in the Missouri Coteau region of North and South Dakota during 1989–2003. Fuelled by demand for starch-based ethanol, development of drought-resistant crops, expiration of conservation contracts, and increasing commodity prices, wetland and grassland conversion has accelerated (Wright & Wimberly 2013). High commodity prices have made farmers less reliant on USDA commodity support programmes and effectively neutralised disincentives for habitat conversion. Specifically, new risk management tools provided by federally-subsidised crop insurance, which protect those farming marginally productive land from economic losses, contradict other policies aimed at conserving grasslands or protecting highly erodible land (Wright & Wimberly 2013). The annual wetland loss rate in the PPR of North and South Dakota (2001–2011) was 0.35% or 15,377 ac/yr (6,223 ha/yr, Johnston 2013). The rate of grassland conversion in the Western Corn Belt of North Dakota, South Dakota, Nebraska, Minnesota and Iowa from 2006 to 2011 ranged between 1 and 5.4% annually with a nearly 1.31 million acres (530,000 ha) net decline in grass-dominated land cover (Wright & Wimberly 2013). Since the mid-1980s, federal and provincial programmes in prairie Canada encouraged conversion of marginal cropland to perennial grassland (typically hay fields and pasture), and removal of grain transportation subsidies in the mid-1990s further encouraged conversion to grass-based agriculture (Riemer 2005). However, despite overall increases in grassland during the past 25 years, the absence of native grassland and wetland protection policies in prairie Canada have resulted in declines in native grassland and wetlands of 10% and 5%, respectively, during 1985–2001 (Watmough & Schmoll 2007).

The lack of effective disincentives for habitat conversion in current U.S. and Canadian agricultural policies, generous risk management tools and ongoing conversion of grasslands and wetlands to croplands pose a significant threat to waterfowl populations in the PPR. For example, 1.4 million temporary and seasonal wetlands of < 1 acre in size, located in crop fields in the eastern Dakotas and northeast Montana, are "at risk" of drainage without effective Swampbuster protections (R.E. Reynolds and C.R. Loesch, unpubl. data). Reynolds and Loesch (unpubl. data) further indicated that loss of these wetlands would reduce the current breeding habitat capacity by about one-third for the five most common breeding ducks in the region. Reversing trends in habitat loss in important waterfowl regions will be extremely challenging and if current rates of habitat conversion to croplands continue and habitat protection rates remain at current levels, regional habitat conservation goals and ultimately waterfowl population goals will need to be reduced (Doherty *et al.* 2013).

The 2014 Farm Bill: reforms, challenges and opportunities

The one-year extension of the 2008 Farm Bill expired on September 30, 2013, resulting in a temporary lapse in funding for farm bill programmes. Passage of a new farm bill was delayed over a year by political gridlock, as Congress debated how to achieve cost savings and streamline programmes to reduce the federal deficit. Finally, on 7 February 2014, the President signed into law a new farm bill called the Agricultural Act of 2014 (hereafter, 2014 Farm Bill) that reauthorised several important conservation programmes and enacted other policy reforms aimed at conserving critical grassland and wetland habitat on private land. In addition to the challenges and delays of getting a new farm bill passed, substantial funding reductions were made to conservation programmes estimated at ~$6 billion over the next 10 years (CBO 2014). The 2014 Farm Bill also included major reforms to commodity programmes, new crop insurance options and consolidated conservation programmes. Given the importance of farm bill programmes and agricultural policy to continental waterfowl populations, resource managers and conservation planners should be prepared to adapt, optimise and deliver targeted conservation programmes much more efficiently with significantly less federal financial resources from 2014–2018.

Re-linking conservation compliance to crop insurance

For nearly 30 years, U.S. agricultural producers have agreed to minimise impacts to HEL and Swampbuster-protected wetlands in exchange for farm programme benefits primarily offered through Title I commodity (*e.g.* direct payments, countercyclical payments, *etc.*) and other farm credit supports. These "conservation compliance" provisions were first established in the 1985 Farm Bill to help reduce adverse effects USDA programmes were having on environmentally-sensitive land by reducing soil erosion on HEL and slowing wetland conversion on agricultural lands. Current law allows agricultural producers to farm through wetlands during dry periods and still retain farm programme benefits provided they do not modify the hydrology of impacted wetlands, or if modifications were undertaken after 23 December 1985 steps must be taken to mitigate for equivalent wetland functions and values. Conservation compliance provisions also disallow USDA loans or payments to producers growing annually-tilled commodities on HEL without a soil conservation plan having first been approved by the Natural Resources Conservation Service (NRCS). According to the USDA, ~100 million acres (40.5 million ha) or 25% of all cropland in the U.S. is considered highly erodible (Claassen 2012). With assistance from USDA, producers have developed conservation plans on over 140 million acres (56.7 million ha) of farmed land and reduced soil erosion on HEL by nearly 40% or 295 million tons of soil per year (Claassen 2005).

From 1985 to 1995, conservation compliance requirements were also tied to federal crop insurance benefits, but Congress decoupled these requirements from crop insurance in the 1996 Farm Bill. A large increase in crop insurance enrolment from 99.7 to 202.6 million acres (40.3 to 82 million ha) occurred in 1994–1995 following the passage of the Federal Crop Insurance Reform Act of 1994; however, ~22 million fewer acres (8.9 million ha) were insured in 1996, suggesting that decoupling conservation compliance had little impact on crop insurance enrolment. For the past three decades, conservation compliance has been very effective at conserving farmed wetlands on private agricultural land (Brady 2005). According to the USDA, up to 3.3 million acres (1.3 million ha) of vulnerable wetlands within or adjacent to cropland were not drained because of conservation compliance policies enacted since the 1985 Farm Bill (Claassen 2012). As USDA works to implement newly authorised farm bill programmes during 2014–2018, it will be important to retain these effective conservation measures.

The 2014 Farm Bill eliminates several Title I (*e.g.* direct payments and counter-cyclical payments) programmes tied to conservation compliance provisions in the farm bill. Many groups within the conservation community advocated the need to reconnect these provisions to federal crop insurance benefits (Title XI). In recent years, many producers have opted out of Title I benefits completely, thereby allowing them to convert wetlands for agriculture, while still receiving federal crop insurance benefits without penalty. Since 1994, federal crop insurance has evolved to become the most important and highest-funded safety net and risk management tool for agricultural producers, particularly in the NGP. Indeed, estimated federal outlays for the crop insurance programme will total nearly $90 billion over the next 10 years (CBO 2014). Re-linking conservation compliance provisions to crop insurance premium subsidies would help ensure that farmers maintain a strong safety net, while ensuring long-standing protections for HEL and farmed wetlands remain in effect. After being decoupled from crop insurance since 1996, the 2014 Farm Bill reconnected conservation compliance provisions for farmed wetlands and HEL to federal crop insurance benefits. These provisions will provide critical protections for millions of farmed wetlands on agricultural land through 2018; however, USDA interpretation and implementation of this new policy will be a key factor in ensuring its effectiveness.

Unlike the U.S., Canada does not maintain similar federal wetland protection policies for wetlands on private land; consequently, current laws vary significantly among provinces and territories (Lynch-Stewart *et al.* 1993). In Canada, provinces have primary jurisdiction over wetland protection policies within their boundaries, whereas the territories generally share authority among federal, territorial and native agencies. However, Canada does maintain fairly robust wetland protection policies on federal Crown land and Environment Canada is the primary agency responsible for coordinating and implementing these policies (Government of Canada 1991). Generally, provincial laws cannot bind the federal Crown, which creates regional differences and geospatial challenges when trying to implement and enforce wetland protection policies on private land across a broad landscape. Current provincial wetland protection policies are being developed and/or implemented in Alberta, Saskatchewan, Manitoba, Ontario, New Brunswick and Nova Scotia. However, these regional and provincial disparities create a significant challenge to wetland conservation for waterfowl on private land in Canada.

Sodsaver: slowing native prairie conversion to croplands

Temperate grasslands are one of the most imperilled ecosystems on the planet, yet maintain one of the lowest habitat protection rates of any major terrestrial biome (Hoekstra *et al.* 2005). Native grasslands that support diverse wildlife populations and grass-based agriculture are being converted to cropland at unprecedented rates across many parts of North America. During 2012, nearly 400,000 acres (161,900 ha) of land with no prior cropping history was converted to crop production across the U.S., including >54,876 acres (22,207 ha) in Nebraska, > 27,128 acres (10,978 ha) in South Dakota, >26,395 acres (10,682 ha) in Texas and >24,961 acres (10,101 ha) in Florida (USDA FSA 2013). At current conversion rates, over half of the native prairie remaining in the U.S. areas of the PPR will be lost in the next 34 years (Stephens *et al.* 2008). Agricultural policies, emerging technologies and economic drivers are fuelling large-scale conversion of these rare and important prairie habitats. Native grasslands provide critical habitat for wildlife, including a globally-significant breeding range for many waterfowl and shorebird species (Ringelman *et al.* 2005). These habitats also support numerous grassland-dependent songbirds, which are experiencing a steeper population decline than any other avian guild in North America (Peterjohn & Sauer 1999). Additionally, native rangelands are fundamentally important for livestock production by providing forage and resilience to drought. Ranching, recreational hunting and ecotourism associated with the native prairie also provide economic diversity and stability to rural economies.

Today, the last remaining grassland-dominated landscapes are largely confined to areas with poor soils, steep topography and climatic conditions largely unsuitable for consistent crop production (Doherty *et al.* 2013). Unfortunately, accelerated grassland conversion is occurring in many of these areas, causing significant ecological and societal impacts. Further loss of native rangeland habitat is also an economically costly proposition, bringing additional disaster-prone land into production, while creating significant taxpayer liabilities through subsidised risk management. Sodsaver legislation enacted in the 2014 Farm Bill, will: 1) limit crop insurance coverage to 65 percent of the applicable transition yield (*i.e.* county average) for the first four years until an actual production history is established on newly broken land; 2) reduce crop insurance subsidies on newly-broken sod by 50 percentage points below the premium subsidy that would otherwise apply for the first four consecutive years of crop production; and 3) make newly-broken acreage ineligible for yield substitution. These provisions were included as a nationwide policy in the 2013 Farm Bill passed by the Senate, but were confined to only the U.S. PPR in the Farm Bill passed by the House of Representatives. As illustrated by the 2008 Farm Bill, a region-only Sodsaver provision is difficult to administer and can create inequities among agricultural producers within and across states. Instead, a national provision would create a more equitable and actuarially sound programme across the country.

The 2014 Farm Bill provides a new regional Sodsaver programme that applies to Montana, South Dakota, North Dakota, Minnesota, Iowa and Nebraska. This provision applies to the entire state, not just the PPR-portion, and is a mandatory requirement, in contrast to the state Governor opt-in programme of the 2008 Farm Bill. This provision will not completely stop native prairie conversion in these six states, but it will provide less financial incentive for converting native prairie, as the crop insurance subsidies have been reduced significantly. Grassland conversion continues to be a national issue that plagues many grassland-dependent species, such as Greater Sage-grouse *Centrocercus urophasianus*, Lesser Prairie-chicken *Tympanuchus pallidicinctus* and many migratory birds that depend on these rare and declining habitats across the U.S. In 2013, 89% of the nearly 400,000 acres (161,900 ha) of perennial cover converted to cropland occurred outside of the U.S. PPR (USDA FSA 2013). Thus, future farm bill policy efforts aimed at grassland protection should focus on enacting a national Sodsaver programme that applies to all states and creates other similar reforms that conserve critical native habitats. Additional policy reforms such as significantly reducing or eliminating crop insurance subsidies on non-arable land (*i.e.* soil classes 6–8) should also be considered.

The future of the Conservation Reserve Program in a changing landscape

The CRP is considered one of the most successful USDA conservation programmes in history and its landscape-level impacts on reducing soil erosion, improving water quality, sequestering carbon and enhancing wildlife habitat are well-documented (see Allen & Vandever 2012). However, growing global demand for commodities, escalating land and cash rent values, stagnant CRP rental rates, biofuel policies, and improved genetics and farming technologies are driving the loss of CRP acreage across much of the U.S. particularly in the PPR. For example, Wright & Wimberly (2013) documented conversion of 1.3 million acres (0.53 million ha) of perennial grasslands (*i.e.* native prairie, tame pasture and CRP) to cropland in 2006–2011, which represents a rate of change in grassland cover not seen since the "Dust Bowl" era of the 1930s. The Farm Service Agency estimates that < 6 million acres (2.43 million ha) of CRP will remain in the U.S. PPR in 2014. This loss represents a substantial decrease (31%) from its peak of 8.3 million acres (3.59 million ha) in 2007 and declining trends are expected to continue over the next 5 years (Fig. 1; USDA FSA 2013). The 2014 Farm Bill reduces the national CRP enrolment cap from 27.5 million acres (12.9 million ha) to 24 million acres (9.71 million ha) by 2017. In order to achieve cost savings, the national enrolment cap on the CRP Farmable Wetland Program (FWP) will also be reduced from 1 to 0.75 million acres (404,690 to 303,500 ha). However, in issuing guidance to USDA for new CRP rule-making, the Manager's report states "overall reduction in the maximum acres enrolled …

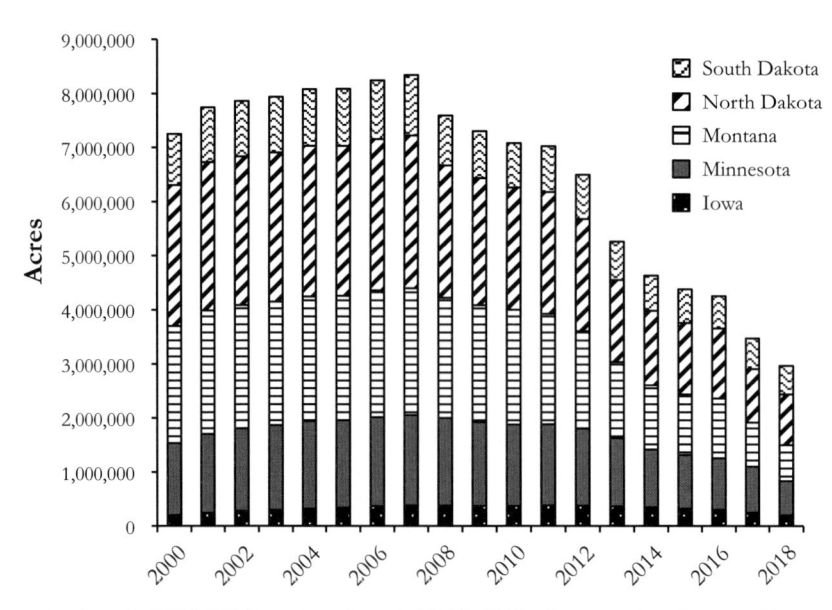

Figure 1. Actual (2000–2013) and projected (2014–2018) Conservation Reserve Program (CRP) enrolment area in the U.S. Prairie Pothole Region. Projected area assumes no new sign-ups and anticipated expirations based on Farm Service Agency reports for 2014–2018 (Ducks Unlimited, unpubl. data).

should not serve as an indicator of declining support for CRP. The Managers intend for CRP to be implemented at authorized levels, using the statutory flexibility, and for the program to continue as one of USDA's key conservation programs in concert with working lands conservation efforts."

Despite a significant reduction in the overall CRP acreage cap, several provisions were included in the 2014 Farm Bill to make the programme more flexible and attractive to producers, while promoting a "working lands" approach. For example, the Secretary of Agriculture will have greater authority to: 1) enrol newly eligible grasslands (up to 2 million acres, or 0.81 million ha); 2) flexibly apply prescribed grazing, burning, haying and other mid-contract management activities outside of the primary nesting season; 3) provide more allowances to use rather than dispose of residue removed from CRP land during contract maintenance and management; and 4) promote expanded use of continuous and Conservation Reserve Enhancement Practices (CREP) sign-up opportunities. Faced with these challenges and opportunities, resource managers will need to focus on making CRP more economically attractive and competitive by updating county rental rates, increasing land-use flexibility and management allowances, maximising continuous sign-up opportunities, and working to modify the national Environmental Benefit Index (EBI) scoring process to elevate the PPR to a national priority area.

Other working land opportunities

The 2014 Farm Bill also consolidates and streamlines 23 conservation programmes authorised under the 2008 Farm Bill into just 13 programmes. For example, former easement programmes such as the Farm and Ranch Lands Protection Program, GRP and WRP were merged into the ACEP. The ACEP establishes two separate tracks for wetland reserve easements (WRE) and agricultural land easements (ALE), while providing > $2 billion of funding for conservation on private land over the 2014–2018 period. It also allows a landowner donation for ALEs as long as another entity matches 50% of the Secretary's contribution and provides a waiver to pay up to 75% USDA cost-share for certain grassland conservation easements. This provision may create new public-private partnership opportunities among state, federal, private and NGO partners to develop easement programmes on private land. The new farm bill also reduces the former 7-year ownership rule to 2 years to become eligible for wetland easement enrolment. This may be an attractive incentive for conservation buyers looking to enrol land into the programme.

The 2014 Farm Bill also consolidates several former regional conservation programmes (*e.g.* the Chesapeake Bay Watershed Program and the Cooperative Conservation Partnership Initiative) into a new Regional Conservation Partnership Program (RCPP). Under RCPP, projects may focus on water quality, erosion, wildlife habitat and other regional resource concerns, and this new programme will create up to eight national critical conservation areas. The RCCP partnership agreements may extend up to 5 years and the programme provides mandatory funding of $100 million per year from 2014–2018. It will facilitate landscape-scale conservation initiatives leverage partnerships and enable managers to direct resources strategically towards priority regions for waterfowl, such as the PPR or the Gulf Coast.

The 2014 Farm Bill also merges EQIP and former Wildlife Habitat Incentives Program (WHIP) into one general EQIP programme, but specifies that "wildlife habitat development" is a defined programme purpose and sets a minimum 5% funding floor for wildlife habitat projects. The programme also requires that at least 60% of the total funds be invested for livestock purposes. The EQIP is one of the highest funded conservation programmes in the new farm bill, providing an average of $1.35 to $1.75 billion per year of conservation funding. The EQIP provides cost-share for a number of wildlife-friendly conservation (wetland development, grassland improvement, *etc.*) and habitat management practices (brush control, weed management, prescribed grazing, forage stand improvement, *etc.*) that may be very compatible with waterfowl and economically attractive to livestock producers, who prefer more short-term working land options as opposed to traditional 10–15 year set-aside programmes such as the CRP.

Habitat potential on agricultural working land

To the extent that waterfowl are able to adapt to habitat changes, or working agricultural lands retain or simulate ecological functions provided by historical habitats, the adverse effects of habitat loss may be dampened. Indeed, exponential growth in Lesser Snow Geese *Chen caerulescens* populations are attributed to behavioural and morphological adjustments that enabled birds to shift from historical to agricultural habitats (Linscombe 1972; Alisauskas 1998). There are numerous other examples of waterfowl using non-traditional or altered habitats, although the demographic consequences of these shifts are generally unknown. Thus, in addition to conserving and restoring traditional habitats, we must identify and work collaboratively to promote working agricultural systems that are both producer- and waterfowl-friendly and provide environmental services in addition to the production of food and fibre. Cultivation of rice *Oryza sativa* and winter cereals represent two such situations.

Agricultural working land and waterfowl: rice agriculture example

Rice agriculture is a major component of the contemporary landscapes of the Gulf Coastal Plain, LMAV, and Central Valley of California. Between 1985 and 2012, 2.3–3.6 million acres (0.93–1.46 million ha) of rice were planted annually nationwide with over half (60%) of this acreage located in the LMAV (Arkansas, Mississippi, Louisiana, and Missouri), 25% in the Gulf Coast region (Louisiana and Texas), and 15% in the Central Valley of California (USDA NASS 2014).

Cultivation practices

Rice is a warm-season crop typically planted in the spring and harvested in summer or autumn. Cultivation practices vary somewhat within and among rice-growing regions as a consequence of differences in climate, geography, soils, topography, surrounding land-uses, water supply, disease-pest issues, rotational cropping opportunities and farming traditions. In California and the LMAV, seeding of rice is similar to seeding practices for other cereal crops. That is, rice seed can be drilled or broadcast under dry to moist conditions in either reduced or conventional tillage systems ("dry seeding"). In southwest Louisiana, rice is most commonly cultivated using a water-seeding system in a 3-year rotation with crawfish (Order: Decapoda) and fallow or soybeans (69% water-seeded, 31% dry seeded; J. Saichuk, pers. comm.). In a water-seeded system, rice is planted aerially into flooded fields in March–June (Blanche *et al.* 2009). Shortly before planting (3–4 days), the seedbed is tilled rough, fertilizer is applied and incorporated, and the field is flooded. Alternatively, rough tillage conducted in autumn or winter may be followed by flooding and, shortly before seeding, water-levelling (tractor pulling a blade through the flooded rice field). Water-levelling agitates the soil and water, producing a thick slurry and level seedbed when the soil settles out of the water. Water-seeded fields typically are dewatered 24 h after seeding.

The principal advantage of water seeding is that it provides an excellent cultural method for control of weeds, especially Red Rice *Oryza punctate* (Webster & Levy 2009). Red Rice is the most troublesome and economically damaging competitor of rice; annually contributing to the loss of tens of thousands of dollars to rice producers in southern states (Webster & Levy 2009). Some producers flood harvested rice fields to facilitate feeding by wintering waterfowl on noxious Red Rice (Smith & Sullivan 1980). Water seeding is also preferred by farmers that plant extensive acreages in areas with high rain and is compatible with other uses of rice fields such as crawfish aquaculture.

Cultivation practices in water- and dry-seeded fields are similar after seeding. Fields are gradually (re)flooded when rice has sprouted 4–6 inches (10–15 cm) and remain flooded throughout the growing season until rice seeds mature. Most of the currently grown rice varieties need ~120 days from seed germination until the grain is ready for harvest. Fields are drained 4 weeks before harvest to allow combine harvesters to operate in the fields.

An assortment of dryland crops are rotated with rice in California and the LMAV, but rotational options are limited in coastal Louisiana and Texas. Along the Gulf Coast, rice typically is not cultivated in the same field during consecutive years because doing so would increase disease and weed prevalence and reduce yields. Management options for rice producers include production of a second or "ratoon" rice crop, preparing fields for winter–spring crawfish production, or idling land for fallow or dryland crop production the following spring–summer. Ratooning is the practice of harvesting grain from tillers originating from the stubble of a previously harvested crop (main crop). The climatic conditions of southwest Louisiana and the early harvest date of commonly grown rice varieties combine to create an opportunity for ratoon crop production, but weather, planting date, quality of the first crop and harvest conditions can all influence ratoon rice development and yield. In general, the first crop should be harvested by 15 August to ensure adequate time for ratoon rice to develop. Harvest of ratoon rice typically occurs in October–November.

The rice-crawfish-fallow (or rice-crawfish-soybean) rotational strategy commonly deployed in the region employs crawfish in a rotational system of rice and sometimes soybeans. Rice is grown and harvested during the summer, and crawfish are grown during autumn, winter and early spring in the same field. Louisiana crawfish producers rely on a forage-based system for providing nourishment to growing crawfish. Rice has become the standard forage crop for the industry because the plant exhibits the desired characteristics under the long-term flooded condition of a crawfish pond and partly because adequate stands of vegetation are achievable and predictable when recommended management practices are followed.

Rice fields managed for crawfish production are commonly fertilised and irrigated to achieve a ratoon crop (re-growth) of forage. Fields are initially flooded in October–December and remain flooded throughout the harvest period, January–June. In southwest Louisiana, fields are typically fallowed following drawdown in May–June, but some producers may drawdown crawfish ponds earlier (April) to plant soybeans (April–June). To control weeds in fields rotating back into rice cultivation, water control structures typically are closed in the autumn (after soybean harvest) to capture available rainfall. Producers may pump water onto fields if fields are leased for waterfowl hunting (November–January) or rainfall is inadequate to completely flood fields by January. Fields are drained in spring so that they may be tilled in preparation for rice planting as described above.

Waterbird use of rice

A wide variety of waterbirds (waterfowl, shorebirds and wading birds) and some landbirds use rice fields (Taft & Elphick 2007). Rice field use by wintering and migrating waterfowl and shorebirds is especially pronounced. Avian use is best documented in Californian rice fields where over 118 species representing 38 families have been recorded during winter (Eadie *et al.* 2008). Densities of non-breeding waterfowl and shorebirds observed in Californian rice fields averaged 730 (peak count = 3,600) and 252 (2,600) birds/km^2, respectively (Eadie *et al.* 2008).

The 2.5–3.75 million acres (1–1.5 million ha) of farmland in coastal Louisiana and Texas operated in rice-crawfish-fallow, rice-fallow, rice-pasture or rice-dryland crop rotational scheme simulate wet, early successional habitats that potentially are highly attractive to wetland-associated wildlife. The close proximity of fields to coastal marshes, their location at the terminus of two major migratory bird flyways, bird-friendly cultivation practices, high annual rainfall, and abundant plant and animal foods further enhance their potential value for waterbirds. Indeed, recent shifts in the distributions of waterbirds from coastal wetlands to inland agricultural wetlands (*e.g.* Fleury & Sherry 1995) coincide with the expansion of crawfish aquaculture and ongoing loss and degradation of coastal wetlands. A minimum of 67 species of waterbirds including 17 waterfowl, 33 shorebirds, 15 wading birds, 2 rail and 1 crane species have been observed using rice fields in coastal Texas and Louisiana (W.L. Hohman, unpubl. data). Peak densities of non-breeding geese, ducks, shorebirds, and wading birds recorded in these rice fields during winters 1996/97 or 1997/98 were 9,300, 4,300, 1,700, and 1100 birds/km^2, respectively (W.L. Hohman, unpubl. data). Estimated seasonal use by waterbirds (excluding geese) from October to May was 72.1 and 125.5 million use-days in 1996/97 and in 1997/98, respectively (W.L. Hohman, unpubl. data). However, because waterbirds use rotational crops (*e.g.* fallow), peak and seasonal use may have been underestimated by ≥50% (W.L. Hohman, unpubl. data).

Use of rice fields by waterbirds is potentially influenced by factors such as field size, timing, duration and extent of flooding, crop rotation, cultivation and harvest practices, grazing, height of vegetation, stubble treatments, frequency of disturbance and surrounding landscape features (*e.g.* land uses, cover types, amount of edge, distance to water, *etc.*). Waterbird richness and density are greater in flooded than unflooded rice fields in California (Eadie *et al.* 2008). Waterbird groups responded differently to water depth, with peak species richness and conservation value (species being indexed by their relative abundance in North America; Elphick & Oring 1998) observed at intermediate water depths (10–20 cm) (Elphick 1998; Eadie *et al.* 2008). In Californian rice fields, however, interpretation of waterbird responses to manipulation of rice straw was confounded by an interaction with the depth of flooding (Eadie *et al.* 2008).

In the Texas and Louisiana rice fields, waterbird species richness/diversity was highest in fallow fields and rice crop cover types, greatly exceeding other crop covers (W.L. Hohman, unpubl. data). Fields in 3-year rice-fallow rotation had higher richness and diversity scores than fields in 3- or 4-year rice rotations with dryland crops. Richness was decreased by grazing and increased by flooding. Duck densities were affected by crop rotation scheme, shorebird densities were affected by crop cover and grazing, and wader densities were affected by both crop cover and rotation scheme. Densities of all three groups increased with flooding.

Louisiana rice fields also provide habitat for breeding waterbirds (Hohman *et al.* 1994), at least one of which (King Rail *Rallus elegans*) has been given special status in 12 states. Increase in the nesting density of King Rails in Louisiana's rice fields is attributed to expansion of crawfish aquaculture. Other common to rare nesting birds include Fulvous Whistling Duck *Dendrocygna bicolor*, Purple Gallinule *Porphyrula martinica*, Common Moorhen *Gallinula chloropus* and Least Bittern *Ixobrychus exilis*.

Opportunities for management of rice fields for waterfowl

Waterbirds are attracted to rice fields because of the abundant foods that occur there. Potential waterbird foods include waste grain, seeds of water tolerant (*i.e.* moist soil) plants, green forage and invertebrates. Rice fields are highly dynamic systems and, although vegetation is highly monotypic, rice fields essentially function like early successional, seasonally-flooded wetlands. That is, they have high detrital (*i.e.* straw) inputs that, when flooded, serve as forage for production of crawfish and other aquatic invertebrates. Further, reduced pesticide use in fields managed for crawfish production may benefit other aquatic invertebrates (McClain *et al.* 2009).

Waterbirds also use rice fields as resting areas. The general openness of the rice agricultural landscape is attractive to many species that during migration and winter must remain vigilant for potential predators (Elphick 2000). Rice agriculture has become especially important for Northern Pintail (*Anas acuta*, hereafter Pintail) wintering in California's Central Valley and along the Texas–Louisiana Gulf Coast (Miller 1987; Cox & Afton 1997). Pintail and other waterbirds may shift to rice field refuges to avoid disturbance in other habitats or, alternatively, hunting disturbance may result in daytime avoidance of rice fields (Rave & Cordes 1993; Cox & Afton 1996).

Agronomic practices typically followed during the 3-year rice-crawfish-fallow or rice-fallow rotational schemes are generally "waterbird friendly." So in most cases, management of Gulf Coast rice fields for wintering and migrating waterbirds involves only minor changes in existing management practices. Because of high annual rainfall, use of flooding for weed control, practice of water-levelling, water-seeding of rice, crawfish aquaculture and the leasing of rice fields for waterfowl hunting, Gulf Coast rice fields tend to be wet and therefore available to waterbirds throughout much of the year. Additionally, many coastal rice fields are left unplanted (*e.g.* pasture rotation) or fallowed every other year. Moist soil plants that grow in fallowed fields produce abundant seeds that are highly preferred foods of wintering waterfowl (Fredrickson & Taylor 1982). Indeed, samples taken at waterfowl feeding sites in Louisiana rice fields indicated biomass of moist soil plant seeds in rice fields may be equivalent to that found in public areas managed specifically for that purpose (Hohman *et al.* 1996). With average rainfall, passive management (*e.g.* simply closing water control structures) is likely to be sufficient to meet the diverse habitat needs of most waterbird species; however, the productivity and attractiveness of Gulf Coast rice fields for waterfowl and other waterbirds may be further enhanced by timely manipulations of rice stubble and flooding, precise control of water levels during rice cultivation, minimising disturbances in fallow fields during March–May, or establishment of some single crop ponds managed solely for crawfish production (W.L. Hohman, unpubl. data).

Following the 2010 Deepwater Horizon Gulf Oil Spill, the USDA NRCS established the Migratory Bird Habitat Initiative (MBHI) to provide inland waterbird habitats to compensate for potential oil impacts on coastal wetlands. Through EQIP, WHIP, and WRP, the MBHI has provided incentives for private landowners in eight states (Alabama, Arkansas, Florida, Georgia, Louisiana, Mississippi, Missouri and Texas) to enhance and increase availability of shallow-water habitats for migrating and wintering waterfowl, shorebirds, and other waterbirds along the Gulf Coast and within the LMAV. The provision of financial assistance and compatibility of management activities with normal agronomic practices and recreational use of sites contributed to the enthusiastic response by landowners who offered almost 1 million acres (> 400,000 ha) for possible enrolment in EQIP or WHIP. To qualify for enrolment the proposed management activity must represent a change from normal agronomic practices (*i.e.* "enhancement"). Approximately half of the offers were accepted into the programme with most of the contracts awarded in the rice growing region of southwest Louisiana. In coastal Louisiana and Texas, the primary management practices implemented through the MBHI entailed manipulations of rice stubble and shallow flooding of rice fields in early autumn or late winter. Stubble manipulations and early flooding were implemented to benefit autumn-migrating shorebirds which pass through the region in August and September; late flooding targeted spring-migrating waterfowl. The net result was that shallow-water habitats were available in coastal regions for an extended duration. Activities undertaken through EQIP and WHIP on agricultural working land were similar, but eligibility differences between the programmes enabled the USDA NRCS to serve a broader clientele.

An evaluation of waterbird responses to MBHI practices by researchers at Mississippi State University is ongoing, but preliminary results further substantiate the importance of rice agriculture for waterbirds. Fields in which a ratoon was produced and subsequently disced were especially important habitat for non-breeding waterbirds, if they were flooded through assistance from MBHI or other means (Marty 2013).

Challenges to management of rice fields for waterfowl

The potential for rice agriculture to provide habitat for waterfowl is substantial on the Gulf Coastal Plain, as it is in other rice growing regions. Challenges to the management of rice fields as waterfowl habitat identified by Eadie *et al.* (2008) include: 1) the provision of habitat for non-target, undesirable or nuisance wildlife (*e.g.* Red-winged Blackbird, *Agelaius phoeniceus*, American Coot *Fulica americana*; Snow Geese, *etc.*); 2) water quality concerns (*e.g.* release of nutrients, particulate matter in water releases from rice fields); 3) additional time and financial costs associated with management (*e.g.* delayed field work and costs of pumping and stubble manipulations); 4) declining rice acreage due to urban growth, farm economics or human disturbance; 5) increased habitat fragmentation; 6) increased harvest efficiency or changes in agronomic practices (*e.g.* straw management, development of glyphosate-tolerant rice varieties) that reduce the availability of waste grain and moist soil plant seeds; 7) decreased availability of water (*e.g.* conflicts caused by increased demand and use by other user groups); and 8) conservation of endangered species.

Additionally, agricultural policy that favours production of other crops or restricts farmer participation in conservation programmes may contribute to a reduction in rice acreage. For example, Louisiana continues to fund MBHI through EQIP, and MBHI was expanded to include activities designed to provide nesting and brood-rearing habitat for resident waterbirds such as the Mottled Duck *Anas fulvigula*. Although this species has adapted to survive on the wet agricultural/coastal marsh interface, it is vulnerable to urban encroachment, coastal land loss and conversion from "wet" agriculture (such as rice and crawfish production) to dry land crops such as soybean, sugarcane and milo (Hohman *et al*. in press). Initial interest in this component of MBHI was constrained by confusion about the level of compensation that was to be provided for various management scenarios. Programme restrictions are also limiting expansion of MBHI. Specifically, the EQIP requirement that only allows for provision of financial and technical assistance for the application of a new practice or activity prevents producers from re-enrolling fields in MBHI. Consequently, the acreage enrolled in MBHI has declined because producers are unwilling to bear the increased costs of management without compensation.

Management of rice fields for recreational activity and income derived from hunting leases can provide a strong motivation for producers to manage rice fields as waterfowl habitat. The value of rice fields for waterbirds in southwest Louisiana was estimated to be > $100 1,028/acre ($247–2,538/ ha) based on the value of hunting leases or the restitution value of waterbirds using rice fields during the breeding and non-breeding periods (W.L. Hohman, unpubl. data). Further, the value of rice fields as waterbird habitat exceeded the return realised by farmers for production of rice and crawfish ($208/acre, or $494/ha; W.L. Hohman, unpubl. data). Knowledge of the value that rice agriculture provides to ecosystem services, and efforts to minimise mismatches between the value of these services and income derived from agricultural production, should further advance stewardship of rice agriculture for the conservation of waterfowl and other wildlife.

Although waterbird use of Californian rice fields is well documented, the extent to which rice fields provide a reasonable substitute for natural wetlands is unclear (Elphick 2000; Eadie *et al*. 2008). The functional equivalence of rice agriculture in comparison with historical wetland habitats along the Texas–Louisiana coast likewise is unknown; nonetheless, Louisiana and Texas have experienced extensive loss and degradation of coastal wetlands. From 1932–2000, coastal Louisiana lost > 4,900 km^2 of land, primarily marsh, with the annual rate of wetland loss estimated to be 43 km^2 between 1985–2010 (Couvillion *et al*. 2011). Continued loss of coastal wetlands, and reductions in rice acreage in coastal Texas–Louisiana, have important implications for waterbird conservation in North America. Enhanced management of agricultural wetlands along the Gulf Coast (*e.g.* as undertaken in response to the Deepwater Horizon oil spill through the NRCS's MBHI) may represent the best opportunity to accommodate waterbirds displaced by wetland loss associated with sea-level rises and other environmental change.

Agricultural working land and waterfowl: autumn cereals example

Opportunities for management of autumn cereals for waterfowl

The glaciated PPR region of central North America serves as the primary breeding area for many of North America's waterfowl and shorebirds (Batt *et al.* 1989; Skagen & Thompson 2007). Historically, extensive native grasslands and diverse wetlands provided ideal habitat for successful waterfowl reproduction in this area (Stephens *et al.* 2005). Since human settlement, however, a majority of the PPR has become an important agricultural production zone for small-grain, oil-seed and row crops. Conversion of grassland to annual cropland, along with drainage and degradation of wetlands, has made significant alterations to the landscapes in which breeding waterfowl and shorebirds nest (Stephens *et al.* 2008). Today, this region of North America is one of the most intensively cropped landscapes in the world, with > 80% of some counties in cropland production (Foley *et al.* 2005; Statistics Canada 2011).

Conversion of grasslands to cropland and associated alteration of predator communities in the PPR are thought to be the primary reason for long-term declines in waterfowl production in this region (Sargeant *et al.* 1993; Greenwood *et al.* 1995; Beauchamp *et al.* 1996; Stephens *et al.* 2008). In addition, the intensity of cropping practices has increased on existing cultivated land in recent history. The largest and most economically and environmentally significant change in agricultural land-use since the 1970s has been the decline in summer fallow, a practice where cropland is left uncropped for alternate growing seasons for moisture accumulation, nitrogen release and weed control (Carlyle 1997). In prairie Canada, the practice of summer fallowing has declined by ~18.8 million acres (7.6 million ha) between 1971–2011 (Statistics Canada 2012). In its place, continuous cropping under minimum and zero-tillage practices with high nutrient and pesticide inputs has prevailed. Podruzny *et al.* (2002) suggested that declines in populations of some bird species, such as Pintail, may have been the result of reduced nest survival as continuous cropping replaced relatively safe nest sites located in summer fallow.

Cropland conversion to grassland began in the U.S. under the CRP in the late 1980s, and in Canada with removal of grain transportation subsidies in 1995. Recent trends and long-term projections of cropland area suggest that conversion of grassland to cropland is again on the rise (Rashford *et al.* 2010; Wright & Wimberly 2013). Biofuel-driven agricultural commodity prices are expected to increase pressure to convert grasslands to croplands in the foreseeable future (Wright & Wimberly 2013). While waterfowl benefited greatly from programmes such as the CRP (Reynolds *et al.* 2006), these benefits are expected to diminish as remaining grasslands are converted to cropland (Stephens *et al.* 2008; Rashford *et al.* 2011). Not all croplands are equal, however, in their potential to affect breeding waterfowl. While many waterfowl species can benefit from croplands as a food resource during non-breeding periods (reviewed in Taft & Elphick 2007), few crops provide relatively safe nesting habitat like the grasslands they replace. Early nesting species, such as Mallard *Anas platyrhynchos* and Pintail, are especially susceptible to nest failure in croplands, but some autumn-seeded cereal crops including winter wheat *Triticum* sp. and autumn rye *Secale cereale* may provide viable nesting habitats (Devries *et al.* 2008).

In North America, wheat has two distinct growing seasons. Winter wheat, accounting for 70–80% of U.S. wheat production, is planted in the autumn, harvested the following summer and is generally grown from the Texas Gulf Coast to prairie Canada (Acquaah 2005). Spring wheat is planted in early spring, harvested in late summer/early autumn and produced primarily in the prairies of the northern U.S. and southern Canada (Acquaah 2005). Within the PPR, the majority of wheat grown is the spring-seeded variety. For example, of 75 million acres (30.4 million ha) of cropland in prairie Canada in 2012, ~21 million acres (8.5 million ha) were wheat, of which only ~1 million acres (0.4 million ha) were winter wheat (Statistics Canada 2012). In North and South Dakota, about 10 million acres (4.1 million ha) out of 40 million cropland acres (16.2 million ha) were wheat in 2012, of which ~2 million acres (0.8 million ha) were winter wheat. Other autumn-seeded cereal grains like autumn rye and triticale (*Triticum* × *Secale* hybrid) generally comprise less than a couple of hundred thousand acres in the PPR.

Croplands are commonly ignored in waterfowl nesting studies despite their documented use by nesting birds (Goelitz 1918; Earl 1950; Milonski 1958; Higgins 1977; Lokemoen & Beiser 1997). This is likely because most waterfowl nesting studies historically avoided searching seeded cropland, or limited timing and frequency of searches relative to other habitats due to crop damage concerns. Hence, our understanding of cropland use by nesting ducks is limited despite the dominance of cropland as potential nest habitat in many landscapes important to breeding waterfowl. Where data are available, nest survival in cropland is typically low due to predation and destruction of nests by machinery during spring-seeding operations (Cowardin *et al.* 1985; Klett *et al.* 1988; Greenwood *et al.* 1995; Richkus 2002).

Autumn-seeded cereal grains, such as winter wheat and autumn rye, however, can provide relatively undisturbed nesting cover for birds during the breeding season and may complement grassland nesting habitats that are available to birds. Several recent studies suggest that autumn cereals may provide high value nesting habitat for breeding waterfowl relative to spring-seeded crops and grasslands. Devries *et al.* (2008) conducted complete nest searches on 4,247 ha of cropland in southern Saskatchewan, including spring-seeded (wheat and barley) and autumn-seeded cereals (winter wheat and autumn rye). Autumn rye and spring-seeded crops were used for nesting by five duck species (Mallard, Pintail, Blue-winged Teal *Anas discors*, Northern Shoveler *A. clypeata*, and Gadwall *A. strepera*), while winter wheat was used by all of the aforementioned species, as well as by Green-winged Teal *A. crecca*, and Lesser Scaup

Aythya affinis). Nest densities were 0.39 and 0.25 nests/ha in winter wheat and autumn rye, respectively, compared to 0.03 nests/ha in spring-seeded cereals. Critically, nest survival was consistently very high throughout the nesting season in winter wheat and autumn rye (~38% and 18%, respectively) whereas survival in spring-seeded crops varied from close to 0% in early nests to close to winter wheat levels for late nests (Devries *et al.* 2008). High nest survival has also been found in winter wheat in comparable studies in North Dakota (Duebbert & Kantrud 1987; B.R. Skone, unpubl. data). Further, nest success rates in autumn-seeded crops are generally greater than those found in grassland habitats throughout much of the PPR (*e.g.* Klett *et al.* 1988; Greenwood *et al.* 1995). Additional research comparing waterfowl nest density and success in winter wheat and adjacent grassland habitat is currently ongoing (B.R. Skone, unpubl. data).

Together, the density and success of waterfowl nests in autumn-seeded cereals suggests that these crops have the potential to recruit many more waterfowl young to breeding populations than spring-seeded cropland, and are comparable to grassland habitats (Devries *et al.* 2008). Providing high nest survival early in the nesting season conveys added value, given the importance of early hatched nests to waterfowl recruitment (*e.g.* Dzus & Clark 1998). The value of autumn-seeded cereals may be most evident in landscapes with high breeding waterfowl populations, many wetlands and extensive croplands. Pintail, especially, could benefit from expansion of autumn-seeded cereal crops, as they nest extensively in cropland stubble (Milonski 1958; Klett *et al.* 1988; Miller & Duncan 1999), initiate nests early in the season prior to spring-seeding operations (Austin & Miller 1995) and re-nest minimally (Austin & Miller 1995; Guyn & Clark 2000). Further, Pintail tend to settle in highly cropped landscapes, especially at high population density (J.H. Devries, unpubl. data). Other priority bird species (*e.g.* Long-billed Curlew *Numenius americanus*) that are known to nest early in cropland stubble are also likely to benefit from autumn cereals (Lokemoen & Beiser 1997; Devries *et al.* 2010).

Challenges for management of autumn cereals for waterfowl

Given the potential benefits of autumn-seeded cereals to nesting waterfowl, Ducks Unlimited (DU), an international non-profit organisation focused on conserving waterfowl habitat, has taken great interest in winter wheat. Efforts by DU include promoting winter cereals in landscapes that have high wetland densities and attract high densities of Pintail and other wetland-dependent birds. As with most agricultural commodities, the primary drivers of winter wheat production are agronomic and in this sense, winter wheat has several advantages. Winter wheat on average provides a 20% yield advantage over spring wheat and generally has lower input costs (Statistics Canada 2013). Further, indirect benefits include: 1) spreading out the annual workload; 2) earlier seeding of spring-seeded crops; 3) decreased exposure to poor spring seeding weather; 4) winter wheat takes full advantage of spring moisture; 5) early growth avoids exposure to certain pests; and 6) winter wheat often outcompetes spring grassy weeds.

Despite the agronomic benefits, barriers to the growth of winter wheat remain. First, extremely cold winters, especially in prairie Canada, challenge existing winter wheat varieties with winter kill. Also in Canada, where hard red spring wheat has been the "gold standard" of the grain market, challenges remain with developing markets for alternate wheat varieties (Mulik & Koo 2006). Finally, changing long-held traditional farming practices remains an impediment, as seeding in September, shallow seeding, seeding into standing stubble and earlier harvests, challenge farmers to make substantial changes to their operations.

To address these challenges, DU has embraced several non-traditional activities for a conservation organisation. Recognising the limitations of available winter wheat varieties, DU provided financial support for the development of new winter wheat varieties at a time when winter wheat variety development in Canada was concluding. Currently, > 90% of winter wheat varieties grown in prairie Canada are those developed with DU support. Further, DU is investing in collaborative research to improve cold-hardiness of winter wheat varieties while providing direct technical assistance to farmers regarding best crop management practices. While incentive payments were initially part of the programme, evaluations have shown that technical expertise provided by agronomists was more attractive and sustainable than cash incentives (DU, unpubl. data). Ducks Unlimited has recently expanded their winter wheat programme in partnership with Bayer Cropscience. A focus of the partnership with Bayer, "Winter Cereals – Sustainability in Action", includes additional extension outreach to increase the acreage of winter wheat planted in the PPR and expansion of winter wheat breeding programmes at several universities across the U.S. and Canada. Since DU initiated the winter cereals programme in 1999, winter wheat acreage in North Dakota has increased over 12-fold from 60,000 acres (24,300 ha) in 1999 to 750,000 acres (303,600 ha) planted in 2012 (USDA NASS 2013). Over the same time period in prairie Canada, winter wheat has grown from 245,000 acres (99,100 ha) to 1.1 million acres (459,500 ha, Statistics Canada 2012).

The remaining barriers to expansion of winter wheat can be overcome. New cold-tolerant varieties are in constant development and additional varietal development is focused on yield, quality and disease resistance. Markets for winter wheat, especially in Canada, are beginning to expand, and realised agronomic benefits should overcome traditional barriers to autumn-seeding. Finally, further research is being conducted to determine whether winter wheat provides landscape-level impacts on duck and shorebird nest survival in addition to the apparent habitat-specific increase in nest survival for nests within winter wheat fields (B. Skone, Montana State University, pers. comm.).

The challenges of strategic conservation targeting on private land

The need to target conservation strategically has long been recognised by policy makers, ecologists and economists. Faced with limited resources (*e.g.* budgets or labour-hours), conservationists cannot select all available projects that produce biological benefits (*e.g.* easement locations or management activities); thus, they seek to select projects that generate the greatest biological benefits possible given resource constraints. What constitutes strategic targeting and how to achieve it, however, can differ substantially across different interest groups and academic disciplines. Moreover, the challenges specific to targeting waterfowl conservation on private land depend on the definition of strategic targeting. These challenges become clear if we begin from the strategic habitat conservation (SHC) framework developed by the USFWS (USFWS 2008).

The SHC framework describes strategic targeting as an iterative process involving biological planning, conservation design and delivery, and monitoring and research that provides feedback to inform the process. Essential to the SHC framework are: 1) defining and measuring specific population objectives (*i.e.* as opposed to simply focusing on habitat area protected, which are inputs to species-specific objectives); 2) using the best scientific information, including population-habitat models and decision support tools, to inform and update iteratively the SHC process; and 3) developing partnerships to design and deliver conservation programmes.

For over a quarter of a century the North American Waterfowl Management Plan (NAWMP; DOI & EC 1986) has served as a transformative model of partnership-based,

landscape-scale conservation delivery. The original plan and subsequent updates in 1994, 1998 and 2004 established abundant waterfowl populations as the plan's ultimate goal. A science-based understanding of waterfowl habitat requirements throughout their annual cycle and population responses to habitat, enabled managers to step-down continental population objectives to important waterfowl regions; regional partnerships between public and private parties (Joint Ventures) were formed to implement management and assess progress towards the achievement of objectives. As in the SHC framework described above, information gathered during monitoring efforts was used to improve population-habitat models and provide a sound science-base for management actions.

The NAWMP was substantially revised in 2012 to reflect, "… the rising challenges presented by a changing climate, social changes, the effects on land-use decisions of global economic pressures, and fiscal restraint faced by agencies …" (DOI *et al.* 2012:iv). Specifically, the 2012 NAWMP seeks to formally integrate objectives for waterfowl populations, habitat conservation, and societal needs and desires. The central thesis of the revised plan is that "… conservation goals can only be achieved with broad public support and by influencing land-use decisions over extensive areas of the continent" (DOI *et al.* 2012:12). The 2012 NAWMP recognises that most of these areas are privately owned "working lands" noting that, "While some conservation outcomes are achieved through regulations and policies, others result from collaborations that lead to voluntary actions. Support from the public and participation by landowners hinges on striking the right balance between conservation outcomes and the socioeconomic drivers that influence land-use decisions. That balance is always shifting, depending on the relative value placed on conservation versus other drivers."

Despite general similarities between the NAWMP and SHC frameworks, many details are lacking, especially with respect to strategic targeting of conservation on private land. The SHC framework is "strategic" only in the sense that conservation activities are based on specific objectives, scientific planning and design, and regular evaluation. In areas dominated by private land, such as the U.S. and Canadian prairies, private landowner incentives, land-use change and agricultural policy significantly complicate the design and delivery of waterfowl conservation. In addition, the SHC framework does not provide sufficient guidance on how to evaluate conservation success. Metrics to evaluate conservation success are especially lacking on private land where conservation delivery can be orders of magnitude more expensive than conservation on public land, spatially targeting conservation is limited by each landowner's willingness to accept conservation, and many conservation activities can focus as much on agricultural activities as on ecological activities (*e.g.* working land conservation; Lewis *et al.* 2011). Measuring success, and thus targeting conservation, in terms of biological benefits as implied by the SHC framework (*e.g.* changes in species-specific populations), can lead to conservation plans that are highly inefficient (*i.e.* waste scarce resources; Duke *et al.* 2013).

There is a general consensus in the economic and ecological literature that three primary factors affect efficiency of conservation delivery: biological benefits, conservation cost and threat of habitat loss or conversion (see Newburn *et al.* 2005 for a review of alternative targeting strategies). Biological benefits, measured in physical units or dollars, refer to the outcomes of conservation. Although some studies focus on targeting biological benefits exclusively (*e.g.* Niemuth *et al.* 2009), the broader literature has consistently demonstrated that benefits must be weighed against conservation costs to generate efficient conservation plans (Naidoo & Iwamura 2007; Duke *et al.* 2013). Plans that maximise benefits only (*e.g.* by selecting sites for protection that have the greatest biological value) often lead to inefficient conservation outcomes because limited budgets are quickly exhausted on high-benefit, high-cost projects (Duke *et al.* 2013). Incorporating costs, using a cost-benefit or return on investment criterion, tends to increase conservation efficiency by maximising the conservation benefit per dollar expended.

More recently, the literature focused on conservation targeting established the important role that threat plays in designing efficient conservation plans (Merenlender *et al.* 2009). Threat refers to the risk that biological benefits will be lost in the absence of conservation. In the case of waterfowl nesting habitat, for example, threat could refer to the probability that dense grassland cover is converted to intensive cropland. Ignoring threats can also result in conservation inefficiencies because limited resources are targeted to areas likely to produce benefits even in the absence of explicit conservation. Incorporating threat, along with benefits and costs, therefore improves efficiency by targeting limited resources towards projects that generate the greatest avoided loss per dollar expended (Newburn *et al.* 2005; Murdoch *et al.* 2007; Withey *et al.* 2012).

Despite the large and growing strategic conservation targeting knowledge base, there has been relatively little research targeted specifically to the design of efficient waterfowl habitat conservation plans. Several studies reported the costs and benefits of specific management treatments (see Williams *et al.* 1999 for a review) and others explored the cost-effectiveness of waterfowl management in hypothetical settings (*e.g.* Rashford & Adams 2007). Several waterfowl studies have considered landscape-level conservation targeting but have focused on conservation benefits only (*e.g.* Reynolds *et al.* 2006; Niemuth *et al.* 2009; Johnson *et al.* 2010), have considered benefits and costs but not threats (Loesch *et al.* 2012), or have considered threats and benefits but not cost (Stephens *et al.* 2008). Rashford *et al.* (2011) demonstrated the cost-benefit-threat tradeoffs associated with targeting grassland conservation in prairie Canada, but their application was for a hypothetical and unrealistic conservation scheme (*i.e.* a fixed payment to all grassland).

Given the consensus in the literature, developing truly strategic plans to target waterfowl conservation on private land will require estimates of benefits for various conservation alternatives (*e.g.* changes in recruitment), conservation costs and measures of threat levels. Much of this information already exists, particularly for the waterfowl breeding grounds of North America. Population-habitat models exist to predict waterfowl distributions and response to landscape-level conservation (Cowardin *et al.* 1995; Reynolds *et al.* 1996; Johnson *et al.* 2010). Conservation costs can also be estimated from existing data (*e.g.* published cropland rental rates to proxy for the cost of conservation easements); however, significant heterogeneity in costs across space and information asymmetry (*i.e.* landowner's private costs are not observable) imply that estimating conservation costs at the landscape level could be complex. Moreover, cost tends to be highly correlated with threat. For example, locations that have a high probability of converting from grassland to intensive cropland will have a high opportunity cost of remaining in grassland, and thus high conservation costs.

Estimating threat or risk of grassland or wetland conversion can also be challenging because conversion in the region is a largely private decision influenced by economic, physical and social factors that are not completely observable. Agricultural policy reform, global commodity markets and stochastic weather patterns are difficult to quantify or predict and can also contribute to uncertainty related to threats. Since these factors are highly heterogeneous across space (*e.g.* soil quality varies considerably across the prairies), conversion risk is likely to be highly heterogeneous (see below).

Given spatially heterogeneous conversion risk, conservation targeting that considers only benefits and cost will be inefficient (Newburn *et al.* 2005). Additionally, effectiveness of voluntary conservation programmes, such as CRP and WRP, are also influenced by threat, *e.g.* land with high conversion threat and thus high opportunity costs of entering conservation programmes is less likely to be enrolled given a fixed payment level (Lewis *et al.* 2011). It is therefore crucial to understand factors affecting private land-use decisions, and thus habitat conversion risk.

Private land-use decisions and habitat conversion risk

Studies in ecology, economics and geography have a long history of modelling private land-use change and its drivers (see *e.g.* Verburg *et al.* 2004). Although theories and approaches differ across disciplines (*e.g.* geographers and ecologists tend to focus on social drivers at the macro-scale, whereas economists tend to focus on private drivers at the micro-scale), there is a general consensus that economic, bio-physical and social/policy factors drive land-use change. A relatively recent and growing body of literature which has specifically examined agricultural land-use change, both in the PPR and in the NGP, suggests several key drivers of land-use change and the risk of habitat conversion (Stephens *et al.* 2008; Rashford *et al.* 2010; Gutzwiller & Flather 2011; Rashford *et al.* 2011; Sohl *et al.* 2012; Feng *et al.* 2013; Wright & Wimberly 2013; Attavanich *et al.* 2014). Although the identified drivers are not mutually exclusive, we categorise and describe them under the broad headings of biophysical, economic and policy drivers.

Bio-physical drivers

The biophysical attributes of land, such as soil quality, hydrology and slope, directly and indirectly affect private land-use. In some cases, biophysical attributes may restrict the set of land uses that are physically possible (*e.g.* land too steep to be tilled). Biophysical attributes also affect yields that can be realised from the land, and thus, the economic returns private landowners can derive from alternative land uses. As a result, land with characteristics that are suited to crop production tends to be placed in crop production. Studies have found strong positive correlations between high soil quality and grassland conversion (Wright & Wimberly 2013). In addition, PPR grassland habitats in land capability Class 1 and 2 (*i.e.* best soils for agricultural production) were found to be 30% to 200% more likely to be converted to cropland than grassland of lower soil capability (Rashford *et al.* 2010). Likewise, hydric soils, when drained, may provide productive farmland, and removal of in-field wetlands can improve the efficiency of tillage operations by removing "obstacles" to farm machinery.

Climatic conditions have also been found to strongly influence land-use decisions. Temperature, precipitation and CO_2 concentrations affect yield and yield variability, and thus the economic returns and risk associated with alternative land uses (Adams *et al.* 1990). Studies in the PPR and NGP have generally found strong positive correlations between climate change and grassland habitat conversion (Sohl *et al.*

2012; Attavanich *et al.* 2014). Warmer and wetter conditions in the PPR are predicted to increase wheat production at the expense of significant pastureland (Fig. 2; B.S. Rashford, unpubl. data). Potential effects of climate change, however, are highly heterogeneous across space and are moderated by other drivers (*e.g.* soil quality). Research in the NGP, where predicted climate changes and soil quality are highly heterogeneous, indicates that grasslands in the central Dakotas will be at increasing risk of conversion, while grassland in the western NGP will remain relatively secure or increase (Fig. 3; B.S. Rashford, unpubl. data). Such climate-induced land-use changes can exacerbate the effect of climate change on waterfowl and must therefore be considered when targeting conservation to mitigate climate change. Attavanich *et al.* (2014), predicted climate and land-use change impacts on waterfowl in the PPR and reported that ignoring land-use response would underestimate the effects of climate change on waterfowl by as much as 10% (300,000 breeding pairs).

Economic drivers (prices)

Economic theory assumes that landowners allocate land to the use that generates the highest discounted stream of returns (Rashford *et al.* 2010). Hence, any factors that affect current or future returns will drive land-use decisions. Many studies in the NGP and PPR concluded that agricultural prices or their derivatives (*e.g.* land rental rates) are important drivers of land-use and habitat conversion (Stephens *et al.* 2008; Rashford *et al.* 2010; Rashford *et al.* 2011; Feng *et al.* 2013). For example, recent research in the NGP indicated that, holding all else constant, a 10% increase in the

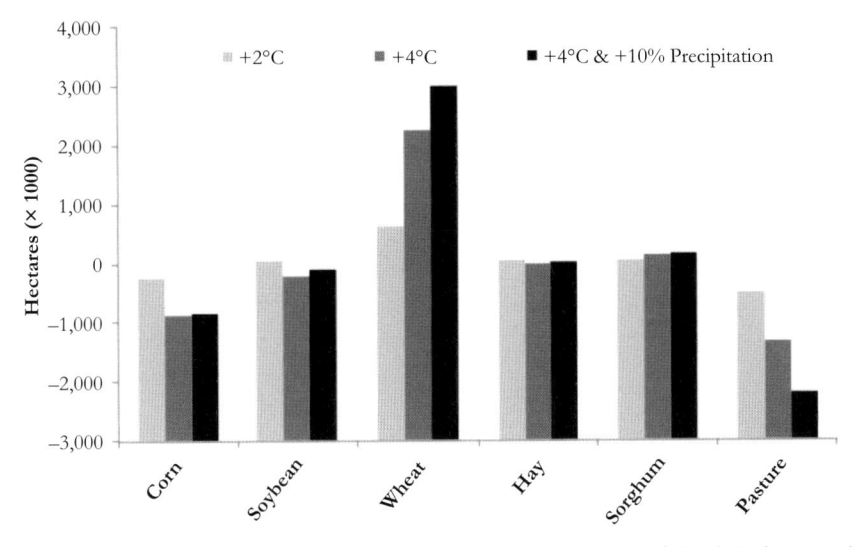

Figure 2. Predicted change in area (1,000s ha) by land use in the North and South Dakota portion of the Prairie Pothole Region for three future climate scenarios (+2°C, +4°C, and +4°C with +10% precipitation) (B.S. Rashford, unpubl. data).

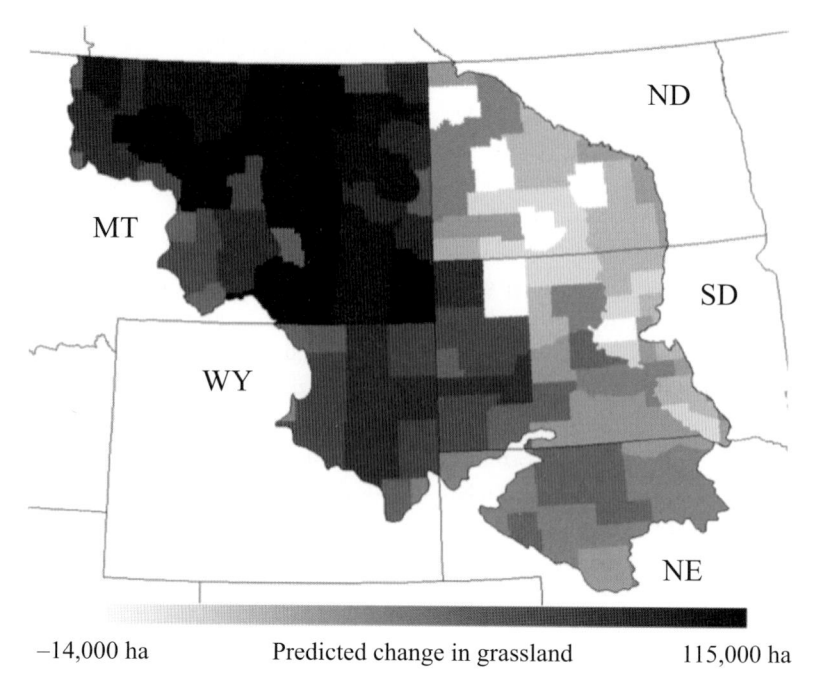

Figure 3. Predicted change in expected grassland area (ha) in 2030, based on the Intergovernmental Panel on Climate Change's A2 scenario (B.S. Rashford, unpubl. data; IPCC 2000). ND = North Dakota, SD = South Dakota, NE = Nebraska, WY = Wyoming and MT = Montana.

returns to cropland would induce ~22,000 ha of grassland to convert to cropland (B.S. Rashford, unpubl. data).

Policy drivers (government payments, crop insurance, conservation payments)

As discussed above, the U.S. farm bill provides a number of programmes that provide incentives for private landowners to choose certain land uses or production practices. Subsidised crop insurance can reduce the financial risk of growing crops on lower quality soils or in areas with less than suitable climates (*i.e.* areas that would be more likely to remain in native covers in the absence of insurance). Thus, increases in crop insurance subsidies have been correlated with increases in crop acreage and decreases in CRP enrolment (Feng *et al.* 2013). Recent research in the NGP suggests that the probability of a field being used for crop production would be as much as 30% lower if there were no direct government payments to agricultural producers, which would imply ~5.4 million additional acres (2.2 million ha) of grassland (B.S. Rashford, unpubl. data).

Future directions for strategically targeting conservation on private land

Strategically targeting conservation in a manner that accounts for economic efficiency, private land-use incentives, and threat of loss may require a slight reconsideration of the current SHC framework. First, conservation agencies will need to consider alternative measures of efficiency when evaluating conservation accomplishments and updating conservation targeting strategies. Simply evaluating programme outcomes in terms of biological benefits will not provide the comprehensive information necessary to decide which programmes should be applied where. Given limited budgets, the cost of achieving outcomes across space must be considered to target conservation cost-effectively. Incorporating costs can imply shifts in conservation focus that are counter to biological targeting, such as focusing resources in regions where the incremental biological benefits are relatively low but the benefits per dollar are high due to low conservation costs. Similarly, the identification of priority areas in the SHC framework could be informed by incorporating costs and conversion threat. For example, prioritising habitat protection for breeding waterfowl based on pair densities may overlook the fact that the costs of achieving population objectives could be reduced by focusing in regions with lower pair densities but relatively less conversion threat (and therefore lower conservation costs).

The decision-support tools critical to the SHC framework may also need to be expanded to make it effective for targeting and delivering conservation on private land. Models of the relationship between habitat and populations may misinform the SHC process if the effects of private land-use decisions are not considered. For example, targeting easements in a particular region may appear to generate a large population response given the current distribution of land-use; however, if probability of land-use change were accounted for, the population response may be substantially different. Additionally, targeting conservation on private land effectively may require wholly different decision support tools than the typical tools that focus on population-habitat relationships. For instance, the use of models that complement traditional population-habitat models, by informing landowners about how different conservation alternatives (*e.g.* working-land conservation or farm bill programmes) can be economically compatible with (or beneficial to) their agricultural production. A site-specific decision support tool, demonstrating the economic and ecological tradeoffs between alternative crop rotations, could thus be used as a conservation delivery mechanism by leading to increased adoption of winter wheat.

Lastly, strategically targeting waterfowl conservation on private land will require recognition that many effective conservation "activities" may have little resemblance to more traditional biological activities. Forming a partnership, itself an emphasis of the SHC framework, to lobby politically for "waterfowl friendly" agricultural policies (*e.g.* conservation compliance) or to invest in agricultural research (*e.g.* new winter wheat varieties) may prove as effective as more traditional direct habitat management. Although many non-profit conservation groups currently use such activities as highlighted in previous sections, incorporating such activities explicitly within the SHC framework would focus the process. Additionally, designing conservation activities for private land should be incorporated specifically (to improve leverage) in farm bill conservation programmes. Farm bill programmes are not particularly well targeted across space, for example, because they largely depend on voluntary participation and opportunistic enrolment. Consequently, it is difficult to control the spatial allocation of such programmes, because of inherent inability to control, or even easily predict, which landowners will choose to participate. A broadened SHC framework that considers existing voluntary enrolment, however, could be used to target other conservation activities to leverage benefits of farm bill programmes, for instance by targeting easements near existing CRP to create larger blocks of waterfowl nesting cover.

Acknowledgements

This paper is the product of a special paper session "Opportunities and Challenges Facing Waterfowl Habitat Conservation on Private Lands" (organisers: W.L. Hohman and E. Lindstrom) conducted at the Ecology and Conservation of North American Waterfowl Conference held in Memphis, TN, 27–31 January 2013. We are especially grateful to the Conference Co-chairs (R.M. Kaminski and J.B Davis, Mississippi State University) and Scientific Committee for the opportunity to participate in the meeting. Contributors to our 2-day special session were *Audubon California*: M. Iglecia and R. Kelsey; *Ducks Unlimited*: M.G. Brasher, J.D. James, M.R. Kaminski, M.T. Merendino, M.J. Petrie and A.J. Wiseman; *Mississippi State University*: A.B. Alford, J.L. Avery, L.W. Burger, L.R. D'Abramo, J.B. Davis, S.C. Grado, R.M. Kaminski, J.R. Marty, M.D. McConnell, W. Schilling and G. Wang; *Point Blue Conservation Science* (formerly Point Reyes Bird Observatory Conservation Science): C. Hickey, M. Reiter, K. Sesser, D. Skalos and K. Strum; *USDA NRCS*: S.J. Brady and R. Weber; and *USFWS*: H.D. Azure, K.E. Doherty, H. Hoistad, C.R. Loesch, W.A. Meeks, N.D. Niemuth, R.E. Reynolds, A.J. Ryba, C.L. Stemler, and J. Tirpak. Skillful editing of the manuscript was provided by L. Webb and E. Rees.

References

Acquaah, G. 2005. *Principles of Crop Production: Theory, Techniques, and Technology*. Pearson Prentice Hall, Upper Saddle River, New Jersey, USA.

Adams, R.M., Rosenzweig, C., Ritchie, J., Peart, P., Glyer, J.D., McCarl, B.A., Curry, B. & Jones, J. 1990. Global Climate Change and Agriculture. *Nature* 345: 219–224.

Alisauskas, R.T. 1998. Winter range expansion and relationships between landscape and morphometrics of midcontinent Lesser Snow Geese. *Auk* 115: 851–862.

Allen, A.W. & Vandever, M.W. 2012. *Conservation Reserve Program (CRP) Contributions to Wildlife Habitat, Management Issues, Challenges and Policy Choices – an Annotated Bibliography*. U.S. Geological Survey Scientific Investigations Report 2012–5066. U.S. Geological Survey, Washington D.C., USA.

Attanavich, W., Rashford, B.S., Adams, R.M. & McCarl, B.A. 2014. Land Use, Climate Change and Ecosystem Services. *In* J.M. Duke and J. Wu (eds.), *Oxford Handbook of Land Economics*, pp. 255–280. Oxford University Press, UK.

Austin, J.E. & Miller, M.R. 1995. Northern Pintail (*Anas acuta*). *In* A. Poole & F. Gill (eds.). *The*

Birds of North America, No. 163. The Birds of North America, Inc., Philadelphia, PA, USA.

Batt, B.D.J., Anderson, M.G., Anderson, C.D. & Caswell, F.D. 1989. The use of Prairie Potholes by North American ducks. In A. van der Valk (ed.), *Northern Prairie Wetlands*, pp. 204–227. Iowa State University Press, Ames, USA.

Beauchamp, W.D., Koford, R.R., Nudds, T.D., Clark, R.G. & Johnson, D.H. 1996. Long-term declines in nest success of prairie ducks. *Journal of Wildlife Management* 60: 247–257.

Blanche, B., Harrell, D. & Saichuk, J. 2009. Weed management. In J. Saichuk (ed.), *Louisiana Rice Production Handbook*, pp. 3–15. Louisiana State University, College of Agriculture, AgCenter, Baton Rouge, USA. Accessible at http://www.lsuagcenter.com/NR/rdonlyres/F7E930BF-346A-45DA-A989-37FE3D0B0F11/97775/pub2331RiceProductionHandbook2014completebook.pdf (last accessed 29 August 2014).

Brady, S.J. 2005. Highly erodible land and Swampbuster provisions of the 2002 Farm Act. In J.B. Haufler (ed.), *Fish and Wildlife Benefits of Farm Bill Conservation Programs: 2000–2005 Update*, pp. 5–15. Technical Review 05-2. The Wildlife Society, Bethesda, Maryland, USA.

Burger, Jr., L.W. 2000. Wildlife responses to the Conservation Reserve Program in the Southeast. In W.L. Hohman & D.J. Halloum (eds.), *A Comprehensive Review of Farm Bill Contributions to Wildlife Conservation, 1985–2000*, pp. 55–73. U.S. Department of Agriculture, Natural Resources Conservation Service, Wildlife Habitat Management Institute, Technical Report USDA/NRCS/WHMI-2000. Madison, Mississippi, USA.

Carlyle, W.J. 1997. The decline of summerfallow on the Canadian Prairies. *The Canadian Geographer/Le Géographe Canadien* 41: 267–280.

Claassen, R. 2005. Has conservation compliance reduced soil erosion on U.S. cropland? In OECD (ed.), *Evaluating Agri-Environmental Policies: Design, Practice and Results*, pp. 293–294. Organization for Economic Co-operation and Development (OECD), Paris, France.

Claassen, R. 2012. *The Future of Environmental Compliance Incentives in U.S. Agriculture: The Role of Commodity, Conservation and Crop Insurance Programs*. U.S. Department of Agriculture, Economic Research Service, Economic Information Bulletin 94. U.S. Department of Agriculture, Economic Research Service, Washington D.C., USA.

Congressional Budget Office (CBO). 2014. *Budget Estimates for H.R. 2642, the Agricultural Act of 2014*. Accessible at http://www.cbo.gov/sites/default/files/cbofiles/attachments/hr2642LucasLtr.pdf (last accessed 14 February 2014).

Couvillion, B.R., Barras, J.A., Steyer, G.D., Sleavin, W., Fischer, M., Beck, H., Trahan, N., Griffin, B. & Heckman, D. 2011. *Land Area Change in Coastal Louisiana from 1932 to 2010*. U.S. Geological Survey Scientific Investigations Map 3164, scale 1:265,000. U.S. Geological Survey, Washington D.C., USA.

Cowardin, L.M., Gilmer, D.S. & Shaiffer, C.W. 1985. Mallard recruitment in the agricultural environment of North Dakota. *Wildlife Monograph* 92: 1–37.

Cowardin, L.M., Shaffer, T.L. & Arnold, P.M. 1995. *Evaluation of Duck Habitat and Estimation of Duck Population Sizes with a Remote-Sensing Based System*. Biological Science Report No. 2. U.S. Department of the Interior, National Biological Service, Washington D.C., USA.

Cox, Jr., R.R. & Afton, A.D. 1996. Evening flights of female Northern Pintails from a major roost site. *Condor* 98: 810–819.

Cox, Jr., R.R. & Afton, A.D. 1997. Use of habitats by female northern pintails wintering in southwestern Louisiana. *Journal of Wildlife Management* 61: 435–443.

Ciuzio, E., Hohman, W.L., Martin, B., Smith, M.D., Stephens, S., Strong, A.M., and Vercauteren, T. 2013. Opportunities and challenges to implementing bird conservation on private lands. *Wildlife Society Bulletin* 37: 267–277.

Dahl, T.E. 1990. *Wetlands Losses in the United States 1780s to 1980s*. Department of the Interior, U.S. Fish and Wildlife Service, Washington, D.C., USA. Accessible at http://www.fws.gov/wetlands/Documents/Wetlands-Losses-in-the-United-States-1780s-to-1980s.pdf (last accessed 19 November 2013).

Dahl, T.E. 2000. *Status and Trends of Wetlands in the Conterminous United States 1986 to 1997*. U.S. Department of the Interior, Fish and Wildlife Service, Washington D.C., USA. Accessible at http://www.fws.gov/wetlands/Documents/Status-and-Trends-of-Wetlands-in-the-Conterminous-United-States-1986-to-1997.pdf (last accessed 19 November 2013).

Devries, J.H., Armstrong, L.M., MacFarlane, R.J., Moats, L. & Thoroughgood, P.T. 2008. Waterfowl nesting in fall-seeded and spring-seeded cropland in Saskatchewan. *Journal of Wildlife Management* 72: 1790–1797.

Devries, J.H., Rimer, S.O. & Walsh, E.M. 2010. Cropland nesting by Long-billed Curlews in southern Alberta. *Prairie Naturalist* 42: 123–129.

Doherty, K.E., Ryba, A.J., Stemler, C.L., Niemuth, N.D. & Meeks, W.A. 2013. Conservation planning in an era of change: State of U.S. Prairie Pothole Region. *Wildlife Society Bulletin* 37: 546–563.

DOI & EC (U.S. Department of Interior & Environment Canada). 1986. *North American Waterfowl Management Plan*. U.S. Department of the Interior, Fish and Wildlife Service, Washington, D.C., USA. Accessible at http://www.fws.gov/birdhabitat/NAWMP/files/NAWMP.pdf (last accessed 27 February 2014).

DOI, EC & Secretario de Desarrollo Social Mexico. 2012. *North American Waterfowl Management Plan 2012: People Conserving Waterfowl and Wetlands*. U.S. Department of the Interior, Fish and Wildlife Service, Washington, D.C., USA. Accessible at http://www.fws.gov/birdhabitat/NAWMP/files/NAWMP-Plan-EN-may23.pdf (last accessed 27 February 2014).

Duebbert, H.F. & Kantrud, H.A. 1987. Use of no-till winter wheat by nesting ducks in North Dakota. *Journal of Soil and Water Conservation* 42: 50–53.

Duke, J.M., Dundas, S.J. & Messer, K.M. 2013. Cost-effective conservation planning: Lessons from economics. *Journal of Environmental Management* 125: 126–133.

Dzus, E.H. & Clark, R.G. 1998. Brood survival and recruitment of Mallards in relation to wetland density and hatching date. *Auk* 115: 311–318.

Eadie, J.M., Elphick, C.S., Reinecke, K.J. & Miller, M.R. 2008. Wildlife values of North American ricelands. *In* S.W. Manley (ed.), *Conservation in Ricelands of North America*, pp. 7–90. The Rice Foundation, Stuttgart, Arkansas, USA.

Earl, J.P. 1950. Production of Mallards on irrigated land in the Sacramento Valley, California. *Journal of Wildlife Management* 14: 332–342.

Elphick, C.S. 2000. Functional equivalency between ricefields and seminatural wetland habitats. *Conservation Biology* 14: 181–191.

Elphick, C.S. & Oring, L.W. 1998. Winter management of Californian rice fields for waterbirds. *Journal of Applied Ecology* 35: 95–108.

Feng, H., Hennessy, D.A. & Miao, R. 2013. The effects of government payments on cropland acreage, conservation reserve program enrollment, and grassland conversion in the Dakotas. *American Journal of Agricultural Economics* 95: 412–418.

Fleury, B.E. & Sherry, T.W. 1995. Long-term population trends of colonial wading birds in the southern United States: The impact of crayfish aquaculture on Louisiana populations. *The Auk* 112: 613–632.

Foley, J.A., Defries, R., Asner, G.P., Barford, C., Bonan, G., Carpenter, S.R., Chapin, F.S., Coe, M.T., Daily, G.C., Gibbs, H.K., Helkowski, J.H., Holloway, T., Howard, E.A., Kucharik, C.J., Monfreda, C., Patz, J.A., Prentice, I.C., Ramankutty, N. & Snyder, P.K. 2005. Global consequences of land use. *Science* 309: 570–574.

Food and Agriculture Organization (FOA). 2010. *The State of Food Insecurity in the World: Addressing Food Insecurity in Protracted Crises.* Food and Agriculture Organization of the United Nations, World Summit on Food Security, November 16–18, 2009, Rome, Italy. Accessible at http://www.fao.org/docrep/013/i1683e/i1683e.pdf (last accessed 19 November 2013).

Frayer, W.E., Monahan, T.J., Bowden, D.C. & Graybill, F.A. 1983. *Status and Trends of Wetlands and Deepwater Habitats in the Conterminous United States, 1950's to 1970's.* Colorado State University, Fort Collins, Colorado, USA.

Fredrickson, L.H. & Taylor, T.S. 1982. *Management of Seasonally Flooded Impoundments for Wildlife.* Resource Publication 148, U.S. Department of the Interior, Fish and Wildlife Service, Washington D.C., USA.

Gerard, P. 1995. *Agricultural Practices, Farm Policy and the Conservation of Biological Diversity.* Biological Science Report No. 4. U.S. Department of the Interior, National Biological Service, Washington D.C., USA.

Goelitz, W.A. 1918. The destruction of nests by farming operations in Saskatchewan. *Auk* 35: 238–240.

Government of Canada. 1991. *The Federal Policy on Wetland Conservation.* Environment Canada. Ottawa, Ontario, Canada.

Greenwood, R.J., Sargeant, A.B., Johnson, D.H., Cowardin, L.M. & Shaffer, T.L. 1995. Factors associated with duck nest success in the Prairie Pothole Region of Canada. *Wildlife Monographs* 128: 1–57.

Gutzwiller, K.J. & Flather, C.H. 2011. Wetland features and landscape context predict the risk of wetland habitat loss. *Ecological Applications* 21: 968–982.

Guyn, K.L. & Clark, R.G. 2000. Nesting effort of Northern Pintails in Alberta. *Condor* 102: 619–628.

Heard, L.P., Allen, A.W., Best, L.B., Brady, S.J., Burger, W., Esser, A.J., Hackett, E., Johnson, D.H., Pederson, R.L., Reynolds, R.E., Rewa, C., Ryan, M.R., Molleur, R.T. & Buck, P. 2000. In W.L. Hohman & D.J. Halloum (eds.), *A Comprehensive Review of Farm Bill Contributions to Wildlife Conservation, 1985–2000.* U.S. Department of Agriculture, Natural Resources Conservation Service, Wildlife Habitat Management Institute, Technical Report USDA/NRCS/WHMI-2000. Madison, Mississippi, USA.

Higgins, K.F. 1977. Duck nesting in intensively farmed areas of North Dakota. *Journal of Wildlife Management* 41: 232–242.

Hoekstra, J.M., Boucher, T.M., Ricketts T.H & Roberts, C. 2005. Confronting a biome crisis: global disparities of habitat loss and protection. *Ecology Letters* 8: 23–29.

Hohman, W.L., Moore, J.L., Stark, T.M., Weisbrich, G.A. & Coon, R.M. 1994. Breeding waterbird use of Louisiana rice fields in relation to planting practices. *Proceedings of the Southeastern Association of Fish and Wildlife Agencies* 48: 31–37.

Hohman, W.L., Kidd, G., Nelson, R. & Pitre, J. In press. Opportunities for crane conservation through U.S. Department of Agriculture Conservation Programs. *In* J.E. Austin & K. Morrison (eds.), *Cranes and Agriculture – a Practical Guide to Conservationists and Land*

Managers, pp. 000–000. International Union for Conservation of Nature, Species Survival Commission, Crane Specialist Group, Gland, Switzerland.

Intergovernmental Panel on Climate Change (IPCC). 2000. *IPCC Special Report: Emissions Scenarios, Summary for Policymakers*. A Special Report of IPCC Working Group III, ISBN: 92-9169-113-5. World Meteorological Organization, United Nations Environment Programme, IPCC. Accessible at http://www.ipcc.ch/pdf/special-reports/spm/sres-en.pdf (last accessed 14 May 2014).

Johnston, C.A. 2013. Wetland losses due to row crop expansion in the Dakota Prairie Pothole Region. *Wetlands* 33: 175–182.

Johnson, D.H. 2000. Grassland bird use of Conservation Reserve Program fields in the Northern Great Plains. *In* W.L. Hohman & D.J. Halloum (eds.), *A Comprehensive Review of Farm Bill Contributions to Wildlife Conservation, 1985–2000*, pp. 19–33. U.S. Department of Agriculture, Natural Resources Conservation Service, Wildlife Habitat Management Institute, Technical Report USDA/NRCS/WHMI-2000. Madison, Mississippi, USA.

Johnson, R.R., Granfors, D.A., Niemuth, N.D., Estey, M.E. & Reynolds, R.E. 2010. Delineating grassland bird conservation areas in the U.S. Prairie Pothole Region. *Journal of Fish and Wildlife Management* 1: 38–42.

Klett, A.T., Shaffer, T.L. & Johnson, D.H. 1988. Duck nest success in the Prairie Pothole Region. *Journal of Wildlife Management* 52: 431–440.

Lewis, D.J., Plantinga, A.J., Nelson, E. & Polasky, S. 2011. The efficiency of voluntary incentives policies for preventing biodiversity loss. *Resource and Energy Economics* 33: 192–211.

Linscombe, R.G. 1972. *Crop damage by waterfowl in southwestern Louisiana*. M.Sc. Thesis, Louisiana State University, Baton Rouge, USA.

Loesch, C.R., Reynolds, R.E. & Hansen, L.T. 2012. An assessment of re-directing breeding waterfowl conservation relative to predictions of climate change. *Journal of Fish and Wildlife Management* 3: 1–22.

Lokemoen, J.T. & Bciscr, J.A. 1997. Bird use and nesting in conventional, minimum-tillage, and organic cropland. *Journal of Wildlife Management* 61: 644–655.

Lynch-Stewart, P., Rubec, C.D.A., Cox, K.W. & Patterson, J.H. 1993. *A Coming of Age: Policy for Wetland Conservation in Canada*. Report No. 93-1. North American Wetlands Conservation Council (Canada), Ottawa, Ontario.

Maizel, M., White, R.D., Gage, S., Osbourne, L., Root, R., Stitt, S. & Muehlbach, G. 1998. Historical Interrelationships between Population Settlement and Farmland in the Conterminous United States, 1790 to 1992. *In* T.D. Sisk (ed.), *Perspectives on the Land-use History of North America: a Context for Understanding or Changing Environment*, pp. 5–13. Biological Science Report USGS/BRD/BSR-1998-0003. U.S. Geological Survey, Biological Resources Division, Washington D.C., USA.

Marty, J.R. 2013. *Seed and waterbird abundances in ricelands in the Gulf Coast Prairies of Louisiana and Texas*. M.Sc. thesis, Mississippi State University, Starkville, USA.

McClain, W.R, Romaire, R.P., Lutz, C.G. & Shirley, M.G. 2009. *Louisiana Crawfish Production Manual*. Publication No. 2637, Louisiana State University, College of Agriculture, AgCenter, Baton Rouge, USA. Accessible at http://www.lsuagcenter.com/en/crops_livestock/aquaculture/crawfish/Crawfish+Production+Manual.htm (last accessed 5 May 2014).

Merenlender, A.M., Newburn, D., Reed, S.E. & Rissman, A.R. 2009. The importance of incorporating threat for efficient targeting and evaluation of conservation investments. *Conservation Letters* 2: 240–41.

Millar, J.B. 1989. Perspectives on the Status of Canadian Prairie Wetlands. *In* R.R. Sharitz & J.W. Gibbons (eds.), *Freshwater Wetlands and Wildlife*, pp. 829–852. U.S. Department of Energy Symposium Series No. 61, U.S. Department of Energy, Oak Ridge, Tennessee, USA.

Miller, M.R. 1987. Fall and winter foods of Northern Pintails in the Sacramento Valley, California. *Journal of Wildlife Management* 51: 405–414.

Miller, M.R. & Duncan, D.C. 1999. The Northern Pintail in North America: status and conservation needs of a struggling population. *Wildlife Society Bulletin* 27: 788–800.

Milonski, M. 1958. The significance of farmland for waterfowl nesting and techniques for reducing losses due to agricultural practices. *Transactions of the North American Wildlife Conference* 23: 215–227.

Mulik, K. & Koo, W.W. 2006. *Substitution Between U.S. and Canadian Wheat by Class*. Center for Agricultural Policy and Trade Studies Department of Agribusiness and Applied Economics, North Dakota State University, Fargo, North Dakota, USA.

Murdoch, W., Polasky, S., Wilson, K.A., Possingham, H.P., Kareiva, P. & Shaw, R. 2007. Maximizing return on investment in conservation. *Biological Conservation* 139: 375–388.

Naidoo, R. & Iwamura, T. 2007. Global-scale mapping of economic benefits from agricultural lands: Implications for conservation priorities. *Biological Conservation* 140: 40–49.

National Fish, Wildlife and Plants Climate Adaptation Strategy Partnership (NFWPCAS). 2012. *Draft National Fish, Wildlife and Plants Climate Adaptation Strategy: Shared Solutions to Protect Shared Values*. Association of Fish and Wildlife Agencies, Council on Environmental Quality, Department of the Interior, Great Lakes Indian Fish and Wildlife Commission, National Oceanic and Atmospheric Administration. Washington, D.C., USA.

Newburn, D., Reed, S., Berck, P. & Merenlender, A. 2005. Economics of land-use change in prioritizing private lands conservation. *Conservation Biology* 19: 1411–1420.

Nickerson, C., Ebel, R., Borchers, A. & Carriazo, F. 2011. *Major Uses of Land in the United States, 2007*, EIB-89. U.S. Department of Agriculture, Economic Research Service, Washington D.C, USA. Accessible at http://www.ers.usda.gov/media/188404/eib89_2_.pdf (last accessed November 2013).

Niemuth, N.D., Reynolds, R.E., Granfors, D.A., Johnson, R.R., Wangler, B. & Estey, M.E. 2009. Landscape-level planning for conservation of wetland birds in the U.S. Prairie Pothole Region. *In* J.J. Millspaugh & F.R. Thompson, III (eds.), *Models for Planning Wildlife Conservation in Large Landscapes*, pp. 533–560. Academic Press, Burlington, Massachusetts, USA.

Noss, R.F., LaRoe III, E.T. & Scott, J.M. 1995. *Endangered Ecosystems of the United States: a Preliminary Assessment of Loss and Degradation*. Biological Report 28. U.S. Department of the Interior, National Biological Service. Washington, D.C., USA.

Peterjohn, B.G. & Sauer, J.R. 1999. Population status of North American grassland birds from the North American Breeding Bird Survey, 1966–1996. *Studies in Avian Biology* 19: 27–44.

Podruzny, K.M. Devries, J.H., Armstrong, L.M. & Rotella, J.J. 2002. Long-term response of Northern Pintails to changes in wetlands and agriculture in the Canadian Prairie Pothole Region. *Journal of Wildlife Management* 66: 993–1010.

Rashford, B.S. & Adams, R.M. 2007. Improving the cost-effectiveness of ecosystem management:

An application to waterfowl production. *American Journal of Agricultural Economics* 89: 755–768.

Rashford, B.S., Walker, J.A. & Bastian, C.T. 2010. Economics of grassland conversion to cropland in the Prairie Pothole Region. *Conservation Biology* 25: 276–284.

Rashford, B.S., Bastian, C.T. & Cole, J.G. 2011. Agricultural land-use change in prairie Canada: Implications for wetland and waterfowl conservation. *Canadian Journal of Agricultural Economics* 59: 185–205.

Rave, D.P. & Cordes, C.L. 1993. Time-activity budget of northern pintails using nonhunted ricefields in southwest Louisiana. *Journal of Field Ornithology* 64: 211–218.

Reynolds, R.E. 2000. Waterfowl responses to the Conservation Reserve Program in the Northern Great Plains. *In* W.L. Hohman & D.J. Halloum (eds.), *A Comprehensive Review of Farm Bill Contributions to Wildlife Conservation, 1985–2000*. pp. 35–43. U.S. Department of Agriculture, Natural Resources Conservation Service, Wildlife Habitat Management Institute, Technical Report USDA/NRCS/WHMI-2000. Madison, Mississippi, USA.

Reynolds, R.E., Cohan, D.R. & Johnson, M.A. 1996. Using landscape information approaches to increase duck recruitment in the Prairie Pothole Region. *Transaction of the North American Wildlife and Natural Resource Conference* 61: 86–93.

Reynolds, R.E., Shaffer, T.L., Loesch, C.R. & Cox, Jr., R.R. 2006. The Farm Bill and duck production in the Prairie Pothole Region. *Wildlife Society Bulletin* 34: 963–974.

Richkus, K.D. 2002. Northern Pintail nest site selection, nest success, renesting ecology and survival in the intensively farmed prairies of southern Saskatchewan: An evaluation of the ecological trap hypothesis. Ph.D. thesis, Louisiana State University, Baton Rouge, Louisiana, USA.

Riemer, G. 2005. Land-use policy change and the ramifications for stewardship and waterfowl conservation in Saskatchewan. *In* T.A. Radenbaugh & G.C. Sutter (eds.), *Managing Changing Prairie Landscapes*, pp. 11–22. University of Regina Press, Regina, Saskatchewan, Canada.

Ringelman, J.K., Forman, K.J., Granfors, D.A., Johnson, R.R., Lively, C.A., Naugle, D.E., Niemuth, N.D. & Reynolds, R.E. 2005. *Prairie Pothole Joint Venture 2005 Implementation Plan*. U.S. Fish and Wildlife Service, Prairie Pothole Joint Venture, Denver, Colorado, USA.

Ryan, M.R. 2000. Impact of the Conservation Reserve Program on wildlife conservation in the Midwest. *In* W.L. Hohman & D.J. Halloum (eds.), *A Comprehensive Review of Farm Bill Contributions to Wildlife Conservation, 1985–2000*. pp. 45–54. U.S. Department of Agriculture, Natural Resources Conservation Service, Wildlife Habitat Management Institute, Technical Report USDA/NRCS/WHMI-2000. Madison, Mississippi, USA.

Samson, F. & Knopf, F. 1994. Prairie conservation in North America. *BioScience* 44: 418–421.

Sargeant, A.B., Greenwood, R.J., Sovada, M.A. & Shaffer, T.L. 1993. *Distribution and Abundance of Predators that Affect Duck Production: the Prairie Pothole Region*. Resource Publication 194. U.S. Department of Interior, Fish and Wildlife Service, Washington D.C., USA.

Skagen, S.K. & Thompson, G. 2007. *Northern Plains/Prairie Potholes Regional Shorebird Conservation Plan (Revised)*. U.S. Department of Interior, Fish and Wildlife Service, Washington D.C., USA.

Smith, R. J. & Sullivan, J.D. 1980. Reducing red rice grain in rice fields by winter feeding of ducks. *Arkansas Farm Research* 29: 3.

Sohl, T.L., Sleeter, B.M., Saylera, K.L., Bouchard, M.A., Reker, R.R., Bennett, S.L., Sleeter, R.R.,

Kanengieter, R.L. & Zhue, Z. 2012. Spatially explicit land-use and land-cover scenarios for the Great Plains of the United States. *Agriculture, Ecosystems and Environment* 153: 1–15.

Statistics Canada. 2011. *2011 Census: Census of Agriculture.* Statistics Canada, Ottawa, Ontario, Canada.

Statistics Canada. 2012. *2012 Census: Census of Agriculture.* Statistics Canada, Ottawa, Ontario, Canada.

Statistics Canada. 2013. *Estimated Areas, Yield, Production, Average Farm Price and Total Farm Value of Principal Field Crops, in Imperial Units, Annual.* Table 001-0017-CANSIM (database). Last updated December 4, 2013. Statistics Canada, Ottawa, Ontario, Canada.

Stephens, S.E., Rotella, J.J., Lindberg, M.S., Taper, M.L. & Ringleman, J.K. 2005. Duck nest survival in the Missouri Coteau of North Dakota: landscape effects at multiple spatial scales. *Ecological Applications* 15: 2137–2149.

Stephens, S.E., Walker, J.A., Blunck, D.R., Jayaraman, A., Naugle, D.E., Ringleman, J.K. & Smith, A.J. 2008. Predicting risk of habitat conversion in native temperate grasslands. *Conservation Biology* 22: 1320–1330.

Taft, O.W. & Elphick, C.S. 2007. *Waterbirds on Working Lands: Literature Review and Bibliography Development.* Technical Report. National Audubon Society, New York, USA.

USDA FSA (U.S. Department of Agriculture, Farm Service Agency). 2013. *U.S. Non-Cropland to Cropland Report.* Washington, D.C., USA. Accessible at http://www.fsa.usda.gov/FSA/webapp?area=newsroom&subject=landing&topic=foi-er-fri-dtc (last accessed 15 January 2014).

USDA NASS (U.S. Department of Agriculture, National Agricultural Statistics Service). 2013. Acreage report. United States Department of Agriculture, Washington D.C., USA.

USDA NASS. 2014. *2012 Census of Agriculture, United States Summary and State Data,* Volume 1, Part 51 (AC-12-A-51). United States Department of Agriculture, Washington D.C., USA. Accessible at http://www.agcensus.usda.gov/Publications/2012/Full_Report/Volume_1,_Chapter_1_US/usv1.pdf (last accessed 5 May 2014).

USDA NRCS (U.S. Department of Agriculture, Natural Resources Conservation Service). 2000. *1997 National Resources Inventory.* U.S. Department of Agriculture, Natural Resources Conservation Service, Resources Inventory Division, Washington, D.C., USA.

USDA NRCS. 2013. *2007 National Resources Inventory.* U.S. Department of Agriculture, Natural Resources Conservation Service, Resources Inventory Division, Washington D.C., USA. Accessible at http://www.nrcs.usda.gov/wps/portal/nrcs/detail/national/technical/nra/nri/?cid=stelprdb1117258 (last accessed 12 February 12 2014).

USFWS (U.S. Fish and Wildlife Service). 2008. *Strategic Habitat Conservation Handbook: A guide to Implementing the Technical Elements of Strategic Habitat Conservation,* Version 1.0. U.S. Department of Interior, Fish and Wildlife Service, Washington D.C., USA.

Verburg, P.H., Schot, P.P. Dijst, M.J. & Veldkamp, A. 2004. Land use change modeling: Current practice and research priorities. *GeoJournal* 61: 309–324.

Watmough, M.D. & Schmoll, M.J. 2007. *Environment Canada's Prairie & Northern Habitat Monitoring Program Phase II: Recent habitat trends in the Prairie Habitat Joint Venture.* Technical Report Series No. 493. Environment Canada, Canadian Wildlife Service, Edmonton, Alberta, Canada.

Webster, E. & Levy, R. 2009. *Weed management. In* R. Saichuk (ed.), *Louisiana Rice Production Handbook,* pp. 46–71. Louisiana State University, College of Agriculture, AgCenter, Baton Rouge, USA. http://www.lsuagcenter.com/en/communications/publications/Publications+Catalog/Crops+and+Livestock/Rice/Rice+Production+Handbook.htm.m (last accessed 5 May 2014).

Williams, B.K., Koneff, M.D. & Smith, D.A. 1999. Evaluation of waterfowl conservation under the North American Waterfowl Management Plan. *Journal of Wildlife Management* 63: 417–440.

Withey, J.C., Lawler, J.J., Polasky, S., Plantinga, A.J., Nelson, E.J., Kareiva, P., Wilsey, C.B., Schloss, C.A., Nogeire, T.M., Ruesch, A., Ramos Jr., J. & Reid, W. 2012. Maximising return on conservation investments in the conterminous USA. *Ecology Letters* 15: 1249–1256.

Wright, C.K. & Wimberly, M.C. 2013. Recent land use change in the Western Corn Belt threatens grasslands and wetlands. *Proceedings of the National Academy of Sciences* 110: 4134–4139.

Photograph: Aerial view of southwest Louisiana rice fields, by John K. Saichuk/Louisiana State University AgCenter.

Estimating habitat carrying capacity for migrating and wintering waterfowl: considerations, pitfalls and improvements

CHRISTOPHER K. WILLIAMS[1]*, BRUCE D. DUGGER[2],
MICHAEL G. BRASHER[3], JOHN M. COLUCCY[4], DANE M. CRAMER[4],
JOHN M. EADIE[5], MATTHEW J. GRAY[6], HEATH M. HAGY[7],
MARK LIVOLSI[1], SCOTT R. McWILLIAMS[8], MARK PETRIE[9],
GREGORY J. SOULLIERE[10], JOHN M. TIRPAK[11] &
ELISABETH B. WEBB[12]

[1]Department of Entomology and Wildlife Ecology, University of Delaware,
250 Townsend Hall, Newark, Delaware 19716, USA.
[2]Department of Fisheries and Wildlife, Oregon State University, Corvallis, Oregon 97331, USA.
[3]Ducks Unlimited, Inc., Gulf Coast Joint Venture, 700 Cajundome Blvd, Lafayette,
Louisiana 70506, USA.
[4]Ducks Unlimited, Inc., Great Lakes/Atlantic Region, 1220 Eisenhower Place, Ann Arbor,
Michigan 48108, USA.
[5]Department of Wildlife, Fish and Conservation Biology, University of California,
1 Shields Avenue, Davis, California 95616, USA.
[6]Department of Forestry, Wildlife and Fisheries, University of Tennessee,
274 Ellington Plant Sciences Building, Knoxville, Tennessee 37996, USA.
[7]Forbes Biological Station – Bellrose Waterfowl Research Center, Illinois
Natural History Survey, University of Illinois at Urbana-Champaign, P.O. Box 590,
Havana, Illinois 62644, USA.
[8]Department of Natural Resources Science, University of Rhode Island,
105 Coastal Institute, Kingston, Rhode Island 02881, USA.
[9]Ducks Unlimited Inc. Western Regional Office, 17800 SE Mill Plain Blvd. Suite 120,
Vancouver, Washington 98683, USA.
[10]U.S. Fish and Wildlife Service, Upper Mississippi River and Great Lakes Region
Joint Venture Science Office, 2651 Coolidge Road, Suite 101, East Lansing,
Michigan 48823, USA.
[11]U.S. Fish and Wildlife Service, Gulf Coastal Plains & Ozarks Landscape Conservation
Cooperative, 700 Cajundome Blvd, Lafayette, Louisiana 70506, USA.
[12]U.S. Geological Survey, Missouri Cooperative Fish and Wildlife Research Unit,
Department of Fisheries and Wildlife Sciences, University of Missouri, Columbia,
Missouri 65211, USA.
*Correspondence author: E-mail: ckwillia@udel.edu

Abstract

Population-based habitat conservation planning for migrating and wintering waterfowl in North America is carried out by habitat Joint Venture (JV) initiatives and is based on the premise that food can limit demography (*i.e.* food limitation hypothesis). Consequently, planners use bioenergetic models to estimate food (energy) availability and population-level energy demands at appropriate spatial and temporal scales, and translate these values into regional habitat objectives. While simple in principle, there are both empirical and theoretical challenges associated with calculating energy supply and demand including: 1) estimating food availability, 2) estimating the energy content of specific foods, 3) extrapolating site-specific estimates of food availability to landscapes for focal species, 4) applicability of estimates from a single species to other species, 5) estimating resting metabolic rate, 6) estimating cost of daily behaviours, and 7) estimating costs of thermoregulation or tissue synthesis. Most models being used are daily ration models (DRMs) whose set of simplifying assumptions are well established and whose use is widely accepted and feasible given the empirical data available to populate such models. However, DRMs do not link habitat objectives to metrics of ultimate ecological importance such as individual body condition or survival, and largely only consider food-producing habitats. Agent-based models (ABMs) provide a possible alternative for creating more biologically realistic models under some conditions; however, ABMs require different types of empirical inputs, many of which have yet to be estimated for key North American waterfowl. Decisions about how JVs can best proceed with habitat conservation would benefit from the use of sensitivity analyses that could identify the empirical and theoretical uncertainties that have the greatest influence on efforts to estimate habitat carrying capacity. Development of ABMs at restricted, yet biologically relevant spatial scales, followed by comparisons of their outputs to those generated from more simplistic, deterministic models can provide a means of assessing degrees of dissimilarity in how alternative models describe desired landscape conditions for migrating and wintering waterfowl.

Key words: agent-based models, bioenergetics, carrying capacity, daily ration models, energy demand, energy supply, waterfowl.

Population-based, habitat conservation of the North American Waterfowl Management Plan (NAWMP 2012) is planned and implemented by regional, collaborative partnerships named Joint Ventures (JVs). Since 1987, JVs have spent approximately US$5 billion to conserve or manage 7.8 million ha of habitat. Nine JVs set waterfowl habitat objectives and deliver programmes in regions that support the majority of ducks wintering in the United States. Conservation plans developed by these JVs are based on the premise that food during the non-breeding period can limit demographics and thus population trends for waterfowl (*i.e.* the food limitation hypothesis). This hypothesis is supported by research indicating that duck body condition correlates with winter habitat conditions (Delnicki & Reinecke 1986; Lovvorn 1994; Thomas 2004; Heitmeyer 2006; Moon *et al.* 2007), which influences diet quality (Loesch & Kaminski 1989), and moreover that body condition influences survival (*e.g.* Moon & Haukos 2006; Bergan & Smith 1993) and the timing of migration phenology (Heitmeyer 1988, 2006). At the population level, winter habitat conditions can influence the distribution of ducks within and across winters (Nichols *et al.* 1983; Hepp & Hines 1991; Lovvorn & Baldwin 1996; Pearse *et al.* 2012). Finally, there is evidence for cross-seasonal influences, with winter and migration habitat conditions influencing subsequent productivity (Heitmeyer & Fredickson 1981; Kaminski & Gluesing 1987; Raveling & Heitmeyer 1989; Guillemain *et al.* 2008; Devries *et al.* 2008; Anteau & Afton 2009).

Most JVs use a bioenergetics model to estimate habitat carrying capacity and project habitat needs to support waterfowl populations at target levels during the non-breeding season (*e.g.* Prince 1979; Reinecke *et al.* 1989; Petrie *et al.* 2011). Bioenergetics models represent a class of resource depletion models and those used by winter habitat JVs often take the form of daily ration models (DRMs; Goss-Custard *et al.* 2003). While DRMs can take different forms, they generally aggregate food energy density across multiple habitat patches (using either habitat-specific values or average values across habitats) and divide by daily energy demands of a target duck species to estimate the theoretical carrying capacity of a given area (Miller & Newton 1999; Goss-Custard *et al.* 2002; Goss-Custard *et al.* 2003). In its simplest form, carrying capacity may be expressed in terms of duck energy-days (DED):

$$\mathrm{DED} = \frac{\text{Food available }(\text{g dry weight})\times}{\text{Daily energy expenditure}} \quad (1)$$
$$\frac{\text{True metabolisable energy }(\text{kcal/g dry weight})}{(\text{kcal/day})}$$

Thus, under the assumption that DRMs reasonably reflect foraging dynamics of free-ranging waterfowl, useful calculations of carrying capacity require estimates of: 1) habitat-specific food production (g dry weight per unit area), 2) functional availability of waterfowl foods (g dry weight per unit area; *e.g.* Greer et al. 2009), 3) true metabolisable energy of available foods (kcal per g dry weight; Miller & Reinecke 1984), 4) daily energy requirements of target species (kcal), and 5) region- and species-specific population targets (Petrie *et al.* 2011). The actual forms of models being used by JVs are more sophisticated than the simple equation depicted above. For example, most JVs model energy supply and demand in time and space (*e.g.* Central Valley JV 2006; Pacific Coast JV 2004) with the understanding that energy supplies may be influenced by natural or intentional flooding of habitats and that demand may vary temporally based on population size, migration chronology, changes in species composition, physiological needs, weather and other endogenous or exogenous factors. Regardless of model sophistication, all DRMs require some estimate of energy supply and demand.

While simple in principle, there are

empirical and theoretical challenges associated with estimating energy supply and demand. When estimating energy supply, bias can occur from inaccurate estimates of: 1) food availability, 2) energy content of specific foods, 3) extrapolating site-specific estimates of food availability to landscapes for focal species, or 4) assuming estimates from a single species of waterfowl apply to other species. Likewise, quantitative challenges exist when estimating energy demand during the non-breeding period. Beyond the challenge of estimating regional population size (Soulliere *et al.* 2013), those planning waterfowl management programmes may also face biased estimates of: 1) resting metabolic rate, 2) cost of daily behaviours, and 3) costs of thermoregulation or tissue synthesis. Finally, bioenergetic models can take various forms ranging from simple DRMs to spatially-explicit, agent-based models that incorporate additional mechanistic details of the systems being modelled. However, the conditions dictating when more complicated models are required is not thoroughly understood (Goss-Custard *et al.* 2003).

To address some of these challenges, a special session was convened at the 6th North American Duck Symposium to consider fundamental aspects of the DRMs most commonly used in conservation planning undertaken by winter habitat JVs in North America. A comprehensive assessment and comparison of the strengths, weaknesses and utility of the full suite of conservation planning models for wintering waterfowl were beyond the scope of this paper. Rather, talks addressed several key elements associated with estimating energy supply and demand, along with alternative model structures and the implications of new model advances and pitfalls, for future conservation planning directed towards ducks at staging and wintering sites.

Energy supply

Calculating energy supply for waterfowl requires an empirical measure of habitat-specific food production and availability to waterfowl. Here we define food availability as the production of food minus an amount not exploitable by waterfowl (*i.e.* the giving-up density, or food availability threshold; Greer *et al.* 2009; Hagy & Kaminski 2012a). When food availability is known or can be reasonably estimated, energy supply can be calculated using energetic values of each food to individual species and extrapolated across the area of interest. Below, we explore each of these components in increased detail.

Estimating food production

Direct and indirect methods have been developed for estimating food availability in the environment. For dabbling ducks, clipping of inflorescences, extracting soil cores and sweep nets are the primary tools used to estimate the biomass of seeds, nektonic and benthic forage produced (Dugger *et al.* 2008; Kross *et al.* 2008a; Evans-Peters *et al.* 2010; Hagy & Kaminski 2012a).

Vegetative food production is often estimated using floristic measurements (*e.g.* Gray *et al.* 1999a, 2009; Naylor *et al.* 2005) or by assessing seed, tuber and plant part biomass at the end of the growing season (Kross *et al.* 2008a). Seeds can be threshed from inflorescences and collected using core samples (Kross *et al.* 2008a), predicted using vegetation morphology indices (Gray *et al.* 1999a) using visual assessments (Naylor *et al.* 2005), or measured directly using other means (Gray *et al.* 2013). Initially, Laubhan & Fredrickson (1992) developed equations that predicted seed production using phytomorphological measurements. These models were not widely used because they required extensive field measurements and predictions outside of the region of development were unreliable (Gray *et al.* 1999a). Gray *et al.* (2009) determined that the area of a seed head was a reliable predictor of seed mass and developed a simplified process of using desktop scanners to predict seed production. Naylor *et al.* (2005) described a process for ranking moist-soil habitat quality for waterfowl in California, USA, based on visual estimates of plant composition and forage quality. Stafford *et al.* (2011) replicated this technique in Illinois and found that the index explained 65% of the variation in moist-soil plant seed biomass collected from core samples. These rapid assessment techniques allow wetland managers to obtain efficiently general estimates of seed biomass for waterfowl in moist-soil wetlands without extensive and costly laboratory or field work. However, Evans-Peters (2010) suggested these visual assessments omit ≤ 30% of seed biomass in the seed bank. Moreover, accuracy of visual assessments of standing vegetation during the growing season or prior to vegetation senescence and inundation by water may not accurately reflect food densities at later points in time useful to managers (Greer *et al.* 2007; Fleming *et al.* 2012).

Traditional core samplers used for sampling duck foods range from 5–10 cm in diameter (Swanson 1983; Stafford *et al.* 2006; Greer *et al.* 2007; Hagy *et al.* 2012b; Smith *et al.* 2012) and 5–10 cm in depth (Greer *et al.* 2007; Kross *et al.* 2008a; Olmstead 2010; Hagy & Kaminski 2012b), with deeper samplers used for larger, longer-necked taxa (*e.g.* 30 cm for swans; Santamaria & Rodriguez-Girones 2002). In relatively shallow water, core samplers may yield simultaneous density estimates for submersed aquatic vegetation (*e.g. Myriophalum* sp., *Ceratophyllum* sp., *Elodia* sp.), nektonic and benthic invertebrates, seeds and tubers (Swanson 1978). In deeper water or in areas with high densities of aquatic vegetation, core samplers may be used in combination with sweep nets (Murkin *et al.* 1996; Tidwell *et al.* 2013), exclusion devices (Straub *et al.* 2012) and box samplers (Synchra & Adamek 2010) to provide better measures of food availability.

Regardless of the direct sampling method used, samples require extensive time to process in the laboratory. Food items typically are sorted from the plant, soil and detritus by hand using a series of graduated sieves, dried to constant mass in a forced air oven, identified and weighed by species or appropriate biological classification. This process is tedious and costly (Stafford *et al.* 2011). Sub-sampling is a well-vetted approach for reducing processing time (*e.g.* Proctor & Marks 1974; Schroth & Kolbe 1993; Murkin *et al.* 1996; Reinecke & Hartke 2005; Smith *et al.* 2012). Waterfowl researchers have recently applied soil core sub-sampling and verified that overall means are similar between sub-sampled and whole-processed samples for moist-soil seeds (Hagy *et al.* 2011; Stafford *et al.* 2011), rice *Oryza* sp. grains (Stafford *et al.* 2006) and macroinvertebrates (M. Livolsi *et al.* University of Delaware, unpubl. data), although variance associated with the estimates may increase (Hagy *et al.* 2011).

Currently, there is little information on the optimal sample size for core samples in waterfowl habitats, and some information suggests optimal sample size may be difficult to predict (Reinecke & Hartke 2005; Marty 2013). While current studies often base sample size on financial and temporal constraints (Sherfy *et al.* 2000; Evans-Peters *et al.* 2012), 20–30 samples per patch have been shown to result in coefficient of variation (C.V.) values of < 10% (Dugger *et al.* 2008; Greer *et al.* 2009; Evans-Peters *et al.* 2012). Preliminary power analysis of saltmarsh systems indicates that *c.* 40 samples per habitat type/location/time period would be reasonable for some habitats, but other habitats such as "high-marsh" and rice fields show greater variability and may require significantly more samples (K. Ringelman, University of Delaware, unpubl. data; Marty 2013). More comprehensive research is needed to provide appropriate sample sizes for quantifying food availability within different habitat types.

Food biomass estimates may also be influenced by improper inclusion of prey seldom consumed or not energetically profitable to waterfowl. For example, 30–70% of seeds sorted from core samples collected in the Mississippi Alluvial Valley, USA, in autumn had little nutritional value or were likely not consumed by most dabbling ducks (Hagy & Kaminski 2012b; Olmstead *et al.* 2013). Additionally, biomass estimates may be biased due to incomplete recovery of seeds during sample processing, through non-detection, loss or destruction. Hagy *et al.* (2011) reported that *c.* 14% of known seeds were not recovered from core samples during sorting, and that recovery rates depended on seed size. Thus, energy information based on incomplete recovery could underestimate food availability by 10–20% (Hagy *et al.* 2011), whereas analyses that do not account for actual diet and food use bias could overestimate food availability by as much as 47% (Hagy *et al.* 2011; Olmstead *et al.* 2013).

Finally, biomass estimates may be variable among locations and geographical regions (*e.g.* Stafford *et al.* 2006a; 2011; Kross *et al.* 2008a; Evans-Peters *et al.* 2012; Hagy & Kaminski 2012b); thus, use of local productivity estimates may be biased when scaled to regional levels. A recent simulation of core sampling indicated that detection probabilities for food items varied by food densities, corer size and the underlying pattern of food distribution (A. Behney, Southern Illinois University, unpubl. data). While biologists can use methodological improvements to reduce local variance, they should acknowledge the possibility of geographic variation, random *versus* clumped food distributions within habitats and the possibility that birds may not follow an ideal free distribution or forage optimally. Therefore, biologists may wish to consider sampling in multiple locations to provide a better representation of values for regional-scale habitat management and conservation (*e.g.* Stafford *et al.* 2006; Kross *et al.* 2008a).

Estimating food availability

Unbiased, precise estimates of food production and its energy density are not sufficient for understanding the energy supply available to migrating and wintering waterfowl, because all foods produced may not be available when waterfowl access habitats, for instance due to the depth or extent (*i.e.* surface area) of flooding (Kross *et al.* 2008b; Foster *et al.* 2010). Assessment of availability includes accounting for physical accessibility and energy acquired given costs of foraging (Hagy *et al.* 2012b).

Use of an area by migrating or wintering waterfowl may lag considerably from the time of seed maturation, and reliance on food production estimates from the end of a growing season may overestimate food available to the birds. Seed abundance in agricultural habitats can decrease substantially between seed maturation and the arrival of waterfowl, especially in mid to southern latitudes of the United States (Manley *et al.* 2004; Stafford *et al.* 2006; Greer *et al.* 2009; Foster *et al.* 2010b). However, Marty (2013) reported an increase in abundance of waste rice in the Gulf Coastal Prairie rice fields of Louisiana and Texas during autumn, due to the production of a second unharvested rice crop (ratoon) in late autumn, following an initial late summer harvest of rice. Seed loss rates vary with hydrology and are attributable to germination, consumption by non-target wildlife and decomposition (Stafford *et al.* 2006; Greer *et al.* 2007; Foster *et al.* 2010a). Hagy *et al.* (2012a) noted that decomposition of moist-soil seeds in flooded emergent wetlands was *c.* 18% per month. Stafford *et al.* (2006) reported that 58% of waste rice in the MAV fields decomposed post-harvest in autumn, compared to 14% and 8% loss from granivory and germination, respectively. Foster *et al.* (2010b) found that monthly rate of loss of seeds for corn *Zea* sp., sorghum *Sorghum* sp. and soybean *Glycine* sp. ranged from 64–84% post-harvest. Decomposition rates likely vary with latitude as warmer temperatures would contribute to higher rates. Thus, production estimates are useful to evaluate management actions, but sampling should either be timed to coincide with waterfowl arrival or appropriate adjustments are needed to initial estimates, to account for seed loss not attributable to waterfowl foraging.

The physical availability of foods for waterfowl depends on the birds' ability to extract foods from wetlands (*e.g.* Nolet *et al.* 2001). Studies of seed and invertebrate biomass have traditionally assumed that every seed or invertebrate captured in the 5–10 cm deep core samples was available to foraging waterfowl. However, species with different foraging behaviours and morphologies (*e.g.* diving ducks *Aythya* sp. *versus* dabbling ducks *Anas* sp.) can affect how much of the production is actually available (Nudds & Kaminski 1984; Murkin *et al.* 1996; Sherfy *et al.* 2000; Evans-Peters 2010; Olmstead 2010). Additionally, foraging efficiency varies with sediment depths and seed type, and it is likely that seeds buried at greater depths may be less profitable energetically (Nolet *et al.* 2001; Smith *et al.* 2011). However, if biologists sample systems with large amounts of macroinvertebrates moving within the soil column, especially in a tidal system, deeper sampling depths may be more appropriate to estimate food availability (although this has yet to be tested).

From behavioural and energetic perspectives, estimates of food availability must be adjusted when food densities are too low for energetically profitable foraging (van Gils *et al.* 2004; Nolet *et al.* 2006; Hagy *et al.* 2012b). There is evidence for a critical food density that remains after individuals either give up foraging (Greer *et al.* 2009) or no longer remove food from patches despite continued foraging effort (Hagy 2010; Hagy *et al.* 2012b). However, the critical food density varies among habitats, regions and potentially even between foraging patches that differ in food composition (Baldassarre & Bolen 1984; Naylor 2002; Greer *et al.* 2009; Hagy 2010). For example, Hagy (2010) noted that foraging thresholds varied widely for moist-soil wetlands, but they were likely at least 200 kg/ha in natural moist-soil wetlands with a wide variety of seed taxa present. Hagy et al (2012b) reported residual seed densities exceeding 250kg/ha in moist-soil wetlands in the MAV after waterfowl ceased to remove additional foods, whereas Naylor (2002) reported residual densities of 30–160 kg/ha in these wetlands in California. Greer *et al.* (2009) and Baldassarre & Bolen (1984) reported residual densities of 50 kg/ha or less in flooded rice and dry corn fields, respectively. Gray *et al.* (2013) provided updated estimates of available food for waterfowl in agricultural fields, moist-soil wetlands and bottomland hardwoods, and incorporated critical food densities into these estimates. Application of fixed critical food densities probably results in inaccurate predictions at the patch level (van Gils *et al.* 2004), but are compatible with daily ration models currently used to predict carrying capacity for large regions (*e.g.* Soulliere *et al.* 2007) and may have some value if patch-specific data are unavailable or impractical to obtain. Therefore, incorporating critical food densities at the patch level will likely increase the accuracy of food availability estimates. Ancillary modelling has indicated that failure to apply foraging thresholds accurately at the patch level could affect food availability estimates by as much as 60%, and this bias varies with values of foraging thresholds and seed density (Pearse & Stafford 2014; H. Hagy, Mississippi State University, unpubl. data).

True metabolisable energy of foods

True metabolisable energy (TME, kcal/g) represents the amount of energy an individual bird receives from a food item, after accounting for metabolic faecal losses and also endogenous urinary losses as metabolised energy (Miller & Reinecke 1984). The TME provides a more accurate estimate of metabolised energy than apparent metabolisable energy (AME), because TME accounts for faecal and urinary losses (Miller & Reinecke 1984; Karasov 1990). The TME of waterfowl foods are important components for accurate assessments of waterfowl bioenergetics and energetic carrying capacity. It may be calculated: a) indirectly, using a regression model of total excretory energy on total food intake, or b) experimentally, determined by feeding birds a controlled diet and measuring excretory energy (Sibbald 1975; Sibbald 1979;

Kaminski & Essig 1992). The TME values can then be used to calculate available energy by multiplying the mass of a food item by its TME value and extrapolating the resulting energy value across an area of interest.

Nonetheless, there is a lack of TME values for common waterfowl foods and species. Current studies have focused on TME values for Mallard *Anas platyrhynchos*, American Black Duck *A. rubripes*, Northern Pintail *A. acuta*, Canada Goose *Branta canadensis*, Blue-winged Teal *A. discors* and Carolina Wood Duck *Aix sponsa* (Hoffman & Bookhout 1985; Jorde & Owne 1988; Petrie 1994; Reinecke *et al.* 1989; Petrie *et al.* 1998; Sherfy 1999; Sherfy *et al.* 2001; Checkett *et al.* 2002; Kaminski *et al.* 2003; Ballard *et al.* 2004; Dugger *et al.* 2007; J. Coluccy *et al.*, Ducks Unlimited, unpubl. data). There is uncertainty associated with applying a TME value to a species other than the one from which it was derived. However, closely related bird species will likely have similar TME values for a given food item due to similarities in gut morphology. While this hypothesis needs to be tested, substituting TME values among similar bird species may suffice when evaluating energy content of food items until species-specific TME values become available.

Another problem with the lack of information on TME values is that existing studies have typically determined TME values for only a few seed or invertebrate species across a select few families. Therefore, researchers are forced to "fill in the gaps" by assigning TME values based on educated guesses with information from few studies. For example, many JVs use a mean seed TME value of 2.5 kcals/g for their projections of moist-soil seeds; however, data from the Lower Klamath National Wildlife Refuge, in northern California, indicate that mean seed TME values can be less than 50% of this estimate (Dugger *et al.* 2008). Most existing TME data are for seed species from the Midwestern United States, and employing TME averages may therefore be inappropriate in regions where seed composition is significantly different. Calculating potential available energy across large areas using variable or biased TME values may result in meaningless estimates of carrying capacity. Thus, additional research efforts focused on deriving TME values experimentally for a wide variety of common food items for waterfowl species will increase precision and accuracy of energetic carrying capacity estimates. Alternatively, because such comprehensive analyses may be impractical, future tests that extrapolate partial knowledge may be of use. For example, researchers could derive TME values for a single plant seed species for a range of duck species that differ in body size and diet (Green-winged Teal *A. crecca*, American Wigeon *A. americana*, Northern Shoveler *A. clypeata*, Mallard) to yield insight into the extrapolation and applicability of TME values for species that have not been included in TME experiments. Alternatively, researchers could relate TME to nutritional composition of the seeds because digestibility of multi-species forage is related to the amount of indigestible fibre. Additionally, researchers may evaluate if known TME values of common waterfowl foods can be predicted from gross energy (GE) estimates of these foods (*e.g.*

Kaminski *et al.* 2003). If GE would explain significant variation in TME (*e.g.* ≥ 70%) such models may be used cost-effectively for habitat conservation planning and implementation.

Extrapolating energy supply to the landscape

Because of the significant potential for biological and sampling error discussed above, the notion of extrapolating estimated useable energy to a landscape level should be approached with caution. Inherently, the potential for multiplication of errors may cause landscape-level variance to be too large to make meaningful management recommendations. However, an equally difficult problem is to quantify correctly what habitat is actually available to waterfowl species.

Various geospatial data have been used to quantify characteristics of important waterfowl habitats (*e.g.* the National Wetlands Inventory (NWI) and National Land Cover Data). However, there are inherent limitations associated with the accuracy of these data. For example, the NWI established by the U.S. Fish and Wildlife Service to conduct a nationwide inventory of wetlands by type is widely used for quantifying the availability of wetlands on staging and wintering areas. Unfortunately, the NWI does not capture and classify all wetlands accurately because NWI maps are derived from aerial photo-interpretation with varying limitations due to scale, photograph quality, inventory techniques and other factors (Federal Geographic Data Committee 2008; Dahl *et al.* 2009). Recent advances in GIS technology, the availability of higher resolution imagery, the integration of NWI data with other geospatial data sources (*e.g.* LiDAR, soil maps, *etc.*) and the development of standardised techniques for wetland identification and delineation have substantially improved the NWI (Dahl *et al.* 2009; Knight *et al.* 2013). However, NWI data are currently only available for 89% of continental United States, and the average date of the NWI for most of the U.S. is from the 1980s. Currently, NWI data are being updated at a rate of ≤ 2% per year due to funding reductions (J. Coluccy, Ducks Unlimited, pers. comm.). Similar limitations exist for quantifying non-wetland habitat types (*e.g.* county crop data or seasonally-flooded cropland). These limitations have hampered efforts to estimate habitat availability accurately for wintering waterfowl, especially at frequencies desired for maintaining a contemporary understanding of the landscape carrying capacity.

In addition to error associated with estimating wetland habitats correctly, biologists need to consider potential indirect impacts of human developments and disturbance that make available habitat avoided and thus reduce carrying capacity (Korschgen & Dahlgren 1992; St. James *et al.* 2013). Often, estimates of available habitat are quickly calculated with the assumption that all wetlands are created equal and waterfowl have unimpeded access to them. However, avoidance behaviour may occur at various temporal and spatial scales, ranging between not settling in an area and/or not utilising a space to its maximum potential (Korschgen & Dahlgren 1992; Laundré *et al.* 2010; Hine *et al.* 2013). While some waterfowl species such as Canada Geese and Mallard may have adapted well to an altered landscape, other species of special concern appear to be particularly sensitive to disturbance (Korschgen & Dahlgren 1992) including Atlantic Brant *Branta bernicla*, American Black Ducks, Canvasback *Aythya valisineria* and Lesser Scaup *Aythya affinis*. Additionally, ducks with a smaller body size and shorter longevity (*e.g.* Green-winged Teal) than larger species (*e.g.* Mallard) may take more risks to forage or otherwise use wetlands that are hunted, and thus exhibit reduced avoidance behaviour (St. James 2011).

For planning purposes, it is important for resource managers to understand how waterfowl separate themselves from anthropogenic development and respond to disturbance, and how these factors influence their ability to extract critical food resources from habitats. For example, on Lake St. Clair between Michigan and Ontario, autumn-staging diving ducks shifted their use of traditional feeding and loafing areas on the U.S. side of the lake to new areas on the Canadian side (Shirkey 2012). Warmer weather and associated increased angling and hunting activity on the U.S. side throughout autumn and early winter is considered to have resulted in a significant shift in habitats important to Canvasback and Lesser Scaup (Shirkey 2012). Diving ducks had options at this location, but obviously disturbance is an important management consideration at staging sites.

Energy demand

In addition to knowing energy availability in the landscape, calculating habitat carrying capacity requires an estimation of the energy needs of individual waterfowl on any given location and day. Energy requirements of a wild vertebrate (or Daily Energy Expenditure, DEE) are usually estimated as the sum of the energy costs of maintenance (or Resting Metabolic Rate, RMR), activity and thermoregulation (King 1973; Servello *et al.* 2005; Fig. 1). If the animal is growing or reproducing, then an additional energy cost associated with this production must be added (Fig. 1). If the animal is not in a steady state and storing energy, then this cost also must be added to the daily energy costs for the animal (Fig. 1). Most bioenergetics approaches currently used by JVs for estimating DEE of waterfowl do not account for the energy costs of thermoregulation, production and storage, thereby reducing DEE to an estimate of energy expenditure through daily activities (Fig. 1). We will summarise current views on estimating resting metabolic rate, the energetic costs of daily activity and the costs of thermoregulation.

Resting metabolic rate

The RMR is usually defined as energy costs of maintaining an animal's physiological systems under a certain, restricted set of conditions including: 1) no activity, 2) no cost of staying warm or cool when measured at ambient temperatures that are within the thermoneutral zone of the animal, 3) no growth or reproduction, and 4) the bird is at steady state and not accumulating any fat or protein reserves or depleting them. Because of these restrictive conditions, there have been few replicated RMR studies for duck species. However,

A simple model of bioenergetics

Daily Energy Expenditure DEE = 3 × RMR

Activity

Resting Metabolic Rate
RMR = 457 × (body mass)$^{0.77}$

+ Thermoregulation

Production (reproduction, growth)

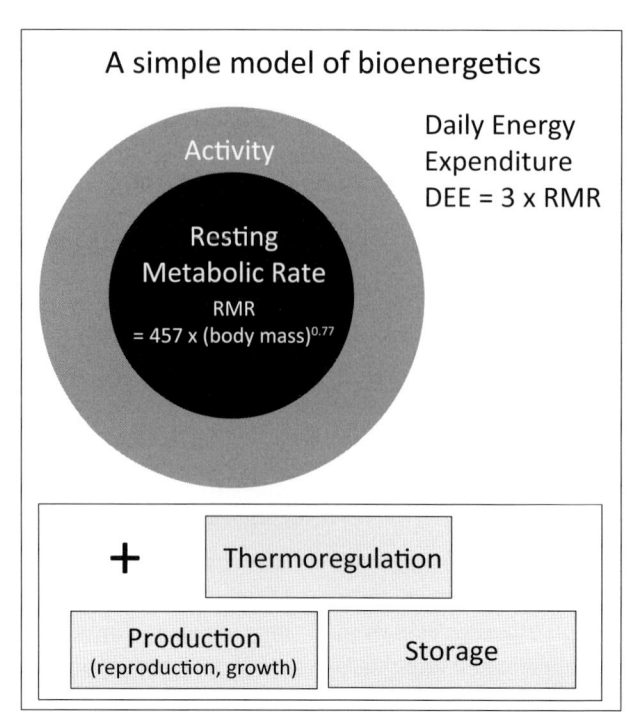

Storage

Figure 1. A conceptual bioenergetics model of the key components that comprise Daily Energy Expenditure (DEE) of a wild vertebrate. DEE is usually estimated as the sum of the energy costs of maintenance (or Resting Metabolic Rate, RMR, calculated from body mass; Miller & Eadie 2006), times 3 to account for activity. Most bioenergetics approaches do not account for the energy costs of thermoregulation, production and storage, thereby reducing DEE to an estimate of energy expenditure through daily activities.

there is established literature on the allometric scaling of RMR of a wide variety of animals (Kleiber 1932, 1961; Prince 1979; Schmidt-Nielsen 1984; Peters 1983; Calder 1984; Brown & West 2000) following:

$$RMR = a(Mass)^b \quad (2)$$

where a = a mass proportionality coefficient, Mass = body mass (kg), and b = slope of the regression line on a log scale. In general, the accuracy of RMR predictions from body size improves substantially when subsets of species such as waterfowl are considered. Miller & Eadie (2006) used all available data from waterfowl RMR to update estimates of the slope (b) and intercept (a) parameters for waterfowl. For JV carrying capacity modelling, this provides a relatively straightforward method to estimate RMR for this component of bioenergetics modelling (but see caveats in Miller & Eadie 2006).

Activity expenditures

Daily Energy Expenditure is based on the previously estimated RMR times the cost of activity. Most bioenergetics models used by JVs calculate DEE as

$$\text{DEE} = \text{RMR} \times 3 \qquad (3)$$

with the multiplier "3" accounting for the average amount of energy expended on activity in any given day (King 1974; Prince 1979; Miller & Eadie 2006). However, if ducks are storing energy, then this adds further energy costs (*e.g.* Heitmeyer 2006 used a 3.4 multiplier for Mallard in winter). While this methodology provides an estimate of DEE, there are questions as to whether this simple approach accurately accounts for variability in behaviour due to external variations (*e.g.* temperature, tide, time of day, month, latitude, harvest pressure, disturbance, *etc.*) that are known to influence both daily activities and DEE (Weathers 1979; Albright *et al.* 1983; Brodsky & Weatherhead 1985; Morton *et al.* 1989). Miller & Eadie (2006) demonstrated that estimates of carrying capacity were highly sensitive to the multiple of RMR used, as well as the mass proportionality coefficient (*a*) from the allometric equation. Thus, the use of a single multiplier for adjusting RMR is likely an oversimplification. Depending on the sophistication of the planning process, JVs could use more refined estimates of activity costs derived from measured time-activity budgets of waterfowl in a given area for which DEE is to be estimated (see Weathers *et al.* 1984; Miller & Eadie 2006). Time-activity budgets rely on extensive behavioural observations to determine a time budget or the percentage of time free-living individuals spend in different behavioural states. Using behaviour-specific factorial increases in energy expenditure over RMR, a time budget can be converted into estimates of energy expenditure or an energy budget (Albright *et al.* 1983; Paulus 1988). Therefore, equation (3) can be expanded to account for multiple activities:

$$\text{DEE} = \sum_{i=1}^{n} \left[(\text{RMR} \times a_i) \times T_i \right] \qquad (4)$$

where a_i = the activity-specific factorial increase in RMR for the *i*th behavioural activity and T_i = the proportion of time engaged in the activity within the 24-hr cycle. For example, Wooley (1976) added heart rate monitors to five Black Ducks and estimated multiplier values of 1.7 for feeding, 1.2 for resting, 2.1 for comfort, 2.2 for swimming, 2.2 for alert, 12.5 for flying, 1.7 for walking, 2.4 for agonistic and 2.4 for courtship. Finally, the activity-specific factorial increase in RMR, a_i, was estimated by Wooley (1976). However, due to the crude nature of this study as compared to a more controlled respirometry study, the validity of the estimates has been queried. For example, how could comfort behaviour be energetically more taxing than feeding? If these estimates are biased, the implications for scaling up to the population level duck-use days could be flawed. Additional studies that measure BMR accurately for many waterfowl species (see McKechnie & Wolf 2004), and consider how environmental factors including cold ambient temperatures affect BMR and DEE (*e.g.* McKechnie 2008; McKinney & McWilliams 2005), can provide more precise estimates of DEE for waterfowl.

Estimating DEE through time-energy budgets is labour intensive, time consuming, assumes random observability, does not account for energetically taxing flight behaviour and is historically limited to diurnal observations (Jorde & Owen 1988). Recently, Jones (2012) conducted a comprehensive analysis of different measures of daily activity energy expenditures using American Black Ducks in coastal New Jersey. Black Duck behaviour (including flying) was quantified during morning crepuscular, diurnal, evening crepuscular and nocturnal periods to create a 24-hr time-energy budget. Behaviours and energy expenditure differed between periods and months, with greatest hourly energy expenditure during the morning crepuscular period and lowest during the nocturnal period. Additionally, precipitation, temperature and tide influenced variation in Black Duck behaviours over the 24-hr period. Moreover, anthropogenic disturbance factors influenced behaviour including increased feeding during diurnal and nocturnal periods on areas open to hunting when the hunting season was closed, and increased resting on areas closed to hunting regardless of whether the hunting season was open. As a result of the detailed time energy budget, DEE was estimated as being 1,218 ± s.e. 19.36 kJ/bird/day, or 2.4 times RMR. The same estimate calculated using RMR times 3 was 21% greater at 1,545 kJ/bird/day, or 21% higher. If one were to apply both estimates to 101,017 ha of New Jersey coastal habitat for which energy supplies were estimated (Cramer et al. 2012), autumn carrying capacity would be ~55,000 ducks while the 24-hr time energy budget estimates would predict a carrying capacity of ~70,000 ducks, closer to the estimated 75,000 Black Ducks estimated by mid-winter surveys (USFWS MBDC 2014).

Cost of thermoregulation

The assumption that waterfowl are not incurring energy costs of thermoregulation, production and storage requires some scrutiny. Waterfowl and all endotherms must increase energy expenditure when ambient temperature is below the lower critical temperature (LCT) or above the upper critical temperature (UCT) of the animal (Fig. 2):

$$m_c * \Delta T_{LCT-Ta} \quad (5)$$

where m_c is the slope of increasing metabolic energy above the lowest critical temperature (LCT) and ΔT_{LCT-Ta} is the difference in ambient temperature from the lowest critical temperature. This allows for an expansion of equation (4):

$$DEE = \sum_{i=1}^{n}\left[\left((RMR \times a_i) + CT\right) \times T_i\right] \quad (6)$$

where CT = the cost of thermoregulation at a specified temperature (kJ/bird/h) in addition to activity-specific increases to RMR. The UCT is rarely considered, as most temperate waterfowl avoid regions with temperatures above the UCT. However, waterfowl during winter may often encounter periods when the ambient temperature is below their LCT. For example, the LCT of brant is 7.5°C (Morehouse 1974); below this ambient temperature, brant would expend additional energy to stay warm. Generally, because thermal conductivity of water is 23 times greater than that of air, waterfowl sitting on

Costs of thermoregulation

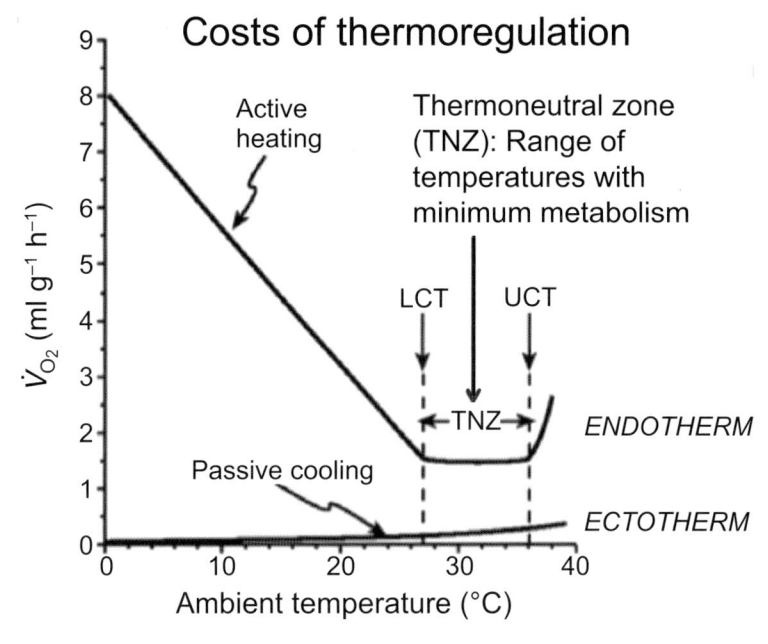

Figure 2. Any endothermic animal incurs additional energy costs when ambient temperatures drop below their lower critical temperature (LCT). These additional costs are linearly related to ambient temperatures below the LCT and are directly a function of the insulative properties of the animal. From Hiebert & Noveral (2007).

water would need to expend much more energy to stay warm compared to birds sitting on shore at the same ambient air temperature. While Richman & Lovvorn (2011) did not find noticeable differences in energy costs for Common Eiders *Somateria mollissima*, in cold water or air, McKinney & McWilliams (2005) estimated that the energy costs of thermoregulation in water could contribute as much as 13–23% of DEE for Bufflehead *Bucephala albeola* during winter in southern New England. Therefore, if DEE is estimated for waterfowl during winter when temperatures are typically below the LCT, including the explicit energy cost of thermoregulation will reduce bias in DEE estimates. However, there is a need for empirical studies of the energy costs of thermoregulation in other waterfowl species to provide a strong foundation for such estimates.

Alternative modelling frameworks

While addressing sources of variation associated with energy supply and demand is fundamental to reliable bioenergetics models, there is also value in considering if alternative modelling frameworks could improve conservation planning for wintering waterfowl. The DRMs have provided a useful approach for estimating bioenergetic needs and landscape carrying capacity for waterfowl in winter for most non-breeding JVs in North America for several reasons: 1) they are well-established and widely accepted, 2) they provide a tool for translating waterfowl population objectives into habitat-based objectives, which is essential to accomplish NAWMP continental goals, 3) they are based on data that can be obtained and validated from field surveys/research (*e.g.* food abundance and daily energy demand), and 4) they allow managers and planners to evaluate the effect of large-scale habitat changes on the availability of food resources and so provide an ability to undertake scenario planning (Central Valley JV Implementation Plan 2006).

However, there are a number of limitations of the DRMs. The DRMs are not spatially explicit, because they assume no cost of travelling between food patches, and food availability is considered to be relatively uniform across the landscape. The set of simplifying assumptions incorporated into most DRMs (*e.g.* ideal free foragers) precludes consideration of how habitat heterogeneity or bird distribution patterns (spatially or temporally) influence carrying capacity. Likewise, in DRM, energy demand is summed over all individuals regardless of sex and age, usually over extended time periods (bi-weekly), and with DEEs that are assumed to be invariant over time (fixed energy costs). In some cases, JVs will sum DEE across all species based on an average or representative body size.

Changes in energy expenditures throughout the non-breeding season, due to changing food availability, temporal differences in thermoregulatory and individual state-strategies (*e.g.* the need to acquire energy reserves in preparation for migration), and interaction with other fitness-maximising strategies (*e.g.* risk aversion, courtship and mate defence), generally are not included in DRMs. A further limitation of current DRMs is that they consider only energy and not other nutrient resource needs from foraging habitat. Yet other habitats, such as roosting or refuge sites with better thermal characteristics, reduced disturbance or fewer predators, may also be important for reliable conservation planning. Finally, the suite of response variables in DRMs is limited, resulting in a conglomerate energy supply and summed population energy demand that yields a surplus or deficit determined on the habitat base available.

One of the challenges for JVs concerned with science-based habitat conservation for non-breeding waterfowl is to develop a measure of how achieving habitat objectives ultimately affects waterfowl demographic parameters. Current DRMs do not make this link, which limits the ability of habitat managers to integrate planning models with demographic models that predict effects of regional actions on waterfowl dynamics at the continental scale. An alternative approach to DRMs is use of agent-based models (hereafter referred to as ABMs). Unlike the top-down population-based approach of DRMs (summed energy demand and supply functions across species and habitats), ABMs instead represent a "bottom-up" approach where systems are modelled as collections of unique individuals or "agents". Unlike more formal mathematical population models, in ABMs the system dynamics emerge from interactions of individuals and their environment. Models are solved by simulation instead of analytical solutions. Given plausible and realistic "rules" of interaction and behaviour, the outcome is determined by simulation of how the agents in the model respond according to the rules and parameters defined.

The ABMs are now beginning to see broad application in conservation fields, although they have been used in ecological research for well over two decades (DeAngelis & Gross 1992; Sutherland & Allport 1994; Goss-Custard *et al.* 2003; Grimm & Railsback 2005; Stillman 2008). McLane *et al.* (2011) recently provided a comprehensive review of use and utility of ABMs in wildlife ecology and management. Several ABMs have been developed that could provide a platform for modification and use by the NAWMP community to model and plan for waterfowl and waterbird use of managed wetlands. Stillman's MORPH programme (Stillman 2008; West *et al.* 2011) provides a foraging model that has been applied to coastal birds in estuarine environments and has broad potential as a platform for extension to other waterbirds and waterfowl. Pettifor *et al.* (2000) developed a spatially explicit, individual-based, behavioural model to examine the annual cycle of migratory geese. Mathevet *et al.* (2003) developed an ABM model as a management tool for waterfowl conservation incorporating farming and hunting practices in France. However, they did not model duck energetics explicitly and instead relied on a spatially-located DRM. Most recently, the Eadie & Shank research group at UC Davis developed a prototype ABM referred to as SWAMP (Spatially-explicit Waterbird Agent-based Modeling Program) to model overwintering waterfowl bioenergetics (Miller *et al.* 2014; K. Ringelman *et al.*, unpubl. data; J. Eadie *et al.*, unpubl. data). This model is intended to provide similar functionality as DRMs, although it is one of the few ABMs specifically designed for use in landscape-based conservation management. The first prototype has been tested, validated and peer-reviewed (Miller *et al.* 2014).

What are the advantages of an agent-based approach?

Why might we need a potentially more complex approach to plan for habitat needs of migrating and wintering waterfowl? Conceptually and pragmatically, exploring and developing an ABM approach includes nine potential advantages. First, ABMs link the behaviour of individuals with population- or community-level processes – "scaling upwards". This provides an opportunity to incorporate individual variation among birds and different populations in different locations or over time. It reveals local effects that could have larger scale impacts (local sinks) that might allow planners to target conservation efforts more effectively. Second, ABMs provide the ability to model individual and population performance metrics of ultimate interest to JVs, such as body condition and survival. Thus, an ABM approach may enable functional integration of habitat changes on wintering and staging areas with regional and continental demographic impacts. Third, ABMs provide a more mechanistic structure for foraging behaviour and dynamics that allows for inclusion of more biologically realistic behaviour (*e.g.* foraging rules, patch choice criteria, flocking, *etc.*) than a generic DRM that subsumes and potentially confounds a large amount of important biology. Fourth, ABMs permit spatially explicit analysis of the effects of alternative management regimes on the area and geographic distribution of wetland habitats. Fifth, ABMs provide a useful tool for cross-boundary JV planning. The condition of birds in one JV region might influence their decision about when to migrate to another region, and the condition of birds when they arrive in a JV region might influence how long they stay. The ABMs provide a method by which to track and link body condition and movement of birds across larger spatial scales. Sixth, the ABMs allow planners to model large numbers (millions) of birds in real time, and on large and small spatial scales, using GIS layers as input. Seventh, ABMs can incorporate other important determinants of habitat use and carrying capacity such as disturbance and dispersion of non-foraging (refuge) habitat. Eighth, ABMs offer the potential to expand the capacity to generalise across taxa, including waterfowl, shorebirds and other wetland-dependent wildlife. Ninth, we can use these models to integrate more directly and completely with existing models of water management, in-stream fish habitat, urban growth, and other spatially-based conservation issues (J. Fleskes, U.S. Geological Survey, pers. comm.).

Challenges of an agent-based approach

Agent-based models are not a panacea for conservation planning. While they offer advantages and potential new insights, we have identified at least three challenges of implementing an ABM approach. First, the richness of possible combinations of deterministic and stochastic processes can make it daunting to simulate through a sufficient range of scenarios to be confident of generating results with a high level of repeatability and generality. Sensitivity analyses are essential, and careful thought and description (and some simplification) are necessary, especially in the early development of a model. Second, ABMs can be complex with a large number of parameters and functions to be estimated. This complexity deterred progress with ABM, because models were specific to a particular situation, not transparent, and not easily communicated or vetted by the scientific and management communities (Grimm & Railsback 2005). However, the field has become sufficiently advanced and well-defined protocols have been established (*e.g.* Grimm *et al.* 2006, 2010). Yet, many parameters are unknown or based on expert opinion. The ABMs are not alone in this regard; traditional DRMs also include a large number of variables and can be extremely complex (*e.g.* the TRUEMET bioenergetic modelling application includes up to 77 time-dependent and 41 time-independent variables). Lastly, ABMs, like DRMs, require a strong foundation of empirical data upon which to base the model and validation. Hence, the sampling and calculation challenges discussed above for DRMs also apply to ABMs. The ABMs allow for heterogeneity in many of these parameters (*e.g.* DEE as a function of daily mass). Although integrating heterogeneity is more realistic, it can also generate increased variation in predicted outcomes. As with any modelling exercise, the quality of output is dictated by the quality and validity of the parameters and functions used as input. Field and laboratory research to refine parameters, estimate unknown parameters and validate outputs are essential.

Given these challenges, do we really need more complex models? Goss-Custard *et al.* (2003) tested carrying capacity predictions between a DRM and a spatial depletion model (SDM) and concluded that predictions were similar under certain circumstances. Comparisons of the output from SWAMP with TRUEMET corroborate this similarity (K. Ringelman *et al.*, unpubl. data). Hence, DRMs may be sufficient for basic determination of foraging habitat needs, with the caveat that DRMs may not reflect energy balance perfectly and uncertainty may be high.

So, when do we need spatially explicit behaviour-based models? We envision several conditions when an ABM approach would be valuable, when: 1) time and energy costs of foraging differ between patches, 2) cost of movement between patches is great and varies over time (*e.g.* with patch depletion close to refuges), 3) sequence of use of patches varies but is important, 4) distance from roosting sites or refuges to foraging patches varies and changes over time, 5) juxtaposition and location of habitats are important, 6) non-foraging habitat is important and managers need to assess consequences and interactions of disturbance, sanctuary, predation and public access, among other non-foraging needs, and 7) there is a need to link habitat conditions to key individual (*e.g.* body condition) and population performance (*e.g.* survival) metrics.

Admittedly, there is a trade-off between complexity and utility, and we need to be vigilant against creating increasingly complex models simply because it is possible. However, the ability to address uncertainty in a more formal and explicit manner, the value of a mechanistic approach that potentially can link habitat to demography, and the ability to consider smaller-scale spatially-explicit planning for waterfowl conservation (*e.g.* by providing non-foraging habitat and refuge areas), provide a strong argument to continue developing and learning from ABM approaches.

Future research and management considerations

Research has made substantial advances to address empirical uncertainties and test assumptions of JV biological planning models over the past several decades. Such research has greatly increased our confidence in the utility of DRMs as a conservation planning tool under certain applications. However, it has also revealed a need to embrace more formal means of identifying model components and parameters and refining them, while also considering alternative modelling frameworks that may produce more reliable conservation planning strategies. Perturbation Analyses or Structural Decision Making are useful tools for considering such uncertainty (Caswell 2001; Hoekman *et al.* 2002, 2006; Coluccy *et al.* 2008), and managers planning the conservation of wintering waterfowl may want to consider expanding their use of such tools.

During our special session and this review, we identified a subset of key simplifying assumptions common to DRMs, some of which deserve additional scrutiny and refinement (*e.g.* constant BMR multipliers). With few exceptions, potential implications of these assumptions and model uncertainties for habitat conservation objectives and priorities have not been quantified (*cf.* Miller & Eadie 2006). Variability, whether originating from natural processes or sampling strategies, is an integral part of any biological system being modelled (*e.g.* Saether *et al.* 2008), yet most DRMs currently being used are deterministic. Even in the few cases where quantitative analyses of variability have occurred, the results have not yet been widely incorporated into conservation recommendations or used to prioritise future investment in science. Thus, a fundamental question is whether habitat objectives arrived at using models based solely on measures of central tendency (*e.g.* mean population abundance, mean TME values) can produce landscapes necessary to achieve NAWMP goals (Straub *et al.* 2012).

Recent progress in our understanding of waterfowl foraging ecology and the successful application of ABMs to inform habitat conservation for coastal waterbirds (*e.g.* Stillman & Goss-Custard 2010) are compelling JV conservation planners to consider the development and use of alternative frameworks for establishing habitat objectives. However, ABMs may be accompanied by even greater uncertainties and requirements for scientific investigations to parameterise, evaluate and refine them. Thus, deliberate considerations must be made to determine if the advantages of more complex models outweigh their greater financial and logistical costs. An initial approach could include developing an ABM at a restricted yet biologically relevant spatial scale, followed by a comparison of their outputs to those generated from more simplistic, deterministic models as a means of assessing degrees of dissimilarity in how alternative models describe desired landscape conditions. Logically, sophisticated models with outputs only marginally different from simpler models would likely not be worthy of adoption (*e.g.* Goss-Custard *et al.* 2003). Moreover, because conservation plans are implemented by the broader JV community, understanding of key models and model-based conservation recommendations is essential for maximum partner engagement and support.

Finally, the challenges associated with developing and refining better biological models are likely to increase as JVs explore strategies to achieve the integrated goals of the current version of NAWMP (NAWMP 2012), which will likely require investments in human dimensions that have until now been at the periphery of JV activities. Skillful evaluation of the costs and benefits to include financial, ecological and evaluation of alternative carrying capacity models and additional model refinements should invoke prudent decisions about their necessity. Just as the waterfowl management community has applied rigorous science to ensure efficient and effective expenditure of limited conservation dollars, so too should it be the foundation for expenditure of science resources, which are often far more limited.

Acknowledgements

We wish to thank Joe Fleskes, Matt Miller, Samantha Richman, Kevin Ringelman and Jeff Schank for reviews and advice regarding the content of this paper. The use of trade, product, industry or firm names or products is for informative purposes only and does not constitute an endorsement by the U.S. Government or other sponsoring or participating agencies.

References

Albright, J.J., Owen, Jr., R.B. & Corr, P.O. 1983. The effects of winter weather on the behavior and energy reserves of black ducks in Maine. *Transactions of the Northeast Section of the Wildlife Society* 40: 118–128.

Anteau, M.J. & Afton, A.D. 2009. Lipid reserves of lesser scaup (*Aythya affinis*) migrating across a large landscape are consistent with the "spring condition" hypothesis. *The Auk* 126: 873–883.

Baldassarre, G.A. & Bolen, E.G. 1984. Field-feeding ecology of waterfowl wintering in the Southern High Plains of Texas. *Journal of Wildlife Management* 48: 63–71.

Ballard, B.M., Thompson, J.E., Petrie, M.J., Checkett, M. & Hewitt, D.G. 2004. Diet and nutrition of northern pintails wintering along the southern coast of Texas. *Journal of Wildlife Management* 68: 371–382.

Bergan, J.F. & Smith, L.M. 1993. Survival rates of female mallards wintering in the Playa Lakes Region. *The Journal of Wildlife Management* 57: 570–577.

Bousquet, F. & Le Page, C. 2004. Multi-agent simulations and ecosystem management: a review. *Ecological Modelling* 176: 313–332.

Brodsky, L.M. & Weatherhead, P.J. 1985. Time and energy constraints on courtship in wintering American black ducks. *Condor* 87: 33–36.

Brown, J.H. & West, G.B. (eds.) 2000. *Scaling in Biology*. Oxford University Press, Oxford, UK.

Calder, W.A., III. 1984. *Size, Function, and Life History*. Harvard University Press, Cambridge, Massachusetts, USA.

Callicutt, J.T., Hagy, H.M. & Schummer, M.L. 2011. The food preference paradigm: a review of autumn-winter food use by North American dabbling ducks (1900–2009). *Journal of Fish and Wildlife Management* 2: 29–40.

Caswell, H. 2001. *Matrix Population Models*. Second edition. Sinauer, Sunderland, Massachusetts, USA.

Central Valley Joint Venture. 2006. *Central Valley Joint Venture Implementation Plan: Conserving Bird Habitat*. U.S. Fish and Wildlife Service, Sacramento, California, USA.

Checkett, J.M., Drobney, R.D., Petrie, M.J. & Graber, D.A. 2002. True metabolizable energy of moist-soil seeds. *Wildlife Society Bulletin* 30: 1113–1119.

Coluccy, J.M., Yerkes, T., Simpson, R., Simpson, J.W., Armstrong, L. & Davis, J.I. 2008. Population dynamics of breeding mallards in the Great Lakes states. *Journal of Wildlife Management* 72: 1181–1187.

Cramer, D.M., Castelli, P.M., Yerkes, T. & Williams, C.K. 2012. Food resource availability for American black ducks wintering in southern New Jersey. *The Journal of Wildlife Management* 76: 214–219.

Dahl, T.E., Dick, J., Swords, J. & Wilen, B.O. 2009. *Data Collection Requirements and Procedures for Mapping Wetland, Deepwater and Related Habitats of the United States*. Division of Habitat Resource Conservation, National Standards and Support Team, Madison, Wisconsin, USA.

DeAngelis D.L. & Gross, L.J. 1992. Individual-based models and approaches in ecology:

populations, communities and ecosystems. *In* D.L. DeAngelis & L.J. Gross (eds.), *Individual-based Models and Approaches in Ecology*. Chapman & Hall, New York, USA.

DeAngelis, D.L. & Mooij, W.M. 2005. Individual-based modeling of ecological and evolutionary processes. *Annual Review of Ecology, Evolution, and Systematics* 36: 147–168.

de Kroon, H., van Groenendael, J. & Ehrlen, J. 2000. Elasticities: A review of methods and model limitations. *Ecology* 81: 607–618.

Delnicki, D. & Reinecke, K.J. 1986. Mid-winter food use and body weights of mallards and wood ducks in Mississippi. *The Journal of Wildlife Management* 50: 43–51.

Devries, J.H., Armstrong, L.M., MacFarlane, R.J., Moats, L. & Thoroughgood, P.T. 2008. Waterfowl nesting in fall-seeded and spring-seeded cropland in Saskatchewan. *Journal of Wildlife Management* 72: 1790–1797.

Dolman, P.M. & Sutherland, W.J. 1995. The response of bird populations to habitat loss. *Ibis* 137: S38–S46.

Dolman, P.M. & Sutherland, W.J. 1997. Spatial patterns of depletion imposed by foraging vertebrates: Theory, review and meta-analysis. *Journal of Animal Ecology* 66: 481–494.

Dugger, B.D., Moore, M.L., Finger, R.S. & Petrie, M.J. 2007. True metabolizable energy for seeds of common moist-soil plant species. *Journal of Wildlife Management* 71: 1964–1967.

Dugger, B.D., Petrie, M.J. & Mauser, D. 2008. *A Bioenergetic Approach to Conservation Planning for Waterfowl at Lower Klamath and Tule Lake National Wildlife Refuge*. Technical Report submitted to United States Fish and Wildlife Service. US Fish & Wildlife Service, Tulelake, California, USA.

Evans-Peters, G.R. 2010. Assessing biological values of Wetlands Reserve Program wetlands for wintering waterfowl. M.Sc. thesis, Oregon State University, Corvallis, USA.

Evans-Peters G.R., Dugger, B.D., Petrie, M.J. 2012. Plant community composition and waterfowl food production on Wetland Reserve Program easements compared to those on managed public lands in western Oregon and Washington. *Wetlands* 32: 391–399.

Federal Geographic Data Committee. 2008. *Federal Geographic Data Committee Wetlands Inventory Mapping Standard*. FGDC Wetland Subcommittee and Wetland Mapping Standard Workgroup. Reston, Virginia, USA. Accessible at http://www.fgdc.gov/participation/working-groups-subcommittees/wsc/ (last accessed 18 August 2014).

Fleming, K.S., Kaminski, R.M., Tietjen, T.E., Schummer, M.L., Ervin, G.N. & Nelms, K.D. 2012. Vegetative forage quality and moist-soil management on Wetlands Reserve Program lands in Mississippi. *Wetlands* 32: 919–929.

Foster, M.A., Gray, M.J., Harper, C.A. & Walls, J.G. 2010a. Comparison of agricultural seed loss in flooded and unflooded fields on the Tennessee National Wildlife Refuge. *Journal of Fish and Wildlife Management* 1: 43–46.

Foster, M.A., Gray, M.J. & Kaminski, R.M. 2010b. Agricultural seed biomass for migrating and wintering waterfowl in the Southeastern United States. *Journal of Wildlife Management* 74: 489–495.

Fredrickson, L.H. & Reid, F.A. 1988. Nutritional values of waterfowl foods. *Waterfowl Management Handbook – Fish and Wildlife Leaflet 13.1.1*. U. S. Fish and Wildlife Service, Washington D.C., USA.

Gill, J.A., Sutherland, W.J. & Norris K. 2001. Depletion models can predict shorebird distribution at different spatial scales. *Proceedings of the Royal Society Biological Sciences Series B* 268: 369–376.

Goss-Custard, J.D., Clarke, R.T., Briggs, K.B., Ens, B.J., Exo, K.M., Smit, C., Beintema, A.J., Caldow, R.W.G., Catt, D.C., Clark, N.A., Dit

Durell, S.E.A.L.V., Harris, M.P., Hulscher, J.B., Meininger, P.L., Picozzi, N., Prys-Jones, R., Safriel, U.N. & West, A.D. 1995. Population consequences of winter habitat loss in a migratory shorebird. I. Estimating model parameters. *Journal of Applied Ecology* 32: 320–336.

Goss-Custard, J.D., West, A.D., Clarke, R.T., Caldow, R.W.G & Dit Durell, S.E.A.L.V. 1996. The carrying capacity of coastal habitats for oystercatchers. *In* J. D. Goss-Custard (ed.), *The Oystercatcher: from Individuals to Populations*. Oxford Ornithology Series, Oxford University Press, Oxford, UK.

Goss-Custard, J.D., Stillman, R.A., West, A.D., Caldow, R.W.G. & McGrorty, S. 2002. Carrying capacity in overwintering migratory birds. *Biological Conservation* 105: 27–41.

Goss-Custard, J.D., Stillman, R.A., Caldow, R.W.G. & West, A.D. 2003. Carrying capacity in overwintering birds: when are spatial models needed? *Journal of Applied Ecology* 40: 176–187.

Gray, M.J., Kaminski, R.M. & Brasher, M.G. 1999a. A new method to predict seed yield of moist-soil plants. *Journal of Wildlife Management* 63: 1269–1272.

Gray, M.J., Kaminski, R.M., & Weerakkody, G. 1999b. Predicting seed yield of moist-soil plants. *Journal of Wildlife Management* 63: 1261–1268.

Gray, M.J., Foster, M.A. & Peña Peniche, L.A. 2009. New technology for estimating seed production of moist-soil plants. *Journal of Wildlife Management* 73: 1229–1232.

Gray, M.J., Hagy, H.A., Nyman, J.A. & Stafford, J.D. 2013. Management of wetlands for wildlife. *In* C.A. Davis & J.T. Anderson (eds), *Wetland Techniques, Volume 3*. Springer, Secaucus, New Jersey, USA.

Greer, A.K., Dugger, B.D., Graber, D.A. & Petrie, M.J. 2007. The effects of seasonal flooding on seed availability for spring migrating waterfowl. *The Journal of Wildlife Management* 71: 1561–1566.

Greer, D.M., Dugger, B.D., Reinecke, K.J. & Petrie, M.J. 2009. Depletion of rice as food of waterfowl wintering in the Mississippi Alluvial Valley. *Journal of Wildlife Management* 73: 1125–1133.

Grimm, V. 1999. Ten years of individual-based modeling in ecology: what have we learned and what could we learn in the future? *Ecological Modeling* 115: 129–148.

Grimm, V. & Railsback, S.F. 2005. *Individual-Based Modeling and Ecology*. Princeton University Press, New Jersey, USA.

Grimm, V., Revilla, E., Berger, U., Jeltsch, F., Mooij, W.M., Railsback, S.F., Thulke, H.H., Weiner, J., Wiegand, T. & DeAngelis, D.L. 2005. Pattern-oriented modeling of agent based complex systems: lessons from ecology. *Science* 310: 987–991.

Grimm, V., Berger, U., Bastiansen, F., Eliassen, S., Ginot, V., Giske, J., Goss-Custard, J., Grand, T., Heinz, S.K., Huse, G., Huth, A., Jepsen, J.U., Jørgensen, C., Mooij, W.M., Müller, B., Pe'er, G., Piou, C., Railsback, S.F., Robbins, A.M., Robbins, M.M., Rossmanith, E., Rüger, N., Strand, E., Souissi, E., Stillman, R.A., Vabø, R., Visser, U. & DeAngelis, D.L. 2006. A standard protocol for describing individual-based and agent-based models. *Ecological Modelling* 198: 115–126.

Grimm, V., Berger, U., DeAngelis, D.L., Polhill, J.G., Giske, J. & Railsback, J.S.F. 2010. The ODD protocol: A review and first update. *Ecological Modeling* 221: 2760–2768.

Guillemain, M., Elmberg, J., Arzel, C., Johnson, A.R. & Simon, G. 2008. The income-capital breeding dichotomy revisited: late winter body condition is related to breeding success in an income breeder. *Ibis* 150: 172–176.

Hagy, H.M. 2010. Winter food and waterfowl dynamics in managed moist-soil wetlands in the Mississippi Alluvial Valley. Ph.D. thesis. Mississippi State University, Mississippi State, USA.

Hagy, H.M. & Kaminski, R.M. 2012a. Winter waterbird and food dynamics in autumn-managed moist-soil wetlands of the Mississippi Alluvial Valley. *Wildlife Society Bulletin* 36: 512–523.

Hagy, H.M. & Kaminski, R.M. 2012b. Apparent seed use by ducks in the Mississippi Alluvial Valley. *Journal of Wildlife Management* 76: 1053–1061.

Hagy, H.M., Straub, J.N. & Kaminski, R.M. 2011. Estimation and correction of seed recovery bias from moist-soil cores. *Journal of Wildlife Management* 75: 959–966.

Heitmeyer, M.E. 1988. Body composition of female mallards in winter in relation to annual cycle events. *Condor* 90: 669–680.

Heitmeyer, M.E. 2006. The importance of winter floods to mallards in the Mississippi Alluvial Valley. *Journal of Wildlife Management* 70: 101–110.

Heitmeyer, M.E. & Fredrickson, L.H. 1981. Do wetland conditions in the Mississippi Delta hardwoods influence mallard recruitment? *Transactions of the North American Wildlife and Natural Resources Conferences* 46: 44–57.

Hepp, G.R. & Hines, J.E. 1991. Factors affecting winter distribution and migration distance of Wood Ducks from southern breeding populations. *Condor* 93: 884–891.

Hiebert, S.M. & Noveral, J. 2007. Are chicken embryos endotherms or ectotherms? A laboratory exercise integrating concepts in thermoregulation and metabolism. *Advances in Physiology Education*. 31:97–109.

Hine, C.S., Hagy, H.M., Yetter, A.P., Horath, M.M., Smith, R.V. & Stafford, J.D. 2013. *Waterbird and Wetland Monitoring at the Emiquon Preserve: Final Report 2007–2013.* Illinois Natural History Survey Technical Report 2013 (20), Champaign, Illinois, USA.

Hoekman, S.T., Mills, L.S., Howerter, D.W., Devries, J.H. & Ball, I.J. 2002. Sensitivity analyses of the life cycle of midcontinent mallards. *Journal of Wildlife Management* 66: 883–900.

Hoekman, S.T., Gabor, T.S., Petrie, M.J., Maher, R., Murkin, H.R. & Lindberg, M.S. 2006. Population dynamics of Mallards breeding in agricultural environments in eastern Canada. *Journal of Wildlife Management* 70: 121–128.

Hoffman, R.D. & Bookhout, T.A. 1985. Metabolizable energy of seeds consumed by ducks in Lake Erie marshes. *Transactions of the North American Wildlife and Natural Resources Conference* 50: 557–565.

Jones III, O.E. 2012. Constructing a 24 hour time-energy budget for American black ducks wintering in coastal New Jersey. M.Sc. thesis, University of Delaware, Newark, USA.

Jorde, D.G. & Owen, Jr., R.B. 1988. Efficiency of nutrient use by American black ducks wintering in Maine. *Journal of Wildlife Management* 59: 209–214.

Kaminski, R.M. & Essig, H.W. 1992. True metabolizable energy estimates for wild and game-farm mallards. *Journal of Wildlife Management* 56: 321–324.

Kaminski, R.M. & Gluesing, E.A. 1987. Density- and habitat-related recruitment in mallards. *The Journal of Wildlife Management* 51: 141–148.

Kaminski, R.M., Davis, J.B., Essig, H.W., Gerard, P.D. & Reinecke, K.J. 2003. True metabolizable energy for wood ducks from acorns compared to other waterfowl foods. *Journal of Wildlife Management* 67: 542–550.

Karasov, W.H. 1990. Digestion in birds: chemical and physiological determinants and ecological implications. *Studies in Avian Biology* 13: 391–415.

King, J.R. 1973. Energetics of reproduction in birds. *In* D.S. Farner (ed.), *Breeding Biology of Birds.* National Academy of Science, Washington, D.C., USA.

King, J.R. 1974. Seasonal allocation of time and energy resources in birds. *Nuttall Ornithological Club Publication* 15: 4–70.

Kleiber, M. 1932. Body size and metabolism. *Hilgardia* 6: 315–353.

Kleiber, M. 1961. *The Fire of Life. An Introduction to Animal Energetics*. Wiley, New York, USA.

Knight, J.F., Tolcser, B.T., Corcoran, J.M. & Rampi, L.P. 2013. The effects of data selection and thematic detail on the accuracy of high spatial resolution wetland classifications. *Photogrammetric Engineering and Remote Sensing* 79: 613–623.

Korschgen, C.E. & Dahlgren, R.B. 1992. Human disturbances of waterfowl: causes, effects, and management. *Waterfowl Management Handbook No. 13*. U.S. Department of the Interior, Washington D.C., USA.

Kross, J., Kaminski, R.M., Reinecke, K.J., Penny, E.J. & Pearse, A.T. 2008a. Moist-soil seed abundance in managed wetlands in the Mississippi Alluvial Valley. *Journal of Wildlife Management* 72: 707–714.

Kross, J., Kaminski, R.M., Reinecke, K.J. & Pearse, A.T. 2008b. Conserving waste rice for wintering waterfowl in the Mississippi Alluvial Valley. *Journal of Wildlife Management* 72: 1383–1387.

Laubhan, M.K. & Fredrickson, L.H. 1992. Estimating seed production of common plants in seasonally flooded wetlands. *Journal of Wildlife Management* 56: 329–337.

Laundré, J.W., Hernandez, L. & Ripple, W.J. 2010. Landscape of fear: ecological implication of being afraid. *The Open Ecology Journal* 3: 1–7.

Loesch, C.R. & Kaminski, K.M. 1989. Winter body-weight patterns of female mallards fed agricultural seeds. *The Journal of Wildlife Management* 53: 1081–1087.

Lovvorn, J.R. 1994. Nutrient reserves, probability of cold spells and the question of reserve regulation in wintering canvasbacks. *Journal of Animal Ecology* 63: 11–23.

Lovvorn, J.R. & Baldwin, J.R. 1996. Intertidal and farmland habitats of ducks in the Puget Sound region: a landscape perspective. *Biological Conservation* 77: 97–114.

Luke S., Cioffi-Revilla, C., Panait, L., Sullivan, K. & Balan, G. 2005. MASON: A Multi-agent Simulation Environment. *Simulation* 81: 517–527.

Marty, J.R. 2013. Seed and waterbird abundances in ricelands in the Gulf Coast Prairies of Louisiana and Texas. M.Sc. Thesis. Mississippi State University, USA.

Mathevet R, Bousquet, F.C., Le Page, C. & Antona, M. 2003. Agent-based simulations of interactions between duck population, farming decisions and leasing of hunting rights in the Camargue (Southern France). *Ecological Modelling* 165: 107–126.

McKechnie, A.E. 2004. The allometry of avian basal metabolic rate: good predictions need good data. *Physiological and Biochemical Zoology* 77: 502–521.

McKechnie, A.E. 2008. Phenotypic flexibility in basal metabolic rate and the changing view of avian physiological diversity: a review. *Journal of Comparative Physiology B* 178: 235–247.

McKinney, R.R.A. & McWilliams, S.T.R. 2005. A new model to estimate daily energy expenditure for wintering waterfowl. *The Wilson Bulletin* 117: 44–55.

McLane A.J., Semeniuk, C., McDermid, G.J. & Marceau, D.J. 2011. The role of agent-based models in wildlife ecology and management. *Ecological Modelling* 222: 1544–1556.

Miller, M.R. & Eadie, J.M. 2006. The allometric relationship between resting metabolic rate and body mass in wild wintering waterfowl (Anatidae) and an application to estimation of winter habitat requirements. *The Condor* 108:166–177.

Miller, M.R. & Newton, W.E. 1999. Population energetics of northern pintails wintering in the Sacramento Valley, California. *Journal of Wildlife Management* 63: 1222–1238.

Miller, M.R. & Reinecke, K.J. 1984. Proper expression of metabolizable energy in avian energetics. *The Condor* 86: 396–400.

Miller M.L., Ringelman, K.M., Schank, J.C. & Eadie, J.M. 2014. SWAMP: An agent-based model for wetland and waterfowl conservation and management. *Simulation* 90: 52–68.

Moon, J.A. & Haukos, D.A. 2006. Survival of female northern pintails wintering in the Playa Lakes Region of northwestern Texas. *Journal of Wildlife Management* 70: 777–783.

Moon, J.A., Haukos, D.A. & Smith, L.M. 2007. Declining body condition of northern pintails wintering in the Playa Lakes region. *Journal of Wildlife Management* 71: 218–221.

Morehouse, K.A. 1974. Development, energetics and nutrition of captive Pacific Brant (*Branta bernicla orientalis*, Tougarinov). Ph.D. thesis, University of Alaska, Fairbanks, Alaska, USA.

Morton, J.M., Fowler, A.C. & Kirkpatrick, R.L. 1989. Time and energy budgets of American black ducks in winter. *Journal of Wildlife Management* 53: 401–410.

Murkin, H.R., Wrubleski, D.A. & Reid, F.A. 1996. Sampling invertebrates in aquatic and terrestrial habitats. *In* T.A. Bookout (ed.), *Research and Management Techniques for Wildlife and Habitats*. Fifth edition. The Wildlife Society, Bethesda, Maryland, USA.

National Ecological Assessment Team. 2006. *Strategic Habitat Conservation – Final Report of the National Ecological Assessment Team*. U.S. Geological Survey and U.S. Fish and Wildlife Service, Washington D.C., USA.

Naylor, L.W. 2002. Evaluating moist-soil seed production and management in Central Valley Wetlands to determine habitat needs for waterfowl. M.Sc. thesis, University of California, Davis, California, USA.

Naylor, L.W., Eadie, J.M., Smith, W.D., Eichholz, M. & Gray, M.J. 2005. A simple method to predict seed yield in moist-soil habitats. *Wildlife Society Bulletin* 33: 1335–1341.

Nichols, J.D., Reinecke, K.J. & Hines, J.E. 1983. Factors affecting the distribution of mallards wintering in the Mississippi Alluvial Valley. *Auk* 100: 932–946.

Nolet, B.A., Langevoord, O., Bevan, R.M., Engelaar, K.R., Klaassen, M., Mulder, R.J.W. & Van Dijks, S. 2001. Spatial variation in tuber depletion by swans explained by difference in net intake rates. *Ecology* 82: 1655–1667.

Nolet, B.A., Gyimesi, A. & Klaassen, R.H. 2006. Prediction of bird-day carrying capacity on a staging site: a test of depletion models. *Journal of Animal Ecology* 75: 1285–1282.

North American Waterfowl Management Plan Committee. 2012. *North American Waterfowl Management Plan: People Conserving Waterfowl and Wetlands*. U.S. Fish and Wildlife Service, Washington D.C., USA.

Nudds, T.D. & Kaminski, R.M. 1984. Sexual size dimorphism in relation to resource partitioning in North American dabbling ducks. *Canadian Journal of Zoology* 62: 2009–2012.

Olmstead, V.G. 2010. Evaluation of management strategies on moist-soil seed availability and depletion on Wetland Reserve Program sites in the Mississippi Alluvial Valley. M.Sc. thesis, Arkansas Tech University, Russellville, USA.

Olmstead, V.G., Webb, E.B. & Johnson, R.W. 2013. Moist-soil seed biomass and species richness on Wetland Reserve Program Easements in the Mississippi Alluvial Valley. *Wetlands* 33: 197–206.

Pacific Coast Joint Venture. 2004. *Pacific Coast Joint Venture: Coastal Northern California Component: Strategic Plan*. U.S. Fish and Wildlife Service, Sacramento, California, USA.

Paulus, S.L. 1988. Time-activity budgets of nonbreeding Anatidae: a review. *In* M. W. Weller (ed.), *Waterfowl in Winter*, pp. 135–165 University of Minnesota Press. Minneapolis, USA.

Pearse, A.T. & Stafford, J.D. 2014. Error propagation in energetic carrying capacity models. *Journal of Conservation Planning* 10: 17–24.

Pearse, A.T., Kaminski, R.M., Reinecke, K.J. & Dinsmore, S.J. 2012. Local and landscape associations between wintering dabbling ducks and wetland complexes in Mississippi. *Wetlands* 32: 859–869.

Peters, R.H. 1983. *The Ecological Implications of Body Size*. Cambridge University Press, Cambridge, UK.

Petrie, M.J. 1994. True metabolizable energy of waterfowl foods. M.Sc. thesis, University of Missouri, Columbia, USA.

Petrie, M.J., Drobney, R.D. & Graber, D.A. 1998. True metabolizable energy estimates of Canada goose foods. *Journal of Wildlife Management* 62: 1147–1152.

Petrie, M. J., Brasher, M.G., Soulliere, G.J., Tirpak, J.M., Pool, D.B. & Reker, R.R. 2011. *Guidelines for Establishing Joint Venture Waterfowl Population Abundance Objectives*. North American Waterfowl Management Plan Science Support Team Technical Report No. 2011-1, U.S. Fish and Wildlife Service, Washington D.C., USA.

Pettifor, R.A., Caldow, R.W.G., Rowcliffe, J.M., Goss-Custard, J.D., Black, J.M., Hodder, K.H., Houston, A.I., Lang, A. & Webb, J. 2000. Spatially explicit, individual-based, behavioural models of the annual cycle of two migratory goose populations. *Journal of Applied Ecology* 37: 103–135.

Prince, H.H. 1979. Bioenergetics of postbreeding dabbling ducks. *In* T. A. Bookhout (ed.), *Waterfowl and Wetlands – An Integrated Review*. Proceedings of the 1977 Symposium of the North Central Section of The Wildlife Society. The Wildlife Society, Madison, Wisconsin, USA.

Proctor, J.R. & Marks, C.F. 1974. The determination of normalizing transformations for nematode count data from soil samples and of efficient sampling schemes. *Nematologica* 20: 395–406.

Railsback, S.F & Grimm, V. 2012. *Agent-based and Individual-based Modeling: A Practical Introduction*. Princeton University Press, New Jersey, USA.

Raveling, D.G. & Heitmeyer, M.E. 1989. Relationships of population size and recruitment of pintails to habitat conditions and harvest. *Journal of Wildlife Management* 53: 1088–1103.

Reinecke, K.J. & Hartke, K.M. 2005. Estimating moist-soil seeds available to waterfowl with double sampling for stratification. *Journal of Wildlife Management* 69: 794–799.

Reinecke, K.J, Kaminski, R.M., Moorhead, D.J., Hodges, J.D. & Nassar, J.R. 1989. Mississippi Alluvial Valley. *In* L.M. Smith, R.L. Pederson & R.M. Kaminski (eds.), *Habitat Management for Migrating and Wntering Waterfowl in North America*. Texas Tech University Press, Lubbock, USA.

Richman, S.E. & Lovvorn, J.R. 2011. Effects of air and water temperatures on resting metabolism of auklets and other diving birds. *Physiological and Biochemical Zoology* 84: 316–332.

Ringelman, K.M. 2014. Predator foraging behavior and patterns of avian nest success: What can we learn from an agent-based model? *Ecological Modelling* 272: 141–149.

Sæther, B.E., Lillegård, M., Grøtan, V., Drever, M.C., Engen, S., Nudds, T. & Podruzny, K.M. 2008. Geographical gradients in the population dynamics of North American prairie ducks. *Journal of Animal Ecology* 77: 869–882.

Santamaría, L. & Rodríguez-Gironés, M.A. 2002. Hiding from swans: optimal burial depth of sago pondweed tubers foraged by Bewick's swans. *Journal of Ecology* 90: 303–315.

Schmidt-Nielsen, K. 1984. *Scaling: Why is Animal Size so Important?* Cambridge University Press, Cambridge, UK.

Schroth, G. & Kolbe, D. 1994. A method of processing soil core samples for root studies by subsampling. *Biology and Fertility of Soils* 18: 60–62.

Servello, F.A., Hellgren, E.C. & McWilliams, S.R. 2005. Techniques for wildlife nutritional ecology. *In* C.E. Braun (ed.), *Research and Management Techniques for Wildlife and Habitats.* The Wildlife Society, Washington D.C., USA.

Sherfy, M.H. 1999. Nutritional value and management of waterfowl and shorebird foods in Atlantic coastal moist-soil impoundments. Ph.D. thesis, Virginia Polytechnic Institute and State University, Blacksburg, Virginia, USA.

Sherfy, M.H., Kirkpatrick, R.L. & Richkus, K.D. 2000. Benthos core sampling and Chironomid vertical distribution: implications for assessing shorebird food availability. *Wildlife Society Bulletin* 28: 124–130.

Sherfy, M.H., Kirkpatrick, R.L. & Webb, Jr., K.E. 2001. Nutritional consequences of gastolith ingestion in blue-winged teal: a test of the hard-seed-for-grit hypothesis. *Journal of Wildlife Management* 65: 406–414.

Shirkey, B.T. 2012. Diving duck abundance and distribution on Lake St. Clair and western Lake Erie. M.Sc. thesis, Michigan State University, East Lansing, USA.

Sibbald, I.R. 1975. The effect of level of feed intake on metabolizable energy values measured with adult roosters. *Poultry Science* 54: 1990–1997.

Sibbald, I.R. 1979. Effects of level of feed input, dilution of test material, and duration of excreta collection on true metabolizable energy values. *Poultry Science* 58: 1325–1329.

Smith, R.V., Stafford, J.D., Yetter, A.P., Whelan, C.J., Hine, C.S. & Horath, M.M. 2011. Foraging thresholds of spring migrating dabbling ducks in central Illinois. *Illinois Natural History Survey Technical Report* 2011(No. 37), Champaign, Illinois, USA.

Smith, R.V., Stafford, J.D., Yetter, A.P., Horath, M.M., Hine, C.S. & Hoover, J.P. 2012. Foraging ecology of fall-migrating shorebirds in the Illinois River valley. *PLoS ONE* 7(9):e45121.

Soulliere, G.J., Potter, B.A., Coluccy, J.M., Gatti, R.C., Roy, C.L., Luukkonen, D.R., Brown, P.W. & Eichholz, M.W. 2007. *Upper Mississippi River and Great Lakes Region Joint Venture Waterfowl Habitat Conservation Strategy.* U.S. Fish and Wildlife Service, Fort Snelling, Minnesota, USA.

Soulliere, G.J., Loges, B.W., Dunton, E.M., Luukkonen, D.R., Eichholz, M.E. & Koch, K.E. 2013. Monitoring Midwest waterfowl during the non-breeding period: priorities, challenges and recommendations. *Journal of Fish and Wildlife Management* 4: 395–405.

St. James, E.A. 2011. Effect of hunting frequency on duck abundance, harvest and hunt quality in Mississippi. M.Sc. thesis, Mississippi State University, USA.

St. James, E.A., Schummer, M.L., Kaminski, R.M., Penny, E.J., Burger, L.W. 2013. Effect of weekly hunting frequency on duck abundances in Mississippi Wildlife Management Areas. *Journal of Fish and Wildlife Management* 4: 144–150.

Stafford, J.D., Kaminski, R.M., Reinecke, K.J. & Manley, S.W. 2006. Waste rice for waterfowl in the Mississippi Alluvial Valley. *Journal of Wildlife Management* 70: 61–69.

Stafford, J.D., Yetter, A.P., Hine, C.S., Smith, R.V. & Horath, M.M. 2011. Seed abundance for waterfowl in wetlands managed by the Illinois Department of Natural Resources. *Journal of Fish and Wildlife Management* 2: 3–11.

Stillman R.A. 2008. MORPH – an individual-based model to predict the effect of environmental change on foraging animal populations. *Ecological Modelling* 216: 265–276.

Stillman R.A. & Goss-Custard, J.D. 2010. Individual-based ecology of coastal birds. *Biological Reviews* 85: 413–434.

Stillman, R.A., Goss-Custard, J.D., West, A.D., Durell, S.E.A.L.V., McGrorty, D.S., Caldow, R.W.G., Norris, K.J., Johnstone, I.G., Ens, B.J., van der Meer, J. & Triplet, P. 2001. Predicting shorebird mortality and population size under different regimes of shellfishery management. *Journal of Applied Ecology* 38: 857–868.

Straub, J.N., Gates, J.L., Schultheis, R.D., Yerkes, T., Coluccy, J.M. & Stafford, J.D. 2012. Wetland food resources for spring-migrating ducks in the upper-Mississippi River and Great Lakes Region. *Journal of Wildlife Management* 76: 768–777.

Sutherland, W.J. 1998. The effect of local change in habitat quality on populations of migratory species. *Journal of Applied Ecology* 35: 418–421.

Sutherland, W.J. & Allport, G.A. 1994. A spatial depletion model of the interaction between bean geese and wigeon with the consequences for habitat management. *Journal of Animal Ecology* 63: 51–59.

Swanson, G.A. 1978. A water column sampler for invertebrates in shallow wetlands. *Journal of Wildlife Management* 42: 670–672.

Swanson, G.A. 1983. Benthic sampling for waterfowl foods in emergent vegetation. *Journal of Wildlife Management* 47: 821–823.

Synchra, J. & Adamek, Z. 2010. Sampling efficiency of the Gerking sampler and sweep net in pond emergent littoral macrophyte beds – a pilot study. *Turkish Journal of Fisheries and Aquatic Sciences* 10: 161–167.

Tidwell, P.R., Webb, E.B., Vrtiska, M.P., Bishop, A.A. 2013. Diets and food selection of female mallards and blue-winged teal during spring migration. *Journal of Fish and Wildlife Management* 4: 63–74.

Thomas, D.R. 2004. Assessment of Waterfowl Body Condition to Evaluate the Effectiveness of the Central Valley Joint Venture. M.Sc. Thesis. University of California, Davis, California, USA.

U.S. Fish and Wildlife Service's Migratory Bird Data Center (USFWS MBDC). 2014. *Midwinter Waterfowl Index Database*. U.S. Fish and Wildlife Service, Patuxent, Maryland, USA. Accessible at http://mbdcapps.fws.gov/ (last accessed 14 March 2014).

van Gils, J.A., Edelaar, P., Escudero, G. & Piersma, T. 2004. Carrying capacity models should not use fixed prey density thresholds: a plea for using more tools of behavioural ecology. *Oikos* 104: 197–204.

Weathers, W.W. 1979. Climate adaptations in avian standard metabolic rate. *Oecologia* 42: 81–89.

Weathers, W.W., Buttemer, W.A., Hayworth, A.M. & Nagy, K.A. 1984. An evaluation of time-budget estimates of daily energy expenditure in birds. *Auk* 101: 459–472.

West A.D., Stillman, R.A., Drewitt, A., Frost, N.J., Mander M. & Miles, C. 2011. MORPH – a user-friendly individual-based model to advise shorebird policy and management. *Methods in Ecology and Evolution* 2: 95–98.

Wooley, Jr., J.B. 1976. Energy expenditure of the black duck under controlled and free-living conditions. M.Sc. thesis, University of Maine, Orono, USA.

Annual variation in food densities and factors affecting wetland use by waterfowl in the Mississippi Alluvial Valley

HEATH M. HAGY[1]*, JACOB N. STRAUB[2],
MICHAEL L. SCHUMMER[3,5] & RICHARD M. KAMINSKI[4]

[1]*Illinois Natural History Survey, Forbes Biological Station – Bellrose Waterfowl Research Center, University of Illinois at Urbana-Champaign, Havana, Illinois 62644, USA.
[2]State University of New York at Plattsburgh, Center for Earth and Environmental Science, Plattsburgh, New York 12901, USA.
[3]Long Point Waterfowl, PO Box 160, Port Rowan, Ontario N0E 1M0, Canada.
[4]Mississippi State University, Department of Wildlife, Fisheries and Aquaculture, Box 9690, Mississippi State, Mississippi 39762, USA.
[5]Current address: State University of New York at Oswego, Department of Biological Sciences, 30 Centennial Drive, Oswego, New York 13126, USA.
*Correspondence author. E-mail: hhagy@illinois.edu

Abstract

Spatial and temporal heterogeneity in habitat quantity and quality, weather and other variables influence the production of food and the distribution of waterfowl, making it difficult to predict carrying capacity accurately. Food densities for waterfowl, which are key parameters of energetic carrying capacity models, were examined in managed moist-soil wetlands and bottomland hardwood forests in or near the Mississippi Alluvial Valley (MAV) of the southern United States of America, to determine variation in those densities across wetlands and years. Secondly, the relationship between migratory waterfowl density in managed wetlands and local and mid-latitude factors north of the study area was examined to identify mechanisms influencing waterfowl density at latitudes used during winter. At individual wetlands and within years, food densities were highly variable, but coefficients of variation (CV) at the scale of the MAV and nearby areas across years were relatively low (moist-soil CV = 21%, bottomland hardwood forest CV = 11%). Local precipitation was inversely related to waterfowl density in managed moist-soil wetlands, and this relationship was stronger than other local and mid-latitude factors including weather severity and temperature. Our data suggest that simplistic daily ration models may reasonably incorporate fixed estimates of food density for managed moist-soil wetlands and bottomland hardwood forests to predict energetic carrying capacity of waterfowl habitat at the scale of the MAV across multiple years. However, substantial variation in food densities among locations and time periods likely limits the utility and accuracy of these models when scaled down temporally or spatially. Therefore, the challenge in predicting annual carrying capacity for waterfowl in the MAV likely depends less on precisely estimating food densities at the scale of individual wetlands and more on determining spatial and temporal availability of habitats that contain food resources for waterfowl.

Key words: bottomland hardwood forest, conservation planning, dabbling duck, daily ration model, migration, moist-soil, weather severity.

Waterfowl ecologists use predictive models to develop habitat conservation objectives sufficient to meet the energy demands of migrating and wintering waterfowl populations in North America and elsewhere (Soulliere *et al.* 2007; Reinecke *et al.* 1989). These models require estimates of food density and foraging demand, with the latter reflecting the number and duration of stay of individuals foraging in a given area (Williams *et al.* 2014). Many previous studies have aimed to measure the density and availability of food in habitats used by waterfowl, but few researchers have examined variability in food densities at multiple temporal and spatial scales (*cf.* Lovvorn & Gillingham 1996). Moreover, variation or changes in regional and habitat use by waterfowl can have significant effects on carrying capacity model predictions of habitat requirements (Hagy *et al.* 2014).

Food availability for waterfowl can be influenced by a range of factors including annual production and seasonal decomposition of plant and animal foods, depletion of food resources by wildlife other than waterfowl, diet selectivity by foragers, ice and snow cover, duration and depth of flooding, disturbance by humans and natural predators and photoperiodic cues triggering migration (Rees 1981; Newton 1998; Schummer *et al.* 2010; Hagy & Kaminski 2012a,b). Moreover, even if available in some parts of the migratory range, foods may not be encountered by waterfowl because of variation within and between years in the timing of migration and regional movements by waterfowl (Bellrose *et al.* 1979; Schummer *et al.* 2010; Krementz *et al.* 2011, 2012; O'Neal *et al.* 2012; Hagy *et al.* 2014). Currently, energetic carrying capacity models used by some Joint Ventures of the North American Waterfowl Management Plan and other conservation partners include fixed parameters (*i.e.* constants) which may not account for spatio-temporal variation in food density or other factors that result in a mismatch of foods becoming available and waterfowl being present to access those foods (Soulliere *et al.* 2007; Williams *et al.* 2014). In order to develop habitat conservation objectives effectively, conservation planners require an understanding of variation in carrying capacity estimates resulting from variable parameter estimates and the mechanisms underlying waterfowl habitat use to determine priority habitats for conservation (Schummer *et al.* 2010; Hagy & Kaminski 2012b; Beatty *et al.* 2014b; Hagy *et al.* 2014; Williams *et al.* 2014).

Recently, information has become

available to help explain waterfowl movements and habitat use during winter in relation to landscape composition (Pearse *et al.* 2012; Beatty *et al.* 2014b). In addition to landscape-scale factors measured close to wintering and stopover sites used by large numbers of waterfowl, factors north of wintering areas could affect the southward movement of individuals and these may be useful in further explaining and predicting wetland use (Schummer *et al.* 2010). For migratory species such as waterfowl, habitat selection is likely a hierarchical process and factors affecting selection may vary temporally and interact spatially (Beatty *et al.* 2014a,b). For example, the cumulative effects of decreasing temperatures, freezing of wetlands and snow cover can cause regional decreases in abundance of waterfowl at autumn staging areas (Schummer *et al.* 2010), but once birds reach their southern wintering grounds where harsh weather conditions are less common, other factors such as precipitation (which influences wetland availability), food availability and intrinsic wetland factors may influence habitat use (Davis *et al.* 2009; Hagy & Kaminski 2012b; Dalby *et al.* 2013). Factors related to migratory movements from mid-latitude areas to more southerly wintering grounds may influence the abundance of birds within southern areas and allow comparison of the relative influence of local and mid-latitude factors on site use by the birds.

Although a number of studies have examined waterfowl movements and abundance in relation to food, habitat juxtaposition and other factors influencing movements along the migration route, there is a need for studies that simultaneously consider factors within wintering areas and those occurring at latitudes north of wintering areas, which may cause movements of birds into wintering areas and influence their access to foods, subsequent fitness and conservation planning (Lovvorn & Baldwin 1996; Haig *et al.* 1998; Pearse *et al.* 2012). The annual variation in food density in managed moist-soil wetlands and bottomland hardwood forests in and near the Mississippi Alluvial Valley (MAV), an important wintering area for North American waterfowl at the continental level (Reinecke *et al.* 1989), was examined to determine variation in parameter estimates used in energetic carrying capacity models at two spatial scales (within and across study wetlands in the MAV) and across years. Secondly, the relative influence of factors measured not only locally but also at a mid-latitude location on migratory waterfowl densities in managed wetlands was investigated. Our objectives were to: 1) describe variation in food densities across wintering areas used by migrating waterfowl during autumn and winter in the MAV, and 2) determine factors that influence waterfowl densities on managed wetlands to better inform the conservation planning process.

Methods

Variation in food abundance

Waterfowl density and associated food densities were estimated in moist-soil wetlands and food density was also estimated in bottomland hardwood forests in or near the MAV (Reinecke *et al.* 1989;

Hagy & Kaminski 2012a,b). Data presented here and related sample collection methods have previously been described in detail by Hagy and Kaminski (2012b) and Straub (2012), but different analyses were conducted to address our novel objectives. Moist-soil and bottomland forest wetlands are used extensively by many species of waterfowl, especially dabbling ducks (*Anas* sp.), for provision of food resources and other life-history needs (Reinecke *et al.* 1989). Moist-soil wetlands provide abundant natural seeds and tubers after they are flooded, which typically occurs in late autumn or early winter, and bottomland hardwood wetlands provide hard mast (*e.g.* acorns) throughout winter which typically become available during periodic bottomland flooding events (Reinecke *et al.* 1989; Hagy & Kaminski 2012b; Straub 2012). To estimate annual densities of sound red oak *Quercus* sp. acorns, we installed and checked 1-m^2 seed traps monthly from November through February 2009–2013 at five study wetlands across the MAV (Straub 2012). Additionally, to estimate moist-soil seed and tuber densities, we collected 10 benthic core subsamples during November or December (*i.e.* before most wintering waterfowl accessed wetlands and depleted foods) in 2006–2008 at each of three wetland plots immediately following flooding that had been either: 1) mown, 2) disced, or 3) not manipulated during the autumn prior to flooding, for 22 wetlands in or near the MAV (2006 = 6 wetlands, 2007 = 9 wetlands, 2008 = 7 wetlands). Laboratory processing followed Hagy and Kaminski (2012b) and Straub (2012); seed and tuber densities were adjusted for seeds lost, missed or destroyed during processing (Hagy *et al.* 2011), seeds and tubers of plant taxa thought to be avoided or infrequently consumed by waterfowl were removed from density estimates (Hagy & Kaminski 2012a) and coefficients of variation (CV) were estimated for each wetland (CV = s.d./mean density for all foods combined in non-manipulated wetland plots), across wetlands for each year, and across years for both moist-soil wetlands and bottomland hardwood forests.

Local and mid-latitude factors affecting waterfowl density

Waterfowl were enumerated by species from elevated hides during crepuscular periods 2–3 times weekly at each wetland plot from first flooding (*i.e.* November–December) through to waterfowl leaving the wetlands and surrounding area (*i.e.* late February), during winters 2006–2009 (Hagy & Kaminski 2012b). We combined densities of all dabbling duck species observed and analysed only this variable as dabbling ducks comprised >90% of observations and their densities were positively correlated with densities of all waterbirds combined (Hagy & Kaminski 2012b).

To accomplish our goal of investigating factors affecting waterfowl densities in wetlands in the study area, we examined the influence of weather and habitat variables measured within or near wetlands in the MAV (*i.e.* local) and weather variables measured *c.* 200 km north of our study area (*i.e.* mid-latitude) where large concentrations of waterfowl may winter if they are not encouraged to migrate further south by weather or other factors (Schummer *et al.*

2010). Concurrent assessment of local wintering area and mid-latitude variables addresses the hypothesis that waterfowl densities might be influenced by extrinsic factors (*e.g.* a weather severity index (WSI); Schummer *et al.* 2010), and these events may exert a greater influence than local conditions in southerly areas (*i.e.* factors near and within the southerly wetlands) to which the birds migrate. Weather data from central Missouri, which is in the northern part of the typical wintering range for Mallard *Anas platyrhynchos* following the Mississippi Flyway, were assumed to provide a reasonable representation of the effects of weather on duck movements from northern staging and wintering areas into the MAV, including our study wetlands. Weather data (*i.e.* WSI, cumulative precipitation during winter (October 1 to observation date), and mean daily temperature) from the closest weather station and water depth gauge readings from within each wetland were used to evaluate local influences on waterfowl densities. Water depths and weather data were recorded on the same day as waterfowl densities. For more northern (mid-latitude) parts of the wintering range, the same weather variables were acquired. Data were acquired from the Historical Climatology Network National Oceanographic Atmospheric Service weather stations at Farmington, Missouri (mid-latitude) and also from weather stations closest to (≤ 50 km from) study wetlands (at Batesville, Corinth, Greenwood, Starkville and Yazoo City in Mississippi; Covington, Jackson and Union City in Tennessee).

Linear mixed models were used in R (nlme; Pinheiro *et al.* 2014) to assess variation in dabbling duck density across managed moist-soil wetlands in relation to various explanatory variables. A set of candidate models (Table 1), each representing a unique biologically-plausible scenario, was built and models were compared for explanatory support using Akaike's information criterion adjusted for small sample size (AIC$_c$; Burnham & Anderson 2002). Because study wetlands were repeatedly sampled within a season, observation date was included as a repeated effect in the model. Additionally, the management category for each wetland plot (*i.e.* mow, disc or no manipulation) nested within wetland was designated as a random effect because evidence (*i.e.* lowest AIC$_c$) suggested that this increased the explanatory power of our global model (Zuur *et al.* 2009). Models were developed to assess support for mid-latitude factors, local factors, a combination of both local and mid-latitude factors, the effects of year and date of surveys, and finally a null model containing only the intercept. Inspection of residual plots and histograms indicated that dabbling duck density (*i.e.* the response variable) was not normally distributed and had heterogeneous variance when plotted against independent variables. Dabbling duck density therefore was natural log transformed prior to analysis. Parameter estimates from the most parsimonious model were back-transformed to describe the size of each effect. We provide marginal and conditional R^2 statistics as a means to assess the fit of each candidate model (Nakagawa & Schielzeth 2013). Marginal R^2 describes the proportion of variance

Table 1. Results of linear mixed models predicting dabbling duck density in managed moist-soil wetlands in or near the Mississippi Alluvial Valley during late autumn and winter 2006–2009, with the difference between each model-specific Akaike Information Criteria adjusted for small sample size (ΔAIC_c) and that of the top model. Model variables include local (LO) and mid-latitude (MO) estimates of cumulative winter precipitation (PrecipW), water depth (Depth), a weather severity index (WSI), temperature (Temp), year, management practice (autumn mowing, discing, or no management; Treat) and Julian day.

Model	AIC_c	ΔAIC_c	R^2_{marg}	R^2_{cond}
LOPrecipW+LODepth	2248.3	0	0.04	0.34
LOPrecipW + LODepth + MOPrecipW	2254.6	6.3	0.05	0.34
LOTemp+LODepth+LOPrecipW	2257.4	9.1	0.04	0.35
LOPrecipW + LODepth + MOWSI	2257.5	9.2	0.04	0.35
LODepth + MOPrecipW	2262.4	14.1	0.00	0.44
LOPrecipW + LOTemp + LODepth + MOWSI	2263.5	15.2	0.04	0.36
LOPrecipW + LOTemp + LODepth + MOPrecipW	2263.6	15.3	0.05	0.34
LOPrecipW + LODepth + MOWSI + MOPrecipW	2263.7	15.4	0.05	0.34
LOTemp + LOPrecipW + LODepth + MOWSI + MOPrecipW	2269.3	21.0	0.05	0.35
LODepth + MOWSI + MOPrecipW	2271.8	23.5	0.00	0.45
LOTemp + LODepth + MOWSI + MOPrecipW	2276.1	27.8	0.00	0.44
LODepth	2286.9	38.6	0.00	0.44
LOTemp+LODepth	2294.9	46.6	0.00	0.41
LODepth + MOWSI	2297.2	48.9	0.00	0.45
LOPrecipW	2407.0	158.7	0.03	0.36
LOPrecipW+LOTemp	2415.7	167.4	0.03	0.36
LOPrecipW + MOWSI	2416.6	168.3	0.03	0.36
MOPrecipW	2417.9	169.6	0.00	0.44
LOPrecipW + MOWSI + MOPrecipW	2420.8	172.5	0.04	0.35
MOPrecipW + MOTemp	2427.3	179.0	0.00	0.45
MOPrecipW + MOWSI	2427.5	179.2	0.00	0.45
Treat	2449.7	201.4	0.07	0.29
Null (Intercept)	2462.0	213.7	0.00	0.43
Year	2468.9	220.6	0.01	0.39
Julian day	2469.9	221.6	0.02	0.36
LOTemp	2470.5	222.2	0.00	0.42
MOTemp	2472.7	224.4	0.00	0.44
MOWSI	2472.8	224.5	0.00	0.45

explained by the fixed factor(s) while conditional R^2 describes the proportion of variance explained by both the fixed and random factors (Nakagawa & Schielzeth 2013).

Additionally, a general linear mixed model was used in SAS 9.3 to evaluate the effects of year and wetland on food density in managed moist-soil wetlands in late autumn (*i.e.* approximately early November, before waterfowl used wetlands) by performing a different analysis of data presented by Hagy and Kaminski (2012b). Year and wetland were included as fixed effects and wetland management practice (*i.e.* discing, mowing or no manipulation of robust moist-soil vegetation; see Hagy & Kaminski 2012b) within each wetland plot was included as a random effect. The response variable (food density) was natural log transformed to normalize residuals and homogeneity of variances among years and wetlands. Results were considered significant at $P < 0.05$.

Results

For managed moist-soil, CVs for seed and tuber densities ranged from 9–77% within wetlands ($\bar{x} = 31\%$, $n = 22$) and from 32–115% across years ($\bar{x} = 69\%$, $n = 3$). Overall, the CV of the annual mean seed density across years and wetlands was less (CV = 21%) than within years or most wetlands (Fig. 1). For bottomland hardwood forests, CVs for red Oak acorn densities ranged from 16–60% within wetlands ($\bar{x} = 33\%$ $n = 5$) and from 11–29% across wetlands ($\bar{x} = 18\%$, $n = 4$). Overall, the CV of the annual mean acorn density across

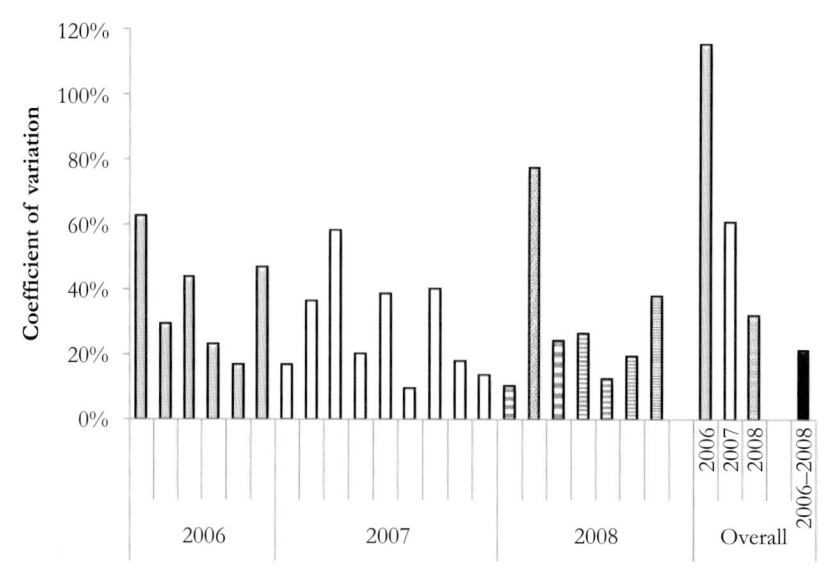

Figure 1. Coefficients of variation (%) for means of seed and tuber density for individual wetland plots, across wetland plots within years, and across wetland plots and years (overall) during late autumn 2006–2008 in managed moist-soil wetlands ($n = 22$ unmanipulated moist-soil wetland plots) in or near the Mississippi Alluvial Valley.

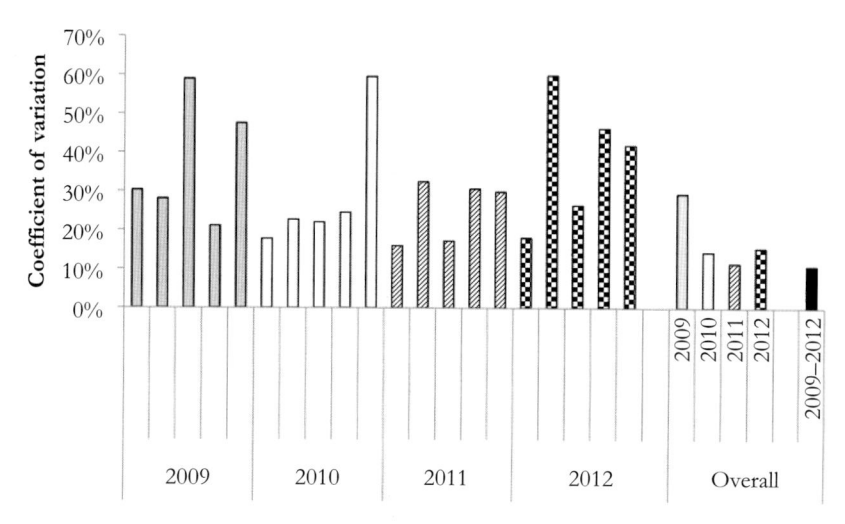

Figure 2. Coefficients of variation (%) for means of red Oak acorn production density for individual wetlands, across wetlands within years, and across wetlands and years (overall) during late autumn and winter 2009–2012 in bottomland hardwood forests (*n* = 5 wetlands surveyed repeatedly) in or near the Mississippi Alluvial Valley.

years and wetlands was less (CV = 11%) than individual years or most wetlands (Fig. 2). Food densities in managed moist-soil wetlands in late autumn varied by wetland ($F_{16,32}$ = 6.56, P < 0.001) but did not differ among years ($F_{2,13}$ = 2.01, P = 0.174) (Fig. 3).

On evaluating the explanatory variables thought to influence dabbling duck density, the top model contained factors measured only at the local scale (Table 1). Dabbling duck densities in managed moist-soil wetlands were negatively associated with local winter precipitation (*i.e.* an assumed correlate of local wetland availability) and positively associated with mean water depth of wetland plots, although confidence intervals associated with the beta estimate overlapped zero for water depth and thus we did not explore that relationship further (Fig. 4). A 9.6 cm increase in local precipitation during winter decreased predicted numbers of dabbling ducks in managed moist-soil wetlands by 1 duck/ha. Models containing only mid-latitude or mid-latitude plus local factors were not competitive (ΔAIC_c > 6.3). Although we had low model uncertainty, the proportion of the variance explained by depth and local winter precipitation (R^2 = 0.04) was less than the variance explained by the combination of fixed effects and random variables (R^2 = 0.34).

Discussion

Small-scale spatial or temporal estimates of food density (*e.g.* among wetlands or years) in managed moist-soil wetlands and bottomland hardwood forests were highly variable and means from these estimates were relatively imprecise, which is consistent with other studies in similar habitats

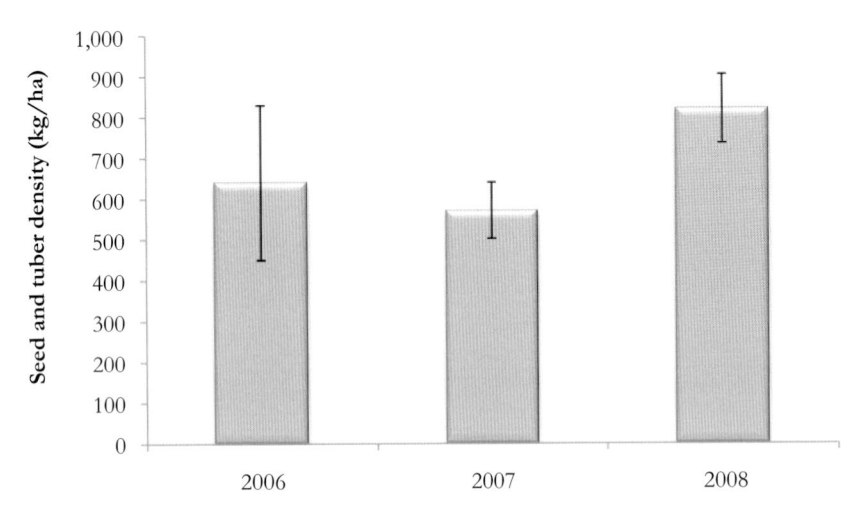

Figure 3. Seed and tuber density (dry weight, in kg/ha ± s.e.) during late autumn 2006–2008 in managed moist-soil wetlands (n = 64 wetland plots) in or near the Mississippi Alluvial Valley (data from Hagy & Kaminski 2012b).

(Stafford et al. 2006; Kross et al. 2008; Evans-Peters et al. 2012; Straub 2012; Olmstead et al. 2013). However, coefficients of variation were relatively low and means were similar when estimated across wetlands and years. Thus, use of fixed food densities in simplistic daily ration models for conservation planning purposes at a large spatial scale appears to be a reasonable practice. At the MAV scale and across the years of the study, food densities in managed moist-soil wetlands and bottomland hardwood forests were relatively constant; however, the spatial distribution of that food within the region varied annually. Because waterfowl are highly mobile and respond to changing habitat conditions, they are likely able to move within the landscape and respond to changing distributions of food and habitat availability. In fact, many factors other than food density likely influence habitat use, and direct relationships between food resources and waterfowl distribution seem to be difficult to detect without incorporating additional local environmental conditions into habitat models (Fleming 2010; Tapp 2013; Weegman 2013).

Hagy and Kaminski (2012b) presented data indicating that early winter food densities varied with management practice in moist-soil wetlands; they considered year and wetland as random effects relative to their research questions, but did not test them explicitly. Herein, re-examination of their data across years indicated that while waterfowl food densities varied by wetland, there was not an apparent annual difference in their study area (Fig. 3). The variation between individual wetlands was much greater than across years. Similarly, Straub (2012) reported that acorn densities in bottomland hardwood forests fluctuated greatly across years for individual wetlands;

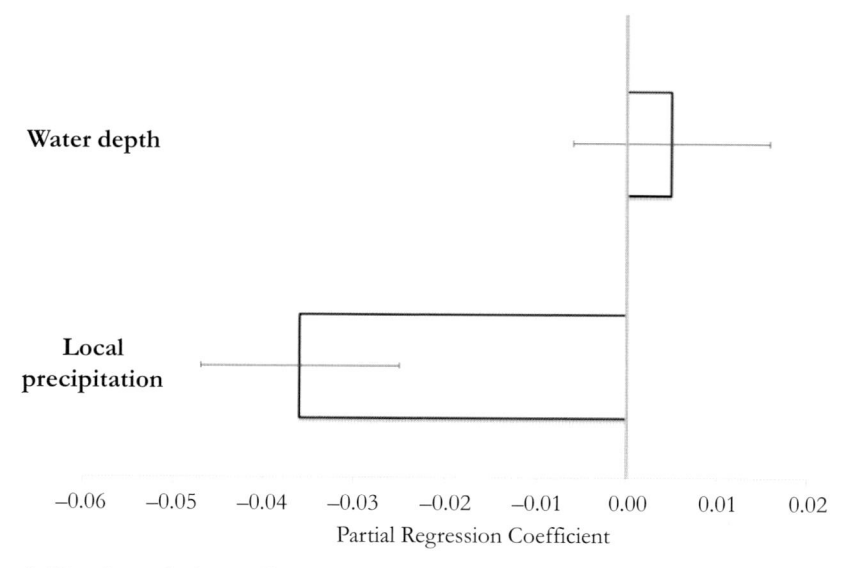

Figure 4. Direction and relative effect size (partial regression coefficient with 95% confidence intervals) for variables in the top model predicting dabbling duck densities during late autumn 2006–2008 in managed moist-soil wetlands (n = 64 wetland plots) in or near the Mississippi Alluvial Valley.

however, at the scale of the MAV, annual estimates were similar. Although some bottomland hardwood forests produced few acorns in some years, low yield never occurred at all wetlands in the same year. Across years, MAV-wide estimates of red Oak acorn abundance were precise (CV = 11%), but variability across wetlands was great (Fig. 2). Stafford *et al.* (2006) and Kross *et al.* (2008) both reached similar conclusions for seeds in rice fields and moist-soil wetlands, respectively, in the MAV. Thus, while daily ration models incorporating fixed food densities may reasonably predict carrying capacity at large spatial scales, with all other parameters being equal, substantial variation among locations and time periods likely limits the predictive accuracy of these models when scaled down spatially.

In managed moist-soil wetlands (Kross *et al.* 2008), bottomland hardwood forests and agricultural rice fields (Stafford *et al.* 2006), food production for waterfowl has been shown to be highly variable between sites but much less variable across years and large regions, such as the MAV. Variation at individual wetlands may be influenced by management practices (Hagy & Kaminski 2012b), management frequency and intensity (Brasher *et al.* 2007; Olmstead *et al.* 2013), and other environmental factors that are difficult to predict accurately. Therefore, the challenge in predicting annual carrying capacity in the managed wetlands and bottomland hardwood forests of the MAV likely depends less on accounting for annual differences in site-specific food densities and more on availability of those wetlands as habitats suitable for waterfowl. However, flooding of foods at the appropriate time

(Greer *et al.* 2007) and to the appropriate depth (Hagy & Kaminski 2012b) may be a challenge in some years. To date, we are aware of few attempts to quantify functionally available habitats at a scale such as a joint venture region (but see Soulliere *et al.* 2007), despite the clear need to determine and quantify food availability.

We examined the relative influences of mid-latitude and local factors on waterfowl densities in managed wetlands in and near the MAV and determined that local precipitation, an assumed surrogate of wetland availability, was most influential in explaining variation in dabbling duck density. Managed wetlands with extensive water control capabilities, such as the wetlands included in our study (see Hagy & Kaminski 2012b), are often flooded before many passively or non-managed wetlands and are available when waterfowl first arrive in the late autumn and early winter. Thus, waterfowl are likely attracted to these managed wetlands initially but may later colonise passively filled or temporary wetlands following periods of sufficient rainfall (Beatty *et al.* 2014a). Given the rapid declines in waterfowl food densities in managed wetlands documented by Hagy and Kaminski (2012b), our results suggest that waterfowl may move to alternative locations, such as agricultural fields (Pearse *et al.* 2012), following precipitation events to acquire newly-available food on flooded farmland.

Interestingly, evidence is accumulating that suggests local habitat availability may be a better predictor of duck density in managed wetlands than food density (Hagy 2010; Tapp 2013) or weather and precipitation in more northerly portions of the Mallard's wintering range (Krementz *et al.* 2012; Beatty *et al.* 2014b). Distributions and duration of stay of waterfowl at autumn staging areas can vary with wetland area, disturbance (Stafford *et al.* 2010), wetland forage quality (O'Neal *et al.* 2012), precipitation (Krementz *et al.* 2011, 2012) and vegetation characteristics (Moon & Haukos 2008), but others have failed to show a relationship between food abundance and waterfowl use of wetlands (Percival *et al.* 1998; Straub 2008; Brasher 2010; Fleming 2010). Weather has been shown to influence regional abundance of ducks (Schummer *et al.* 2010), but Krementz *et al.* (2012) reported that temperature and the onset of freezing conditions were not significant determinants of departure date during autumn migration. Cumulative research suggests that there is a significant degree of plasticity in the autumn migration of duck species to the wintering grounds (Bellrose *et al.* 1979; Krementz *et al.* 2012). We posit that a suite of factors including photoperiod, the location and timing of severe weather events and habitat availability and quality interact with considerable inter- and intraspecific variation to determine migratory patterns, but continued research is necessary to differentiate their relative contribution to the timing of waterfowl migration and habitat use.

At a continental or flyway scale, a suite of factors influences waterfowl habitat selection, including wetland availability (Beatty *et al.* 2014b). Krementz *et al.* (2011) and Hagy *et al.* (2014) anecdotally noted that, during spring- and autumn-migration, Mallard stopover use and duration of stay may have been related to precipitation and the availability of local wetlands, respectively. Interestingly, we identified a relationship between local precipitation and duck densities in the wintering region of the MAV, which might also suggest that once ducks migrate to latitudes where influences of photoperiod, weather severity and other factors decrease (*i.e.* wintering areas; Schummer *et al.* 2010), wetland habitat availability is also an important driver of habitat selection (Webb *et al.* 2010; Pearse *et al.* 2012). Thus, conservation planning models can benefit from considering timing and extent of flooding (*i.e.* wetland inundation) when determining habitat objectives at large spatial scales (*e.g.* Joint Ventures).

Although challenging to build and parameterise, spatially explicit models that incorporate variables (*e.g.* precipitation) that account for spatial and temporal variability in habitat availability may be needed for more accurate predictions of food availability and site use by ducks. However, relatively simplistic daily ration models that incorporate fixed estimates of food density are likely adequate for predicting energetic carrying capacity at the MAV scale for waterfowl that are highly mobile and can respond rapidly to changing habitat conditions and availability. A critical next step in improving the accuracy of energetic carrying capacity models is estimating the spatial and temporal extent and variability of habitats by modelling the wetland areas suitable for exploitation of food resources by waterfowl (Williams *et al.* 2014). Future modelling attempts could incorporate the spatial arrangement of patches (*i.e.* costs of food acquisition), temporal availability of patches within and among years and individual patch value rather than aggregate food availability to improve accuracy and utility at smaller scales. In reality, extensive inter- and intraspecific variation in migration timing and life-history strategies add considerable uncertainty to energetic models (Hagy *et al.* 2014) and efforts aimed at reducing these uncertainties or quantifying the relative effects on energetic carrying capacity models at annual and longer-term timescales would be beneficial (see Notaro *et al.* 2014; Schummer *et al.* 2014).

In summary, our data indicate that annual food densities in managed wetlands and bottomland hardwood forests were generally stable across the 3-year study at a regional scale used by a Joint Venture for conservation planning. It may be beneficial for conservation planners to quantify longer-term variability in food resources and examine factors influencing the availability of these resources to waterfowl in other regions used by wintering waterfowl. Furthermore, if the results are extended to other wintering areas and habitat types, quantifying and facilitating habitat availability within the landscape for waterfowl, in sufficient quantities to meet their energetic demands at the appropriate time and location, is a challenge worthy of additional exploration.

Acknowledgements

We thank the many individuals who assisted during collection of the data used in this manuscript. Many thanks are due to two anonymous reviewers, L. Webb and E. Rees for their helpful suggestions which greatly improved this manuscript. We thank the Illinois Natural History Survey, the Forest and Wildlife Research Center (FWRC) of the Mississippi State University (MSU) Department of Wildlife, Fisheries and Aquaculture, the United States Forest Service, Ducks Unlimited and other organisations for funding and support.

References

Beatty, W.S., Kesler, D.C., Webb, E.B., Raedeke, A.H., Naylor, L.W. & Humburg, D.D. 2014a. The role of protected area wetlands in waterfowl habitat conservation: implications for protected are network design. *Biological Conservation* 176: 144–152.

Beatty, W.S., Webb, E.B., Kesler, D.C., Raedeke, A.H., Naylor, L.W. & Humburg, D.D. 2014b. Landscape effects on mallard habitat selection at multiple spatial scales during the non-breeding period. *Landscape Ecology* 29: 989–1000.

Bellrose, F.C., Paveglio, Jr. F.L. & Steffeck, D.W. 1979. Waterfowl populations and the changing environment of the Illinois River valley. *Illinois Natural History Survey Bulletin* 32: 1–54.

Brasher, M.G. 2010. Duck use and energetic carrying capacity of actively and passively managed wetlands in Ohio during autumn and spring migration. Ph.D. thesis, The Ohio State University, Columbus, USA.

Brasher, M.G., Steckel, J.D. & Gates, R.J. 2007. Energetic carrying capacity of actively and passively managed wetlands for migrating ducks in Ohio. *Journal of Wildlife Management* 71: 2532–2541.

Burnham, K.P. & Anderson, D.R. 2002. *Model Selection and Multimodel Inference: a Practical Information-Theoretic Approach. Second Edition.* Springer-Verlag, New York, USA.

Dalby, L., Fox, A.D., Petersen, I.K., Delany, S. & Svenning, J.C. 2013. Temperature does not dictate the wintering distributions of European dabbling duck species. *Ibis* 155: 80–88.

Davis, B.E., Afton, A.D. & Cox, R.R., Jr. 2009. Habitat use by female mallards in the lower Mississippi Alluvial Valley. *The Journal of Wildlife Management* 73: 701–709.

Evans-Peters, G.R., Dugger, B.D. & Petrie, M.J. 2013. Plant community composition and waterfowl food production on Wetland Reserve Program Easements compared to those on managed public lands in western Oregon. *Wetlands* 32: 391–399.

Fleming, K.S. 2010. Effects of management and hydrology on vegetation, winter waterbird use, and water quality on Wetlands Reserve Program lands, Mississippi. M.Sc. thesis, Mississippi State University, Mississippi State, USA.

Greer, A.K., Dugger, B.D., Graber, D.A. & Petrie, M.J. 2007. The effects of seasonal flooding on seed availability for spring migrating waterfowl. *The Journal of Wildlife Management* 71: 1561–1566.

Haig, S.M., Mehlman, D.W. & Oring, L.W. 1998. Avian movements and wetland connectivity in landscape conservation. *Conservation Biology* 12: 749–758.

Hagy, H.M. 2010. Winter food and waterfowl dynamics in managed moist-soil wetlands in the Mississippi Alluvial Valley. Ph.D. Dissertation, Mississippi State University, Mississippi State, Mississippi, USA.

Hagy, H.M. & Kaminski, R.M. 2012a. Apparent seed use by ducks in the Mississippi Alluvial Valley. *Journal of Wildlife Management* 76: 1053–1061.

Hagy, H.M. & Kaminski, R.M. 2012b. Winter waterbird and food dynamics in autumn-managed moist-soil wetlands of the Mississippi Alluvial Valley. *Wildlife Society Bulletin* 36: 512–523.

Hagy, H.M., Straub, J.N. & Kaminski, R.M. 2011. Estimation and correction of seed recovery bias from moist-soil cores. *Journal of Wildlife Management* 75: 959–966.

Hagy, H.M., Yetter, A.P., Stodola, K.W., Horath, M.M., Hine, C.S., Ward, M.P., Benson, T.J., Smith, R.V. & Stafford, J.D. 2014. Stopover duration of mallards during autumn in the Illinois River Valley. *Journal of Wildlife Management* 78: 747–752.

Krementz, D.G., Asanti, K. & Naylor, L.W. 2011. Spring migration of mallards from Arkansas as determined by satellite telemetry. *Journal of Fish and Wildlife Management* 2: 156–168.

Krementz, D.G., Asanti, K. & Naylor, L.W. 2012. Autumn migration of Mississippi Flyway mallards as determined by satellite telemetry. *Journal of Fish and Wildlife Management* 3: 238–251.

Kross, J., Kaminski, R.M., Reinecke, K.J., Penny, E.J. & Pearse, A.T. 2008. Moist-soil seed abundance in managed wetlands in the Mississippi Alluvial Valley. *Journal of Wildlife Management* 72: 707–714.

Lovvorn, J.R. & Baldwin, J.R. 1996. Intertidal and farmland habitats of ducks in the Puget Sound region: a landscape perspective. *Biological Conservation* 77: 97–114.

Lovvorn, J.R. & Gillingham, M.P. 1996. Food dispersion and foraging energetics: a mechanistic synthesis for field studies of avian benthivores. *Ecology* 77: 435–451.

Moon, J.A. & Haukos, D.A. 2008. Habitat use of female northern pintails in the playa lakes region of Texas. *Proceedings of the Annual Conference of the Southeastern Association of Fish and Wildlife Agencies* 62: 32–87.

Nakagawa, S. & Schielzeth, H. 2013. A general and simple method for obtaining R^2 from generalized linear-mixed effects models. *Methods in Ecology and Evolution* 4: 133–142.

Notaro, M., Lorenz, D., Hoving, C. & Schummer, M. 2014. 21st Century projections of snowfall and winter severity across central-eastern North America. *Journal of Climate* doi: http://dx.doi.org/10.1175/JCLI-D-13-00520.1.

Newton, I. 1998. *Population Limitation in Birds*. Academic Press, San Diego, California, USA.

O'Neal, B.J., Stafford, J.D. & Larkin, R.P. 2012. Stopover duration of fall-migrating dabbling ducks. *Journal of Wildlife Management* 76: 285–293.

Olmstead, V.G., Webb, E.B. & Johnson, R.W. 2013. Moist-soil seed biomass and species richness on Wetland Reserve Program Easements in the Mississippi Alluvial Valley. *Wetlands* 33: 197–206.

Pearse, A.T., Kaminski, R.M., Reinecke, K.J. & Dinsmore, S.J. 2012. Local and landscape associations between wintering dabbling ducks and wetland complexes in Missisppi. *Wetlands* 32: 859–869.

Percival, S.M., Sutherland, W.J. & Evans, P.R. 1998. Intertidal habitat loss and wildfowl numbers, applications of a spatial depletion model. *Journal of Applied Ecology* 35: 57–63.

Pinheiro, J., Bates, D., DebRoy, S., Sarkar, D. & R Core Team. 2014. *Nlme: Linear and Nonlinear Mixed Effects Models*. R package version 3.1-117. Accessible at http://CRAN.R-project.org/package=nlme (last accessed 08.07.14).

Rees, E.C. 1982. The effect of photoperiod on the timing of spring migration in the Bewick's Swan. *Wildfowl* 33: 119–132.

Reinecke, K.J., Kaminski, R.M., Moorhead, D.J., Hodges, J.D. & Nassar, J.R. 1989. Mississippi Alluvial Valley. In L.M. Smith, R.L. Pederson & R.M. Kaminski (eds.), *Habitat Management for Migrating and Wintering Waterfowl in North America*, pp. 203–247. Texas Tech University Press, Lubbock, USA.

Schummer, M.L., Kaminski, R.M., Raedeke, A.H. & Graber, D.A. 2010. Weather-related indices of autumn-winter dabbling duck abundance in middle North America. *Journal of Wildlife Management* 74: 94–101.

Schummer, M.L., Cohen, J., Kaminski, R.M., Brown, M.E. & Wax, C.L. 2014. Atmospheric teleconnections and Eurasian snow cover as predictors of a weather severity index in relation to Mallard *Anas platyrhynchos* autumn–winter migration. *Wildfowl* (Special Issue No. 4): 451–469.

Soulliere, G.J., Potter, B.A., Coluccy, J.M., Gatti, R.C., Roy, C.L., Luukkonen, D.R., Brown, P.W. & Eichholz, M.W. 2007. *Upper Mississippi River and Great Lakes Region Joint Venture Waterfowl Habitat Conservation Strategy*. U.S. Fish and Wildlife Service, Fort Snelling, Minnesota, USA.

Stafford, J.D., Kaminski, R.M., Reinecke, K.J. & Manley, S.W. 2006. Waste rice for waterfowl in the Mississippi Alluvial Valley. *Journal of Wildlife Management* 70: 61–69.

Stafford, J.D., Horath, M.M., Smith, R.V., Yetter, A.P. & Hine, C.S. 2010. Historical and contemporary characteristics and waterfowl use of Illinois River valley wetlands. *Wetlands* 30: 565–576.

Straub, J.N. 2008. Energetic carrying capacity of habitats used by spring migrating waterfowl in the Upper Mississippi and Great Lakes Region during spring migration. M.Sc. thesis, Ohio State University, Columbus, USA.

Straub, J.N. 2012. Estimating and modelling red oak acorn production and abundance in the Mississippi Alluvial Valley. Ph.D. thesis, Mississippi State University, Mississippi State, USA.

Straub, J.N., Gates, R.J., Schultheis, R.D., Yerkes, T., Coluccy, J.M. & Stafford, J.D. 2012. Wetland food resources for spring-migrating ducks in the upper Mississippi River and Great Lakes region. *Journal of Wildlife Management* 76: 768–777.

Tapp, J.L. 2013. Waterbird use and food availability on Wetland Reserve Program Easements enrolled in the Migratory Bird Habitat Initiative. M.Sc. thesis, University of Missouri, Columbia, USA.

Webb, E.B., Smith, L.M., Vrtiska, M.P. & Lagrange, T.G. 2010. Effects of local and landscape variables on wetland bird habitat use during migration through the rainwater basin. *Journal of Wildlife Management* 74: 109–111.

Weegman, M.M. 2013. Waterbird and seed abundances in Migratory Bird Habitat Initiative and non-managed wetlands in Mississippi and Louisiana. M.Sc. thesis, Mississippi State University, Mississippi State, USA.

Williams, C.K., Dugger, B., Brasher, M., Collucy, J., Cramer, D.M., Eadie, J.M., Gray, M., Hagy, H.M., Livolsi, M., McWilliams, S.R., Petrie, M., Soulliere, G.J., Tirpak, J. & Webb, L. 2014. Estimating habitat carrying capacity for migrating and wintering waterfowl: considerations, pitfalls and improvements. *Wildfowl* (Special Issue No. 4): 407–435.

Zuur, A.F., Ieno, E.N., Walker, N.J., Saveliev, A.A. & Smith, G.M. 2009. *Mixed Effects Models and Extensions in Ecology with R*. Springer, New York, USA.

Atmospheric teleconnections and Eurasian snow cover as predictors of a weather severity index in relation to Mallard *Anas platyrhynchos* autumn–winter migration

MICHAEL L. SCHUMMER[1,5]*, JUDAH COHEN[2], RICHARD M. KAMINSKI[3], MICHAEL E. BROWN[4] & CHARLES L. WAX[4]

[1]Long Point Waterfowl, PO Box 160, Port Rowan, Ontario N0E 1M0, Canada.
[2]Atmospheric and Environmental Research, Lexington, Massachusetts USA.
[3]Mississippi State University, Department of Wildlife, Fisheries, and Aquaculture, Mississippi State, Mississippi 39762 USA.
[4]Department of Geosciences, Mississippi State University, Mississippi State, Mississippi 39762, USA.
[5]Current address. SUNY-Oswego, Department of Biological Sciences, 30 Centennial Drive, Oswego, New York 13126, USA.
*Correspondence author. E-mail: mschummer@longpointwaterfowl.org

Abstract

Research on long-term trends in annual weather severity known to influence migration and winter distributions of Mallard *Anas platyrhynchos* and other migratory birds is needed to predict effects of changing climate on: 1) annual distributions and vital rates of these birds, 2) timing of habitat use by migratory birds, and 3) demographics of the hunters of these species. Weather severity thresholds developed previously for Mallard were used to calculate weather severity and spatially-depicted Weather Severity Index Anomalies (± km^2, WSIA), in comparison with normal conditions, for Mallard in eastern North America from November–January 1950–2008. We determined whether WSIA differed among decades and analysed the effects of atmospheric teleconnections and Eurasian snow cover on annual variation in WSIA. Weather severity was mildest (+ WSIA) during the 2000s compared to other decades and differed substantially from the 1960s and 1970s (– WSIA). The Arctic Oscillation Index explained substantial variation in WSIA during El Niño and La Niña episodes, but not when the Oceanic Niño Index was neutral. Eurasian snow cover models accurately predicted if the WSIA would be greater or less than normal for 75% of the years studied. Our results may provide a partial explanation for recent observations of interrupted or reduced migration to southern latitudes by Mallard and other migratory birds during autumn–winter. Our models also provide ecologists with teleconnection models to help predict future distributions of Mallard and potentially other migratory birds in eastern North America. Future investigations could include testing the influence of WSIA on Mallard survival and on annual movements and distributions for other migratory birds, to provide a better understanding of the influences of climate and changes in climate on population dynamics and the need to conserve particular habitats.

Key words: Arctic Oscillation, climate, El Niño, migration, waterfowl, weather, winter.

The winter distribution of migratory birds is influenced by a range of variables including food and habitat availability, weather, evolutionary and ecological mechanisms, body condition and anthropogenic factors (Miller *et al.* 2005; Newton 2007, 2008; Schummer *et al.* 2010). Evidence suggests that global climate change has lengthened growing seasons, increased winter temperatures and decreased snow accumulation at many locations worldwide (Field *et al.* 2007). Concurrently, delays in autumn migration and changes in the birds' winter distributions have been documented in North America and Eurasia (Sokolov *et al.* 1999; Cotton 2003; La Sorte & Thompson 2007; Brook *et al.* 2009; Sauter *et al.* 2010). Many migratory birds distribute annually along a latitudinal gradient as a function of physiological tolerances to the severity of winter weather (Root & Schneider 1995). Although food and habitat availability, refugia, flooding and other exogenous factors can mitigate influences of weather severity on bird energy budgets, climate remains a primary determinant of winter distributions of many species (Root 1988). Thus, climate warming is often cited as the mechanism underlying a shift in range towards the pole (Crick 2004; Gordo 2007; La Sorte & Thompson 2007; Nevin *et al.* 2010).

Northward shifts in winter range and changes in autumn migration phenology can arise from changes in weather severity, food and habitat availability, or a combination of these (Gordo 2007; Newton 2008; Schummer *et al.* 2010). Concurrent with milder weather during winter, habitat has become increasingly available to waterfowl through wetland conservation programmes in northern and mid-latitude regions of North America (NAWMP 2004; Dahl 2006). This has led to considerable debate regarding the mechanism(s) determining current winter distributions of Nearctic waterfowl (National Flyway Council and Wildlife Management Institute 2006, Greene and Krementz 2008; Brook *et al.* 2009). Research into the long-term changes in weather conditions known to elicit southerly migrations by waterfowl therefore is needed to understand influences of weather severity and habitat availability on the birds' winter distribution and migration phenology (*e.g.* Schummer *et al.* 2010). Further, success of wetland restoration is often measured by monitoring the recolonisation and use of habitats by wildlife (Jordan *et al.* 1987; Scodari 1997), though confirming the outcome of restoration efforts is difficult without concurrent quantification of the weather factors that influence distribution.

Increasingly, studies are identifying changes in the frequency and amplitude of atmospheric teleconnections as influencing temporal and spatial distributions of populations and species (Stenseth et al. 2003; Wang & Schimel 2003). Atmospheric teleconnnections are defined as recurring, persistent, large-scale patterns of atmospheric pressure, circulation and temperature anomalies occurring over thousands of square kilometres (Bridgman & Oliver 2006). Prominent teleconnections dominant in the North American sector include the El Niño Southern Oscillation (ENSO), the Pacific North America teleconnection pattern (PNA), the Arctic Oscillation (AO) and the North Atlantic Oscillation (NAO; Bridgman & Oliver 2006). A substantial portion of the warming trends in North America over the past 20 years have been attributed to sustained positive phases in atmospheric oscillations (Serreze & Barry 2005; Hurrell & Deser 2009). Further, some models indicate increased frequency and amplitude of positive AO and NAO phases as a result of global climate change (Corti et al. 1999; Gillet et al. 2003). Positive phase teleconnections may result in increased frequency of mild winters and changes in winter distributions and migration phenology of birds (Cotton 2003). The AO and NAO can only be forecast accurately c. 7 days in advance; thus we sought additional indices of winter weather used to produce long-term seasonal forecasts for eastern North America (see Fig. 1). Cohen and Jones (2011) detected a strong correlation between the advance of Eurasian snow cover during October and the winter AO, which influences winter temperatures in eastern North America (i.e. the Snow Advance Index; SAI). Thus, we reasoned that the SAI may predict severe winter weather months in advance of traditional atmospheric teleconnections and provide seasonal forecasts of waterfowl migration timing and intensity.

Using temperature and snow data, Schummer et al. (2010) developed several competing models to explain rates of change in relative abundance of Mallard *Anas platyrhynchos* at mid-latitude staging areas in North America (i.e. Missouri) during autumn–winter. The model which best explained annual variation in the rate of change in Mallard abundance was calculated daily for November–January inclusive as the mean daily temperature $-(°C) +$ the number of consecutive days where the mean temperature was $\leq 0°C +$ snow depth $+$ the number of consecutive days with snow cover (i.e. the Weather Severity Index, WSI; Schummer et al. 2010). Temperatures of $< 0°C$ were given a positive algebraic sign (i.e. indicating more severe weather and accumulating positively when added to other variables in the model), and temperatures $> 0°C$ were given a negative sign. Here we use the WSI to rank severity of winter weather for nearly six decades and determine differences in mean decadal rank for the winters 1950/51–2008/09 (hereafter 1950–2008) in eastern North America (see Fig. 1). Eastern North America was selected because it corresponded with the geographic area in which the WSI was developed (Schummer et al. 2010) and ecoregions used in conservation planning for Nearctic waterfowl (i.e. Mississippi and Atlantic Flyways: Bellrose 1980; NAWMP

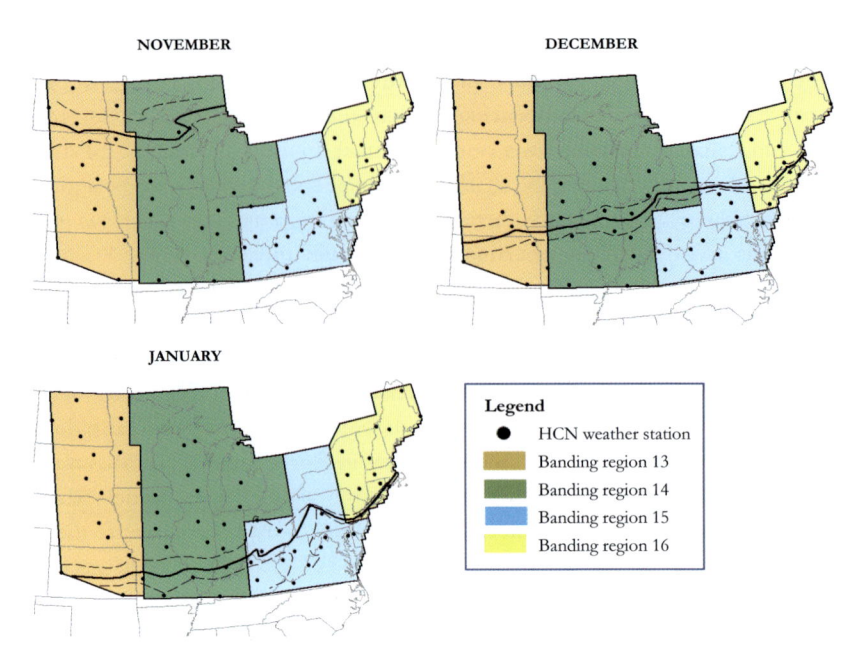

Figure 1. Long-term mean location (solid line) ± 95% C.I. (dashed line) of the threshold weather severity index (WSI = 8) for Mallard in the Mississippi Flyway (USFWS banding regions 13–14) and the Atlantic Flyway (USFWS banding regions 15–16), for winters 1950–2008, calculated using Historical Climatology Network (HCN) data (black dots).

2012). The period 1950–2008 was used because monitoring long-term metrics of Mallard breeding population and habitat dynamics (*i.e.* the Breeding Waterfowl and Habitat Survey: Baldassarre & Bolen 2006) commenced during the 1950s. We used data through to 2008, which was the end of the initial funding period for data compilation and analysis (Zimmerman 2009). As the overall aim of the study was to investigate if indices used to forecast winter weather (*sensu* Cohen & Jones 2011) could be used to predict WSI and potentially provide longer-term projections of Mallard distributions associated with climate change, we conclude by relating our findings to global climate and climate change models.

Methods

Weather data

Sixty two weather stations in the United States Historical Climatology Network (HCN) at *c.* ≥ 35°N were selected to provide long-term state-wide weather data for North Dakota, South Dakota, Nebraska, Kansas, Minnesota, Iowa, Missouri, Wisconsin, Illinois, Michigan, Indiana, Kentucky, Ohio, New York, Pennsylvania, West Virginia, Maryland, Virginia, Vermont, Delaware, New Jersey, New Hampshire, Massachusetts, Connecticut, Rhode Island and Maine (Quinlan *et al.* 1987; Williams *et al.* 2006; Fig. 1). The median latitude HCN station was selected within each state and inspected for missing temperature and snow data. If > 10% of daily temperature or snow data were missing, we alternately selected stations immediately north and south until a weather station that contained < 10% missing data was identified. For states ≥ 28,500 km² in area (*i.e.* larger than Maryland), two additional weather stations were also included (one in the north and the other in the south of the state), starting with the 25% and 75% quartile latitude HCN weather stations and using the aforementioned criterion for missing data. We also selected a station in Michigan's upper peninsula which met this criterion. Only two HCN weather stations in Nebraska were found that met our quality criteria.

Daily mean temperature and snow data were obtained for November–January 1950–2008 from selected HCN weather stations to calculate a daily WSI for each station (n = 330,832 values), as follows: WSI = Temp + Temp days + Snow depth + Snow days (following Schummer *et al.* 2010), where Temp = mean daily temperature (°C; with temperatures < 0°C given a positive algebraic sign to indicate more severe weather, and those > 0°C a negative sign, as described above); Temp days = consecutive days with mean temperature ≤ 0°C; Snow depth = (snow depth, in cm) × 0.394; and Snow days = consecutive days ≥ 2.54 cm of snow. Values for snow depth and snow days convert to 1 inch of snow, a measurement used in weather forecasting and reporting in the United States. For days when temperature and snow depth data were missing, we interpolated missing values as being the mean of data recorded on days next before and after these date(s) (see Schummer *et al.* 2010); however, months with ≥ 7 consecutive days of missing WSI values were omitted from the analyses. Annual monthly mean WSI values were calculated from the daily WSI values for each HCN station, for November–January 1950–2008 (n = 10,788; PROC MEANS in SAS Institute 2009). For each month, we also calculated a long-term (1950–2008) monthly mean (± 95% C. I.s) from annual monthly mean WSI values for each HCN station. Monthly PNA, AO and NAO (November–January) and three-month mean ENSO indices (September–November, SON; October–December, OND; and November–January, NDJ) were obtained from the National Oceanic and Atmospheric Administration/Climate Prediction Center at College Park, Maryland, USA (NOAA 2010).

Spatial analyses

Interpolation analyses were conducted using the natural neighbour method in ArcView Spatial Analyst, with a cell output size of 0.07 degrees (*c*. 7.7 km; ESRI 2009), to depict the long-term average WSI spatially, and to calculate ± 95% C.I. for monthly mean WSI values (for November–January,1950–2008) from the 62 HCN weather stations (Fig. 1). Schummer *et al.* (2010) reported that daily WSI values of ≥ 7.2 coincided with an increased likelihood of Mallard leaving mid-latitude locations in North America. We assumed that Mallard reacted similarly to weather severity throughout the range of our present study as in our earlier study area (at *c.* 33°–49°N), and a conservative threshold WSI value of 8 therefore was used to demarcate by month and year the geographic extent of the WSI threshold value. Three polylines were digitised as layers representing the long-term average (LTA) WSI and the upper and lower 95% C.I.s for each month of the study (Fig. 1). We conducted the same interpolation analyses to produce a monthly WSI threshold demarcation line for November–January 1950–2008 inclusive (n = 232 maps). The areas above and below the 95% C.I.s were digitised as polygon layers to determine the area (km^2) assumed to be available or unavailable to Mallard, based on the LTA and 95% C.I.s (Fig. 1). The "extract by mask" function in ArcView Spatial Analyst (ESRI 2009) was then used to develop polygon areas (*i.e.* km^2) above and below the 95% C.I. within the U.S. Fish and Wildlife Service waterfowl banding regions 13–16 in eastern North America (hereafter, banding regions; Fig. 1). Banding regions were modified by extending the northern border to capture areas of southern Canada where data interpolation of WSI occurred. Within the banding regions, we subtracted polygon areas (*i.e.* km^2) below the lower 95% C.I. from polygon areas above the upper 95% C.I. to determine changes to net potential habitat for Mallard (± km^2; hereafter the weather severity index anomaly, WSIA) for each region. Thus, the WSIA is the area assumed available or unavailable for use by Mallard compared to the LTA by banding region on a monthly basis (November–January) for each year (1950–2008). Data were summarised by banding regions because they represent different Mallard populations and migratory flyways (*i.e.* the Mississippi and Atlantic Flyways).

Decadal analysis of winter severity index anomalies (WSIA)

We summed WSIA from banding regions 13 and 14 (hereafter, Mississippi Flyway) for each month and year, likewise summed data for regions 15 and 16 (Atlantic Flyway) to follow regional scales at which Mallard migrate, and thus conservation planning decisions are made for Nearctic waterfowl (Baldassarre & Bolen 2006), and because data from individual banding regions were not normally distributed. We also calculated mean WSIA within the Mississippi Flyway (MSF), the Atlantic Flyway (AF) and the combined MSF and AF means for the three-month period (November–January; NDJ). Data passed normality and equal variance tests (P > 0.05). A one-way Analysis of Variance (ANOVA) therefore was used to test the null hypothesis that there is no difference between the six decadal periods (1950s–2000s inclusive) in the WSIA values recorded (α = 0.05; Sokal & Rohlf 1981; Systat Software Inc. 2008). We predicted that WSIA during the 2000s would be greater in comparison with other decades because of observations of interrupted or delayed (*i.e.* "short-stopping") migration by Mallard to southern latitudes in eastern North America in the last decade (Greene & Krementz 2008; Brook *et al.* 2009; Nevin *et al.* 2010).

Effects of atmospheric teleconnections

We used an information-theoretic approach (Burnham & Anderson 2002) to investigate climatic factors potentially influencing variation in WSIA. Five candidate teleconnection-based models, associated with weather patterns in eastern North America, were developed *a priori* for this analysis as listed below.

El Niño Southern Oscillation (ENSO)

Climate variability resulting from ENSO can produce wide-ranging ecological consequences (Stenseth *et al.* 2003). ENSO is most often characterised as El Niño, La Niña or Neutral. El Niño occurs when the equatorial Pacific Ocean sea surface temperatures (SSTs) are unusually warm, whereas La Niña occurs when the same region is dominated by unusually cold SSTs, and Neutral occurs when SSTs are near normal. The Oceanic Niño Index (ONI) is the principal tool for monitoring and assessing ENSO. To be considered a full El Niño or La Niña episode, the ONI must exceed ± 0.5 (°C) for ≥ 5 consecutive overlapping 3-month seasons (Tozuka *et al.* 2005; NOAA 2010). During winter months, a positive ONI (El Niño, ≥ +0.5) is associated with increased precipitation and colder than normal temperatures in the southeast United States, but warmer than normal temperatures and reduced ice coverage in the Great Lakes region and eastern Canada (Ropelewski & Halpert 1987; Halpert & Ropelewski 1992; Assel *et al.* 2000). A negative ONI (La Niña, ≤ –0.5) commonly results in warmer and drier conditions over the southeast United States, but colder than normal air penetrating into the northern Great Plains of Canada and the United States (Ropelewski & Halpert 1987; Halpert & Ropelewski 1992). During ONI neutral phases (–0.5 to 0.5), equatorial Pacific Ocean SSTs appear to have much less influence on global climates and forecasting long-term regional weather patterns becomes difficult (Bridgman & Oliver 2006).

North Atlantic Oscillation (NAO)

The NAO is often associated with interdecadal and interannual shifts in ecological processes in marine and terrestrial systems of Europe and North America (Hurrell *et al.* 2003). A positive NAO index increases the likelihood of mild, wet winters in the eastern United States while increased cold accompanies a negative NAO index over the same area (Hurrell 1995). Because the NAO is also correlated with the Arctic Oscillation (AO), there is considerable debate as to whether the NAO is merely a regional expression of the AO, which is a hemispheric mode of climatic variability (Ambaum *et al.* 2001; Hurrell *et al.* 2003). Moreover, scientists are uncertain if the NAO or the AO has the greatest influence on winter climates and related ecological processes in North America (Aanes *et al.* 2002; Stenseth *et al.* 2003).

Arctic Oscillation (AO)

Effects of the AO on North American climate are most pronounced from December through March (Serreze & Barry 2005). A positive AO index reduces the number of cold air intrusions east of the Rocky Mountains, causing much of the eastern United States to experience warmer winters than normal. A negative AO index produces the opposite effect over the same area (Serreze & Barry 2005; Bridgman & Oliver 2006).

Pacific North American Oscillation (PNA)

Like the AO, the PNA tends to be most pronounced during the winter months. The positive phase of the PNA is associated with above average temperatures in western Canada and the Pacific coast of the United States, but below average temperatures across the south-central and southeast United States. The PNA has also been associated with moisture variability in the same region (Coleman & Rogers 1995; Rogers & Coleman 2003). Anomalies in regional temperatures, ice cover and snow cover have been related to the PNA phase (Assel 1992; Serreze *et al.* 1998). Also, positive PNA index values are more frequent during El Niño episodes (Wang & Fu 2000).

Plausible interactions and combined influences

Several studies have investigated modulation of the ONI by other atmospheric teleconnections (Gershunov & Barnett 1998, Bridgman & Oliver 2006) and the combined influences of atmospheric teleconnections on winter temperature, precipitation and snowfall (Serreze *et al.* 1998). For example, during El Niño or La Niña episodes, inclusion of other atmospheric teleconnections as covariates improves explained variation in precipitation and temperature throughout much of North America (Higgins *et al.* 2000). However, several atmospheric teleconnections are often correlated and their combination in models could create statistical bias. We calculated Pearson's correlation coefficients for relationships between continuous variables (NAO, AO, PNA) and did not include combinations of model predictors that were correlated ($r \geq 0.70$; Dormann *et al.* 2013). We also used ANOVA to test the null hypothesis of no difference in NAO, AO and PNA among the ONI categories of El Niño, La Niña and Neutral ($P < 0.05$). Data passed normality and equal variance tests ($P > 0.05$, Sokal & Rohlf 1981). If an ANOVA approached significance ($P < 0.10$), we did not include that atmospheric teleconnection in models containing ONI categories. Positive associations were detected between the AO and NAO ($0.64 \leq r_{58} \leq 0.76$, $P < 0.001$) and the ONI and PNA ($0.37 \leq r_{58} \leq 0.40$, $P < 0.01$). The PNA also varied with ONICAT ($3.49 \leq F_{2,55} \leq 8.16$, $P < 0.05$) but not with the AO and NAO ($0.64 \leq F_{2,55} \leq 0.09$, $P > 0.05$). Because of the latter results, we did not include the AO and PNA within models containing the NAO and ONICAT, respectively. Thus, we evaluated 18 candidate models for explaining variation in WSIA: 1) AO, 2) NAO, 3) PNA, 4) ONI (categorical, ONICAT), 5) ONI (continuous, ONI), 6) ONICAT AO, 7) ONICAT NAO, 8) ONICAT × AO, 9) ONICAT × NAO, 10) ONI AO, 11) ONI NAO, 12) ONI × AO, 13) ONI × NAO, 14) PNA AO, 15) PNA NAO, 16) PNA × AO, 17) PNA × NAO, and 18) the NULL. In addition to appropriate multivariate models, we modelled interactions of continuous variables and ONI category to determine if including the relationship among slopes improved the model fit.

An Akaike's Information Criterion (AIC*c*) was calculated for each model (PROC MIXED; SAS Institute 2009). Competing models were ranked according to Δ AIC*c* values and selection was based on the lowest Δ AIC*c* value (Burnham & Anderson 2002; Littell *et al.* 2007). We considered models competitive when Δ AIC*c* values were ≤ 2 units of the best model (*i.e.* with Δ AIC*c* =

0; Burnham & Anderson 2002). Year was included as a random variable and variance components (VC) were used from a suite of tested covariance structures (*i.e.* CS, UN, TOEP, AR(1), ARH(1), VC), because VC produced the best fit models (lowest AIC*c* values: Littell *et al.* 2007). We calculated Akaike weights (w_i) to assess relative support for each atmospheric teleconnection-based model in explaining variation in WSIA. When top and competing models contained an interaction effect, results were interpreted from the slopes and intercepts of the relationship between the dependent and interacting explanatory variables (Gutzwiller & Riffell 2007; SAS Institute 2009).

Effects of the snow advance index (SAI)

To evaluate whether the SAI was better than random at predicting WSIA during November–January, AIC*c* values for the SAI model were compared with those for the null models (PROC MIXED; SAS Institute 2009 to determine whether SAI was better than random chance of it predicting WSIA. SAI is a strong predictor for AO (Cohen & Jones 2011) and NAO and AO are correlated (Hurrell *et al.* 2003) so we only included SAI and the null in our analysis. Analyses were conducted on 1972–2008 data because the SAI values were available only for 1972 onwards. Whether removing ONI-neutral years improved the utility of the model for predicting WSIA was also investigated.

Results

Decadal analysis of WSIA

The WSIA rank recorded for NDJ each year did not differ significantly across the decades ($F_{5,52} = 1.31$, $P = 0.26$, n.s.; Table 1). The 2000s had the highest mean WSIA but also greatest range in values (Table 1; Appendix 1). Inspection of data indicated that winter 2000/01 was a statistical outlier (with WSIA > 200 units in comparison with the next

Table 1. Descriptive statistics for Weather Severity Index Anomalies (WSIA, ± thousands of km²), by decade, for November–January in eastern North America.

Decade	n	Mean (± s.e.) WSIA (× 1,000 km²)	Median WSIA (× 1,000 km²)	25% Quartile WSIA (× 1,000 km²)	75% Quartile WSIA (× 1,000 km²)
1950s	9	69.0 (50.5)	72.5	–77.4	210.0
1960s	10	–4.9 (25.1)	–15.8	–57.6	52.9
1970s	10	–33.5 (32.9)	–16.2	–123.5	74.0
1980s	10	68.7 (57.9)	60.2	–51.9	252.5
1990s	10	14.3 (43.9)	8.5	–113.3	110.3
2000s	9	129.2 (85.4)	129.4	49.6	272.1

coldest winters, whereas WSIA values for the two warmest winters differed by *c.* 10 units) and ranked as the most negative WSIA in the 58 years of the study (Appendix 1). On removing 2000/01 data, the mean WSIAs differed significantly among decades (ANOVA: $F_{5,51}$ = 2.99, P = 0.02). *Post-hoc* Tukey pair-wise comparisons with 2000/01 data removed detected that WSIA was greater in the 2000s ($\bar{x}$ = 197.7 ± 57.9) than the 1970s (P = 0.01) and 1960s (P = 0.04) but no other decadal differences were detected for WSIA ($P \geq 0.30$, n.s.; Table 1).

Effects of atmospheric teleconnections

Models containing interactions of AO or NAO with ONICAT were the only models that were ≤ 2 Δ AICc units from the best model for all locations and time periods (Table 2). The AO and NAO explained a greater portion of variation (R^2) in WSIA during El Niño and La Niña episodes than during Neutral conditions for all locations and periods. Models derived using 3-month means generally were better fit models (*i.e.* by w_i and R^2) than those using monthly data. In the Mississippi Flyway, AO had a greater influence on WSIA during NDJ, whereas the NAO was more influential in the Atlantic Flyway (Table 2). Greatest weight of evidence explaining variation in WSIA for Mississippi and Atlantic Flyways combined was associated with the interaction of AO with ONICAT (Table 2). However, AICc values for the interaction of NAO and ONICAT were < 2 units from the top model (ONICAT × AO) and received substantial weight of evidence (Table 2). For eastern North America, a substantial portion of the variation in WSIA was explained by the AO during El Niño and La Niña episodes but not during Neutral conditions (Table 2, Fig. 2).

Effects of SAI

The SAI (AICc = 477.3) was a better predictor of WSIA than the null model (AICc = 481.9), but utility of the model for predicting WSIA was relatively poor (Fig. 3). Including the ONI category (El Nino, La Nina or Neutral) did not improve the predictive ability of our model (AICc = 480.7). Despite poor model fit, the SAI did predict correctly whether WSIA would be positive (mild winter compared to normal) or negative (severe winter compared to normal) in 27 (75%) of 36 years (Fig 3).

Discussion

So far as we are aware, this study is the first to assess the influence of long-term trends in annual weather severity (*i.e.* ± WSIA) on the autumn–winter distribution of Mallard in the Nearctic (Schummer *et al.* 2010). We detected a trend toward less severe weather (+ve WSIA) during recent decades (1990s and 2000s) but also detected that severe events (–ve WSIA) occurred during this period (*e.g.* winter 2000/2001). The results identified that WSIA was positively related to the Arctic and North Atlantic Oscillation Indices during El Niño and La Niña episodes but not during ONI Neutral conditions. Similar to other studies, Arctic and North Atlantic Oscillation Indices were found to be strongly correlated (Ambaum *et al.* 2001; Hurrell *et al.* 2003). Our models indicate that weather severity, which is known to influence autumn–winter

Table 2. Akaike's Information Criteria (AICc) for linear relationships between Weather Severity Index Anomalies (WSIA, ± thousands km²) and atmospheric teleconnection models in the Mississippi and Atlantic Flyways, in November–January 1950–2008. Eighteen candidate models for explaining variation in WSIA were considered for each time period (including combinations of AO, NAO, PNA, continuous ONI (ONI), categorical ONI (ONICAT) and year; Methods section provides further details); the top two models in each case are presented here.

Location[a]	Period[b]	Models[c]	Δ AICc	AICc	w_i	R² by Category[d]
Mississippi Flyway	November (N)	ONICAT × NAO	1404.5	0.00	0.92	EL (0.02), LA (0.23), NA (0.06)
		ONICAT × AO	1409.3	4.80	0.08	EL (0.00), LA (0.30), NA (0.00)
	December (D)	ONICAT × AO	1449.3	0.00	0.51	EL (0.16), LA (0.51), NA (0.12)
		ONICAT × NAO	1449.4	0.01	0.49	EL (0.31), LA (0.21), NA (0.08)
	January (J)	ONICAT × AO	1415.4	0.00	0.76	EL (0.25), LA (0.46), NA (0.03)
		ONICAT × NAO	1417.7	2.30	0.24	EL (0.01), LA (0.17), NA (0.10)
	NDJ	ONICAT × AO	1374.9	0.00	0.67	EL (0.14), LA (0.55), NA (<0.00)
		ONICAT × NAO	1376.3	1.40	0.33	EL (0.18), LA (0.32), NA (<0.00)
Atlantic Flyway	December (D)	ONICAT × AO	1378.3	0.00	0.51	EL (0.41), LA (0.60), NA (0.04)
		ONICAT × NAO	1378.4	0.01	0.49	EL (0.47), LA (0.49), NA (0.01)
	January (J)	ONICAT × AO	1316.4	0.00	0.93	EL (0.34), LA (0.43), NA (0.06)
		ONICAT × NAO	1321.6	5.20	0.07	EL (0.20), LA (0.25), NA (0.01)
	NDJ	ONICAT × NAO	1292.3	0.00	0.66	EL (0.66), LA (0.48), NA (0.00)
		ONICAT × AO	1293.6	1.30	0.34	EL (0.60), LA (0.55), NA (0.04)
Mississippi and Atlantic Flyways	NDJ	ONICAT × AO	1401.0	0.00	0.66	EL (0.32), LA (0.64), NA (0.01)
		ONICAT × NAO	1402.3	1.30	0.34	EL (0.28), LA (0.27), NA (0.01)

[a] Mississippi Flyway = USFWS banding regions 13 and 14; Atlantic Flyway = USFWS banding regions 15 and 16 (see Fig. 1).
[b] November is not included in the Atlantic Flyway because most observations were outside of the study area (see Fig. 1).
[c] ONICAT = Oceanic Niño Index category (e.g. El Niño, La Niña and Neutral), AO = Arctic Oscillation Index, NAO = North Atlantic Oscillation Index. Only models with weight of evidence (w_i) are included.
[d] derived R² by category: EL = El Niño, LA = La Niña and NA = Neutral.

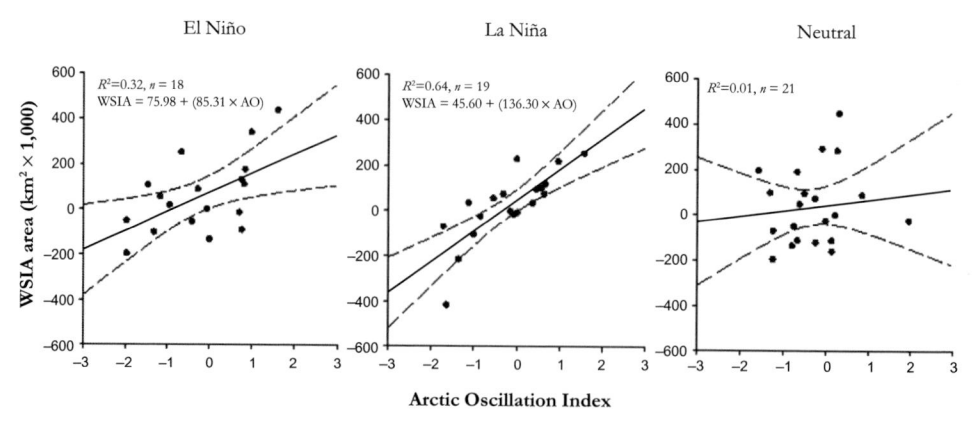

Figure 2. Relationships between the Arctic Oscillation Index (AO) and Weather Severity Index Anomalies (WSIA, ± thousands of km^2) in eastern North America by Oceanic Niño Index category (El Niño, La Niña and Neutral) for November–January, 1950–2008.

Mallard migration (Bellrose 1980; Schummer et al. 2010), is reduced during El Niño and La Niña episodes when the Arctic and North Atlantic Oscillation Indices are in a positive phase. Thus, we think Mallard migration may be interrupted or delayed (i.e. "short-stopping") during these conditions. Investigation of the usefulness of the SAI, as a long-term seasonal predictor of weather influencing Mallard migration, provided mixed results. The SAI predicted accurately the coming winter (NDJ) as being either less or more severe than normal 75% of the time, but its capacity to predict WSIA was limited. We suggest continued investigation of the SAI with additional weather indices. Long-term forecasting of weather known to influence Mallard and other waterfowl migrations would be helpful for managers charged with providing waterfowl habitat at key times of migration throughout the non-breeding season and for managing timing of hunting seasons (Schummer et al. 2010).

The results of the study provides a potential explanation for recent observations of delayed autumn–winter migration in Nearctic Mallard and other migratory birds (National Flyway Council and Wildlife Management Institute 2006; Nevin et al. 2010) and may be helpful in modelling future autumn–winter distributions of Mallard (and possibly other migratory birds) in eastern North America and more widely (see Notaro et al. 2014). Colleagues are encouraged to test for the influence of WSIA on the movements and annual distribution of Mallard and other waterfowl, its influence on waterfowl hunter demographics and behaviour, and whether WSIA affects survival rates and trends in population size. To facilitate these analyses we suggest development of weather indices for other waterfowl and web-based tools for distribution of these data to scientists and managers. Overall, the results suggest that recent observations of delayed waterfowl migration (National Flyway Council and Wildlife Management Institute 2006; Brook et al. 2009) may be related to reduced weather

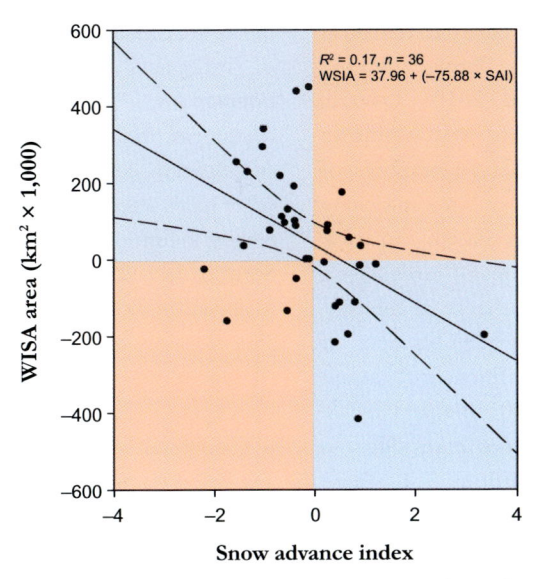

Figure 3. Relationship between the Snow Advance Index (SAI) and Weather Severity Index Anomalies (WSIA, ± thousands of km^2) in eastern North America for November–January 1972–2008. Data within blue areas were correctly designated as WSIA greater or lesser than normal by SAI (27 of 36 years, 75%), whereas red areas were incorrectly designated.

severity at northern and mid-latitudes in eastern North America.

The results can also be used for determining the relative contributions of weather severity and habitat availability to the winter distribution and migration phenology of waterfowl and other migratory wildlife. WSIA was quantified here for Mallard in eastern North America, but the availability and quality of wetland habitat, human-related disturbance, waterfowl population sizes, and other factors potentially influencing habitat use, migration and population dynamics should also be taken in account (Bellrose 1980; Kaminski & Gluesing 1987; Newton 2008). Such an evaluation would add clarity to the debate regarding the mechanism(s) that influence the current winter distributions of Nearctic waterfowl (Greene & Krementz 2008). In addition, such information could potentially aid conservationists in predicting future winter distributions of Nearctic waterfowl and habitat protection, management and restoration needs under various climate change models (e.g. Ruosteenoja et al. 2003; La Sorte & Jetz 2010). Further, WSIA could be included in models used to determine benefits of wetland restoration because habitat use by birds (i.e. a metric of restoration success) can be highly dynamic and may be related to the severity of winter weather in addition to habitat quality and other metrics (Newton 1998).

A sustained positive Arctic Oscillation Index could result in increased frequency of mild winters (+ve WSIA), during which decreased ice cover, snowfall and increased temperatures may allow Mallard to remain at more northern latitudes during autumn–winter (Schummer *et al.* 2010). Overall, reduced weather severity may result in delayed migration or reduced numbers of Mallard and possibly other birds migrating to southern latitudes in North America. Nearctic waterfowl (millions of birds), use a diversity of aquatic and terrestrial foraging niches, and can feed at rates capable of causing strong trophic influences (Newton 1998; Abraham *et al.* 2005; Baldassaree & Bolen 2006). Sustained northern shifts in autumn–winter distributions of these abundant species could increase foraging pressure at northern latitudes while reducing such effects at southerly locations and cause changes in trophic relationships (*i.e.* trophic cascades) throughout eastern North America (Crick 2004; Inkley *et al.* 2004). An increase in foraging intensity at more northern latitudes during autumn and winter may also deplete the food available for waterfowl during spring migration in some locations (Straub *et al.* 2012; Greer *et al.* 2007; Long Point Waterfowl, unpubl. data). Predicting future distributions of waterfowl using forecasts of WSIA may help conservationists develop adaptive plans to meet the habitat needs of waterfowl and other migratory birds in a changing climate (Lehikoinen *et al.* 2006; Seavy *et al.* 2008; La Sorte & Jetz 2010, Notaro *et al.* 2014).

We used a simplistic WSI developed for Mallard and other dabbling ducks (Schummer *et al.* 2010) to examine their potential past and future distributions. Results from our study corroborate those for European Mallard which showed reduced winter migration distance with long-term warming (Sauter *et al.* 2010). We encourage including a broader suite of influences such as potential species interactions, physiology and energy-dependent "bottle-necks" at different stages of migration, concurrent habitat changes, and other biotic interactions, to increase biological realism in future analyses (Seavy *et al.* 2008; La Sorte & Jetz 2010). Further examination of Mallard distribution using satellite telemetry and volunteer observation programmes (*e.g.* Christmas Bird Count, Mallard Migration Network) in relation to the WSI and WSIA would provide further validation of temporal and spatial distributions of these birds. Other Nearctic waterfowl and migratory species may react differently to annual variation in winter severity and changing climate (Crick 2004; Sauter *et al.* 2010), for instance because habitat generalists (*e.g.* Mallard) often respond to climate change more readily than habitat specialists (La Sorte & Jetz 2010). Thus, we also encourage continued research aimed at understanding the effects and threats of a changing climate for a variety of migratory species (Thomas *et al.* 2001; Walther *et al.* 2002).

Acknowledgements

Finances were provided by the Forest and Wildlife Research Center (FWRC), Mississippi State University (MSU); the James C. Kennedy Endowed Chair in Waterfowl and Wetlands Conservation; the United States Forest Service, Center for Bottomland Hardwoods Research, Stoneville, Mississippi; and the Department of Geosciences, MSU. Statistical advice was provided by G. Wang. We thank D. Krementz and B. Cooke for providing helpful comments during the development of the manuscript. We also thank C. Zimmerman for completing a pilot project (Zimmerman 2009) and for assisting with compilation of weather data. L. Tucker assisted with weather data collection and error checking. J. Cohen is supported by the National Science Foundation grant BCS-1060323. Our manuscript has been approved for publication as MSU-FWRC journal article WFA-310.

References

Aanes, R., Saether, B.E., Smith, F.M. Cooper, E.J., Wookey, P.A. & Øritsland, N.A. 2002. The Arctic Oscillation predicts effects of climate change in two trophic levels in a high-arctic ecosystem. *Ecology Letters* 5: 445–453.

Abraham, K.F., Jefferies, R.L. & Rockwell, R.F. 2005. Goose-induced changes in vegetation and land cover between 1976 and 1997 in an Arctic coastal marsh. *Arctic, Antarctic, and Alpine Research* 37: 269–275.

Assel, R.A. 1992. Great Lakes winter-weather 700-hPa PNA teleconnections. *Monthly Weather Review* 120: 2156–2163.

Assel R.A., Janowiak, J., Boyce, D., O'Connors, C., Quinn, F.H. & Norton, D.C. 2000. Laurentian Great Lakes ice and weather conditions for the 1998 El Niño winter. *Bulletin of the American Meteorological Society* 81: 703–718.

Ambaum, M.H.P., Hoskins, B.J. & Stephenson, D.B. 2001. Arctic Oscillation or North Atlantic Oscillation? *Journal of Climate* 14: 3495–3507.

Baldassarre, G.A. & Bolen, E.G. 2006. *Waterfowl Ecology and Management*. Krieger, Malabar, Florida, USA.

Bellrose, F.C. 1980. *Ducks, Geese and Swans of North America*. Stackpole Books, Mechanicsburg, Pennsylvania, USA.

Brook, R.W., Ross, R.K., Abraham, K.F., Fronczak, D.L. & Davies, J.C. 2009. Evidence for black duck winter distribution change. *Journal of Wildlife Management* 73: 98–103.

Burnham, K.P. & Anderson, D.R. 2002. *Model Selection and Multimodel Inference: a Practical Information-Theoretic Approach. Second Edition.* Springer-Verlag, New York, New York, USA.

Bridgman, H.A. & Oliver, J.E. 2006. *The global climate system: patterns, processes and teleconnections.* Cambridge University Press, Cambridge, UK.

Cohen, J. & Jones, J. 2011. A new index for more accurate winter predictions. *Geophysical Research Letters* 38: L21701.

Coleman, J.S.M. & Rogers, J.C. 1995. Ohio River Valley winter moisture condition associated with the Pacific-North American teleconnection pattern. *Journal of Climate* 16: 969–981.

Corti, S., Molteni, F. & Palmer, T.N. 1999. Signature of recent climate change in frequencies of natural atmospheric circulation regimes. *Nature* 398: 799–802.

Cotton, P.A. 2003. Avian migration phenology and global climate change. *Proceedings of Nature and Science* 100: 12219–12222.

Crick, H.Q.P. 2004. The impact of climate change on birds. *Ibis* 146:48–56.

Dahl, T.E. 2006. *Status and Trends of Wetlands in the Conterminous United States 1998 to 2004.* U.S. Department of the Interior, Fish and Wildlife Service, Washington D.C., USA.

Dormann, C.F., Elith, J., Bacher, S., Buchmann, C., Carl, G., Carré, G., Diekötter, T., García Marquéz, J., Gruber, B., Lafourcade, B., Leitão, P.J., Münkemüller, T., McClean, C., Osborne, P., Reineking, B., Schröder, B., Skidmore, A.K., Zurell, D., Lautenbach, S. 2013. Collinearity: a review of methods to deal with it and a simulation study evaluating their performance. *Ecography* 36: 27–46.

Environmental Systems Research Institute (ESRI). 2009. *ArcGIS 9.2 Desktop Help.* Accessible at www.webhelp.esri.com/arcgisdesktop/9.2/index.cfm? TopicName=welcome (last accessed 15 June 2014).

Field, C. B., Mortsch, L.D., Brklacich, M., Forbes, D.L., Kovacs, P., Patz, J.A., Running, S.W., Scott, M.J., Andrey, J., Cayan, D., Demuth, M., Hamlet, A., Jones, G., Mills, E., Mills, S., Minns, M.K., Sailor, D., Saunders, M., Scott, D., Solecki, W., and MacCracken, M. 2007. North America: Climate change 2007: impacts, adaptation and vulnerability – contribution of working group II to the fourth assessment report of the Intergovernmental Panel on Climate Change. *In* M.L. Parry, F. Canziani, J.P. Palutikof, P.J. van der Linden & C.E. Hanson (eds.), *North America: Climate Change 2007*, pp. 617– 620. Cambridge University Press, Cambridge, UK.

Gershunov, A. & Barnett, T.P. 1998. Interdecadal modulation of ENSO teleconnections. *Bulletin of the American Meteorological Society* 80: 2715–2725.

Gillet, N.P., Baldwin, M.P. & Allen, M.R. 2003. Climate change and the North Atlantic Oscillation. The North Atlantic Oscillation: Climate Significance and Environmental Impact. *Geophysical Monograph* 134: 193–199.

Gordo, O. 2007. Why are bird migration dates shifting? A review of weather and climate effects on avian migration phenology. *Climate Research* 35: 37–58.

Greene, A.W. & Krementz, D.G. 2008. Mallard harvest distributions in the Mississippi and Central Flyways. *Journal of Wildlife Management* 72: 1328–1334.

Greer, A.K., Dugger, B.D., Graber, D.A. & Petrie, M.J. 2007. The effects of seasonal flooding on seed availability for spring migrating waterfowl. *Journal of Wildlife Management* 71: 1561–1566.

Gutzwiller, K.J. & Riffell, S.K. 2007. Using statistical models to study temporal dynamics of animal-landscape relations. *In* J.A. Bissonette & I. Storch (eds.), *Temporal Dimensions of Landscape Ecology: Wildlife Responses to Variable Resources*, pp. 93–118. Springer Inc., New York, New York, USA.

Halpert, M.S. & Ropelewski, C.F. 1992. Surface temperature patterns associated with the Southern Oscillation. *Journal of Climate* 5: 577–593.

Higgins, R.W., Leetmaa, A., Xue, Y. & Barston, A. 2000. Dominant factors influencing the seasonal predictability of U.S. precipitation and surface air temperature. *Journal of Climate* 13: 3994–4017.

Hurrell, J.R. 1995. Decadal trends in the North Atlantic Oscillation and relationships to regional temperature and precipitation. *Science* 269: 676– 679.

Hurrell, J.R. & Deser, C. 2009. North Atlantic climate variability: The role of the North Atlantic Oscillation. *Journal of Marine Systems* 78: 28–41.

Hurrell, J.R., Kushnir, Y., Visbeck, M. & Ottersen, G. 2003. An overview of the North Atlantic Oscillation. The North Atlantic Oscillation: Climate Significance and Environmental Impact. *Geophysical Monograph Series* 134: 1–35.

Inkley, D.B., Anderson, M.G., Blaustein, A.R., Burkett, V.R., Felzer, B., Griffith, B., Price, J. & Root, T.L. 2004. *Global climate change and wildlife in North America, Wildlife Society Technical Review 04-2*. The Wildlife Society, Bethesda, Maryland, USA.

Jordan, W.R., Gilpin, M.E., & Aber, J.D. 1987. *Restoration ecology: A Synthetic Approach to Ecological Research*. Cambridge University Press, Cambridge, UK.

Kaminski, R.M. & Gluesing, E.A. 1987. Density-and habitat-related recruitment in mallards. *Journal of Wildlife Management* 51: 141–148.

La Sorte, F.A. & Jetz, W. 2010. Avian distributions under climate change: towards improving projections. *Journal of Experimental Biology* 213: 862–869.

La Sorte, F.A. & Thompson, F.R. 2007. Poleward shifts in winter ranges of North American birds. *Ecology* 88: 1803–1812.

Lehikoinen, A., Kilpi, M. & Öst, M. 2006. Winter climate affects subsequent breeding success of common eiders. *Global Change Biology* 12: 1355–1365.

Littell, R.C., Milliken, G.A., Stroup, W.W., Wolfinger, R.D. & Schabenberger, O. 2007. *SAS for Mixed Models. Second edition.* SAS Institute Inc., Cary, North Carolina, USA.

Miller, R.M., Takekawa, J.Y., Fleskes, J.P., Orthmeyer, D.L., Casazza, M.L. & Perry, W.M. 2005. Spring migration of Northern Pintails from California's Central Valley wintering area tracked with satellite telemetry: routes, timing and destinations. *Canadian Journal of Zoology* 83: 1314–1332.

National Flyway Council and Wildlife Management Institute (NFC and WMI). 2006. *National Duck Hunter Survey 2005: National Report.* Wildlife Management Institute, Washington D.C., USA.

National Oceanic and Atmospheric Administration (NOAA). 2010. *Climate Prediction Center: El Niño-Southern Oscillation Index.* Available at www.cpc.noaa.gov/products/precip/CWlink/MJO/enso.shtml#history (last accessed 6 May 2012).

North American Waterfowl Management Plan (NAWMP) Committee. 2004. *North American Waterfowl Management Plan. 2004. Implementation Framework: Strengthening the Biological Foundation.* Canadian Wildlife Service, U.S. Fish & Wildlife Service and Secretaria de Medio Ambiente y Recursos Naturales. Accessible at http://www.fws.gov/birdhabitat/NAWMP/Planstrategy.shtm (last accessed 15 June 2014).

Newton, I. 1998. *Population Limitation in Birds.* Academic Press, San Diego, USA.

Newton, I. 2007. Weather-related mass mortality events in migrants. *Ibis* 149: 453–467.

Newton, I. 2008. *The Migration Ecology of Birds.* Academic Press, San Diego, USA.

Nevin, D.K., Butcher, G.S. & Bancroft, G.T. 2010. Christmas bird counts and climate change: northward shifts in early winter abundance. *American Birds* 109: 10–15.

Notaro, M., Lorenz, D., Hoving, C., & Schummer, M.L. 2014. 21st Century projections of snowfall and winter severity across central-eastern North America. *Journal of Climate* online early: doi: http://dx.doi.org/10.1175/JCLI-D-13-00520.1.

Quinlan, F.T., Karl, T.R. & Williams, C.N., Jr. 1987. *United States Historical Climatology Network (HCN) Serial Temperature and Precipitation Data.* Carbon Dioxide Information Analysis Center, Oak Ridge National Laboratory NDP-019, Oak Ridge, Tennessee, USA.

Rogers, J.C. & Coleman, J.S.M. 2003. Interactions between the Atlantic/Multidecadal Oscillation, El Niño/La Niña, and the PNA in winter Mississippi Valley stream flow. *Geophysical Research Letters* 30: 1518–1522.

Ropelewski, C.F. &. Halpert, M.S. 1987. Global and regional scale precipitation patterns associated with the El Niño/Southern Oscillation. *Monthly Weather Review* 115: 1606–1626.

Root, T. 1988. Energy constraints on avian distributions and abundances. *Ecology* 69: 330–339.

Root, T. & Schneider, S.H. 1995. Ecology and climate: research strategies and implications. *Science* 269: 334–341.

Ruosteenoja, K., Carter, T.R., Jylha, K., & Tuomenvirta, H. 2003. *Future climate in World Regions: an Intercomparison of Model-based Projections for the New IPCC Emissions Scenarios.* Finnish Environment Institute, Helsinki, Finland.

SAS Institute. 2009. *SAS/STAT User's Guide.* SAS Institute, Cary, North Carolina, USA.

Sauter, A., Korner-Nievergelt, F. & Jenni, L. 2010. Evidence of climate change effects on within-winter movements of European Mallards *Anas platyrhynchos. Ibis* 152: 600–609.

Schummer, M.L., Kaminski, R.M., Raedeke, A.H. & Graber, D.A. 2010. Weather-related indices of autumn–winter dabbling duck abundance in middle North America. *Journal of Wildlife Management* 74: 94–101.

Scodari, P.F. 1997. *Measuring the Benefits of Federal Wetland Programs*. Environmental Law Institute, Washington D.C., USA.

Seavy, N.E.,. Dybala, K.E. & Snyder, M.A. 2008. Perspectives in ornithology: climate models and ornithology. *Auk* 125: 1–10.

Serreze M.C. & Barry, R.G. 2005. *The Arctic Climate System*. Cambridge University Press, Cambridge, UK.

Serreze, M.C., Clark, M.P., McGinnis, D.L., & Robinson, D.A. 1998. Characteristics of snowfall over the eastern half of the United States and relationships with principal modes of low-frequency atmospheric variability. *Journal of Climate* 11: 234–250.

Sokal, R.R., & Rohlf, F.J. 1981. *Biometry*. W.H. Freeman and Company, San Francisco, USA.

Sokolov, L.V., Markovets, M.Y. & Morozov, Y.G. 1999. Long-term dynamics of the mean date of autumn migration in passerines on the Courish Spit of the Baltic Sea. *Avian Ecology and Behaviour* 2: 1–18.

Stenseth, N.C., Ottersen, G., Hurrell, J.W., Mysterud, A., Lima, M., Chan, K.-S., Yoccoz, N.G., & Ådlandsvik, B. 2003. Studying climate effects on ecology through the use of climate indices, the North Atlantic Oscillation, El Niño Southern Oscillation and beyond. *Proceeding of the Royal Society of London, B Biological Science* 270: 2087–2096.

Straub, J.N., Gates, R.J., Schultheir, R.D., Yerkes, T., Coluccy, J.M. & Stafford, J.D. 2012. Wetland food resources for spring-migrating ducks in the Upper Mississippi River and Great Lakes region. *Journal of Wildlife Management* 76: 768–777.

Systat Software, Inc. 2008. *Systat 11.0*. Richmond, California, USA.

Thomas, D.E., Blondel, J., Perret, R., Lambrechts, M.M. & Speakman, J.R. 2001. Energetic and fitness costs of mismatching resource supply and demand in seasonally breeding birds. *Science* 291: 2598–2600.

Tozuka, T, Luo, J.-J., Masson, S., Behera, S.K. & Yamagata, T. 2005. Annual ENSO simulated in a coupled ocean atmosphere model. *Dynamics of Atmospheres and Oceans* 39: 41–60.

US Fish and Wildlife Service (USFWS). 2010. *Upper Mississippi River National Wildlife Refuge Fall Flight Surveys: How are Fall Flights Conducted?* Accessible at http://www.fws.gov/midwest/UpperMississippiRiver/Documents/InfoFF.pdf (last accessed 20 October 2010).

Walther, G., Post, E., Convey, P., Menzel, A., Parmesan, C., Beebee, T.C.J., Fromentin, J., Hugh-Guldberg, O. & Bairlein, F. 2002. Ecological response to recent climate change. *Nature* 416: 389–395.

Wang, G. & Schimel, D. 2003. Climate change, climate modes, and climate impacts. *Annual Review of Environmental Resources* 28: 1–28.

Wang, H. & Fu, R. 2000. Winter monthly mean atmospheric anomalies over the North Pacific and North America associated with El Niño SSTs. *Journal of Climate* 13: 3435–3447.

Williams, C.N., Jr., Menne, M.J., Vose, R.S. & Easterling, D.R. 2006. *United States Historical Climatology Network Daily Temperature, Precipitation, and Snow Data*. Oak Ridge National Laboratory/Carbon Dioxide Information Analysis Center-118, NDP-070, Oak Ridge, Tennessee, USA.

Zimmerman, C.E. 2009. Long-term trend analysis of climatic factors influencing autumn-winter migration of Mallards in the Mississippi Flyway. M.Sc. thesis, Mississippi State University, Mississippi State, Mississippi, USA.

Appendix 1. Rank of Weather Severity Index Anomalies (WSIA, ± thousands of km^2) for eastern North America (Atlantic and Mississippi Flyway combined) in November–January 1950–2008.

Winter	WSIA (× 1,000 km^2)	Rank*	Winter	WSIA (× 1,000 km^2)	Rank*
1950/51	−26.8	21	1979/80	292.0	55
1951/52	−93.5	13	1980/81	93.8	39
1952/53	197.0	49	1981/82	−51.9	18
1953/54	283.4	54	1982/83	339.2	56
1954/55	72.5	35	1983/84	−161.5	5
1955/56	−106.9	11	1984/85	33.1	29
1956/57	117.9	45	1985/86	−197.1	4
1957/58	249.0	52	1986/87	−0.4	27
1958/59	−72.0	15	1987/88	87.3	38
1959/60	96.8	41	1988/89	252.5	53
1960/61	−137.5	6	1989/90	−114.4	9
1961/62	−27.8	19	1990/91	86.1	37
1962/63	−72.5	14	1991/92	110.3	44
1963/64	−57.6	16	1992/93	−27.1	20
1964/65	52.9	32	1993/94	−0.8	26
1965/66	106.5	43	1994/95	173.8	47
1966/67	47.0	31	1995/96	−217.4	2
1967/68	−3.7	25	1996/97	−113.3	10
1968/69	−53.0	17	1997/98	−17.8	28
1969/70	−103.3	12	1998/99	227.6	51
1970/71	−17.5	22	1999/00	217.1	50
1971/72	94.8	40	2000/01	−418.5	1
1972/73	−14.8	23	2001/02	447.7	58
1973/74	−9.0	24	2002/03	54.8	33
1974/75	99.2	42	2003/04	71.7	34
1975/76	74.0	36	2004/05	129.4	46
1976/77	−198.9	3	2005/06	189.8	48
1977/78	−135.6	7	2006/07	437.1	57
1978/79	−123.5	8	2007/08	33.9	30

* Rank 1 = most severe winter; 58 = least severe.

Waterfowl populations of conservation concern: learning from diverse challenges, models and conservation strategies

JANE AUSTIN[1]*, STUART SLATTERY[2] & ROBERT G. CLARK[3]

[1]US Geological Survey, Northern Prairie Wildlife Research Center, Jamestown, North Dakota 58401, USA.
[2]Ducks Unlimited Canada, Institute for Waterfowl and Wetland Research, Oak Hammock, Manitoba MB R0C 2Z0, Canada.
[3]Environment Canada, Prairie and Northern Wildlife Research Center, Saskatoon, Saskatchewan S7N 0X4, Canada.
*Correspondence author. E-mail jaustin@usgs.gov

Abstract

There are 30 threatened or endangered species of waterfowl worldwide, and several sub-populations are also threatened. Some of these species occur in North America, and others there are also of conservation concern due to declining population trends and their importance to hunters. Here we review conservation initiatives being undertaken for several of these latter species, along with conservation measures in place in Europe, to seek common themes and approaches that could be useful in developing broad conservation guidelines. While focal species may vary in their life-histories, population threats and geopolitical context, most conservation efforts have used a systematic approach to understand factors limiting populations and to identify possible management or policy actions. This approach generally includes *a priori* identification of plausible hypotheses about population declines or status, incorporation of hypotheses into conceptual or quantitative planning models, and the use of some form of structured decision making and adaptive management to develop and implement conservation actions in the face of many uncertainties. A climate of collaboration among jurisdictions sharing these birds is important to the success of a conservation or management programme. The structured conservation approach exemplified herein provides an opportunity to involve stakeholders at all planning stages, allows for all views to be examined and incorporated into model structures, and yields a format for improved communication, cooperation and learning, which may ultimately be one of the greatest benefits of this strategy.

Key words: Anatidae, conservation strategy, decision framework, population model, status and trends.

More than 20 species or populations of waterfowl in North America, with diverse life-histories, have experienced substantial declines over the past 25 years, or their numbers remain well below conservation goals (Table 1). Duck species of conservation concern range from the non-migratory Mottled Duck *Anas fulvigula*, which has small populations of limited distribution, to migratory scaup (Greater Scaup *Aythya marila* and Lesser Scaup *A. affinis*, combined hereafter as scaup) and sea ducks (Tribe: Mergini) with continental distributions. While some species share traits, such as geographic overlap of scaup and scoter *Melanitta* sp. breeding ranges in the boreal forests of North America, others seem to have little in common (*e.g.* Northern Pintail *Anas acuta* and sea ducks). These declines and persistent low populations have concerned biologists, managers and hunters alike (Miller & Duncan 1999; Austin *et al.* 2000). One aspect shared across species is considerable uncertainty about the factors that may be limiting populations, which creates substantial challenges for developing effective conservation strategies.

A wide range of environmental factors pose threats to the persistence of many duck, sea duck and goose populations globally. Of 228 waterfowl taxa (sub-species level) investigated by Green (1996), 48 vulnerable or endangered taxa (37 ducks and sea ducks; 11 geese) were threatened mainly by habitat loss, hunting and predation by invasive species. These same threats were also the most common among the 29 threatened or endangered duck and goose species recently assessed by the International Union for Conservation of Nature (IUCN; IUCN 2013), although high degrees of uncertainty were noted regarding limiting factors. Problems associated with habitat loss, fragmentation and degradation (*e.g.* reduced water quality) have continued unabated since Green's (1996) work (An *et al.* 2007; Dahl & Stedman 2013; Junk *et al.* 2013). Hence, many challenges faced by North American waterfowl have relevance globally, even if those populations are not considered threatened by global standards.

In this paper we examine approaches to addressing contemporary challenges faced by several duck species of special management concern in North America: Mottled Duck, American Black Duck *A. rubripes* (hereafter Black Duck), Northern Pintail, scaup and sea ducks. We also examine conservation challenges facing the Common Eider *Somateria mollissima* in western Europe, where collaborative research has developed but eider monitoring and management depends in large part on agreement among many countries. Although all of these species are designated as being of "least concern" by international nature conservation agencies, they have become focal species for several reasons. First, in the case of the North American dabbling and diving ducks, all are numerically important harvested species valued by hunters (Raftovich & Wilkins 2013). For instance, Northern Pintail, Lesser Scaup and Black Duck are highly prized by hunters in the Pacific, Mississippi and Atlantic Flyways, respectively, for a variety of cultural reasons. Second, all of these species have experienced substantial population declines at some point in the past 30 years, with no evidence of strong recoveries

Table 1. Summary of North American waterfowl species that were below conservation objectives of the North American Waterfowl Management Plan (NAWMP) in 2011, and species listed by federal agencies or International Union for the Conservation of Nature (IUCN) as being "at-risk" in 2013.

Below NAWMP objective[a]	Species of special concern[a,b]	At-risk species[b]
Northern Pintail *Anas acuta*	Mottled Duck *Anas fulvigula*	Steller's Eider (Alaska) *Polysticta stelleri*
American Black Duck *Anas rubripes*	Barrow's Goldeneye (eastern) *Bucephala islandica*	Spectacled Eider *Sometaria fischeri*
American Wigeon *Anas americana*	Black Scoter *Melanitta americana*	Laysan Duck *Anas laysanensis*
Greater Scaup *Aythya marila*	White-winged Scoter *Melanitta fusca*	Hawaiian Duck *Anas wyvilliana*
Lesser Scaup *Aythya affinis*	Long-tailed Duck *Clangula hyemalis*	
Atlantic Flyway Canada Goose (resident) *Branta canadensis*	Harlequin Duck (eastern) *Histrionicus histrionicus*	
Dusky Canada Goose *Branta c. occidentalis*	King Eider *Sometaria spectabilis*	
	Tule Goose *Anser albifrons elgasi*	
	Cackling Goose *Branta hutchinsonii*	
	Western High Arctic Brant *Branta bernicla*	
	Emperor Goose *Chen canagica*	
	Hawaiian Goose *Branta sandvicensis*	

[a]NAWMP 2012, Appendix A (Tables 1, 2, 3).
[b]Canadian Wildlife Service Waterfowl Committee 2012, USFWS 2008, IUCN 2013.

(Zimpfer *et al.* 2013; Ekroos *et al.* 2012). Population sizes of Northern Pintail, scaup and Black Duck remain below goals established by the North American Waterfowl Management Plan (NAWMP; NAWMP 2012). Third, population management objectives achieved through harvest regulations should be guided by science, and the manifold reasons for persistently low populations have not been adequately resolved. This situation can create debates between advocates of conservative harvest regulations or season closures and proponents of liberal harvest quotas who may question the lack of evidence for adverse harvest effects on populations. And, fourth, these species provide unique opportunities to learn about the application of formal decision analysis (*e.g.* Clemen 1996; Conroy & Peterson 2012; Gregory *et al.* 2012) to address concerns surrounding the management and conservation of harvested duck populations, while these species remain relatively common. These taxa represent a range of conservation goals, geographic scope, confidence in survey results, availability of data to inform hypotheses and models, modelling approaches and organisational history. Additionally, planning efforts within each taxon generally follow a robust conservation framework (*sensu* strategic habitat conservation; Johnson *et al.* 2009) that facilitates systematic and collaborative planning, typically under the auspices of NAWMP infrastructure (NAWMP 2012). The goals of this paper are to review the conceptual framework, demonstrate how the framework was applied in case studies and highlight the value of planning models for making decisions when much uncertainty is involved. We believe this approach is applicable whether the population of concern is the Lesser Scaup, which is still common in North America, or a globally threatened species.

Conceptual framework

A conceptual framework is here defined as an organisation of ideas into a set of logical steps to solve a problem and develop strategies to achieve desired goals. For North American waterfowl, those goals are population levels sufficient to meet conservation and societal demands and are implicit in subsequent discussions. Conservation efforts generally follow a framework that begins with a broad approach to the formulation of hypotheses about why populations either decline or remain below conservation goals. The strength of this approach lies in proposing plausible hypotheses to explain low populations, a process that typically involves a thorough evaluation of existing evidence and debate about defensible and sometimes speculative explanations for population patterns. One way of visualising this is with a decision tree (adapted from Platt's (1964) logical tree), as was used to summarise and illustrate explanations for low populations of Northern Pintail (J. Eadie, University of California-Davis, cited in Miller *et al.* 2003) and scaup (Fig. 1). Mechanisms that could produce observed population changes allow for explicit predictions about the expected demographic responses to specific management or policy actions. This approach provides a stronger conceptual

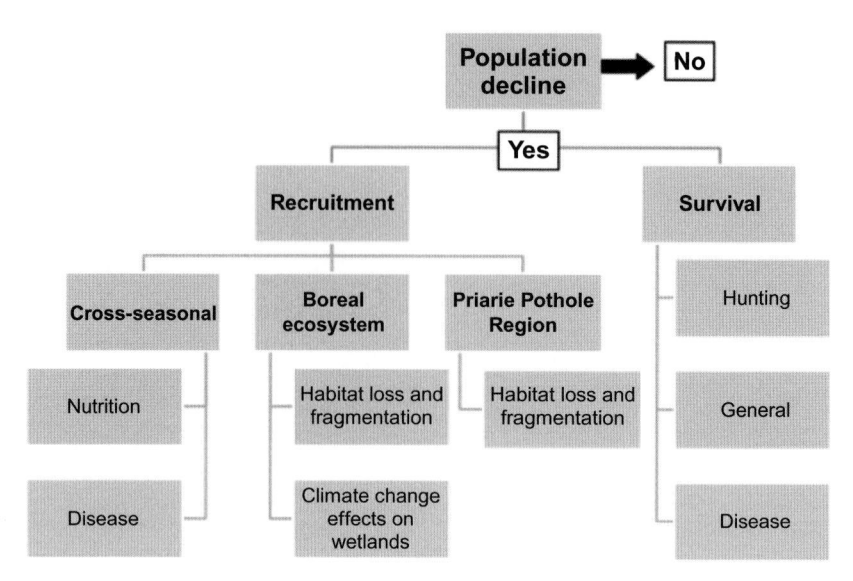

Figure 1. Graphical representation of a decision tree (modified from a logical tree; Platt 1964), designed to represent main working hypotheses proposed for scaup population status or declines (*e.g.* recruitment, survival) and putative mechanisms responsible for such changes (see Table 2). Management or policy alternatives would be implemented to improve demographic rates (survival or reproductive success), and then evaluated for effectiveness with targeted monitoring, research or adaptive management programmes. Factors affecting recruitment in the boreal ecosystem differ from those affecting recruitment in the Prairie Pothole Region; other factors affecting recruitment cross seasons for both breeding regions. Hypotheses were generated during waterfowl community workshops (Austin *et al.* 2000), as well as via research, monitoring and modelling studies.

framework for integrating critical steps by pinpointing: 1) likely bottlenecks to positive population growth rates; 2) suites of management or policy actions with the potential to alleviate these bottlenecks; 3) predicted demographic and population responses to these actions; and 4) monitoring required to evaluate the effectiveness of management interventions.

Population models, whether qualitative or quantitative, are at the core of implementing this conceptual framework. The use of population models to inform conservation actions has a long history (Shaffer 1981; Caswell 2000). The role and sophistication of models have greatly expanded over the last few decades, enabling biologists to integrate and simultaneously to model potential drivers of demographic variation encountered on breeding, staging and wintering areas, in order to predict population change (Mattsson *et al.* 2012; Osnas *et al.* 2014). Model objectives and structure reflect existing hypotheses or primary issues of concern, the availability of data and potential management actions. In addition, integrated models are increasingly able to leverage multiple sources of limited data (Schaub & Abadi 2011). Models therefore are fundamental in our case studies because they codify the decision tree and management actions, and thus allow measurable predictions about expected demographic outcomes under different management and conservation scenarios. Additionally, this structure can be applied to other management goals (*e.g.* hunter recruitment and retention; NAWMP 2012), and used to identify key uncertainties. There is considerable optimism, indeed expectation, that these model-based approaches will be pivotal in setting new, integrated objectives for NAWMP in the next several years (NAWMP 2012; Osnas *et al.* 2014). In the following case studies, we examine how application and outcomes of the framework evolved under the unique life history, data availability and socio-political settings for each duck species.

Case studies

Mottled Duck

Background – The Mottled Duck is a non-migratory species with two genetically distinct sub-populations, one in Florida, the other occurring along the Western Gulf Coast (WGC) portions of Alabama, Mississippi, Louisiana, Texas and northeast Mexico. The combined population estimate is *c.* 172,000 birds (M. Brasher, U.S. Fish and Wildlife Service, and R. Bielefeld, Florida Fish and Wildlife Commission, pers. comm.). Acquiring reliable status and trends information has been hampered by the lack of long-term, range-wide surveys corrected for visibility bias. Spring surveys in Florida suggest that the sub-population there has been stable (at *c.* 53,300) since 1984, but local surveys and indices suggest declines in coastal Texas during 1994–2005 (Johnson 2009) and relatively stable trends elsewhere (Bielefeld *et al.* 2010). It is also one of the least studied Anatini in North America. The species is considered to be of conservation concern because of its restricted distribution, relatively small population sizes, loss and degradation of key coastal habitats in the WGC (Wilson 2007) and introgressive hybridization with Mallard *Anas platyrhynchos* in the Florida sub-population (Table 2).

In the WGC, the primary conservation concerns are the degradation and loss of critical habitats, notably coastal and inland palustrine marshes, rice fields and native prairie and pastureland important for nesting (Wilson 2007). Highly variable breeding propensity, which affects population growth rates, may be tied to wetland conditions (Rigby & Haukos 2012). In Florida, similar concerns about the loss and degradation of wetland habitats have raised questions about the duck's nutritional status, which can affect reproductive success (Florida Fish and Wildlife Conservation Commission 2011). Because of small home ranges, the species can be sensitive to local habitat changes and harvest pressure. Hybridization with feral Mallard is however the main threat to Mottled Duck in Florida. Both regions share concerns about harvest rates and potential impacts of climate change on habitat conditions, with likely increased frequency of severe weather and further habitat loss (Florida Fish and Wildlife Conservation Commission 2011).

Approach – Conservation plans were developed based on expert opinion and limited existing data, and implemented for both Florida (Florida Fish and Wildlife

Table 2. Hypothesised factors causing declines or low populations of selected waterfowl species of conservation concern. Vital rates: R = recruitment, S = survival, P_b = probability of breeding. Occurrence of hybridization is noted with an X.

Presumed cause(s)	Mottled Duck	Black Duck	Northern Pintail	Scaup	North American sea ducks	Baltic Common Eider
Habitat loss and degradation on breeding grounds	R, S, P_b	R, S, P_b	R, S, P_b	R, S, P_b		
Habitat loss and degradation on wintering grounds		R, S, P_b	S	S	S	
Habitat changes on spring migration areas				R, P_b		
Habitat changes due to climate change	R, S, P_b			R, P_b	R, S, P_b	
Habitat changes related to agriculture	R, S, P_b		R, S, P_b			
Interspecific hybridization with Mallard	X	X				
Human disturbance during breeding	R, P_b					
Lead poisoning from spent shotgun pellets	S	S			S	
Over-harvest	S	S	S	S	S	S
Diseases and parasites		S	S		S	S
Contaminants (lead poisoning from spent shotgun pellets)		S			S	S
Bioaccumulation of contaminants from foods on migration and wintering areas		S		S, P_b	S, P_b	
Altered predator-prey relations on breeding grounds			R, S	R, S		R, S
Disturbance on migration and wintering areas from shipping, wind-power development					S	
Commercial exploitation of food resources on wintering grounds						R, S, P_b

Conservation Commission 2011) and WGC sub-populations (Wilson 2007). Experts identified factors most likely to limit sub-population growth, which stimulated research to elucidate how those factors were affecting vital rates, such as breeding distribution, nesting effort and survival.

Florida's plan focused on addressing uncertainties related to hybridization with Mallard by developing tools to identify species and hybrids more accurately, and on assessing the impact of wetland quality on productivity and the energy demands of this species. Results from studies on habitat use patterns for urban and rural Mottled Ducks in Florida improved predictions of Mottled Duck distribution and habitat use during multiple periods of the annual cycle and under contrasting water conditions (Bielefeld & Cox 2006; Varner 2013, 2014). The findings should improve the targeting of habitat conservation actions, and also improve the efficiency and effectiveness of the annual population surveys. Recent information on temporal and spatial patterns of survival in Florida (Bielefeld & Cox 2006) have improved predictions of how future habitat loss and alteration (including continued urbanisation and wetland creation associated with urban development and the Comprehensive Everglades Restoration Plan; Anonymous 1999), will affect the Mottled Duck sub-population. New techniques based on plumage characteristics (R. Bielefeld, pers. comm.) will be valuable for assessing the extent and distributional aspects of Mottled Duck x Mallard hybridization, and thus for identifying the most appropriate conservation actions.

In the WGC, a sex-specific matrix model identified female annual survival as an important factor in Mottled Duck population dynamics and potential target for management actions (Johnson 2009). A pattern of high breeding season survival and low breeding incidence suggested a trade-off between nesting effort and female survival (Rigby & Haukos 2012). Combined, these models indicate that improved habitat quality will be critical for conserving this species in the WGC region. Two main conservation actions identified by the model are the enhancement and restoration of coastal marshes (primarily for creating suitable (*i.e.* low salinity) brood habitat), and the restoration of coastal prairie and associated wetlands to enhance nesting propensity, nest success and brood survival. Partners have developed a spatially-explicit decision support tool to aid delivery of Mottled Duck habitat conservation in locations where demographic responses are likely to be more favourable. Finally, implementation of an annual range-wide, visibility-corrected survey of Mottled Ducks in the WGC will likely reduce uncertainties about population sizes and trends.

American Black Duck

Background – The American Black Duck is distributed in eastern North America from Ontario to the Maritime Provinces and south through states of the Mississippi Flyway and the Atlantic Flyway. Historically, it was the most abundant dabbling duck in eastern North America and also the most heavily harvested (Rusch *et al.* 1989). Estimates from the Mid-Winter Waterfowl

Inventory, conducted annually across most key wintering areas in the U.S., indicated the population experienced a rapid and sustained decline of > 50% between 1955 and the 1990s (Conroy *et al.* 2002). Christmas Bird Count data, a citizen-science survey programme conducted annually in selected areas, suggest that Black Duck numbers declined in the southern and central portion of wintering range during 1966–2003 but that populations in the northeast were stable (Link *et al.* 2006). While these wintering surveys provide the longest time period to assess trends, they both suffer from substantial shortcomings, such as temporal and spatial variation in survey effort and methodology. Breeding ground surveys conducted with more rigorous methods since 1990 indicate stable or slightly increasing trends (Zimpfer *et al.* 2013). Contrasting population trends among these three surveys could be related to counting methods or temporal shifts in winter distributions. However, some have raised questions about possible regional differences in population demographics (Conroy *et al.* 2002; Black Duck Joint Venture 2008).

Researchers and managers have proposed several hypotheses to explain the historic decline of Black Duck populations (Table 2), including over-harvest, competition and hybridization with Mallard, decrease in quality and quantity of wintering and breeding habitat, parasites and disease (*e.g.* duck viral enteritis) and environmental contaminants (*e.g.* lead shot, mercury, DDT). Conroy *et al.* (2002) found support for four major, continental-scope factors that may influence Black Duck populations: 1) loss in the quantity or quality of breeding habitats; 2) loss in the quantity or quality of wintering habitats; 3) harvest; and 4) competitive interactions or hybridization with Mallards. They concluded that no single factor could explain the Black Duck decline. A common theme across these issues and trends is uncertainty about the role of density dependence on reproduction and survival, and potential cross-seasonal influences of putative density-dependent effects. Also unclear is the degree to which competition and hybridization with Mallards may have interacted with other factors such as harvest and habitat changes (Nudds *et al.* 1996; Petrie *et al.* 2000). Although numerous investigations have addressed these issues, there is a lack of consensus about the role these factors play in limiting the population.

Approach – The Black Duck Joint Venture (BDJV) was established in 1989 as the first "species joint venture" (JV) to implement and coordinate a cooperative population monitoring, research and communications programme to provide information required to manage Black Duck populations and restore numbers to the NAWMP goal of 640,000 breeding birds (NAWMP 2012). Initial priorities included development and implementation of improved surveys to monitor breeding populations and harvest, directed projects to provide estimates of vital rates and habitat requirements, research to incorporate spatial information into the breeding ground survey to identify habitat features affecting Black Duck abundance, and development of a life-cycle model (Conroy *et al.* 2002) and a model to estimate autumn age-ratios.

The annual life-cycle model provides a

mechanistic description of population growth, assesses hypotheses concerning factors potentially limiting Black Duck population growth, and links hypotheses to parameters that could be estimated from available data. Sensitivity analyses were used to explore effects of statistical uncertainty in parameter values on population growth rates. Results indicated that reproductive rates were positively influenced by breeding habitat quantity and negatively influenced by Black Duck and Mallard densities, and that the proportion of Black Ducks harvested also declined with increasing densities of both species (Conroy *et al.* 2002). Conroy *et al.*'s (2002) modelling work formalised uncertainties about factors that influence the Black Duck population and provided the foundation for an adaptive management framework.

The BDJV, in partnership with the Eastern Habitat and Atlantic Coast JVs, is developing a decision framework that integrates habitat and population management that will enable the JV to produce an objective, science-based estimate of carrying capacity and make recommendations for revising the NAWMP population goal (Black Duck Joint Venture 2008; Devers *et al.* 2011). The framework focuses on area of habitat restored or protected at the Bird Conservation Region (BCR; North American Bird Conservation Initiative 2000) level, framed as a resource allocation issue. Decision framework objectives include: 1) achieving the NAWMP population goal under a harvest strategy of 98% maximum sustainable yield; 2) maintaining current distribution of breeding and wintering Black Ducks corresponding to the 1990–2012 period; 3) maintaining habitat to support desired abundance, distribution and harvest opportunity; and 4) increasing understanding of the density-dependence mechanism and of factors limiting the species to make increasingly more informed decisions. Underlying the framework is the Conroy *et al.* (2002) model of Black Duck population dynamics and habitat, with competing hypotheses on the role of density dependence on reproduction on the breeding grounds, survival on wintering grounds (post-hunting season), carry-over effects of wintering habitat conditions and changes in movement patterns from breeding to wintering areas. Drivers of vital rates include weather and carrying capacity as affected by habitat loss and habitat management. Strength of density dependence in winter is assumed to be related to energy intake and expenditure (*e.g.* per capita food supply and weather conditions). Harvest is included in the decision framework but is not a focus of management actions. Research is underway to address key uncertainties related to energetic capacities on the wintering grounds, return on investment for winter habitat restoration and to improve parameter estimates such as post-season survival rates in relation to variation in weather and food (Osnas *et al.* 2014).

Northern Pintail

Background – The Northern Pintail is one of the most abundant dabbling ducks in North America. The main breeding habitats are in Alaska and the Prairie Pothole Region of southern Canada and the northern U.S.

Great Plains, and winter habitats are along the coasts and throughout the southern U.S.A. The species is closely associated with temporary and seasonal wetlands, and historically pintail numbers have tracked wetland conditions on the prairies (Miller *et al.* 2003). Population levels were high during the 1950s and 1970s (5.5–9.9 million), periodically fell below 4 million birds during short-term droughts on the prairies during the 1960s–1980s, then fell to record lows during an extensive prairie drought in 1988–1991 (1.8–2.3 million). Despite greatly improved wetland conditions in the prairies since the mid-1990s, pintail numbers over the last decade have averaged 3.2 million, 43% below the NAWMP goal of 5.6 million (Zimpfer *et al.* 2013). Most of the recent decline occurred in Prairie Canada, and the once-close relationship between numbers of breeding pintail and number of prairie wetlands has weakened substantially since the 1990s (Podruzny *et al.* 2002).

Three main biological hypotheses have been suggested to account for the pintail decline. The most plausible is that conversion of prairie to cropland and changing cropping practices on the breeding grounds, especially in prairie Canada, has reduced nest success (Table 2). But it was also speculated that fewer females nested (persistently) due to cross-seasonal effects from reduced habitat quality during winter and spring migration. Finally, over-harvest and higher mortality of adult females during the breeding season due to diseases (primarily Avian Botulism *Clostridium botulinum* and Avian Cholera *Pasturella multocida*) and predation have also been suggested. Meanwhile, biologists have also expressed uncertainty about the ability of the traditional waterfowl survey (survey strata 1–50; Zimpfer *et al.* 2013) to count breeding Northern Pintail reliably during dry years on the prairies, when the pintails may overfly the region to settle in unsurveyed areas further north.

Empirical research since the 2001 Pintail Workshop (Miller *et al.* 2003), undertaken both at breeding sites and on the wintering grounds, has helped to fill many information gaps and reduced some uncertainties, such as those relating agricultural practices to nest survival (Podruzny *et al.* 2002; Kowalchuck 2012; J. Devries, Ducks Unlimited Canada, unpubl. data) and migration chronology relative to timing of surveys within traditional survey areas (Miller *et al.* 2005). For the migration and winter periods, research findings have generally downplayed the importance of low survival rates (Miller *et al.* 2005; Haukos *et al.* 2006; Fleskes *et al.* 2007; Rice *et al.* 2010). However, studies of Northern Pintail wintering on the Texas Gulf Coast identified new concerns about low overwinter survival associated with the loss of wetlands and rice agriculture (Moon & Haukos 2006; Anderson 2008). These unexpected results and the implementation of a national harvest strategy for Northern Pintail in 1997 (USFWS 2010) were among the factors elevating the importance of harvest rates in population dynamics models.

Approach – The Pintail Action Group (PAG) was created in 2003, operating as a working group under NAWMP, with a mission to advocate and support the coordination and evaluation of Northern

Pintail management and research among JVs, North American Flyways, government agencies, and organisations (Duncan et al. 2003). JVs have since pursued large-scale habitat programmes on key breeding areas and maintenance of key migration and wintering areas. For example, the Prairie Habitat JV, which encompasses the Prairie Pothole Region in Canada, has developed programmes to encourage conversion of spring-seeded cropland to more pintail-friendly uses (Devries et al. 2008), such as Winter Wheat *Triticum aestivum* and forage crops. Other conservation efforts include direct land protection and enhancement, agricultural partnerships, and policy initiatives. The PAG coordinated work to construct an empirically based meta-population model that integrates the effects of habitat and harvest on vital rates, and provides a platform to link habitat change and regional management actions to key demographic rates and population responses (Mattson et al. 2012). The model approach and structure is described below and by Osnas et al. (2014).

The predictive life-cycle model (Mattson et al. 2012) enables evaluation of how alternative habitat and harvest management strategies simultaneously influence continental-scale pintail population dynamics. This was the first model to integrate habitat and harvest explicitly into a modelling framework, the goal of the NAWMP Joint Task Group (Anderson et al. 2007). Mattson et al. (2012) discuss the general assumptions, common to most other species models, that population dynamics are regulated by external (i.e. habitat and harvest management) and internal (i.e. density-dependent) mechanisms, and that those mechanisms may interact. These assumptions in turn lead to the dual assumptions that habitat management by JVs (or JVs encompassing main pintail regions of North America) has a direct influence, and that harvest management has an indirect influence on population-specific vital rates through density-dependent mechanisms. This linkage allows simultaneous prediction of the effects of harvest and habitat management on continental pintail population dynamics (Table 3).

Greater and Lesser Scaup

Background – Greater and Lesser Scaup cannot be distinguished in aerial surveys so the species are usually combined and identified as 'scaup' in population estimates. Their combined range is the most widespread of North American diving ducks. Greater Scaup breed primarily in tundra regions from western Alaska to eastern Canada, with some also breeding in the boreal forest, whereas Lesser Scaup breed largely in boreal and prairie regions. In winter most scaup are found along the coasts of the Pacific and Atlantic Oceans, the Gulf of Mexico and the Great Lakes, although Lesser Scaup also winter on inland waters. The combined breeding populations of scaup declined from 5.7–7.6 million birds in the 1970s to a record low of 3.25 million birds in 2006 before showing signs of partial recovery; 4.2 million scaup were reported in 2013 (Zimpfer et al. 2013). The current population estimate remains 33% below the NAWMP goal of 6.3 million. The prolonged decline and uncertainties about

Table 3. Examples of conceptual and predictive models developed for species of conservation concern in North America, and their uses in guiding research and conservation programmes. References that describe details of these models are given in respective case-studies.

Species	Model	Conservation applications
Mottled Duck	*Florida*: Model to differentiate Mottled Duck age and sex, and to identify Mottled Duck-Mallard hybrids based on plumage characteristics	Accurate identification of species and hybrids in surveys and where hybrids are most prevalent. Improve population and harvest estimates; research.
	Western Gulf Coast: Sex-specific annual life-cycle model	Pinpoint critical vital rates (female annual survival) in population dynamics. Clarify relationship between female survival and nesting effort under different habitat conditions. Guide habitat actions (protection and restoration of critical wetlands).
Black Duck[a]	Integrated life-cycle and habitat model	Guide habitat actions; identify best areas or actions to achieve conservation goals.
Northern Pintail[a]	Integrated life-cycle and habitat model	Simultaneously evaluate effects of harvest and habitat management on continental-scale population dynamics. Framework for cost-effective allocation of conservation resources. Identify research and management priorities. Evaluate estimated costs and benefits of management actions.
Scaup[a]	Integrated life-cycle, habitat and hunter model	Simultaneously evaluate effects of harvest and habitat management on continental-scale scaup population dynamics and hunter recruitment and retention. Identify best areas or actions to manage species.

Table 3 (*continued*).

Species	Model	Conservation applications
Sea ducks		
Atlantic Common Eider	Demographic model	Sustainability of harvest and recommendations for harvest levels. Impact of periodic mortality events from avian cholera on population dynamics.
Pacific Common Eider	Stochastic, stage-based, matrix life-cycle model	Identify research and management priorities.
Long-tailed Duck	Matrix population model	Determine impacts of lead poisoning and subsistence harvest on population dynamics.
Spectacled Eider	Energy balance simulation model for wintering in the Bering Sea	Assess implications of changing climate, changing food resource availability and design of marine protected areas.

[a]See Osnas *et al.* (2014) for details.

factors contributing to the low numbers resulted in both species being listed as "focal species of concern" by the U.S. Fish and Wildlife Service (USFWS 2011). The largest decline occurred in the boreal forest, the core breeding region for Lesser Scaup, but numbers also declined in prairie Canada. Numbers of scaup in the tundra survey strata, presumed to be Greater Scaup, have been stable or slightly increasing since the 1970s. Hence, the main focus of concern is on Lesser Scaup.

Specific hypotheses explaining the population decline were first put forward by Austin *et al.* (2000) and Afton & Anderson (2001) and with further debate evolved into six key hypotheses (Table 2, Fig. 1). The *Disease Hypothesis* (*i.e.* contaminants) proposes that environmental contaminants have had a negative effect on scaup survival and productivity; this was based on known environmental contamination of wintering and staging areas (primarily selenium and PCBs) and on high levels of contaminants recorded in some preferred scaup foods such as the exotic Zebra Mussels *Dreissena polymorpha* (Custer and Custer 2000; Petrie *et al.* 2007). The *Spring Condition Hypothesis* posits that body condition during migration and pre-breeding has declined compared to historic levels due to reduced food abundance or quality on spring migration areas, and subsequently reduced body condition has negatively affected scaup survival and productivity (*e.g.* through lower breeding propensity, smaller clutch sizes, and later nest initiation dates). Original concerns about widespread habitat changes on the breeding grounds have been refocused to two inter-related hypotheses.

The *Climate Change-Habitat Hypothesis* suggests that warming climate in northern breeding regions has reduced the abundance or quality of wetland habitats for scaup at large scales, potentially reducing food resources, availability of nesting or brood-rearing habitat, and breeding propensity; altered habitat conditions may also have altered scaup's exposure to predators, reducing nest success or adult female survival. This hypothesis is founded on data indicating that the greatest change in annual mean temperatures coincides with the location of core Lesser Scaup breeding habitats in the western boreal forest, and evidence for substantial long-term declines in wetland areas in Alaska's boreal region (Riordan *et al.* 2006). The *Climate Change-Mismatch Hypothesis* asserts that earlier spring phenology and warmer water temperatures in northern breeding wetlands has caused invertebrates (the scaup's main food resource) to advance their reproductive cycles, possibly reducing their availability to scaup later in the season (see Drever *et al.* 2012). The *Predation Hypothesis* postulates that fluctuations in predators and alternate prey indirectly affect waterfowl productivity (Brook *et al.* 2005). A *Harvest Impact Hypothesis* was put forward to acknowledge possible links between harvest management and scaup population size, but was not considered a strong contributor to the scaup decline (Afton & Anderson 2001).

Approach – A Scaup Action Team (SAT) was created in 2008, also as a special working group under the auspices of the NAWMP, to help strengthen the biological foundations of conservation programmes. The interests of the SAT and the listing of scaup as a "Migratory Birds of Management Concern" (USFWS 2011) led to development of a conservation action plan. The SAT is using a structured decision-making process, starting by framing the problem (resource allocation among alternative management actions) and identifying fundamental objectives of scaup conservation planning not only in terms of objectives for scaup populations and their habitats but also for scaup hunter populations. The foundation of the decision framework is based on predictive models for both scaup and hunter populations, linked via harvest rate, the former building on work of Flint *et al.* (2006) and Koons *et al.* (2006).

The prototype predictive model for scaup is designed as a top-down decision framework to address three objectives: 1) achieve landscape conditions (continental carrying capacity, *i.e.* habitat) capable of supporting target populations; 2) ensure desired levels of sustainable harvest; and 3) sustain the diving duck hunting tradition (*i.e.* diving duck hunter population). The framework provides a means to identify the best areas or actions to be targeted for managing scaup and for learning (reducing uncertainty). The framework explicitly links two life-cycle models, one for scaup populations and a second for diving duck hunters, and identifies alternative management actions and their (inter-) relationships to scaup and/or hunter vital rates.

The scaup life-cycle model incorporates separate population estimates and respective vital rates for three breeding regions: prairie and boreal regions (Lesser Scaup) and tundra (Greater Scaup). In workshops (*e.g.* Austin *et al.* 2010), experts formulated competing hypotheses about causes of population changes (Fig. 1) and identified measurable features (attributes), such as wetland density or percent of the landscape in cropland, that likely influenced vital rates (Table 2) and that could be influenced through management (or policy) actions. Functional relationships were then developed for each vital rate and measurable attribute and incorporated into the scaup model (Austin *et al.* 2010). Density dependence is incorporated in two parts of the life-cycle model. During breeding, the mechanism of density dependence is via probability of breeding (habitat and/or food limitation in boreal and tundra regions). For birds in late winter (*i.e.* after the hunting season), density dependence may operate through survival, with food limitation as the primary mechanism. The model relates the number of ducks in the post-hunting population to survival during the following season (here, late winter–early spring) with either compensatory or additive harvest mortality. The process allowed many issues of uncertainty to be identified (*e.g.* interactions among alternative actions, lag effects of environmental change and reliability of vital rate estimates).

The scaup life-cycle model is explicitly linked to a simple model of diving duck hunters, which identifies putative factors and their functional relationships affecting hunter recruitment and retention. The two models are linked via an empirically based harvest rate parameter, and are both projected forward through time to estimate numbers of scaup, number of scaup harvested, and numbers of diving duck hunters under different habitat and harvest regulations scenarios. The model also can be used to explore potential impacts of large-scale ecosystem change on scaup reproduction and carrying capacity. Ultimately, the model will provide the necessary framework to perform decision analyses and evaluate estimated costs and benefits of specific management actions as well as to make transparent, informed trade-offs among multiple objectives.

North American sea ducks

Background – Among the least studied of North American waterfowl are 15 species of sea ducks (Mergini). Their distributions fall largely in remote arctic or marine areas, outside of traditional survey areas, so reliable indices of their populations and productivity have been lacking. Moreover, some groups of sea ducks have not been differentiated to species during surveys (three species of scoters *Melanitta* sp.; Common Goldeneye *Bucephala clangula* and Barrow's Goldeneye *B. islandica*; and Red-breasted Merganser *Mergus serrator* and Common Merganser *M. merganser*). Consequently, abundance, relative densities and population trends cannot be accurately estimated for most sea duck populations. Eight of 22 species or populations are thought to be below historic levels and 5 are thought to be at or above historic levels; the status of remaining species remains unknown (Bowman *et al.* 2015). Since 1986, Barrow's Goldeneye and the eastern population of Harlequin Ducks *Histrionicus histrionicus* have been listed as species of concern in Canada and as threatened in Maine (Table 1). Spectacled Eider *Sometaria fisheri* and Alaskan-breeding population of Steller's Eiders *Polysticta stelleri* are listed as threatened in the U.S. Where population data do exist, trends of several sea duck species were correlated with large-scale oceanic regime shifts, although the direction of relationships varied within and among species, and these populations appear to have been stable or increasing for the last 20 years (Flint 2013).

For many species, ecological knowledge in the early 1990s was insufficient to identify priority threats or factors contributing to apparent declines. Threats related to loss and degradation of breeding and wintering habitats, and the implications to long-term health and security of populations, are shared by multiple sea duck populations (Table 2). Habitat-related threats include oil, gas and wind power development, shellfish aquaculture on staging and wintering areas, and effects of changing climate on critical habitat. Harvest threatens several populations (SDJV Management Board 2008). Other issues of concern include bioaccumulation of contaminants, effects of disease and parasites (*e.g.* Avian Cholera die-offs affecting Common Eiders), consumption of spent lead shot and disturbances from shipping lanes and offshore wind power development.

Approach – Evolving awareness and concerns surrounding habitat, contaminants and harvest for all sea duck species led to the establishment of the Sea Duck JV (SDJV) in 1998 as a multi-species JV to advance sea duck conservation. The focus of the SDJV to date has primarily been to fill key information gaps on population trends, vital rates, habitat use, and delineation of functional populations (SDJV Management Board 2008). Conservation efforts involve a coordinated international approach (mainly U.S. and Canada, but also Russia and Greenland). Partners used existing data and expert opinion to rank species and research priorities for species known or believed to be facing significant threats. The SDJV has supported programmes to develop and improve population monitoring and delineation, such as winter sea duck surveys off the Atlantic Coast and on the Great Lakes, counting Black Scoters *Melanitta americana* molting in James Bay, and delineating functional populations using satellite telemetry and genetic markers. Because of the diversity of species and issues, biologists have pursued targeted research projects rather than broad conceptual models more generally applicable to seaducks.

The targeted sea duck projects have led to development of at least eight different population models that have or can aid decision-makers (see Table 3 for examples). Model types included stage-based matrix projections (for Common Eider: Gilliland *et al.* 2009; Iles 2012; Wilson *et al.* 2012; for King Eider *Sometaria spectabilis*: Bentzen & Powell 2012; for Long-tailed Duck *Clangula hyemalis*: Schamber *et al.* 2009), reverse-time capture-recapture (White-winged Scoter *Melanitta fusca*: Alisauskas *et al.* 2004), individual-based models (Harlequin Duck: Harwell *et al.* 2012), and spatially-explicit simulations of energy balance (Spectacled Eider: Lovvorn *et al.* 2009). While these models are too numerous to review here, they have been used to assess and guide regulations towards sustainable harvest levels, identify vital rates most responsive to management action or requiring further research, quantify population-level risk to the Exxon Valdez oil spill, and assess size requirements for marine protected areas. Most models generally demonstrated high and stable annual female survival, the vital rate to which population changes were most sensitive. However, given these patterns in adult survival, fecundity parameters (nest success and especially duckling survival) were more often indicated as potential targets for management actions.

Common Eider in the Baltic

Background – The Common Eiders of the Baltic/Wadden Sea flyway breed in Sweden, Finland, Denmark, Norway, Estonia, the Netherlands and Germany and winter mainly in Denmark, Germany, the Netherlands, Sweden, Norway and Poland. This population has been well-studied and has been the subject of long-term international monitoring programmes because of its status in the European harvest. Recent evidence from mid-winter surveys suggests the population may have experienced a substantial decline. Coordinated aerial surveys in the Dutch, German and Danish Wadden Sea show numbers halved from *c.* 320,000 in 1993 to *c.* 160,000 in 2007; coordination of counts in other winter regions is weaker, leading to substantial uncertainties in overall trends (Ekroos *et al.* 2012). Ability to assess population trends across the entire winter range is further compromised by changes in count methodology from "total counts" to aerial survey methods that rely on distance sampling and spatial modelling. The "best" estimates of winter totals for the years 1991, 2000, and 2009 were of 1,181,000, 760,000 and 976,000 Common Eiders, respectively. Ekroos *et al.* (2012) questioned whether the apparent increase between 2000 and 2009 was real or due to: 1) changes in survey methods; 2) the generation of mid-winter counts from some states using data collected over several winters during 2006–2010; or 3) birds short-stopping further east in response to milder winters, where they may be less well counted. These negative population trends contrast with breeding-ground surveys that show no consistent trends in breeding abundance (Desholm *et al.* 2002; BirdLife International 2004). Desholm *et al.* (2002) suggested the decline may represent a decline in numbers of non-breeding "floaters", which would not be represented in breeding ground counts but would be included in winter counts. There is also evidence of a decline in the adult sex ratio among Common Eiders harvested in Denmark (Ekroos *et al.* 2012). Hence, there are substantial underlying uncertainties about winter survey data and population demographics within different breeding regions.

The most immediate threat to the Baltic/Wadden Sea population is commercial exploitation of shellfish, which has likely reduced food availability to eiders and is linked to mass starvation of Common Eiders in some years (Table 2) and regional reductions in other years (Camphuysen *et al.* 2002). Furthermore, declines may be related to unknown factors causing delays among females in first year of breeding combined with reduced breeding propensity. Decline in some nesting colonies have been attributed to varying causes, usually associated with changes in predation, including greater predation of incubating females and eggs by White-tailed Sea Eagles *Haliaeetus albicilla* in Finland and invasive American Mink *Neovison vison* that have reduced reproductive success and female survival in Sweden (Desholm *et al.* 2002). Declines in other colonies have been linked to lower duckling survival (related to density dependent regulation and viral disease), competition with other waterbirds, and poor pre-nesting body condition in spring (Desholm *et al.* 2002). Other issues of concern include disease and parasite infestations affecting survival and reproduction, pollutants generally, lead poisoning in Finland, avian cholera in Denmark, bycatch in gill nets, and offshore collisions with high-speed boats and offshore structures such as wind turbines and bridges. The impact of harvest on the population is also unclear (Gilliland *et al.* 2009).

Approach – Because of its global importance to many waterbirds in the flyway, the Danish Wadden Sea has been recognised as a Ramsar site, a Natura-2000 site, an Important Bird Area, a Man and Biosphere Reserve, and a World Heritage Site. It is encompassed under the Western Palearctic Anatidae Agreement (WPAA), Trilateral Governmental Conference (The Netherlands, Germany and Denmark), and the 1995 Agreement on the Conservation of African-Eurasian Migratory Waterbirds (AEWA; Boere & Piersma 2012). The latter provides the best legal, intergovernmental instrument for collaborative management of the Wadden Sea. However, these international conventions and collaborative partnerships have not been entirely effective in protecting waterbirds and their habitats in the Wadden Sea (Boere & Piersma 2012). Conflicting economic and political interests of the multiple nations continue to challenge conservation planning and implementation in the Baltic/Wadden Sea region. Conservation and management of the Common Eider, and other sea ducks in Europe, would be greatly enhanced by the development of a conservation plan to help prioritise, coordinate and implement research, monitoring and management actions (also see Elmberg *et al.* 2006).

Decision making in the face of uncertainty

We have outlined a basic framework that integrates critical steps for defining actions to conserve populations of concern, ranging from identifying plausible hypotheses about factors influencing demographic parameters and population status to determining suites of potentially effective management or policy actions to monitoring the outcomes of those actions. We also provided examples of how this framework has been used for several waterfowl taxa, including the development of sophisticated planning models that quantify key relationships between stressors, demography and desired management outcomes (Table 3). These are essential steps towards addressing population concerns and revealing critical research needs. Uncertainties exist at each stage of the framework, which generally can be grouped into the following categories:

(1) *Population Assessments*. While many species are of concern because of low numbers or perceptions of substantial population decline, robust data for assessing population trajectories are often lacking. As well, spatial variation in demographic rates, where such data exist, suggests that population trajectories might be driven by sub-populations. However, demographically distinct sub-populations are not clearly identified for many species.

(2) *Demographic trends and relationships*. We have little information on spatial and temporal patterns in survival and reproduction for many species of conservation concern. Furthermore, functional relationships between demographic parameters and habitat quality or other limiting factors, plus underlying biological mechanisms driving those patterns, are often unknown. These include harvest, cross-seasonal and density-dependent effects.

(3) *Status and trends of key limiting factors*. Models assume linkages between demography and habitat quantity and quality or the presence of other stressors, *e.g.* "invasive" species (genetic competitors), contaminants, or predators. Key to targeting conservation action and evaluating the outcome of management actions is an understanding of how environmental conditions change due to and in spite of management actions. However, such information is often absent at spatial or temporal scales consistent with the scale of conservation concerns.

(4) *Predicting future relationships in a changing world.* Modelled relationships between limiting factors and waterfowl demography are built on expert opinion and/or existing data. However, managers cannot assume that systems they are trying to manage are static (*i.e.* constant through time and space), and therefore that known current values are useful for predicting future patterns. For example, climate change may induce changes in the ecological processes that drive patterns of waterfowl distribution and demography, which may alter those patterns (Nichols *et al* 2011). This potential change in system dynamics through time may be difficult to predict, but is valuable to explore (Drever *et al.* 2012).

(5) *Predicting outcomes and cost effectiveness of management or policy actions.* Uncertainty in the above categories can hinder the identification and implementation of appropriate conservation actions. For example, competing hypotheses about relationships between limiting factors and demography may lead to different management strategies. Moreover, limited ability to control how management actions are deployed (*e.g.* due to unplanned financial constraints), and also the effectiveness of actions given environmental variation, make predicting and also realising desired outcomes challenging. Uncertainty regarding the success of conservation outcomes confounds estimating return on investment, an essential component in determining how best to allocate the limited finances allocated to conservation.

Despite these many uncertainties, many conservation decisions must be made now. Often these decisions are time sensitive and cannot wait for perfect information. As Nichols *et al.* (2011) highlight, such decisions are regular occurrences for population managers. There is a large field of adaptive management and structured decision-making that describes an active, transparent and defensible process for arriving at decisions and reducing uncertainty to inform future decisions (*e.g.* Williams *et al.* 2002; Conroy *et al.* 2012). It is not our intent to repeat this information here, but rather to focus on the use of population models for advancing adaptive decision-making.

We recognise that models are an over-simplification of complex relationships, with inherent errors, uncertainties and assumptions. However, models are key components to adaptive management because they provide a defensible structure from which to communicate, make decisions, and learn about population dynamics and the impacts of our decisions. Both conceptual and quantitative models articulate contrasting views about how the systems we are trying to influence operate, allowing us to predict potential outcomes of alternate conservation actions. Further, by specifying key relationships, parameterizing equations and conducting sensitivity analyses, we bring key information gaps and debate into greater focus. This focus can inform research agendas, strengthen fundraising efforts, and guide development of conservation programmes. Finally, competing hypotheses about relationships between limiting factors and demography can be weighted based upon the degree of confidence we have in the probability that they are correct (Nichols *et al.* 2011). These weights can be modified as we learn through directed research or implementation of conservation programmes arising from strategic decision-making, thereby improving future decisions. Thus, while not a panacea for every situation, models provide a mechanism for structured, long-term learning; the most crucial research questions and monitoring needs typically emerge during this process and can be integrated with conservation action plans.

Conclusions

The structure of conservation efforts has evolved somewhat differently for each of the waterfowl species of concern, reflecting different issues, histories, geopolitical context and associated uncertainties about current and future system dynamics and management effectiveness. However, the examples we present share common components: formulation of hypotheses at initial stages; application of conceptual and quantitative models that integrate hypotheses with conservation actions; development of formal conservation frameworks and plans based on adaptive management; and use of collaborations and partnerships, largely through the NAWMP's JVs. We believe this approach provides the most defensible, and perhaps repeatable, method for allocating limited resources and advancing learning.

Review of IUCN threats for threatened and endangered waterfowl worldwide indicated great uncertainty in fully understanding the limiting factors and actions required to alleviate their impacts (Green 1996). While the approach we described herein has been used for several data-rich species, we argue that it is equally applicable for data-poor species. This benefit is due, in part, to the ability of conceptual models to help shape and communicate biological reasoning. However, we fully recognise the challenges associated with conserving species that migrate across multiple countries that potentially have different levels of resources (people and financial resources) and perspectives on goals and collaboration for conservation. The approach we outlined can be useful for rapidly assessing risks and guiding conservation efforts for diverse waterfowl species of conservation concern.

Acknowledgements

This paper is based on a special session at the 2013 Ecology and Conservation of North American Waterfowl Symposium (ECNAW). We would like to thank the ECNAW scientific committee for accepting our session proposal and guiding development of content. We are also greatly indebted to the following presenters and their coauthors; without your hard work there would have been no session: R.R. Bielefeld, T. Bowman, P. Devers, J.-M. DeVink, K. Guyn, A.D. Fox and J. Lovvorn. Finally, we thank A. Green, B. Mattsson, E. Rees and L. Webb for their valuable comments on earlier versions of this paper.

References

Afton, A.D. & Anderson, M.G. 2001. Declining scaup populations: a retrospective analysis of long-term population and harvest survey data. *Journal of Wildlife Management* 65: 781–796.

Alisauskas, R.T., Traylor, J.J., Swoboda, C.J. & Kehoe, F.P. 2004. Components of population growth rate for white-winged scoters in Saskatchewan, Canada. *Animal Biodiversity and Conservation* 27: 1–12.

Anderson, J.T. 2008. Survival, habitat use and movements of female Northern Pintails wintering along the Texas coast. M.Sc. thesis, Texas A & M University, Kingsville, USA.

Anderson, M.G., Caswell, D., Eadie, J.M., Herbert, J.T., Huang, M., Humburg, D.D., Johnson, F.A., Koneff, M.D., Mott, S.E., Nudds, T.D., Reed, E.T., Ringelman, J.K., Runge, M.C. & Wilson, B.C. 2007. *Report from the Joint Task Group for Clarifying North American Waterfowl Management Plan Population Objectives and Their Use in Harvest Management*. NAWMP Joint Task Group unpublished report, U.S. Fish & Wildlife Service and U.S. Geological Survey, Washington D.C., USA. Accessible at http://nawmprevision.org/sites/default/files/jtg_final_report.pdf.

An, S., Li, H., Guan, B., Zhou, C., Wang, Z., Deng, Z., Zhik, Y., Lui, Y., Xu, C., Fang, S., Jiang, J. & Li, H. 2007. China's natural wetlands: past problems, current status and future challenges. *Ambio* 36: 355–342.

Anonymous. 1999. *Final Feasibility Report and Program Environmental Impact Statement. Central and Southern Florida Project Comprehensive Review Study*. U.S. Army Corps of Engineers, Southern Florida Water Management District. Accessible at http://www.evergladesplan.org.

Austin, J.E., Afton, A.D., Anderson, M.G., Clark, R.G., Custer, C.M., Lawrence, J.S., Pollard, J.B. & Ringelman, J.K. 2000. Declining scaup populations: issues, hypotheses and research needs. *Wildlife Society Bulletin* 28: 254–263.

Austin, J.E., Boomer, G.S., Brasher, M., Clark, R.G., Cordts, S.D., Eggeman, D., Eichholz, M.W., Hindman, L., Humburg, D.D., Kelley, J.R., Jr., Kraege, D., Lyons, J.E., Raedeke, A.H., Soulliere, G.J. & Slattery, S.M. 2010. *Development of Scaup Conservation Plan*. Unpublished report from the Third Scaup Workshop – Migration, Winter and Human Dimensions, 21–25 September 2009, Memphis Tennessee, USA. U.S. Geological Survey, Jamestown, North Dakota, USA.

Bentzen, R.L. & Powell, A.N. 2012. Population dynamics of King Eiders breeding in northern Alaska. *Journal of Wildlife Management* 76: 1011–1020.

Bielefeld, R.R. &. Cox, R.R., Jr. 2006. Survival and cause-specific mortality of adult female mottled ducks in east-central Florida. *Wildlife Society Bulletin* 34: 388–394.

Bielefeld, R.R., Brasher, M.G., Moorman, T.E. & Gray, P.N. 2010. Mottled Duck (*Anas fulvigula*). *In* A. Poole (ed.), *The Birds of North America Online*. Cornell Lab of Ornithology, Ithaca, New York, USA. Accessible at http://bna.birds.cornell.edu/bna/.

BirdLife International. 2004. *Birds in Europe: Population Estimates, Trends and Conservation Status*. BirdLife Conservation Series No. 12. BirdLife International, Wageningen, the Netherlands.

Black Duck Joint Venture. 2008. *Triennial report to the North American Waterfowl Management Plan Committee, July, 2008*. Unpublished report, U.S. Fish and Wildlife Service, Laurel, Maryland, USA.

Boere, G.C. & Piersma, T. 2012. Flyway protection and the predicament of our migratory birds: a look at international conservation policies and the Dutch Wadden Sea. *Oceans and Coastal Management* 68: 157–168.

Bowman, T.D, Silverman, E.D, Gilliland, S.G. & Leirness, J.B. 2015. Status and trends of North American sea ducks: reinforcing the need for better monitoring. *In* J.-P.L. Savard, D.V. Derksen, D. Esler & J.M. Eadie (eds.), *Ecology and Conservation of North American Sea*

Ducks, pp. 1–27. Studies in Avian Biology (in press), CRP Press, New York, USA.

Brook, R.W., Duncan, D.C., Hines, J.E., Carrière, S. & Clark, R.G. 2005. Effects of small mammal cycles on productivity of boreal ducks. *Wildlife Biology* 11: 3–11.

Camphuysen, C.J., Berrevoets, C.M., Cremers, H.J.W.M., Dekinga, A., Dekker, R., Ens, B.J., van der Have, T.M., Kats, R.K.H., Kuiken, T., Leopold, M.F., van der Meer, J. & Piersma, T. 2002. Mass mortality of Common Eiders (*Somateria mollissima*) in the Dutch Wadden Sea, winter 1999/2000: starvation in a commercially exploited wetland of international importance. *Biological Conservation* 106: 303–317.

Canadian Wildlife Service Waterfowl Committee. 2012. *Population Status of Migratory Game Birds in Canada: November 2011*. Canadian Wildlife Service Migratory Birds Regulatory Report Number 37. Canadian Wildlife Service, Gatineau, Quebec, Canada.

Caswell, H. 2000. Prospective and retrospective perturbation analyses: Their roles in conservation biology. *Ecology* 81: 619–627.

Clemen, R.T. 1996. *Making Hard Decisions: an Introduction to Decision Analysis. Second Edition*. Duxbury Press, Belmont, California, USA.

Conroy, M.J. & Peterson, J.T. 2012. *Decision Making in Natural Resource Management: A Structured, Adaptive Approach*. John Wiley & Sons, New York, USA.

Conroy, M.J., Miller, M.W. & Hines, J. E. 2002. Identification and synthetic modelling of factors affecting American Black Duck populations. *Wildlife Monographs* 150: 1–64.

Custer, C.M. & Custer, T.W. 2000. Organochlorine and trace element contamination in wintering and migrating diving ducks in the southern Great Lakes, USA, since the zebra mussel invasion. *Environmental Toxicology and Chemistry* 19: 2821–2829.

Dahl, T.E. & Stedman, S.M. 2013. *Status and trends of wetlands in the coastal watersheds of the Conterminous United States 2004 to 2009*. U.S. Department of the Interior, Fish and Wildlife Service and National Oceanic and Atmospheric Administration, National Marine Fisheries Service, Washington D.C., USA.

Desholm, M., Christensen, T.K., Scheiffarth, G., Hario, M., Andersson, Å., Ens, B., Camphuysen, C.J., Nilsson, L., Waltho, C.M., Lorentsen, S.-H., Kuresoo, A., Kats, R.K.H., Fleet, D.M. & Fox, A.D. 2002. Status of the Baltic/Wadden Sea population of the Common Eider *Somateria m. mollissima*. *Wildfowl* 53: 167–204.

Devers, P.K., McGowan, C., Mattson, B., Brook, R., Huang, M., Jones, T., McAuley, D. & Zimmerman, G. 2011. *American Black Duck Adaptive Management – Preliminary Integrated Habitat and Population Dynamics Framework. A Case Study from the Structured Decision Making Workshop, 13–17 September 2010*. National Conservation Training Center, Shepherdstown, West Virginia, USA.

Devries, J.H., Armstrong, L.M., MacFarlane, R.J., Moats, L. & Thoroughgood, P.T. 2008. Waterfowl nesting in fall seeded and spring seeded cropland in Saskatchewan. *Journal of Wildlife Management* 72: 1790–1797.

Drever, M.C., Clark, R.G., Derksen, C., Slattery, S.M., Toose, P. & Nudds, T.D. 2012. Population vulnerability to climate change linked to timing of breeding in boreal ducks. *Global Change Biology* 18: 480–492.

Duncan, D.C., Miller, M.R., Guyn, K.L., Flint, P.L. & Austin, J.E. 2003. Part 2. A management and research plan for Northern Pintails. *In* M.R. Miller, D.C. Duncan, K.L. Guyn, P.L. Flint, & J.E. Austin (eds.), *The Northern Pintail in North America: The Problem and a Prescription for Recovery*, pp. 28–38. Ducks Unlimited, Sacramento, California, USA.

Ekroos, J., Fox, A.D., Christensen, T.K., Petersen, I.K., Kilpi, M., Jónsson, J.E., Green, M., Laursen, K., Cervencl, A., de Boer, P., Nilsson, L., Meissner, W., Garthe, S. & Öst, M. 2012. Declines amongst breeding Eider *Somateria mollissima* numbers in the Baltic/Wadden Sea flyway. *Ornis Fennica* 89: 81–90.

Elmberg, J., Nummi, P., Pöysä, H., Sjöberg, K., Gunnarsson, G., Clausen, P., Guillemain, M., Rodrigues, D. & Väänänen, V.M. 2006. The scientific basis for new and sustainable management of European migratory ducks. *Wildlife Biology* 12: 121–127.

Fleskes, J.P., Yee, J.L., Yarris, G.S., Miller, M.R. & Casazza, M.L. 2007. Northern Pintail and Mallard survival in California relative to habitat, abundance and hunting. *Journal of Wildlife Management* 71: 2238–2248.

Flint, P.L. 2013. Changes in size and trends of North American sea duck populations associated with North Pacific oceanic regime shifts. *Marine Biology* 160: 59–65.

Flint, P.L., Grand, J.B., Fondell, J.A. & Morse, J.A. 2006. Population dynamics of Greater Scaup breeding on the Yukon-Kuskokwim Delta, Alaska. *Wildlife Monographs* 162: 1–22.

Florida Fish and Wildlife Commission. 2011. *A Conservation Plan for the Florida Mottled Duck*. Florida Fish & Wildlife Commission, Tallahassee, Florida, USA.

Gilchrist, H.G., Gilliland, S., Rockwell, R., Savard, J.-P., Robertson, G.J. & Merkel, F.R. 2001. *Population Dynamics of the Northern Common Eider in Canada and Greenland: Results of a Computer Simulation Model*. Canadian Wildlife Service Unpublished Report. National Wildlife Research Centre, Carleton University, Ottawa, Ontario, Canada.

Gilliland, S.G., Gilchrist, H.G., Rockwell, R.F., Robertson, G.J., Savard, J-P.L., Merkel, F. & Mosbech, A. 2009. Evaluating the sustainability of harvest among northern Common Eiders *Sometaria mollissima borealis* in Greenland and Canada. *Wildlife Biology* 15: 24–36.

Green, A.J. 1996. Analyses of globally threatened Anatidae in relation to threats distribution, migration patterns, and habitat use. *Conservation Biology* 10: 1435–1445.

Gregory, R., Failing, L., Harstone, M., Long, G., McDaniels, T. & Ohlson, D. 2012. *Structured Decision Making: A Practical Guide to Environmental Management Choices*. John Wiley & Sons, New York, USA.

Harwell, M.A., Gentile, J.H. & Parker, K.R. 2012. Quantifying population-level risks using and individual based model: Sea otters, Harlequin Ducks, and Exxon Valdez oil spill. *Integrated Environmental Assessment and Management* 8: 503–522.

Haukos, D.A., Miller, M.R., Orthmeyer, D.L., Takekawa, J.Y., Fleskes, J.P., Casazza, M.L., Perry, W.M. & Moon, J.A. 2006. Spring migration of Northern Pintails from Texas and New Mexico, USA. *Waterbirds* 29: 127–136.

Iles, D.T. 2012. Drivers of nest success and stochastic population dynamics of the common eider (*Somateria mollissima*). Unpublished M.Sc. thesis, Utah State University, Logan, Utah, USA.

International Union for Conservation of Nature (IUCN). 2013. *IUCN Red List of Threatened Species. Version 2013.2*. IUCN, Gland, Switzerland. Accessible at www.iucnredlist.org (last accessed 11 December 2013).

Johnson, F.A. 2009. Variation in population growth rates of Mottled Ducks in Texas and Louisiana. U.S. Geological Survey Administrative Report to the U.S. Fish and Wildlife Service. U.S. Geological Survey, Gainesville, Florida, USA.

Junk, W.J., An, S., Finlayson, C.M., Gopal, B., Kvêt, J., Mitchell, S.A., Mitsch, W.J. & Robarts, R.D. 2013. Current state of knowledge regarding the world's wetlands and their future under global climate change: a synthesis. *Aquatic Science* 75: 151–167.

Koons, D.N., Rotella, J.J., Willey, D.W., Taper, M., Clark, R.G., Slattery, S., Brook, R.W, Corcoran, R.M. & Lovvorn, J.R. 2006. Lesser Scaup population dynamics: what can be learned from available data? *Avian Conservation and Ecology* 1(3): 6.

Kowalchuk, T.A. 2012. *Breeding ecology of Northern Pintails in prairie landscapes: tests of habitat selection and reproductive trade-off models.* Unpublished Ph.D. thesis, University of Saskatchewan, Saskatoon, Canada.

Link, W.A., Sauer, J.R. & Niven, D.K. 2006. A hierarchical model for regional analysis of population change using Christmas Bird Count Data, with application to the American Black Duck. *Condor* 108: 13–24.

Lovvorn, J.R., Grebmeier, J.M., Cooper, L.W., Bump, J.K. & Richman, S.E. 2009. Modeling marine protected areas for threatened eiders in a climatically changing Bering Sea. *Ecological Applications* 19: 1596–1613.

Mattsson, B.J., Runge, M.C., Devries, J.H., Boomer, G.S., Eadie, J.M., Haukos, D.A., Fleskes, J.P., Koons, D.N., Thogmartin, W.E., & Clark, R.G. 2012. A modeling framework for integrated harvest and habitat management of North American waterfowl: case-study of Northern Pintail metapopulation dynamics. *Ecological Modelling* 225: 146–158.

Miller, M.R. & Duncan, D.C. 1999. The Northern Pintail in North America: status and conservation needs of a struggling population. *Wildlife Society Bulletin* 27: 788–800.

Miller, M.R., Duncan, D.C., Guyn, K.L., Flint, P.L. & Austin, J.E. (eds.). 2003. Part 1. Proceedings of the Northern Pintail Workshop, 25 March 2001, Sacramento, California, USA. *In* M.R. Miller, D.C. Duncan, K.L. Guyn, P.L. Flint & J.E. Austin (eds.), *The Northern Pintail in North America: The problem and a Prescription for Recovery*, pp. 6–25. Ducks Unlimited, Sacramento, California, USA.

Miller, M.R., Takekawa, J.Y., Fleskes, J.P., Orthmeyer, D.L., Casazza, M.L. & Perry, W.M. 2005. Spring migration of Northern Pintails from California's Central Valley wintering area tracked with satellite telemetry: routes, timing and destinations. *Canadian Journal of Zoology* 83: 1314–1332.

Moon, J.A. & Haukos, D.A. 2006. Survival of female Northern Pintails wintering in the Playa Lakes Region of northwestern Texas. *Journal of Wildlife Management* 70: 777–783.

Nichols, J.D., Koneff, M.D., Heglund, P.J., Knutson, M.G., Seamans, M.E., Lyons, J.E., Morton, J.M., Jones, M.T., Boomer, G.S. & Williams, B.K. 2011. Climate change, uncertainty and natural resource management. *Journal of Wildlife Management* 75: 6–18.

North American Bird Conservation Initiative. 2000. *Bird Conservation Region Descriptions*. U.S. Fish and Wildlife Service, Division of Bird Habitat Conservation, Arlington, Virginia, USA. Accessible at http://www.nabci-us.org/aboutnabci/bcrdescrip.pdf.

North American Waterfowl Management Plan (NAWMP) Committee. 2012. North American Waterfowl Management Plan 2012: People conserving waterfowl and wetlands. North American Waterfowl Management Plan Committee Report to the Canadian Wildlife Service and the U.S. Fish and Wildlife Service. U.S. Department of the Interior, Washington D.C., USA.

Nudds, T.D., Miller, M.W. & Ankney, C.D. 1996. Black ducks: harvest, mallards or habitat? *In* J. T. Ratti (ed.), *Seventh International Waterfowl Symposium*, pp. 50–60. Institute for Wetland and Waterfowl Research, Memphis, Tennessee, USA.

Osnas, E.E., Runge, M.C., Mattsson, B.J., Austin, J.E., Boomer, G.S., Clark, R.G., Devers, P., Eadie, J.M., Lonsdorf, E.V. & Tavernia, B.G. 2014. Managing harvest and habitat as integrated components. *Wildfowl* (Special Issue No. 4): 305–328.

Petrie, M.J., Drobney, R.D. & Sears, D.T. 2000. Mallard and Black Duck breeding parameters in New Brunswick: a test of the reproductive rate hypothesis. *Journal of Wildlife Management* 64: 832–838.

Petrie, S.A, Badzinski, S.S. & Drouillard, K.G. 2007. Contaminants in lesser and greater scaup staging on the lower Great Lakes. *Archives for Environmental Contaminants and Toxicology* 52: 580–589.

Platt, J. 1964. Strong inference. *Science* 146: 347–353.

Podruzny, K.M., Devries, J.H., Armstrong, L.M. & Rotella, J.J. 2002. Long-term response of Northern Pintails to changes in wetlands and agriculture in the Canadian Prairie Pothole Region. *Journal of Wildlife Management* 66: 993–1010.

Raftovich, R.V. & Wilkins, K.A. 2013. *Migratory bird hunting activity and harvest during the 2011–12 and 2012–13 hunting seasons*. U.S. Fish and Wildlife Service, Laurel, Maryland, USA.

Rice, M.B., Haukos, D.A., Dubovsky, J.A. & Runge, M.C. 2010. Continental survival and recovery rates of Northern Pintails using band-recovery data. *Journal of Wildlife Management* 74: 778–787.

Rigby, E.A. & Haukos, D.A. 2012. Breeding season survival and breeding incidence of female Mottled Ducks on the Upper Texas Gulf Coast. *Waterbirds* 35: 260–269.

Riordan, B., Verbyla, D. & McGuire, A.D. 2006. Shrinking ponds in subarctic Alaska based on 1950–2002 remotely sensed images. *Journal of Geophysical Research*: (2005–2012)111.G4.

Rusch, D.H., Ankney, C.D., Boyd, H., Longcore, J.R., Montalbano, F., Ringelman, J.K. & Stotts, V.D. 1989. Population ecology and harvest of the American black duck: a review. *Wildlife Society Bulletin* 17: 379–406.

Schamber, J.L., Flint, P.L., Grand, J.B., Wilson, H.M. & Morse, J.A. 2009. Population dynamics of Long-tailed Ducks breeding on the Yukon-Kuskokwim Delta, Alaska. *Arctic* 62: 190–200.

Schaub, M. & Abadi, F. 2011. Integrated population models: a novel analysis framework for deeper insight into population dynamics. *Journal of Ornithology* 152: S227–S237.

Sea Duck Joint Venture (SDJV) Management Board. 2008. *Sea Duck Joint Venture Strategic Plan 2008–2012*. U.S. Fish and Wildlife Service, Anchorage, Alaska, USA, and Canadian Wildlife Service, Sackville, New Brunswick, Canada. Accessible at http://seaduckjv.org.

Shaffer, M.L. 1981. Minimum population sizes for species conservation. *BioScience* 31: 131–134.

U.S. Fish and Wildlife Service. 2010. *Northern Pintail Harvest Strategy 2010*. U.S. Department of the Interior, Fish and Wildlife Service, Division of Migratory Bird Management, Washington D.C., USA.

U.S. Fish and Wildlife Service. 2011. *Migratory Bird Program, Focal Species Strategy*. U.S. Department of Interior, Fish and Wildlife Service, Division of Migratory Bird Management, Arlington, Virginia, USA. Accessible at www.fws.gov/migratorybirds/currentbirdissues/management/FocalSpecies.html.

Varner, D.M., Bielefeld, R.R. & Hepp, G.R. 2013. Nesting ecology of Florida Mottled Ducks using altered habitats. *Journal of Wildlife Management* 77: 1002–1009.

Varner, D.M., Hepp, G.R. & Bielefeld, R.R. 2014. Annual and seasonal survival of adult female

Mottled Ducks in southern Florida, USA. *The Condor* 116: 134–143.

Williams, B., Nichols, J.D. & Conroy, M.J. 2002. *Analysis and Management of Animal Populations: Modeling, Estimation and Decision Making.* Academic Press, San Diego, California, USA.

Wilson, B.C. 2007. *North American Waterfowl Management Plan, Gulf Coast Joint Venture: Mottled Duck Conservation Plan.* North American Waterfowl Management Plan, Albuquerque, New Mexico, USA.

Wilson, H.M., Flint, P.L., Powell, A.N., Grand, J.B. & Moran, C.L. 2012. Population ecology of breeding Pacific Common Eiders on the Yukon-Kuskokwim Delta, Alaska. *Wildlife Monographs* 182: 1–28.

Zimpfer, N.L., Rhodes, W.E., Silverman, E.D., Zimmerman, G.S. & Richkus, K.D. 2013. *Trends in waterfowl breeding populations, 1955–2013.* U.S. Fish and Wildlife Service Administrative Report, 12 July 2013. U.S. Fish and Wildlife Service, Laurel, Maryland, USA.

Photograph: Common Eider males and females at Joekulsarlon, Iceland, by Imagebroker/FLPA.

Waterfowl in Cuba: current status and distribution

PEDRO BLANCO RODRÍGUEZ[1], FRANCISCO J. VILELLA[2]* & BÁRBARA SÁNCHEZ ORIA[3]

[1]Instituto de Ecología y Sistemática, Ministerio de Ciencia, Tecnología y Medio Ambiente, Apartado Postal 8010, La Habana 10800, Cuba.
[2]U.S. Geological Survey, Mississippi Cooperative Fish and Wildlife Research Unit, Department of Wildlife, Fisheries and Aquaculture, Mississippi State University, Mississippi 39762, USA.
[3]Instituto de Ecología y Sistemática, Ministerio de Ciencia, Tecnología y Medio Ambiente, Apartado Postal 8010, La Habana 10800, Cuba.
*Correspondence author. E-mail: fvilella@cfr.msstate.edu

Abstract

Cuba and its satellite islands represent the largest landmass in the Caribbean archipelago and a major repository of the region's biodiversity. Approximately 13.4% of the Cuban territory is covered by wetlands, encompassing approximately 1.48 million ha which includes mangroves, flooded savannas, peatlands, freshwater swamp forests and various types of managed wetlands. Here, we synthesise information on the distribution and abundance of waterfowl on the main island of Cuba, excluding the numerous surrounding cays and the Isla de la Juventud (Isle of Youth), and report on band recoveries from wintering waterfowl harvested in Cuba by species and location. Twenty-nine species of waterfowl occur in Cuba, 24 of which are North American migrants. Of the five resident Anatid species, three are of conservation concern: the West Indian Whistling-duck *Dendrocygna arborea* (globally vulnerable), White-cheeked Pintail *Anas bahamensis* (regional concern) and Masked Duck *Nomonyx dominicus* (regional concern). The most abundant species of waterfowl wintering in Cuba include Blue-winged Teal *A. discors*, Northern Pintail *A. acuta*, and Northern Shoveler *A. clypeata*. Waterfowl banded in Canada and the United States and recovered in Cuba included predominantly Blue-winged Teal, American Wigeon and Northern Pintail. Banding sites of recovered birds suggest that most of the waterfowl moving through and wintering in Cuba are from the Atlantic and Mississippi flyways. Threats to wetlands and waterfowl in Cuba include: 1) egg poaching of resident species, 2) illegal hunting of migratory and protected resident species, 3) mangrove deforestation, 4) reservoirs for irrigation, 5) periods of pronounced droughts, and 6) hurricanes. Wetland and waterfowl conservation efforts continue across Cuba's extensive system of protected areas. Expanding collaborations with international conservation organisations, researchers and governments in North America will enhance protection of waterfowl and wetlands in Cuba.

Key words: Anatidae, Caribbean, Cuba, conservation, habitat, management.

The Caribbean islands are a priority area globally for biodiversity conservation because of the high rate of habitat loss in the region (Brooks *et al.* 2006; Shi *et al.* 2005). The archipelago straddles the boundary of the Neotropical and Nearctic regions with tropical and subtropical climates. Rainfall patterns in the insular Caribbean are highly variable, and on many islands precipitation exceeds potential evapotranspiration, a condition that provides ample water to sustain wetland environments (Lugo 2002). With a total land mass of 110,860 km^2 and over 1,600 offshore islands and cays, Cuba represents the largest and most diverse island group in the West Indies (Fig. 1). It harbours the greatest biological diversity and degree of endemism in the West Indies; over 50% of its flowering plants and 32% of its vertebrates are unique to the country (ACC-ICGC 1978; Woods 1989; González 2002; Rodríguez-Schettino 2003; Borroto & Mancina 2011). Despite its regional importance, very little published information on waterfowl in Cuba, including on their distribution and general ecology, has become available to scientists working on these species in other parts of their range (Scott & Carbonell 1986; Santana 1991). Given the scarcity of publications on Cuban waterfowl

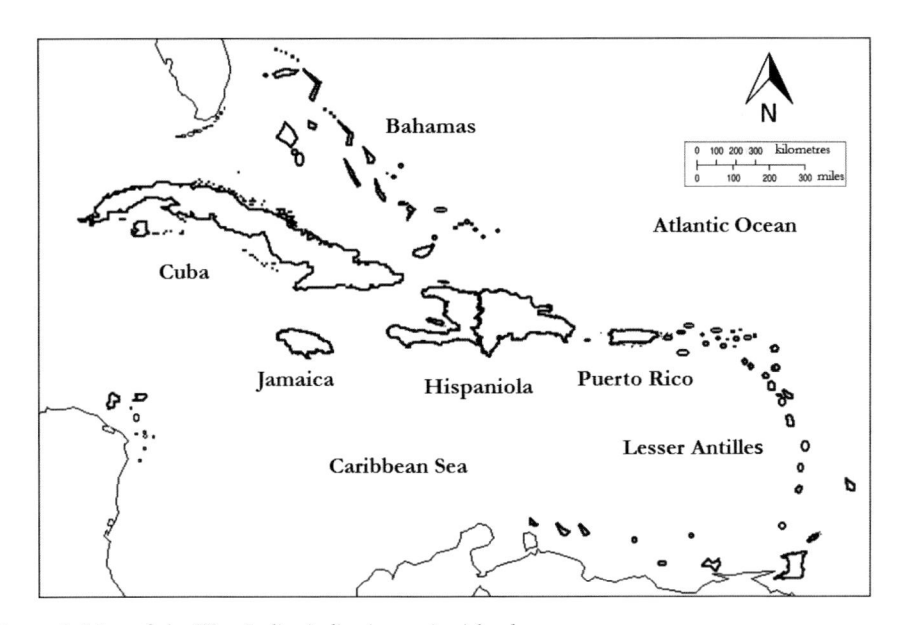

Figure 1. Map of the West Indies indicating major island groups.

and threats to wetland conservation in the region, it is important to summarise available information for the benefit of the broader scientific community. Furthermore, condensing the available literature on waterfowl and wetlands of Cuba into a single document may also be useful for researchers and managers interested in the region.

Wetland types in Cuba include mangrove forest (riparian and estuarine), freshwater marsh, seasonally flooded savanna, swamp forest, riverine wetlands and managed wetlands such as salt pans and rice fields (Borhidi *et al.* 1993). Coastal regions of Cuba feature large expanses of mangrove forest characterised by the four tree species common in the Caribbean (Red Mangrove *Rhyzophora mangle*, Black Mangrove *Avicennia germinans*, White Mangrove *Laguncularia racemosa* and Buttonwood Mangrove *Conocarpus erectus*). The interior regions of the main island of Cuba are characterised by flat topography where seasonally flooded savannas are found. These savannas represent the most floristically diverse wetlands of the Caribbean and include a great number of endemic palm species (Armenteros *et al.* 2007). Dominant species of savanna wetlands include *Eleocharris interstincta, Claudium jamaicense, Paspalum giganteum, Cyperus* sp., *Isoetes palustris, Erianthus giganteus, Thalia geniculata, Nymphaea odorata* and *Brasenia scheberi*. Swamp forests are characterised by arboreal elements and epiphytes with canopy heights of up to 20 m. Here the dominant species include *Tabebuia angustata, Fraxinus cubensis, Annona glabra, Gueltarda combiri, Bucida palustris, Hibiscus tiliaceus* and *Chrysobalanus icaco* (Armenteros *et al.* 2007).

Approximately 30% of the 1.48 million ha of Cuban wetlands are included in the national system of protected areas. Some of the most important wetlands include: the complex of lagoons south of Pinar del Río province, the Birama marshes in Granma province, the Río Máximo wildlife refuge in Camagüey province and the Lanier Swamp in the Isla de la Juventud. With a total area of 450 km^2, the Zapata Swamp (22°01'–22°40' N, 80°33'–82°09' W) is the largest and most complex drainage system in the Caribbean (Kirkconnell *et al.* 2005). Some 625,354 ha of this large wetland complex are protected as a Biosphere Reserve. Here we present information on the distribution and abundance of waterfowl on the main island of Cuba, excluding the numerous surrounding cays and the Isle of Youth. We also report on band recoveries, by species and location, for waterfowl caught and ringed in Canada and the United States that were harvested in mainland Cuba. There have been no formal waterfowl banding programmes to date within Cuba.

Methods

We reviewed and summarised information on abundance patterns and geographic distribution from a large number of unpublished reports and publications for the period 1975–2010 (most notably from Garrido & Schwartz 1968; Garrido 1980; Llanes *et al.* 1987; Sánchez *et al.* 1991; Torres & Solana 1994; Acosta & Mugica 1994; Goossen *et al.* 1994; Morales & Garrido 1996; Melián 2000, Rodríguez 2000; Peña *et al.* 2000; Wiley *et al.* 2002; Barrios *et al.* 2003). Presence of duck species and location coordinates were transferred to

118 cartographic quadrangles (37 × 18.5 km, scale 1:100,000) covering the entire main island. Residence categories for each species in Cuba followed Garrido and Kirkconnell (2000). For instance, bimodal resident (BR) refers to species (*e.g.* Wood Duck) that include both permanent breeding residents and also a small number of transient migrants; winter resident and transient (WR-T) refers to species that mostly winter in Cuba but with some individuals that occur as transients as they move through Cuba to and from wintering sites on the mainland (Central and South America); mostly transient winter residents (T-WR) are species that only occur as transients during migration peaks; introduced breeding residents (I-BR) are introduced species (e.g. the Muscovy Duck) known to breed in Cuba; and accidental (Ac) species occur only occasionally in Cuba.

Band recovery information of waterfowl harvested in Cuba from 1930–2010 was obtained from the U.S. Geological Survey's Bird Banding Laboratory and the Canadian Bird Banding Office (Blanco & Sánchez 2005). Location information from field surveys and band recoveries were georeferenced and incorporated in a Geographic Information System using ArcView 3.1 (ESRI 2001).

Results

Waterfowl in Cuba are represented by 29 species in 14 genera (Garrido & Kirkconnell 2011; Raffaele *et al.* 1998), of which 24 are migratory species with varying degrees of residence (Table 1). Species recorded in Cuba represent 93.5% (29 of 31) of waterfowl reported for the West Indies (Raffaele *et al.* 1998), highlighting the regional importance of Cuba for waterfowl in the Caribbean. North American migratory waterfowl contribute greatly to the widespread distribution of ducks in Cuba and were registered in 98 (83%) of the 118 topographic quadrangles (Fig. 2). The migratory species are almost exclusively from North America, with the possible exception of the White-faced Whistling Duck *Dendrocygna viduata* which comes from Central and/or South America and is considered an "accidental" species in Cuba. Migrant waterfowl most frequently recorded in topographic quadrangles included Blue-winged Teal *Anas discors*, American Wigeon *A. americana*, Northern Shoveler *A. clypeata*, Northern Pintail *A. acuta* and Fulvous Whistling-duck *Dendrocygna bicolor*. A total of 1,842 bands from 11 waterfowl species were recovered in Cuba during 1930–2010 (Table 2). Of these, 91.5% were recovered from Blue-winged Teal, 2.2% from American Wigeon and 2.1% from Northern Pintail. A small number of Wood Duck banded in Florida and Georgia were recovered in Cuba, suggesting that, in addition to the permanent breeding residents, occasional transient individuals arrive from North America; the species is therefore classed as a bimodal resident (Blanco & Sánchez 2005; Garrido & Kirkconnell 2011).

Arrival dates and presence of migratory waterfowl have been reported by various Cuban researchers working in natural wetlands and fields with rice *Oryza* sp. cultivation. Unfortunately, much of this information is only available in scientific journals published in Cuba or in regional

Table 1. Waterfowl species in Cuba by residence type according to Garrido and Kirkconnell (2011). Residence categories include: accidental (Ac), permanent resident (PR), bimodal resident (BR), bimodal resident and transient (BR-T) winter resident and transient (WR-T), mostly transient winter resident (T-WR), introduced breeding resident (I-BR).

Common name	Scientific name	Category of residence
White-faced Whistling-Duck	*Dendrocygna viduata*	Ac
Black-bellied Whistling-Duck	*Dendrocygna autumnalis*	PR
West Indian Whistling-Duck	*Dendrocygna arborea*	PR
Fulvous Whistling-Duck	*Dendrocygna bicolor*	I-BR
Greater White-fronted Goose	*Anser albifrons*	Ac
Snow Goose	*Chen caerulescens*	Ac
Canada Goose	*Branta canadensis*	Ac
Tundra Swan	*Cygnus columbianus*	Ac
Muscovy Duck	*Cairina moschata*	I-BR
Wood Duck	*Aix sponsa*	BR-T
Gadwall	*Anas strepera*	Ac
American Wigeon	*Anas americana*	WR-T
Mallard	*Anas platyrhynchos*	T-WR
Blue-winged Teal	*Anas discors*	WR-T
Cinnamon Teal	*Anas cyanoptera*	Ac
Northern Shoveler	*Anas clypeata*	WR-T
White-cheeked Pintail	*Anas bahamensis*	PR
Northern Pintail	*Anas acuta*	WR-T
Green-winged Teal	*Anas crecca*	WR-T
Canvasback	*Aythya valisineria*	Ac
Redhead	*Aythya americana*	Ac
Ring-necked Duck	*Aythya collaris*	WR-T
Greater Scaup	*Aythya marila*	Ac
Lesser Scaup	*Aythya affinis*	WR-T
Bufflehead	*Bucephala albeola*	Ac
Hooded Merganser	*Lophodytes cucullatus*	Ac
Red-breasted Merganser	*Mergus serrator*	WR-T
Masked Duck	*Nomonyx dominicus*	PR
Ruddy Duck	*Oxyura jamaicensis*	BR-T

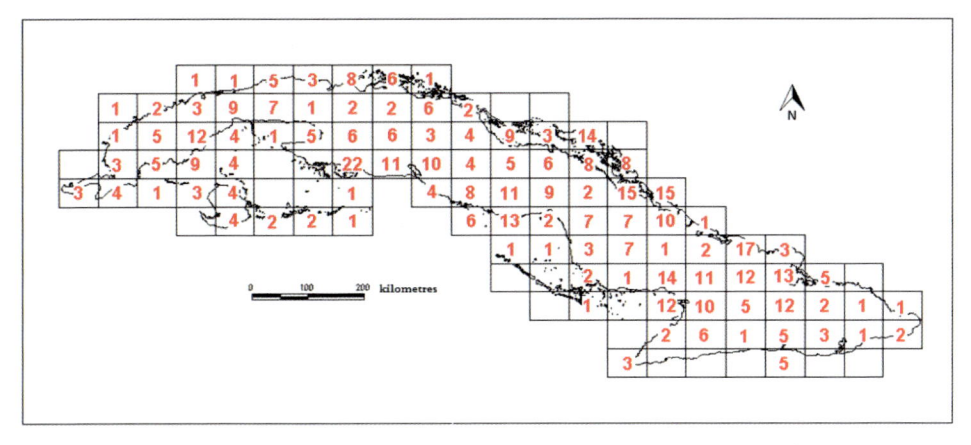

Figure 2. Number of waterfowl species recorded per topographic quadrangle in Cuba (2007–2013).

publications, which may be difficult to access by researchers outside the Caribbean (Acosta *et al.* 1992; Blanco 1996; Blanco *et al.* 1996; Sánchez & Rodríguez 2000; Sánchez *et al.* 2008). Published reports indicate the major entry points for migrant waterfowl include the Zapata Swamp lagoons, the Río Máximo refuge and mangrove wetlands in cays of the Sabana-Camagüey archipelago. Coastal lagoons in the south-central part of the island adjacent to the rice producing regions (*e.g.* Sur del Jíbaro) are also important arrival sites for the birds (Mugica 2000; Mugica *et al.* 2001; Acosta and Mugica 2006).

Band recovery data indicated that waterfowl recovered in Cuba originated in 10 Canadian provinces and 34 states of the USA. Overall, *c.* 84.6% of the 1,842 banded waterfowl recovered in Cuba originated from 24 U.S. states and Canadian provinces in the eastern and central regions of North America. The remaining recoveries include a small number of birds originating in states of the western United States (*e.g.* California and Idaho). The greatest number of duck species and individuals recovered in 10 of 15 Cuban provinces originated in three Canadian provinces (Manitoba, Saskatchewan and Ontario) as well as the states of Illinois, Iowa, Louisiana, Michigan, Minnesota, Missouri, North Dakota and South Dakota (Fig. 3). Bands were recovered mostly from September–April, coinciding with the months when migratory waterfowl are most commonly found in the Caribbean (Raffaele *et al.* 1998). These results suggest a prominent role for both the Mississippi and Atlantic Flyways in the stopover and migration patterns of waterfowl in Cuba.

Migratory waterfowl begin to arrive in Cuba during August and into early September. Small flocks comprised of Blue-winged Teal, Northern Pintail and Northern Shoveler containing 20–100 birds are common in coastal wetlands during this period. Numbers of these and other species gradually increase until they reach a peak in November, then decrease in coastal areas as wintering waterfowl move to interior wetlands on the island (Fong *et al.* 2005; Kirkconnell *et al.* 2005). During February and March, numbers of Blue-winged Teal

Table 2. Number of waterfowl in North America (U.S.A. and Canada) recovered in Cuba in the years 1930–2010. Cuban provinces include: Isla de la Juventud (IJ), Pinar del Río (PR), Habana (H), Ciudad de la Habana (CH), Matanzas (Mtz), Cienfuegos (Cfg), Villa Clara (VC), Sancti Spíritus (SSp), Ciego de Ávila (CAv), Camagüey (Cam), Las Tunas (Tun), Holguín (Hol), Granma (Gra), Santiago de Cuba (Stgo), Guantánamo (Gmo).

Species	IJ	PR	H	CH	Mtz	Cfg	VC	SSp	CAv	Cam	Tun	Hol	Gra	Stgo	Gmo
A. sponsa	1		1							1	2				
A. acuta		4	3	1	4	4	2	7	4	1	2	1	2	2	
A. Americana		4			3	3	3	2	4	5	3	6	5	1	1
A. clypeata	1	2						1	2		1				
A. crecca			1					5	3						1
A. cyanoptera										1					
A. discors	12	403	290	28	96	52	51	219	128	111	63	83	91	33	26
A. strepera	1														
A. affinis		1			1	1	3		1	3	1			1	
A. collaris		3	1		1	2	1	1	1	3		1	1		
D. bicolour	1	9			2		1	10	2	2			1		

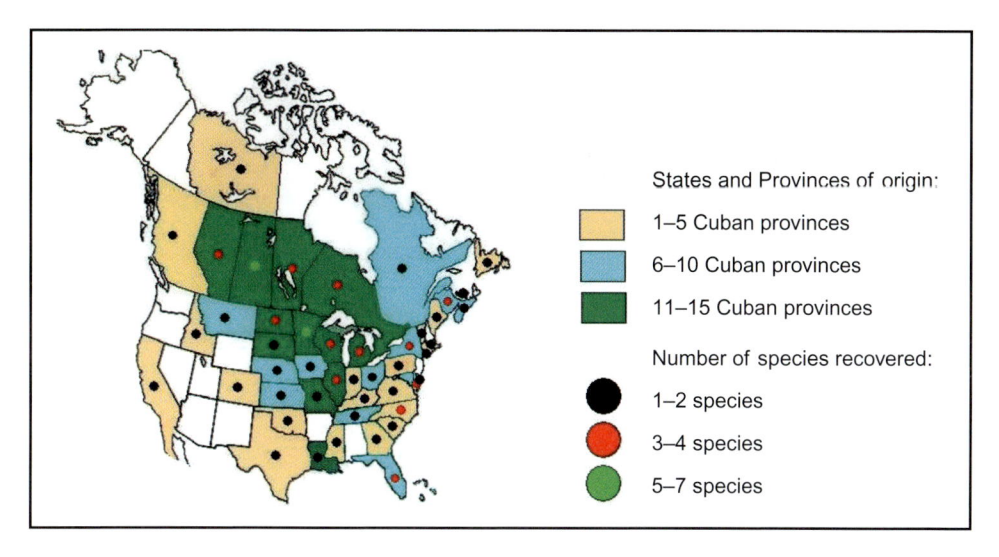

Figure 3. Origin of bands recovered in Cuba (1930–2010) indicating number of Cuban provinces represented and duck species recovered in US states and Canadian provinces. Information provided by the U.S. Geological Survey's Bird Banding Laboratory and the Canadian Bird Banding Office.

and Northern Pintail gradually increase in coastal wetlands as birds prepare for northern migration back to their breeding grounds (Blanco & Sánchez 2005).

Cuba harbours six species of resident waterfowl, including introduced breeding species. Nesting normally occurs from April–October though many species have extended nesting seasons (Garrido & Kirkconnell 2011). The Fulvous Whistling-duck was introduced to Cuba in 1931 and occurs at a limited number of sites on the island (Garrido & García 1975). The species nests in flooded forests surrounding some of Cuba's major reservoirs such as Mampostón (Mayabeque province) and Leonero (Granma province). Nests have been reported in cavities of various tree species including the Cuban Royal Palm *Roystonea regia*. Although information is available on the presence and distribution of waterfowl resident in Cuba (*e.g.* for White-cheeked Pintail *Anas bahamensis* and West Indian Whistling-duck *Dendrocygna arborea*; Fig. 4), less is known regarding their nesting ecology and productivity. Yet such data are important for species conservation, particularly as three resident species are classified as being of global and regional conservation concern (González *et al.* 2012; BirdLife International 2013): the West Indian Whistling-duck (globally vulnerable), White-cheeked Pintail (regional concern) and Masked Duck *Nomonyx dominicus* (regional concern). Moreover, while Cuba harbours the largest numbers of these resident species in the Caribbean (*e.g.* around 10,000 West Indian Whistling-ducks), information suggests that Cuban populations of resident ducks are declining (Acosta & Mugica 2006).

Cuba boasts a vast network of protected

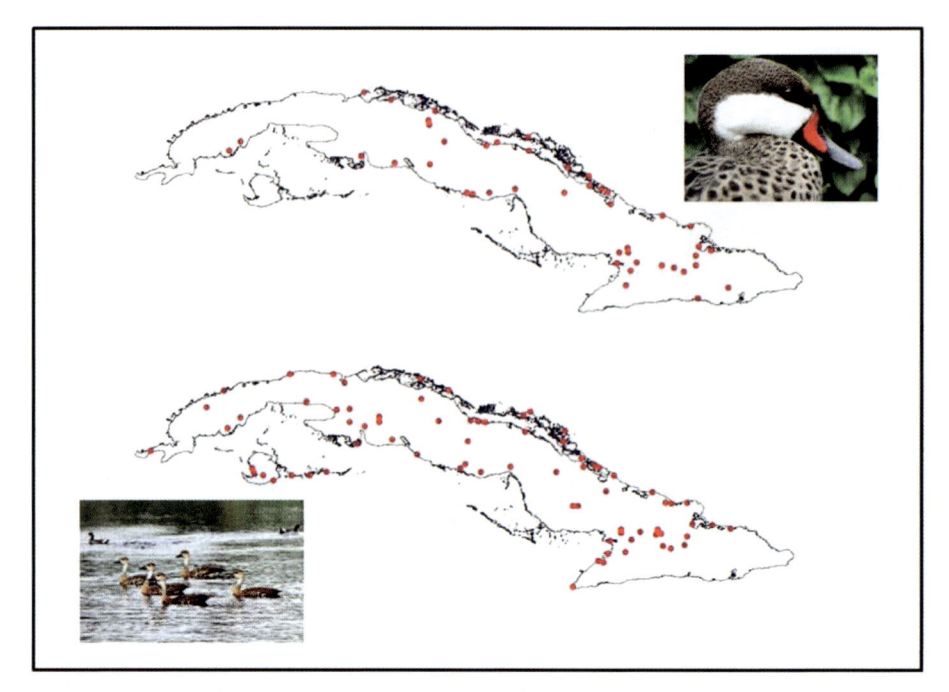

Figure 4. Location records for White-cheeked Pintail (top, photo: Alberto Puente) and West Indian Whistling-Duck (bottom, photo: Mike Morel) in Cuba and on the Isle of Youth.

areas including marine and terrestrial ecosystems. Approximately 16.9% of the terrestrial surface of Cuba (10.98 million ha) is protected by 253 different conservation units (CNAP 2009). Within this network of conservation sites, 27 of the protected areas harbour some of the most important locations for Cuba's threatened resident waterfowl (Fig. 5). These include: four biosphere reserves, five national parks, four ecological reserves, seven wildlife refuges and all six Ramsar sites designated within Cuba (CNAP 2009; Aguilar 2010).

Discussion

The broad geographic distribution of waterfowl across the main island of Cuba and their diverse taxonomic representation is largely due to the contribution of migratory species from North America. While Cuba is much larger in size than other Caribbean islands, the diversity of waterfowl present likely reflects the relatively undisturbed condition of most wetland ecosystems, including interior as well as coastal regions containing extensive areas of mangrove forest (Giri *et al.* 2011). Moreover, the proximity and interspersion of many of the principal wetlands of Cuba to rice production areas likely benefits not only resident but also migratory waterfowl.

Rice cultivation has long been a component of Cuban agriculture, and it currently represents the second most important crop (in terms both of the area planted and in yield) after sugarcane

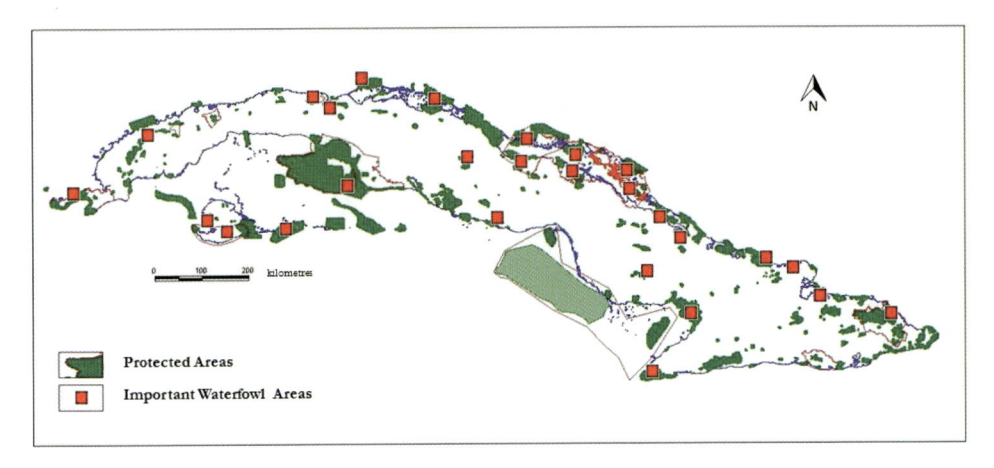

Figure 5. Important waterfowl areas and the protected areas network in Cuba (CNAP 2009).

Saccharum sp. Areas in rice production total *c.* 150,000 ha and are concentrated along the southern part of the island bordering coastal wetlands in the provinces of Pinar del Río, Matanzas, Sancti Spiritus, Camaguey and Granma. Following the dissolution of the Soviet Union in 1991, Cuba was faced with major economic challenges. The island nation responded with a large-scale food production programme and experimented with alternatives to industrialised farming due to lack of chemical fertilisers and pesticides (FAO 2002). At present, the most extensive rice production regions are in close proximity to natural wetlands, facilitating the waterbirds' use of the rice fields as feeding areas. Further, the general lack of pesticide and herbicide use promotes high levels of vertebrate and invertebrate biodiversity (Mugica *et al.* 2006). Consequently, waterbird populations thrive in the rice-producing areas of Cuba. Waterbirds are an important biotic component of the rice agro-ecosystem. Most waterbirds feed on invertebrates and weed seeds rather than rice seeds, and their waste adds nutrients to the soil, promoting an energy flow between the rice paddies and the nearby wetlands (Elphick 2000; Mugica *et al.* 2006). Ongoing research and outreach by the avian ecology group of the University of Havana has helped greatly in changing attitudes of farmers, and has encouraged them to manage rice fields to support biodiversity at these sites (Mugica *et al.* 2006).

Band recoveries suggest that the Mississippi and Atlantic Flyways contribute greatly to the diversity of waterfowl species wintering in and migrating through Cuba (Fig. 3). Recent advances in the study of migratory strategies for terrestrial birds suggest that North American warblers *Parulidae* sp. exhibit similar overwintering patterns. For instance, stable isotope analysis indicated that warblers wintering in western Cuba originate from New England states, while some warblers which winter further east in Cuba and in the rest of the Caribbean are derived from southern Appalachians populations (Faaborg *et al.* 2010). These new techniques may be useful to provide further insights into the biogeography of waterfowl in Cuba and the rest of the Caribbean, and would greatly enhance information derived from the banding data.

Efforts to increase communication between Cuban waterfowl biologists and banding laboratories in the United States and Canada should be expanded (Blanco & Sánchez 2005), not least because waterfowl species resident in the southeastern United States (*e.g.* Wood Duck and Fulvous Whistling-duck) may move regularly between Cuba and the continent (Turnbull *et al.* 1989). The links between mainland and insular populations of these species and the functional role of Cuban wetlands in their annual cycle are still unknown. Quantitative studies on population ecology and habitat relationships of breeding resident species are also considered a priority by Cuban biologists, for informing the conservation and management of waterfowl and wetlands in the region (Acosta & Mugica 2006).

Approximately 1.19 million ha of wetlands are currently protected under various conservation categories in Cuba (ACC-ICGC 1993; Aguilar 2010). Despite the extensive network of protected areas and environmental legislation aimed at expanding protection of mangrove forest, Cuban wetlands and waterfowl face numerous threats. Although Cuban legislation prohibits the harvest of duck species classified as threatened, subsistence hunting of resident waterfowl persists across several regions of Cuba. Similarly, illegal harvest of eggs from threatened waterfowl species and other waterbirds occurs in Cuba, as it does in other islands of the Caribbean (Erwin *et al.* 1984). Illegal harvest of mangrove for charcoal production continues in remote coastal regions of Cuba. Further, illegal logging is ongoing in areas of palm forest and swamp forest where resident species such as the West Indian Whistling-duck nest. Fires also degrade these savannas and seasonally flooded forests. In recent years Cuba has experienced periods of pronounced droughts resulting in lower water levels and, consequently, reduced productivity of wetlands (Sims & Vogelmann 2002). Finally, hurricanes can impact coastal wetlands due to storm timing, frequency and intensity, which in turn can alter coastal wetland hydrology, geomorphology, biotic structure, productivity and nutrient cycling (Michener *et al.* 1997). Hurricane impact on waterbirds highlights the importance of establishing long-term studies for identifying complex environmental and ecological interactions that may otherwise be dismissed as stochastic processes (Green *et al.* 2011).

Cuba is considered a high priority country for biodiversity conservation within the Caribbean basin region, yet it remains largely ignored by most conservation organisations in North America, and few long-term conservation programmes have been established by international NGOs. The state of U.S.-Cuba relations should not exclude the island-nation from regional conservation programmes (*e.g.* the Atlantic Coast Joint Venture), given the prominent role of Cuba's wetland resources compared to the rest of the Caribbean. International cooperation with Cuban scientists, universities and environmental organisations should be expanded if an integrated and effective conservation strategy for wetlands and waterfowl in the Caribbean region is to be achieved (Margulis & Kunz 1984; Santana 1991).

Acknowledgements

B.S.O and F.J.V. dedicate this paper to the memory of our dear friend and collaborator, Pedro Blanco Rodríguez, who passed away before publication of this manuscript. We are grateful to our colleagues at the Institute of Ecology and Systematics for support and assistance during field expeditions. L. Webb and J.M. Wunderle provided helpful and constructive comments to an earlier version of the manuscript. Any use of trade, firm or product names is for descriptive purposes only and does not imply endorsement by the U.S. Government.

References

ACC-ICGC 1993. *Estudio Geográfico Integral. Ciénaga de Zapata.* Publicaciones del Servicio de Información y Traducciones, La Habana, Cuba.

Acosta, M. & Mugica, L. 1994. Notas sobre la comunidad de aves del Embalse Leonero, provincia Granma. *Ciencias Biológicas* 27: 169–171.

Acosta, M., Morales, J., González, M. & Mugica, L. 1992. Dinámica de la comunidad de aves de la playa La Tinaja, Ciego de Ávila, Cuba. *Ciencias Biológicas* 24: 44–58.

Acosta, M. & Mugica, L. 2006. Reporte final de aves acuáticas en Cuba. Facultad de Biología, Universidad de La Habana, La Habana, Cuba.

Aguilar, S. (ed.). 2010. *Áreas Importantes para la Conservación de las Aves en Cuba.* Editorial Academia, La Habana, Cuba.

Armenteros, M., Williams, J.P., Hidalgo, G. & González-Sansón, G. 2007. Community structure of meio- and macrofauna in seagrass meadows and mangroves from NW shelf of Cuba (Gulf of Mexico). *Revista de Investigaciones Marinas* 28: 139–150.

American Ornithologists' Union. 2009. *49th Supplement to the American Ornithologists' Union's Check-list of North American birds.* Accessible at http://www.aou.org/checklist (last accessed on 3 March 2014).

Barrios, O., Blanco, P. & Soriano, R. 2003. Nuevos registros de aves en Cayo Sabinal, Camagüey, Cuba. *Journal of Caribbean Ornithology* 16: 22–23.

Bellrose, F.C. 1976. *Duck, Geese and Swans of North America.* Stackpole Books, Harrisburg, Pennsylvania, USA.

BirdLife International. 2013. *Threatened Birds of the World.* Accessible at http:www.birdlife.org (last accessed on 3 March 2014).

Blanco, P. 1996. Censos de aves acuáticas en un humedal costero de la Ciénaga de Zapata, Cuba. *Avicennia* 4–5: 51–55.

Blanco, P. & Sánchez, B. 2005. Recuperación de aves migratorias neárticas del orden Anseriformes en Cuba. *Journal of Caribbean Ornithology* 18: 1–6.

Blanco, P., Zuñiga D., Gómez, R., Socarras, E., Suárez M. & Morera, F. 1996. Aves del sistema insular Los Cayos de Piedra, Sancti Spíritus, Cuba. *Oceanides* 11: 49–56.

Borroto, R. & Mancina, C. (eds.). 2011. *Mamíferos en Cuba.* UPC Print, Vaasa, Finland.

Borhidi, A., Muñoz, O. & Del Risco, E. 1993. Plant communities of Cuba. I. Fresh and salt water, swamp and coastal vegetation. *Acta Botanica Hungarica* 29: 337–376.

Brooks, T.M., Mittermeier, R.A., da Fonseca, G.A.B., Gerlach, J., Hoffman, M., Lamoreux, J.F., Mittermeier, C.G., Pilgrim, J.D. & Rodrigues, A.S.L. 2006. Global biodiversity conservation priorities. *Science* 313: 58–61.

Centro Nacional de Áreas Protegidas (CNAP). 2009. *Plan del Sistema Nacional de Áreas*

Protegidas 2009–2013. Centro Nacional de Áreas Protegidas, Havana, Cuba.

Elphick, C.S. 2000. Functional equivalency between rice fields and the semi-natural wetland habitat. *Conservation Biology* 14: 181–191.

Erwin, R.M., Kushlan, J.A., Luthin, C., Price, I.M. & Sprunt, A. 1984. Conservation of colonial waterbirds in the Caribbean basin: summary of a panel discussion. *Colonial Waterbirds* 7: 139–142.

Environmental Systems Research Institute (ESRI). 2001. Getting to know ArcGIS desktop. Environmental Systems Research Institute, Inc., Redlands, California, USA.

Faaborg, J., Holmes, R.T., Anders, A.D., Bildstein, K.L., Dugger, K.M., Gauthreaux, S.A., Heglund, P., Hobson, K.A., Jahn, A.E., Johnson, D.H., Latta, S.C., Levey, D.T., Marra, P.P., Merkord, C.L., Nol, E., Rothstein, S.I., Sherry, T.W., Sillet, T.S., Thompson, F. & Warnock, N. 2010. Recent advances in understanding migration systems of New World land birds. *Ecological Monographs* 80: 3–48.

Food and Agriculture Organization (FOA). 2002. Cuba. FAO Rice Information Bulletin, FAO, Rome, Italy.

Fong, A., Maceira, D., Alverson, W.S. & Wachter, T. (eds.). 2005. Cuba: Parque Nacional "Alejandro de Humboldt". Rapid Biological Inventories Report No. 14. The Field Museum, Chicago, Illinois, USA.

Garrido, O.H. 1980. Los vertebrados terrestres de la Península de Zapata. *Poeyana* 203: 1–49.

Garrido, O.H. & García, F. 1975. *Catálogo de las Aves de Cuba*. Editorial Academia, La Habana, Cuba.

Garrido, O.H. & Kirkconnell, A. 2000. *Birds of Cuba (Helm Field Guides)*. Christopher Helm Publishers, London, UK.

Garrido, O.H. & Kirkconnell, A. 2011. *Aves de Cuba*. Comstock Publishing Associates, Ithaca, New York, USA.

Garrido, O.H. & Schwartz, A. 1968. Anfibios, reptiles y aves de la Península de Guanahacabibes, Cuba. *Poeyana* 53: 1–68.

Giri, C., Ochieng, E., Tieszen, L.L., Zhu, Z., Singh, A., Loveland, T., Masek, J. & Duke, N. 2011. Status and distribution of mangrove forests of the world using earth observation satellite data. *Global Ecology and Biogeography* 20: 154–159.

González, H. (ed.). 2002. *Aves de Cuba*. UPC Print, Vaasa, Finland.

González, H., Rodríguez Shettino, L., Rodríguez, A., Mancina, C.A. & Ramos, I. 2012. *Libro Rojo de los Vertebrados de Cuba*. Editorial Academia, La Habana, Cuba.

Goossen, J. P., Blanco, P., Sirois, J. & Gonzalez, H. 1994. Waterbird and shorebird count in the province of Matanzas, Cuba. *Canadian Wildlife Service Technical Report Series* 170: 1–18.

Green, M.C., Hill, A., Troy, J.R., Holderby, Z. & Geary B. 2011. Status of breeding reddish egrets on Great Inagua, Bahamas with comments on breeding territoriality and the effects of hurricanes. *Waterbirds* 34: 213–217.

Kirkconnell, A., Stotz, D.F. & Shoplaand, M. (eds.). 2005. Cuba: Peninsula de Zapata. *Rapid Biological Inventories Report* No. 7. The Field Museum, Chicago, Illinois, USA.

Llanes, A., Kirkconnell, A., Posada, R.M. & Cubillas, S. 1987. Aves de Cayo Saetía, Archipiélago de Camagüey, Cuba. *Misceláneas Zoológicas, Instituto Zoología, Academia de Ciencias de Cuba* 35: 3–4.

Lugo, A.E. 2002. Conserving Latin American and Caribbean mangroves: issues and challenges. *Madera y Bosques* 8 (Supplement): 5–25.

Margulis, L. & Kunz, T.H. 1984. Glimpses of biological research and education in Cuba. *BioScience* 34: 634–639.

Melián, L.O. 2000. Inventario de las aves en zonas húmedas de San Miguel de Parada. *Biodiversidad de Cuba Oriental* IV: 90–93.

Michener, W.K., Blood, E.R., Bildstein, K.L., Brinson, M.M. & Gardner, L.R. 1997. Climate change, hurricanes and tropical storms, and rising sea level in coastal wetlands. *Ecological Applications* 7: 770–801.

Morales, J. & Garrido, O.H. 1996. Aves y reptiles de cayo Sabinal, Archipiélago Sabana Camagüey, Cuba. *El Pitirre* 9: 9–11.

Mugica, L. 2000. Estructura espacio temporal y relaciones energéticas en la comunidad de aves de la arrocera Sur del Jíbaro, Sancti Spíritus, Cuba. Ph.D. thesis, Universidad de La Habana, La Habana, Cuba.

Mugica, L., Acosta, M. & Dennis, D. 2001. Dinámica temporal de la comunidad de aves asociada a la arrocera Sur del Jíbaro. *Biología* 15: 86–97.

Mugica, L., Acosta, M., Denis, D., Jiménez, A., Rodríguez, A. & Ruiz, X. 2006. Rice culture in Cuba as an important wintering site for migrant waterbirds from North America. *In* G.C. Boere, C.A. Galbraith & D.A. Stroud (eds.), *Waterbirds Around the World*, pp. 172–176. The Stationery Office, Edinburgh, U.K.

Peña, C., Fernández, A., Navarro, N., Reyes, E. & Sigarreta, S. 2000. Avifauna asociada al sector costero de Playa Corinthia, Holguín, Cuba. *El Pitirre* 13: 31–34.

Raffaele, H., Wiley, J., Garrido, O.H., Keith, A. & Raffaele, J. 1998. *Birds of the West Indies*. Princeton University Press, Princeton, New Jersey, USA.

Rodríguez, F. 2000. Lista de los vertebrados del Parque Natural bahía de Naranjo, Cuba. *Biodiversidad de Cuba Oriental V*: 143–146.

Rodríguez Shettino, L. (ed.). 2003. *Anfibios y Reptiles de Cuba*. UPC Print, Vaasa, Finland.

Sánchez, B. & Rodríguez, D. 2000. Avifauna associated with the aquatic and coastal ecosystems of Cayo Coco, Cuba. *El Pitirre* 13: 68–75.

Sánchez, B., García, M.E. & Rodríguez, D. 1991. Aves de Cayo Levisa, Archipiélago de los Colorados, Pinar del Río, Cuba. Inv. Mar. *Cicimar* 6: 247–249.

Sánchez, B., Blanco, P., Oviedo, R., Hernández, A., del Pozo, P., Lamela, W., Torres, M. & Rodríguez, R. 2008. Composición y abundancia de las comunidades de aves en localidades de la provincia de Cienfuegos, Cuba. *Poeyana* 496: 10–19.

Santana, E.C. 1991. Nature conservation and sustainable development in Cuba. *Conservation Biology* 5: 13–17.

Scott, D.A. & Carbonell M. 1986. *A Directory of Neotropical Wetlands*. International Union for Conservation of Nature and Natural Resources (IUCN), Conservation Monitoring Centre. Cambridge, U.K.

Shi, H., Singh, A., Kant, S., Zhu, Z. & Waller, E. 2005. Integrating habitat status, human population pressure, and protection status into biodiversity conservation priority setting. *Conservation Biology* 19: 1273–1285.

Sims, H. & Vogelmann, K. 2002. Popular mobilization and disaster management in Cuba. *Public Administration and Development*, 22: 389–400.

Torres, A. & Solana, E. 1994. Listado de las aves observadas dentro del corredor migratorio de Gibara, provincia Holguín; Cuba. *Garciana* 22: 1–4.

Turnbull, R.E., Johnson, F.A. & Brakhage, D.H. 1989. Status, distribution and foods of Fulvous Whistling-Ducks in south Florida. *Journal of Wildlife Management* 53: 1046–1051.

Wiley, J.W., Ruiz, A., Pérez, E., Faife, M., Díaz, L., González, M., Rivero, Y., Chirino, G., Soto, O., Morejón, R., Vales, A. & Ibarra, M.E. 2002. Bird survey in the mogote vegetational complex in the Sierra del Infierno, Pinar del Río, Cuba. *El Pitirre* 15: 7–15.

Woods, C.A. 1989. Biogeography of the West Indies past, present and future. Sandhill Crane Press, Gainesville, Florida, USA.

Assessment of the 6th North American Duck Symposium (2013): "Ecology and Conservation of North American Waterfowl"

LUCIEN P. LABORDE, JR.[1]*, RICHARD M. KAMINSKI[2] & J. BRIAN DAVIS[2]

[1]School of Renewable Natural Resources, Louisiana State University, Baton Rouge, Louisiana 70803, USA.
[2]Department of Wildlife, Fisheries and Aquaculture, Mississippi State University, Mississippi State, Mississippi 39762, USA.
*Correspondence author. E-mail: llabor2@tigers.lsu.edu

Abstract

Using web-based technology, we e-mailed a survey link to all 450 conferees of the 6th North American Duck Symposium (NADS 6), "Ecology and Conservation of North American Waterfowl" (ECNAW), seeking feedback from attendees in order to guide the organisation of future waterfowl and other wildlife symposia. Twelve questions were posed to evaluate the 2013 all-waterfowl symposium and a further 18 questions to assist planning future similar meetings. A total of 284 responses (63%) were received; the feedback suggested that NADS 6 was well organised, that it presented relevant information and that it was valuable to conferees. Perceptions of respondents on the structure of NADS 6 (*i.e.* whether presentation on all waterfowl should be included, or only on ducks) may not be representative of attendees of previous NADS meetings, as these focused on duck species. Nevertheless, respondents suggested that future symposia should continue on a 3-year rotation and retain its 4-day format with four morning plenary sessions, concurrent afternoon oral presentations and evening poster and mentee-mentor sessions. Respondents also recommended a maximum of three concurrent afternoon sessions and indicated that future symposia might continue to embrace geese and swans as well as ducks. The results suggest a need for officials of NADS to determine future meeting frameworks and venues. Web-based surveys provide a useful tool for conference evaluation and can promote effective design and relevance of future meetings and related events.

Key words: conference evaluation, ducks, NADS, survey, symposium, waterfowl.

Waterfowl (Anatidae: ducks, geese and swans) are important birds ecologically, environmentally and economically (Baldassarre & Bolen 2006; Grado *et al.* 2011; Green & Elmberg 2014). They have been foci of continual research, conservation and recreational endeavours in North America since the early 20th century (Bellrose 1976; Baldassarre & Bolen 2006). To help sustain waterfowl resources in North America, scientists and managers have convened conferences and symposia periodically to communicate contemporary knowledge about these birds and their habitats, particularly for species and populations of conservation concern (*e.g.* Canadian Wildlife Service 1969; Bookhout 1979; Boyd 1983; Whitman & Meredith 1987; Weller 1988; Smith *et al.* 1989; Fredrickson *et al.* 1990; Batt *et al.* 1992; Rusch *et al.* 1998). The inaugural North American Duck Symposium and Workshop (NADS 1) was convened in Baton Rouge, Louisiana, USA in 1997. This seminal event attracted professionals and students from North America and Europe to address the ecology and management of wild ducks, to synthesise acquired knowledge and to convey future needs and directions for research, management and conservation. An important objective of NADS from inception has been to attract students to present their research and promote their professionalism among colleagues. The founders of NADS believed that addressing research questions and management issues related to sustaining duck populations, maintaining the wildfowling tradition and advancing ecological studies, as well as the involvement of the next generation of students, were of paramount importance. Additionally, the founders considered that waterfowl ecologists had led major advances in avian ecology, analytical procedures and conservation, and they therefore sought to perpetuate and develop this legacy, ultimately through the creation of NADS, Inc., a non-profit organisation established to facilitate future symposia and workshops.

Six NADS have been convened to date, in: 1) Baton Rouge, Louisiana, USA (1997); 2) Saskatoon, Saskatchewan, Canada (2000); 3) Sacramento, California, USA (2003); 4) Bismarck, North Dakota, USA (2006); 5) Toronto, Ontario, Canada (2009); and 6) Memphis, Tennessee, USA (2013). Locations generally have rotated between the United States and Canada, among administrative waterfowl flyways, and generally in northern and southern locations of North America. Each NADS was organised under the direction of a scientific committee, which had discretion regarding the theme, content and venue. The Science Programme Committee for NADS 6 agreed the symposium would be expanded to include all taxa of waterfowl. Thus, NADS 6 was subtitled "Ecology and Conservation of North American Waterfowl" (ECNAW), with the North American Arctic Goose Conference and Workshop and the International Sea Duck Conference included as joint partners in NADS 6/ECNAW (http://www.northamericanducksymposium.org).

Feedback and evaluation are crucial for improving wildlife science and conservation programmes, and also for stakeholder engagement (Sholtes 1988; Jacobson 2012; Lauber *et al.* 2012). Although five NADS have been held previously, none were evaluated by surveying the conferees. Following NADS 6/ECNAW, its lead organisers (R.M. Kaminski and J.B. Davis) decided to conduct a post-symposium assessment of the meeting and asked the senior author (L.P. Laborde, Jr.), with human-dimensions and survey-sampling skills, to develop a questionnaire. The primary objective was to poll participants on evaluative and planning criteria for future NADS and similar meetings. We believe results from this survey may also benefit wildlife and natural resources professionals in planning and implementing other large conferences and symposia.

Methods

The NADS 6/ECNAW post-symposium survey was developed to address 12 evaluative, 18 planning and three demographic-related questions. We evaluated conference sessions using an ordinal scale of 1 = "not valuable", 2 = "marginally valuable", 3 = "moderately valuable", and 4 = "highly valuable", and likewise used an ordinal scale of 1 = "strongly disagree", 2 = "disagree", 3 = "neither agree nor disagree", 4 = "agree", and 5 = "strongly agree" to rank agreement with statements addressing the symposium venue, scheduling, programme and costs (Dillman et al. 2009). Additionally, we invited open comments from conferees. Confirmed e-mail addresses were obtained from all 450 participants and the survey was distributed on 21 February 2013, three weeks after the symposium. Each conferee was asked to complete the survey, and each e-mail contained an embedded link to the survey using Qualtrics™ v. 12000 (Qualtrics Labs, Inc., Provo, Utah, USA; Vaske 2008). We contacted conferees up to three times at 5-day intervals to elicit their response and then thanked all respondents. An alternative response system – via electronic document, post or e-mail – was also provided (Vaske 2008). Responses were limited to one per Internet Protocol (IP) address to minimise poll crashing (i.e. multiple responses per attendee; Dillman et al. 2009). Survey protocols ensured anonymity and confidentiality and were approved by the Louisiana State University Agriculture Center Institutional Review Board (Protocol Number HE 13-7). We collected responses through to 21 March 2013. We calculated the margin of error as the 95% confidence interval for the true population value of responses, following Dillman et al. (2009). Chi-square tests ($\alpha = 0.05$) were used to assess non-response bias and to test frequencies of response among demographic classes. Simple descriptive statistics are presented to analyse evaluative and planning questions.

Results

We received 284 (63%) responses to the survey. Based on this response rate and the population of 450 conferees, we report results within a margin of error of ± 4%, indicating that 19 out of 20 times (i.e. 95% of occasions) the true population value will be within 4 percentage points of our reported sample estimate. We used three demographic variables to evaluate non-response bias. Respondents were 85% male (15% female), but gender proportions of respondents did not differ significantly from non-respondents ($\chi^2_1 = 2.82, P = 0.093$, n.s.). Age distribution of respondents was ≤ 25 (11%), 26–35 (30%), 36–45 (24%), 46–55 (19%), 56–65 (13%), and > 65 years (3%), but ages of non-respondents were not available for comparison. By occupation, 23% of respondents were students, 15% were academicians and 62% were grouped as professionals, including biologists, managers, administrators and retirees. A higher proportion of the academicians who attended the meeting responded to the survey (89%) than did students (62%) or professionals (57%; $\chi^2_2 = 11.96$, $P = 0.002$).

Respondents evaluated each of five conference sessions separately. The four daily plenary sessions of the symposium were attended by ≥ 88% of respondents, and their ratings averaged 3.2–3.3 (s.d. = 1.1–1.3), indicating that their assessment of plenaries ranged from moderate to high value. One day before the grand opening of the symposium, there was a special session on the 2012 revision of the North American Waterfowl Management Plan; it was attended by 28% of the respondents who arrived early to the meeting and rated it, on average, moderately valuable (mean = 3.1, s.d. = 1.5). The remaining seven evaluative statements considered the relevancy of information presented at the symposium, logistics, the host hotel and nearby venues. Mean ratings ranged from 3.3–4.4 (s.d. = 0.7–1.1), indicating their assessment ranged from moderate to strong agreement (Table 1).

Eighteen questions addressed respondents' preferences for future NADS, of which five specifically addressed the format of future symposia. For NADS 7 (scheduled to be held in Annapolis, Maryland, USA; February 2016), morning plenary and afternoon oral presentations, of the same length as NADS 6, were favoured by 63% and 77% of respondents respectively. During NADS 6, 6–7 concurrent sessions were held during three afternoons, but feedback indicated that this was too many, with 58% of respondents preferring only 2–3 concurrent sessions, and 32% suggesting 4–5 concurrent sessions. Only 2% wanted to continue the NADS 6 format of 6–7 sessions being held at the same time. Given options for convening

Table 1. Level of agreement with statements evaluating the North American Duck Symposium and Workshop 6, Ecology and Conservation of North American Waterfowl (2013), held at Memphis, Tennessee, USA.

Statement	Mean[a]	s.d.	n
Information presented was directly relevant to my work	4.4	0.7	281
The registration cost was a fair value	4.1	0.8	283
The symposium was well organised	4.2	0.9	280
Adequate time was allowed for breaks	4.2	0.8	280
Adequate time was devoted to issues of waterfowl management	4.0	0.8	281
The hotel cost was fair relative to amenities received	3.3	1.1	279
Service from the hotel staff was excellent	4.3	0.8	280

[a]Rated on a scale of 1 = strongly disagree, to 5 = strongly agree.

more frequent symposia, 66% of respondents preferred the current format of a 4-day symposium every three years. Given options of integrated plenary topics in a 4-day or two consecutive 2-day formats with different registration options and costs, 73% preferred the current format of 4 days with four different daily plenary sessions and a single registration fee. Cross-tabulation of the above five questions about the format of future symposia confirmed that ≥ 50% of the three major occupational groups (*i.e.* students, academicians and professionals) ranked alternatives identically as described above.

Eleven statements addressing the symposium venue, scheduling, programme and costs were rated as described previously (Table 2). Responses to eight statements ranged from neutral to agreeable (means = 3.1–4.0, s.d. = 0.7–1.1; Table 2), indicating that evenings were preferred for poster sessions, that the student mentee-mentor session should be continued, door prizes should be given to students and professionals during breaks between sessions, and that the cost of public transport between airports and hotels should be considered on choosing the hotel. Respondents neither agreed nor disagreed that breakfasts should be provided as part of registration costs, and there was no consensus that speakers and entertainment during lunch enhanced the symposium. Although nearly all respondents did not wish to continue the NADS 6 format of

Table 2. Level of agreement with statements for planning the North American Duck Symposium and Workshop 7 (scheduled to be held at Annapolis, Maryland, USA in February, 2016).

Statement	Mean[a]	s.d.	n
Professional and student presentations should be intermingled	4.0	0.9	280
Evening receptions were a good time for poster sessions	3.9	0.8	279
The Student Mentor-Mentee session should be continued	3.9	0.9	273
Door prizes for students should be continued	4.0	0.8	278
Door prizes should be availed to all conferees	3.1	1.0	279
Speakers and entertainment during lunch enhanced the programme	2.8	1.1	279
Breakfast should not be included to reduce registration costs	2.3	1.1	280
Alcoholic beverages should be included during evening receptions	3.4	1.1	281
Restaurants and amenities should be available within walking distance	4.3	0.7	281
The conference should be within a $30 cab ride of an airport	3.7	0.9	279
Future symposiums should address "ducks only" and not all waterfowl	2.1	1.1	282

[a]Rated on a scale of 1 = strongly disagree, to 5 = strongly agree.

6–7 concurrent afternoon sessions, which was necessary to accommodate many presenters at the all-waterfowl symposium, 73% of respondents "strongly disagree" or "disagree" that future NADS should address ducks only.

Two questions addressed participation in NADS 7. A total of 172 respondents (65%) indicated they were "likely" or "very likely" to attend NADS 7, and 35 respondents (12%) volunteered to serve on an organising committee for NADS 7. One hundred and two respondents (36%) offered comments, of which 47 were congratulatory in nature. Twenty-nine comments indicated there were too many concurrent afternoon sessions, and 26 comments stated that the daily cost of the host hotel exceeded federal and some state expenses limits. For additional details, the complete survey and its summarised results and comments are available on the NADS 6/ECNAW website (http://www.northamericanducksymposium.org/index.cfm?page=survey) or from the senior author.

Discussion

We surveyed conferees of NADS 6/ECNAW, a symposium that embraced all waterfowl (*i.e.* ducks, geese and swans), unlike previous NADS which focused on ducks alone. The results and interpretations therefore reflect data from respondents attending this all-waterfowl conference and may not be the perception of attendees of NADS 1–5. Nonetheless, the results likely will be useful for planning future NADS and similar large meetings. Because the overall response rate was > 60%, with > 55% of conferees in each of the occupation classes responding, respondents representing age classes from ≤ 25 to > 65 years, and there being no significant difference in the gender of responding and non-responding conferees, we believe that the responses were reasonably representative of those attending the conference. A representative sample from a majority of the surveyed population is considered more relevant than a high response rate for generalisations from survey results (Vaske 2008).

Survey results suggested that NADS 6/ECNAW was well organised, that it presented relevant content and was valuable to conferees. Responses and comments suggested that the host hotel rates should fall within federal and other expense guidelines, and that the location of the host hotel should be within walking distance of restaurants and other amenities. Respondents also suggested that future NADS should continue on a 3-year rotation and retain its single registration, 4-day format with 4 morning plenary sessions. There was an overwhelming preference to reduce the number of concurrent sessions in future NADS, likely because previous NADS featured non-concurrent sessions enabling possible holistic attendance of sessions by conferees. Nonetheless, > 70% of respondents expressed the preference that future symposia embrace all taxa of waterfowl. Because this opinion reflected perceptions only of respondents to the NADS 6/ECNAW survey, we conclude there is a need for the scientific committee of future NADS to work with NADS, Inc. to determine if subsequent NADS should focus on ducks or be inclusive of all waterfowl taxa. Indeed, there are major trade-offs between holding an all-waterfowl symposium with the number of concurrent sessions in large meetings, during which conferees would be unable to attend all sessions, presenters (notably students) would not have an opportunity to address most conferees (if speaking in one of several concurrent sessions), and there may also be competition for attendance and fund-raising between NADS and other waterfowl, ornithological or wildlife conferences. Multiple concurrent sessions at all-waterfowl symposia may thus lessen opportunities for students to gain knowledge and receive expert feedback on their work, which has been identified as an important objective for NADS by NADS, Inc.

Access to web-based survey tools and to the conferees' e-mail addresses make electronic post-symposia surveys an inexpensive and relatively efficient method for evaluating meetings and planning future events. While we did not incur any direct costs to administer the survey, future surveyors may experience charges for development and use of a survey instrument. The NADS 6 post-symposia survey implied the relevance and value of symposia presentations, the frequency and possible format of future symposia, and general guidelines for the location and cost of host hotels. The survey was also able to identify volunteers for organising committees of the next symposium.

As far as we are aware, the survey of the NADS 6/ECNAW conferees was the first formal evaluation of an international waterfowl conference. We recommend using similar electronic survey methods for evaluating future NADS so that the data are comparable and not confounded by survey methodology. Additionally, we recommend that other wildlife and natural resources conferences and symposia conduct similar post-meeting surveys. When combined with early programme and hotel planning, and with the involvement of experienced committee volunteers and fundraisers from previous symposia, post-conference surveys can promote the effective design and relevance of future events.

Acknowledgements

We are indebted to the public- and private-sector sponsors of NADS 6/ECNAW. We also thank the scientists and managers who composed the Scientific Program Committee and the Graduate Student Scholarship and Awards Committee, the graduate students at Mississippi State University, especially Justyn Foth who helped to ensure success of the symposium, the session chairs and moderators, and the staff of the Peabody Hotel, Memphis, Tennessee, USA. We thank all conferees of NADS/ECNAW, especially those who completed the post-symposia survey and enabled this evaluation. We thank A. Afton, D. Haukos, A.D. Fox and E.C. Rees for reviewing an earlier draft of this paper. Finally, we thank the School of Renewable Natural Resources at Louisiana State University for supporting design, implementation and evaluation of the survey, and the Forest and Wildlife Research Center (FWRC), Mississippi State University, for supporting J.B. Davis and R.M. Kaminski as symposia co-chairs. This manuscript has been approved for publication by the FWRC and the Department of Wildlife, Fisheries, and Aquaculture (WFA) at Mississippi

State University as FWRC/WFA publication WF379.

References

Baldassarre, G.A. & Bolen, E.G. 2006. *Waterfowl Ecology and Management, 2nd Edition*. Krieger Publishing Company, Malabar, Florida, USA.

Batt, B.D.J., Afton, A.D., Anderson, M.G., Ankney, C.D., Johnson, D.H., Kadlec, J.A. & Krapu, G.L. (eds.) 1992. *Ecology and Management of Breeding Waterfowl*. University of Minnesota Press, Minneapolis, Minnesota, USA.

Bellrose, F.C. 1976. *Ducks, Geese and Swans of North America, 2nd Edition*. Stackpole Books, Harrisburg, Pennsylvania, USA.

Bookhout, T.A. (ed.) 1979. *Waterfowl and Wetlands – an Integrated Review*. Proceedings of the 39th Mid-west Fish and Wildlife Conference, Madison, Wisconsin. LaCrosse Printing Company, LaCrosse, Wisconsin, USA.

Boyd, H. (ed.) 1983. First western hemisphere waterfowl and waterbird symposium. Canadian Wildlife Service, Catalogue No. CW66-63/1983E. Canadian Wildlife Service, Ottawa, Ontario, Canada.

Canadian Wildlife Service. 1969. Saskatoon wetlands seminar. Canadian Wildlife Service Report Series, No. 6. Canadian Wildlife Service, Ottawa, Ontario, Canada.

Dillman, D.A., Smyth, J.D. & Christian, L.M. 2009. *Internet, Mail, and Mixed-mode Surveys: The Tailored-design Method, 3rd Edition*. John Wiley & Sons, Inc., Hoboken, New Jersey, USA.

Fredrickson, L.H., Burger, G.V., Havera, S.P., Graber, D.A., Kirby, R.E. & Taylor, T.S. (eds.) 1990. *Proceedings of the 1988 North American Wood Duck Symposium*. St. Louis, Missouri, USA.

Grado, S.C., Hunt, K.M., Hutt, C.P., Santos, X. & Kaminski, R.M. 2011. Economic impacts of waterfowl hunting in Mississippi derived from a state-based mail survey. *Human Dimensions of Wildlife* 16: 100–113.

Green, A.J. & Elmberg, J. 2014. Ecosystem services provided by waterbirds. *Biological Reviews* 89: 105–122.

Jacobson, S.K. 2012. Communications and Outreach. *In* N. J. Silva (ed.), *The Wildlife Techniques Manual, Volume 2, 7th Edition*, pp. 2–42. The Johns Hopkins University Press, Baltimore, Maryland, USA.

Lauber, T.B., Decker, D.J., Leong, K.M., Chase, L.C. & Schusler, T.M. 2012. Stakeholder engagement in wildlife management. *In* D.J. Decker, S.J. Riley & W.F. Siemer (eds.), *Human Dimensions of Wildlife Management*, pp. 139–156. The Johns Hopkins University Press, Baltimore, Maryland, USA.

Rusch, D.H., Samuel, M.D., Humburg, D.D. & Sullivan, B.D. 1998. *Biology and Management of Canada Geese*. Proceedings of the International Canada Goose Symposium, Milwaukee, Wisconsin, USA.

Scholtes, P.R. 1988. *The Team Handbook: How to use Teams to Improve Quality*. Joiner Associates Inc. Madison, Wisconsin, USA.

Smith, L.M., Pederson, R.L. & Kaminski, R.M. 1989. *Habitat Management for Migrating and Wintering Waterfowl in North America*. Texas Tech University Press, Lubbock, Texas, USA.

Vaske, J.J. 2008. *Survey Research and Analysis: Applications in Parks, Recreation and Human Dimensions*. Venture Publishing, Inc., State College, Pennsylvania, USA.

Weller, M.W. (ed.) 1988. *Waterfowl in Winter*. University of Minnesota Press, Minneapolis, Minnesota, USA.

Whitman, W.R. & Meredith, W.H. (eds.) 1987. *Waterfowl and Wetlands Symposium: Proceedings of a Symposium on Waterfowl and Wetlands Management in the Coastal Zone of the Atlantic Flyway*. Delaware Department of Natural Resources and Environmental Control Document No. 40-05/87/07/01. Delaware Coastal Management Program, Dover, Delaware, USA.

Photograph: Flushing Mallard (and a likely male Mallard x American Black Duck hybrid) in an interspersed bottomland-hardwood and moist-soil wetland in the Mississippi Alluvial Valley of Mississippi, by James C. Kennedy.